# *The*
# HUMAN BRAIN *during the*
# LATE FIRST
# TRIMESTER

# ATLAS OF
# HUMAN CENTRAL NERVOUS SYSTEM DEVELOPMENT
## SERIES

## Shirley A. Bayer *and* Joseph Altman

### VOLUME 1
*The Spinal Cord from Gestational Week 4 to the 4th Postnatal Month*

### VOLUME 2
*The Human Brain During the Third Trimester*

### VOLUME 3
*The Human Brain During the Second Trimester*

### VOLUME 4
*The Human Brain During the Late First Trimester*

### VOLUME 5
*The Human Brain During the Early First Trimester*

*The*
# HUMAN BRAIN *during the*
# LATE FIRST
# TRIMESTER

## Shirley A. Bayer *and* Joseph Altman

**CRC Press**
Taylor & Francis Group
Boca Raton  London  New York

CRC Press is an imprint of the
Taylor & Francis Group, an **informa** business

A TAYLOR & FRANCIS BOOK

CRC Press
Taylor & Francis Group
6000 Broken Sound Parkway NW, Suite 300
Boca Raton, FL 33487-2742

First issued in paperback 2019

© 2006 by Taylor & Francis Group, LLC
CRC Press is an imprint of Taylor & Francis Group

No claim to original U.S. Government works

ISBN-13: 978-0-8493-1423-0 (hbk)
ISBN-13: 978-0-367-39094-5 (pbk)

**Library of Congress Cataloging-in-Publication Data**

Catalog record is available from the Library of Congress

**Visit the Taylor & Francis Web site at
http://www.taylorandfrancis.com**

**and the CRC Press Web site at
http://www.crcpress.com**

# DEDICATION

We dedicate this volume to the memory of Prof. Dr. Ferdinand Hochstetter (1861-1954) for his pioneering description of the development of the human brain during the first trimester. To this day, his publication that appeared in Vienna and Leipzig in 1919 is the best source for systematically showing the external features of the human brain at this stage, and the current work is built on the foundation he provided.

# ACKNOWLEDGMENTS

We thank Dr. William DeMyer, pediatric neurologist at Indiana University Medical Center, for access to his personal library on human central nervous system development. We also thank the staff of the National Museum of Health and Medicine at the Armed Forces Institute of Pathology, Walter Reed Hospital, Washington, D.C.: Dr. Adrianne Noe, Director; Archibald J. Fobbs, Curator of the Yakovlev Collection; Elizabeth C. Lockett; and William Discher. We are most grateful to Dr. James M. Petras at the Walter Reed Institute of Research who made his darkroom facilities available so that we could develop all the photomicrographs on location rather than in our laboratory in Indiana. Finally, we thank publisher Barbara Norwitz, project manager Jim McGovern, proofreader Samar Haddad, Randy Brehm, and Kari Budyk at CRC Press/Taylor and Francis for their personal attention to us and for expert help during production of the manuscript.

# CONTENTS

# CONTENTS

# CONTENTS

# CONTENTS

# CONTENTS

# PART I

# INTRODUCTION

## A. Organization of the Atlas

This Atlas focuses on the development of the human brain during the late first trimester, and is Volume 4 in the *Atlas of Human Central Nervous System Development* series. Volume 1 (Bayer and Altman, 2002) provides a record of the development of the spinal cord from the 4th gestational week (GW4) to the 4th postnatal month. Volume 2 (Bayer and Altman, 2004a) records brain development during the third trimester, with the specimens ranging in age from GW37 to GW26. Volume 3 (Bayer and Altman, 2005) presents brain development during the second trimester, from GW24 to GW13.5. The specimens dealt with in the present volume cover the period from GW11 to GW7.5. The major theme of Volume 2 is the *maturation* of the brain's *settled* and *enduring* neuron populations. The major theme of Volume 3 is the *migration, sojourning,* and *settling* of neuronal populations. Prominent migratory streams and sojourning structures are evident by GW13.5 and many of these are still present at GW24. The major theme of the present volume is *late neurogenesis* and *early neuronal differentiation.* The structures of central interest are the visually distinguishable divisions, or mosaics, of the germinal neuroepithelium that generate different populations of neurons and neuroglia. The major theme of the forthcoming Volume 5, which will deal with brain development during the early first trimester (GW3-GW7), will be *neuroepithelial expansion* along the *expanding brain ventricles,* and *early neurogenesis.* As in Volumes 2 and 3, the specimens are presented in a reverse, older-to-younger order. We do that for both heuristic and pedagogical reasons: proceeding from the familiar and better known to what is less familiar and often uncertain or hypothetical.

The present volume contains grayscale photographs of nine sets of Nissl-stained, and one set of Bodian-stained sections, of normal brains ranging in age from early fetuses,

crown-rump (CR) length 57-60 mm and estimated age GW11, to late embryos, CR length 21-23 mm and estimated age GW7.5. As in the previous volumes, we sought to select for each age group, brains sectioned in the coronal, horizontal, and sagittal planes. However, this posed a problem as the terms "coronal" and "horizontal" (but not "sagittal") become more and more ambiguous when applied to younger and younger fetal brains and then embryonic brains. The conventional neuroanatomical designation of "coronal" comes from the practice of placing the dissected brain on a horizontal surface, and with the base of the cerebral cortex and the cerebellar cortex lying flat on that surface, making transverse (left-to-right) cuts perpendicular to the supporting surface. In a similar manner, "horizontal" sections are made by making the cuts parallel to the surface. With this convention, there is little ambiguity, for instance, that the frontal lobe is situated anteriorly and the occipital lobe posteriorly. However, this practice has not been consistently followed by embryologists; indeed, it could not be followed when the brains are not shelled out and the embryo is sectioned *de toto*. Moreover, since the whole embryo, and more particularly the medulla, pons and brainstem (as described in this volume) undergo a series of flexures, a coronal cut made at one point becomes an oblique or horizontal cut at another point. The same applies to horizontal cuts. Indeed, perusal of embryological textbooks reveals little consistency in the positioning of an embryonic brain or series of brains in the illustrations. In order to produce consistency in presenting the human brains of different fetal and embryonic ages, and facilitate comparisons of the changes across all the ages, we have therefore adopted the following procedure. Each brain, irrespective whether it was designated in the original protocol as "coronal" or "horizontal," has been placed into an X-Y coordinate frame as if it were a mature brain; i.e., the ventral surface of the cortex and the cerebellum lie flat on an imaginary line that we call the "cardinal horizontal plane." The line perpendicular to it, we call the "cardinal coronal plane." Given this coordinate system we then indicate the deviation of a given brain from the cardinal coronal or cardinal horizontal. This method allows a consistent positioning of the telencephalon, diencephalon, mesencephalon, and cerebellum (but not of the pons, medulla and spinal cord) into an age-independent stereotactic framework, and designations like anterior and posterior, and dorsal and ventral pointing in the same direction for easy comparisons across ages.

Each specimen is presented in a separate part of the Atlas: GW11 in the coronal plane in **Part II**; GW11 in the horizontal plane in **Part III**; GW9 in the sagittal plane in **Part IV**; GW9 in the horizontal plane in **Part V**; GW9 in the coronal plane in **Part VI**; GW8 in the sagittal plane in **Part VII**; GW8 in the coronal plane in **Part VIII**; GW8 in the horizontal plane in **Part IX**; GW7.5 in the coronal plane in **Part X**; and GW7.5 in the sagittal plane in **Part XI**. Selected "coronal" plates are presented from rostral to caudal in portrait orientation. The dorsal part of each section is toward the top of the page, the ventral part at the bottom, and the midline is in the vertical center of each section. Sagittal plates are presented from medial to lateral in portrait orientation. The anterior part of each section is facing left, posterior right, dorsal top, and ventral bottom. "Horizontal" plates are presented from rostral to caudal in landscape orientation. The anterior part of each section is facing to the left (bottom of page), posterior to the right (top of page), and the midline is in the horizontal center of each section. Each part contains *companion plates*, designated as **A** and **B** on facing pages; some of the plates are expanded into **C** and **D** parts. Parts **A** and **C** of each plate on the left page show the full contrast photograph without labels; parts **B** and **D** show low contrast copies of the same photograph on the right page with superimposed outlines of structures and unabbreviated labels. The *low magnification plates* show entire sections to identify the large structures of the brain, such as the various lobes and gyri of the cerebral cortex, and large subdivisions of the brain core, such as the basal ganglia, thalamus, hypothalamus, midbrain, pons, cerebellum, and medulla. The *high magnification plates* feature enlarged views of the brain core to identify smaller structures. For ease of interpretation in all plates, the ventricles are labeled in **CAPITALS**, the neuroepithelium and other germinal zones in **Helvetica bold**, transient structures in *Times bold italic*, and permanent structures in Times Roman or **Times bold**. Fixation artifacts and processing damage are usually outlined with *dashed lines* in parts **B** or **D** of each plate. Finally, an alphabetized **Glossary** gives brief definitions of most labels used in the plates with expanded definitions of all transient developmental structures.

## B. Specimens

A total of 10 specimens are illustrated and annotated in this Volume. All of them are from the Collections of human embryos or fetal brains currently kept at the National Museum of Health and Medicine, Armed Forces Institute of Pathology, Washington, D.C. One brain, estimated age GW11, is from the Yakovlev Collection and is designated, using the prefix Y, as Y1-59. We give a brief description of the Yakovlev Collection in Volume 2 (Bayer and Altman, 2004a) and more detail is provided in Haleem (1990). Eight specimens are from the Carnegie Collection and are designated by their respective numbers with the added prefix C. The Carnegie Collection was started by Franklin P. Mall (1862-1917) and expanded at the Carnegie Institution of Washington under the direction of George L. Streeter (1873-1948) and George W. Corner (1889-1981). One specimen, estimated age GW9, is from the Minot Collection and is designated, using the prefix M, as M841. The Minot Collection is named after Charles S. Minot (1852-1914), who collected and prepared more than 1900 embryos of different animal species, and about 100 human embryos, close to a century ago. We made use of models of the embryonic and fetal brains prepared in the Carnegie Institution, and by Hochstetter (1919).

## C. Photography and Computer Processing

Selected sections of the Yakovlev specimen (Y1-59) were photographed at low magnification, using a macro lens (Vivitar, 55mm 1:2.8 auto macro) attached to a Nikkormat 35mm camera that was mounted on a stand over a fluorescent light board. The Carnegie and Minot specimens and all higher magnification views were photographed using either an Olympus photomicroscope or a Wild phototmakroskop. The magnification varied for each specimen according to the size of the dissected brain or entire fetal head: the section with the largest area that could be accommodated within the field of view set the magnification for all sections of a particular specimen. All photographs were taken with a green filter to increase the contrast of the black and white film (Kodak technical pan #TP442). The film was developed at 20°C for 6 to 7 minutes in Kodak HC110 developer (dilution F), followed by Kodak stop bath for 30 seconds, Kodak fixer for 5 minutes, Kodak hypo clearing agent for 1 minute, running water rinse for 10 minutes, and a brief rinse in Kodak photoflo before drying.

The negatives were scanned at 2700 dots-per-inch (dpi) with a Nikon Coolscan-1000 35mm film scanner which was interfaced to a PowerPC G3 Macintosh computer running Adobe Photoshop (version 5.02) with a plug-in Nikon driver. To capture the subtle shades of gray, the negatives were scanned as color positives, inverted, and converted to grayscale. Using the enhancement features built into Adobe Photoshop and the additional features of Extensis

Intellihance, adjustments were made to increase contrast and sharpness. When the image resolution was set to 300 dpi, a full-size photographic file printed at approximately 12-10 inches. Images are shown at slightly reduced full size on separate pages. Adobe Illustrator was used to superimpose labels and outline structural details on low contrast copies of the Adobe Photoshop files. The Plates were placed into a book-form layout using Adobe InDesign. Finally, camera-ready files were provided to Taylor and Francis in Adobe portable document format.

The entire brain and upper cervical spinal cord of C966 was three-dimensionally reconstructed to show the large superventricles and the surface features of the neuroepithelium in the telencephalon, diencephalon, and rhombencephalon (**Figures 10-19, Part X**). The brain reconstructions (**Figures 10-15**) involved five steps. *First*, photographs of serial sections were made throughout the entire brain; the negatives were scanned and converted to computer files as described in the preceding paragraph. *Second*, all the files of sections selected for the reconstruction were placed into one large Photoshop file that contained a separate photograph in each layer. By altering the visibility and transparency of these layers the sections were aligned to each other as they were before sectioning. Then each layer was saved as a separate file. *Third*, Adobe Illustrator was used to outline the brain surface and the edge of the ventricles, and the outlines of each section were saved in separate Adobe Illustrator eps (encapsulated postscript) files. *Fourth*, the eps files were imported into 3D space (x, y, and z coordinates) using Cinema 4DXL (C4D, Maxon Computer, Inc.), a modeling and animation software package. For each section, points on the outlines have unique x-y coordinates and the same z coordinate. By calculating the distance between sections, the entire array of outlines was stretched out in the z axis. The outlines were segregated into two groups, one for the brain surface, the other for the ventricular surface. The C4D loft tool builds a "skin" for each group of outlines by creating a spline mesh of polygons. The polygons start from the x-y points on the first outline with the most anterior z coordinate, to the x-y points on the next outline behind it, and finish with the x-y points on the last outline at the most posterior z coordinate. The spline meshes were rendered either as completely opaque or partly transparent surfaces using the C4D ray-tracing engine. Selected surfaces can be made either invisible or visible using the various options in C4D. The complex structure of the brain requires that the surface be built in several different lofts to avoid twisting the surface in the regions of the brain flexures. When these different loft segments are shown together, there are a few unavoidable artifacts, such as surface indentations and changes in the way light reflects from the surface; some of these are labeled in **Figures 10-19**. *Fifth*, the rendered images were converted to Adobe Photoshop files, and Adobe Illustrator was used to label the structures and to draw thin lines on some of the surfaces to make the images easier to under-

stand. The reconstruction of the parts of the neuroepithelium (**Figures 16-19**) followed the same five steps except that only the selected parts of the neuroepithelium in each section were outlined, and these were always rendered as completely opaque blocks of tissue.

## D. Identification of Immature Brain Structures

In contrast to the mature human brain, there are no comprehensive textbooks or atlases available on the entire course of the human brain development. The few publications that cover the late first trimester give overviews that identify the largest brain structures in chapters included in embryology textbooks (e.g., Patten, 1953; Hamilton et al., 1964) and atlases (Gasser, 1975; O'Rahilly and Müller, 1994). Some facets of human brain development during this period are part of an overview in an edited publication (Sidman and Rakic, 1982), and there are a few research papers available on the development of specific brain regions (e.g., Gilbert, 1935; Pearson, 1941; Kappers, 1958; Humphrey, 1960, 1968; Richter, 1965; Kahle, 1956; Hewitt, 1958; Shuangshoti and Netsky, 1966; Rakic and Sidman, 1970; Moore et al., 1999; Forutan et al., 2001; Koutcherov et al., 2002). There is no single publication that labels both large and small structures in the developing human brain in serially sectioned specimens at several different ages during the late first trimester. Many of the transient features of the early fetal/late embryonic human brain have never been studied. Therefore, we rely to a great extent on our experimental work in the developing rat brain to identify formative brain structures, migratory streams, and components of the germinal neuroepithelium. In those studies, we labeled proliferating cells in the germinal matrices of the brain with [3]H-thymidine administered at daily intervals through the entire prenatal period, and every other day up to weaning. By varying survival times after [3]H-thymidine exposure from 1 hour, to days, we used the autoradiographic technique to establish quantitative timetables of neurogenesis, trace the speed and route of neuronal migrations, and document the settling patterns of neurons in the maturing brain. The results of these studies were published over a period of three decades in journal articles (see below) and reviewed in chapters contributed to edited books (Altman and Bayer, 1975, 1995, 2004; Altman, 1992; Bayer and Altman, 1995a, 1995b). We have also presented evidence for many parallels in the sequential order of brain development in rat and man (Bayer et al., 1993). Readers interested in details of those experimental results and the rationale of some of the anatomic and morphogenetic identifications made in this Atlas may consult the following publications.

*Amygdala*: Bayer (1980c).
*Basal Ganglia*: Bayer (1984, 1985b, 1987).
*Cerebellum*: Altman and Bayer (1978a, 1982, 1985a, 1985b, 1985c, 1997).

*Cerebral Cortex*:  Altman and Bayer (1990a, 1990b); Bayer and Altman (1990, 1991a, Bayer et al., 1991a).
*Cranial Nerve Nuclei*:  Altman and Bayer (1980a, 1980b, 1980c, 1982b).
*Hippocampus*:  Altman and Bayer  (1975, 1990c, 1990d, 1990e); Bayer (1980a, 1980b).
*Hypothalamus*:  Altman and Bayer  (1978c, 1978d, 1978e, 1986).
*Medulla*:  Altman and Bayer (1980a, 1980b, 1980c, 1982b).
*Midbrain*:  Altman and Bayer  (1981a, 1981b, 1981c).
*Olfactory Bulb*:  Altman (1969); Bayer (1983).
*Pontine Area*:  Altman and Bayer  (1978b, 1980d, 1987a, 1987b, 1987c, 1987d).
*Preoptic Area*:   Bayer and Altman (1987).
*Rhinencephalon*:   Bayer (1985a, 1986a, 1986b); Bayer and Altman (1991b).
*Septal Area*:  Bayer (1979a, 1979b).
*Spinal Cord*:  Altman and Bayer (1984, 2001).
*Thalamus*:  Altman and Bayer (1979a, 1979b, 1979c, 1988a, 1988b, 1988c, 1989a, 1989b, 1989c).

## E.  Major Transitional Brain Structures in the Late First Trimester

Transitional brain structures in late first trimester fetuses include the following:  *(i)* The neuroepithelium (NEP), the primary germinal matrix of neural stem cells that are the ultimate source of all the neurons and neuroglia of the central nervous system.  *(ii)* The rhombencephalic, mesencephalic, diencephalic, and telencephalic superventricles that provide the greatly expanded cerebrospinal fluid (CSF) interface for the shuttling nuclei of the pseudostratified NEP cells that undergo mitotic division near the lumen.  *(iii)* The expanding fetal choroid plexus (CP), and its later shrinkage and transformations, which appears to play a role in neurogenesis and neuronal differentiation.  *(iv)* The spreading glioepithelia (GEP) along several fiber tracts, and the glioepithelia associated with the formative ependymal layer that lines the enduring portions of the brain ventricles (G/EP).  *(v)* The secondary germinal matrices abutting the neuroepithelium, such as the subventricular zone (SVZ) of the cerebral cortex, and others situated some distance from the ventricles. *(vi)* Short-distance migratory routes, and long-distance migratory streams, such as the telencephalic rostral migratory and lateral migratory streams and several precerebellar migratory streams. *(vii)* The transient sojourn zones and stratified transitional fields, such as the cortical stratified transitional field (STF) and the cerebellar transitional field (CTF).

**The *Neuroepithelium and its Mosaic Organization.*** The NEP is a cell-dense pseudostratified proliferative matrix that lines the extensive ventricular system of the developing brain and spinal cord.  It used to be called the ependymal layer, but that term is no longer used because the ependyma is a highly specialized tissue that lines the

shrunken ventricles of the maturing and adult central nervous system. The NEP is composed of stem cells that either differentiate as specialized populations of neurons and neuroglia or give rise to secondary germinal matrices some distance from the ventricular system.  The latter include the SVZ of the cerebral cortex, the external germinal layer of the cerebellar cortex, and the subgranular zone of the hippocampal dentate gyrus.  At the outset of its development, the NEP stem cells are the sole constituents of the brain (to be described and illustrated in Volume 5).  The bulk of NEP cells either keep proliferating or generate differentiating (postmitotic) neurons that, following a precise spatio-temporal order, migrate over a short or long distance to form various brain structures.  While NEP cells all look alike in Nissl-stained preparations, discontinuous NEP "stretches" and "patches," or *mosaics*, can be distinguished along the ventricles in terms of matrix thickness (pseudostratified cell depth), cell packing density, and as growing or shrinking ventricular protuberances, invaginations and evaginations. Using experimental techniques in rats (short-survival, sequential survival, and long-survival $^3$H-thymidine autoradiography, respectively), we dated the changing proliferative dynamics of these NEP mosaics, tracked their migratory paths and settling patterns, and related that information to the time of origin of different neuronal populations in various brain regions and structures.  On the assumption that the neurons that form discrete parenchymal regions and structures (different lobes, ganglia, nuclei, etc.) in the neighborhood of these NEP mosaics are progeny of those mosaics, we name the latter as putative NEPs of those structures.  Examples of broader identifications of NEP "stretches" are frontal cortical NEP, hypothalamic NEP, thalamic NEP, and hippocampal NEP. An example of more specific identification of "patches" along the hippocampal NEP are subicular NEP (the mosaic that generates neurons of the subiculum), ammonic NEP (the mosaic that generates neurons the pyramidal cells of Ammon's horn), and dentate  NEP (the mosaic that generates the granule cells of the dentate gyrus).  In cases where a germinal matrix abuts a fiber tract, we identify that as a putative glioepithelium (GEP), e.g, the fornical GEP.  In addition to the cells that migrate short distances from their germinal sites of origin, there are others that migrate over long distances.  Examples are the cells of the rostral migratory stream in the anterior telencephalon that migrate to the olfactory bulb (Altman, 1969), and the intramural (parenchymal) and extramural (subpial) migratory streams of the rhombencephalon that form the precerebellar nuclei in the medulla and pons (Altman and Bayer, 1987a-d).  In such instances, we identify these NEP mosaics on the basis of the available experimental evidence of their ultimate destinations.  Finally, there are many sites where we have no information at all on the destination of the neurons leaving particular NEP mosaics and indicate that uncertainty either by a question mark following a tentative identification or by omitting any identification.

***Neurogenesis and its relation to the Expansion and Shrinkage of the Superventricles***. The NEP originates as an open proliferative sheet, the neural plate, without its own fluid environment. As we shall illustrate in Volume 5, the ventricles begin to form at about GW3, after the closure of the neural tube (the future spinal cord) and the cephalic vesicles (the future brain) is completed. Following that momentous morphogenetic event, the ventricles expand enormously during the rest of the first trimester and, with differences among the different ventricles, then begin to shrink during the second or third trimester, and gradually assume their mature size and configuration. The first lumen to expand is the rhombencephalic ventricle. Beginning as a shallow invagination beneath a thin membrane (the medullary velum) at about GW4, it expands greatly as a cavernous cistern is formed by the flexures of the medulla, pons and brainstem, underneath the cover of the medullary velum. As illustrated in this Volume, the expansion of the rhombencephalic ventricle continues until about GW9, and then it starts to shrink and gradually assumes the form of the familiar 4th ventricle. Next, the mesencephalic ventricle (the future aqueduct) and the diencephalic ventricle (the future 3rd ventricle) start to expand and, after a shorter developmental period, begin to shrink. Finally, at about GW5.5, the lateral ventricles begin to form as symmetrical balloon-like fluid compartments of the midline diencephalic ventricle and expand enormously up to GW11; thereafter the lumen of the lateral ventricles begin to shrink as the basal ganglia and the cerebral cortex develop. We name these greatly expanded embryonic cisterns *superventricles* to distinguish them from the greatly diminished and transformed mature ventricles for the following reasons.

(1) Unlike the shrunken mature ventricles, the cavernous embryonic ventricles are lined by proliferating NEP cells rather than differentiated ependymal cells. This raises the possibility that the embryonic and mature ventricles and the CSF they contain serve different functions.

(2) We know since the pioneering studies of Sauer (1936) that the NEP is a pseudostratified epithelium and that the nuclei of NEP cells shuttle to the ventricular lumen before undergoing mitosis (Sauer called the process interkinetic nuclear migration). This shuttling to the lumen as a prerequisite for cell division means that the rate of NEP cell proliferation is limited by the length of the ventricular shoreline because the elongated cells straddling the depth of the pseudostratified NEP cannot divide unless there is room for them to move to the NEP/CSF interface. An interesting point in this context is the spatial orientation of the dividing cells near the lumen. It has been noted in the past (e.g., Bayer and Altman, 1991a) that the cleavage plane of these mitotic cells may be radial, or vertical (tangential to the ventricular lining) or horizontal (parallel to the ventricular lining). Vertically cleaving cells occupy double of the NEP/CSF interface when compared with hor-

izontally cleaving cells. It has been hypothesized recently that vertical cleavage results in symmetrical cell division, and horizontal cleavage in asymmetrical cell division (e.g., Chenn and McConnell, 1995; Kornack and Rakic, 1995). Symmetric cell division is assumed to produce two proliferative NEP cells lying next to one another along the ventricle; collectively that should result in an expansion of the ventricular shoreline. Asymmetric cell division is assumed to produce a postmitotic daughter cell, presumably the one farther from the ventricle, a differentiating cell that is ready to leave the NEP. Some evidence has been presented recently of an increase in asymmetric division in the cortical NEP of mice as a function of increasing fetal age (Estivill-Torrus et al., 2002). Asymmetric division should not affect the length of NEP/CSF interface. The shrinkage of the NEP during fetal development would take place if more and more NEP cells lost their mitotic potency and their nuclei no longer returned to the ventricular lumen.

(3) Why is it obligatory for NEP cell nuclei to move to the ventricular lumen to divide when the stem cells of secondary germinal matrices, like the subpial external germinal layer of the cerebellum or the subgranular zone of the hippocampus, divide some distance from the ventricles? One possibility is that the embryonic CSF contains some factor or factors necessary for primary (pluripotent) NEP cell division. This is supported by reports that embryonic CSF promotes NEP cell survival, proliferation, and neurogenesis in experimental animals (Mashayekhi et al., 2002; Gato et al., 2005).

(4) The distinctiveness of the two ventricular systems is suggested by a different functional relationship between the *superventricles* and the *fetal CP*, and the enduring *ventricles* and the *mature CP*. To begin with, the initial expansion of the rhombencephalic and the telencephalic superventricles starts before the CP forms (this will be documented in Volume 5 of this series). Hence, unlike the CSF of the mature ventricular system, the CSF of the embryonic superventricles must originate from some other source than CP secretion.

(5) Moreover, the fetal CP that forms later and rapidly expands to fill the rhombencephalic and telencephalic superventricles (as illustrated in this Volume) has a different cellular organization than the mature CP (Kappers, 1958; Tennyson and Pappas, 1964; Shuangshoti and Netsky, 1966; Dohrmann, 1970; Dziegielewska et al., 2001; Johansson et al., 2005). The adult choroid plexus is a distinctive frond-like tissue composed of a monolayer of differentiated cuboidal cells that surround a capillary core. The exposed surface of these cuboidal cells is covered by a rich meshwork of microvilli and some cilia, and the cell interior is filled with mitochondria. These features, and added evidence, is the basis of the widely held view that the mature CP is a secretory tissue involved in CSF production. In contrast, the fetal CP is a smooth, multilayered

(pseudostratified) epithelium composed of spindle-shaped cells that have a simple exposed surface and contain few mitochondria. Unlike the mature CP cells, these fetal CP cells are filled with glycogen. Hence, it has been suggested that the principal function of the embryonic CP is the glycolytic (anaerobic) support of NEP cell proliferation and neurogenesis rather than the production of CSF.

(6) Finally, and most importantly, the chronological differences in the expansion, configurational transformations, and persistence of the different superventricles can be related to regional differences in the kinetics, date of origin, and time span of cell proliferation in stretches and patches of the NEP lining that produce neurons and neuroglia for different brain structures. For instance, the dorsal roof and lateral wall of the telencephalic superventricle is formed by an extensive, continuous dome-like stretch of NEP that generates an immense number and homogeneous set of cortical neurons in a precise sequential order (infragranular layer cells first, granular layer cells next, and supragranular layer cells last) through its entire extent and over an extended period of time. These neurogenetic features account for the smooth, spherical configuration of the NEP lining of the telencephalic superventricle over an extended period during embryonic and fetal development. In contrast, the ventral and medial shores of the telencephalic superventricles are much less regular because discrete NEP stretches or patches generate neurons here for diverse brain structures – septum, basal ganglia, nucleus accumbens, basal telencephalon, amygdala, hippocampus, etc. – each with different neurogenetic timetables, cell compositions, and population sizes. At these sites, the expanding and shrinking NEP mosaics (and the differentiating parenchymal structures associated with them) produce variably shaped ventricular pools, recesses, and narrows. Still more complex are the configurational changes over time in the diencephalic, mesencephalic, and rhombencephalic superventricles. This is so because at these locations short NEP patches give rise to a multitude of structurally and functionally different brain nuclei, many of which are composed of a relatively small number of neurons generated over a short time span. For instance, we can distinguish along the diencephalic superventricle not only NEP stretches that give rise to the thalamus, epithalamus, subthalamus and hypothalamus, but also patches – temporary evaginations and invaginations – that generate neurons, for instance, for the distinct thalamic nuclei. These expanded and short-lived shorelines create an expanded NEP/CSF interface for the optimal generation of neural stem cells of a particular type.

*Glioepithelia and the Ependymal Linings of the Enduring Ventricles.* Glioepithelia (GEP) are fate-restricted tissues of proliferative cells that produce neuroglia. On the basis of experimental studies in rats, in which proliferative cells are tagged with $^3$H-thymidine, we distinguish four types of GEP. The first is difficult to distinguish from the primary NEP lining the ventricles except that these patches tend to be thin and display proliferative activity for some time after local neurogenesis has ended. These patches begin to appear caudally in the older fetuses of the age group covered in this Volume, and because at most of these sites an ependymal lining will be forming around the enduring ventricles, they are designated uncertainly as glioepithelium/ependyma (G/EP). Examples are the medullary, pontine, hypothalamic, and thalamic G/EP. The second type lines fiber tracts that are devoid of neuronal cell bodies. In the older fetuses of this age group this type is exemplified by the fornical GEP, a continuation of the hippocampal NEP. The fornical GEP probably contains precursors of oligodendrocytes. The third type is the perifascicular GEP that surrounds and penetrates large fiber tracts. In the older specimens of this age group these are seen around the olfactory tract and the optic nerve, chiasm and tract. The fourth type, the subpial GEP, is found as a covering of the telencephalon, known as the subpial granular layer (Brun, 1965). The perifascicular and subpial GEPs may be continuous and of placodal rather than neuroepithelial origin.

*Secondary Germinal Matrices, Migratory Streams, and Sojourn Zones.* Prominent transitional structures of the developing brain during the second trimester, as illustrated in Volume 3 of this series, are the secondary germinal matrices, the prominent sojourn zones, and the large migratory streams. The principal secondary germinal matrices are the subventricular zone (SVZ) abutting the cortical NEP, the SVZ of the basal ganglionic (anterolateral, anteromedial and posterior) eminences, the subgranular zone of the hippocampal dentate gyrus, and the external germinal layer of the cerebellar cortex. All of these secondary matrices are late-generated, fate-restricted stem cells that divide some distance from the ventricles and produce microneurons (neurons with small cell bodies and locally ramifying axons); many of which are known as "granule cells." In the specimens we use, the time of emergence of the cortical SVZ could not be determined because of the difficulty of distinguishing it from the NEP, but it is clearly present by GW11. The basal ganglionic SVZ begins to form as early as GW7.5 and is prominent between the NEP and the parenchyma by GW8. The external germinal layer that originates in the germinal trigone of the cerebellar NEP is not recognizable until GW9. The hippocampal dentate NEP, which is the source of the dentate migration (and which will later form the subgranular zone) is present by GW9; however, the dentate gyrus is only a miniscule structure at GW11. In general, there is no evidence for the onset of microneuron production during this period.

There are long and short migratory streams that carry neural stem cells and/or immature neurons from the NEP to their destination. A prominent long-distance migration during the second trimester is the rostral migratory stream of the telencephalon which, among others, contains cells that differentiate as olfactory bulb granule cells. The rostral migratory stream is not clearly distinguishable at GW9 from the olfactory NEP, which generates the output neurons (mitral cells) of the olfactory bulb; it becomes recognizable as a distinct entity by GW11, and expands greatly during the second trimester (Volume 3). Among the long-distance precerebellar migratory streams, the posterior intramural migratory stream, which contains the neurons that migrate to form the prominent inferior olive, is evident by GW7.5, and remains so throughout the late first trimester. Also present in these specimens are the posterior extramural migratory stream that contains neurons of the lateral reticular and external cuneate nuclei, and the anterior extramural migratory stream that contains pontine gray neurons. However, the pontine gray nucleus is not evident until GW11 when the earliest corticofugal fibers begin to traverse it and the earliest fibers of the middle cerebellar peduncle begin to form.

In Volume 3 during the second trimester, we have illustrated the prominence of the cortical stratified transitional field (STF) sandwiched between the NEP and the cortical plate, the future gray matter. We have identified six distinct cellular and fibrous layers within the STF, where cortical neurons sojourn for some time and mingle with afferent, efferent, and commissural fibers before they resume their migration and settle in the cortical plate. We postulated that the STF is a staging area where connections are formed between unspecified cortical neurons and the functionally and topographically specified thalamocortical fiber systems that provide input to them. As seen in the present Volume, the cortical STF is developing during the late first trimester. STF1 and STF5 are first evident at GW8 in the earlier-maturing anterolateral cortical region spreading into the later-maturing dorsomedial cortical region. The formation of STF5 appears to coincide with the growth of thalamocortical fibers through the diencephalic-telencephalic junction and their penetration into the formative cerebral cortex. Between GW9 and GW11 an additional layer, STF4, begins to emerge slowly and uncertainly in the earlier maturing regions of the cortex (i.e., the lateral aspect of the anterior hemisphere). The emergence of STF4 may be associated with the onset of the descent of corticofugal fibers. The other STF layers (STF3, STF2, and STF6) do not start to form until the beginning of the second trimester (Volume 3).

Several other transitional fields are seen in the developing brain during this period. In the GW7.5-GW8 specimens, large cell aggregates surround the thalamic NEP that we interpret to be sojourning neuronal populations of the anterior, dorsal and ventral thalamic complexes. And, based on observations in rats, we assume that the posterior thalamic complex that will produce neurons for the lateral geniculate body and the pulvinar is initially in a dorsomedial position and then migrates gradually ventrolaterally toward the growing optic tract. The morphogenetic significance of the transitional cell columns and fiber bands seen in the GW9 thalamus, before the thalamic neurons assume their "nuclear" configuration, remains to be elucidated. Another region with alternating bands of sojourning cells and growing fibers is the cerebellar transitional field (CTF). Six layers are distinguished in the formative cerebellum in GW7.5 specimens from the surface toward the NEP: the fibrous CTF1, the cellular CTF2, CTF3 with cells and fibers, the cellular CTF4, CTF5 with cells and fibers, and the cellular CTF6. On the basis of experimental evidence in rats, we interpret the upper cellular layers (CTF2-4) as the earlier generated deep neurons that sojourn for a while superficially, and the lower cellular layers (CTF5-6) as consisting mainly of the later generated Purkinje cells. This stratification becomes blurred by GW8 and GW9 as the Purkinje cells migrate toward the surface and the deep neurons move back toward the core of the cerebellum. The upward migration of Purkinje cells is associated with the spreading of the external germinal layer over the surface of the formative cerebellar cortex. By GW11 most of the Purkinje cells are in a superficial position, where they form parasagittal bands beneath the continuous canopy of the external germinal layer.

## F. A Note on Genetic Analyses of NEP Mosaicism

As we noted earlier, the NEP cells that line the banks of the superventricles look alike in Nissl-stained preparations. The idea of NEP mosaicism is based on the site- and age-dependent distinctiveness of stretches and patches of the NEP in terms of their thickness, the temporary formation of larger protuberances and smaller evaginations and invaginations into the ventricle, and their differential proliferative dynamics, as ascertained with $^3$H-thymidine autoradiography. These spatiotemporal differences in the regional appearance and proliferative dynamics of NEP mosaics, and the fact that they give rise to different brain structures and cell types, raise the possibility that the stem cells composing them are different genetically before they start to differentiate. There is, indeed, emerging experimental evidence in embryonic and fetal mammals (mostly mice) for the genetic distinctiveness of some of the stretches and patches of NEP that are homologous with those we describe here in the first trimester human brain. The experiments are based mostly on *in situ* hybridization of homeodomain-containing transcription factors that are visualized in the developing brain of normal and mutant mice with immunohistochemical markers. Several laboratories, for instance, have reported a pronounced expression of *Pax6*, in combination with other transcription factors, in the dorsal neocortical NEP (e.g., Walther and Gruss, 1991; Stoykova and Gruss, 1994; Warren et al., 1999; Estivill-Torrus et al., 2002; Kimura et al., 2005). At the embryonic age when *Pax6* expression is limited to the dorsal telencephalon, *Nkx2.1* and *Gsh2* are expressed in the ventral telencephalon and the medial telencephalon (Corbin et al.

2003). Some differences were noted in the expression of these factors in the lateral, medial and posterior components of the ganglionic ventral telencephalon. According to another study, different markers highlight different components of the medial telencephalon and the diencephalon in 12.5 day-old mouse embryos (Kimura et al., 2005). Using a different terminology from that used by the authors, *Wnt8b* appears to demarcate the entire hippocampal primordium in the medial cortex, and the epithalamus and the mammillary body in the diencephalon. Within the domain of the medial telencephalon, *Ephb1* demarcates the Ammonic NEP, *Wnt3a* the dentate NEP, and *TTR* the primordium of the telencephalic choroid plexus. Although as yet this genetic approach is limited to a single species in mammals, it is expected that it will shed considerable light on NEP cell heterogeneity and the regional differentiation of the embryonic and fetal brain in other animals and man.

## G. A Note on Functional Maturation

There is currently considerable scientific interest in the physiological maturation of the prenatal human brain, and the correlated issue of the mental status of the embryo and fetus is receiving much public attention. Studies in the first half of the twentieth century with aborted fetuses have indicated that embryos of about 20-21 mm CR length (corresponding to the GW7.5 specimens in the present volume) begin to reliably respond to tactile stimulation with *holokinetic* ("total pattern") body movements (Fitzgerald and Windle, 1942; Hooker, 1942). In GW10 fetuses (CR 48.5 mm) *ideokinetic* or isolated movements are also elicited, such as partial closure of the fingers (though not effective grasping), when the palm of the hand was stimulated (Humphrey, 1964). The more recent introduction of ultrasonic recording techniques has permitted the observation of the emergence of "spontaneous" fetal behavior in normal embryos and fetuses *in utero*. A pioneering study showed that the holokinetic "startle" response emerges as early as GW6, isolated arm and leg movements by GW7, and head rotation and hand and face contact beginning at about GW8 (de Vries, 1982, 1985). According to a more recent study with improved ultrasonic recording methods, isolated arm movements are more frequent than isolated leg movements during the first trimester, and head turning and hand to head contact do not occur with high frequency until the second trimester (Kurjak et al., 2005).

Are the late-embryonic and early-fetal movements reflex reactions mediated by lower-level spinal cord and brain stem mechanisms, or are they emerging voluntary activities carried out under higher-level cortical guidance, reflecting sentience? We have raised this question earlier in the context of our study of the development of the reflex circuitry of the human spinal cord (Altman and Bayer, 2001). Because the spinal cord substrate of the sensorimotor reflex arc begins to form between GW7 and GW8 – the collateral branches of dorsal root sensory nerves sprout and then reach the ventral horn motor neurons during this period – we proposed that the isolated limb movements displayed by embryos of that age are reflex reactions. The morphogenetic evidence presented in this Volume extends that inference by showing that the first trimester embryonic movements cannot be cortically mediated voluntary activities. First, the higher-level sensory channel, the medial lemniscal system relayed in the thalamus, does not reach the cortical plate (the future cortical gray matter) during this period. The internal capsule that contains the thalamocortical fibers is not yet evident at GW7.5 either in the lateral border of the thalamus or the narrow parenchymal bridge that links the diencephalon and the telencephalon. A large collection of thalamocortical fibers "funnel" through the internal capsule by GW8, passing through the basal ganglia and approaching the base of the formative cerebral cortex. These fibers begin to penetrate STF4 of the paracentral lobule by GW9 and that process continues through GW11. It is probable, but this needs to be experimentally verified, that the neurons of layer IV of the cortex, the principal target of thalamocortical fibers, are either still being generated or are still in the STF5 sojourn zone. Thus, there is as yet no functional connection between the thalamus and cells of the cortical plate. The second consideration concerns corticofugal output to the brain stem and spinal cord, which is a prerequisite for cortically mediated motor control. The pontine gray, through which the corticospinal tract descends to the spinal cord, is not evident at GW9; it begins to form about GW11. In one such specimen illustrated (Y1-59; CR 60 mm), but not in the other (C1500; CR 57 mm), a bundle of corticofugal fibers traverses the pons. Although the earliest contact between thalamocortical fibers and differentiating cortical neurons may be established during this period, we have shown earlier that the lateral and ventral corticospinal tracts do not penetrate the spinal cord until the second trimester (Altman and Bayer, 2001: Figs. 7-23, 7-24, 7-35) and are still unmyelinated at birth (Altman and Bayer, 2001: Fig. 8-14).

## H. References

Altman, J. (1969) Autoradiographic and histological studies of postnatal neurogenesis. IV. Cell proliferation and migration in the anterior forebrain, with special reference to persisting neurogenesis in the olfactory bulb. *Journal of Comparative Neurology*, 137:433-458.

Altman, J. (1992) The early stages of nervous system development: Neurogenesis and neuronal migration. In: A. Björklund, T. Hökfelt and M. Tohyama (eds.) *Handbook of Chemical Neuroanatomy*. Volume 10. *Ontogeny of Transmitters and Peptides in the CNS*, pp. 1-31. Amsterdam: Elsevier.

Altman, J. and S. A. Bayer. (1975) Postnatal development of the hippocampal dentate gyrus under normal and experimental conditions. In: R. L. Isaacson and K. H. Pribram (eds.), *The Hippocampus.* Vol. 1. *Structure and Development*, pp. 95-122. New York: Plenum Press.

Altman, J. and S. A. Bayer (1978a) Prenatal development of the cerebellar system in the rat. I. Cytogenesis and histogenesis of the deep nuclei and the cortex of the cerebellum. *Journal of Comparative Neurology,* 179:23-48.

Altman, J. and S. A. Bayer (1978b) Prenatal development of the cerebellar system in the rat. II. Cytogenesis and histogenesis of the inferior olive, pontine gray, and the precerebellar reticular nuclei. *Journal of Comparative Neurology,* 179:49-76.

Altman, J. and S. A. Bayer (1978c) Development of the diencephalon in the rat. I. Autoradiographic study of the time of origin and settling patterns of neurons of the hypothalamus. *Journal of Comparative Neurology,* 182:945-972.

Altman, J. and S. A. Bayer (1978d) Development of the diencephalon in the rat. II. Correlation of the embryonic development of the hypothalamus with the time of origin of its neurons. *Journal of Comparative Neurology,* 182:973-994.

Altman, J. and S. A. Bayer (1978e) Development of the diencephalon in the rat. III. Ontogeny of the specialized ventricular linings of the hypothalamic third ventricle. *Journal of Comparative Neurology,* 182:995-1016.

Altman, J. and S. A. Bayer (1979a) Development of the diencephalon in the rat. IV. Quantitative study of the time of origin of neurons and the internuclear chronological gradients in the thalamus. *Journal of Comparative Neurology,* 188:455-472.

Altman, J. and S. A. Bayer (1979b) Development of the diencephalon in the rat. V. Thymidine-radiographic observations on internuclear and intranuclear gradients in the thalamus. *Journal of Comparative Neurology,* 188:473-500.

Altman, J. and S. A. Bayer (1979c) Development of the diencephalon in the rat. VI. Re-evaluation of the embryonic development of the thalamus on the basis of thymidine-radiographic datings. *Journal of Comparative Neurology,* 188:501-524.

Altman, J. and S. A. Bayer (1980a) Development of the brain stem in the rat. I. Thymidine-radiographic study of the time of origin of neurons of the lower medulla. *Journal of Comparative Neurology,* 194:1-35.

Altman, J. and S. A. Bayer (1980b) Development of the brain stem in the rat. II. Thymidine-radiographic study of the time of origin of neurons of the upper medulla, excluding the vestibular and auditory nuclei. *Journal of Comparative Neurology,* 194:37-56.

Altman, J. and S. A. Bayer (1980c) Development of the brain stem in the rat. III. Thymidine-radiographic study of the time of origin of neurons of the vestibular and auditory nuclei of the upper medulla. *Journal of Comparative Neurology,* 194:877-904.

Altman, J. and S. A. Bayer (1980d) Development of the brain stem in the rat. IV. Thymidine-radiographic study of the time of origin of neurons in the pontine region. *Journal of Comparative Neurology,* 194:905-929.

Altman, J. and S. A. Bayer (1981a) Time of origin of neurons of the rat inferior colliculus and the relations between cytogenesis and tonotopic order in the auditory pathway. *Experimental Brain Research,* 42:411-423.

Altman, J. and S. A. Bayer (1981b) Time of origin of neurons of the rat superior colliculus in relation to other components of the visual and visuomotor pathways. *Experimental Brain Research,* 42:424-434.

Altman, J. and S. A. Bayer (1981c) Development of the brain stem in the rat. V. Thymidine-radiographic study of the time of origin of neurons of the midbrain tegmentum. *Journal of Comparative Neurology,* 198:677-716.

Altman, J. and S. A. Bayer (1982a) Morphological development of the rat cerebellum and some of its mechanisms. In: S. L. Palay and V. Chan-Palay (eds.). *The Cerebellum: New Vistas*, pp. 8-49. Berlin: Springer-Verlag.

Altman, J. and S. A. Bayer (1982b) *Development of the Cranial Nerve Ganglia and Related Nuclei in the Rat.* (*Advances in Anatomy, Embryology and Cell Biology,* Vol. 74). Berlin: Springer-Verlag.

Altman, J. and S. A. Bayer (1984) *The Development of the Rat Spinal Cord.* (*Advances in Anatomy, Embryology and Cell Biology,* Vol. 85). Berlin: Springer-Verlag.

Altman, J. and S. A. Bayer (1985a) Embryonic development of the rat cerebellum. I. Delineation of the rat cerebellum and early cell movements. *Journal of Comparative Neurology,* 231:1-26.

Altman, J. and S. A. Bayer (1985b) Embryonic development of the rat cerebellum. II. Translocation and regional distribution of the deep neurons. *Journal of Comparative Neurology,* 231:27-41.

Altman, J. and S. A. Bayer (1985c) Embryonic development of the rat cerebellum. III. Regional differences in the time of origin, migration and settling of Purkinje cells. *Journal of Comparative Neurology,* 231:42-65.

Altman, J. and S. A. Bayer (1986) *The Development of the Rat Hypothalamus. (Advances in Anatomy, Embryology and Cell Biology,* Vol.100). Berlin: Springer-Verlag.

Altman, J. and S. A. Bayer (1987a) Development of the precerebellar nuclei in the rat. I. The precerebellar neuroepithelium of the rhombencephalon. *Journal of Comparative Neurology,* 257:477-489.

Altman, J. and S. A. Bayer (1987b) Development of the precerebellar nuclei in the rat. II. The intramural olivary migratory stream and the neurogenetic organization of the inferior olive. *Journal of Comparative Neurology,* 257:490-512.

Altman, J. and S. A. Bayer (1987c) Development of the precerebellar nuclei in the rat. III. The posterior precerebellar extramural migratory stream and the lateral reticular and external cuneate nuclei. *Journal of Comparative Neurology,* 257:513-528.

Altman, J. and S. A. Bayer (1987d) Development of the precerebellar nuclei in the rat. IV. The anterior precerebellar extramural migratory stream and the nucleus tegmenti pontis and the basal pontine gray. *Journal of Comparative Neurology,* 257:529-552.

Altman, J. and S. A. Bayer (1988a) Development of the rat thalamus. I. Mosaic organization of the thalamic neuroepithelium. *Journal of Comparative Neurology,* 275:346-377.

Altman, J. and S. A. Bayer (1988b) Development of the rat thalamus. II. Time and site of origin and settling pattern of neurons derived from the anterior lobule of the thalamic neuroepithelium. *Journal of Comparative Neurology,* 275:378-405.

Altman, J. and S. A. Bayer (1988c) Development of the rat thalamus. III. Time and site of origin and settling pattern of neurons of the reticular nucleus. *Journal of Comparative Neurology,* 275:406-428.

Altman, J. and S. A. Bayer (1989a) Development of the rat thalamus. IV. The intermediate lobule of the thalamic neuroepithelium, and the time and site of origin and settling pattern of neurons of the ventral nuclear complex. *Journal of Comparative Neurology,* 284:534-566.

Altman, J. and S. A. Bayer (1989b) Development of the rat thalamus. V. The posterior lobule of the thalamic neuroepithelium and the time and site of origin and settling pattern of neurons of the medial geniculate body. *Journal of Comparative Neurology,* 284:567-580.

Altman, J. and S. A. Bayer (1989c) Development of the rat thalamus. VI. The posterior lobule of the thalamic neuroepithelium and the time and site of origin and settling pattern of neurons of the lateral geniculate and lateral posterior nuclei. *Journal of Comparative Neurology,* 284:581-601.

Altman, J. and S. A. Bayer (1990a) Vertical compartmentation and cellular transformations in the germinal matrices of the embryonic rat cerebral cortex. *Experimental Neurology,* 107:23-35.

Altman, J. and S. A. Bayer (1990b) Horizontal compartmentation in the germinal matrices and intermediate zone of the embryonic rat cerebral cortex. *Experimental Neurology,* 107:36-47.

Altman, J. and S. A. Bayer (1990c) Mosaic organization of the hippocampal neuroepithelium and the multiple germinal sources of dentate granule cells. *Journal of Comparative Neurology,* 301:325-342.

Altman, J. and S. A. Bayer (1990d) The prolonged sojourn of developing pyramidal cells in the intermediate zone of the hippocampus and their settling in the stratum pyramidale. *Journal of Comparative Neurology,* 301:343-364.

Altman, J. and S. A. Bayer (1990e) The migration and distribution of two populations of hippocampal granule cell precursors during the perinatal and postnatal periods. *Journal of Comparative Neurology,* 301:365-381.

Altman, J. and S. A. Bayer (1990f) Mosaic organization of the hippocampal neuroepithelium and the multiple germinal sources of dentate granule cells. *Journal of Comparative Neurology,* 301:325-342.

Altman, J. and S. A. Bayer (1995) *Atlas of Prenatal Rat Brain Development.* Boca Raton, FL: CRC Press

Altman, J. and S. A. Bayer (1997) *Development of the Cerebellar System in Relation to Its Evolution, Structure, and Functions.* Boca Raton, FL: CRC Press.

Altman, J. and S. A. Bayer (2001) *Development of the Human Spinal Cord: An Interpretation Based on Experimental Studies in Animals.* New York, NY: Oxford University Press.

Altman, J. and S. A. Bayer (2002) Regional differences in the stratified transitional field and the honeycomb matrix of the developing human cerebral cortex. *Journal of Neurocytology,* 31:613-632.

Altman, J. and S. A. Bayer (2004) Neuroembryology. In: G. Adelman and B. H. Smith (eds.) *Encyclopedia of Neuroscience.* (Third edition) Amsterdam: Elsevier. (CD ROM)

Bayer, S. A. (1980a) Development of the hippocampal region in the rat. I. Neurogenesis examined with [³H]thymidine autoradiography. *Journal of Comparative Neurology,* 190:87-114.

Bayer, S. A. (1980b) Development of the hippocampal region in the rat. II. Morphogenesis during embryonic and early postnatal life. *Journal of Comparative Neurology,* 190:115-134.

Bayer, S. A. (1980c) Quantitative [³H]thymidine radiographic analyses of the neurogenesis in the rat amygdala. *Journal of Comparative Neurology,* 194:845-875.

Bayer, S. A. (1983) [³H]thymidine radiographic studies of neurogenesis in the rat olfactory bulb. *Experimental Brain Research,* 50:329-340.

Bayer, S. A. (1984) Neurogenesis in the rat neostriatum. *International Journal of Developmental Neuroscience,* 2:163-175.

Bayer, S. A. (1985a) Neurogenesis in the olfactory tubercle and islands of Calleja in the rat. *International Journal of Developmental Neuroscience,* 3:135-147.

Bayer, S. A. (1985b) Neurogenesis in the magnocellular basal telencephalic nuclei in the rat. *International Journal of Developmental Neuroscience,* 3:229-243.

Bayer, S. A. (1986a) Neurogenesis in the anterior olfactory nucleus and its associated transition areas in the rat brain. *International Journal of Developmental Neuroscience,* 4:225-249.

Bayer, S. A. (1986b) Neurogenesis in the rat primary olfactory cortex. *International Journal of Developmental Neuroscience,* 4:251-271.

Bayer, S. A. (1987) Neurogenetic and morphogenetic heterogeneity in the bed nucleus of the stria terminalis. *Journal of Comparative Neurology,* 265:47-64.

Bayer, S. A. and J. Altman (1987) Development of the preoptic area: Time and site of origin, migratory routes, and settling patterns of its neurons. *Journal of Comparative Neurology,* 265:65-95.

Bayer, S. A. and J. Altman (1990) Development of layer I and the subplate in the rat neocortex. *Experimental Neurology,* 107:48-62.

Bayer, S. A. and J. Altman (1991a) *Neocortical Development.* New York, NY: Raven Press.

Bayer, S. A. and J. Altman (1991b) Development of the endopiriform nucleus and the claustrum in the rat brain. *Neuroscience,* 45:391-412.

Bayer, S. A and J. Altman (1995a) Neurogenesis and neuronal migration. In: G. Paxinos (ed.) *The Rat Nervous System* (Second edition), pp. 1041-1078. San Diego, CA: Academic Press.

Bayer, S. A and J. Altman (1995b) Principles of neurogenesis, neuronal migration, and neural circuit formation. In: G. Paxinos (ed.) *The Rat Nervous System* (Second edition), pp. 1079-1098. San Diego, CA: Academic Press.

Bayer, S. A. and J. Altman (2002) *Atlas of Human Central Nervous System Development.* Volume 1: *The Spinal Cord from Gestational Week 4 to the 4th. Postnatal Month.* Boca Raton, FL: CRC Press.

Bayer, S. A. and J. Altman (2004a) *Atlas of Human Central Nervous System Development.* Volume 2: *The Human Brain during the Third Trimester.* Boca Raton, FL: CRC Press.

Bayer, S. A. and J. Altman (2004b) Development of the telencephalon: Neural stem cells, neurogenesis and neuronal migration. In: G. Paxinos (ed.) *The Rat Nervous System* (Third edition), pp. 27-73. San Diego, CA: Academic Press.

Bayer, S. A. and J. Altman (2005) *Atlas of Human Central Nervous System Development.* Volume 3. *The Human Brain during the Second Trimester.* Boca Raton, FL: CRC Press.

Bayer, S. A., J. Altman, X. Dai and L. Humphreys (1991b) Planar differences in nuclear area and orientation in the subventricular and intermediate zones of the rat embryonic neocortex. *Journal of Comparative Neurology,* 307:487-498.

Bayer, S. A., J. Altman, R. J. Russo, X. Dai and J. A. Simmons. (1991c) Cell migration in the rat embryonic neocortex. *Experimental Neurology,* 307:499-516.

Bayer, S. A., J. Altman, R. J. Russo, and X. Zhang (1993) Timetables of neurogenesis in the human brain based on experimentally determined patterns in the rat. *Neurotoxicology*, 14:83-144.

Bayer, S. A., J. Altman, R. J. Russo, and X. Zhang (1995) Embryology. In: S. Duckett (ed.) *Pediatric Neuropathology*, pp. 54-107. Baltimore, MD: Williams & Wilkins.

Brun, A. (1965) The subpial granular layer of the foetal cerebral cortex in man. *Acta Pathologica et Microbiologica Scandinavica*, Suppl. 179:1-98.

Chenn, A, and S. K. McConnell. (1995) Cleavage orientation and the asymmetric inheritance of Notch 1 immunoreactivity in mammalian neurogenesis. *Cell*, 82:631-641.

Corbin, J. G., M. Rutlin, N. Gaiano and G. Fischell (2003) Combinatorial function of the homeodomain proteins Nkx2.1 and Gsh2 in ventral telencephalic patterning. *Development*, 130:4895-4906.

De Vries, J. I. P., G. H. A. Visser and H. F. R. Prechtl (1982) The emergence of fetal behavior. I. Qualitative Aspects. *Early Human Development*, 7:302-322.

De Vries, J. I. P., G. H. A. Visser and H. F. R. Prechtl (1985) The emergence of fetal behavior. II. Quantitative aspects. *Early Human Development*, 12:99-120.

Dohrmann, G. J. (1970) The choroid plexus: A historical review. *Brain Research*, 18:197-218.

Dziegielewska, K. M., J. Ek, M. D. Habgood and N. R. Saunders. (2001) Development of the choroid plexus. *Microscopic Research Techniques*, 52:5-20.

Estivill-Torrus, G., H. Pearson, V. van Heyningen and D. J. Price (2002) *Pax6* is required to regulate the cell cycle and the rate of progression from symmetrical to asymmetrical division in mammalian cortical progenitors. *Development*, 129:455-466.

Fitzgerald, G. E. and W. Windle (1942) Some observations on early human fetal movements. *Journal of Comparative Neurology*, 76:159-167.

Forutan, F., J. K. Mai, K. W. S. Ashwell, S. Lensing-Höhn, D. Nohr, T. Voss, J. Bohl, and C. Andressen (2001) Organisation and maturation of the human thalamus as revealed by CD15. *Journal of Comparative Neurology*, 437:476-495.

Gasser, R. F. (1975) *Atlas of Human Embryos*. New York: Harper & Row.

Gato, A., J. A. Moro, M. I. Alonso, D. Bueno, A. De La Mano and C. Martin. (2005) Embryonic cerebrospinal fluid regulates neuroepithelial survival, proliferation, and neurogenesis in chick embryos. *Anatomical Record*, A, 284:475-484.

Gilbert, M. (1935) The early development of the human diencephalon. *Journal of Comparative Neurology*, 62:81-115.

Haleem, M. (1990) *Diagnostic Categories of the Yakovlev Collection of Normal and Pathological Anatomy and Development of the Brain*. Washington, D.C.: Armed Forces Institute of Pathology.

Hamilton, W. J., J. D. Boyd and H. W. Mossman (1964) *Human Embryology*. (3rd ed., revised) Baltimore, MD: Williams & Wilkins.

Hewitt, W. (1958) The development of the human caudate and amygdaloid nuclei. *Journal of Anatomy*, 92:377-382.

Hochstetter, F. (1919) *Beiträge zur Entwicklungsgeschichte des menschlichen Gehirns*. Vol. 1. Leipzig und Wien: Deuticke.

Hooker, D. (1942) *The Origin of Overt Behavior*. Ann Arbor, MI: University of Michigan Press.

Humphrey, T. (1960) The development of the pyramidal tracts in human fetuses, correlated with cortical differentiation. In: D. B. Tower and J. P. Schadé (eds.), *Structure and Functions of the Cerebral Cortex*. pp. 93-103. Amsterdam: Elsevier.

Humphrey, T. (1964) Some correlations between the appearance of human fetal reflexes and the development of the nervous system. *Progress in Brain Research*, 4:93-135.

Humphrey, T. (1968) The development of the human amygdala during early embryonic life. *Journal of Comparative Neurology*, 132:135-166.

Johansson, P. A., K. M. Dziegielewska, C. J. Ek, M. D. Habgood, K. Mollgard, A. Potter, M. Schuliga and N. R. Saunders (2005) Aquaporin-1 in the choroid plexuses of developing mammalian brain. *Cell and Tissue Research*, 322:353-364.

Kahle, W. (1956) Zur Entwicklung des menschlichen Zwischenhirnes: Studien über die Matrixphasen und die örtlichen Reifungsunterschiede im embryonalen meschlichen Gehirn. II. Mitteilung. *Deutsche Zeitschrift für Nervenheilkunde*, 175:259-318.

Kappers, J. Ariëns (1958) Structural and functional changes in the telencephalic choroid plexus during human ontogenesis. In: G. E. W. Wolstenholme and C. M. O'Connor (eds.) *The CIBA Foundation Symposium on the Cerebrospinal Fluid*, pp. 3-31. Boston, MA: Little, Brown.

Kimura, J., Y. Suda, D. Kurokawa, Z. M. Hossain, M. Nakamura and M. Takahashi (2005) *Emx2* and *Pax6* function in cooperation with *Otx2* and *Otx1* to develop caudal forebrain primordium that includes future archipallium. *Journal of Neuroscience*, 25:5097-5108.

Kornack, D. R. and P. Rakic. (1995) Radial and horizontal deployment of clonally related cells in the primate neocortex: Relationship to distinct mitotic lineages. *Neuron*, 15:311-321.

Koutcherov, Y., J. K. Mai, K. W. S. Ashwell, and G. F. Paxinos (2002) Organisation of human hypothalamus in fetal development. *Journal of Comparative Neurology*, 446:301-324.

Kurjak, A., M. Stanojevic, W. Andonotopo, E. Scazzocchio-Duenas, G. Azumendi and J. M. Carrera. (2005) Fetal behavior assessed in all three trimesters of normal pregnancy by four-dimensional ultrasonography. *Croatian Medical Journal*, 46:772-780.

Mashayekhi, F., C. E. Draper, C. M. Bannister, M. Pourghasem, P. J. Owen-Lynch and J. A. Miyan. (2002) Deficient cortical development in the hydrocephalic Texas (H-Tx) rat: A role for CSF. *Brain*, 125:1859-1874.

Moore, J. K., D. D. Simmons, and Y-L. Guan (1999) The human olivocochlear system: Organization and development. *Audiology and Neuro-otology*, 4:311-325.

O'Rahilly, R. and F. Müller (1994) *The Embryonic Human Brain. An Atlas of Developmental Stages*. New York: Wiley-Liss.

Patten, B. M. (1953) *Human Embryology*. 2nd ed. New York: McGraw-Hill.

Pearson, A. A. (1941) The development of the olfactory nerve in man. *Journal of Comparative Neurology*, 75:199-217.

Rakic, P. and R. L. Sidman (1970) Histogenesis of cortical layers in human cerebellum, particularly the lamina dissecans. *Journal of Comparative Neurology*, 139:473-500.

Richter, E. (1965) *Die Entwicklung des Globus Pallidus und des Corpus Subthalamicum*. Berlin: Springer-Verlag.

Sauer, F. C. (1936) The interkinetic migration of embryonic epithelial nuclei. *Journal of Morphology*, 60:1-11.

Sidman, R. L. and P. Rakic (1982) Development of the human central nervous system. In: W. Haymaker and R. D. Adams (eds.) *Histology and Histopathology of the Nervous System*, pp. 3-145. Springfield, IL: Thomas.

Shuangshoti, S. and M. G. Netsky (1966) Histogenesis of the choroid plexus in man. *American Journal of Anatomy*, 118:283-316.

Stoykova, A. and P. Gruss (1994) Roles of Pax-genes in developing and adult brain as suggested by expression patterns. *Journal of Neuroscience*, 14:1395-1412.

Tennyson, V. M. and G. D. Pappas (1964) Fine structure of the developing telencephalic and myelencephalic choroid plexus in the rabbit. *Journal of Comparative-Neurology*, 123:379-412.

Van Hartesveldt, C., B. Moore and B. K. Hartman (1986) Transient midline raphe glial structure in the developing rat. *Journal of Comparative Neurology*, 253:174-184.

Walther, C. and P. Gruss (1991) *Pax-6*, a murine paired box gene, is expressed in the developing CNS. *Development*, 113:1435-1449.

Warren, N., D. Caric, T. Pratt, J. A. Clausen, P. Asavaritikrai, J. O. Mason, R. E. Hill and D. J. Price (1999) The transcription factor, *Pax6*, is required for cell proliferation in the developing cerebral cortex. *Cerebral Cortex*, 9:627-635.

# PART II: GW11 CORONAL

This specimen is a stillborn female fetus with a crown-rump length (CR) of 60 mm estimated to be at gestational week (GW) 11 (Yakovlev case number RPSL-WX-1-59, referred to here as Y1-59). The brain was cut in the coronal (frontal) plane in 35-μm thick sections and is classified as a Normative Control in the Yakovlev Collection (Haleem, 1990). Since there is no photograph of this brain before it was embedded and cut, a specimen from Hochstetter (1919) that is comparable in age to Y1-59 is used to show the approximate section plane and external features of a GW11 brain (**Figure 1**). Photographs of 22 Nissl-stained sections (**Levels 1-22**) are shown at low magnification in **Plates 1-21**. **Plates 22-35** show high-magnification views of various parts of the brain from the cerebral cortex (**Plates 22-23**) to the midbrain, pons, and medulla (**Plates 33-35**). Several high-magnification plates are rotated 90° (landscape orientation) to more efficiently use page space. Y1-59 has more mature brain structures in the diencephalon, midbrain, pons, and medulla than any other specimen in this Volume. Immature brain structures predominate in the telencephalon and the cerebellum.

Throughout the cerebral cortex, the **neuroepithelium and subventricular zone** are prominent. The **stratified transitional field (STF)** contains mainly **STF1** and **STF5** throughout; with **STF4** only in lateral areas and a questionable **STF2** in a few areas; **STF6** and **STF3** are not present. The **STF** is filled with migrating and sojourning neurons and, unlike any specimen in Volume 3, *has no regional heterogeneity*. The cerebral cortex is completely smooth except for a questionable calcarine sulcus in the left cerebral hemisphere (**Plate 15**); "lobes" are identified as future lobes. The most prominent developmental feature of the cerebral cortex is that both the **STF** layers and the cortical plate have a pronounced lateral (thicker) to medial (thinner) maturation gradient. There is no corpus callosum. The olfactory bulb is beneath the anterior septum and striatum; it contains the **rostral migratory stream** in its core. In anterolateral parts of the cerebral cortex, streams of neurons and glia appear to leave **STF4** and enter the **lateral migratory stream**. The hippocampus is in an immature position dorsal to the thalamus and medial to the temporal lobe. Cells are entering Ammon's horn pyramidal layer in the **ammonic migration**, and granule cells and their precursors are migrating to the hilus of the presumptive dentate gyrus in the **dentate migration**; there is no granular layer. A massive **neuroepithelium/subventricular zone** overlies the amygdala, nucleus accumbens, and striatum (caudate and putamen) where neurons (and glia) are being generated. The caudate, the putamen, and basolateral parts of the amygdala are smaller than in the GW13.5 specimen in Volume 3, but are similar to that specimen because the **striatal neuroepithelium and subventricular zone** have indistinct subdivisions. The **strionuclear glioepithelium** forms definite continuities with the fornical glioepithelium in the telencephalon. Unlike the GW13.5 specimen, the septum in Y1-59 has a **neuroepithelium** at the ventricular surface instead of a **glioepithelium/ependyma**.

The cerebellum is a thick, smooth plate overlying the posterior pons and medulla. However, there is only a thin **glioepithelium/ependyma** at the ventricular surface, indicating that all deep neurons and Purkinje cells have been generated. The deep neurons are in place beneath the cortex, but have indefinite nuclear subdivisions. The cortical surface is covered by an **external germinal layer (egl)** that is actively producing neuronal stem cells, granule, stellate, and basket cells of the cerebellar cortex. Lamination in the cortex is nearly absent, except for a thin molecular layer beneath the **egl**. Nearly all Purkinje cells are migrating, some in discrete clumps. Lobulation has barely begun in the vermis and is nearly absent in the hemispheres. The **germinal trigone** is prominent in the dorsal rhombic lip.

The third ventricle, aqueduct, and fourth ventricle are lined by a thin **glioepithelium/ependyma** indicating that neurogenesis in the primary neuroepithelium is complete. In the medulla, there are two active germinal sites in anterior and posterior parts of the ventral rhombic lip. 1) The **auditory neuroepithelium** generates cochlear nucleus neurons. 2) A large **precerebellar neuroepithelium** generates precerebellar (mainly pontine gray) neurons.

Neurons throughout the diencephalon are settling in fairly well-defined nuclear divisions; the major exceptions are the immature appearance of the lateral and medial geniculate bodies in the posterior thalamus and the hypothalamic medial mammillary body. Neurons are settled in the midbrain tegmentum, pons, and medulla. But the pontine gray is nearly absent, and neurons are still migrating into it from the large **anterior extramural migratory stream**. **Posterior extramural and intramural migratory streams** contain lateral reticular neurons, external cuneate nucleus neurons, and inferior olive neurons. The corticospinal tract forms a small cerebral peduncle in the midbrain, but has not grown into the pons.

15

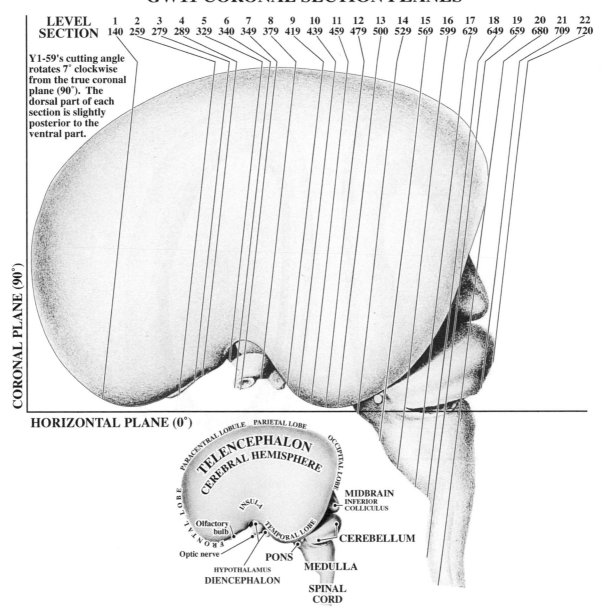

# GW11 CORONAL SECTION PLANES

| LEVEL | 1 | 2 | 3 | 4 | 5 | 6 | 7 | 8 | 9 | 10 | 11 | 12 | 13 | 14 | 15 | 16 | 17 | 18 | 19 | 20 | 21 | 22 |
|---|---|---|---|---|---|---|---|---|---|---|---|---|---|---|---|---|---|---|---|---|---|---|
| SECTION | 140 | 259 | 279 | 289 | 329 | 340 | 349 | 379 | 419 | 439 | 459 | 479 | 500 | 529 | 569 | 599 | 629 | 649 | 659 | 680 | 709 | 720 |

Y1-59's cutting angle rotates 7° clockwise from the true coronal plane (90°). The dorsal part of each section is slightly posterior to the ventral part.

CORONAL PLANE (90°)

HORIZONTAL PLANE (0°)

PARACENTRAL LOBULE — PARIETAL LOBE — OCCIPITAL LOBE

TELENCEPHALON
CEREBRAL HEMISPHERE

FRONTAL LOBE

INSULA — TEMPORAL LOBE

Olfactory bulb

Optic nerve

HYPOTHALAMUS
DIENCEPHALON

PONS

MEDULLA

SPINAL CORD

MIDBRAIN
INFERIOR COLLICULUS

CEREBELLUM

**Figure 1.** The lateral view of the brain and upper cervical spinal cord from a specimen with a crown-rump length of 68 mm (modified from Figure 47, Table VIII, Hochstetter, 1919) serves to show the approximate locations and cutting angles of the illustrated sections of Y1-59 in the following pages. The small inset identifies the major structural features. The cut beneath the cerebellum is the edge of the medullary velum.

## PLATE 1A

**GW11 Coronal
CR 60 mm, Y1-59
Level 1: Section 140**

### LAYERS OF THE CORTICAL *STRATIFIED TRANSITIONAL FIELD (STF)*

*STF1* Superficial fibrous layer with an early developmental stage *(t1)* when many cells are migrating through it, followed by a late stage *(t2)* with sparse cells. Endures as the subcortical white matter.

*STF2* Upper cellular layer, the most superficial sojourn zone where cells translocate to the cortical plate.

*STF3* Honeycomb trilaminar matrix *(3a, 3b, 3c)* of cells and fibers that is not present during the first trimester and is found in granular (sensory) cortices.

*STF4* Complex middle layer where sojourning and migrating cortical neurons grow corticofugal axons and intermingle with corticopetal axons.

*STF5* Deep cellular layer that is prominent during the first trimester, the first sojourn zone to appear outside the germinal matrix.

*STF6* Late-forming deep layer of callosal fibers outside the germinal matrix that is not present during the first trimester.

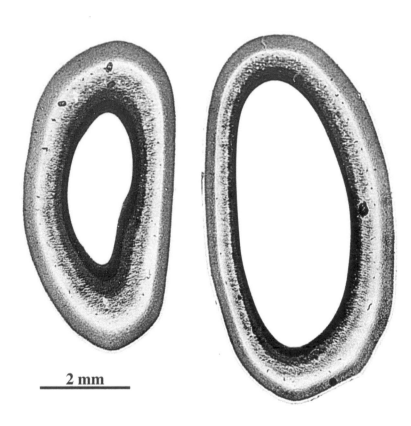

2 mm

FONT KEY:
VENTRICULAR DIVISIONS – CAPITALS
Germinal zone - Helvetica bold
*Transient structure - Times bold italic*
Permanent structure - Times Roman or **Bold**

ABBREVIATIONS:
NEP - Neuroepithelium
SVZ - Subventricular zone

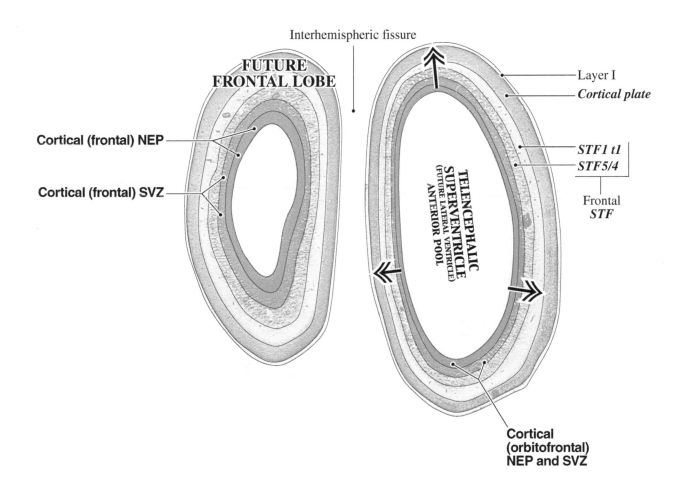

Interhemispheric fissure

**FUTURE
FRONTAL LOBE**

Layer I

*Cortical plate*

**Cortical (frontal) NEP**

**Cortical (frontal) SVZ**

TELENCEPHALIC
SUPERVENTRICLE
(FUTURE LATERAL VENTRICLE)
ANTERIOR POOL

*STF1 t1*

*STF5/4*

Frontal
*STF*

**Cortical
(orbitofrontal)
NEP and SVZ**

**Arrows** indicate the
presumed *direction of
neuron migration* from
neuroepithelial sources.

18

**GW11 Coronal
CR 60 mm, Y1-59
Level 2: Section 259**

## LAYERS OF THE CORTICAL *STRATIFIED TRANSITIONAL FIELD (STF)*

*STF1* Superficial fibrous layer with an early developmental stage *(t1)* when many cells are migrating through it, followed by a late stage *(t2)* with sparse cells. Endures as the subcortical white matter.

*STF2* Upper cellular layer, the most superficial sojourn zone where cells translocate to the cortical plate.

*STF3* Honeycomb trilaminar matrix *(3a, 3b, 3c)* of cells and fibers that is not present during the first trimester and is found in granular (sensory) cortices.

*STF4* Complex middle layer where sojourning and migrating cortical neurons grow corticofugal axons and intermingle with corticopetal axons.

*STF5* Deep cellular layer that is prominent during the first trimester, the first sojourn zone to appear outside the germinal matrix.

*STF6* Late-forming deep layer of callosal fibers outside the germinal matrix that is not present during the first trimester.

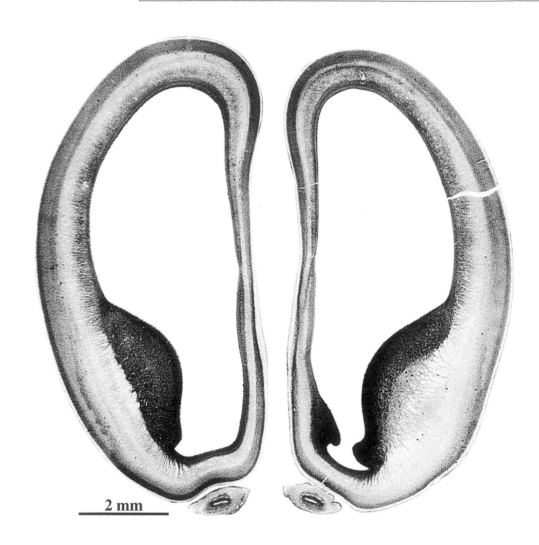

2 mm

**See a high-magnification view of the frontal cortex
from section 269 in Plates 22A and B.**

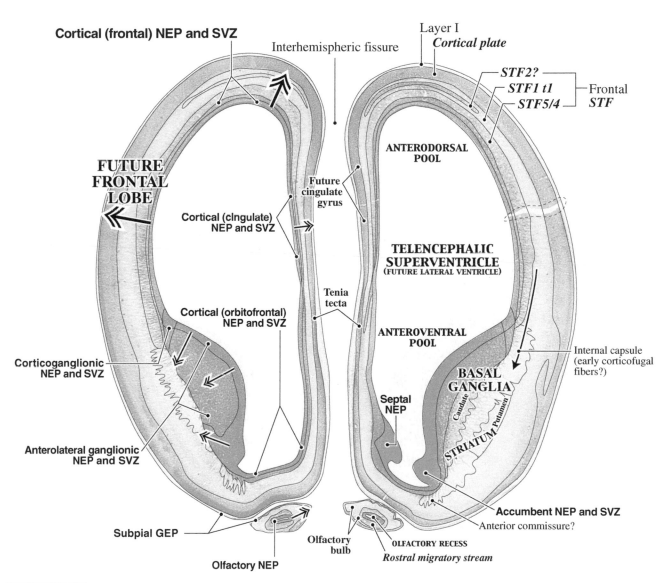

FONT KEY:
VENTRICULAR DIVISIONS – CAPITALS
Germinal zone - Helvetica bold
*Transient structure - Times bold italic*
Permanent structure - Times Roman or **Bold**

ABBREVIATIONS:
GEP - Glioepithelium
NEP - Neuroepithelium
SVZ - Subventricular zone

Cortical (frontal) NEP and SVZ

Interhemispheric fissure

Layer I
*Cortical plate*

*STF2?*
*STF1 t1*   Frontal
*STF5/4*   *STF*

ANTERODORSAL
POOL

FUTURE
FRONTAL
LOBE

Future
cingulate
gyrus

Cortical (cIngulate)
NEP and SVZ

TELENCEPHALIC
SUPERVENTRICLE
(FUTURE LATERAL VENTRICLE)

Tenia
tecta

Cortical (orbitofrontal)
NEP and SVZ

ANTEROVENTRAL
POOL

Internal capsule
(early corticofugal
fibers?)

Corticoganglionic
NEP and SVZ

BASAL
GANGLIA

Septal
NEP

Caudate

Putamen

STRIATUM

Anterolateral ganglionic
NEP and SVZ

Accumbent NEP and SVZ

Subpial GEP

Anterior commissure?

Olfactory
bulb

OLFACTORY RECESS
*Rostral migratory stream*

Olfactory NEP

**Arrows** indicate the
presumed *direction of
neuron migration* from
neuroepithelial sources.

**Arrows** indicate the
presumed *direction of
axon growth* in brain
fiber tracts.

Dashed lines indicate staining
and/or sectioning artifacts.

# PLATE 3A

## GW11 Coronal
## CR 60 mm, Y1-59
## Level 3: Section 279

### LAYERS OF THE CORTICAL *STRATIFIED TRANSITIONAL FIELD (STF)*

| | |
|---|---|
| *STF1* | Superficial fibrous layer with an early developmental stage *(t1)* when many cells are migrating through it, followed by a late stage *(t2)* with sparse cells. Endures as the subcortical white matter. |
| *STF2* | Upper cellular layer, the most superficial sojourn zone where cells translocate to the cortical plate. |
| *STF3* | Honeycomb trilaminar matrix *(3a, 3b, 3c)* of cells and fibers that is not present during the first trimester and is found in granular (sensory) cortices. |
| *STF4* | Complex middle layer where sojourning and migrating cortical neurons grow corticofugal axons and intermingle with corticopetal axons. |
| *STF5* | Deep cellular layer that is prominent during the first trimester, the first sojourn zone to appear outside the germinal matrix. |
| *STF6* | Late-forming deep layer of callosal fibers outside the germinal matrix that is not present during the first trimester. |

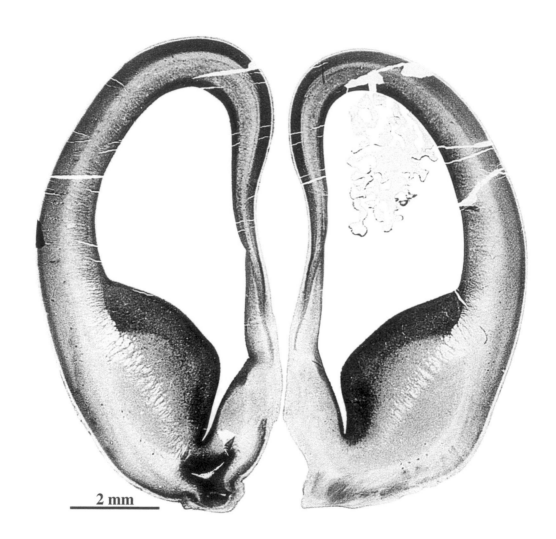

2 mm

**See a high-magnification view of the frontal cortex
from section 269 in Plates 22A and B.**

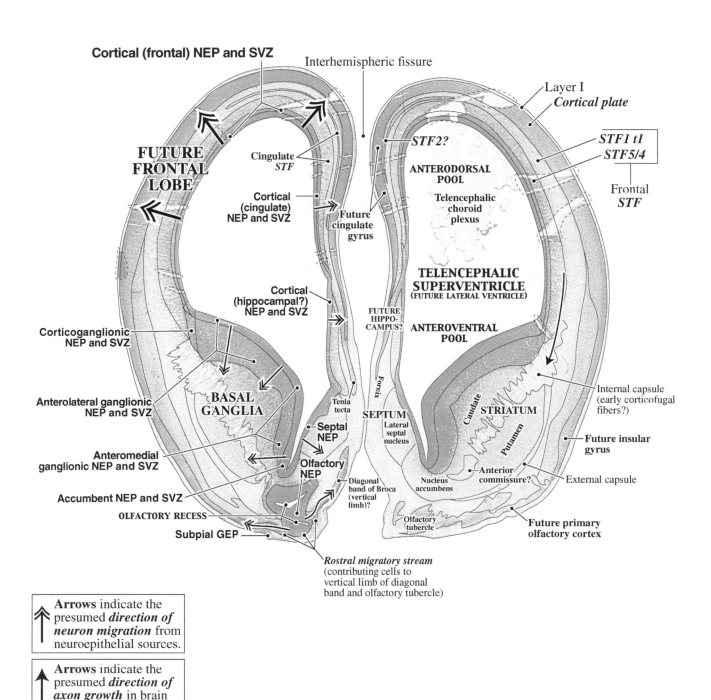

FONT KEY:
VENTRICULAR DIVISIONS – CAPITALS
Germinal zone - Helvetica bold
*Transient structure - Times bold italic*
Permanent structure - Times Roman or **Bold**

ABBREVIATIONS:
GEP - Glioepithelium
NEP - Neuroepithelium
SVZ - Subventricular zone

**Cortical (frontal) NEP and SVZ**

Interhemispheric fissure

Layer I
*Cortical plate*

**FUTURE FRONTAL LOBE**

Cingulate *STF*

*STF2?*

**ANTERODORSAL POOL**

**Telencephalic choroid plexus**

*STF1 t1*
*STF5/4*

Frontal *STF*

**Cortical (cingulate) NEP and SVZ**

*Future cingulate gyrus*

**TELENCEPHALIC SUPERVENTRICLE**
(FUTURE LATERAL VENTRICLE)

**Cortical (hippocampal?) NEP and SVZ**

*FUTURE HIPPO-CAMPUS?*

**ANTEROVENTRAL POOL**

**Corticoganglionic NEP and SVZ**

*Fornix*

Internal capsule (early corticofugal fibers?)

**Anterolateral ganglionic NEP and SVZ**

**BASAL GANGLIA**

*Tenia tecta*

**SEPTUM**

Caudate

**STRIATUM**

**Future insular gyrus**

**Anteromedial ganglionic NEP and SVZ**

**Septal NEP**

Lateral septal nucleus

Putamen

External capsule

**Accumbent NEP and SVZ**

**Olfactory NEP**

Diagonal band of Broca (vertical limb)?

Nucleus accumbens

Anterior commissure?

**Future primary olfactory cortex**

**OLFACTORY RECESS**

Olfactory tubercle

**Subpial GEP**

*Rostral migratory stream*
(contributing cells to vertical limb of diagonal band and olfactory tubercle)

**Arrows** indicate the presumed *direction of neuron migration* from neuroepithelial sources.

**Arrows** indicate the presumed *direction of axon growth* in brain fiber tracts.

Dashed lines indicate staining and/or sectioning artifacts.

# PLATE 4A

**GW11 Coronal
CR 60 mm, Y1-59
Level 4: Section 289**

## LAYERS OF THE CORTICAL *STRATIFIED TRANSITIONAL FIELD (STF)*

*STF1* Superficial fibrous layer with an early developmental stage *(t1)* when many cells are migrating through it, followed by a late stage *(t2)* with sparse cells. Endures as the subcortical white matter.

*STF2* Upper cellular layer, the most superficial sojourn zone where cells translocate to the cortical plate.

*STF3* Honeycomb trilaminar matrix *(3a, 3b, 3c)* of cells and fibers that is not present during the first trimester and is found in granular (sensory) cortices.

*STF4* Complex middle layer where sojourning and migrating cortical neurons grow corticofugal axons and intermingle with corticopetal axons.

*STF5* Deep cellular layer that is prominent during the first trimester, the first sojourn zone to appear outside the germinal matrix.

*STF6* Late-forming deep layer of callosal fibers outside the germinal matrix that is not present during the first trimester.

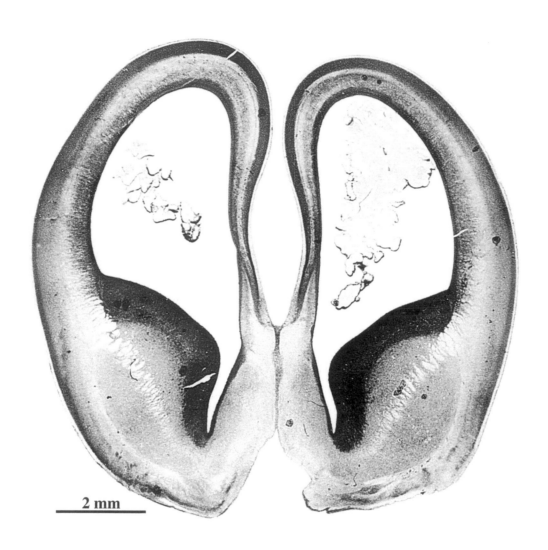

2 mm

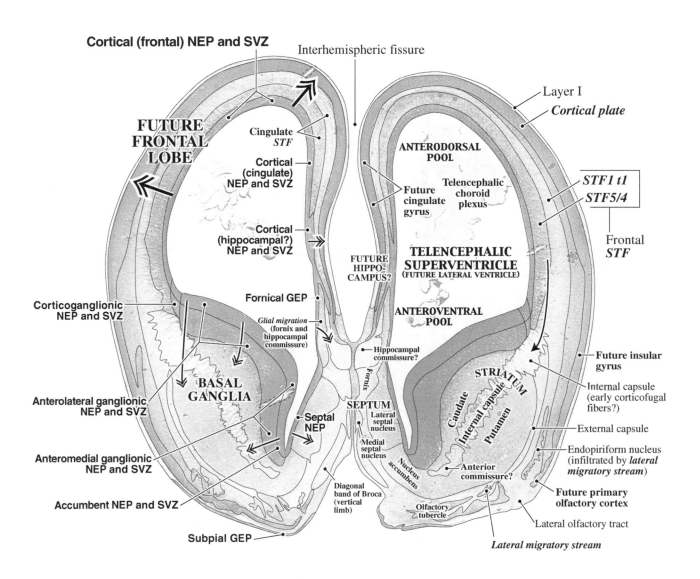

Cortical (frontal) NEP and SVZ

Interhemispheric fissure

Layer I
*Cortical plate*

**FUTURE
FRONTAL
LOBE**

Cingulate
*STF*

Cortical
(cingulate)
NEP and SVZ

Cortical
(hippocampal?)
NEP and SVZ

**ANTERODORSAL
POOL**

*Future
cingulate
gyrus*

Telencephalic
choroid
plexus

*STF1 t1*
*STF5/4*

Frontal
*STF*

*FUTURE
HIPPO-
CAMPUS?*

**TELENCEPHALIC
SUPERVENTRICLE**
**(FUTURE LATERAL VENTRICLE)**

Fornical GEP

*Glial migration
(fornix and
hippocampal
commissure)*

**ANTEROVENTRAL
POOL**

Corticoganglionic
NEP and SVZ

Hippocampal
commissure?

**Future insular
gyrus**

**STRIATUM**

Internal capsule
(early corticofugal
fibers?)

**BASAL
GANGLIA**

Fornix

Caudate

Internal capsule

Putamen

External capsule

Anterolateral ganglionic
NEP and SVZ

**SEPTUM**

Lateral
septal
nucleus

Endopiriform nucleus
(infiltrated by *lateral
migratory stream*)

Anteromedial ganglionic
NEP and SVZ

Medial
septal
nucleus

Septal
NEP

Nucleus
accumbens

Anterior
commissure?

**Future primary
olfactory cortex**

Accumbent NEP and SVZ

Diagonal
band of Broca
(vertical
limb)

Olfactory
tubercle

Lateral olfactory tract

Subpial GEP

*Lateral migratory stream*

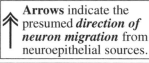

**Arrows** indicate the
presumed *direction of
neuron migration* from
neuroepithelial sources.

**Arrows** indicate the
presumed *direction of
axon growth* in brain
fiber tracts.

Dashed lines indicate staining
and/or sectioning artifacts.

## PLATE 5A

**GW11 Coronal**
**CR 60 mm, Y1-59**
**Level 5: Section 329**

## LAYERS OF THE CORTICAL *STRATIFIED TRANSITIONAL FIELD (STF)*

*STF1* Superficial fibrous layer with an early developmental stage *(t1)* when many cells are migrating through it, followed by a late stage *(t2)* with sparse cells. Endures as the subcortical white matter.

*STF2* Upper cellular layer, the most superficial sojourn zone where cells translocate to the cortical plate.

*STF3* Honeycomb trilaminar matrix *(3a, 3b, 3c)* of cells and fibers that is not present during the first trimester and is found in granular (sensory) cortices.

*STF4* Complex middle layer where sojourning and migrating cortical neurons grow corticofugal axons and intermingle with corticopetal axons.

*STF5* Deep cellular layer that is prominent during the first trimester, the first sojourn zone to appear outside the germinal matrix.

*STF6* Late-forming deep layer of callosal fibers outside the germinal matrix that is not present during the first trimester.

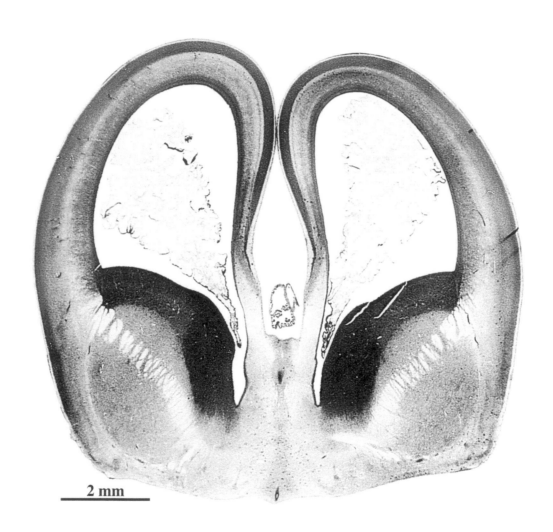

2 mm

**FONT KEY:**
**VENTRICULAR DIVISIONS – CAPITALS**
**Germinal zone - Helvetica bold**
*Transient structure - Times bold italic*
Permanent structure - Times Roman or **Bold**

**ABBREVIATIONS:**
**GEP - Glioepithelium**
**G/EP - Glioepithelium/ependyma**
**NEP - Neuroepithelium**
**SVZ - Subventricular zone**

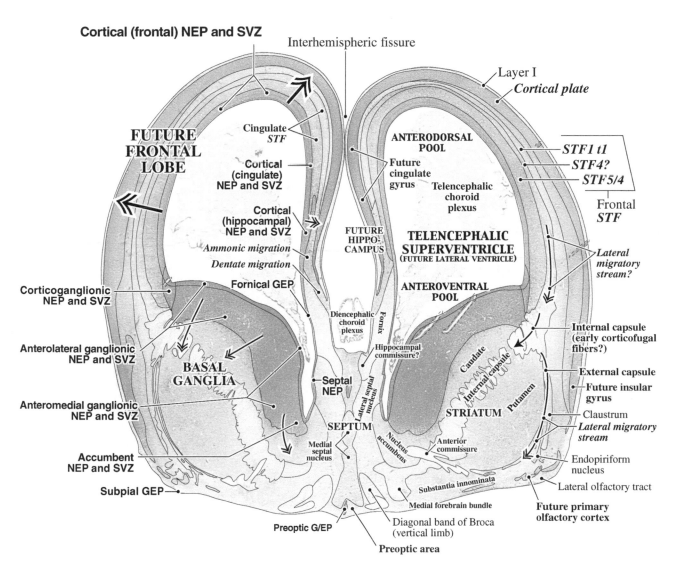

**Cortical (frontal) NEP and SVZ**

Interhemispheric fissure

Layer I
*Cortical plate*

Cingulate
*STF*

**FUTURE**
**FRONTAL**
**LOBE**

**ANTERODORSAL**
**POOL**

*STF1 t1*
*STF4?*
*STF5/4*

Cortical
(cingulate)
**NEP and SVZ**

Future
cingulate
gyrus

Telencephalic
choroid
plexus

Frontal
*STF*

Cortical
(hippocampal)
**NEP and SVZ**

**TELENCEPHALIC**
**SUPERVENTRICLE**
**(FUTURE LATERAL VENTRICLE)**

*Lateral*
*migratory*
*stream?*

*Ammonic migration*

*Dentate migration*

**FUTURE**
**HIPPO-**
**CAMPUS**

**Fornical GEP**

**ANTEROVENTRAL**
**POOL**

**Corticoganglionic**
**NEP and SVZ**

Diencephalic
choroid
plexus

Fornix

**Internal capsule**
**(early corticofugal**
**fibers?)**

**Anterolateral ganglionic**
**NEP and SVZ**

Hippocampal
commissure?

Caudate

Internal capsule

**External capsule**
**Future insular**
**gyrus**

**BASAL**
**GANGLIA**

**Septal**
**NEP**

Lateral septal nucleus

Putamen

Claustrum

*Lateral migratory*
*stream*

**Anteromedial ganglionic**
**NEP and SVZ**

**STRIATUM**

**SEPTUM**

Nucleus
accumbens

Anterior
commissure

**Endopiriform**
**nucleus**

**Accumbent**
**NEP and SVZ**

Medial
septal
nucleus

Lateral olfactory tract

**Subpial GEP**

Substantia innominata

**Future primary**
**olfactory cortex**

Medial forebrain bundle

**Preoptic G/EP**

Diagonal band of Broca
(vertical limb)

**Preoptic area**

**Arrows** indicate the
presumed *direction of*
*neuron migration* from
neuroepithelial sources.

**Arrows** indicate the
presumed *direction of*
*axon growth* in brain
fiber tracts.

Dashed lines indicate staining
and/or sectioning artifacts.

## PLATE 6A

**GW11 Coronal
CR 60 mm, Y1-59
Level 6: Section 340**

### LAYERS OF THE CORTICAL *STRATIFIED TRANSITIONAL FIELD (STF)*

*STF1* Superficial fibrous layer with an early developmental stage *(t1)* when many cells are migrating through it, followed by a late stage *(t2)* with sparse cells. Endures as the subcortical white matter.

*STF2* Upper cellular layer, the most superficial sojourn zone where cells translocate to the cortical plate.

*STF3* Honeycomb trilaminar matrix *(3a, 3b, 3c)* of cells and fibers that is not present during the first trimester and is found in granular (sensory) cortices.

*STF4* Complex middle layer where sojourning and migrating cortical neurons grow corticofugal axons and intermingle with corticopetal axons.

*STF5* Deep cellular layer that is prominent during the first trimester, the first sojourn zone to appear outside the germinal matrix.

*STF6* Late-forming deep layer of callosal fibers outside the germinal matrix that is not present during the first trimester.

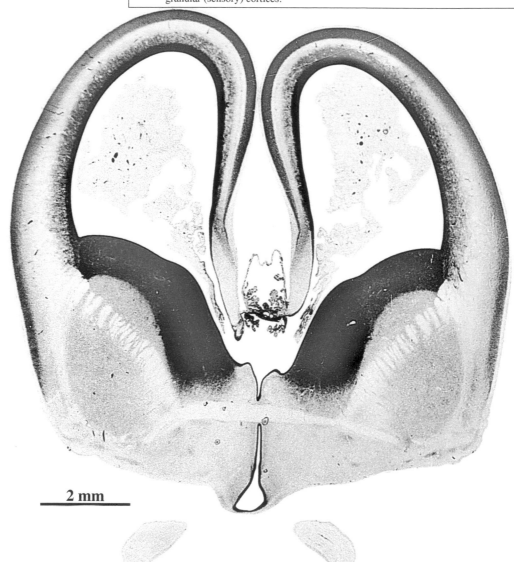

2 mm

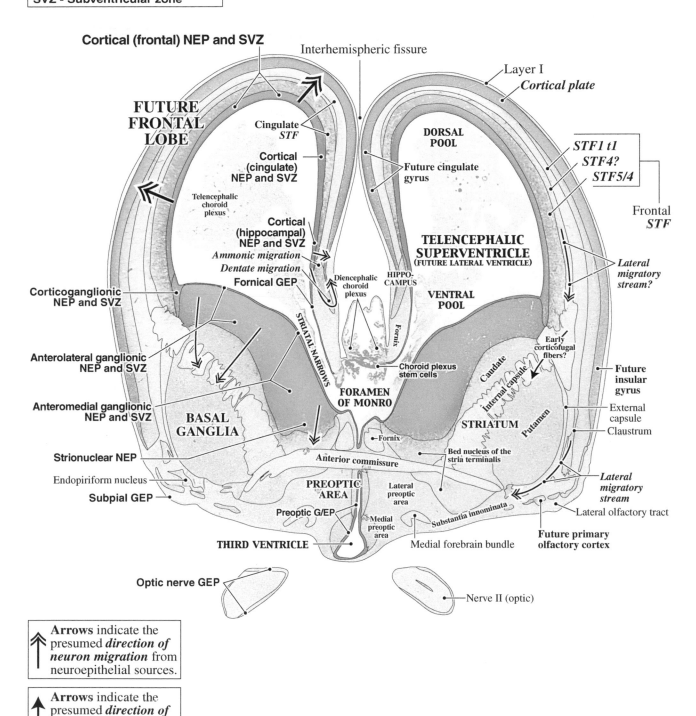

FONT KEY:
VENTRICULAR DIVISIONS – CAPITALS
Germinal zone - Helvetica bold
*Transient structure - Times bold italic*
Permanent structure - Times Roman or **Bold**

ABBREVIATIONS:
GEP - Glioepithelium
G/EP - Glioepithelium/ependyma
NEP - Neuroepithelium
SVZ - Subventricular zone

**Cortical (frontal) NEP and SVZ**

Interhemispheric fissure

Layer I
*Cortical plate*

**FUTURE FRONTAL LOBE**

Cingulate *STF*

**Cortical (cingulate) NEP and SVZ**

Telencephalic choroid plexus

**DORSAL POOL**

Future cingulate gyrus

*STF1 t1*
*STF4?*
*STF5/4*

Frontal *STF*

**Cortical (hippocampal) NEP and SVZ**

*Ammonic migration*
*Dentate migration*

**Fornical GEP**

Diencephalic choroid plexus

HIPPO-CAMPUS

**TELENCEPHALIC SUPERVENTRICLE**
(FUTURE LATERAL VENTRICLE)

*Lateral migratory stream?*

**VENTRAL POOL**

Fornix

**Corticoganglionic NEP and SVZ**

STRIATAL NARROWS

Choroid plexus stem cells

Caudate

Internal capsule

Early corticofugal fibers?

**Future insular gyrus**

**Anterolateral ganglionic NEP and SVZ**

Putamen

External capsule

Claustrum

**Anteromedial ganglionic NEP and SVZ**

**BASAL GANGLIA**

**FORAMEN OF MONRO**

Fornix

**STRIATUM**

Bed nucleus of the stria terminalis

**Strionuclear NEP**

Anterior commissure

Endopiriform nucleus

**PREOPTIC AREA**

Lateral preoptic area

*Lateral migratory stream*

**Subpial GEP**

Preoptic G/EP

Medial preoptic area

Substantia innominata

Lateral olfactory tract

**THIRD VENTRICLE**

Medial forebrain bundle

**Future primary olfactory cortex**

Optic nerve GEP

Nerve II (optic)

**Arrows** indicate the presumed *direction of neuron migration* from neuroepithelial sources.

**Arrows** indicate the presumed *direction of axon growth* in brain fiber tracts.

Dashed lines indicate staining and/or sectioning artifacts.

## PLATE 7A

**GW11 Coronal
CR 60 mm, Y1-59
Level 7: Section 349**

## LAYERS OF THE CORTICAL *STRATIFIED TRANSITIONAL FIELD (STF)*

*STF1* Superficial fibrous layer with an early developmental stage *(t1)* when many cells are migrating through it, followed by a late stage *(t2)* with sparse cells. Endures as the subcortical white matter.

*STF2* Upper cellular layer, the most superficial sojourn zone where cells translocate to the cortical plate.

*STF3* Honeycomb trilaminar matrix *(3a, 3b, 3c)* of cells and fibers that is not present during the first trimester and is found in granular (sensory) cortices.

*STF4* Complex middle layer where sojourning and migrating cortical neurons grow corticofugal axons and intermingle with corticopetal axons.

*STF5* Deep cellular layer that is prominent during the first trimester, the first sojourn zone to appear outside the germinal matrix.

*STF6* Late-forming deep layer of callosal fibers outside the germinal matrix that is not present during the first trimester.

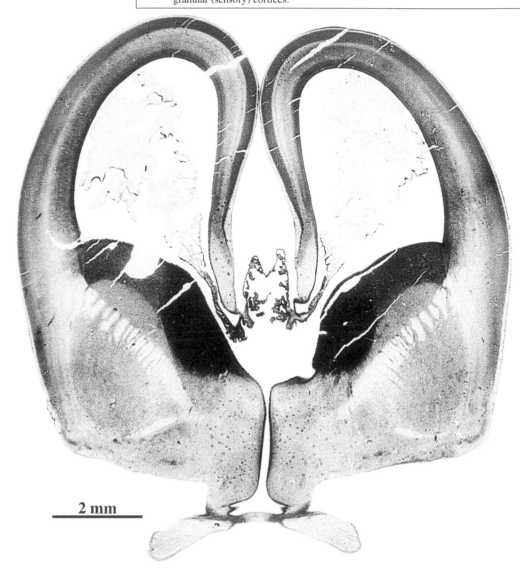

2 mm

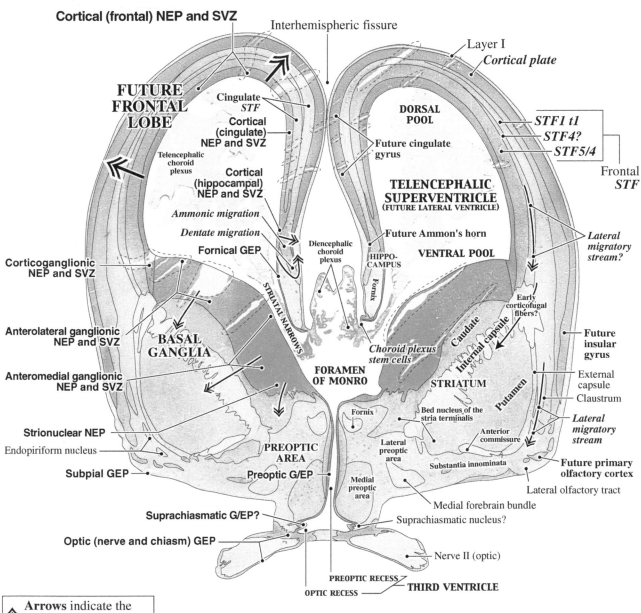

Cortical (frontal) NEP and SVZ

Interhemispheric fissure

Layer I
*Cortical plate*

**FUTURE
FRONTAL
LOBE**

Cingulate
*STF*

Cortical
(cingulate)
NEP and SVZ

**DORSAL
POOL**

*STF1 t1*
*STF4?*
*STF5/4*

Frontal
*STF*

*Telencephalic
choroid
plexus*

Future cingulate
gyrus

Cortical
(hippocampal)
NEP and SVZ

**TELENCEPHALIC
SUPERVENTRICLE**
(FUTURE LATERAL VENTRICLE)

*Ammonic migration*

*Dentate migration*

*Diencephalic
choroid
plexus*

Future Ammon's horn

*Lateral
migratory
stream?*

Fornical GEP

**HIPPO-
CAMPUS**

**VENTRAL POOL**

Corticoganglionic
NEP and SVZ

*Fornix*

*Early
corticofugal
fibers?*

STRIATAL NARROWS

*Choroid plexus
stem cells*

Caudate

Internal capsule

**Future
insular
gyrus**

Anterolateral ganglionic
NEP and SVZ

**BASAL
GANGLIA**

**FORAMEN
OF MONRO**

**STRIATUM**

Putamen

External
capsule

Claustrum

Anteromedial ganglionic
NEP and SVZ

Fornix

Bed nucleus of the
stria terminalis

Anterior
commissure

*Lateral
migratory
stream*

Strionuclear NEP

Endopiriform nucleus

**PREOPTIC
AREA**

Lateral
preoptic
area

Substantia innominata

**Future primary
olfactory cortex**

Subpial GEP

Preoptic G/EP

Medial
preoptic
area

Lateral olfactory tract

Medial forebrain bundle

Suprachiasmatic G/EP?

Suprachiasmatic nucleus?

Optic (nerve and chiasm) GEP

Nerve II (optic)

**PREOPTIC RECESS**

**THIRD VENTRICLE**

**OPTIC RECESS**

# PLATE 8A

**GW11 Coronal
CR 60 mm, Y1-59
Level 8: Section 379**

## LAYERS OF THE CORTICAL *STRATIFIED TRANSITIONAL FIELD (STF)*

*STF1* Superficial fibrous layer with an early developmental stage *(t1)* when many cells are migrating through it, followed by a late stage *(t2)* with sparse cells. Endures as the subcortical white matter.

*STF2* Upper cellular layer, the most superficial sojourn zone where cells translocate to the cortical plate.

*STF3* Honeycomb trilaminar matrix *(3a, 3b, 3c)* of cells and fibers that is not present during the first trimester and is found in granular (sensory) cortices.

*STF4* Complex middle layer where sojourning and migrating cortical neurons grow corticofugal axons and intermingle with corticopetal axons.

*STF5* Deep cellular layer that is prominent during the first trimester, the first sojourn zone to appear outside the germinal matrix.

*STF6* Late-forming deep layer of callosal fibers outside the germinal matrix that is not present during the first trimester.

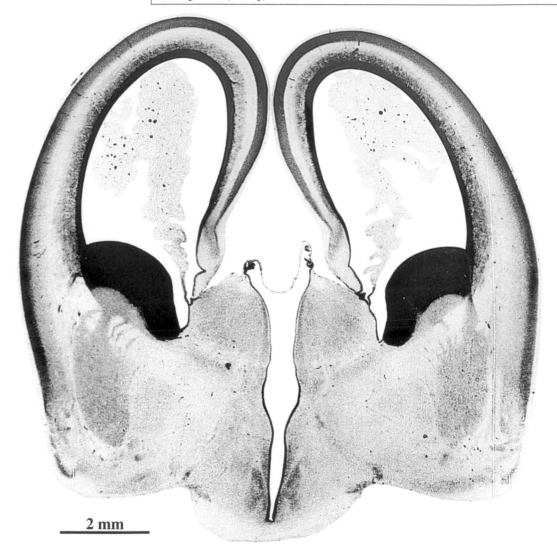

2 mm

**See high-magnification views of the right paracentral cortex, thalamus, and basal ganglia from section 399 in Plates 24A and B to 25A and B.**

**See a high-magnification view of the diencephalon and basal telencephalon from section 389 in Plates 26A and B.**

FONT KEY:
VENTRICULAR DIVISIONS - CAPITALS
Germinal zone - Helvetica bold
*Transient structure - Times bold italic*
Permanent structure - Times Roman or **Bold**

ABBREVIATIONS:
GEP - Glioepithelium
G/EP - Glioepithelium/ependyma
NEP - Neuroepithelium
SVZ - Subventricular zone

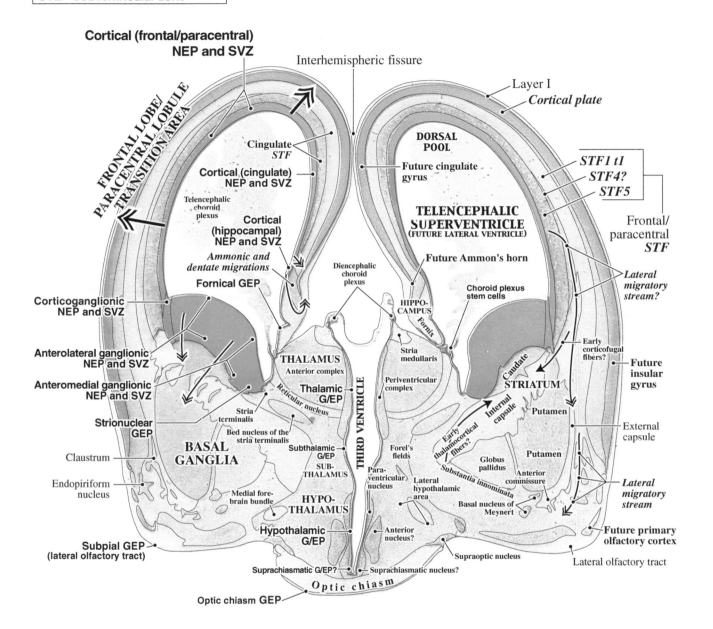

**Cortical (frontal/paracentral) NEP and SVZ**

Interhemispheric fissure

Layer I
*Cortical plate*

*FRONTAL LOBE/ PARACENTRAL LOBULE TRANSITION AREA*

**Cingulate** *STF*

**DORSAL POOL**

Future cingulate gyrus

*STF1 t1*
*STF4?*
*STF5*

Frontal/ paracentral *STF*

**Cortical (cingulate) NEP and SVZ**

Telencephalic choroid plexus

**Cortical (hippocampal) NEP and SVZ**

*Ammonic and dentate migrations*

**Fornical GEP**

**TELENCEPHALIC SUPERVENTRICLE**
(FUTURE LATERAL VENTRICLE)

**Future Ammon's horn**

Diencephalic choroid plexus

Choroid plexus stem cells

*Lateral migratory stream?*

**Corticoganglionic NEP and SVZ**

HIPPO-CAMPUS

*Fornix*

Early corticofugal fibers?

**Future insular gyrus**

**Anterolateral ganglionic NEP and SVZ**

**THALAMUS**
Anterior complex

Stria medullaris

Caudate

**STRIATUM**

**Anteromedial ganglionic NEP and SVZ**

**Thalamic G/EP**

*Reticular nucleus*

Periventricular complex

*Early thalamocortical fibers?*

Internal capsule

Putamen

External capsule

**Strionuclear GEP**

Stria terminalis

Bed nucleus of the stria terminalis

**THIRD VENTRICLE**

Forel's fields

Putamen

*Lateral migratory stream*

Claustrum

**BASAL GANGLIA**

**Subthalamic G/EP**

**SUB-THALAMUS**

Para-ventricular nucleus

Globus pallidus

Substantia innominata

Anterior commissure

Endopiriform nucleus

Medial fore-brain bundle

**HYPO-THALAMUS**

Lateral hypothalamic area

Basal nucleus of Meynert

*Future primary olfactory cortex*

**Subpial GEP**
(lateral olfactory tract)

**Hypothalamic G/EP**

Anterior nucleus?

Supraoptic nucleus

Lateral olfactory tract

**Suprachiasmatic G/EP?**

Suprachiasmatic nucleus?

*Optic chiasm*

**Optic chiasm GEP**

**Arrows** indicate the presumed *direction of neuron migration* from neuroepithelial sources.

**Arrows** indicate the presumed *direction of axon growth* in brain fiber tracts.

# PLATE 9A

## GW11 Coronal
## CR 60 mm, Y1-59
## Level 9: Section 419

### LAYERS OF THE CORTICAL *STRATIFIED TRANSITIONAL FIELD (STF)*

*STF1*  Superficial fibrous layer with an early developmental stage *(t1)* when many cells are migrating through it, followed by a late stage *(t2)* with sparse cells. Endures as the subcortical white matter.

*STF2*  Upper cellular layer, the most superficial sojourn zone where cells translocate to the cortical plate.

*STF3*  Honeycomb trilaminar matrix *(3a, 3b, 3c)* of cells and fibers that is not present during the first trimester and is found in granular (sensory) cortices.

*STF4*  Complex middle layer where sojourning and migrating cortical neurons grow corticofugal axons and intermingle with corticopetal axons.

*STF5*  Deep cellular layer that is prominent during the first trimester, the first sojourn zone to appear outside the germinal matrix.

*STF6*  Late-forming deep layer of callosal fibers outside the germinal matrix that is not present during the first trimester.

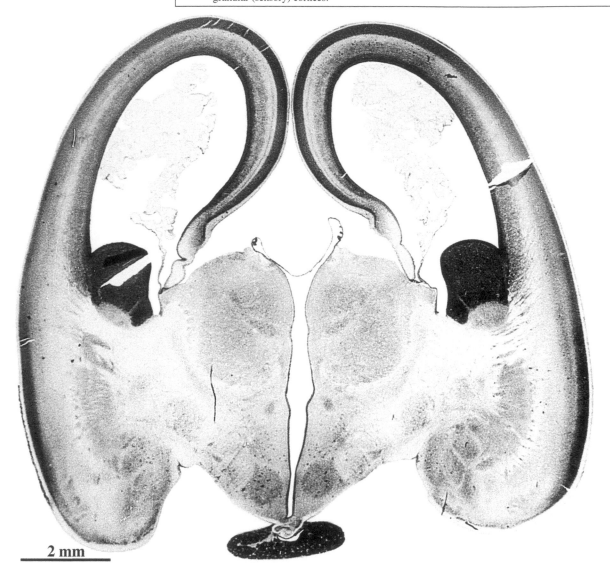

2 mm

**See a high-magnification view of the diencephalon and basal telencephalon from level 9 in Plates 27A and B.**

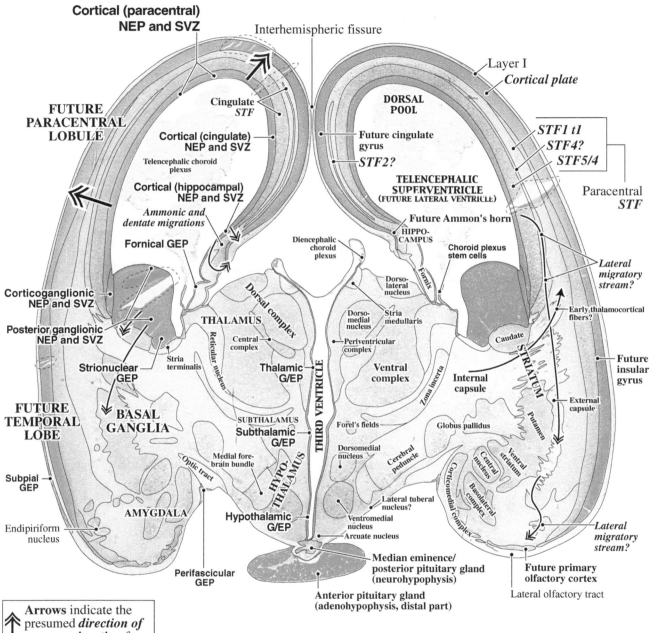

**Cortical (paracentral) NEP and SVZ**

Interhemispheric fissure

Layer I
*Cortical plate*

**Cingulate** *STF*

**DORSAL POOL**

**FUTURE PARACENTRAL LOBULE**

**Cortical (cingulate) NEP and SVZ**

Future cingulate gyrus

*STF2?*

*STF1 t1*
*STF4?*
*STF5/4*

*Telencephalic choroid plexus*

**TELENCEPHALIC SUPERVENTRICLE**
**(FUTURE LATERAL VENTRICLE)**

Paracentral *STF*

**Cortical (hippocampal) NEP and SVZ**

*Ammonic and dentate migrations*

Future Ammon's horn

**Fornical GEP**

*Diencephalic choroid plexus*

**HIPPO-CAMPUS**

Choroid plexus stem cells

*Lateral migratory stream?*

*Fornix*

**Corticoganglionic NEP and SVZ**

Dorsolateral nucleus

Early thalamocortical fibers?

*Dorsal complex*

**THALAMUS**

Dorso-medial nucleus

Stria medullaris

Caudate

**Posterior ganglionic NEP and SVZ**

*Reticular nucleus*

Central complex

Periventricular complex

**STRIATUM**

**Future insular gyrus**

**Strionuclear GEP**

Stria terminalis

**Thalamic G/EP**

Ventral complex

**Internal capsule**

*Zona incerta*

External capsule

*Putamen*

**FUTURE TEMPORAL LOBE**

**BASAL GANGLIA**

**SUBTHALAMUS**

**Subthalamic G/EP**

Forel's fields

Globus pallidus

*Ventral striatum*

*Central nucleus*

**Subpial GEP**

*Optic tract*

Medial fore-brain bundle

**HYPO-THALAMUS**

**THIRD VENTRICLE**

Dorsomedial nucleus

Cerebral peduncle

*Corticomedial complex*

*Basolateral complex*

**AMYGDALA**

**Hypothalamic G/EP**

Lateral tuberal nucleus?

Ventromedial nucleus

*Lateral migratory stream?*

Endipiriform nucleus

Arcuate nucleus

**Median eminence/ posterior pituitary gland (neurohypophysis)**

**Future primary olfactory cortex**

**Perifascicular GEP**

Anterior pituitary gland (adenohypophysis, distal part)

Lateral olfactory tract

**Arrows** indicate the presumed *direction of neuron migration* from neuroepithelial sources.

**Arrows** indicate the presumed *direction of axon growth* in brain fiber tracts.

Dashed lines indicate staining and/or sectioning artifacts.

## PLATE 10A

**GW11 Coronal
CR 60 mm, Y1-59
Level 10: Section 439**

## LAYERS OF THE CORTICAL *STRATIFIED TRANSITIONAL FIELD (STF)*

| | | | |
|---|---|---|---|
| **STF1** | Superficial fibrous layer with an early developmental stage *(t1)* when many cells are migrating through it, followed by a late stage *(t2)* with sparse cells. Endures as the subcortical white matter. | **STF4** | Complex middle layer where sojourning and migrating cortical neurons grow corticofugal axons and intermingle with corticopetal axons. |
| **STF2** | Upper cellular layer, the most superficial sojourn zone where cells translocate to the cortical plate. | **STF5** | Deep cellular layer that is prominent during the first trimester, the first sojourn zone to appear outside the germinal matrix. |
| **STF3** | Honeycomb trilaminar matrix *(3a, 3b, 3c)* of cells and fibers that is not present during the first trimester and is found in granular (sensory) cortices. | **STF6** | Late-forming deep layer of callosal fibers outside the germinal matrix that is not present during the first trimester. |

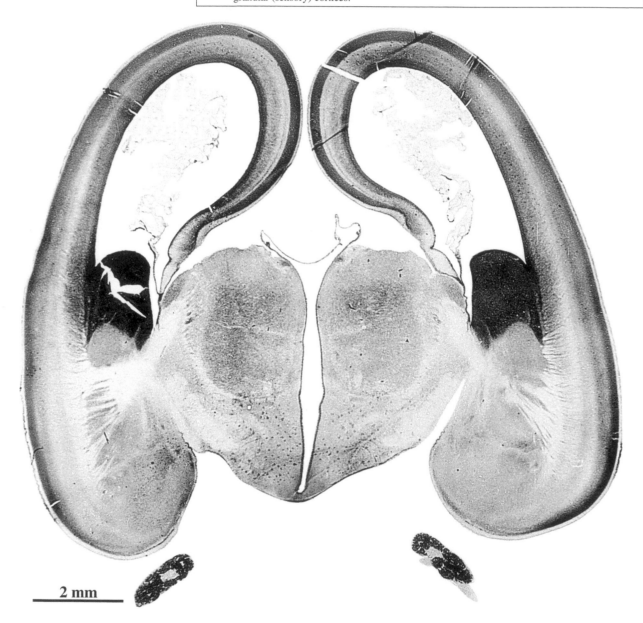

2 mm

**See a high-magnification view of the diencephalon
from section 449 in Plates 28A and B.**

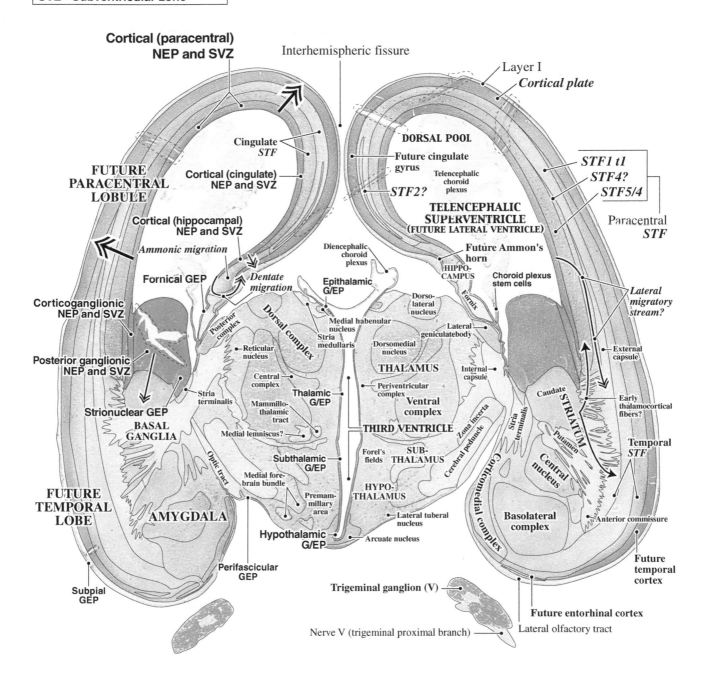

**Arrows** indicate the presumed *direction of axon growth* in brain fiber tracts.

**Arrows** indicate the presumed *direction of neuron migration* from neuroepithelial sources.

Dashed lines indicate staining and/or sectioning artifacts.

# PLATE 11A

## GW11 Coronal
## CR 60 mm, Y1-59
## Level 11: Section 459

### LAYERS OF THE CORTICAL *STRATIFIED TRANSITIONAL FIELD (STF)*

*STF1* Superficial fibrous layer with an early developmental stage *(t1)* when many cells are migrating through it, followed by a late stage *(t2)* with sparse cells. Endures as the subcortical white matter.

*STF2* Upper cellular layer, the most superficial sojourn zone where cells translocate to the cortical plate.

*STF3* Honeycomb trilaminar matrix *(3a, 3b, 3c)* of cells and fibers that is not present during the first trimester and is found in granular (sensory) cortices.

*STF4* Complex middle layer where sojourning and migrating cortical neurons grow corticofugal axons and intermingle with corticopetal axons.

*STF5* Deep cellular layer that is prominent during the first trimester, the first sojourn zone to appear outside the germinal matrix.

*STF6* Late-forming deep layer of callosal fibers outside the germinal matrix that is not present during the first trimester.

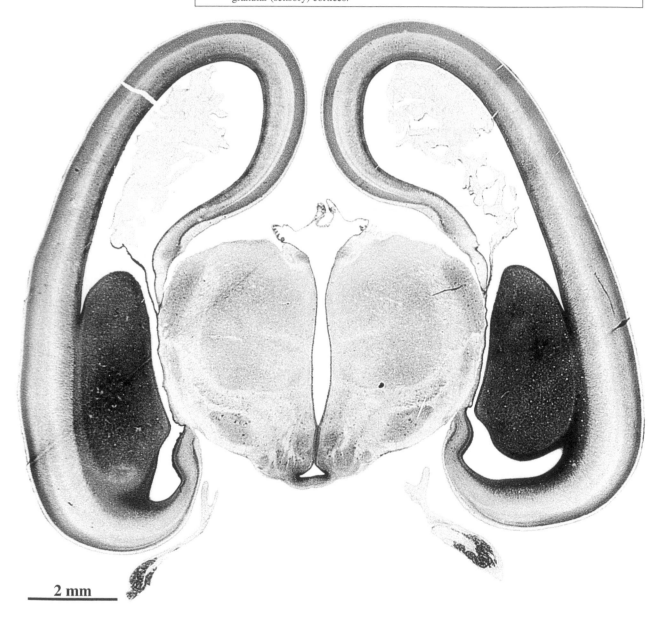

2 mm

**See a high-magnification view of the diencephalon from section 449 in Plates 28A and B, from section 469 in Plates 29A and B.**

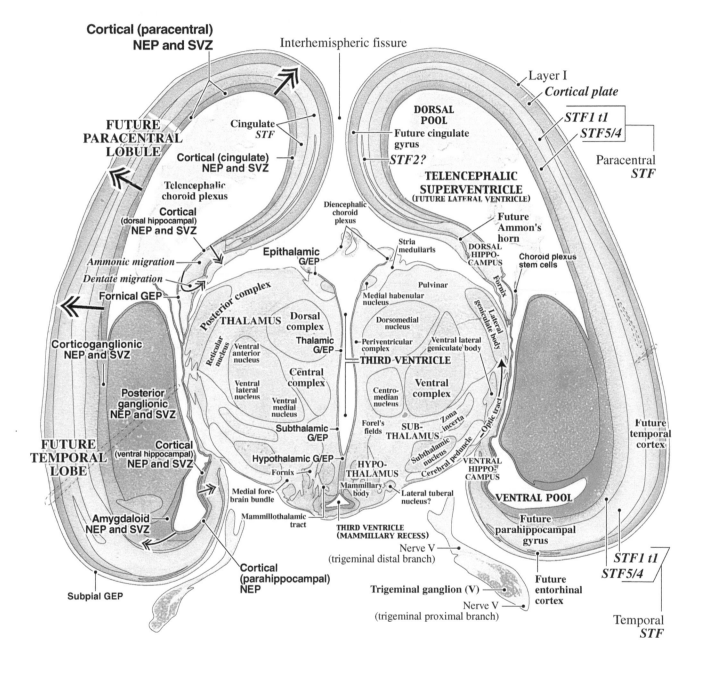

**Arrows** indicate the presumed *direction of axon growth* in brain fiber tracts.

**Arrows** indicate the presumed *direction of neuron migration* from neuroepithelial sources.

Dashed lines indicate staining and/or sectioning artifacts.

## PLATE 12A

**GW11 Coronal
CR 60 mm, Y1-59
Level 12: Section 479**

### LAYERS OF THE CORTICAL *STRATIFIED TRANSITIONAL FIELD (STF)*

*STF1*  Superficial fibrous layer with an early developmental stage *(t1)* when many cells are migrating through it, followed by a late stage *(t2)* with sparse cells. Endures as the subcortical white matter.

*STF2*  Upper cellular layer, the most superficial sojourn zone where cells translocate to the cortical plate.

*STF3*  Honeycomb trilaminar matrix *(3a, 3b, 3c)* of cells and fibers that is not present during the first trimester and is found in granular (sensory) cortices.

*STF4*  Complex middle layer where sojourning and migrating cortical neurons grow corticofugal axons and intermingle with corticopetal axons.

*STF5*  Deep cellular layer that is prominent during the first trimester, the first sojourn zone to appear outside the germinal matrix.

*STF6*  Late-forming deep layer of callosal fibers outside the germinal matrix that is not present during the first trimester.

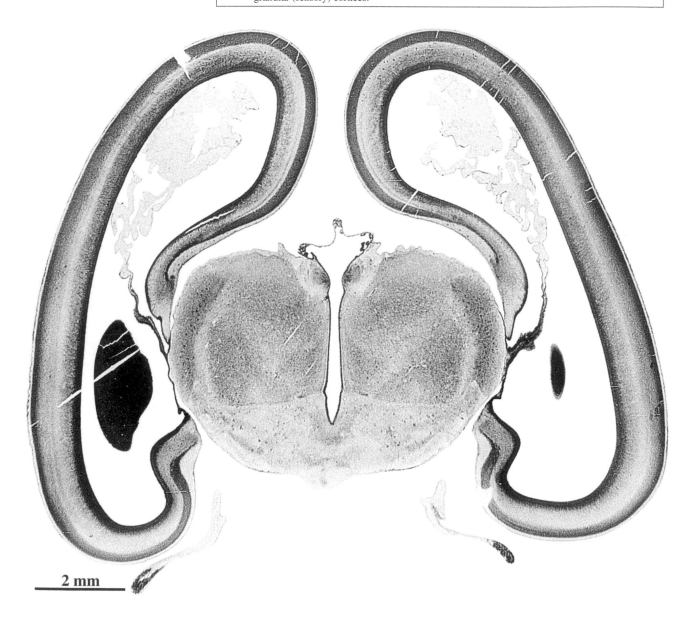

2 mm

**See a high-magnification view of the diencephalon
from section 469 in Plates 29A and B.**

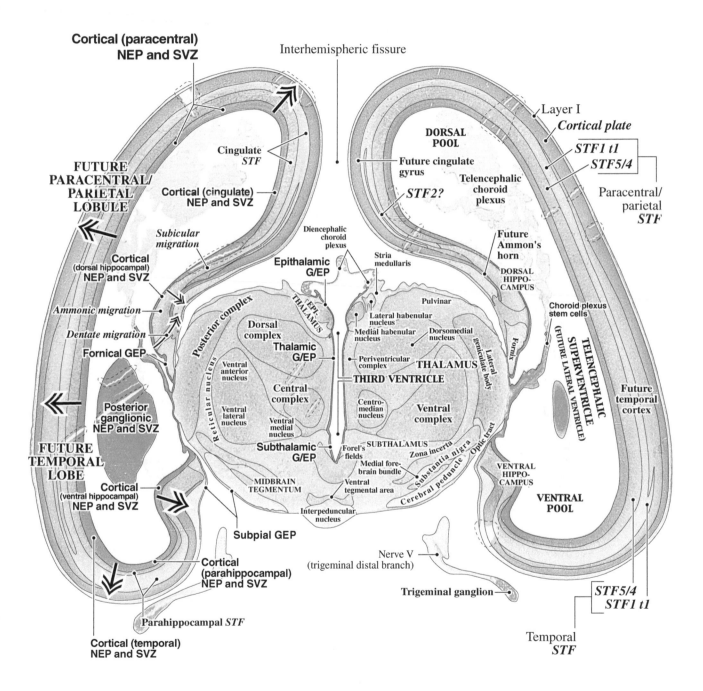

FONT KEY:
VENTRICULAR DIVISIONS – CAPITALS
Germinal zone - Helvetica bold
*Transient structure - Times bold italic*
Permanent structure - Times Roman or **Bold**

ABBREVIATIONS:
GEP - Glioepithelium
G/EP - Glioepithelium/ependyma
NEP - Neuroepithelium
SVZ - Subventricular zone

**Cortical (paracentral) NEP and SVZ**

Interhemispheric fissure

Layer I
*Cortical plate*

**DORSAL POOL**

*STF1 t1*
*STF5/4*

Cingulate *STF*

**Future cingulate gyrus**

Telencephalic choroid plexus

Paracentral/ parietal *STF*

**FUTURE PARACENTRAL/ PARIETAL LOBULE**

**Cortical (cingulate) NEP and SVZ**

*STF2?*

*Subicular migration*

Diencephalic choroid plexus

**Future Ammon's horn**

**DORSAL HIPPO- CAMPUS**

**Cortical (dorsal hippocampal) NEP and SVZ**

Epithalamic **G/EP**

Stria medullaris

*Ammonic migration*

Pulvinar

Choroid plexus stem cells

*Dentate migration*

Posterior complex

Dorsal complex

Lateral habenular nucleus
Medial habenular nucleus
Dorsomedial nucleus

**Fornical GEP**

Thalamic **G/EP**

Periventricular complex

Fornix

Lateral geniculate body

**TELENCEPHALIC SUPERVENTRICLE (FUTURE LATERAL VENTRICLE)**

**Posterior ganglionic NEP and SVZ**

Reticular nucleus

Ventral anterior nucleus

Central complex

**THALAMUS**

**THIRD VENTRICLE**

**Future temporal cortex**

Ventral lateral nucleus

Centro- median nucleus

Ventral complex

**FUTURE TEMPORAL LOBE**

Ventral medial nucleus

Subthalamic **G/EP**

Forel's fields
**SUBTHALAMUS**
Zona incerta

**Cortical (ventral hippocampal) NEP and SVZ**

MIDBRAIN TEGMENTUM

Medial fore- brain bundle
Ventral tegmental area

Substantia nigra
Cerebral peduncle
Optic tract

**VENTRAL HIPPO- CAMPUS**

Subpial GEP

Interpeduncular nucleus

**VENTRAL POOL**

Nerve V
(trigeminal distal branch)

**Cortical (parahippocampal) NEP and SVZ**

*Parahippocampal STF*

Trigeminal ganglion

*STF5/4*
*STF1 t1*

**Cortical (temporal) NEP and SVZ**

Temporal *STF*

**Arrows** indicate the presumed *direction of neuron migration* from neuroepithelial sources.

Dashed lines indicate staining and/or sectioning artifacts.

## PLATE 13A

**GW11 Coronal
CR 60 mm, Y1-59
Level 13: Section 500**

### LAYERS OF THE CORTICAL *STRATIFIED TRANSITIONAL FIELD (STF)*

*STF1*  Superficial fibrous layer with an early developmental stage *(t1)* when many cells are migrating through it, followed by a late stage *(t2)* with sparse cells. Endures as the subcortical white matter.

*STF2*  Upper cellular layer, the most superficial sojourn zone where cells translocate to the cortical plate.

*STF3*  Honeycomb trilaminar matrix *(3a, 3b, 3c)* of cells and fibers that is not present during the first trimester and is found in granular (sensory) cortices.

*STF4*  Complex middle layer where sojourning and migrating cortical neurons grow corticofugal axons and intermingle with corticopetal axons.

*STF5*  Deep cellular layer that is prominent during the first trimester, the first sojourn zone to appear outside the germinal matrix.

*STF6*  Late-forming deep layer of callosal fibers outside the germinal matrix that is not present during the first trimester.

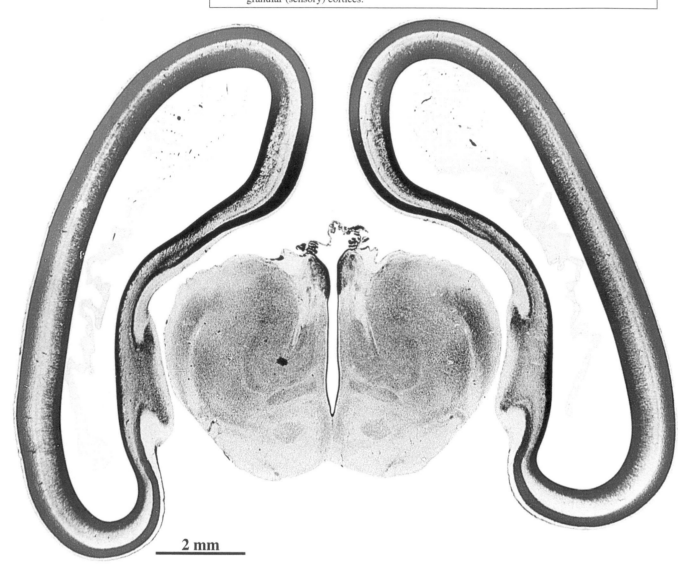

2 mm

**See a high-magnification view of the midbrain and thalamus from section 499 in Plates 30A and B.**

**FONT KEY:**
**VENTRICULAR DIVISIONS – CAPITALS**
Germinal zone - Helvetica bold
*Transient structure - Times bold italic*
Permanent structure - Times Roman or **Bold**

**ABBREVIATIONS:**
**GEP** - Glioepithelium
**G/EP** - Glioepithelium/ependyma
**NEP** - Neuroepithelium
**SVZ** - Subventricular zone

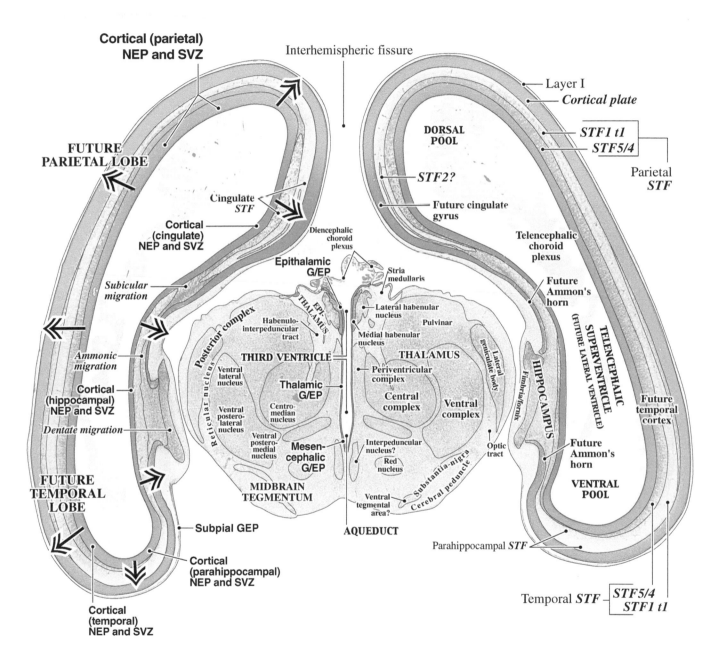

**Cortical (parietal) NEP and SVZ**

Interhemispheric fissure

Layer I

*Cortical plate*

**DORSAL POOL**

*STF1 t1*
*STF5/4*

*STF2?*

Parietal *STF*

**FUTURE PARIETAL LOBE**

Cingulate *STF*

**Cortical (cingulate) NEP and SVZ**

*Future cingulate gyrus*

Telencephalic choroid plexus

Diencephalic choroid plexus

*Subicular migration*

Epithalamic G/EP

EPI-THALAMUS

Stria medullaris

**Future Ammon's horn**

Posterior complex

Habenulo-interpeduncular tract

Lateral habenular nucleus

Pulvinar

Medial habenular nucleus

**THIRD VENTRICLE**

**THALAMUS**

Lateral geniculate body

*Ammonic migration*

Reticular nucleus

Ventral lateral nucleus

Thalamic G/EP

Periventricular complex

**Cortical (hippocampal) NEP and SVZ**

Ventral postero-lateral nucleus

Centro-median nucleus

Central complex

Ventral complex

**TELENCEPHALIC SUPERVENTRICLE (FUTURE LATERAL VENTRICLE)**

**HIPPOCAMPUS**

Fimbria/fornix

*Dentate migration*

Ventral postero-medial nucleus

Mesencephalic G/EP

Interpeduncular nucleus?

Red nucleus

Optic tract

**Future temporal cortex**

**FUTURE TEMPORAL LOBE**

**MIDBRAIN TEGMENTUM**

Substantia nigra

Cerebral peduncle

**Future Ammon's horn**

**VENTRAL POOL**

Ventral tegmental area?

**AQUEDUCT**

**Subpial GEP**

Parahippocampal *STF*

**Cortical (parahippocampal) NEP and SVZ**

Temporal *STF*

*STF5/4*
*STF1 t1*

**Cortical (temporal) NEP and SVZ**

**Arrows** indicate the presumed *direction of neuron migration* from neuroepithelial sources.

Dashed lines indicate staining and/or sectioning artifacts.

## PLATE 14A

**GW11 Coronal
CR 60 mm, Y1-59
Level 14: Section 529**

### LAYERS OF THE CORTICAL *STRATIFIED TRANSITIONAL FIELD (STF)*

*STF1* Superficial fibrous layer with an early developmental stage *(t1)* when many cells are migrating through it, followed by a late stage *(t2)* with sparse cells. Endures as the subcortical white matter.

*STF2* Upper cellular layer, the most superficial sojourn zone where cells translocate to the cortical plate.

*STF3* Honeycomb trilaminar matrix *(3a, 3b, 3c)* of cells and fibers that is not present during the first trimester and is found in granular (sensory) cortices.

*STF4* Complex middle layer where sojourning and migrating cortical neurons grow corticofugal axons and intermingle with corticopetal axons.

*STF5* Deep cellular layer that is prominent during the first trimester, the first sojourn zone to appear outside the germinal matrix.

*STF6* Late-forming deep layer of callosal fibers outside the germinal matrix that is not present during the first trimester.

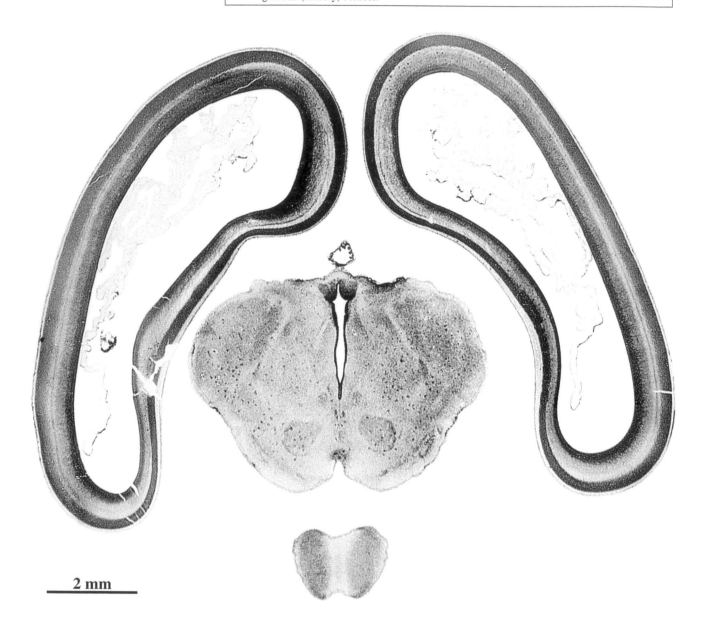

2 mm

**See a high-magnification view of the parietal cortex from section 519 in Plates 23A and B, and of the thalamus and midbrain from this section in Plates 31A and B.**

FONT KEY:
VENTRICULAR DIVISIONS – CAPITALS
Germinal zone - Helvetica bold
*Transient structure - Times bold italic*
Permanent structure - Times Roman or **Bold**

ABBREVIATIONS:
GEP - Glioepithelium
G/EP - Glioepithelium/ependyma
NEP - Neuroepithelium
SVZ - Subventricular zone

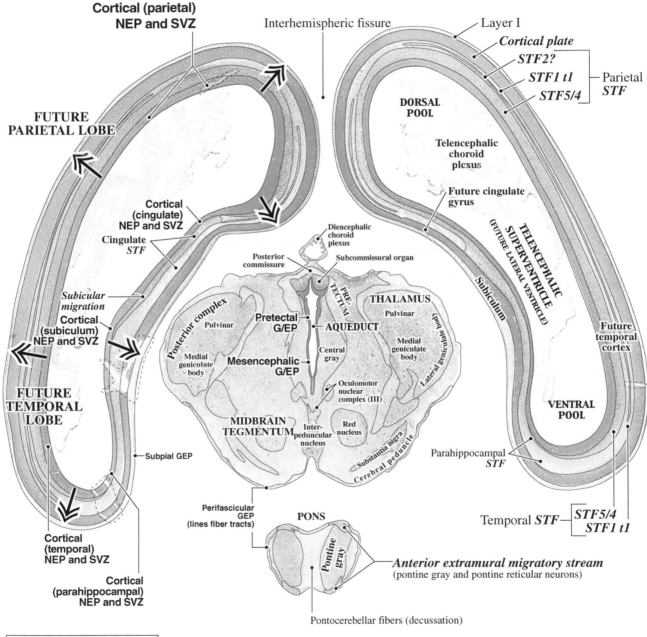

**Cortical (parietal) NEP and SVZ**

Interhemispheric fissure

Layer I

*Cortical plate*

*STF2?*

*STF1 t1*

*STF5/4*

Parietal *STF*

**FUTURE PARIETAL LOBE**

**DORSAL POOL**

Telencephalic choroid plexus

**Cortical (cingulate) NEP and SVZ**

**Future cingulate gyrus**

Cingulate *STF*

*Subicular migration*

**Cortical (subiculum) NEP and SVZ**

**TELENCEPHALIC SUPERVENTRICLE** (FUTURE LATERAL VENTRICLE)

Subiculum

**Future temporal cortex**

**FUTURE TEMPORAL LOBE**

Diencephalic choroid plexus

Subcommissural organ

Posterior commissure

**PRE-TECTUM**

**THALAMUS**

Pulvinar

Posterior complex

Pulvinar

**Pretectal G/EP**

**AQUEDUCT**

Central gray

Medial geniculate body

**Mesencephalic G/EP**

Medial geniculate body

Oculomotor nuclear complex (III)

Lateral geniculate body

**VENTRAL POOL**

**MIDBRAIN TEGMENTUM**

Inter-peduncular nucleus

Red nucleus

Substantia nigra

Cerebral peduncle

Parahippocampal *STF*

Subpial GEP

**Cortical (temporal) NEP and SVZ**

Perifascicular GEP (lines fiber tracts)

**PONS**

Pontine gray

Temporal *STF*

*STF5/4*

*STF1 t1*

*Anterior extramural migratory stream* (pontine gray and pontine reticular neurons)

**Cortical (parahippocampal) NEP and SVZ**

Pontocerebellar fibers (decussation)

**Arrows** indicate the presumed *direction of neuron migration* from neuroepithelial sources.

Dashed lines indicate staining and/or sectioning artifacts.

## PLATE 15A

**GW11 Coronal**
**CR 60 mm, Y1-59**
**Level 15: Section 569**

### LAYERS OF THE CORTICAL *STRATIFIED TRANSITIONAL FIELD (STF)*

*STF1* Superficial fibrous layer with an early developmental stage *(t1)* when many cells are migrating through it, followed by a late stage *(t2)* with sparse cells. Endures as the subcortical white matter.

*STF2* Upper cellular layer, the most superficial sojourn zone where cells translocate to the cortical plate.

*STF3* Honeycomb trilaminar matrix *(3a, 3b, 3c)* of cells and fibers that is not present during the first trimester and is found in granular (sensory) cortices.

*STF4* Complex middle layer where sojourning and migrating cortical neurons grow corticofugal axons and intermingle with corticopetal axons.

*STF5* Deep cellular layer that is prominent during the first trimester, the first sojourn zone to appear outside the germinal matrix.

*STF6* Late-forming deep layer of callosal fibers outside the germinal matrix that is not present during the first trimester.

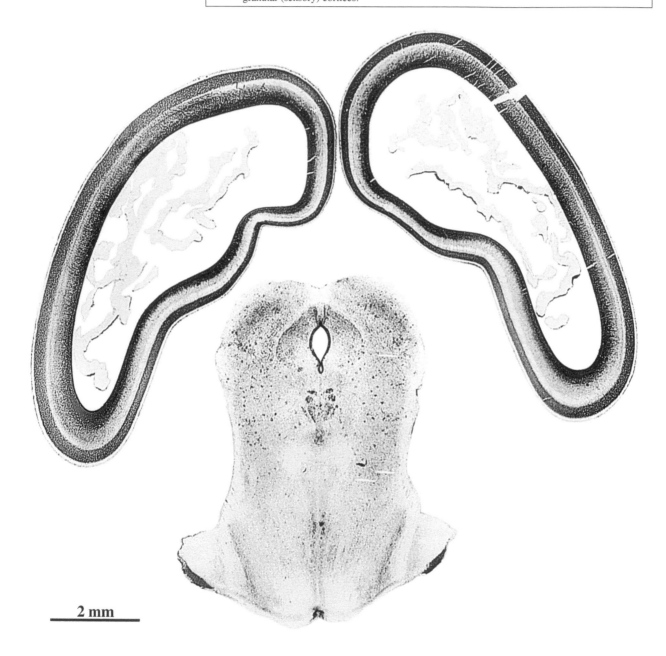

2 mm

**See a high-magnification view of the midbrain and pons from section 549 in Plates 32A and B, and from this section in Plates 33A and B.**

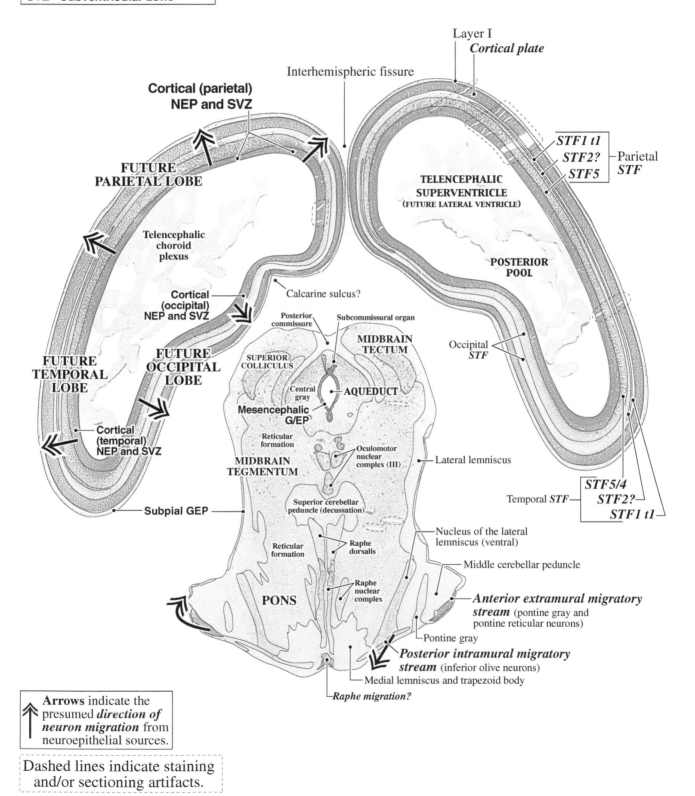

FONT KEY:
VENTRICULAR DIVISIONS – CAPITALS
Germinal zone - Helvetica bold
*Transient structure - Times bold italic*
Permanent structure - Times Roman or **Bold**

ABBREVIATIONS:
GEP - Glioepithelium
G/EP - Glioepithelium/ependyma
NEP - Neuroepithelium
SVZ - Subventricular zone

Layer I
*Cortical plate*

Interhemispheric fissure

**Cortical (parietal) NEP and SVZ**

*STF1 t1*
*STF2?*
*STF5*
— Parietal *STF*

**FUTURE PARIETAL LOBE**

**TELENCEPHALIC SUPERVENTRICLE**
(FUTURE LATERAL VENTRICLE)

**Telencephalic choroid plexus**

**POSTERIOR POOL**

**Cortical (occipital) NEP and SVZ**

Calcarine sulcus?

Posterior commissure
Subcommissural organ

**MIDBRAIN TECTUM**

SUPERIOR COLLICULUS

Occipital *STF*

**FUTURE TEMPORAL LOBE**

**FUTURE OCCIPITAL LOBE**

Central gray
**AQUEDUCT**
Mesencephalic **G/EP**

Reticular formation

Oculomotor nuclear complex (III)

Lateral lemniscus

**MIDBRAIN TEGMENTUM**

**Cortical (temporal) NEP and SVZ**

*STF5/4*
*STF2?*
*STF1 t1*

Temporal *STF*

Superior cerebellar peduncle (decussation)

Nucleus of the lateral lemniscus (ventral)

Middle cerebellar peduncle

**Subpial GEP**

Reticular formation

Raphe dorsalis

Raphe nuclear complex

**PONS**

*Anterior extramural migratory stream* (pontine gray and pontine reticular neurons)

Pontine gray

*Posterior intramural migratory stream* (inferior olive neurons)

Medial lemniscus and trapezoid body

*Raphe migration?*

**Arrows** indicate the presumed *direction of neuron migration* from neuroepithelial sources.

Dashed lines indicate staining and/or sectioning artifacts.

## PLATE 16A

**GW11 Coronal
CR 60 mm, Y1-59
Level 16: Section 599**

### LAYERS OF THE CORTICAL *STRATIFIED TRANSITIONAL FIELD (STF)*

*STF1* Superficial fibrous layer with an early developmental stage *(t1)* when many cells are migrating through it, followed by a late stage *(t2)* with sparse cells. Endures as the subcortical white matter.

*STF2* Upper cellular layer, the most superficial sojourn zone where cells translocate to the cortical plate.

*STF3* Honeycomb trilaminar matrix *(3a, 3b, 3c)* of cells and fibers that is not present during the first trimester and is found in granular (sensory) cortices.

*STF4* Complex middle layer where sojourning and migrating cortical neurons grow corticofugal axons and intermingle with corticopetal axons.

*STF5* Deep cellular layer that is prominent during the first trimester, the first sojourn zone to appear outside the germinal matrix.

*STF6* Late-forming deep layer of callosal fibers outside the germinal matrix that is not present during the first trimester.

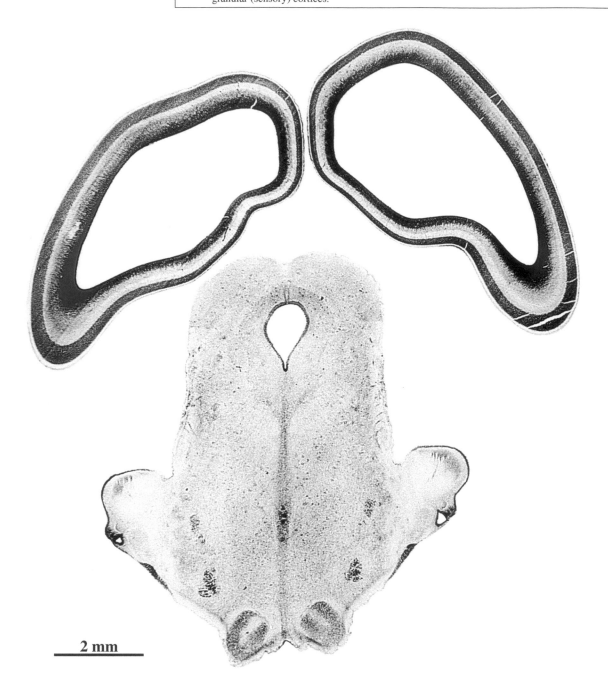

2 mm

**See a high-magnification view of the midbrain and pons from section 589 in Plates 34A and B and from this section in Plates 35A and B.**

FONT KEY:
**VENTRICULAR DIVISIONS – CAPITALS**
**Germinal zone - Helvetica bold**
*Transient structure - Times bold italic*
Permanent structure - Times Roman or **Bold**

ABBREVIATIONS:
**GEP - Glioepithelium**
**G/EP - Glioepithelium/ependyma**
**NEP - Neuroepithelium**
**SVZ - Subventricular zone**

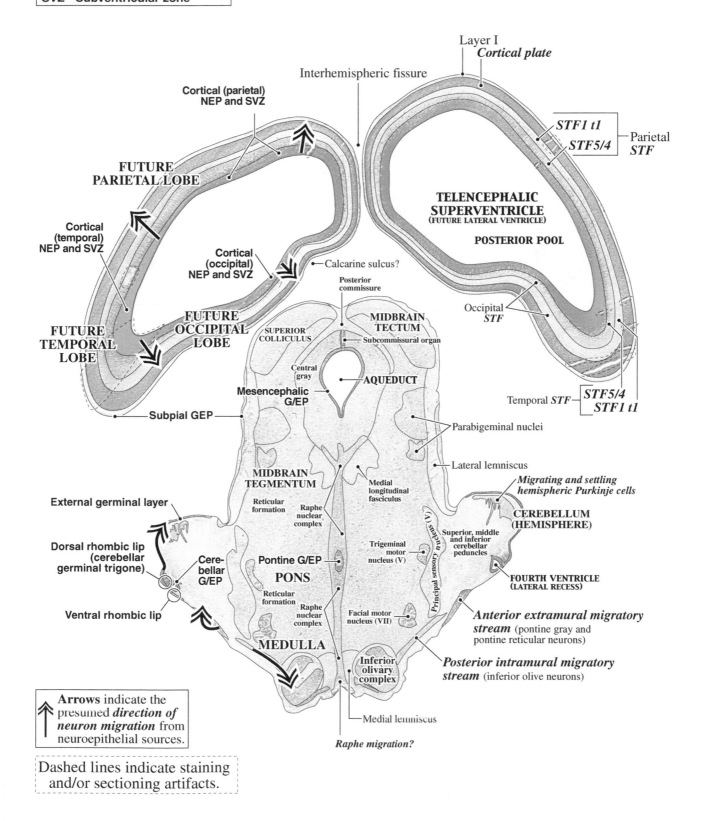

Layer I
*Cortical plate*

Interhemispheric fissure

Cortical (parietal)
NEP and SVZ

*STF1 t1*
*STF5/4*

Parietal
*STF*

**FUTURE
PARIETAL LOBE**

**TELENCEPHALIC
SUPERVENTRICLE**
(FUTURE LATERAL VENTRICLE)

**POSTERIOR POOL**

Cortical
(temporal)
NEP and SVZ

Cortical
(occipital)
NEP and SVZ

Calcarine sulcus?

**FUTURE
OCCIPITAL
LOBE**

Posterior
commissure

**MIDBRAIN
TECTUM**

Occipital
*STF*

**FUTURE
TEMPORAL
LOBE**

SUPERIOR
COLLICULUS

Subcommissural organ

Central
gray

**AQUEDUCT**

Temporal *STF*

*STF5/4*
*STF1 t1*

Mesencephalic
**G/EP**

**Subpial GEP**

Parabigeminal nuclei

Lateral lemniscus

**MIDBRAIN
TEGMENTUM**

Medial
longitudinal
fasciculus

*Migrating and settling
hemispheric Purkinje cells*

Reticular
formation

Raphe
nuclear
complex

**External germinal layer**

**Dorsal rhombic lip
(cerebellar
germinal trigone)**

Trigeminal
motor
nucleus (V)

**CEREBELLUM
(HEMISPHERE)**

Superior, middle
and inferior
cerebellar
peduncles

**Cere-
bellar
G/EP**

**Pontine G/EP**

**PONS**

Reticular
formation

**Ventral rhombic lip**

Raphe
nuclear
complex

Facial motor
nucleus (VII)

**FOURTH VENTRICLE**
(LATERAL RECESS)

*Anterior extramural migratory
stream* (pontine gray and
pontine reticular neurons)

**MEDULLA**

Inferior
olivary
complex

*Posterior intramural migratory
stream* (inferior olive neurons)

↑ **Arrows** indicate the
presumed *direction of
neuron migration* from
neuroepithelial sources.

Medial lemniscus

*Raphe migration?*

Dashed lines indicate staining
and/or sectioning artifacts.

Principal sensory nucleus (V)

# PLATE 17A

**GW11 Coronal
CR 60 mm, Y1-59
Level 17: Section 629**

## LAYERS OF THE CORTICAL *STRATIFIED TRANSITIONAL FIELD (STF)*

*STF1* Superficial fibrous layer with an early developmental stage *(t1)* when many cells are migrating through it, followed by a late stage *(t2)* with sparse cells. Endures as the subcortical white matter.

*STF2* Upper cellular layer, the most superficial sojourn zone where cells translocate to the cortical plate.

*STF3* Honeycomb trilaminar matrix *(3a, 3b, 3c)* of cells and fibers that is not present during the first trimester and is found in granular (sensory) cortices.

*STF4* Complex middle layer where sojourning and migrating cortical neurons grow corticofugal axons and intermingle with corticopetal axons.

*STF5* Deep cellular layer that is prominent during the first trimester, the first sojourn zone to appear outside the germinal matrix.

*STF6* Late-forming deep layer of callosal fibers outside the germinal matrix that is not present during the first trimester.

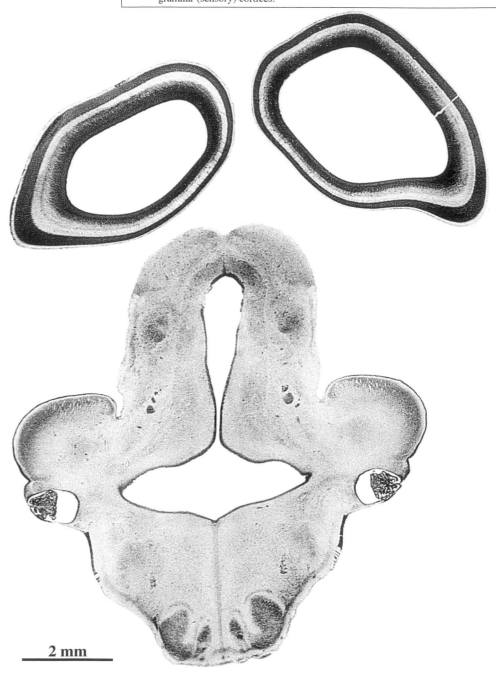

2 mm

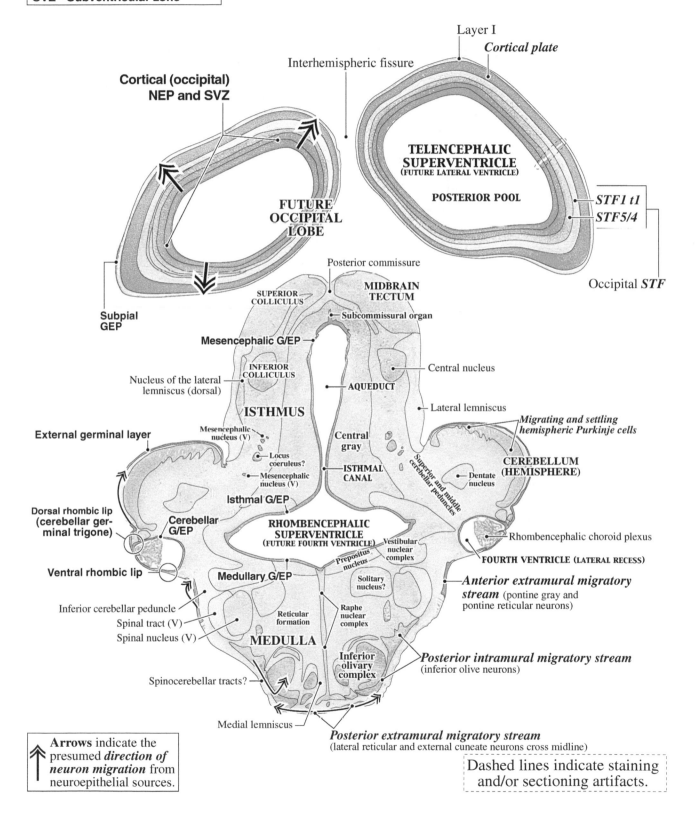

Layer I
*Cortical plate*

Interhemispheric fissure

**Cortical (occipital)
NEP and SVZ**

**TELENCEPHALIC
SUPERVENTRICLE**
(FUTURE LATERAL VENTRICLE)

**POSTERIOR POOL**

*STF1 t1*
*STF5/4*

**FUTURE
OCCIPITAL
LOBE**

**Subpial
GEP**

Occipital *STF*

Posterior commissure

**MIDBRAIN
TECTUM**

**SUPERIOR
COLLICULUS**

Subcommissural organ

**Mesencephalic G/EP**

Central nucleus

Nucleus of the lateral
lemniscus (dorsal)

**INFERIOR
COLLICULUS**

**AQUEDUCT**

Lateral lemniscus

**ISTHMUS**

Mesencephalic
nucleus (V)

*Migrating and settling
hemispheric Purkinje cells*

**External germinal layer**

Locus
coeruleus?

Central
gray

**CEREBELLUM
(HEMISPHERE)**

Mesencephalic
nucleus (V)

**ISTHMAL
CANAL**

Dentate
nucleus

Superior
and middle
cerebellar
peduncles

**Isthmal G/EP**

**Dorsal rhombic lip
(cerebellar ger-
minal trigone)**

**Cerebellar
G/EP**

**RHOMBENCEPHALIC
SUPERVENTRICLE**
(FUTURE FOURTH VENTRICLE)

Vestibular
nuclear
complex

Rhombencephalic choroid plexus

Prepositus
nucleus

**FOURTH VENTRICLE** (LATERAL RECESS)

**Ventral rhombic lip**

**Medullary G/EP**

Solitary
nucleus?

*Anterior extramural migratory
stream* (pontine gray and
pontine reticular neurons)

Inferior cerebellar peduncle

Raphe
nuclear
complex

Reticular
formation

Spinal tract (V)
Spinal nucleus (V)

**MEDULLA**

Inferior
olivary
complex

*Posterior intramural migratory stream*
(inferior olive neurons)

Spinocerebellar tracts?

Medial lemniscus

*Posterior extramural migratory stream*
(lateral reticular and external cuneate neurons cross midline)

**Arrows** indicate the
presumed *direction of
neuron migration* from
neuroepithelial sources.

Dashed lines indicate staining
and/or sectioning artifacts.

# PLATE 18A

**GW11 Coronal**
**CR 60 mm, Y1-59**
**Level 18: Section 649**

## LAYERS OF THE CORTICAL *STRATIFIED TRANSITIONAL FIELD (STF)*

**STF1** Superficial fibrous layer with an early developmental stage *(t1)* when many cells are migrating through it, followed by a late stage *(t2)* with sparse cells. Endures as the subcortical white matter.

**STF2** Upper cellular layer, the most superficial sojourn zone where cells translocate to the cortical plate.

**STF3** Honeycomb trilaminar matrix *(3a, 3b, 3c)* of cells and fibers that is not present during the first trimester and is found in granular (sensory) cortices.

**STF4** Complex middle layer where sojourning and migrating cortical neurons grow corticofugal axons and intermingle with corticopetal axons.

**STF5** Deep cellular layer that is prominent during the first trimester, the first sojourn zone to appear outside the germinal matrix.

**STF6** Late-forming deep layer of callosal fibers outside the germinal matrix that is not present during the first trimester.

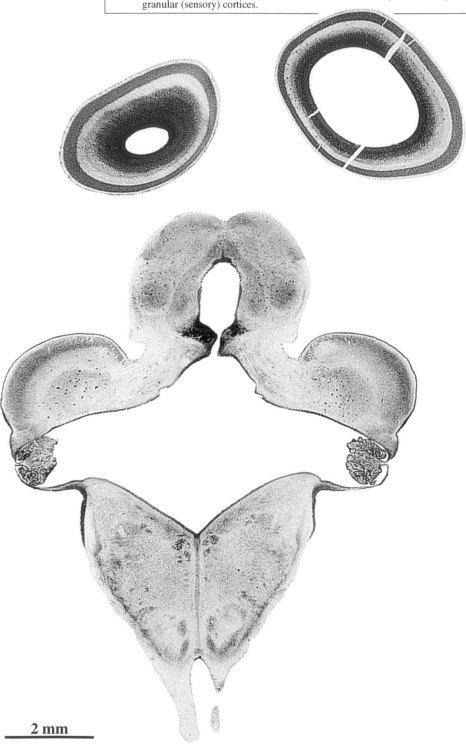

2 mm

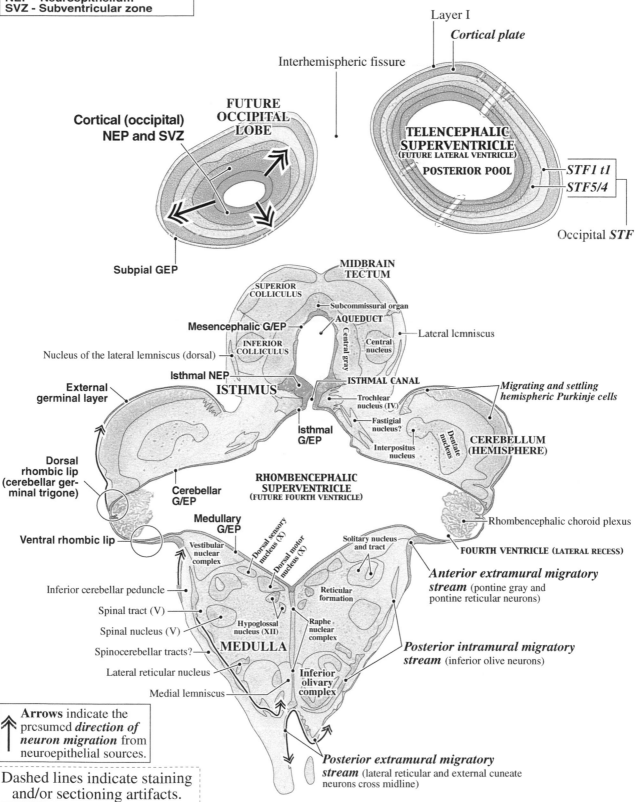

FONT KEY:
VENTRICULAR DIVISIONS – CAPITALS
Germinal zone - Helvetica bold
*Transient structure - Times bold italic*
Permanent structure - Times Roman or **Bold**

ABBREVIATIONS:
GEP - Glioepithelium
G/EP - Glioepithelium/ependyma
NEP - Neuroepithelium
SVZ - Subventricular zone

Interhemispheric fissure

Layer I
*Cortical plate*

FUTURE
OCCIPITAL
LOBE

Cortical (occipital)
NEP and SVZ

TELENCEPHALIC
SUPERVENTRICLE
(FUTURE LATERAL VENTRICLE)

POSTERIOR POOL

*STF1 t1*
*STF5/4*

Occipital *STF*

Subpial GEP

MIDBRAIN
TECTUM

SUPERIOR
COLLICULUS

Subcommissural organ

AQUEDUCT

Central
nucleus

Mesencephalic G/EP

INFERIOR
COLLICULUS

Central gray

Lateral lemniscus

Nucleus of the lateral lemniscus (dorsal)

Isthmal NEP

ISTHMUS

ISTHMAL CANAL

Trochlear
nucleus (IV)

*Migrating and settling
hemispheric Purkinje cells*

External
germinal layer

Isthmal
G/EP

Fastigial
nucleus?

Interpositus
nucleus

Dentate
nucleus

CEREBELLUM
(HEMISPHERE)

Dorsal
rhombic lip
(cerebellar ger-
minal trigone)

Cerebellar
G/EP

RHOMBENCEPHALIC
SUPERVENTRICLE
(FUTURE FOURTH VENTRICLE)

Rhombencephalic choroid plexus

Ventral rhombic lip

Medullary
G/EP

Vestibular
nuclear
complex

Dorsal sensory
nucleus (X)

Dorsal motor
nucleus (X)

Solitary nucleus
and tract

FOURTH VENTRICLE (LATERAL RECESS)

Inferior cerebellar peduncle

Reticular
formation

*Anterior extramural migratory
stream* (pontine gray and
pontine reticular neurons)

Spinal tract (V)

Spinal nucleus (V)

Hypoglossal
nucleus (XII)

Raphe
nuclear
complex

Spinocerebellar tracts?

MEDULLA

*Posterior intramural migratory
stream* (inferior olive neurons)

Lateral reticular nucleus

Medial lemniscus

Inferior
olivary
complex

**Arrows** indicate the
presumed *direction of
neuron migration* from
neuroepithelial sources.

*Posterior extramural migratory
stream* (lateral reticular and external cuneate
neurons cross midline)

Dashed lines indicate staining
and/or sectioning artifacts.

# PLATE 19A

**GW11 Coronal
CR 60 mm, Y1-59
Level 19: Section 659**

## LAYERS OF THE CORTICAL *STRATIFIED TRANSITIONAL FIELD (STF)*

*STF1*  Superficial fibrous layer with an early developmental stage *(t1)* when many cells are migrating through it, followed by a late stage *(t2)* with sparse cells. Endures as the subcortical white matter.

*STF2*  Upper cellular layer, the most superficial sojourn zone where cells translocate to the cortical plate.

*STF3*  Honeycomb trilaminar matrix *(3a, 3b, 3c)* of cells and fibers that is not present during the first trimester and is found in granular (sensory) cortices.

*STF4*  Complex middle layer where sojourning and migrating cortical neurons grow corticofugal axons and intermingle with corticopetal axons.

*STF5*  Deep cellular layer that is prominent during the first trimester, the first sojourn zone to appear outside the germinal matrix.

*STF6*  Late-forming deep layer of callosal fibers outside the germinal matrix that is not present during the first trimester.

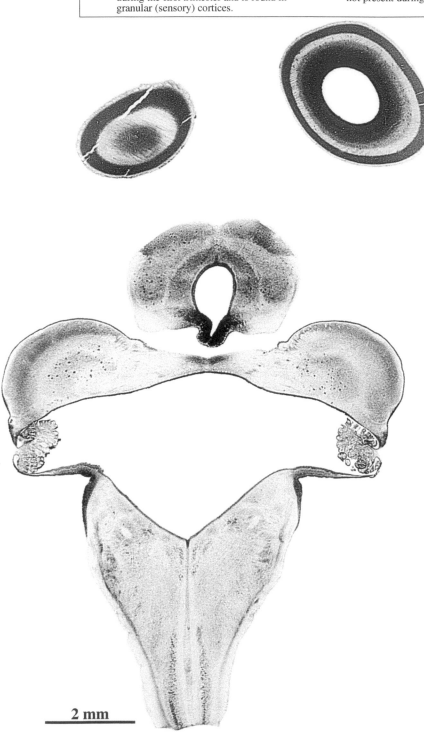

2 mm

FONT KEY:
VENTRICULAR DIVISIONS – CAPITALS
Germinal zone - Helvetica bold
*Transient structure - Times bold italic*
Permanent structure - Times Roman or **Bold**

ABBREVIATIONS:
GEP - Glioepithelium
G/EP - Glioepithelium/ependyma
NEP - Neuroepithelium
SVZ - Subventricular zone

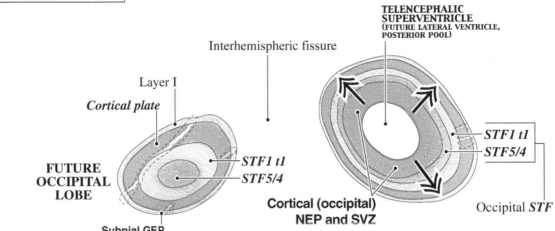

Interhemispheric fissure

Layer I
*Cortical plate*

**TELENCEPHALIC
SUPERVENTRICLE**
(FUTURE LATERAL VENTRICLE,
POSTERIOR POOL)

**FUTURE
OCCIPITAL
LOBE**

*STF1 t1*
*STF5/4*

*STF1 t1*
*STF5/4*

Cortical (occipital)
NEP and SVZ

Occipital *STF*

Subpial GEP

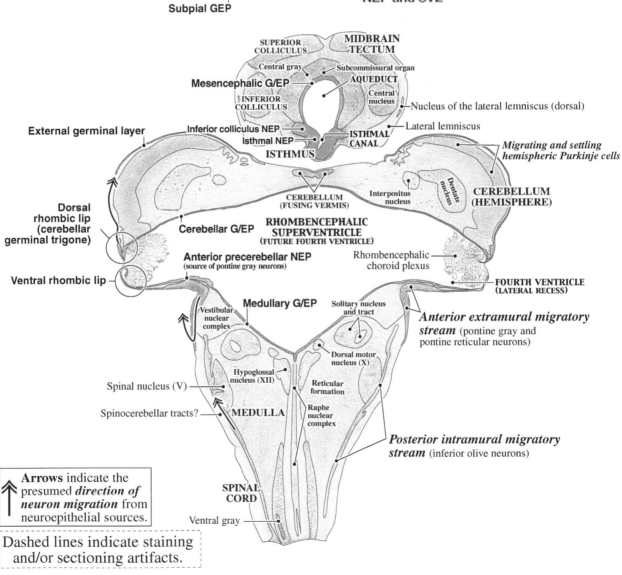

SUPERIOR
COLLICULUS

**MIDBRAIN
TECTUM**

Central gray
Subcommissural organ

Mesencephalic G/EP
**AQUEDUCT**

INFERIOR
COLLICULUS

Central
nucleus

Nucleus of the lateral lemniscus (dorsal)

**External germinal layer**

Inferior colliculus NEP
Isthmal NEP

**ISTHMAL
CANAL**

Lateral lemniscus

*Migrating and settling
hemispheric Purkinje cells*

**ISTHMUS**

CEREBELLUM
(FUSING VERMIS)

Interpositus
nucleus

Dentate
nucleus

**CEREBELLUM
(HEMISPHERE)**

**Dorsal
rhombic lip
(cerebellar
germinal trigone)**

Cerebellar G/EP

**RHOMBENCEPHALIC
SUPERVENTRICLE**
(FUTURE FOURTH VENTRICLE)

**Anterior precerebellar NEP**
(source of pontine gray neurons)

Rhombencephalic
choroid plexus

**Ventral rhombic lip**

**Medullary G/EP**

Solitary nucleus
and tract

**FOURTH VENTRICLE**
(LATERAL RECESS)

*Anterior extramural migratory
stream* (pontine gray and
pontine reticular neurons)

Vestibular
nuclear
complex

Dorsal motor
nucleus (X)

Hypoglossal
nucleus (XII)

Reticular
formation

Spinal nucleus (V)

Raphe
nuclear
complex

Spinocerebellar tracts?

**MEDULLA**

*Posterior intramural migratory
stream* (inferior olive neurons)

**SPINAL
CORD**

Ventral gray

**Arrows** indicate the
presumed *direction of
neuron migration* from
neuroepithelial sources.

Dashed lines indicate staining
and/or sectioning artifacts.

**PLATE 20A**

**GW11 Coronal**
**CR 60 mm, Y1-59**
**Level 20: Section 680**

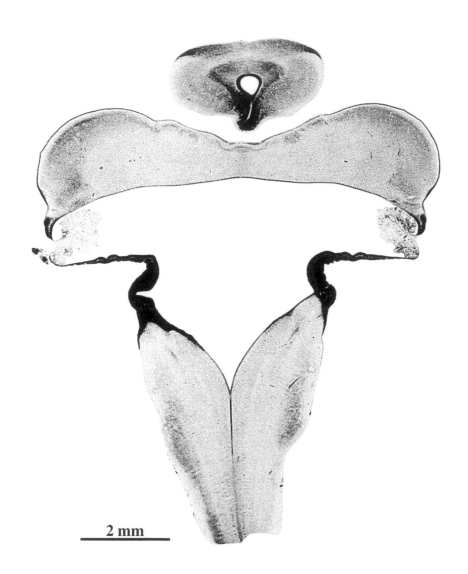

2 mm

FONT KEY:
**VENTRICULAR DIVISIONS – CAPITALS**
**Germinal zone - Helvetica bold**
*Transient structure - Times bold italic*
Permanent structure - Times Roman or **Bold**

ABBREVIATIONS:
**G/EP - Glioepithelium/ependyma**
**NEP - Neuroepithelium**

**Arrows** indicate the
presumed *direction of*
*neuron migration* from
neuroepithelial sources.

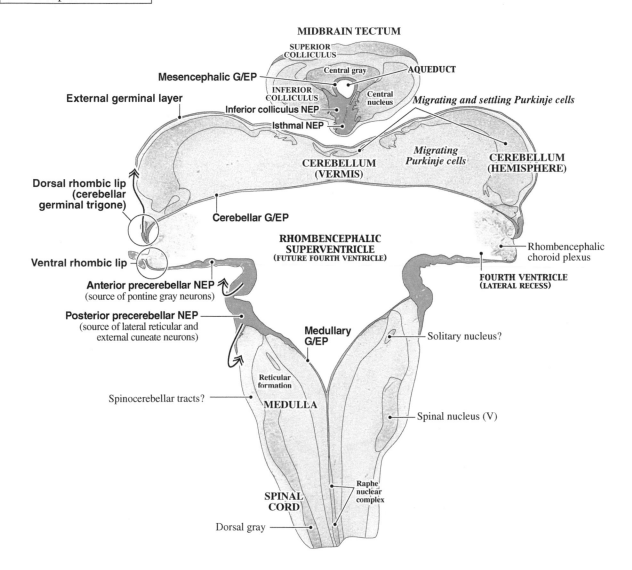

**MIDBRAIN TECTUM**

**SUPERIOR COLLICULUS**

Central gray

**AQUEDUCT**

Mesencephalic G/EP

**INFERIOR COLLICULUS**

Central nucleus

External germinal layer

Inferior colliculus NEP

*Migrating and settling Purkinje cells*

Isthmal NEP

*Migrating Purkinje cells*

**CEREBELLUM (VERMIS)**

**CEREBELLUM (HEMISPHERE)**

Dorsal rhombic lip
(cerebellar germinal trigone)

Cerebellar G/EP

**RHOMBENCEPHALIC SUPERVENTRICLE**
**(FUTURE FOURTH VENTRICLE)**

Rhombencephalic choroid plexus

Ventral rhombic lip

**Anterior precerebellar NEP**
(source of pontine gray neurons)

**FOURTH VENTRICLE**
**(LATERAL RECESS)**

**Posterior precerebellar NEP**
(source of lateral reticular and external cuneate neurons)

**Medullary G/EP**

Solitary nucleus?

*Reticular formation*

Spinocerebellar tracts?

**MEDULLA**

Spinal nucleus (V)

**SPINAL CORD**

*Raphe nuclear complex*

Dorsal gray

**PLATE 21A**

**GW11 Coronal
CR 60 mm, Y1-59**

**Level 21: Section 709**

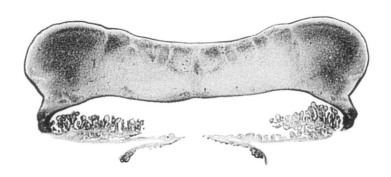

**Level 22: Section 720**

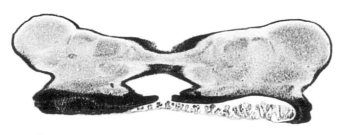

2 mm

## Level 21: Section 709

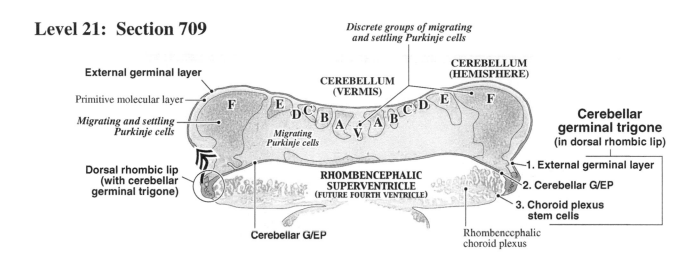

## Level 22: Section 720

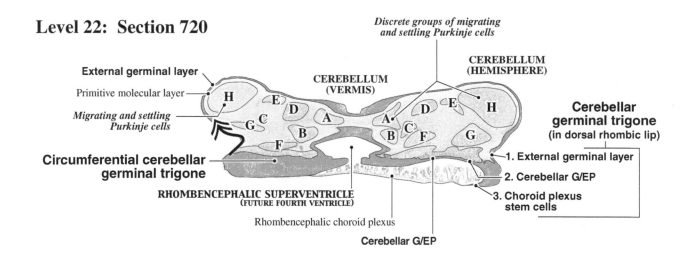

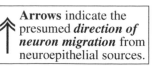

**Arrows** indicate the presumed *direction of neuron migration* from neuroepithelial sources.

58

GW11 Coronal
CR 60 mm, Y1-59
Between levels 2 and 3:
Section 269
FRONTAL
CORTEX

**PLATE 22A**

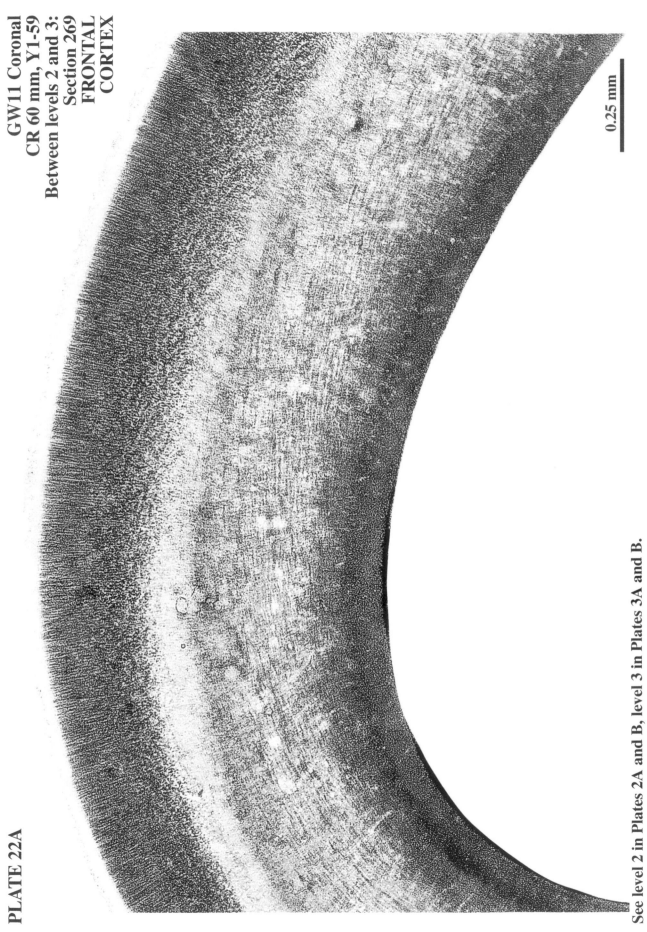

0.25 mm

See level 2 in Plates 2A and B, level 3 in Plates 3A and B.

## PLATE 22B
## FRONTAL CORTEX

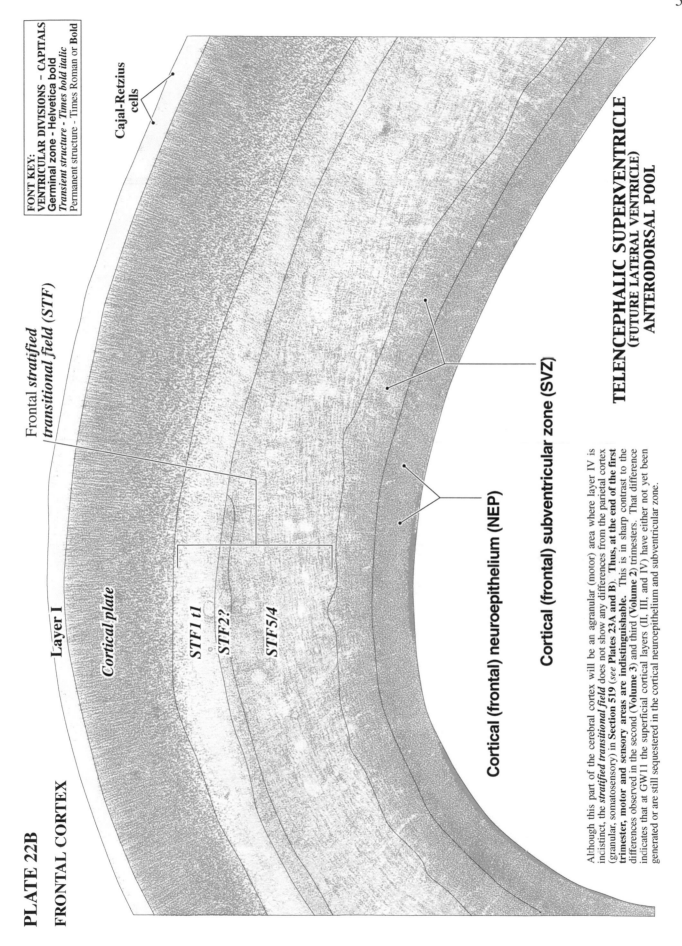

Cajal-Retzius cells

Layer I

Frontal *stratified transitional field (STF)*

*Cortical plate*

*STF1 t1*

*STF2?*

*STF5/4*

**Cortical (frontal) neuroepithelium (NEP)**

**Cortical (frontal) subventricular zone (SVZ)**

**TELENCEPHALIC SUPERVENTRICLE**
**(FUTURE LATERAL VENTRICLE)**
**ANTERODORSAL POOL**

Although this part of the cerebral cortex will be an agranular (motor) area where layer IV is incistinct, the *stratifed transitional field* does not show any differences from the parietal cortex (granular, somatosensory) in Section 519 (*see* **Plates 23A and B**). **Thus, at the end of the first trimester, motor and sensory areas are indistinguishable.** This is in sharp contrast to the differences observed in the second (**Volume 3**) and third (**Volume 2**) trimesters. That difference indicates that at GW11 the superficial cortical layers (II, III, and IV) have either not yet been generated or are still sequestered in the cortical neuroepithelium and subventricular zone.

GW11 Coronal
CR 60 mm, Y1-59
Near level 14:
Section 519
PARIETAL
CORTEX

0.25 mm

PLATE 23A

See level 14 in Plates 14A and B.

61

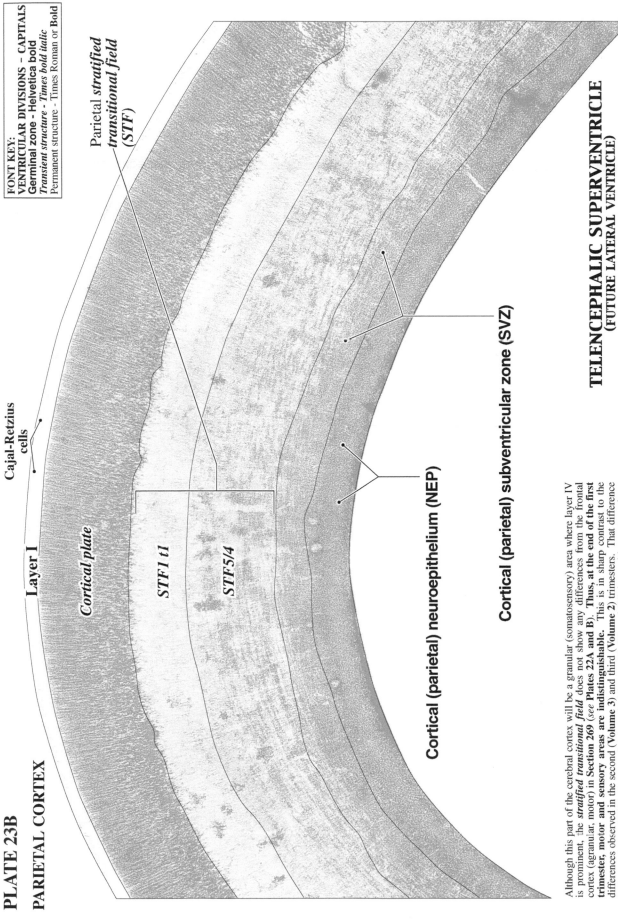

PLATE 23B
PARIETAL CORTEX

FONT KEY:
VENTRICULAR DIVISIONS – CAPITALS
Germinal zone - Helvetica bold
Transient structure - Times bold italic
Permanent structure - Times Roman or Bold

Parietal *stratified transitional field (STF)*

Cajal-Retzius cells

Layer I

Cortical plate

STF1 t1

STF5/4

Cortical (parietal) neuroepithelium (NEP)

Cortical (parietal) subventricular zone (SVZ)

TELENCEPHALIC SUPERVENTRICLE
(FUTURE LATERAL VENTRICLE)
DORSAL POOL

Although this part of the cerebral cortex will be a granular (somatosensory) area where layer IV is prominent, the *stratified transitional field* does not show any differences from the frontal cortex (agranular, motor) in **Section 269** (*see* **Plates 22A and B**). **Thus, at the end of the first trimester, motor and sensory areas are indistinguishable.** This is in sharp contrast to the differences observed in the second (Volume 3) and third (Volume 2) trimesters. That difference indicates that at GW11 the superficial cortical layers (II, III, and IV) have either not yet been generated or are still sequestered in the cortical neuroepithelium and subventricular zone.

**PLATE 24A**

**PARACENTRAL
CORTEX**

See levels 8 and 9
in Plates
8 to 9A and B.

Enlarged in
Plates 25A and B.

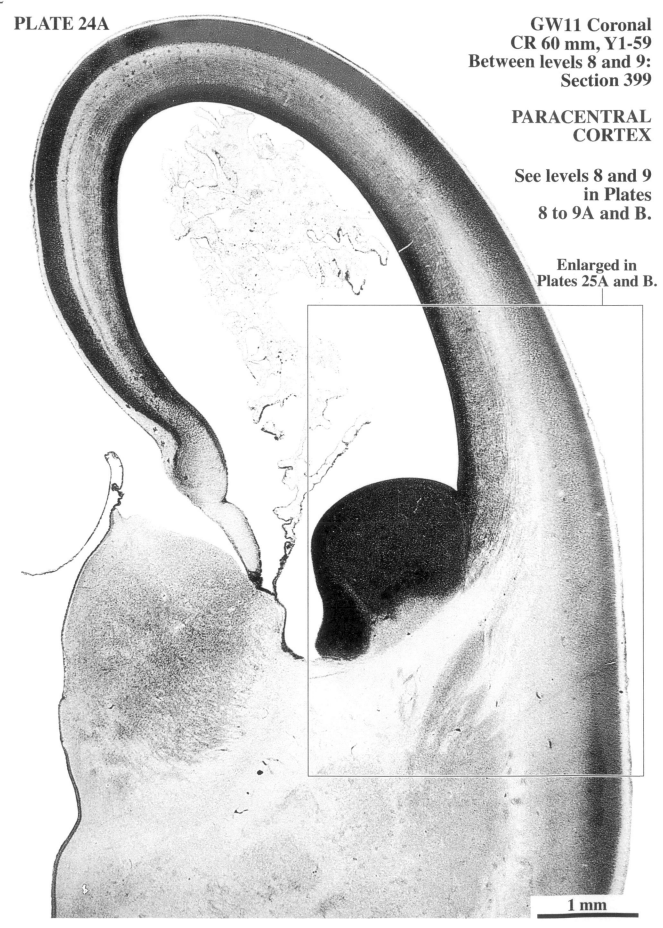

1 mm

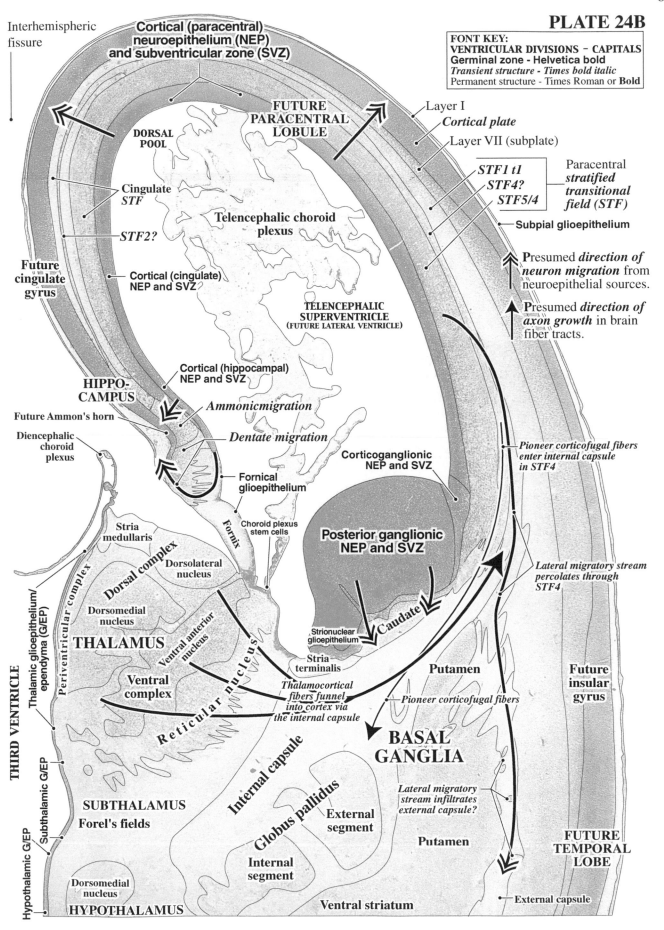

Interhemispheric fissure

Cortical (paracentral) neuroepithelium (NEP) and subventricular zone (SVZ)

**PLATE 24B**

**FUTURE PARACENTRAL LOBULE**

Layer I
*Cortical plate*
Layer VII (subplate)

*STF1 t1*
*STF4?*
*STF5/4*

Paracentral *stratified transitional field (STF)*

**DORSAL POOL**

Cingulate *STF*

*STF2?*

Telencephalic choroid plexus

Subpial glioepithelium

Presumed *direction of neuron migration* from neuroepithelial sources.

**Future cingulate gyrus**

Cortical (cingulate) NEP and SVZ

Presumed *direction of axon growth* in brain fiber tracts.

TELENCEPHALIC SUPERVENTRICLE
(FUTURE LATERAL VENTRICLE)

Cortical (hippocampal) NEP and SVZ

**HIPPO-CAMPUS**

*Ammonicmigration*

Future Ammon's horn

*Dentate migration*

Diencephalic choroid plexus

Corticoganglionic NEP and SVZ

*Pioneer corticofugal fibers enter internal capsule in STF4*

Fornical glioepithelium

Choroid plexus stem cells

Stria medullaris

Posterior ganglionic NEP and SVZ

*Lateral migratory stream percolates through STF4*

Dorsal complex

Dorsolateral nucleus

Dorsomedial nucleus

Caudate

**THALAMUS**

Ventral anterior nucleus

Strionuclear glioepithelium

**Future insular gyrus**

Ventral complex

Stria terminalis

Putamen

*Thalamocortical fibers funnel into cortex via the internal capsule*

*Reticular nucleus*

*Pioneer corticofugal fibers*

**THIRD VENTRICLE**

Thalamic glioepithelium/ependyma (G/EP)

Periventricular complex

**BASAL GANGLIA**

Internal capsule

**SUBTHALAMUS**
Forel's fields

Globus pallidus

External segment

*Lateral migratory stream infiltrates external capsule?*

**FUTURE TEMPORAL LOBE**

Subthalamic G/EP

Putamen

Internal segment

Hypothalamic G/EP

Dorsomedial nucleus

**HYPOTHALAMUS**

Ventral striatum

External capsule

**PLATE 25A**

**GW11 Coronal**
**CR 60 mm, Y1-59**
**Between levels 8 and 9:**
**Section 399**

**ENTRY/EXIT ZONE IN LATERAL**
**PARACENTRAL CORTEX**

**See levels 8 and 9 in**
**Plates 8 to 9A and B.**

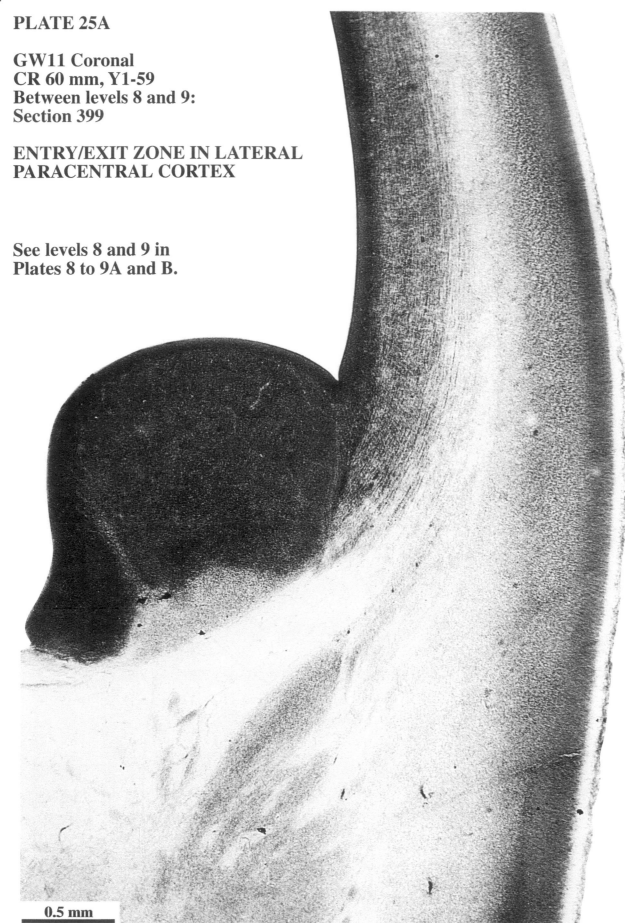

0.5 mm

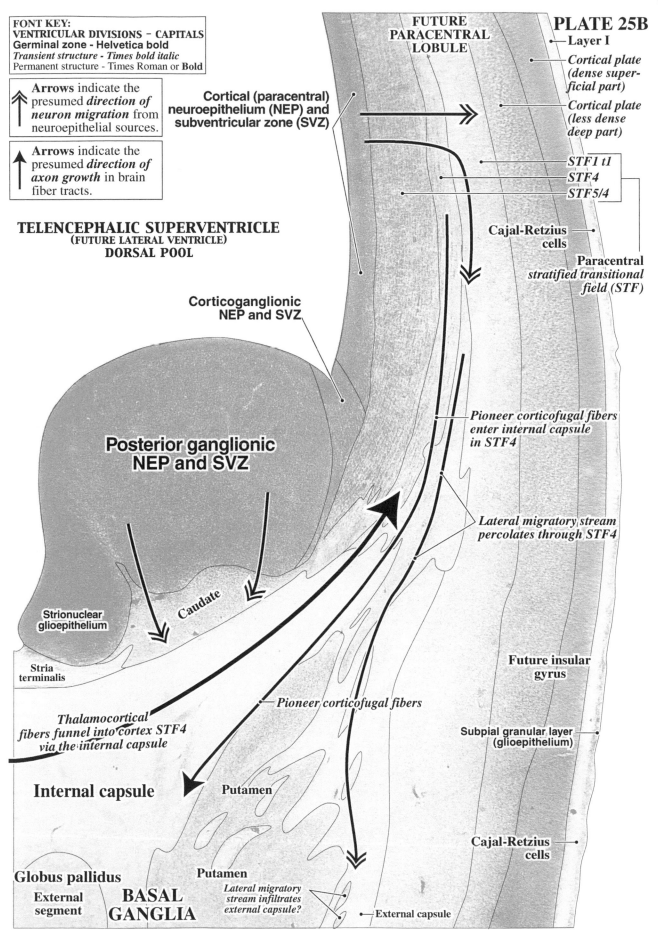

65

PLATE 25B

FUTURE PARACENTRAL LOBULE

Layer I

Cortical plate (dense super-ficial part)

Cortical plate (less dense deep part)

STF1 t1
STF4
STF5/4

Cajal-Retzius cells

Paracentral stratified transitional field (STF)

FONT KEY:
VENTRICULAR DIVISIONS – CAPITALS
Germinal zone - Helvetica bold
Transient structure - Times bold italic
Permanent structure - Times Roman or **Bold**

**Arrows** indicate the presumed *direction of neuron migration* from neuroepithelial sources.

**Arrows** indicate the presumed *direction of axon growth* in brain fiber tracts.

Cortical (paracentral) neuroepithelium (NEP) and subventricular zone (SVZ)

TELENCEPHALIC SUPERVENTRICLE
(FUTURE LATERAL VENTRICLE)
DORSAL POOL

Corticoganglionic NEP and SVZ

**Posterior ganglionic NEP and SVZ**

*Pioneer corticofugal fibers enter internal capsule in STF4*

*Lateral migratory stream percolates through STF4*

Strionuclear glioepithelium

Caudate

Stria terminalis

*Thalamocortical fibers funnel into cortex STF4 via the internal capsule*

*Pioneer corticofugal fibers*

**Future insular gyrus**

**Subpial granular layer** (glioepithelium)

**Internal capsule**

Putamen

**Cajal-Retzius cells**

**Globus pallidus**

Putamen

**External segment**

**BASAL GANGLIA**

*Lateral migratory stream infiltrates external capsule?*

External capsule

PLATE 26A
GW11 Coronal, CR 60 mm, Y1-59, Near level 8: Section 389

DIENCEPHALON, BASAL GANGLIA,
and BASAL TELENCEPHALON

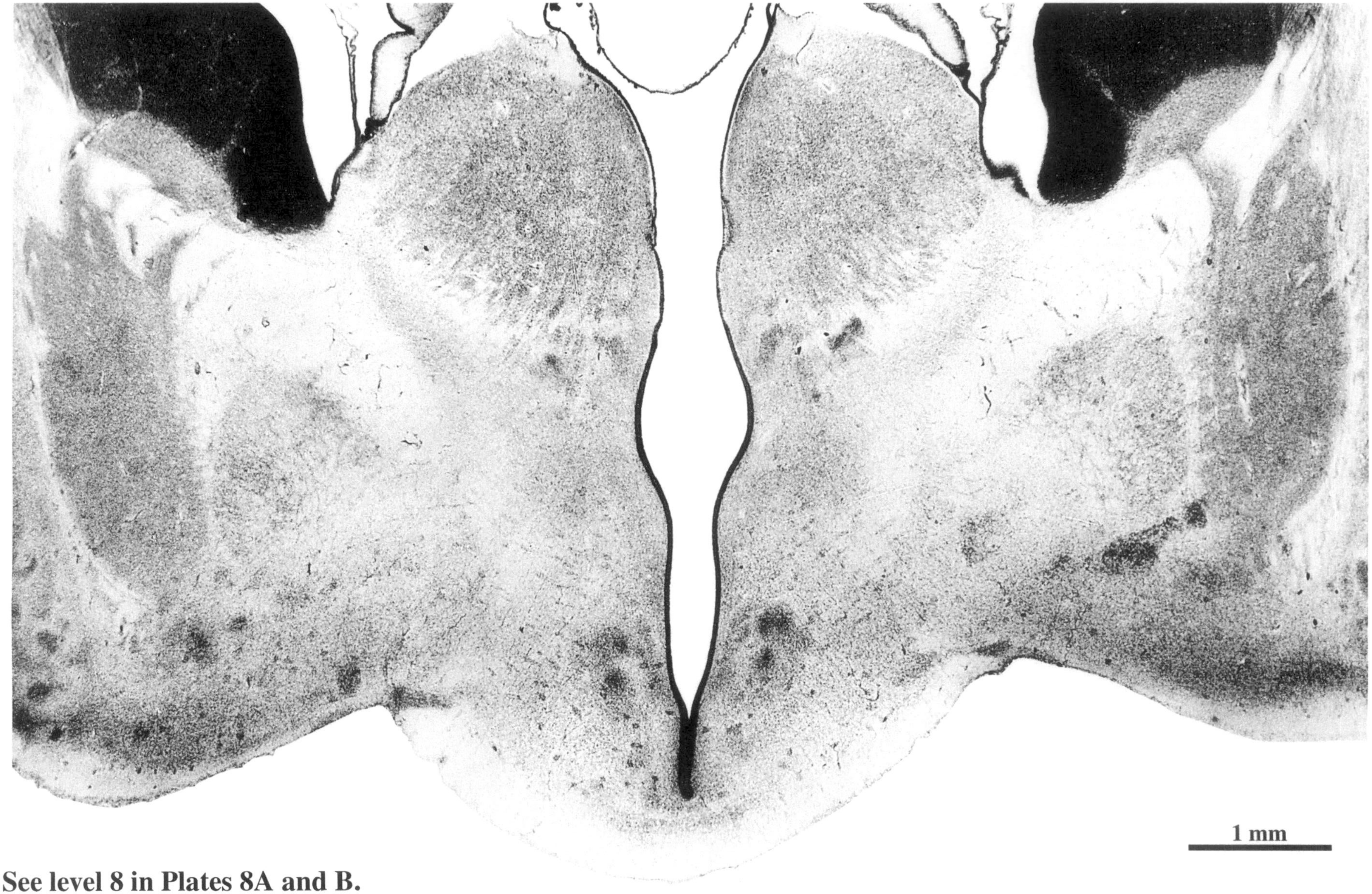

1 mm

See level 8 in Plates 8A and B.

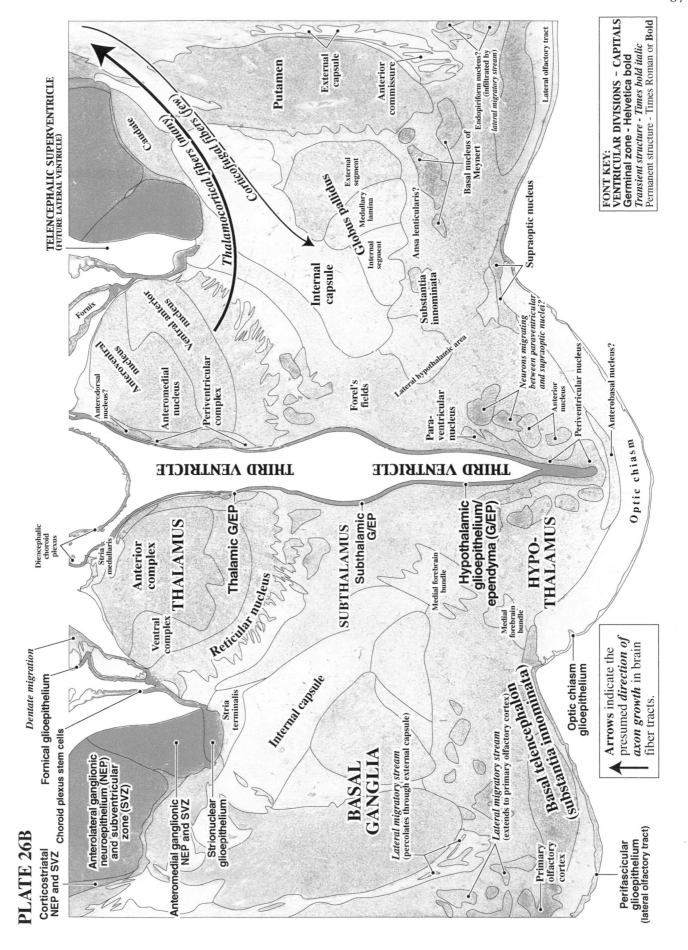

PLATE 26B

67

FONT KEY:
VENTRICULAR DIVISIONS – CAPITALS
Germinal zone - Helvetica bold
Transient structure - Times bold italic
Permanent structure - Times Roman or Bold

Arrows indicate the presumed direction of axon growth in brain fiber tracts.

Corticostriatal NEP and SVZ

Fornical glioepithelium
Choroid plexus stem cells

Dentate migration

Anterolateral ganglionic neuroepithelium (NEP) and subventricular zone (SVZ)

Anteromedial ganglionic NEP and SVZ

Strionuclear glioepithelium

Diencephalic choroid plexus

Stria t. medullaris

THALAMUS

Anterior complex

Ventral complex

Thalamic G/EP

Reticular nucleus

Stria terminalis

Internal capsule

SUBTHALAMUS

Subthalamic G/EP

Medial forebrain bundle

BASAL GANGLIA

Lateral migratory stream (percolates through external capsule)

Lateral migratory stream (extends to primary olfactory cortex)

Hypothalamic glioepithelium/ependyma (G/EP)

HYPO-THALAMUS

Basal telencephalon (substantia innominata)

Primary olfactory cortex

Perifascicular glioepithelium (lateral olfactory tract)

Optic chiasm glioepithelium

THIRD VENTRICLE

THIRD VENTRICLE

TELENCEPHALIC SUPERVENTRICLE (FUTURE LATERAL VENTRICLE)

Caudate

Fornix

Thalamocortical fibers (many)
Corticofugal fibers (few)

Anteroventral nucleus

Anterodorsal nucleus?

Ventral anterior nucleus

Anteromedial nucleus

Periventricular complex

Forel's fields

Lateral hypothalamic area

Paraventricular nucleus

Neurons migrating between paraventricular and supraoptic nuclei?

Anterior nucleus

Periventricular nucleus

Anterobasal nucleus?

Optic chiasm

Supraoptic nucleus

Substantia innominata

Ansa lenticularis?

Basal nucleus of Meynert

Endopiriform nucleus? (infiltrated by lateral migratory stream)

Lateral olfactory tract

Anterior commissure

External capsule

Putamen

Globus Pallidus
External segment
Medullary lamina
Internal segment

Internal capsule

Medial forebrain bundle

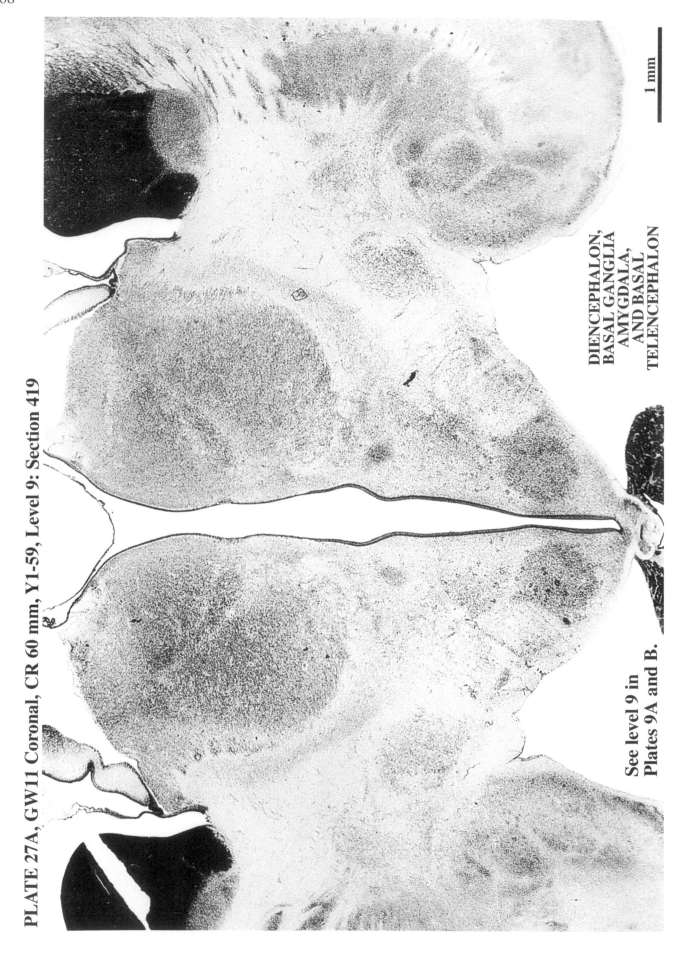

PLATE 27A, GW11 Coronal, CR 60 mm, Y1-59, Level 9: Section 419

DIENCEPHALON, BASAL GANGLIA, AMYGDALA, AND BASAL TELENCEPHALON

See level 9 in Plates 9A and B.

1 mm

# PLATE 27B

**Dentate migration**

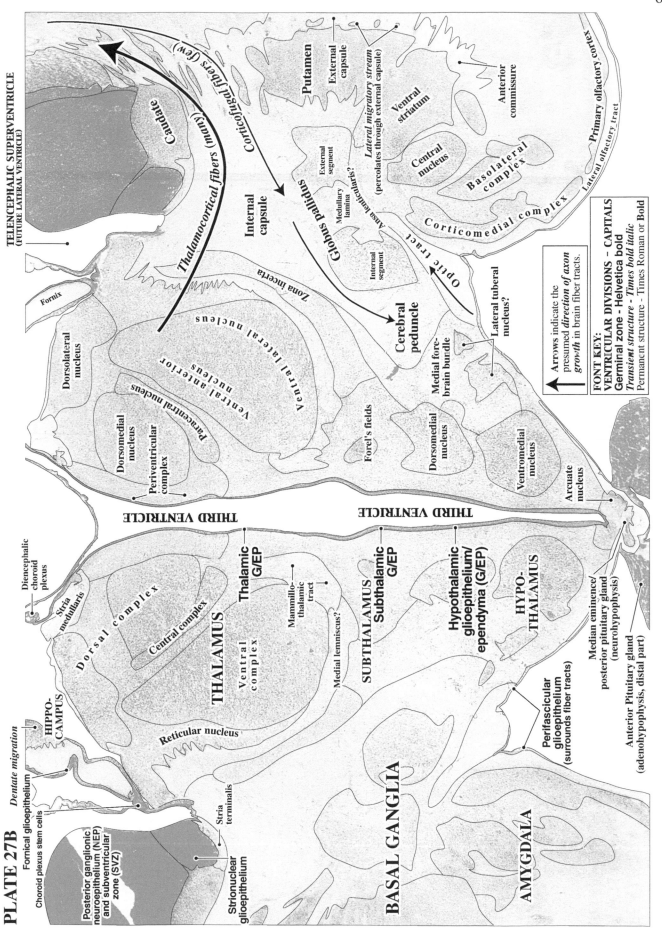

Fornical glioepithelium

Choroid plexus stem cells

**Posterior ganglionic neuroepithelium (NEP) and subventricular zone (SVZ)**

Strionuclear glioepithelium

Stria terminalis

**HIPPO-CAMPUS**

Diencephalic choroid plexus

Stria medullaris

Dorsal complex

Central complex

**THALAMUS**

Ventral complex

Reticular nucleus

**Thalamic G/EP**

Mammillo-thalamic tract

Medial lemniscus?

**SUBTHALAMUS**
**Subthalamic G/EP**

**Hypothalamic glioepithelium/ ependyma (G/EP)**

**HYPO-THALAMUS**

Perifascicular glioepithelium (surrounds fiber tracts)

Median eminence/ posterior pituitary gland (neurohypophysis)

Anterior Pituitary gland (adenohypophysis, distal part)

**BASAL GANGLIA**

**AMYGDALA**

**THIRD VENTRICLE**

**THIRD VENTRICLE**

**TELENCEPHALIC SUPERVENTRICLE**
(FUTURE LATERAL VENTRICLE)

Caudate

Thalamocortical fibers (many)

Corticofugal fibers (few)

Corticofugal fibers (many)

Internal capsule

Fornix

Zona Incerta

Dorsolateral nucleus

Ventral anterior nucleus

Ventral lateral nucleus

Dorsomedial nucleus

Paracentral nucleus

Periventricular complex

Forel's fields

Dorsomedial nucleus

Ventromedial nucleus

Arcuate nucleus

Putamen

External capsule

Lateral migratory stream (percolates through external capsule)

Ventral striatum

External segment

Medullary lamina

Ansa lenticularis?

Internal segment

**Globus Pallidus**

Cerebral peduncle

Optic tract

Medial fore-brain bundle

Lateral tuberal nucleus?

Central nucleus

Basolateral complex

Anterior commissure

Corticomedial complex

Primary olfactory cortex

Lateral olfactory tract

Arrows indicate the presumed *direction of axon growth* in brain fiber tracts.

FONT KEY:
VENTRICULAR DIVISIONS - CAPITALS
Germinal zone - Helvetica bold
*Transient structure - Times bold italic*
Permanent structure - Times Roman or Bold

DIENCEPHALON

1 mm

PLATE 28A
GW11 Coronal, CR 60 mm, Y1-59,
Between levels 10 and 11: Section 449

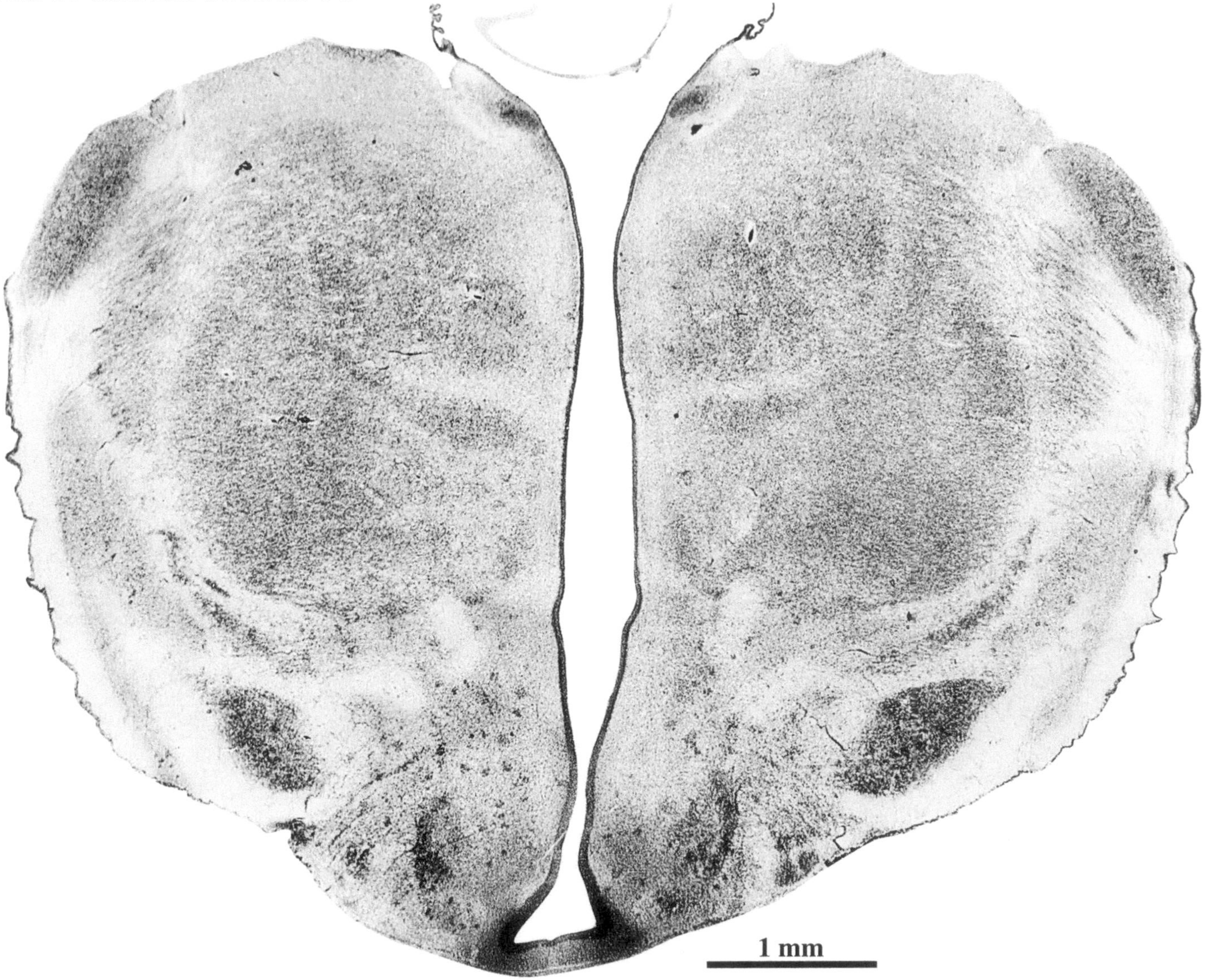

See levels 10 and 11 in Plates 10A and B to 11A and B.

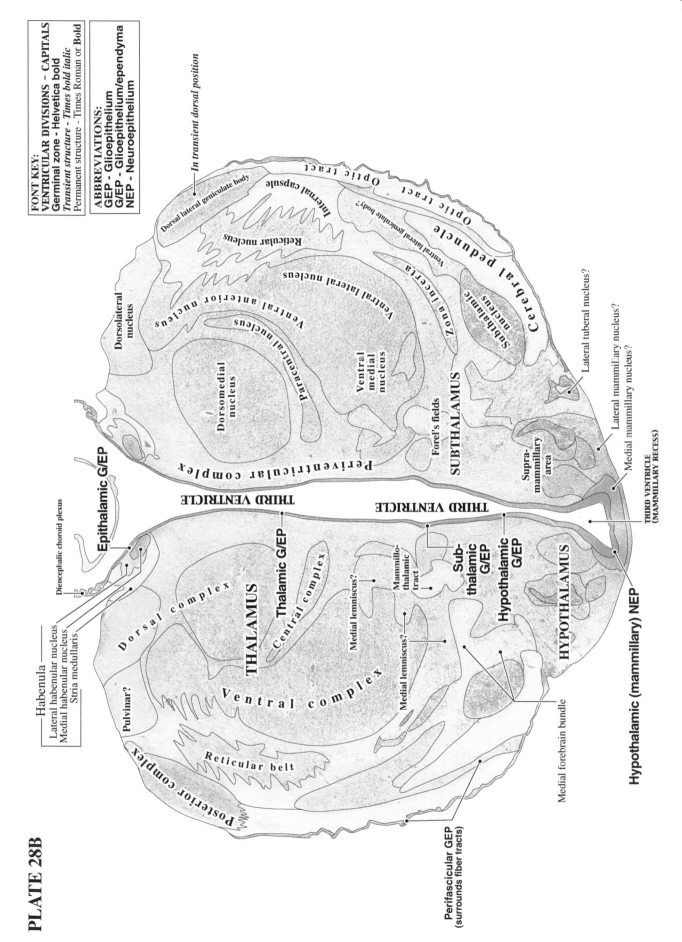

**PLATE 28B**

FONT KEY:
VENTRICULAR DIVISIONS - CAPITALS
Germinal zone - Helvetica bold
*Transient structure - Times bold italic*
Permanent structure - Times Roman or Bold

ABBREVIATIONS:
GEP - Glioepithelium
G/EP - Glioepithelium/ependyma
NEP - Neuroepithelium

*In transient dorsal position*

*Dorsal lateral geniculate body*

**Internal capsule**

**Optic tract**

*Reticular nucleus*

**Optic tract**

Dorsolateral nucleus

*Ventral anterior nucleus*

*Ventral lateral nucleus*

*Ventral lateral geniculate body?*

**Cerebral peduncle**

*Zona incerta*

*Paracentral nucleus*

Dorsomedial nucleus

Ventral medial nucleus

Subthalamic nucleus

Lateral tuberal nucleus?

Forel's fields

**SUBTHALAMUS**

Lateral mammillary nucleus?

Medial mammillary nucleus?

Supra-mammillary area

THIRD VENTRICLE (MAMMILLARY RECESS)

Epithalamic G/EP

*Periventricular complex*

**THIRD VENTRICLE**

**THIRD VENTRICLE**

Diencephalic choroid plexus

Habenula
Lateral habenular nucleus
Medial habenular nucleus
Stria medullaris

Pulvinar?

*Dorsal complex*

**THALAMUS**

Thalamic G/EP

*Central complex*

Medial lemniscus?

*Mammillo-thalamic tract*

Sub-thalamic G/EP

Hypothalamic G/EP

**HYPOTHALAMUS**

Medial lemniscus?

*Ventral complex*

Medial lemniscus?

*Reticular belt*

*Posterior complex*

Perifascicular GEP (surrounds fiber tracts)

Medial forebrain bundle

**Hypothalamic (mammillary) NEP**

72

DIENCEPHALON

1 mm

PLATE 29A
GW11 Coronal, CR 60 mm, Y1-59,
Between levels 11 and 12: Section 469

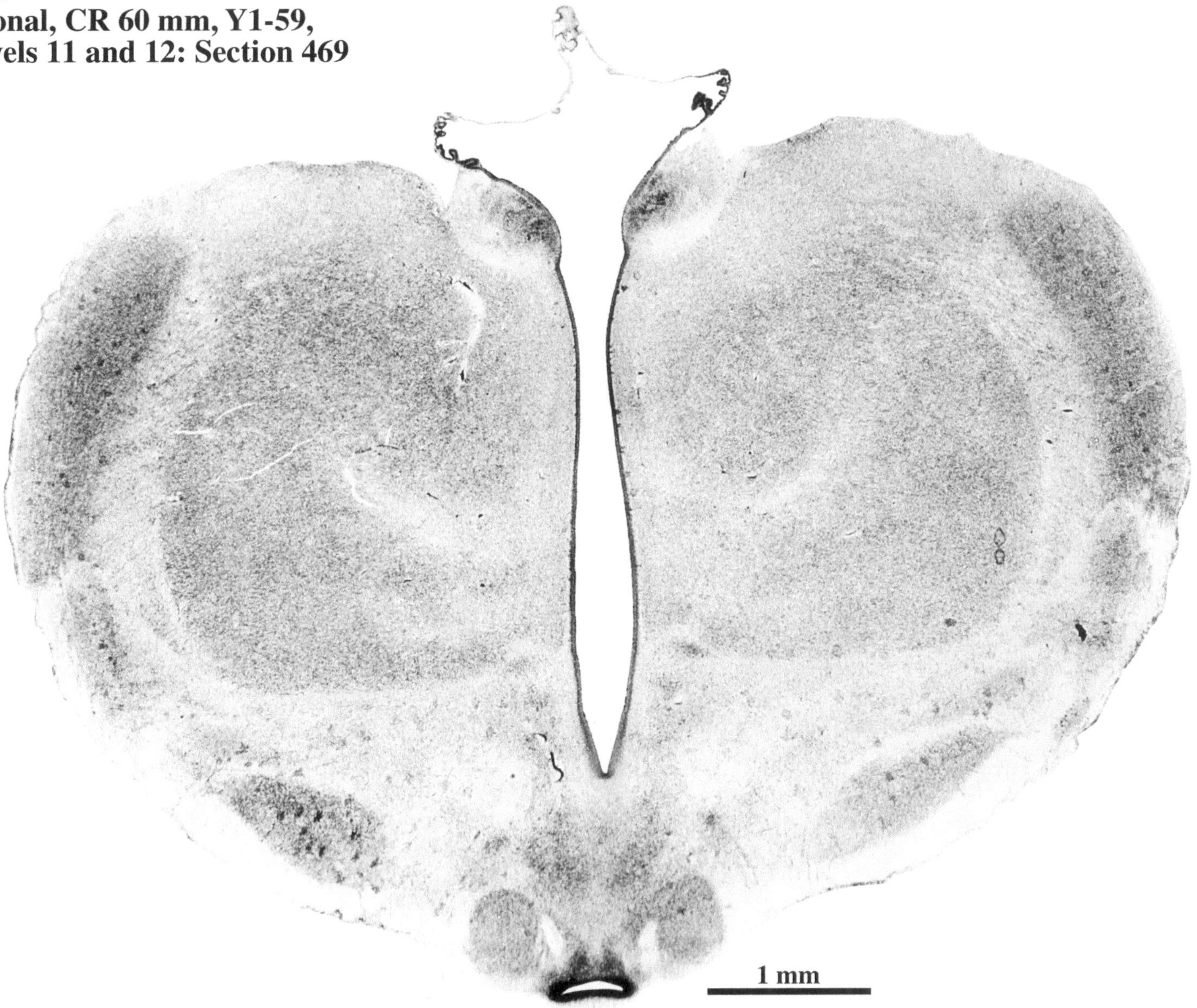

See levels 11 and 12 in Plates 11A and B to 12A and B.

**PLATE 29B**

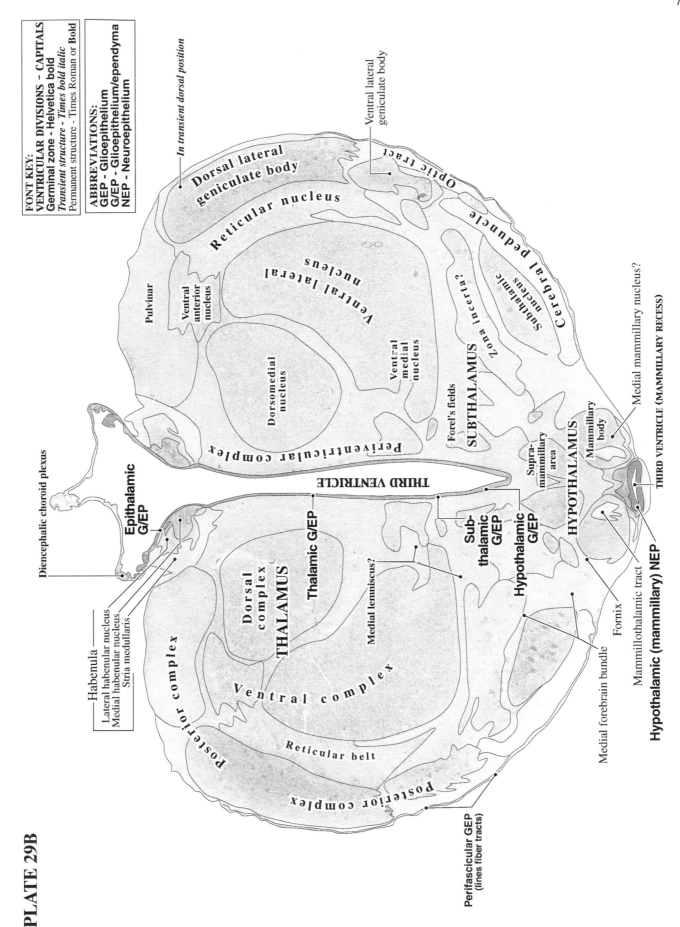

FONT KEY:
VENTRICULAR DIVISIONS – CAPITALS
Germinal zone - Helvetica bold
*Transient structure - Times bold italic*
Permanent structure - Times Roman or **Bold**

ABBREVIATIONS:
GEP - Glioepithelium
G/EP - Glioepithelium/ependyma
NEP - Neuroepithelium

*In transient dorsal position*

Ventral lateral geniculate body

Dorsal lateral geniculate body

Optic tract

Reticular nucleus

Cerebral peduncle

Pulvinar

Ventral anterior nucleus

Ventral lateral nucleus

Subthalamic nucleus

Dorsomedial nucleus

Ventral medial nucleus

Zona incerta?

Forel's fields

SUBTHALAMUS

Periventricular complex

Mammillary body

Medial mammillary nucleus?

Supra-mammillary area

HYPOTHALAMUS

THIRD VENTRICLE (MAMMILLARY RECESS)

Diencephalic choroid plexus

THIRD VENTRICLE

**Epithalamic G/EP**

Habenula
Lateral habenular nucleus
Medial habenular nucleus
Stria medullaris

Posterior complex

Dorsal complex

**THALAMUS**

**Thalamic G/EP**

Medial lemniscus?

**Sub-thalamic G/EP**

**Hypothalamic G/EP**

**HYPOTHALAMUS**

Fornix

Mammillothalamic tract

**Hypothalamic (mammillary) NEP**

Ventral complex

Reticular belt

Posterior complex

Medial forebrain bundle

**Perifascicular GEP**
(lines fiber tracts)

**DIENCEPHALON**

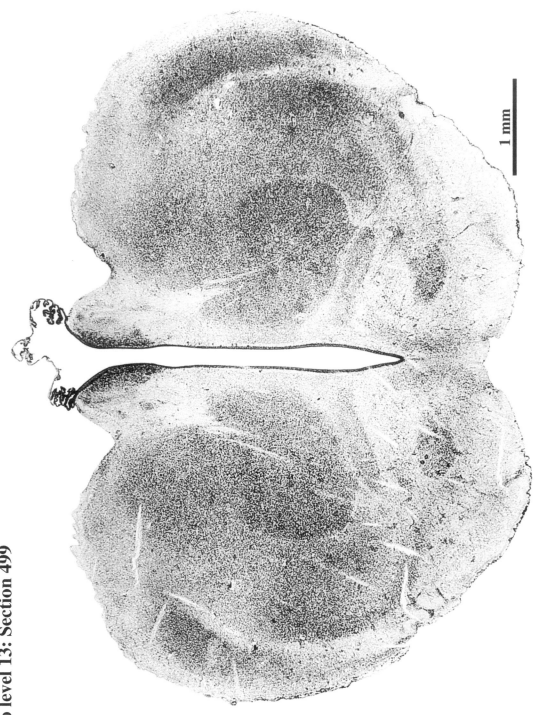

**PLATE 30A**
**GW11 Coronal, CR 60 mm, Y1-59,**
**Adjacent to level 13: Section 499**

1 mm

See level 13 in Plates 13A and B.

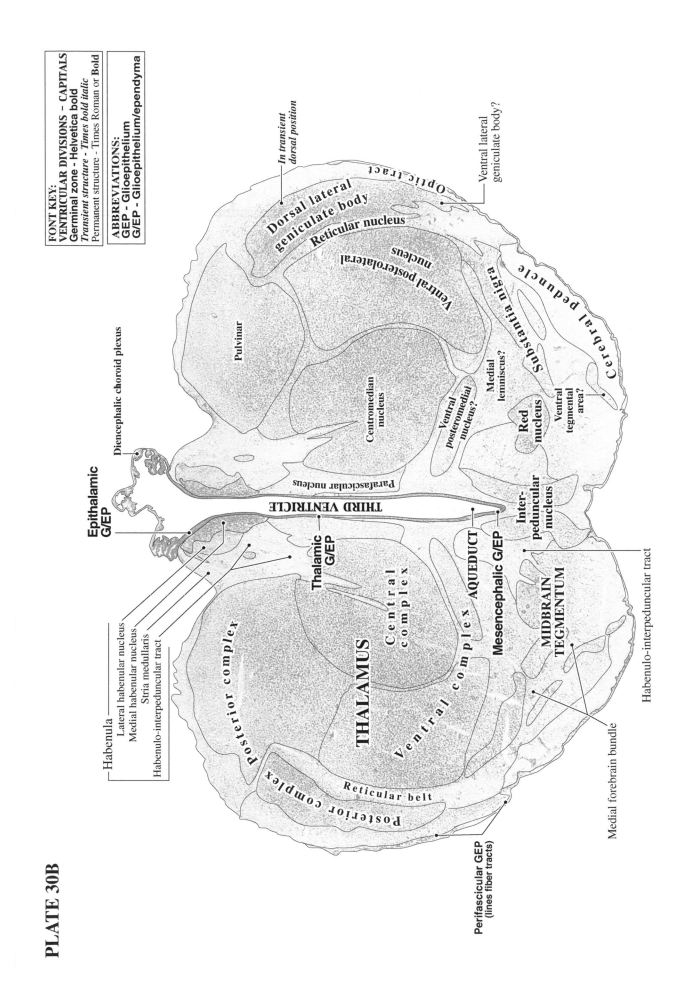

PLATE 30B

FONT KEY:
VENTRICULAR DIVISIONS - CAPITALS
Germinal zone - Helvetica bold
*Transient structure - Times bold italic*
Permanent structure - Times Roman or Bold

ABBREVIATIONS:
GEP - Glioepithelium
G/EP - Glioepithelium/ependyma

*In transient dorsal position*

Optic tract

Dorsal lateral geniculate body
Reticular nucleus
Ventral posterolateral nucleus

Ventral lateral geniculate body?

Diencephalic choroid plexus

Pulvinar

Substantia nigra

Cerebral peduncle

Centromedian nucleus

Medial lemniscus?

Ventral posteromedial nucleus?

Red nucleus

Ventral tegmental area?

Parafascicular nucleus

Epithalamic G/EP

THIRD VENTRICLE

Interpeduncular nucleus

Thalamic G/EP

Central complex

AQUEDUCT

Mesencephalic G/EP

Habenulo-interpeduncular tract

Habenula
Lateral habenular nucleus
Medial habenular nucleus
Stria medullaris
Habenulo-interpeduncular tract

Posterior complex

THALAMUS

Ventral complex

MIDBRAIN TEGMENTUM

Medial forebrain bundle

Posterior complex

Reticular belt

Perifascicular GEP
(lines fiber tracts)

**DIENCEPHALON AND MIDBRAIN TEGMENTUM**

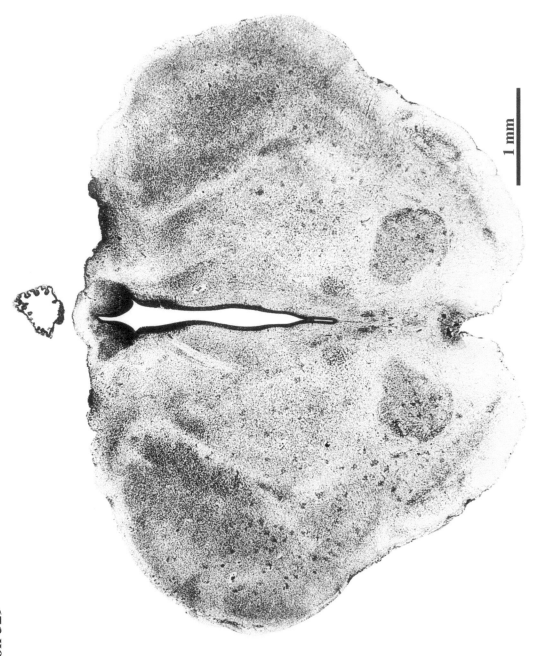

1 mm

**PLATE 31A**
**GW11 Coronal, CR 60 mm, Y1-59,**
**Level 14: Section 529**

See level 14 in Plates 14A and B.

77

PLATE 31B

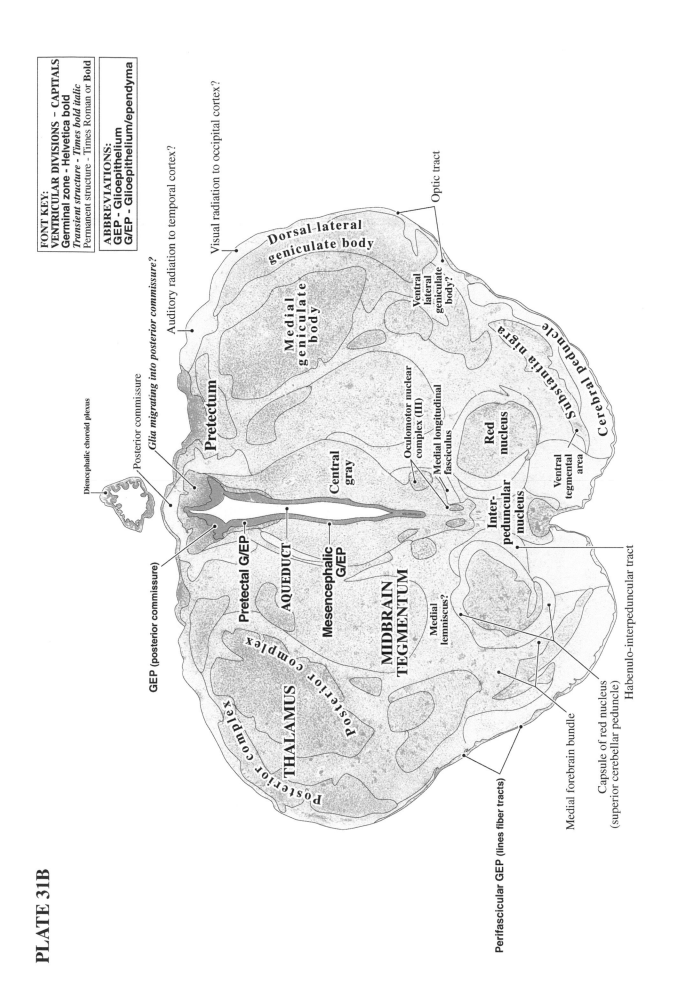

FONT KEY:
VENTRICULAR DIVISIONS – CAPITALS
Germinal zone – Helvetica bold
*Transient structure – Times bold italic*
Permanent structure – Times Roman or **Bold**

ABBREVIATIONS:
**GEP** - Glioepithelium
**G/EP** - Glioepithelium/ependyma

Dorsal-lateral geniculate body

Visual radiation to occipital cortex?

Optic tract

Medial geniculate body

Ventral lateral geniculate body?

Auditory radiation to temporal cortex?

*Glia migrating into posterior commissure?*

Posterior commissure

Diencephalic choroid plexus

**GEP (posterior commissure)**

**Pretectum**

Central gray

Oculomotor nuclear complex (III)

Medial longitudinal fasciculus

Red nucleus

Substantia nigra

Cerebral peduncle

Ventral tegmental area

Inter-peduncular nucleus

**Pretectal G/EP**

**AQUEDUCT**

**Mesencephalic G/EP**

**MIDBRAIN TEGMENTUM**

*Medial lemniscus?*

*Posterior complex*

**THALAMUS**

**Perifascicular GEP (lines fiber tracts)**

Medial forebrain bundle

Capsule of red nucleus (superior cerebellar peduncle)

Habenulo-interpeduncular tract

**GW11 Coronal
CR 60 mm, Y1-59
Near level 15:
Section 549**

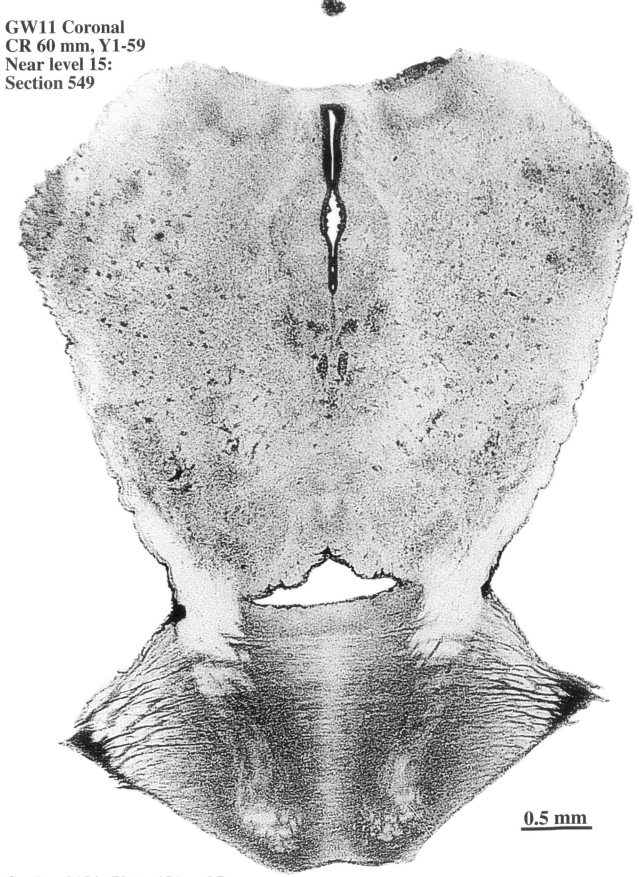

0.5 mm

**See Level 15 in Plates 15A and B.**

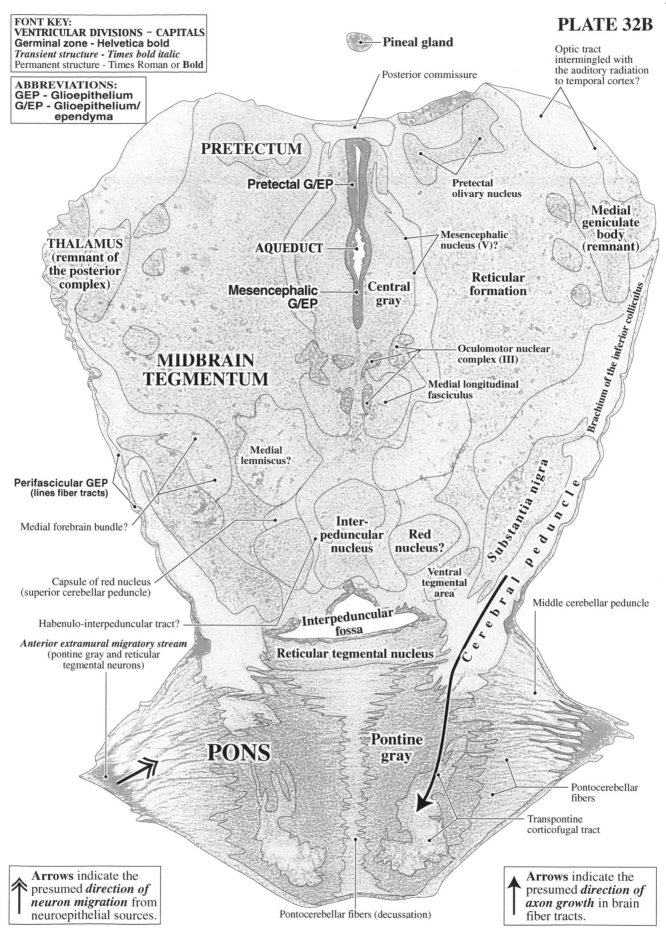

FONT KEY:
VENTRICULAR DIVISIONS – CAPITALS
Germinal zone - Helvetica bold
*Transient structure - Times bold italic*
Permanent structure - Times Roman or **Bold**

ABBREVIATIONS:
GEP - Glioepithelium
G/EP - Glioepithelium/
ependyma

Pineal gland

Optic tract
intermingled with
the auditory radiation
to temporal cortex?

Posterior commissure

**PRETECTUM**

**Pretectal G/EP**

Pretectal
olivary nucleus

**Medial
geniculate
body
(remnant)**

Mesencephalic
nucleus (V)?

**THALAMUS
(remnant of
the posterior
complex)**

**AQUEDUCT**

**Reticular
formation**

**Mesencephalic
G/EP**

**Central
gray**

**MIDBRAIN
TEGMENTUM**

Oculomotor nuclear
complex (III)

Medial longitudinal
fasciculus

Medial
lemniscus?

**Perifascicular GEP**
(lines fiber tracts)

Medial forebrain bundle?

**Inter-
peduncular
nucleus**

**Red
nucleus?**

Ventral
tegmental
area

Middle cerebellar peduncle

Capsule of red nucleus
(superior cerebellar peduncle)

Habenulo-interpeduncular tract?

**Interpeduncular
fossa**

*Anterior extramural migratory stream*
(pontine gray and reticular
tegmental neurons)

**Reticular tegmental nucleus**

**PONS**

**Pontine
gray**

Pontocerebellar
fibers

Transpontine
corticofugal tract

**Arrows** indicate the
presumed *direction of
neuron migration* from
neuroepithelial sources.

Pontocerebellar fibers (decussation)

**Arrows** indicate the
presumed *direction of
axon growth* in brain
fiber tracts.

Brachium of the inferior colliculus

Substantia nigra

Cerebral peduncle

GW11 Coronal
CR 60 mm, Y1-59
Level 15:
Section 569

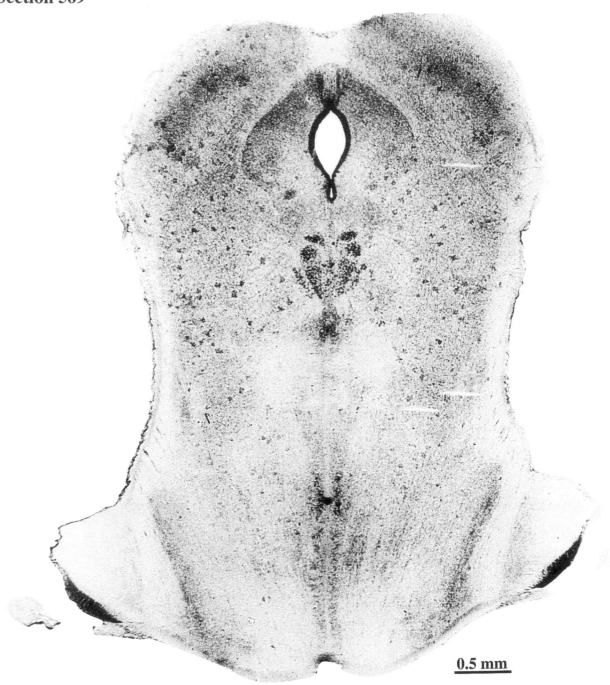

0.5 mm

See Level 15 in Plates 15A and B.

**PLATE 33B**

FONT KEY:
VENTRICULAR DIVISIONS - CAPITALS
**Germinal zone - Helvetica bold**
***Transient structure - Times bold italic***
Permanent structure - Times Roman or **Bold**

ABBREVIATIONS:
**GEP - Glioepithelium**
**G/EP - Glioepithelium/ependyma**

Posterior commissure
Subcommissural organ
Optic tract
SUPERIOR COLLICULUS
MIDBRAIN TECTUM
Migrating superior colliculus neurons sorting into various gray layers
Central gray
AQUEDUCT
Mesencephalic G/EP
Brachium of the inferior colliculus
Mesencephalic nucleus (V)?
Reticular formation
Oculomotor nuclear complex (III)
MIDBRAIN TEGMENTUM
Medial longitudinal fasciculus
Perifascicular GEP (lines fiber tracts)
Superior cerebellar peduncle (decussation)
Reticular formation
Lateral lemniscus
Raphe nuclear complex
Middle cerebellar peduncle
PONS
Pontine gray
Medial lemniscus and trapezoid body
Anterior extramural migratory stream
(pontine gray and reticular tegmental neurons)
Nerve VIII (vestibulocochlear)?
Nucleus of the lateral lemniscus (ventral)


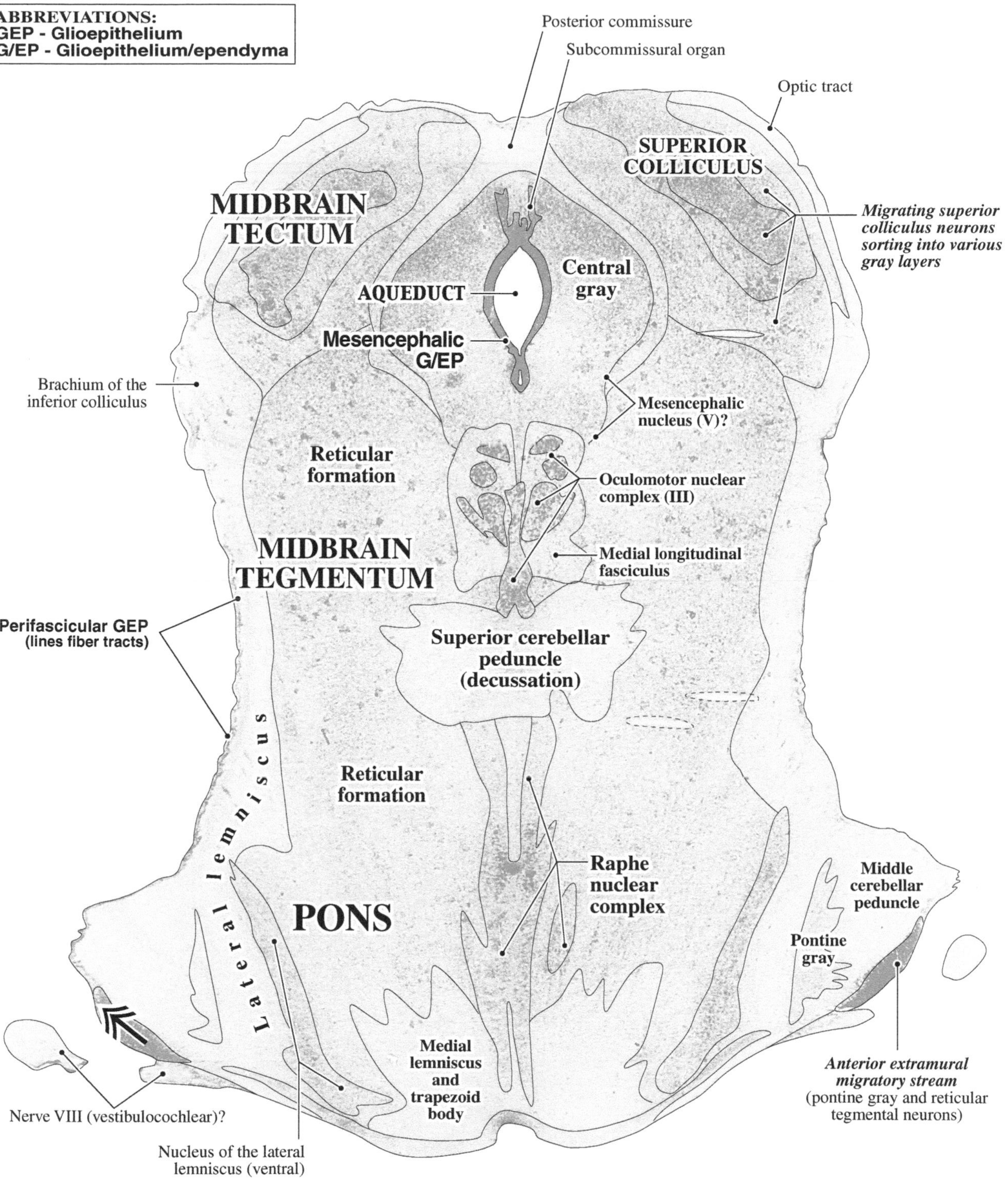

Dashed lines indicate staining and/or sectioning artifacts.

**Arrows** indicate the presumed ***direction of neuron migration*** from neuroepithelial sources.

**PLATE 34A**

**GW11 Coronal**
**CR 60 mm, Y1-59**
**Near level 16:**
**Section 589**

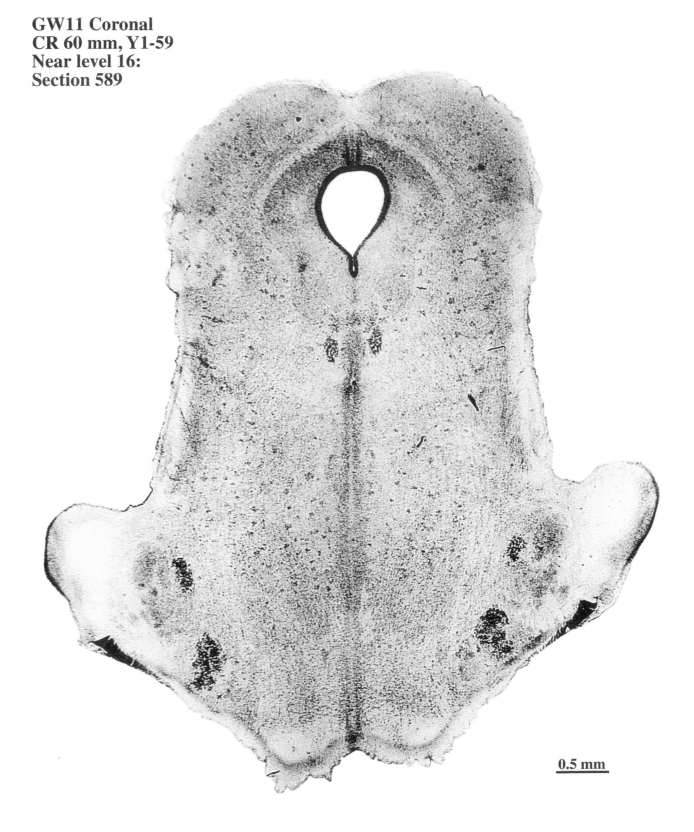

0.5 mm

**See Level 16 in Plates 16A and B.**

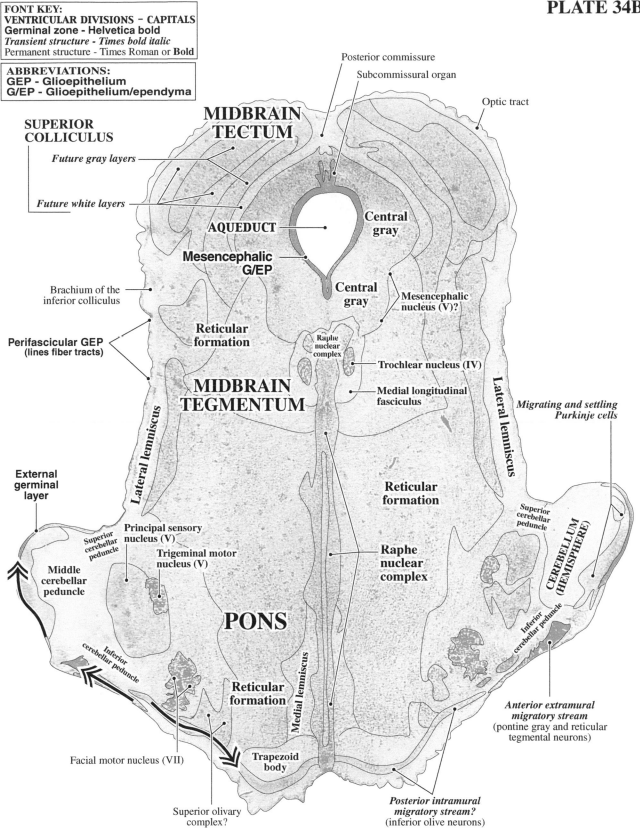

Posterior commissure
Subcommissural organ
Optic tract

**MIDBRAIN TECTUM**

**SUPERIOR COLLICULUS**

*Future gray layers*

*Future white layers*

**AQUEDUCT**

**Central gray**

**Mesencephalic G/EP**

**Central gray**

**Mesencephalic nucleus (V)?**

Brachium of the inferior colliculus

**Reticular formation**

*Raphe nuclear complex*

**Trochlear nucleus (IV)**

**Perifascicular GEP**
**(lines fiber tracts)**

**MIDBRAIN TEGMENTUM**

**Medial longitudinal fasciculus**

*Migrating and settling Purkinje cells*

Lateral lemniscus

Lateral lemniscus

**External germinal layer**

**Reticular formation**

Superior cerebellar peduncle

**Principal sensory nucleus (V)**

*Superior cerebellar peduncle*

**CEREBELLUM (HEMISPHERE)**

**Trigeminal motor nucleus (V)**

**Raphe nuclear complex**

**Middle cerebellar peduncle**

*Inferior cerebellar peduncle*

*Inferior cerebellar peduncle*

**PONS**

**Reticular formation**

Medial lemniscus

*Anterior extramural migratory stream*
(pontine gray and reticular tegmental neurons)

**Reticular formation**

**Trapezoid body**

Facial motor nucleus (VII)

Superior olivary complex?

*Posterior intramural migratory stream?*
(inferior olive neurons)

**PLATE 35A**

**GW11 Coronal**
**CR 60 mm, Y1-59**
**Level 16:**
**Section 599**

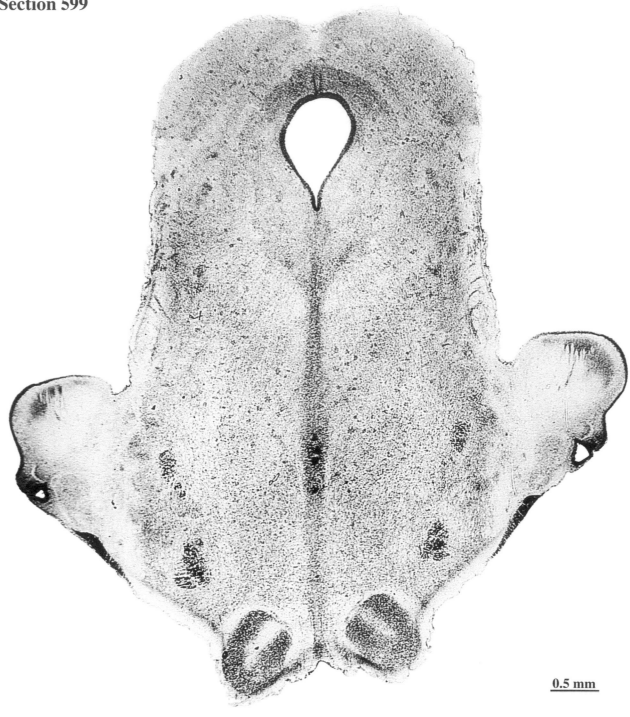

0.5 mm

See Level 16 in Plates 16A and B.

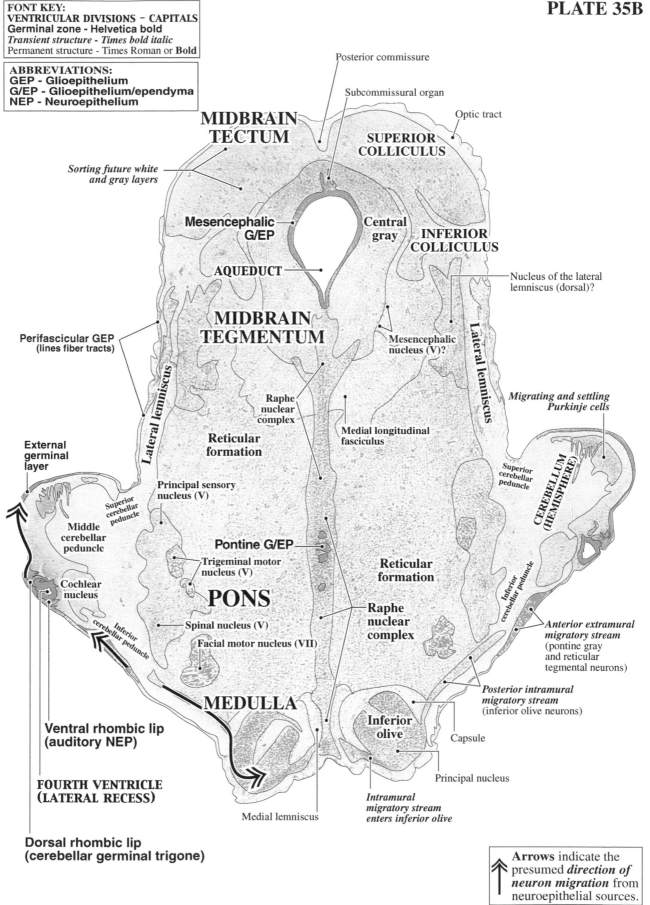

Posterior commissure

Subcommissural organ

Optic tract

**MIDBRAIN TECTUM**

**SUPERIOR COLLICULUS**

*Sorting future white and gray layers*

**Mesencephalic G/EP**

**Central gray**

**INFERIOR COLLICULUS**

**AQUEDUCT**

Nucleus of the lateral lemniscus (dorsal)?

**MIDBRAIN TEGMENTUM**

Mesencephalic nucleus (V)?

Lateral lemniscus

Perifascicular GEP (lines fiber tracts)

Lateral lemniscus

Raphe nuclear complex

Medial longitudinal fasciculus

*Migrating and settling Purkinje cells*

**Reticular formation**

Superior cerebellar peduncle

**External germinal layer**

**CEREBELLUM (HEMISPHERE)**

Superior cerebellar peduncle

Principal sensory nucleus (V)

**Middle cerebellar peduncle**

**Pontine G/EP**

**Reticular formation**

Inferior cerebellar peduncle

**Cochlear nucleus**

Trigeminal motor nucleus (V)

**PONS**

**Raphe nuclear complex**

*Anterior extramural migratory stream* (pontine gray and reticular tegmental neurons)

Inferior cerebellar peduncle

Spinal nucleus (V)

Facial motor nucleus (VII)

*Posterior intramural migratory stream* (inferior olive neurons)

**Ventral rhombic lip (auditory NEP)**

**MEDULLA**

**Inferior olive**

Capsule

**FOURTH VENTRICLE (LATERAL RECESS)**

Principal nucleus

Medial lemniscus

*Intramural migratory stream enters inferior olive*

**Dorsal rhombic lip (cerebellar germinal trigone)**

**Arrows** indicate the presumed *direction of neuron migration* from neuroepithelial sources.

# PART III: GW11 HORIZONTAL

This is specimen number 1500 in the Carnegie Collection, designated here as C1500. A normal fetus with a crown-rump length (CR) of 56 mm was collected in 1916. The fetus is estimated to be early in gestational week (GW) 11. The entire fetus was fixed in formalin and was cut in the horizontal plane. The records are not clear regarding section thickness, but the brain is present in 2,400 sections, more than three times the number of 35 μm sections in the GW17 horizontally sectioned brain in the Yakovlev Collection (Bayer and Altman, 2005, Volume 3, Part IX). We estimate that the sections are between 8-10 μm thick. All sections were stained with Bodian's method to show developing fiber tracts. Since there is no photograph of C1500's brain before it was embedded and cut, a specimen from Hochstetter (1919) that is comparable in age is used to show the approximate section plane and external features in early GW11 (**Figure 2**). **Levels 1-10**, large sections containing the cerebral hemispheres, are shown at low magnification in **Plates 36-45**. **Levels 11-21**, small sections containing only the brainstem, are shown at a higher magnification in **Plates 46-56**. To maximize image size within page space, C1500's sections are rotated 90° (landscape orientation). The anterior part of each section is on the left (page bottom), and the posterior part of each section is on the right (page top).

C1500 has many of the same features as Y1-59, except that it is slightly less mature. Throughout the cerebral cortex, the *neuroepithelium and subventricular zone* are prominent. The *stratified transitional field (STF)* contains *STF1* and *STF5* throughout; with *STF4* only in lateral areas. The most prominent developmental feature of the cerebral cortex is that both the *STF* layers and the cortical plate have a pronounced lateral (thicker) to medial (thinner) maturation gradient. The olfactory bulb beneath the anterior septum and striatum contains a small *rostral migratory stream* in its core. In anterolateral parts of the cerebral cortex, streams of neurons and glia appear to leave *STF4* and enter the *lateral migratory stream*. The hippocampus is in an immature position dorsal to the thalamus and medial to the temporal lobe. Cells are entering Ammon's horn pyramidal layer in the *ammonic migration*, and granule cells and their precursors are migrating to the hilus of the presumptive dentate gyrus in the *dentate migration*; there is no granular layer. A massive *neuroepithelium/subventricular zone* overlies the amygdala, nucleus accumbens, and striatum (caudate and putamen) where neurons (and glia) are being generated.

The cerebellum is a thick, smooth plate overlying the posterior pons and medulla. However, there is only a thin *glioepithelium/ependyma* at the ventricular surface, indicating that all deep neurons and Purkinje cells have been generated. The deep neurons are in place beneath the cortex, but have indefinite nuclear subdivisions. The cortical surface is covered by an *external germinal layer (egl)* that is actively producing neuronal stem cells, granule, stellate, and basket cells of the cerebellar cortex. Lamination in the cortex is nearly absent, except for a thin molecular layer beneath the *egl*. Nearly all Purkinje cells are migrating, and settling. In contrast to Y1-59, there is no evidence of lobulation in the cerebellar cortex.

The third ventricle, aqueduct, and fourth ventricle are lined by a thin *glioepithelium/ependyma* indicating that neurogenesis in the primary neuroepithelium is complete. In the medulla there are two active germinal sites in anterior and posterior parts of the ventral rhombic lip. 1) The *auditory neuroepithelium* generates cochlear nucleus neurons. 2) A large *precerebellar neuroepithelium* generates precerebellar (mainly pontine gray) neurons.

Neurons throughout the diencephalon, midbrain tegmentum, pons, and medulla are settling. Because C1500 is not Nissl-stained, nuclear divisions are very indistinct. The large *anterior extramural*, *posterior extramural, and intramural migratory streams* are prominent in the medulla and pons. The Bodian stain clearly shows several fiber tracts and nerves throughout the brainstem. The optic nerve and tract are well defined, along with the medial forebrain bundle. Unlike Y1-59, there is no sure evidence of a cerebral peduncle in the midbrain tegmentum or pontine gray. However, pontine gray fibers cross the midline and a distinct middle cerebellar peduncle is present. There is also a distinct superior and inferior cerebellar peduncle. There is definite staining in the trigeminal nerve and tract, the facial nerve, the abducens nerve, and the glossopharyngeal nerve.

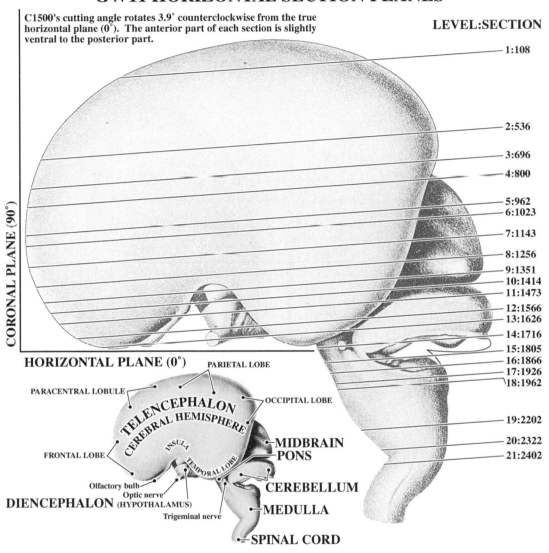

# GW11 HORIZONTAL SECTION PLANES

C1500's cutting angle rotates 3.9° counterclockwise from the true horizontal plane (0°). The anterior part of each section is slightly ventral to the posterior part.

LEVEL:SECTION

1:108
2:536
3:696
4:800
5:962
6:1023
7:1143
8:1256
9:1351
10:1414
11:1473
12:1566
13:1626
14:1716
15:1805
16:1866
17:1926
18:1962
19:2202
20:2322
21:2402

CORONAL PLANE (90°)

HORIZONTAL PLANE (0°)

PARIETAL LOBE
PARACENTRAL LOBULE
TELENCEPHALON
CEREBRAL HEMISPHERE
OCCIPITAL LOBE
INSULA
FRONTAL LOBE
TEMPORAL LOBE
MIDBRAIN
PONS
Olfactory bulb
Optic nerve
DIENCEPHALON (HYPOTHALAMUS)
Trigeminal nerve
CEREBELLUM
MEDULLA
SPINAL CORD

**Figure 2.** The lateral view of the brain and upper cervical spinal cord from a specimen with a crown-rump length of 53 mm (modified from Figure 46, Table VIII, Hochstetter, 1919) serves to show the approximate locations and cutting angles of the illustrated sections of C1500 in the following pages. The small inset identifies the major structural features. The cut beneath the cerebellum is the edge of the medullary velum.

87

88

**PLATE 36A**

**GW11 Horizontal**
**CR 57 mm**
**C1500**
**Level 1: Section 108**

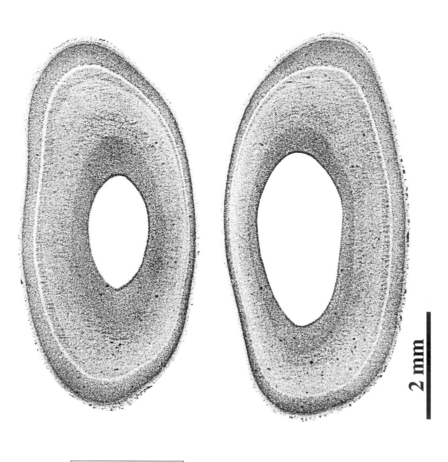

2 mm

**LAYERS OF THE CORTICAL**
**STRATIFIED TRANSITIONAL FIELD (STF)**

*STF1*  Superficial fibrous layer with an early
developmental stage (*t1*) when many
cells are migrating through it, followed
by a late stage (*t2*) with sparse cells.
Endures as the subcortical white matter.

*STF5*  Deep cellular layer that is prominent
during the first trimester, the first sojourn
zone to appear outside the germinal
matrix.

# PLATE 36B

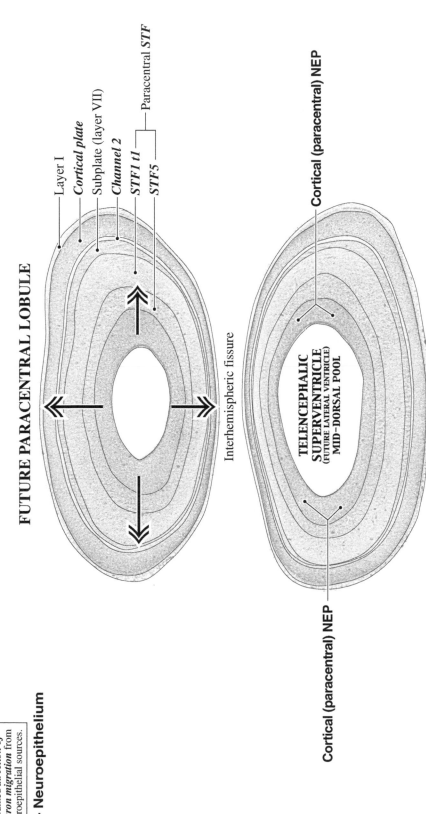

**FUTURE PARACENTRAL LOBULE**

Layer I

*Cortical plate*

Subplate (layer VII)

*Channel 2*

*STF1 t1*

*STF5*

Paracentral *STF*

Interhemispheric fissure

**Cortical (paracentral) NEP**

TELENCEPHALIC
SUPERVENTRICLE
(FUTURE LATERAL VENTRICLE)
MID–DORSAL POOL

**Cortical (paracentral) NEP**

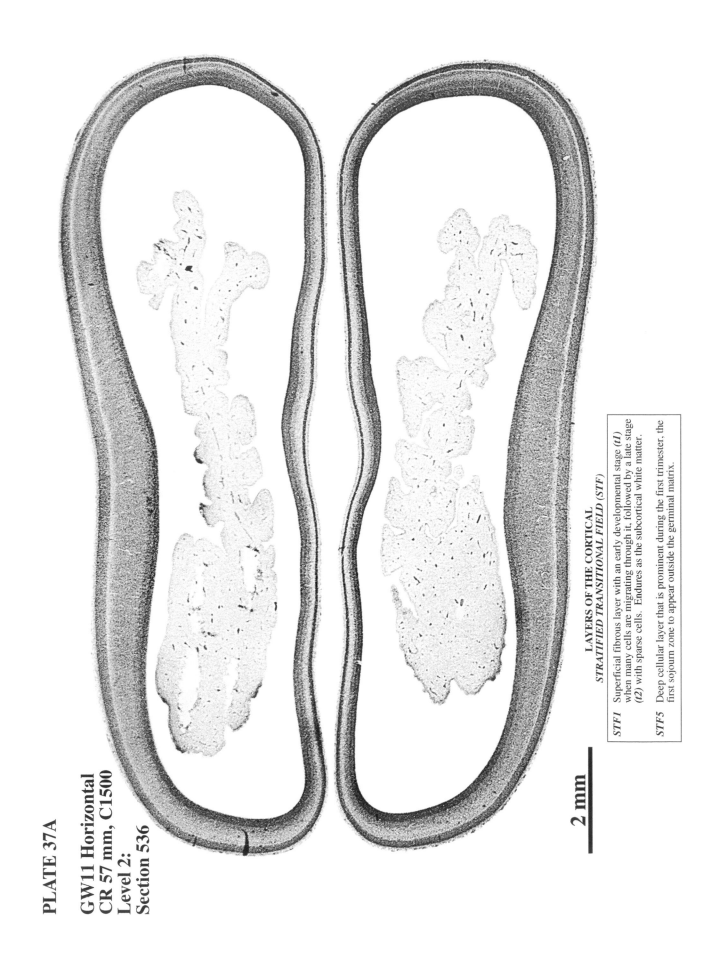

**PLATE 37A**

**GW11 Horizontal**
**CR 57 mm, C1500**
**Level 2:**
**Section 536**

**LAYERS OF THE CORTICAL**
*STRATIFIED TRANSITIONAL FIELD (STF)*

*STF1*  Superficial fibrous layer with an early developmental stage (*t1*) when many cells are migrating through it, followed by a late stage (*t2*) with sparse cells. Endures as the subcortical white matter.

*STF5*  Deep cellular layer that is prominent during the first trimester, the first sojourn zone to appear outside the germinal matrix.

**2 mm**

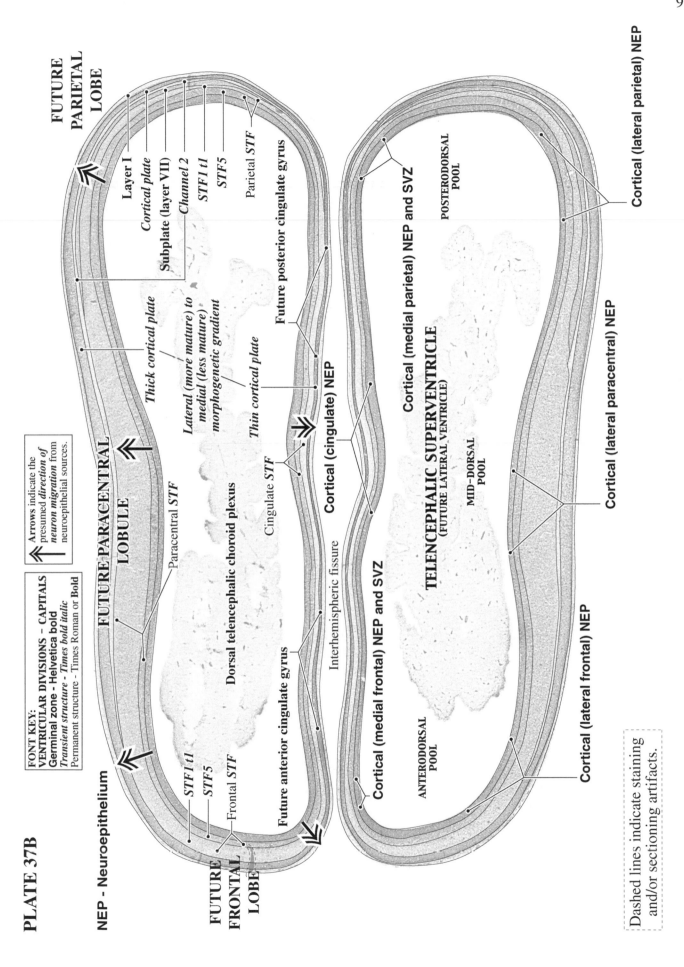

PLATE 37B

FONT KEY:
VENTRICULAR DIVISIONS - CAPITALS
Germinal zone - Helvetica bold
*Transient structure - Times bold italic*
Permanent structure - Times Roman or Bold

NEP - Neuroepithelium

Arrows indicate the presumed *direction of neuron migration* from neuroepithelial sources.

FUTURE PARIETAL LOBE

Layer I
*Cortical plate*
Subplate (layer VII)
*Channel 2*
*STF1 t1*
*STF5*
*Parietal STF*

FUTURE PARACENTRAL LOBULE

*Thick cortical plate*

*Lateral (more mature) to medial (less mature) morphogenetic gradient*

*Thin cortical plate*

Future posterior cingulate gyrus

Cortical (cingulate) NEP

*Cingulate STF*

*Paracentral STF*

Dorsal telencephalic choroid plexus

Interhemispheric fissure

Future anterior cingulate gyrus

*STF1 t1*
*STF5*
*Frontal STF*

FUTURE FRONTAL LOBE

Cortical (medial frontal) NEP and SVZ

Cortical (lateral frontal) NEP

ANTERODORSAL POOL

Cortical (medial parietal) NEP and SVZ

POSTERODORSAL POOL

Cortical (lateral parietal) NEP

TELENCEPHALIC SUPERVENTRICLE
(FUTURE LATERAL VENTRICLE)

MID-DORSAL POOL

Cortical (medial parietal) NEP

Cortical (lateral paracentral) NEP

Dashed lines indicate staining and/or sectioning artifacts.

91

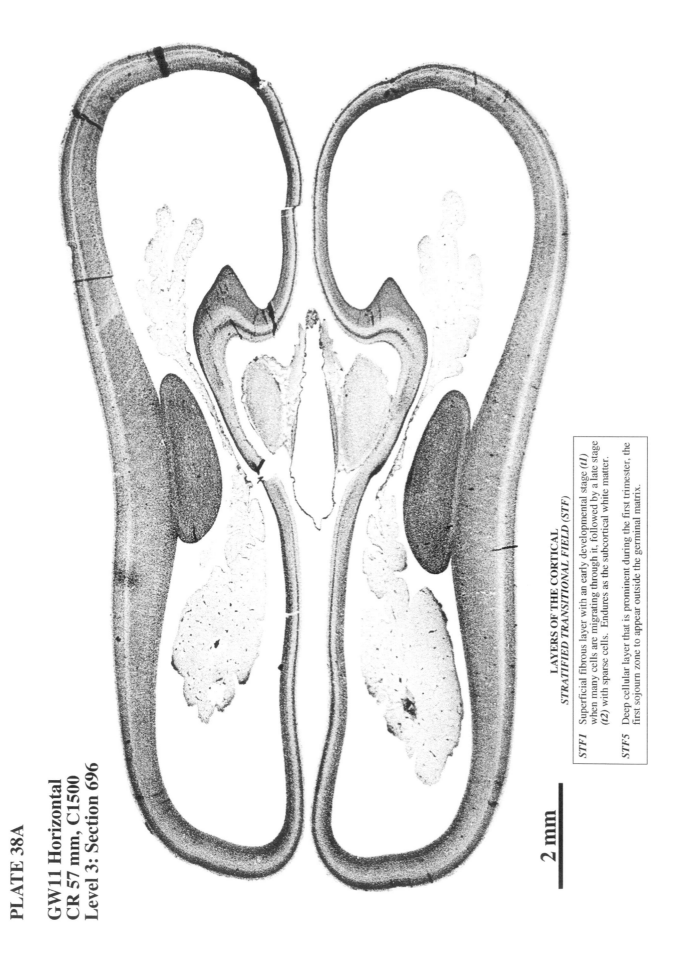

**PLATE 38A**

**GW11 Horizontal**
**CR 57 mm, C1500**
**Level 3: Section 696**

**LAYERS OF THE CORTICAL**
*STRATIFIED TRANSITIONAL FIELD (STF)*

*STF1* Superficial fibrous layer with an early developmental stage (*t1*) when many cells are migrating through it, followed by a late stage (*t2*) with sparse cells. Endures as the subcortical white matter.

*STF5* Deep cellular layer that is prominent during the first trimester, the first sojourn zone to appear outside the germinal matrix.

**2 mm**

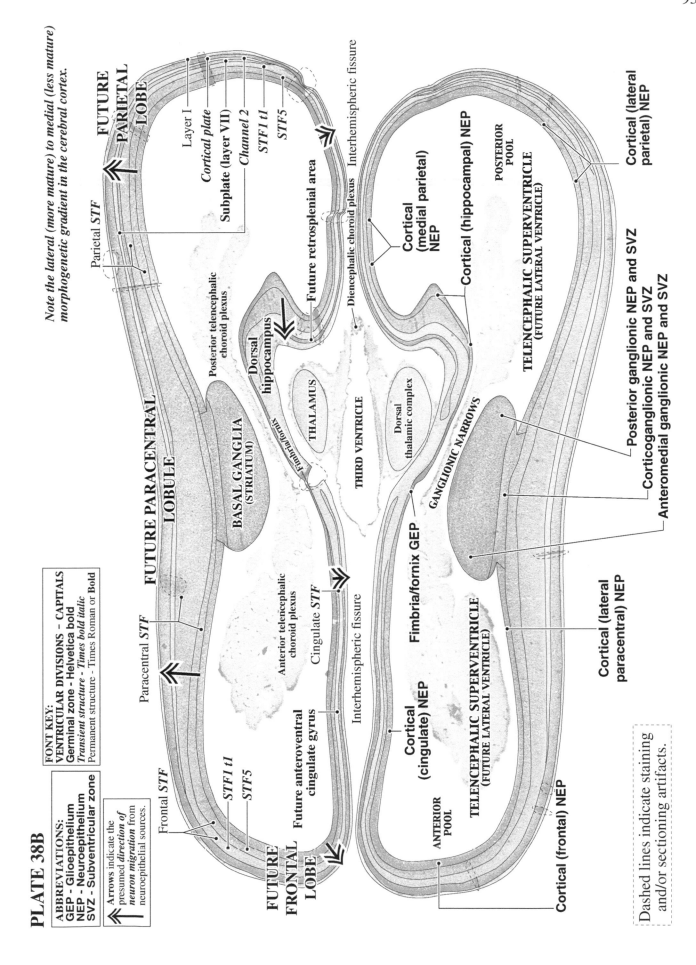

PLATE 38B

**ABBREVIATIONS:**
GEP - Glioepithelium
NEP - Neuroepithelium
SVZ - Subventricular zone

Arrows indicate the presumed *direction of neuron migration* from neuroepithelial sources.

FONT KEY:
VENTRICULAR DIVISIONS – CAPITALS
Germinal zone - Helvetica bold
*Transient structure - Times bold italic*
Permanent structure - Times Roman or Bold

*Note the lateral (more mature) to medial (less mature) morphogenetic gradient in the cerebral cortex.*

93

Dashed lines indicate staining and/or sectioning artifacts.

FUTURE PARIETAL LOBE

Parietal *STF*

Layer I
*Cortical plate*
**Subplate (layer VII)**
*Channel 2*
*STF1 t1*
*STF5*

Posterior telencephalic choroid plexus

*Dorsal hippocampus*

*Future retrosplenial area*

Interhemispheric fissure

Diencephalic choroid plexus

Cortical (medial parietal) NEP

Cortical (hippocampal) NEP

POSTERIOR POOL

Cortical (lateral parietal) NEP

TELENCEPHALIC SUPERVENTRICLE (FUTURE LATERAL VENTRICLE)

FUTURE PARACENTRAL LOBULE

Paracentral *STF*

**BASAL GANGLIA (STRIATUM)**

*Fimbria/fornix*

**THALAMUS**

Anterior telencephalic choroid plexus

Cingulate *STF*

**THIRD VENTRICLE**

Dorsal thalamic complex

GANGLIONIC NARROWS

Posterior ganglionic NEP and SVZ
Corticoganglionic NEP and SVZ
Anteromedial ganglionic NEP and SVZ

Interhemispheric fissure

Frontal *STF*

*STF1 t1*
*STF5*

**Future anteroventral cingulate gyrus**

FUTURE FRONTAL LOBE

**Fimbria/fornix GEP**

**Cortical (cingulate) NEP**

ANTERIOR POOL

TELENCEPHALIC SUPERVENTRICLE (FUTURE LATERAL VENTRICLE)

Cortical (lateral paracentral) NEP

**Cortical (frontal) NEP**

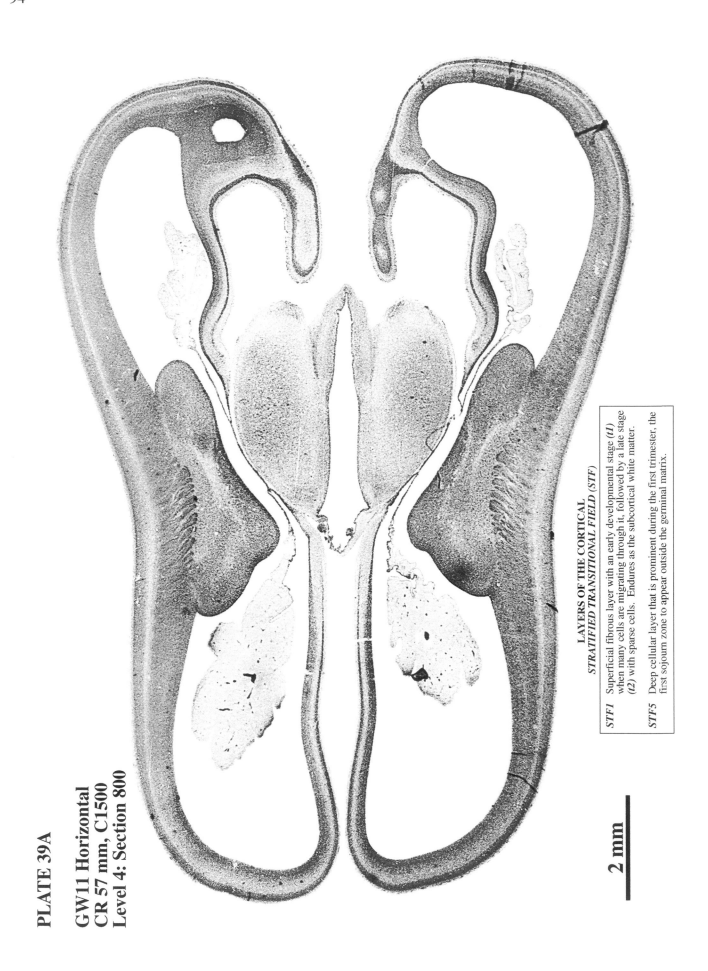

**PLATE 39A**

**GW11 Horizontal**
**CR 57 mm, C1500**
**Level 4: Section 800**

**LAYERS OF THE CORTICAL**
*STRATIFIED TRANSITIONAL FIELD (STF)*

*STF1* Superficial fibrous layer with an early developmental stage (*t1*) when many cells are migrating through it, followed by a late stage (*t2*) with sparse cells. Endures as the subcortical white matter.

*STF5* Deep cellular layer that is prominent during the first trimester, the first sojourn zone to appear outside the germinal matrix.

**2 mm**

# PLATE 39B

ABBREVIATIONS:
GEP - Glioepithelium
G/EP - Glioepithelium/ependyma
NEP - Neuroepithelium
SVZ - Subventricular zone

FONT KEY:
VENTRICULAR DIVISIONS – CAPITALS
Germinal zone - Helvetica bold
*Transient structure - Times bold italic*
Permanent structure - Times Roman or Bold

*Note the lateral (more mature) to medial (less mature)
morphogenetic gradient in the cerebral cortex.*

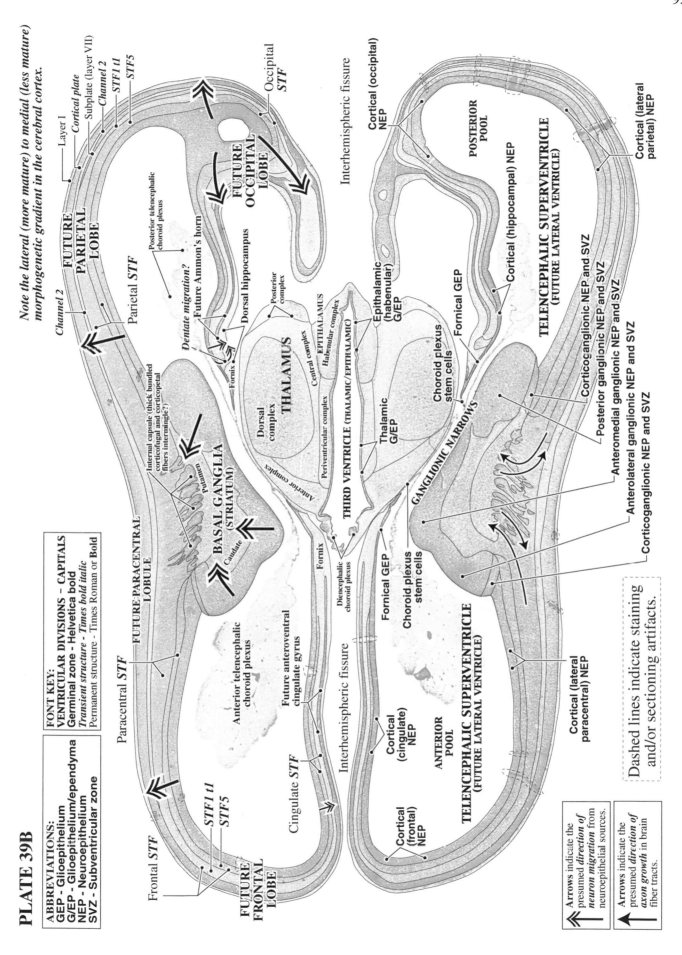

Dashed lines indicate staining
and/or sectioning artifacts.

**Arrows** indicate the
presumed *direction of
neuron migration* from
neuroepithelial sources.

**Arrows** indicate the
presumed *direction of
axon growth* in brain
fiber tracts.

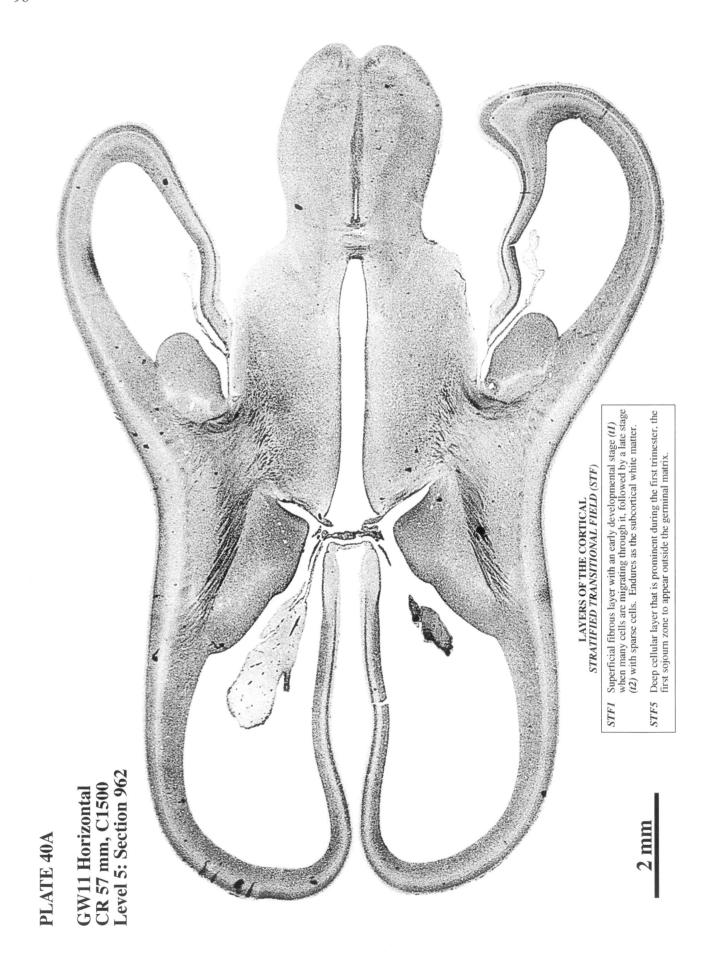

**LAYERS OF THE CORTICAL**
*STRATIFIED TRANSITIONAL FIELD (STF)*

*STF1* Superficial fibrous layer with an early developmental stage (*t1*) when many cells are migrating through it, followed by a late stage (*t2*) with sparse cells. Endures as the subcortical white matter.

*STF5* Deep cellular layer that is prominent during the first trimester, the first sojourn zone to appear outside the germinal matrix.

2 mm

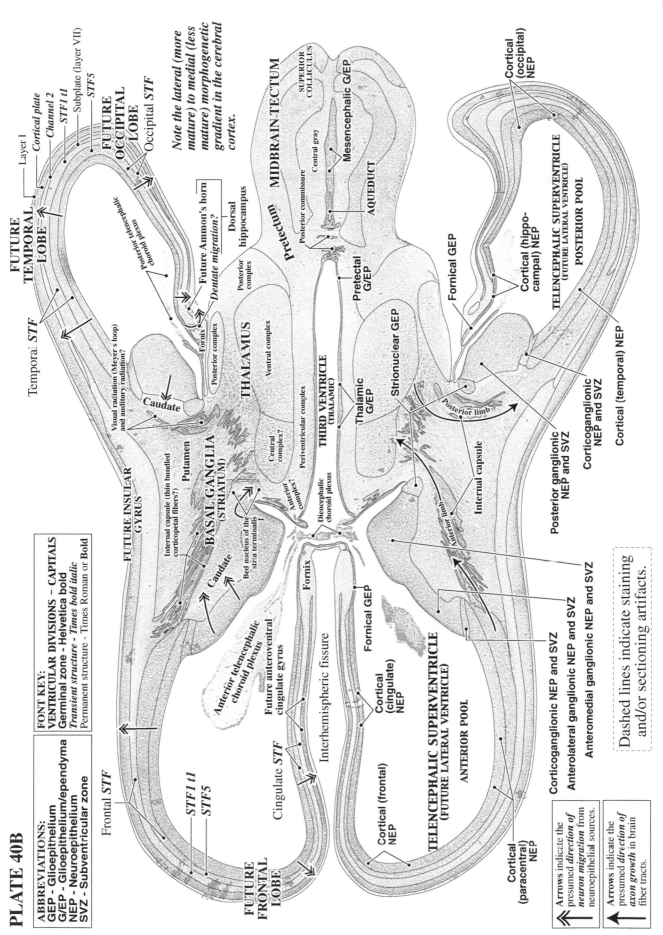

## PLATE 40B

**FONT KEY:**
VENTRICULAR DIVISIONS – CAPITALS
Germinal zone - Helvetica bold
*Transient structure - Times bold italic*
Permanent structure - Times Roman or Bold

**ABBREVIATIONS:**
GEP - Glioepithelium
G/EP - Glioepithelium/ependyma
NEP - Neuroepithelium
SVZ - Subventricular zone

97

*Note the lateral (more mature) to medial (less mature) morphogenetic gradient in the cerebral cortex.*

Layer 1
Cortical plate
Channel 2
*STF1 t1*
Subplate (layer VII)
*STF5*

**FUTURE OCCIPITAL LOBE**
Occipital *STF*

**FUTURE TEMPORAL LOBE**

Temporal *STF*

Posterior telencephalic choroid plexus

Future Ammon's horn
*Dentate migration?*

Dorsal hippocampus

Fornix

Posterior complex

Visual radiation (Meyer's loop) and auditory radiation?

**Caudate**

**FUTURE INSULAR GYRUS**

Internal capsule (thin bundled corticopetal fibers?)

**Putamen**

**BASAL GANGLIA (STRIATUM)**

**Caudate**

Bed nucleus of the stria terminalis

Anterior telencephalic choroid plexus

Future anteroventral cingulate gyrus

Interhemispheric fissure

**Fornix**

**Cortical (cingulate) NEP**

**Fornical GEP**

**Cortical (frontal) NEP**

**TELENCEPHALIC SUPERVENTRICLE (FUTURE LATERAL VENTRICLE)**

**ANTERIOR POOL**

**Cortical (paracentral) NEP**

**FUTURE FRONTAL LOBE**

*STF1 t1*
*STF5*

Cingulate *STF*

Frontal *STF*

Posterior complex

Ventral complex

Central complex?

Periventricular complex

**THIRD VENTRICLE (THALAMIC)**

Anterior complex?

Diencephalic choroid plexus

**THALAMUS**

**MIDBRAIN-TECTUM**

SUPERIOR COLLICULUS

Central gray

**Mesencephalic G/EP**

**AQUEDUCT**

Posterior commissure

**Pretectum**

Pretectal G/EP

Thalamic G/EP

**Strionuclear GEP**

Posterior limb

Internal capsule

Anterior limb

Fornical GEP

Corticoganglionic NEP and SVZ

Anterolateral ganglionic NEP and SVZ

Anteromedial ganglionic NEP and SVZ

**Fornical GEP**

**Cortical (occipital) NEP**

**TELENCEPHALIC SUPERVENTRICLE (FUTURE LATERAL VENTRICLE)**

**POSTERIOR POOL**

Cortical (hippo-campal) NEP

Posterior ganglionic NEP and SVZ

Corticoganglionic NEP and SVZ

Cortical (temporal) NEP

Dashed lines indicate staining and/or sectioning artifacts.

Arrows indicate the presumed *direction of neuron migration* from neuroepithelial sources.

Arrows indicate the presumed *direction of axon growth* in brain fiber tracts.

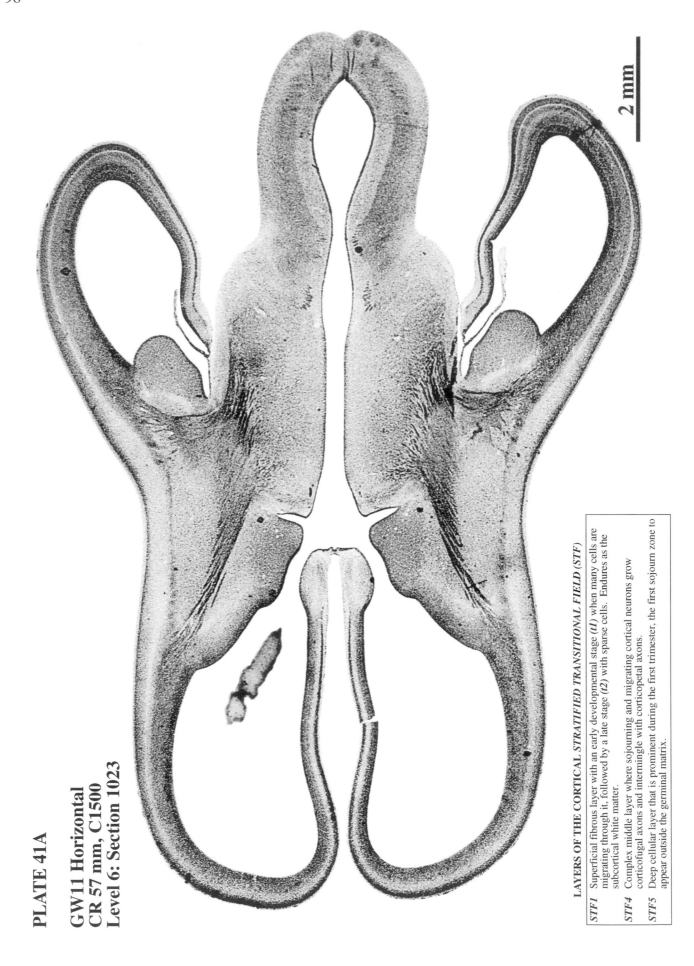

98

**PLATE 41A**

**GW11 Horizontal**
**CR 57 mm, C1500**
**Level 6: Section 1023**

2 mm

*LAYERS OF THE CORTICAL STRATIFIED TRANSITIONAL FIELD (STF)*

*STF1*  Superficial fibrous layer with an early developmental stage (*t1*) when many cells are migrating through it, followed by a late stage (*t2*) with sparse cells. Endures as the subcortical white matter.

*STF4*  Complex middle layer where sojourning and migrating cortical neurons grow corticofugal axons and intermingle with corticopetal axons.

*STF5*  Deep cellular layer that is prominent during the first trimester, the first sojourn zone to appear outside the germinal matrix.

# PLATE 41B

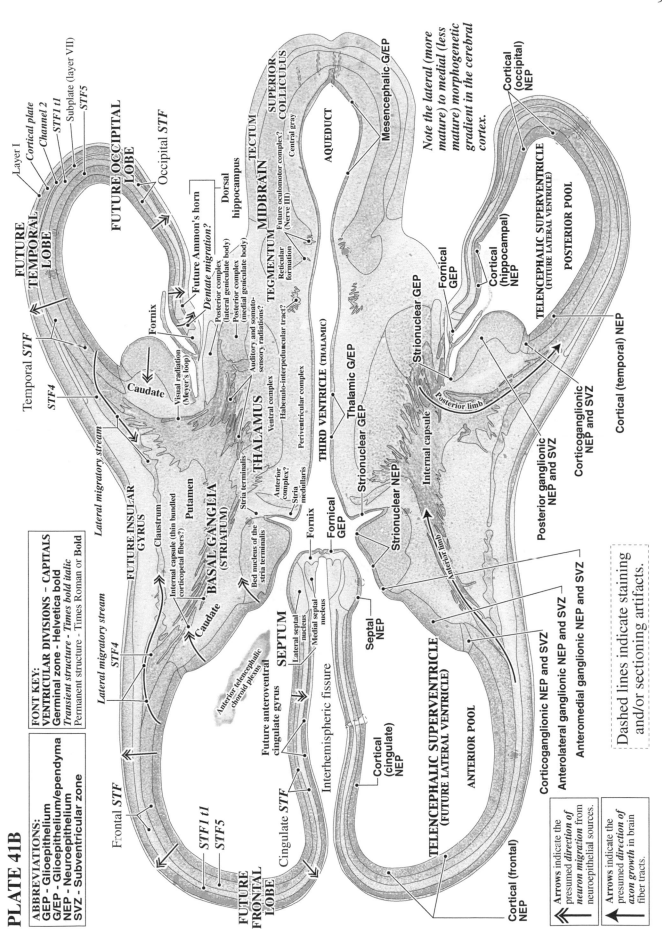

99

**ABBREVIATIONS:**
GEP - Glioepithelium
G/EP - Glioepithelium/ependyma
NEP - Neuroepithelium
SVZ - Subventricular zone

**FONT KEY:**
VENTRICULAR DIVISIONS – CAPITALS
Germinal zone - Helvetica bold
*Transient structure - Times bold italic*
Permanent structure - Times Roman or Bold

Layer I
*Cortical plate*
*Channel 2*
*STF1 t1*
Subplate (layer VII)
*STF5*

FUTURE
TEMPORAL
LOBE

FUTURE OCCIPITAL
LOBE

Occipital *STF*

Temporal *STF*

*STF4*

Fornix

Future Ammon's horn
*Dentate migration?*

Dorsal
hippocampus

Posterior complex
(lateral geniculate body)
Posterior complex
(medial geniculate body)

Visual radiation
(Meyer's loop)

Auditory and somato-
sensory radiations?

**TECTUM**

**SUPERIOR
COLLICULUS**

Mesencephalic G/EP

Central gray

Future oculomotor complex?
(Nerve III)

**MIDBRAIN**

**TEGMENTUM**

Reticular
formation

AQUEDUCT

Note the lateral (more
mature) to medial (less
mature) morphogenetic
gradient in the cerebral
cortex.

Cortical
(occipital)
NEP

Caudate

Lateral migratory stream

**THALAMUS**

Ventral complex

Habenulo-interpeduncular tract?

Periventricular complex

**THIRD VENTRICLE** (THALAMIC)

Thalamic G/EP

Strionuclear GEP

Fornical
GEP

Cortical
(hippocampal)
NEP

**TELENCEPHALIC SUPERVENTRICLE**
(FUTURE LATERAL VENTRICLE)

POSTERIOR POOL

**FUTURE INSULAR
GYRUS**

Claustrum

Internal capsule (thin bundled
corticopetal fibers?)

Putamen

Stria terminalis

Anterior
complex?
Stria
medullaris

**BASAL GANGLIA
(STRIATUM)**

Bed nucleus of the
stria terminalis

Caudate

Strionuclear GEP

Strionuclear NEP

Posterior limb

Internal capsule

Strionuclear GEP

Corticoganglionic
NEP and SVZ

Cortical (temporal) NEP

*STF4*

Lateral migratory stream

*STF5*

*STF1 t1*

Frontal *STF*

Fornix

Fornical
GEP

**SEPTUM**

Lateral septal
nucleus

Medial septal
nucleus

Anterior telencephalic
choroid plexus

Future anteroventral
cingulate gyrus

Interhemispheric fissure

Cortical
(cingulate)
NEP

Cingulate *STF*

**FUTURE
FRONTAL
LOBE**

Septal
NEP

Anterior limb

Posterior ganglionic
NEP and SVZ

Corticoganglionic
NEP and SVZ

**TELENCEPHALIC SUPERVENTRICLE**
(FUTURE LATERAL VENTRICLE)

ANTERIOR POOL

Cortical (frontal)
NEP

Corticoganglionic NEP and SVZ

Anterolateral ganglionic NEP and SVZ

Anteromedial ganglionic NEP and SVZ

Dashed lines indicate staining
and/or sectioning artifacts.

⇐ Arrows indicate the
presumed *direction of
neuron migration* from
neuroepithelial sources.

← Arrows indicate the
presumed *direction of
axon growth* in brain
fiber tracts.

**PLATE 42A**

**GW11 Horizontal**
**CR 57 mm, C1500**
**Level 7: Section 1143**

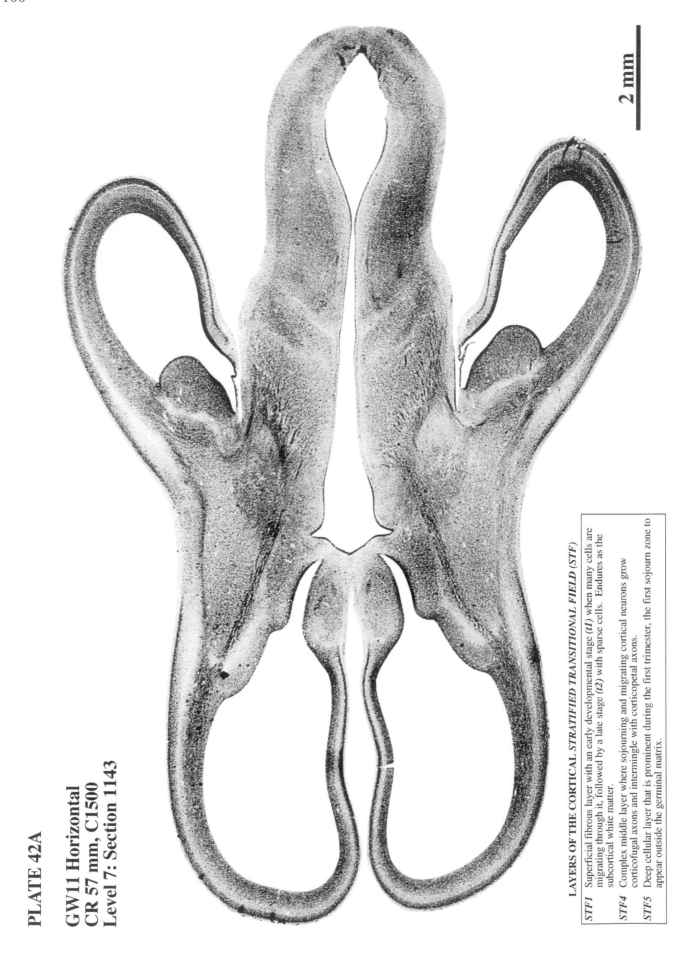

**LAYERS OF THE CORTICAL *STRATIFIED TRANSITIONAL FIELD (STF)***

*STF1* Superficial fibrous layer with an early developmental stage (*t1*) when many cells are migrating through it, followed by a late stage (*t2*) with sparse cells. Endures as the subcortical white matter.

*STF4* Complex middle layer where sojourning and migrating cortical neurons grow corticofugal axons and intermingle with corticopetal axons.

*STF5* Deep cellular layer that is prominent during the first trimester, the first sojourn zone to appear outside the germinal matrix.

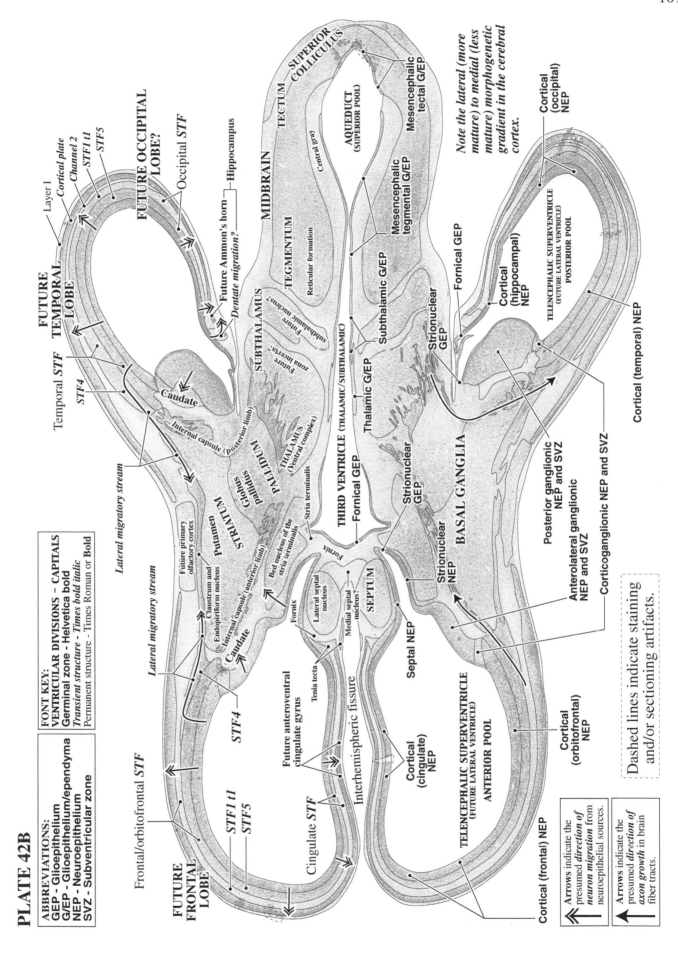

## PLATE 42B

**ABBREVIATIONS:**
GEP - Glioepithelium
G/EP - Glioepithelium/ependyma
NEP - Neuroepithelium
SVZ - Subventricular zone

**FONT KEY:**
VENTRICULAR DIVISIONS – CAPITALS
Germinal zone - Helvetica bold
*Transient structure - Times bold italic*
Permanent structure - Times Roman or Bold

Arrows indicate the presumed *direction of neuron migration* from neuroepithelial sources.

Arrows indicate the presumed *direction of axon growth* in brain fiber tracts.

Dashed lines indicate staining and/or sectioning artifacts.

Note the lateral (more mature) to medial (less mature) morphogenetic gradient in the cerebral cortex.

SUPERIOR COLLICULUS

TECTUM

MIDBRAIN

TEGMENTUM

AQUEDUCT (SUPERIOR POOL)

Central gray

Reticular formation

Mesencephalic tectal G/EP

Mesencephalic tegmental G/EP

Subthalamic G/EP

Fornical GEP

Cortical (occipital) NEP

Cortical (hippocampal) NEP

TELENCEPHALIC SUPERVENTRICLE (FUTURE LATERAL VENTRICLE) POSTERIOR POOL

Cortical (temporal) NEP

SUBTHALAMUS

*Future subthalamic nucleus?*

*Future zona incerta?*

THIRD VENTRICLE (THALAMIC/SUBTHALAMIC)

Fornical GEP

Thalamic G/EP

Strionuclear GEP

Strionuclear GEP

Strionuclear NEP

BASAL GANGLIA

Posterior ganglionic NEP and SVZ

Anterolateral ganglionic NEP and SVZ

Corticoganglionic NEP and SVZ

THALAMUS (Ventral complex)

Stria terminalis

PALLIDUM

Globus pallidus

STRIATUM

Putamen

Bed nucleus of the stria terminalis

Internal capsule (anterior limb)

*Caudate*

Fornix

FORNIX

Septal NEP

SEPTUM

Medial septal nucleus?

Lateral septal nucleus

Tenia tecta

Cortical (cingulate) NEP

Future anteroventral cingulate gyrus

Interhemispheric fissure

Cortical (orbitofrontal) NEP

TELENCEPHALIC SUPERVENTRICLE (FUTURE LATERAL VENTRICLE) ANTERIOR POOL

Cortical (frontal) NEP

FUTURE FRONTAL LOBE

*Cingulate STF*

*STF1 t1*
*STF5*

*Frontal/orbitofrontal STF*

*Lateral migratory stream*

*STF4*

Endopiriform nucleus
Claustrum and
Internal capsule (posterior limb)

Future primary olfactory cortex

*Caudate*

*Lateral migratory stream*

*Lateral migratory stream*

*STF4*

Temporal *STF*

FUTURE TEMPORAL LOBE

FUTURE OCCIPITAL LOBE?

Occipital *STF*

*Dentate migration?*

Future Ammon's horn
Hippocampus

Layer 1
Cortical plate
*Channel 2*
*STF1 t1*
*STF5*

**PLATE 43A**

**GW11 Horizontal**
**CR 57 mm, C1500**
**Level 8: Section 1256**

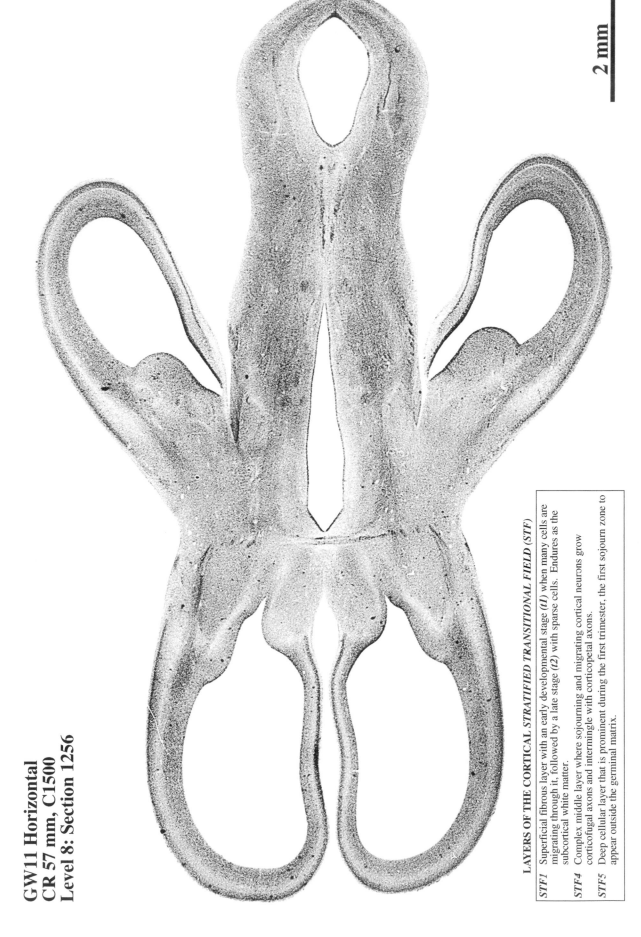

**2 mm**

**LAYERS OF THE CORTICAL *STRATIFIED TRANSITIONAL FIELD (STF)***

*STF1* Superficial fibrous layer with an early developmental stage (*t1*) when many cells are migrating through it, followed by a late stage (*t2*) with sparse cells. Endures as the subcortical white matter.

*STF4* Complex middle layer where sojourning and migrating cortical neurons grow corticofugal axons and intermingle with corticopetal axons.

*STF5* Deep cellular layer that is prominent during the first trimester, the first sojourn zone to appear outside the germinal matrix.

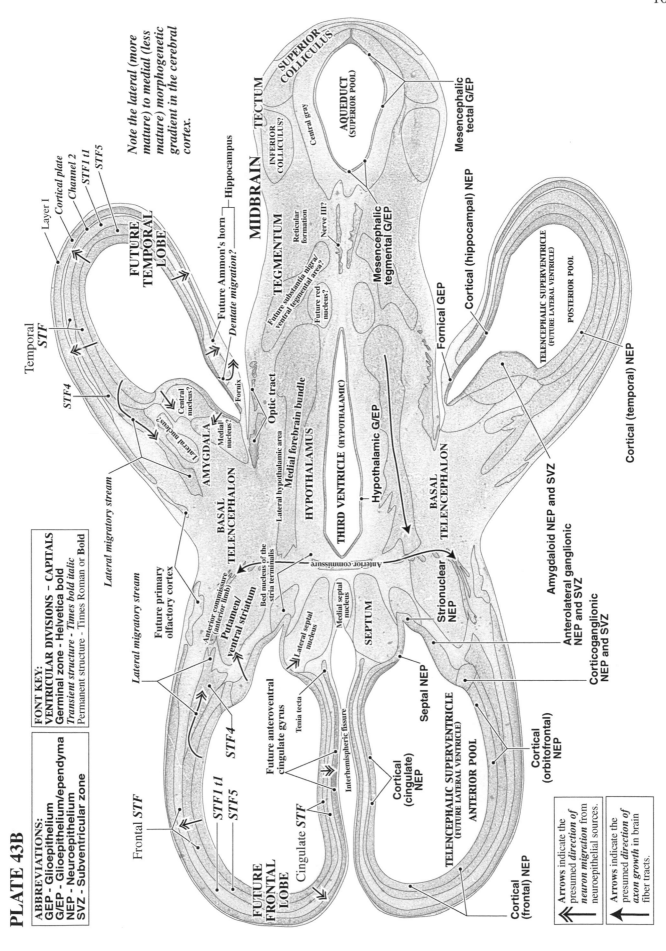

## PLATE 43B

**FONT KEY:**
VENTRICULAR DIVISIONS – CAPITALS
Germinal zone - **Helvetica bold**
*Transient structure - Times bold italic*
Permanent structure - Times Roman or Bold

**ABBREVIATIONS:**
GEP - Glioepithelium
G/EP - Glioepithelium/ependyma
NEP - Neuroepithelium
SVZ - Subventricular zone

*Note the lateral (more mature) to medial (less mature) morphogenetic gradient in the cerebral cortex.*

Layer 1
Cortical plate
*Channel 2*
*STF1 t1*
*STF5*

**FUTURE TEMPORAL LOBE**

Future Ammon's horn ⎤ Hippocampus
*Dentate migration?* ⎦

Temporal *STF*

*STF4*

AMYGDALA
Central nucleus?
Medial nucleus?
Lateral nucleus?

BASAL TELENCEPHALON

*Lateral migratory stream*

Future primary olfactory cortex

*Lateral migratory stream*

Anterior commissure (anterior limb)
Bed nucleus of the stria terminalis

**Putamen/ ventral striatum**

Tenia tecta

Interhemispheric fissure

Future anteroventral cingulate gyrus

Frontal *STF*

*STF1 t1*
*STF5*
*STF4*

**FUTURE FRONTAL LOBE**

Cingulate *STF*

**Cortical (frontal) NEP**

**TELENCEPHALIC SUPERVENTRICLE (FUTURE LATERAL VENTRICLE) ANTERIOR POOL**

**Cortical (orbitofrontal) NEP**

**Cortical (cingulate) NEP**

**Septal NEP**

Lateral septal nucleus
Medial septal nucleus

**SEPTUM**

Fornix

Optic tract

Lateral hypothalamic area
Medial forebrain bundle

**HYPOTHALAMUS**

**THIRD VENTRICLE (HYPOTHALAMIC)**

Future substantia nigra/ ventral tegmental area?
Future red nucleus?

**TEGMENTUM**

Reticular formation
Nerve III?
Central gray

**MIDBRAIN**

INFERIOR COLLICULUS?

**TECTUM**

SUPERIOR COLLICULUS

**AQUEDUCT (SUPERIOR POOL)**

Mesencephalic tectal G/EP

**Mesencephalic tegmental G/EP**

**Hypothalamic G/EP**

Anterior commissure

**BASAL TELENCEPHALON**

**Strionuclear NEP**

**Amygdaloid NEP and SVZ**

**Anterolateral ganglionic NEP and SVZ**

**Corticoganglionic NEP and SVZ**

**Fornical GEP**

**Cortical (hippocampal) NEP**

**TELENCEPHALIC SUPERVENTRICLE (FUTURE LATERAL VENTRICLE) POSTERIOR POOL**

**Cortical (temporal) NEP**

⇐ Arrows indicate the presumed *direction of neuron migration* from neuroepithelial sources.

◀ Arrows indicate the presumed *direction of axon growth* in brain fiber tracts.

103

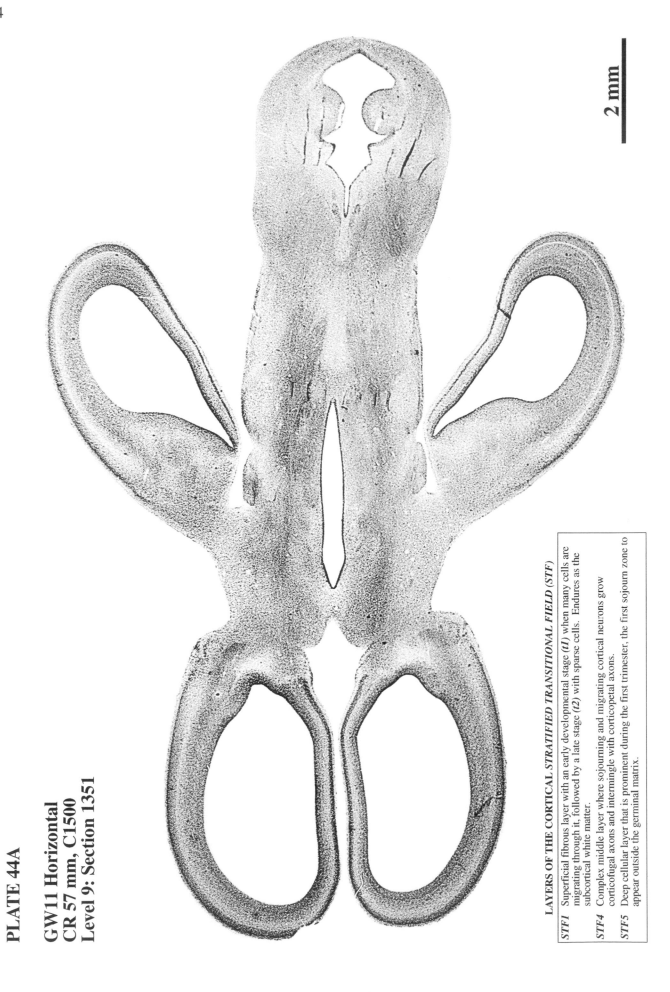

PLATE 44A

**GW11 Horizontal**
**CR 57 mm, C1500**
**Level 9: Section 1351**

2 mm

**LAYERS OF THE CORTICAL *STRATIFIED TRANSITIONAL FIELD* (STF)**

*STF1*  Superficial fibrous layer with an early developmental stage (*t1*) when many cells are migrating through it, followed by a late stage (*t2*) with sparse cells. Endures as the subcortical white matter.

*STF4*  Complex middle layer where sojourning and migrating cortical neurons grow corticofugal axons and intermingle with corticopetal axons.

*STF5*  Deep cellular layer that is prominent during the first trimester, the first sojourn zone to appear outside the germinal matrix.

105

PLATE 44B

**ABBREVIATIONS:**
G/EP - Glioepithelium/ependyma
NEP - Neuroepithelium
SVZ - Subventricular zone

**FONT KEY:**
VENTRICULAR DIVISIONS – CAPITALS
Germinal zone - Helvetica bold
*Transient structure - Times bold italic*
Permanent structure - Times Roman or Bold

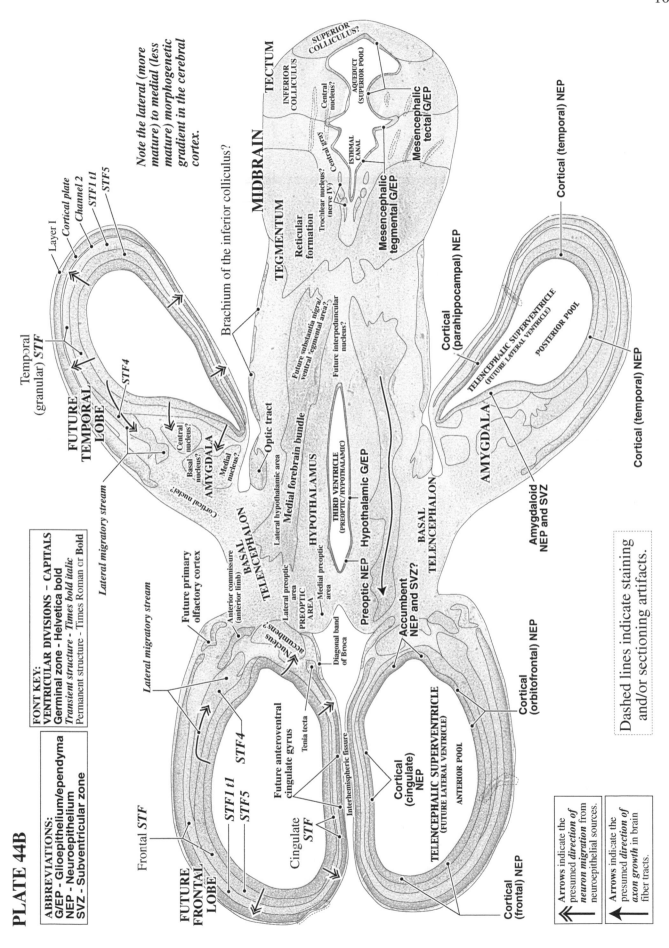

*Note the lateral (more mature) to medial (less mature) morphogenetic gradient in the cerebral cortex.*

Dashed lines indicate staining and/or sectioning artifacts.

Arrows indicate the presumed *direction of neuron migration* from neuroepithelial sources.

Arrows indicate the presumed *direction of axon growth* in brain fiber tracts.

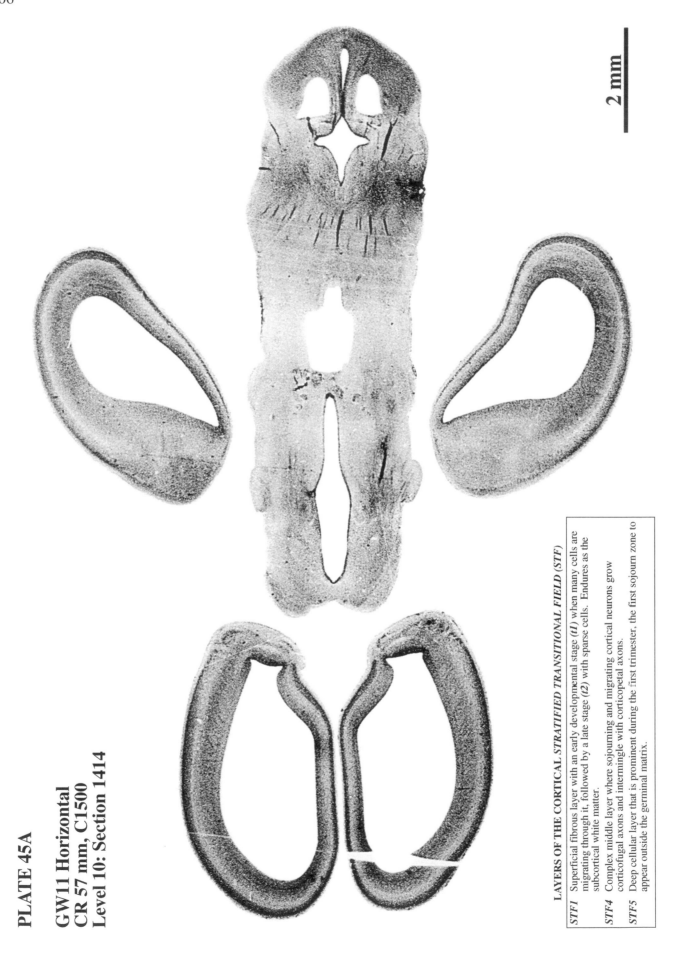

**PLATE 45A**

**GW11 Horizontal**
**CR 57 mm, C1500**
**Level 10: Section 1414**

2 mm

**LAYERS OF THE CORTICAL STRATIFIED TRANSITIONAL FIELD (STF)**

*STF1* Superficial fibrous layer with an early developmental stage (*t1*) when many cells are migrating through it, followed by a late stage (*t2*) with sparse cells. Endures as the subcortical white matter.

*STF4* Complex middle layer where sojourning and migrating cortical neurons grow corticofugal axons and intermingle with corticopetal axons.

*STF5* Deep cellular layer that is prominent during the first trimester, the first sojourn zone to appear outside the germinal matrix.

107

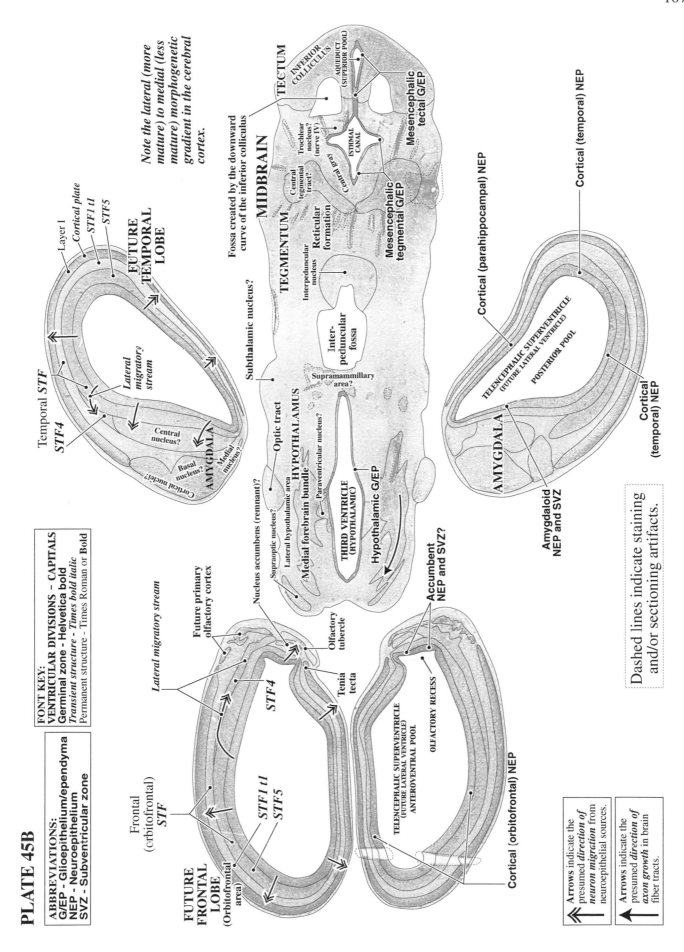

PLATE 45B

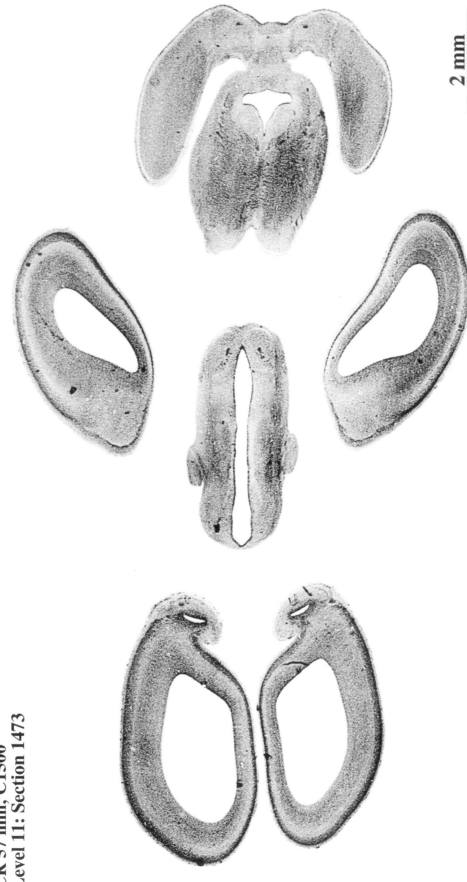

**PLATE 46A**

**GW11 Horizontal**
**CR 57 mm, C1500**
**Level 11: Section 1473**

---

**LAYERS OF THE CORTICAL *STRATIFIED TRANSITIONAL FIELD* (*STF*)**

*STF1*  Superficial fibrous layer with an early developmental stage (*t1*) when many cells are migrating through it, followed by a late stage (*t2*) with sparse cells.  Endures as the subcortical white matter.

*STF4*  Complex middle layer where sojourning and migrating cortical neurons grow corticofugal axons and intermingle with corticopetal axons.

*STF5*  Deep cellular layer that is prominent during the first trimester, the first sojourn zone to appear outside the germinal matrix.

# PLATE 46B

**ABBREVIATIONS:**
G/EP - Glioepithelium/ependyma
NEP - Neuroepithelium
SVZ - Subventricular zone

**FONT KEY:**
VENTRICULAR DIVISIONS – CAPITALS
Germinal zone - **Helvetica bold**
*Transient structure - Times bold italic*
Permanent structure - Times Roman or **Bold**

*Note the lateral (more mature) to medial (less mature) morphogenetic gradient in the cerebral cortex.*

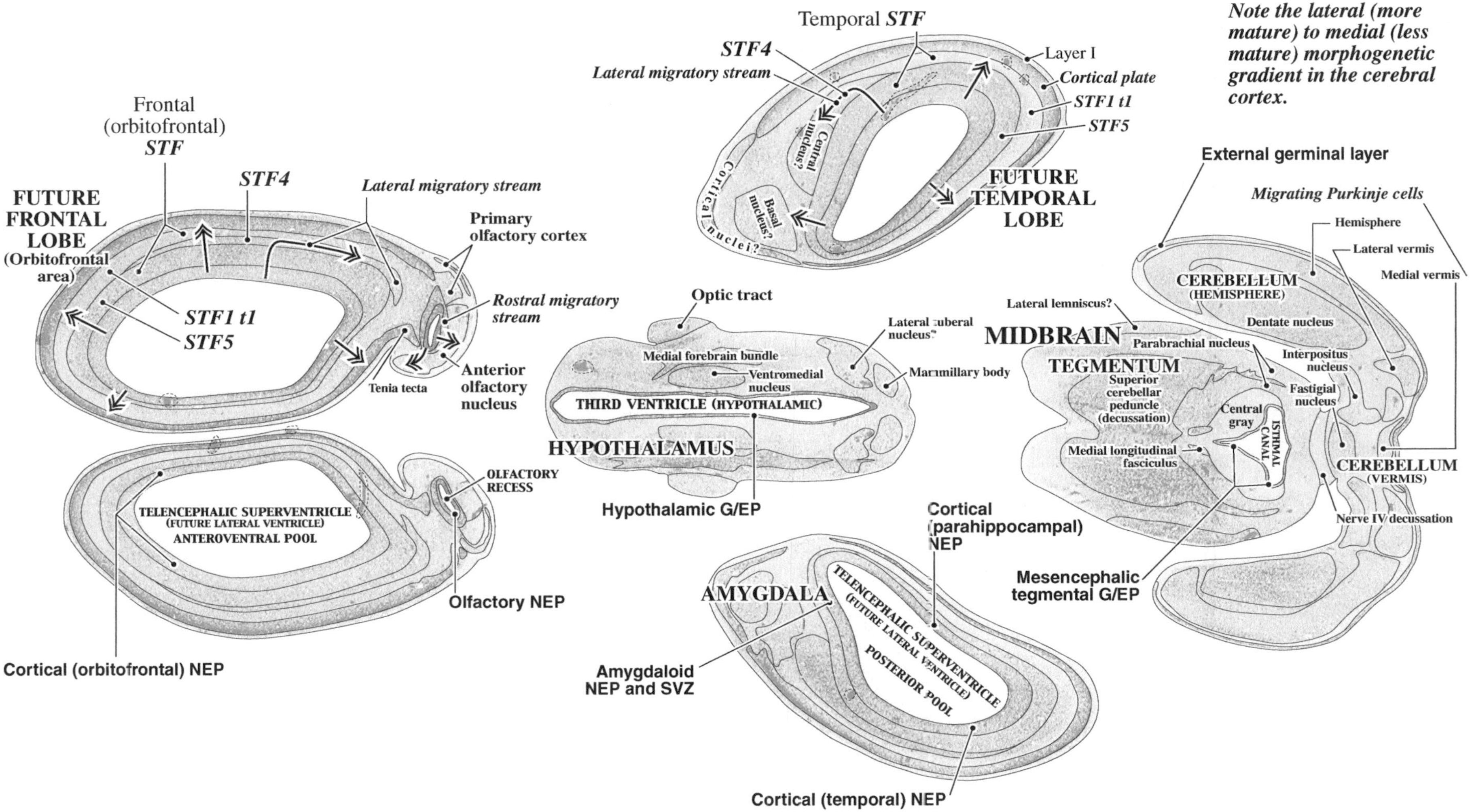

External germinal layer

*Migrating Purkinje cells*

Lateral vermis

Medial vermis

Hemisphere

**CEREBELLUM (HEMISPHERE)**

**CEREBELLUM (VERMIS)**

Dentate nucleus

Interpositus nucleus

Fastigial nucleus

Nerve IV decussation

Parabrachial nucleus

ISTHMAL CANAL

Central gray

Superior cerebellar peduncle (decussation)

Medial longitudinal fasciculus

**TEGMENTUM**

**MIDBRAIN**

Lateral lemniscus?

Mesencephalic tegmental G/EP

Layer I

*Cortical plate*

*STF1 t1*

*STF5*

**FUTURE TEMPORAL LOBE**

Temporal *STF*

*STF4*

*Lateral migratory stream*

*Central nucleus?*

*Cortical nuclei?*

*Basal nucleus?*

Mammillary body

Lateral tuberal nucleus?

Cortical (parahippocampal) NEP

**TELENCEPHALIC SUPERVENTRICLE (FUTURE LATERAL VENTRICLE)**

**POSTERIOR POOL**

**AMYGDALA**

Cortical (temporal) NEP

Amygdaloid NEP and SVZ

Optic tract

Ventromedial nucleus

Medial forebrain bundle

**THIRD VENTRICLE (HYPOTHALAMIC)**

**HYPOTHALAMUS**

Hypothalamic G/EP

*Lateral migratory stream*

Primary olfactory cortex

*Rostral migratory stream*

Anterior olfactory nucleus

Tenia tecta

OLFACTORY RECESS

Olfactory NEP

Frontal (orbitofrontal) *STF*

*STF4*

*STF1 t1*

*STF5*

**FUTURE FRONTAL LOBE**
(Orbitofrontal area)

**TELENCEPHALIC SUPERVENTRICLE**
**(FUTURE LATERAL VENTRICLE)**
**ANTEROVENTRAL POOL**

Cortical (orbitofrontal) NEP

Dashed lines indicate staining and/or sectioning artifacts.

Arrows indicate the presumed *direction of neuron migration* from neuroepithelial sources.

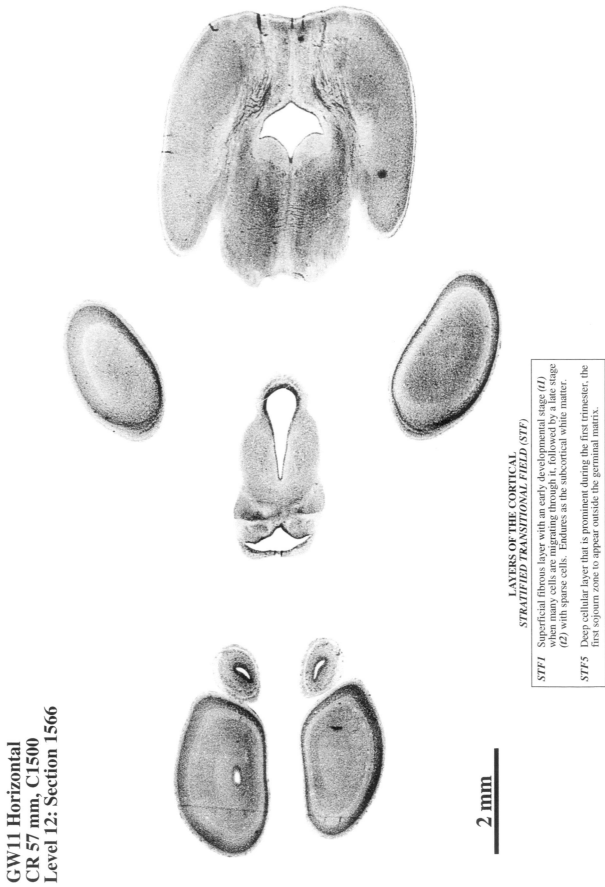

**PLATE 47A**

**GW11 Horizontal**
**CR 57 mm, C1500**
**Level 12: Section 1566**

**2 mm**

**LAYERS OF THE CORTICAL**
*STRATIFIED TRANSITIONAL FIELD (STF)*

*STF1* Superficial fibrous layer with an early developmental stage (*t1*) when many cells are migrating through it, followed by a late stage (*t2*) with sparse cells. Endures as the subcortical white matter.

*STF5* Deep cellular layer that is prominent during the first trimester, the first sojourn zone to appear outside the germinal matrix.

# PLATE 47B

ABBREVIATIONS:
GEP - Glioepithelium
G/EP - Glioepithelium/ependyma
NEP - Neuroepithelium
SVZ - Subventricular zone

FONT KEY:
VENTRICULAR DIVISIONS – CAPITALS
Germinal zone - Helvetica bold
Transient structure - Times bold italic
Permanent structure - Times Roman or Bold

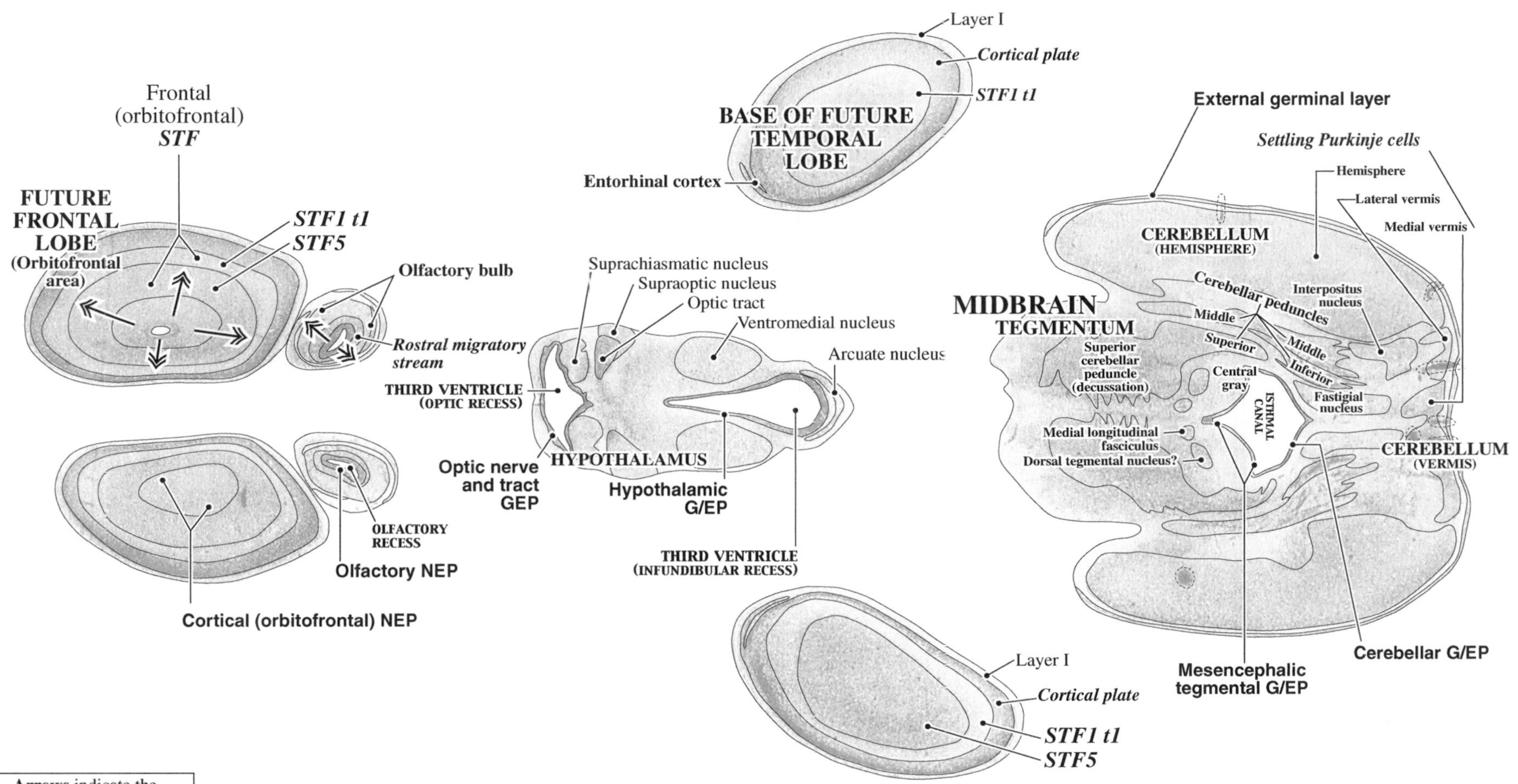

Arrows indicate the presumed *direction of neuron migration* from neuroepithelial sources.

Dashed lines indicate staining and/or sectioning artifacts.

**PLATE 48A**

**GW11 Horizontal**
**CR 57 mm, C1500**
**Level 13: Section 1626**

2 mm

# PLATE 48B

**FONT KEY:**
VENTRICULAR DIVISIONS – CAPITALS
Germinal zone - Helvetica bold
*Transient structure - Times bold italic*
Permanent structure - Times Roman or Bold

## G/EP - Glioepithelium/ependyma

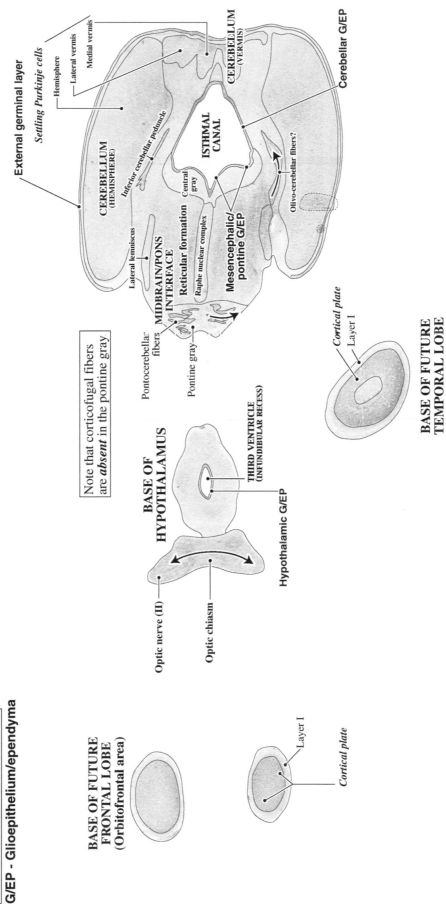

External germinal layer

*Settling Purkinje cells*

Medial vermis
Lateral vermis
Hemisphere

CEREBELLUM
(VERMIS)

Cerebellar G/EP

CEREBELLUM
(HEMISPHERE)

Inferior cerebellar peduncle

ISTHMAL
CANAL

Olivo-cerebellar fibers?

Central
gray

Cerebellar G/EP

Lateral lemniscus

MIDBRAIN/PONS
INTERFACE

Reticular formation

Raphe nuclear complex

Mesencephalic/
pontine G/EP

Pontocerebellar
fibers

Pontine gray

Note that corticofugal fibers
are *absent* in the pontine gray.

BASE OF
HYPOTHALAMUS

THIRD VENTRICLE
(INFUNDIBULAR RECESS)

Hypothalamic G/EP

Optic nerve (II)

Optic chiasm

*Cortical plate*
Layer I

BASE OF FUTURE
TEMPORAL LOBE

BASE OF FUTURE
FRONTAL LOBE
(Orbitofrontal area)

Layer I

*Cortical plate*

Arrows indicate the
presumed *direction of
axon growth* in brain
fiber tracts.

Dashed lines indicate staining
and/or sectioning artifacts.

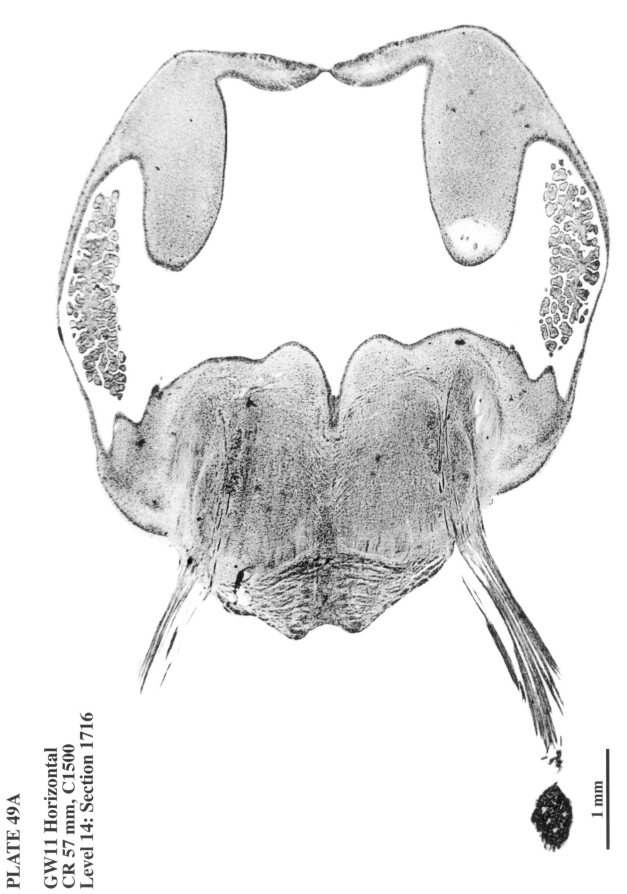

**PLATE 49A**

**GW11 Horizontal**
**CR 57 mm, C1500**
**Level 14: Section 1716**

1 mm

Levels 14 to 21 are only shown at high magnification.

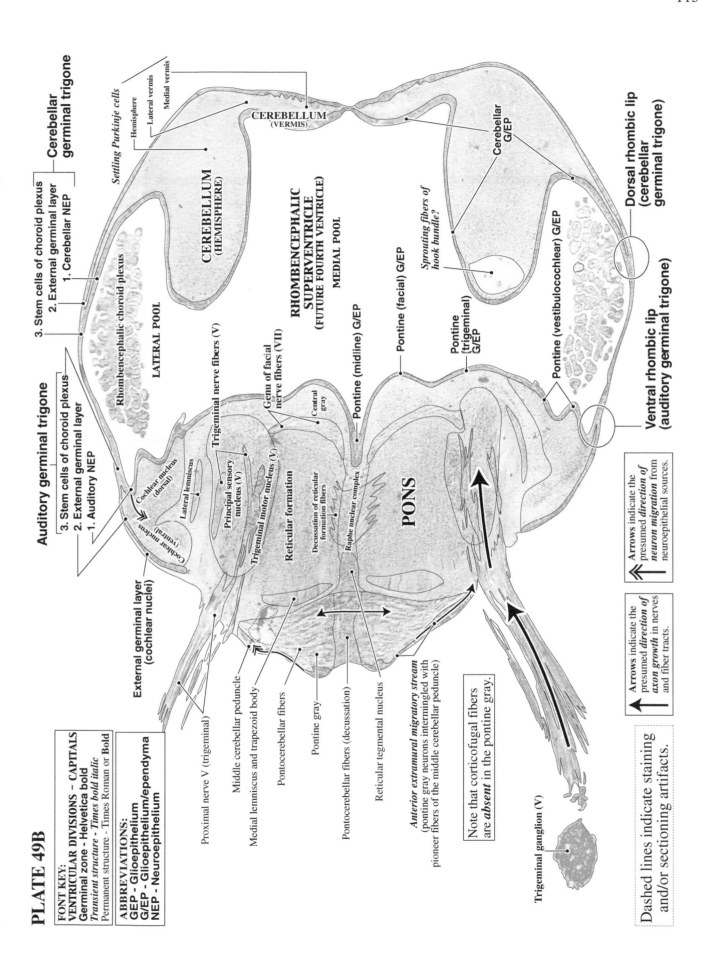

## PLATE 49B

FONT KEY:
VENTRICULAR DIVISIONS - **CAPITALS**
Germinal zone - **Helvetica bold**
*Transient structure - Times bold italic*
Permanent structure - Times Roman or **Bold**

ABBREVIATIONS:
GEP - Glioepithelium
G/EP - Glioepithelium/ependyma
NEP - Neuroepithelium

Cerebellar germinal trigone

3. Stem cells of choroid plexus
2. External germinal layer
1. Cerebellar NEP

Auditory germinal trigone

3. Stem cells of choroid plexus
2. External germinal layer
1. Auditory NEP

Dorsal rhombic lip (cerebellar germinal trigone)

Ventral rhombic lip (auditory germinal trigone)

*Settling Purkinje cells*

Medial vermis
Lateral vermis
Hemisphere

**CEREBELLUM (VERMIS)**

Cerebellar G/EP

*Sprouting fibers of hook bundle?*

**CEREBELLUM (HEMISPHERE)**

*Rhombencephalic choroid plexus*

**LATERAL POOL**

**RHOMBENCEPHALIC SUPERVENTRICLE (FUTURE FOURTH VENTRICLE)**

**MEDIAL POOL**

Pontine (facial) G/EP

Pontine (trigeminal) G/EP

Pontine (vestibulocochlear) G/EP

Pontine (midline) G/EP

Trigeminal nerve fibers (V)

Genu of facial nerve fibers (VII)

Central gray

Cochlear nucleus (dorsal)

Cochlear nucleus (ventral)

Lateral lemniscus

Principal sensory nucleus (V)

Trigeminal motor nucleus (V)

Reticular formation

Decussation of reticular formation fibers

Raphe nuclear complex

**PONS**

External germinal layer (cochlear nuclei)

Proximal nerve V (trigeminal)

Middle cerebellar peduncle

Medial lemniscus and trapezoid body

Pontocerebellar fibers

Pontine gray

Pontocerebellar fibers (decussation)

Reticular tegmental nucleus

*Anterior extramural migratory stream (pontine gray neurons intermingled with pioneer fibers of the middle cerebellar peduncle)*

Note that corticofugal fibers are *absent* in the pontine gray.

Trigeminal ganglion (V)

Arrows indicate the presumed *direction of neuron migration* from neuroepithelial sources.

Arrows indicate the presumed *direction of axon growth* in nerves and fiber tracts.

Dashed lines indicate staining and/or sectioning artifacts.

116

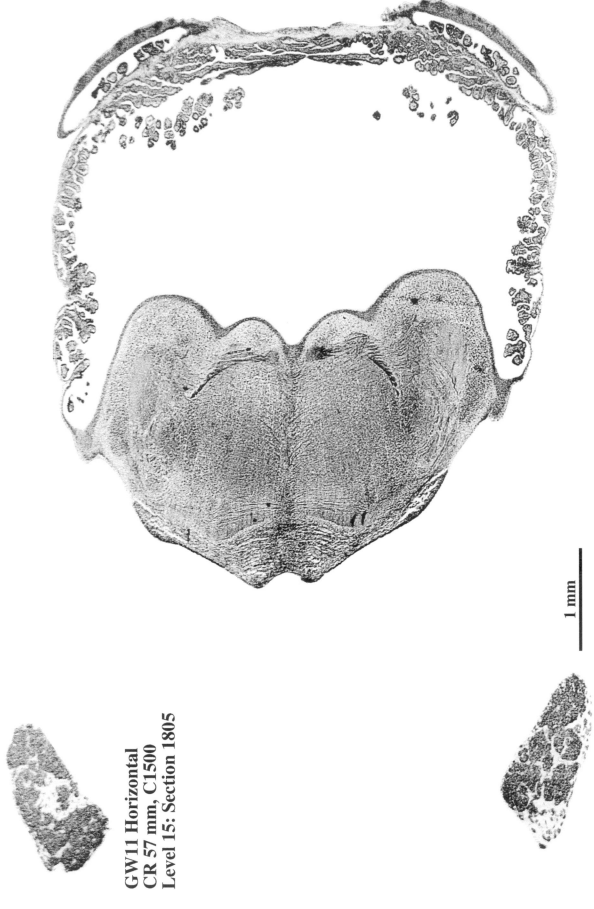

PLATE 50A

**GW11 Horizontal**
**CR 57 mm, C1500**
**Level 15: Section 1805**

1 mm

Levels 14 to 21 are only shown at high magnification.

**PLATE 50B**

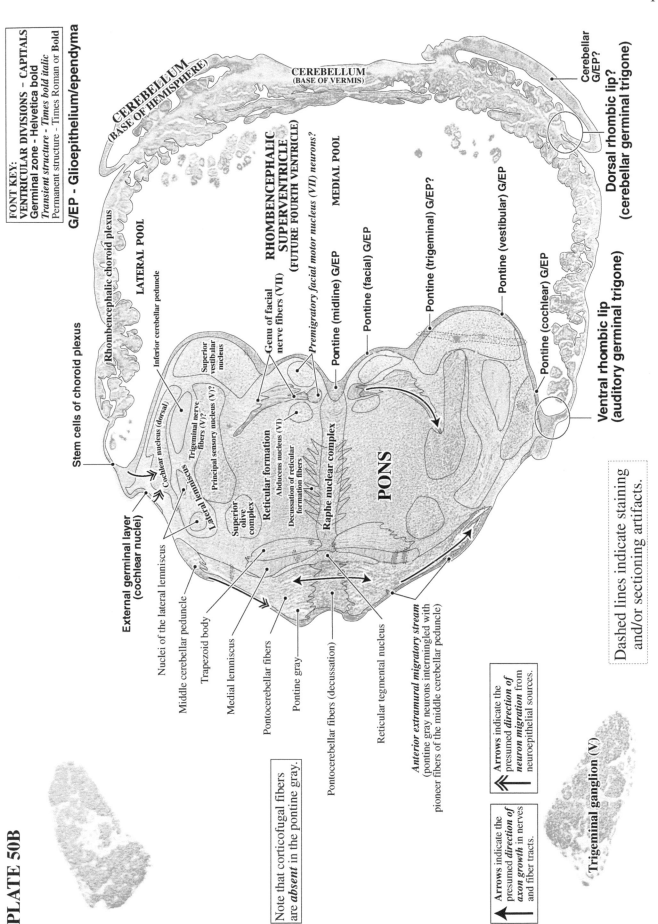

CEREBELLUM
(BASE OF HEMISPHERE)

CEREBELLUM
(BASE OF VERMIS)

Cerebellar
G/EP?

**Dorsal rhombic lip?**
**(cerebellar germinal trigone)**

Rhombencephalic choroid plexus

**LATERAL POOL**

Inferior cerebellar peduncle

Superior
vestibular
nucleus

Genu of facial
nerve fibers (VII)

**RHOMBENCEPHALIC**
**SUPERVENTRICLE**
**(FUTURE FOURTH VENTRICLE)**

*Premigratory facial motor nucleus (VII) neurons?*

**MEDIAL POOL**

**Pontine (midline) G/EP**

**Pontine (facial) G/EP**

**Pontine (trigeminal) G/EP?**

**Pontine (vestibular) G/EP**

**Pontine (cochlear) G/EP**

Stem cells of choroid plexus

Cochlear nucleus (dorsal)

Trigeminal nerve
fibers (V)?

Principal sensory nucleus (V)?

**Reticular formation**

Abducens nucleus (VI)

*Decussation of reticular
formation fibers*

**Raphe nuclear complex**

**PONS**

**Ventral rhombic lip**
**(auditory germinal trigone)**

**External germinal layer**
**(cochlear nuclei)**

Lateral lemniscus

Superior
olive
complex

Nuclei of the lateral lemniscus

Middle cerebellar peduncle

Trapezoid body

Medial lemniscus

Pontocerebellar fibers

Pontine gray

Pontocerebellar fibers (decussation)

Reticular tegmental nucleus

*Anterior extramural migratory stream*
(pontine gray neurons intermingled with
pioneer fibers of the middle cerebellar peduncle)

Note that corticofugal fibers
are *absent* in the pontine gray.

Arrows indicate the
presumed *direction of
axon growth* in nerves
and fiber tracts.

Arrows indicate the
presumed *direction of
neuron migration* from
neuroepithelial sources.

Dashed lines indicate staining
and/or sectioning artifacts.

**Trigeminal ganglion (V)**

118

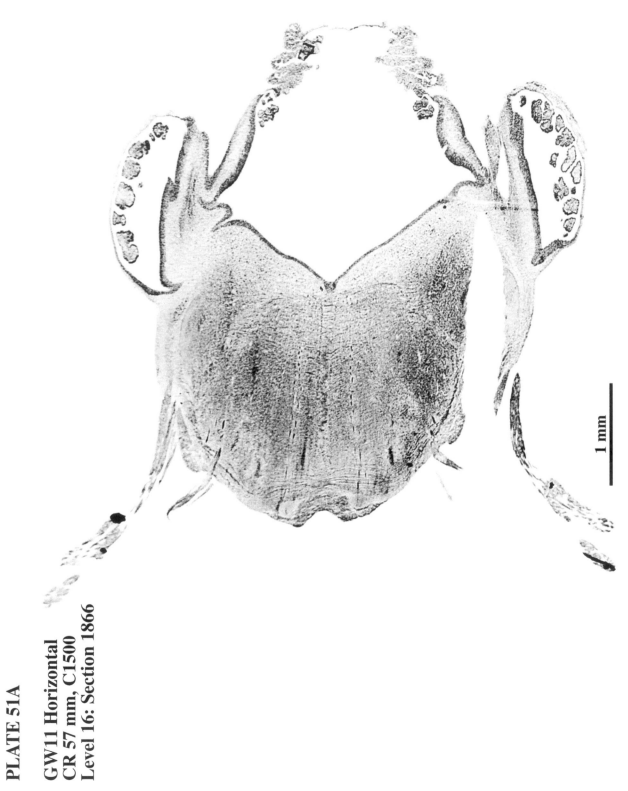

PLATE 51A

**GW11 Horizontal**
**CR 57 mm, C1500**
**Level 16: Section 1866**

1 mm

Levels 14 to 21 are only shown at high magnification.

119

PLATE 51B

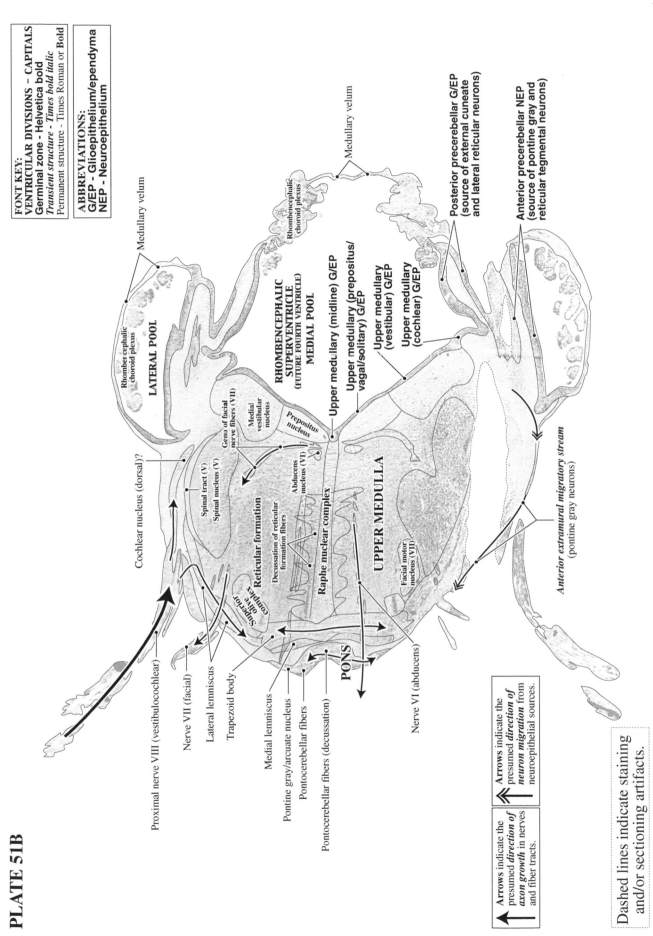

FONT KEY:
VENTRICULAR DIVISIONS - CAPITALS
Germinal zone - Helvetica bold
*Transient structure - Times bold italic*
Permanent structure - Times Roman or Bold

ABBREVIATIONS:
G/EP - Glioepithelium/ependyma
NEP - Neuroepithelium

Arrows indicate direction of *axon growth* in nerves and fiber tracts.

Arrows indicate the presumed *direction of neuron migration* from neuroepithelial sources.

Dashed lines indicate staining and/or sectioning artifacts.

**PLATE 52A**

**GW11 Horizontal
CR 57 mm, C1500
Level 17: Section 1926**

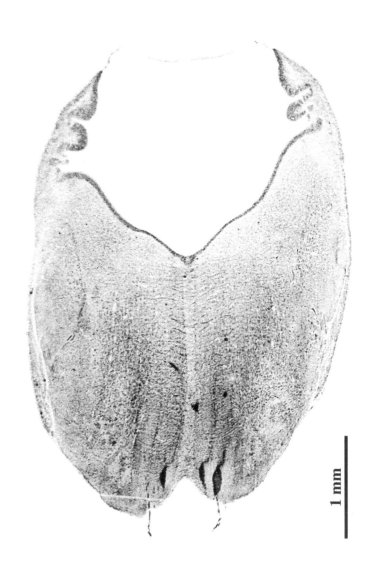

1 mm

Levels 14 to 21 are only shown at high magnification.

# PLATE 52B

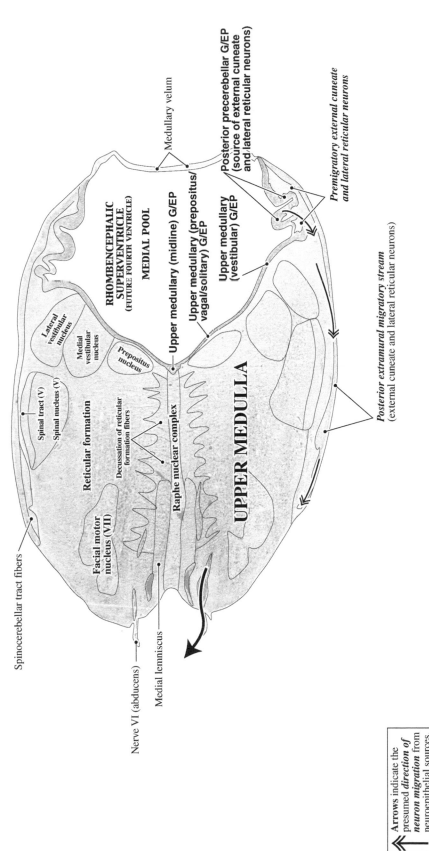

Medullary velum

Posterior precerebellar G/EP
(source of external cuneate
and lateral reticular neurons)

*Premigratory external cuneate
and lateral reticular neurons*

RHOMBENCEPHALIC
SUPERVENTRICLE
(FUTURE FOURTH VENTRICLE)

MEDIAL POOL

Upper medullary (midline) G/EP

Upper medullary (prepositus/
vagal/solitary) G/EP

Upper medullary
(vestibular) G/EP

*Lateral
vestibular
nucleus*

*Medial
vestibular
nucleus*

*Prepositus
nucleus*

**Spinal tract (V)**

**Spinal nucleus (V)**

**Reticular formation**

*Decussation of reticular
formation fibers*

**Raphe nuclear complex**

**Facial motor
nucleus (VII)**

# UPPER MEDULLA

*Posterior extramural migratory stream
(external cuneate and lateral reticular neurons)*

Spinocerebellar tract fibers

Nerve VI (abducens)

Medial lemniscus

Arrows indicate the
presumed *direction of
neuron migration* from
neuroepithelial sources.

Arrows indicate the
presumed *direction of
axon growth* in nerves
and fiber tracts.

**PLATE 53A**

**GW11 Horizontal
CR 57 mm, C1500
Level 18: Section 1962**

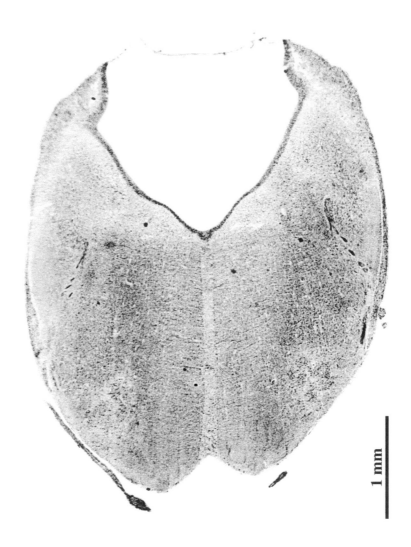

1 mm

Levels 14 to 21 are only shown at high magnification.

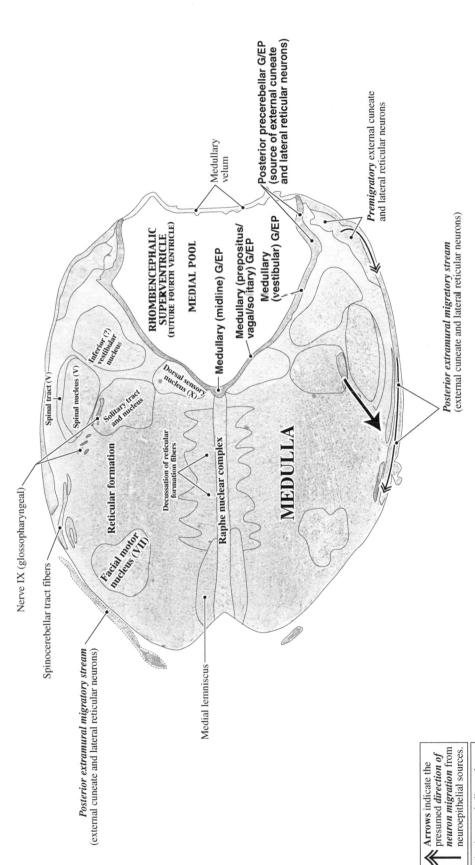

PLATE 53B

FONT KEY:
VENTRICULAR DIVISIONS – CAPITALS
Germinal zone - Helvetica bold
*Transient structure - Times bold italic*
Permanent structure - Times Roman or Bold

**G/EP - Glioepithelium/ependyma**

Medullary
velum

Posterior precerebellar G/EP
(source of external cuneate
and lateral reticular neurons)

*Premigratory* external cuneate
and lateral reticular neurons

*Posterior extramural migratory stream*
(external cuneate and lateral reticular neurons)

RHOMBENCEPHALIC
SUPERVENTRICLE
(FUTURE FOURTH VENTRICLE)

MEDIAL POOL

Medullary (midline) G/EP

Medullary (prepositus/
vagal/solitary) G/EP

Medullary
(vestibular) G/EP

*Inferior (?)
vestibular
nucleus*

*Dorsal sensory
nucleus (X)*

*Spinal tract (V)*

*Spinal nucleus (V)*

*Solitary tract
and nucleus*

**Reticular formation**

Decussation of reticular
formation fibers

**Raphe nuclear complex**

Nerve IX (glossopharyngeal)

**MEDULLA**

**Facial motor
nucleus (VII)**

Spinocerebellar tract fibers

*Posterior extramural migratory stream*
(external cuneate and lateral reticular neurons)

Medial lemniscus

Dashed lines indicate staining
and/or sectioning artifacts.

Arrows indicate the
presumed *direction of
neuron migration* from
neuroepithelial sources.

Arrows indicate the
presumed *direction of
axon growth* in nerves
and fiber tracts.

124

PLATE 54A

GW11 Horizontal
CR 57 mm, C1500
Level 19: Section 2202

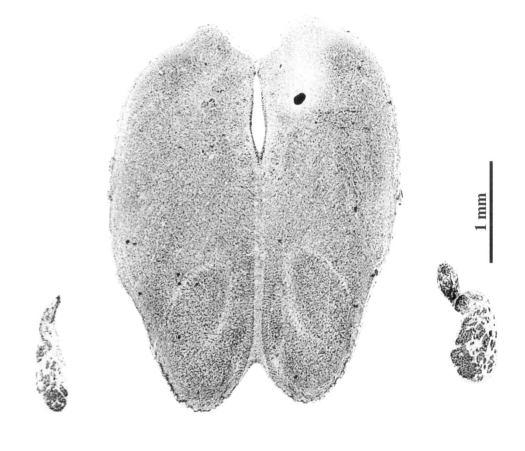

1 mm

Levels 14 to 21 are only shown at high magnification.

## PLATE 54B

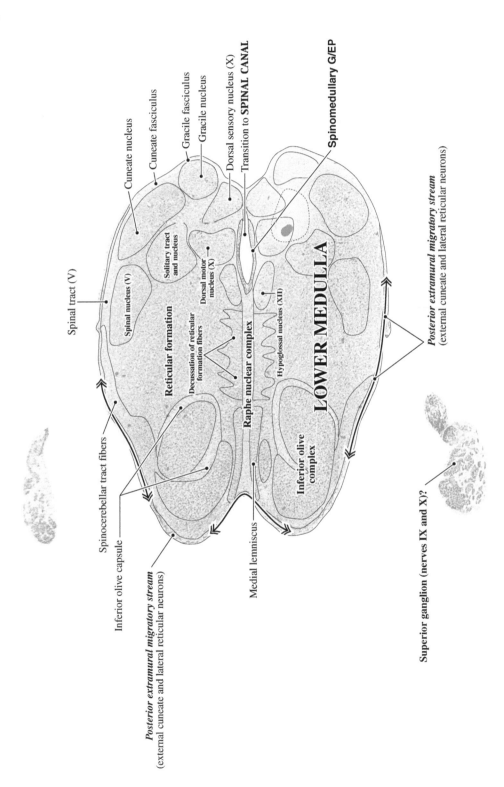

Cuneate nucleus

Cuneate fasciculus

Gracile fasciculus

Gracile nucleus

Dorsal sensory nucleus (X)

Transition to **SPINAL CANAL**

**Spinomedullary G/EP**

*Posterior extramural migratory stream*
(external cuneate and lateral reticular neurons)

Spinal tract (V)

*Solitary tract
and nucleus*

Spinal nucleus (V)

*Dorsal motor
nucleus (X)*

**Reticular formation**

*Decussation of reticular
formation fibers*

**Raphe nuclear complex**

*Hypoglossal nucleus (XII)*

**LOWER MEDULLA**

**Inferior olive
complex**

Spinocerebellar tract fibers

Inferior olive capsule

*Posterior extramural migratory stream*
(external cuneate and lateral reticular neurons)

Medial lemniscus

Superior ganglion (nerves IX and X)?

Dashed lines indicate staining
and/or sectioning artifacts.

Arrows indicate the
presumed *direction of
neuron migration* from
neuroepithelial sources.

**PLATE 55A**

**GW11 Horizontal**
**CR 57 mm, C1500**
**Level 20: Section 2322**

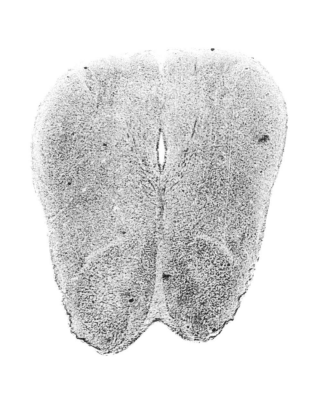

1 mm

Levels 14 to 21 are only shown at high magnification.

# PLATE 55B

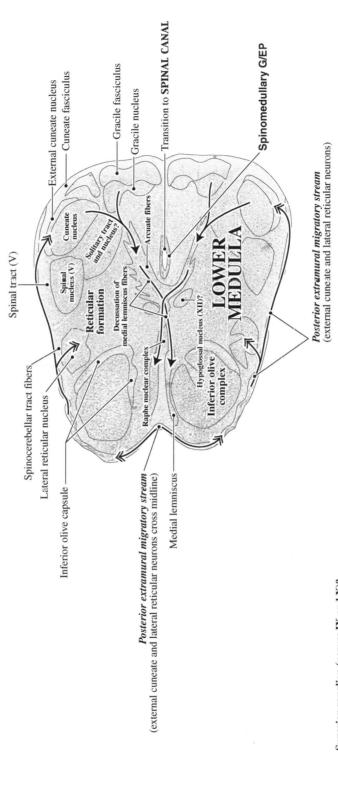

External cuneate nucleus

Cuneate fasciculus

Gracile fasciculus

Gracile nucleus

Transition to **SPINAL CANAL**

**Spinomedullary G/EP**

Cuneate nucleus

*Solitary tract and nucleus?*

Arcuate fibers

Spinal tract (V)

Spinal nucleus (V)

**Reticular formation**

*Decussation of medial lemniscus fibers*

*Posterior extramural migratory stream*
(external cuneate and lateral reticular neurons)

**LOWER MEDULLA**

Hypoglossal nucleus (XII)?

Spinocerebellar tract fibers

Lateral reticular nucleus

Inferior olive capsule

*Posterior extramural migratory stream*
(external cuneate and lateral reticular neurons cross midline)

Medial lemniscus

Raphe nuclear complex

Inferior olive complex

Superior ganglion (nerves IX and X)?

Inferior ganglion (nerves IX and X)?

⇐ **Arrows** indicate the presumed *direction of neuron migration* from neuroepithelial sources.

← **Arrows** indicate the presumed *direction of axon growth* in brain fiber tracts.

128

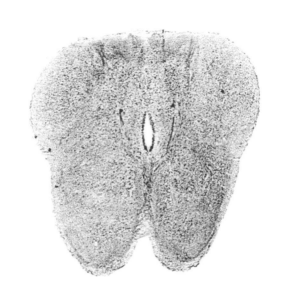

1 mm

Levels 14 to 21 are only shown at high magnification.

**PLATE 56B**

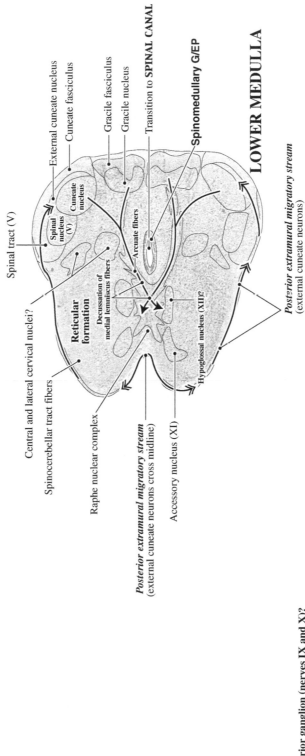

External cuneate nucleus

Cuneate fasciculus

Gracile fasciculus

Gracile nucleus

Transition to **SPINAL CANAL**

**Spinomedullary G/EP**

**LOWER MEDULLA**

*Posterior extramural migratory stream*
(external cuneate neurons)

Spinal tract (V)

Cuneate
nucleus

**Spinal
nucleus
(V)**

*Arcuate fibers*

Central and lateral cervical nuclei?

**Reticular
formation**

*Decussation of
medial lemniscus fibers*

Hypoglossal nucleus (XII)?

Spinocerebellar tract fibers

Raphe nuclear complex

*Posterior extramural migratory stream*
(external cuneate neurons cross midline)

Accessory nucleus (XI)

Superior ganglion (nerves IX and X)?

Inferior ganglion (nerves IX and X)?

Arrows indicate the
presumed *direction of
neuron migration* from
neuroepithelial sources.

Arrows indicate the
presumed *direction of
axon growth* in brain
fiber tracts.

# PART IV: GW9 SAGITTAL

This is specimen number 6658 in the Carnegie Collection, designated here as C6658, a female with a crown-rump length (CR) of 40 mm estimated to be at gestational week (GW) 9. The entire fetus was cut in the sagittal plane in 40 μm sections and stained with hematoxylin and eosin. Information on the date of specimen collection, fixative, and embedding medium (appears to be celloidin) was not available to us. The histology is excellent, and this is one of the best preserved specimens in any of the Collections at the National Museum of Health and Medicine. Since there is no photograph of this specimen's brain before histological processing, a specimen from Hochstetter (1919) that is comparable in age to C6658 is used to show external brain features at GW9 (**A**, **Figure 3**). C6658's brain structures are more difficult to understand because the sections are not cut parallel to the midline; **Figure 3** shows the approximate rotations in horizontal (**B**) and vertical (**C**) dimensions. Photographs of 11 sections (**Levels 1-11**) are illustrated at low magnification in four-parts (**Plates 57A-D** through **67A-D**). The **A/B** parts show the brain in place in the skull; the **C/D** parts show only the brain (and some peripheral ganglia) at slightly higher magnification. **Plates 68-83** show high-magnification views of various parts of the brain at different levels from the cerebral cortex (**Plate 68**) to the pons and sensory ganglia (**Plates 82-83**). All of the high-magnification plates are rotated 90° (landscape orientation) to more efficiently use page space.

C6658 is considerably less mature than the GW11 specimens. Throughout the cerebral cortex, the *neuroepithelium* is prominent and appears to be without a subventricular zone. The *stratified transitional field (STF)* contains *STF1* and *STF5* throughout; with *STF4* only in lateral areas. The most prominent developmental feature of the cerebral cortex is that both the *STF* layers and the cortical plate have a pronounced anterolateral (thicker) to dorsomedial (thinner) maturation gradient. The olfactory bulb is just beginning to evaginate in front of the basal telencephalic neuroepithelium. In anterolateral parts of the cerebral cortex, streams of neurons and glia appear to leave *STF4* and enter the *lateral migratory stream*. The hippocampus contains *ammonic and dentate migrations*, but there is no evidence of a pyramidal in Ammon's horn or a dentate gyrus. A massive *neuroepithelium/subventricular zone* overlies the amygdala, nucleus accumbens, and striatum (caudate and putamen) where neurons (and glia) are being generated.

The cerebellum is a thick, smooth plate overlying the posterior pons and medulla, and a definite *neuroepithelium* at the ventricular surface, indicating some Purkinje cells are still being generated. Many Purkinje cells are sojourning in a dense layer outside the neuroepithelium, and others are migrating upward. Many of the deep neurons are superficial in the cerebellum, but some are migrating downward to intermingle with upwardly migrating Purkinje cells. The cortical surface is partially covered by an *external germinal layer (egl)* that is actively producing neuronal stem cells, as it grows over the surface of the cerebellar cortex.

The third ventricle, aqueduct, and fourth ventricle are lined by thin *neuroepithelia*. The midbrain tegmentum, pons, and medulla have the thinnest neuroepithelia indicating that only the latest generated neurons are being produced at this time. The thick precerebellar neuroepithelium is an exception in the medulla. Thicker neuroepithelia are in the cerebellum (see above) and midbrain tectum, indicating many neurons are still being generated, although the majority of the neurons in these sites are already postmitotic. The neuroepithelium is still thicker in the hypothalamus and thalamus, in accordance with the later maturation of the diencephalon compared to the rest of the brainstem.

Neurons throughout the diencephalon, midbrain tegmentum, pons, and medulla are migrating and settling. Nuclear divisions are very indistinct throughout the diencephalon. More definition is seen in the midbrain tegmentum, pons, and medulla. The large *anterior extramural*, *posterior extramural, and intramural migratory streams* are prominent in the medulla and pons.

131

## GW9 SAGITTAL

A perfect sagittal cut through the brain bisects the cerebral cortex into two separate hemispheres by passing through the interhemispheric fissure, and does the same in the brainstem by passing through the midline of the ventricles.

Sections of C6658's brain are not parallel to the midline either horizontally (-11.71°, top view) or vertically (-6.64°, back view). In each of the illustrated sections on the following pages, the anterior edge of the cortex (top right) is tilted away from the observer, while the medulla and upper spinal cord (bottom) are tilted toward the observer.

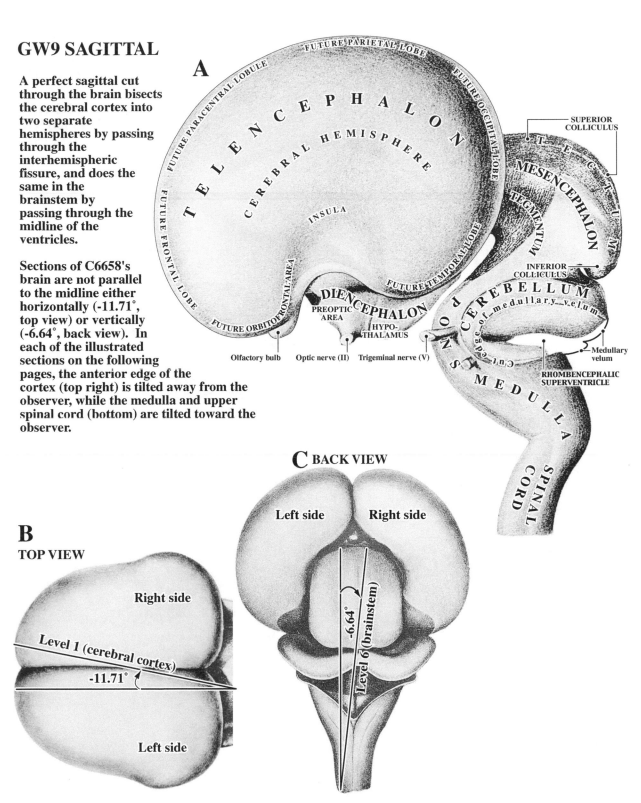

**Figure 3. A**, Lateral view of the brain and upper cervical spinal cord from a specimen with a crown-rump length of 38 mm (modified from Figure 43, Table VII, Hochstetter, 1919) identifies external features of a brain similar to C6658 (CR 40 mm). **B**, Top view of the brain in **A** (modified from Figure 45, Table VIII, Hochstetter, 1919) shows how C6658's sections rotate from a line parallel to the horizontal midline in the interhemispheric fissure. **C**, Back view of the brain in **A** (modified from Figure 44, Table VIII, Hochstetter, 1919) shows how C6658's sections rotate from a line parallel to the vertical midline in the brainstem and upper cervical spinal cord.

**PLATE 57A**

**GW9 Sagittal**
**CR 40 mm, C6658**
**Level 1: Slide 53, Section 1**
**HEAD STRUCTURES,**
**MAJOR BRAIN REGIONS,**
**AND VENTRICULAR**
**DIVISIONS**

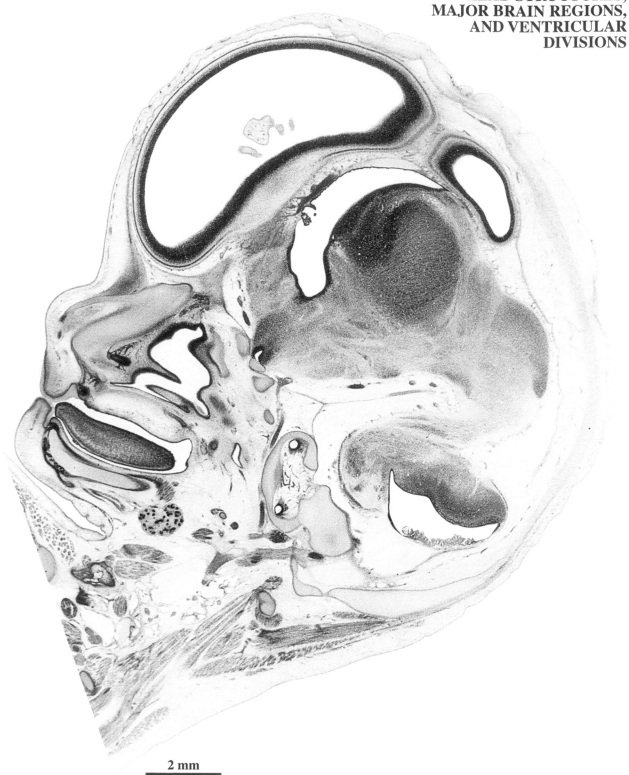

2 mm

**Neuroepithelial divisions, glioepithelial divisions, and differentiating structures are labeled in Parts C and D of this plate on the following pages.**

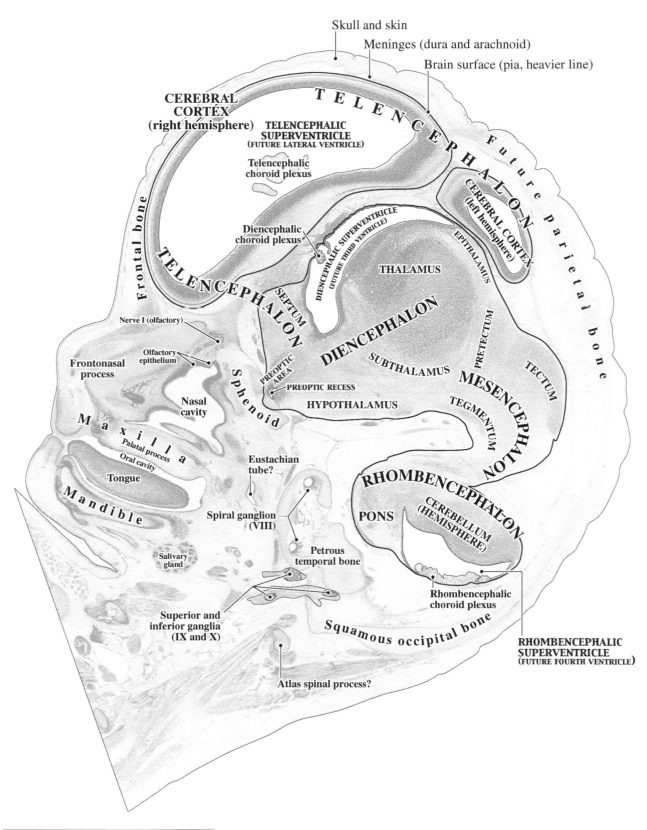

Skull and skin

Meninges (dura and arachnoid)

Brain surface (pia, heavier line)

**CEREBRAL CORTEX**
**(right hemisphere)**

**TELENCEPHALIC**
**SUPERVENTRICLE**
**(FUTURE LATERAL VENTRICLE)**

*T E L E N C E P H A L O N*

*F u t u r e  p a r i e t a l  b o n e*

Telencephalic
choroid plexus

*Frontal bone*

Diencephalic
choroid plexus

**DIENCEPHALIC SUPERVENTRICLE**
**(FUTURE THIRD VENTRICLE)**

**CEREBRAL CORTEX**
**(left hemisphere)**

**EPITHALAMUS**

**THALAMUS**

**TELENCEPHALON**

**SEPTUM**

**DIENCEPHALON**

Nerve I (olfactory)

Olfactory
epithelium

**Frontonasal**
**process**

**PREOPTIC**
**AREA**

**SUBTHALAMUS**

**PRETECTUM**

**TECTUM**

**MESENCEPHALON**

**PREOPTIC RECESS**

**HYPOTHALAMUS**

**TEGMENTUM**

*Sphenoid*

**Nasal**
**cavity**

*M a x i l l a*

*Palatal process*

*Oral cavity*

**Tongue**

**Eustachian**
**tube?**

**RHOMBENCEPHALON**

*M a n d i b l e*

Spiral ganglion
(VIII)

**PONS**

**CEREBELLUM**
**(HEMISPHERE)**

**Salivary**
**gland**

Petrous
temporal bone

Rhombencephalic
choroid plexus

**Superior and**
**inferior ganglia**
**(IX and X)**

*Squamous occipital bone*

**RHOMBENCEPHALIC**
**SUPERVENTRICLE**
**(FUTURE FOURTH VENTRICLE)**

Atlas spinal process?

FONT KEY:
**VENTRICULAR DIVISIONS - CAPITALS**
Major brain structure - Times **Bold CAPITALS**
All other structures - Times Roman or **Bold**

**PLATE 57C**

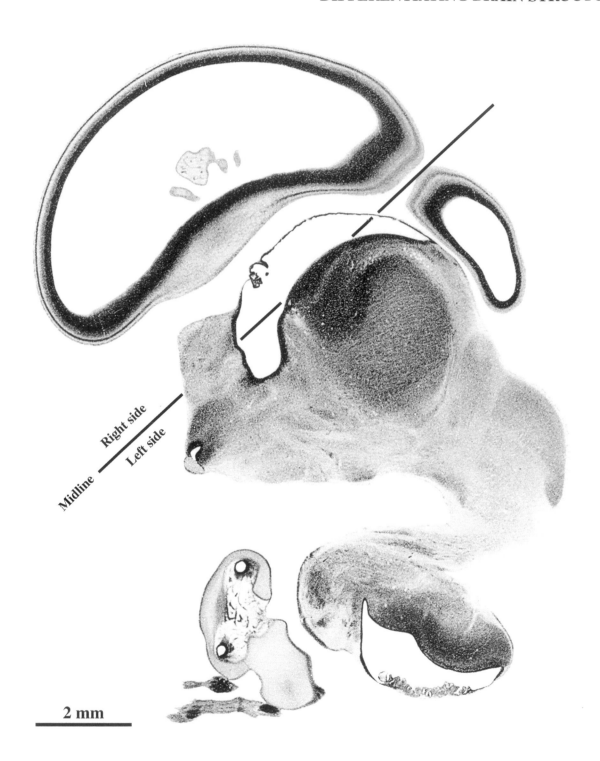

Right side

Left side

Midline

2 mm

The head, major brain structures, and ventricular divisions are
labeled in Parts A and B of this plate on the preceding pages.

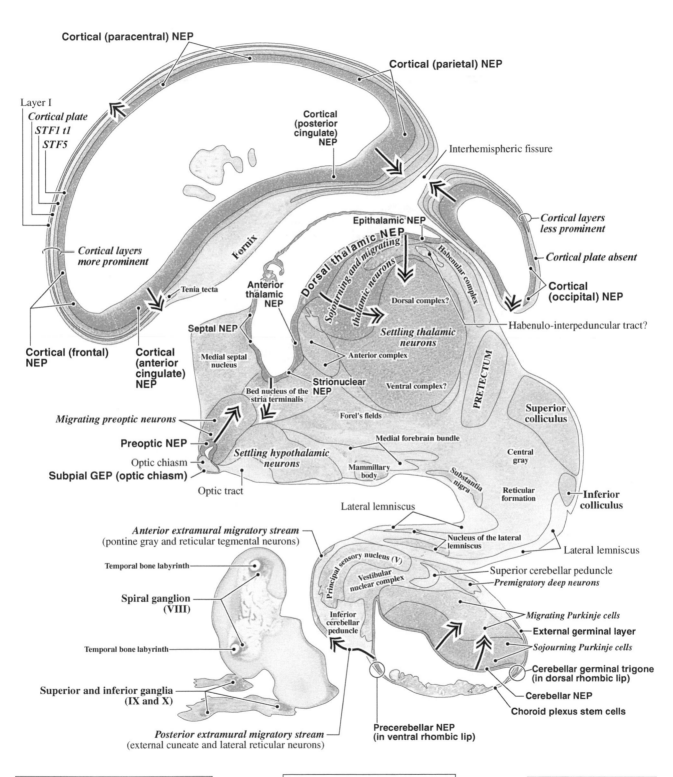

Cortical (paracentral) NEP

Cortical (parietal) NEP

Layer I
*Cortical plate*
*STF1 t1*
*STF5*

Cortical
(posterior
cingulate)
NEP

Interhemispheric fissure

Epithalamic NEP

*Cortical layers
less prominent*

Dorsal thalamic NEP

Habenular complex

*Cortical plate absent*

*Sojourning and migrating
thalamic neurons*

**Cortical
(occipital) NEP**

*Cortical layers
more prominent*

Fornix

Tenia tecta

Anterior
thalamic
NEP

Dorsal complex?

Habenulo-interpeduncular tract?

Septal NEP

*Settling thalamic
neurons*

**Cortical (frontal)
NEP**

**Cortical
(anterior
cingulate)
NEP**

Medial septal
nucleus

Anterior complex

**PRETECTUM**

**Superior
colliculus**

Strionuclear
NEP

Ventral complex?

Bed nucleus of the
stria terminalis

Forel's fields

Central
gray

*Migrating preoptic neurons*

Medial forebrain bundle

**Preoptic NEP**

*Settling hypothalamic
neurons*

Mammillary
body

Substantia
nigra

Reticular
formation

Optic chiasm

**Subpial GEP (optic chiasm)**

Optic tract

**Inferior
colliculus**

Lateral lemniscus

*Anterior extramural migratory stream*
(pontine gray and reticular tegmental neurons)

Nucleus of the lateral
lemniscus

Lateral lemniscus

Temporal bone labyrinth

Principal sensory nucleus (V)

Superior cerebellar peduncle

*Premigratory deep neurons*

**Spiral ganglion
(VIII)**

Vestibular
nuclear complex

*Migrating Purkinje cells*

Temporal bone labyrinth

Inferior
cerebellar
peduncle

**External germinal layer**

*Sojourning Purkinje cells*

**Cerebellar germinal trigone
(in dorsal rhombic lip)**

**Superior and inferior ganglia
(IX and X)**

**Cerebellar NEP**

**Choroid plexus stem cells**

*Posterior extramural migratory stream*
(external cuneate and lateral reticular neurons)

**Precerebellar NEP
(in ventral rhombic lip)**

**FONT KEY:**
**Germinal zone - Helvetica bold**
*Transient structure - Times bold italic*
Permanent structure - Times Roman or **Bold**

**ABBREVIATIONS:**
**GEP - Glioepithelium**
**NEP - Neuroepithelium**
*STF - Stratified transitional field*

**Arrows** indicate the
presumed *direction of
neuron migration* from
neuroepithelial sources.

**PLATE 58A**

**GW9 Sagittal**
**CR 40 mm, C6658**
**Level 2: Slide 63, Section 1**
**HEAD STRUCTURES,**
**MAJOR BRAIN REGIONS,**
**AND VENTRICULAR**
**DIVISIONS**

2 mm

**Neuroepithelial divisions, glioepithelial divisions, and differentiating structures are labeled in Parts C and D of this plate on the following pages.**

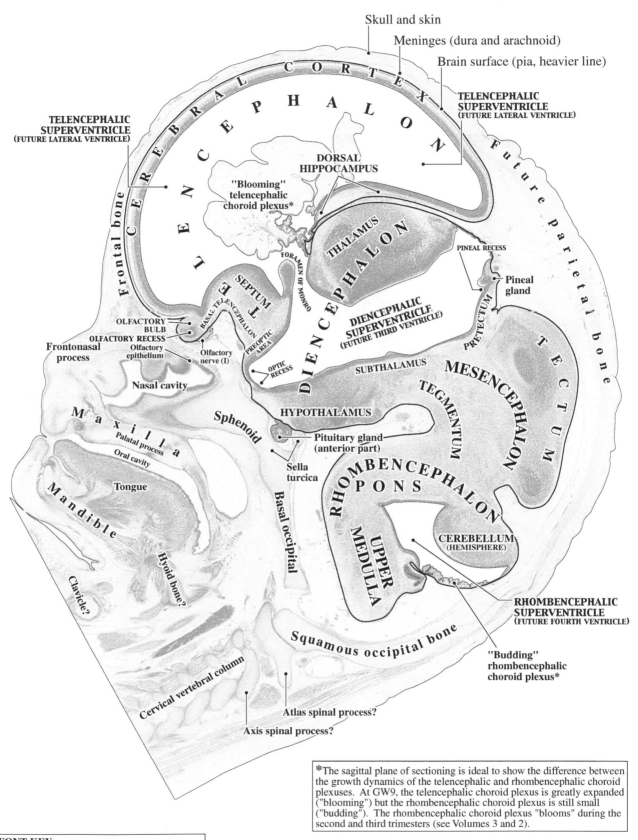

Skull and skin

Meninges (dura and arachnoid)

Brain surface (pia, heavier line)

**TELENCEPHALIC SUPERVENTRICLE**
(FUTURE LATERAL VENTRICLE)

**TELENCEPHALIC SUPERVENTRICLE**
(FUTURE LATERAL VENTRICLE)

CEREBRAL CORTEX

TELENCEPHALON

Future parietal bone

Frontal bone

**DORSAL HIPPOCAMPUS**

"Blooming" telencephalic choroid plexus*

THALAMUS

PINEAL RECESS

Pineal gland

SEPTUM

FORAMEN OF MONRO

BASAL TELENCEPHALON

PREOPTIC AREA

DIENCEPHALON

**DIENCEPHALIC SUPERVENTRICLE**
(FUTURE THIRD VENTRICLE)

PRETECTUM

OLFACTORY BULB

**OLFACTORY RECESS**

Olfactory epithelium

Olfactory nerve (I)

OPTIC RECESS

Frontonasal process

Nasal cavity

SUBTHALAMUS

**MESENCEPHALON**

TEGMENTUM

TECTUM

Maxilla

Palatal process

Oral cavity

Sphenoid

**HYPOTHALAMUS**

Pituitary gland (anterior part)

Sella turcica

Tongue

Mandible

Basal occipital

**RHOMBENCEPHALON PONS**

UPPER MEDULLA

**CEREBELLUM**
(HEMISPHERE)

Hyoid bone?

Clavicle?

Squamous occipital bone

**RHOMBENCEPHALIC SUPERVENTRICLE**
(FUTURE FOURTH VENTRICLE)

"Budding" rhombencephalic choroid plexus*

Cervical vertebral column

Atlas spinal process?

Axis spinal process?

*The sagittal plane of sectioning is ideal to show the difference between the growth dynamics of the telencephalic and rhombencephalic choroid plexuses. At GW9, the telencephalic choroid plexus is greatly expanded ("blooming") but the rhombencephalic choroid plexus is still small ("budding"). The rhombencephalic choroid plexus "blooms" during the second and third trimesters (see Volumes 3 and 2).

**FONT KEY:**
**VENTRICULAR DIVISIONS – CAPITALS**
Major brain structure - Times **Bold CAPITALS**
All other structures - Times Roman or **Bold**

**PLATE 58C**

NEUROEPITHELIAL/GLIOEPITHELIAL DIVISIONS AND
DIFFERENTIATING BRAIN STRUCTURES

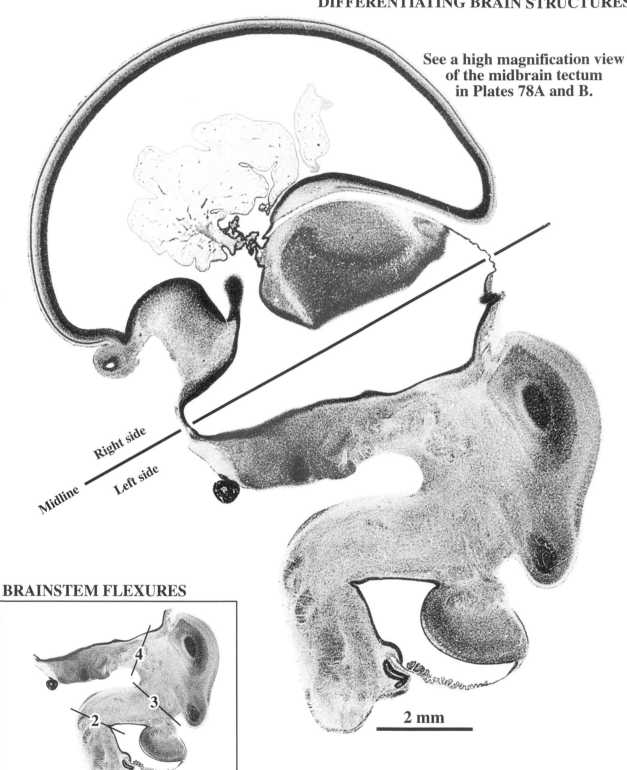

See a high magnification view
of the midbrain tectum
in Plates 78A and B.

Right side

Midline — Left side

**BRAINSTEM FLEXURES**

4

3

2

2. Pontine
3. Mesencephalic
4. Diencephalic

2 mm

The head, major brain structures, and ventricular
divisions are labeled in Parts A and B of this plate on
the preceding pages.

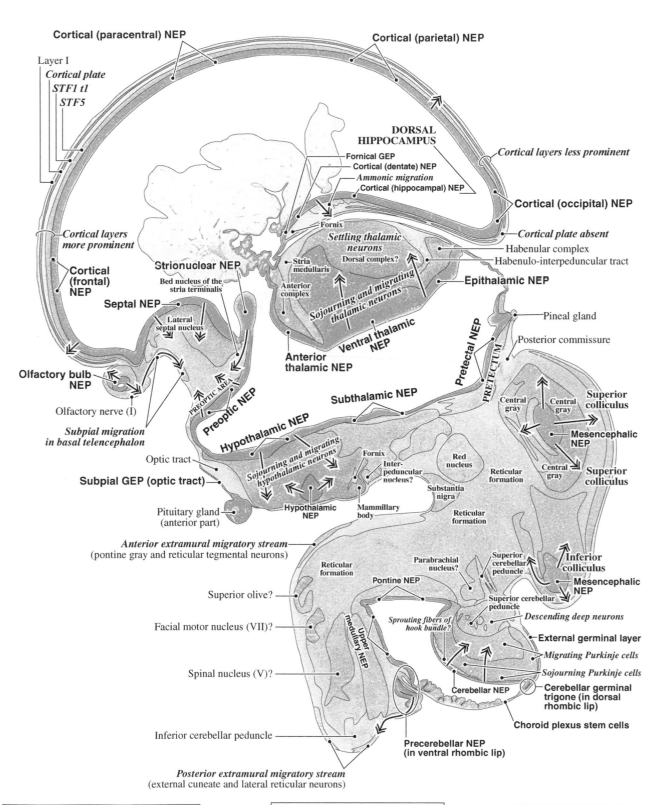

**PLATE 59A**

GW9 Sagittal
CR 40 mm, C6658
Level 3: Slide 67, Section 1
HEAD STRUCTURES,
MAJOR BRAIN REGIONS,
AND VENTRICULAR
DIVISIONS

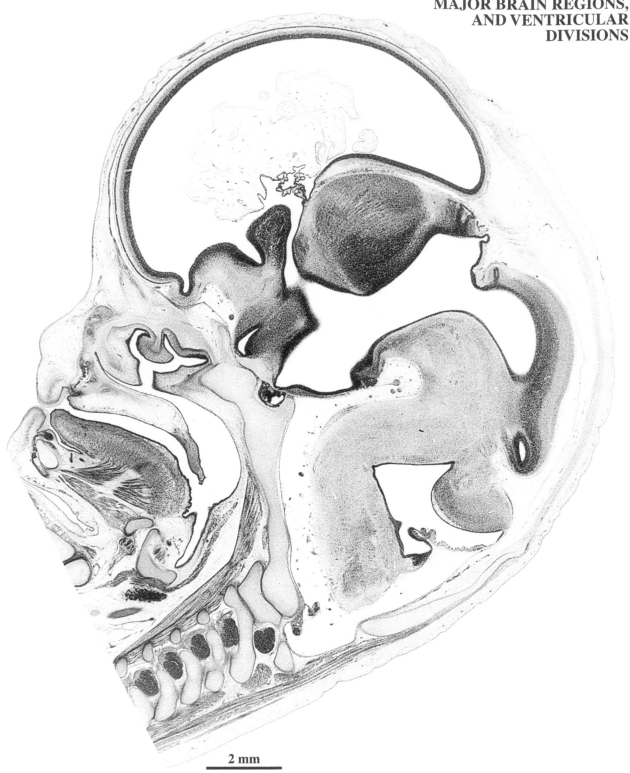

2 mm

Neuroepithelial divisions, glioepithelial divisions, and differentiating
structures are labeled in Parts C and D of this plate on the following pages.

**PLATE 59B**

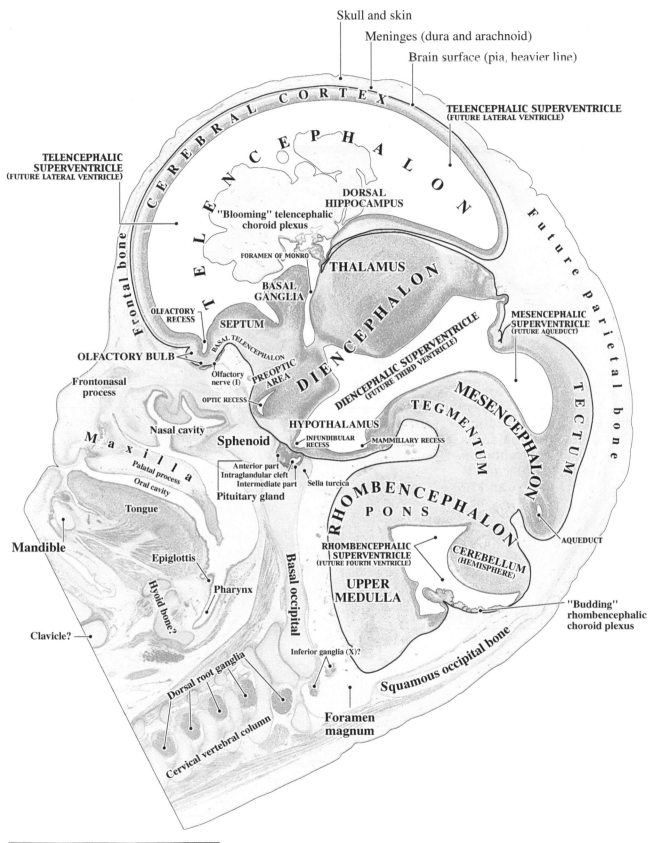

Skull and skin

Meninges (dura and arachnoid)

Brain surface (pia, heavier line)

C E R E B R A L   C O R T E X

**TELENCEPHALIC SUPERVENTRICLE**
(FUTURE LATERAL VENTRICLE)

**TELENCEPHALIC SUPERVENTRICLE**
(FUTURE LATERAL VENTRICLE)

T E L E N C E P H A L O N

Future parietal bone

**DORSAL HIPPOCAMPUS**

"Blooming" telencephalic choroid plexus

FORAMEN OF MONRO

**THALAMUS**

**BASAL GANGLIA**

Frontal bone

OLFACTORY RECESS

**SEPTUM**

BASAL TELENCEPHALON

D I E N C E P H A L O N

**MESENCEPHALIC SUPERVENTRICLE**
(FUTURE AQUEDUCT)

**OLFACTORY BULB**

Olfactory nerve (I)

PREOPTIC AREA

**DIENCEPHALIC SUPERVENTRICLE**
(FUTURE THIRD VENTRICLE)

**M E S E N C E P H A L O N**

**T E C T U M**

Frontonasal process

OPTIC RECESS

**HYPOTHALAMUS**

T E G M E N T U M

Nasal cavity

M a x i l l a

**Sphenoid**

INFUNDIBULAR RECESS

MAMMILLARY RECESS

Palatal process

Oral cavity

Anterior part
Intraglandular cleft
Intermediate part
**Pituitary gland**

Sella turcica

R H O M B E N C E P H A L O N

P O N S

AQUEDUCT

**Tongue**

**CEREBELLUM**
(HEMISPHERE)

**Mandible**

**Epiglottis**

**Pharynx**

Hyoid bone?

Basal occipital

**RHOMBENCEPHALIC SUPERVENTRICLE**
(FUTURE FOURTH VENTRICLE)

**UPPER MEDULLA**

"Budding" rhombencephalic choroid plexus

Clavicle?

Inferior ganglia (X)?

Squamous occipital bone

Dorsal root ganglia

Cervical vertebral column

**Foramen magnum**

**FONT KEY:**
**VENTRICULAR DIVISIONS – CAPITALS**
Major brain structure - Times **Bold CAPITALS**
All other structures - Times Roman or **Bold**

**PLATE 59C**

**GW9 Sagittal**
**CR 40 mm, C6658, Level 3: Slide 67, Section 1**
**NEUROEPITHELIAL/**
**GLIOEPITHELIAL DIVISIONS**
**AND DIFFERENTIATING**
**BRAIN STRUCTURES**

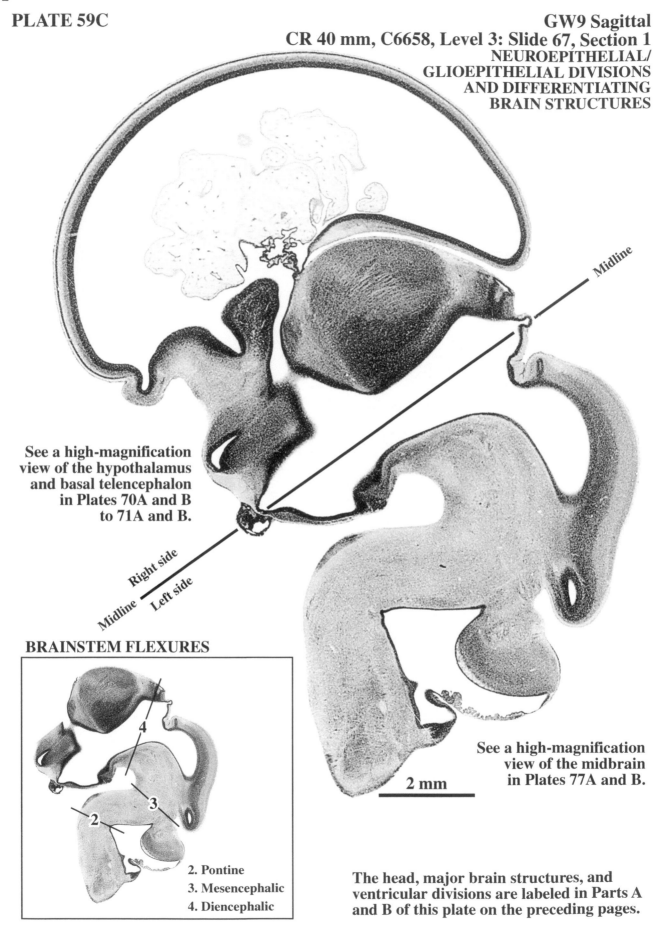

Midline

See a high-magnification
view of the hypothalamus
and basal telencephalon
in Plates 70A and B
to 71A and B.

Right side

Midline  Left side

See a high-magnification
view of the midbrain
in Plates 77A and B.

2 mm

**BRAINSTEM FLEXURES**

4

2

3

2. Pontine
3. Mesencephalic
4. Diencephalic

The head, major brain structures, and
ventricular divisions are labeled in Parts A
and B of this plate on the preceding pages.

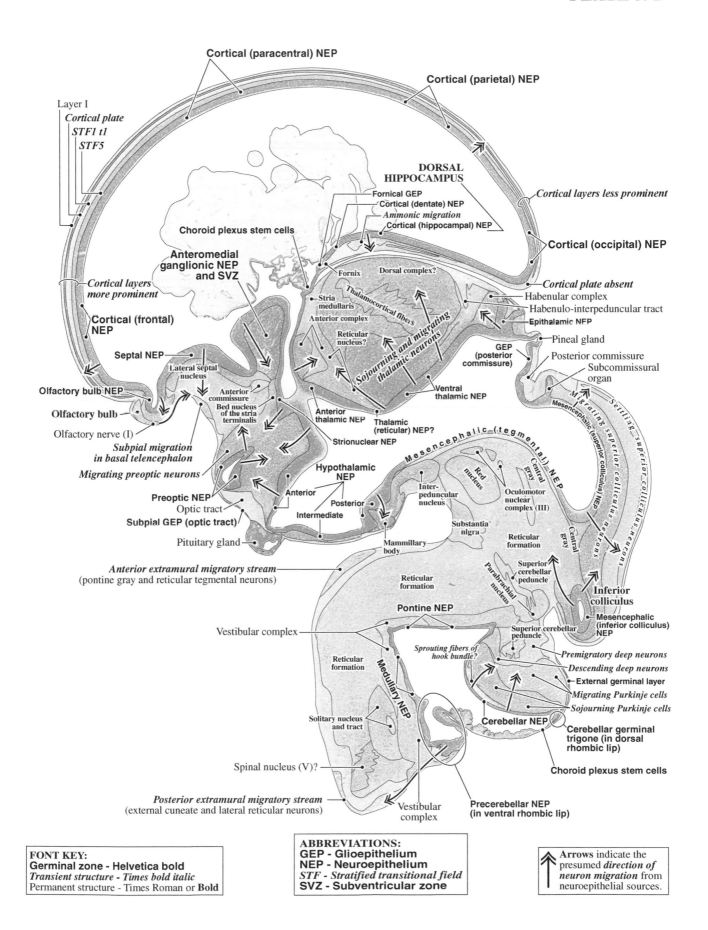

Layer I
*Cortical plate*
*STF1 t1*
*STF5*

Cortical (paracentral) NEP

Cortical (parietal) NEP

**DORSAL HIPPOCAMPUS**

Fornical GEP
Cortical (dentate) NEP
*Ammonic migration*
Cortical (hippocampal) NEP

Choroid plexus stem cells

*Cortical layers less prominent*

**Cortical (occipital) NEP**

**Anteromedial ganglionic NEP and SVZ**

Fornix
Dorsal complex?
*Cortical plate absent*
Habenular complex
Habenulo-interpeduncular tract
Epithalamic NEP

*Cortical layers more prominent*

Stria medullaris
*Thalamocortical fibers*
Anterior complex

Pineal gland
Posterior commissure
Subcommissural organ

**Cortical (frontal) NEP**

Reticular nucleus?
*Sojourning and migrating thalamic neurons*

GEP (posterior commissure)

*Migrating superior colliculus neurons*
*Settling superior colliculus neurons*

**Septal NEP**

**Lateral septal nucleus**
Anterior commissure
Anterior thalamic NEP
Ventral thalamic NEP

*Mesencephalic (superior colliculus) NEP*

**Olfactory bulb NEP**

Bed nucleus of the stria terminalis
Thalamic (reticular) NEP?

**Olfactory bulb**

Strionuclear NEP

*Mesencephalic (tegmental) NEP*

Olfactory nerve (I)

**Hypothalamic NEP**

Red nucleus
Central gray
Oculomotor nuclear complex (III)

*Subpial migration in basal telencephalon*

Inter-peduncular nucleus

*Migrating preoptic neurons*

Anterior

Central gray

**Preoptic NEP**
Optic tract
Posterior
Substantia nigra
Reticular formation

**Subpial GEP (optic tract)**

Intermediate

Pituitary gland

Mammillary body

Parabrachial nucleus
Superior cerebellar peduncle

**Inferior colliculus**

*Anterior extramural migratory stream*
(pontine gray and reticular tegmental neurons)

Reticular formation

Reticular formation

Mesencephalic (inferior colliculus) NEP

Vestibular complex

**Pontine NEP**

Superior cerebellar peduncle

*Premigratory deep neurons*
*Descending deep neurons*

Reticular formation

*Sprouting fibers of hook bundle?*

**External germinal layer**
*Migrating Purkinje cells*
*Sojourning Purkinje cells*

**Medullary NEP**

**Cerebellar NEP**

Solitary nucleus and tract

**Cerebellar germinal trigone (in dorsal rhombic lip)**

Spinal nucleus (V)?

Vestibular complex

**Choroid plexus stem cells**

*Posterior extramural migratory stream*
(external cuneate and lateral reticular neurons)

**Precerebellar NEP (in ventral rhombic lip)**

**FONT KEY:**
Germinal zone - Helvetica bold
*Transient structure - Times bold italic*
Permanent structure - Times Roman or **Bold**

**ABBREVIATIONS:**
**GEP** - Glioepithelium
**NEP** - Neuroepithelium
*STF* - Stratified transitional field
**SVZ** - Subventricular zone

**Arrows** indicate the presumed *direction of neuron migration* from neuroepithelial sources.

**PLATE 60A**

**GW9 Sagittal, CR 40 mm, C6658**
**Level 4: Slide 71, Section 2**
**HEAD STRUCTURES,**
**MAJOR BRAIN REGIONS,**
**AND VENTRICULAR**
**DIVISIONS**

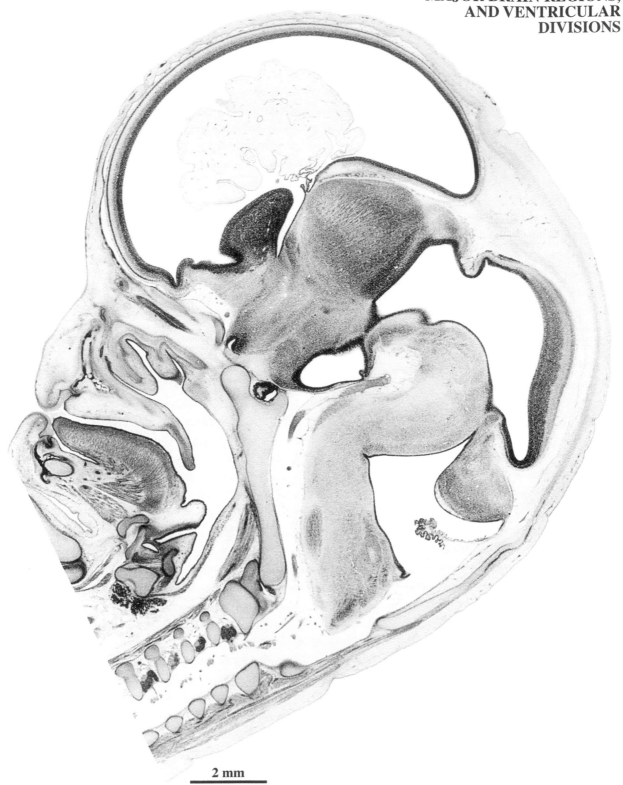

2 mm

**Neuroepithelial divisions, glioepithelial divisions, and differentiating structures are labeled in Parts C and D of this plate on the following pages.**

**PLATE 60B**

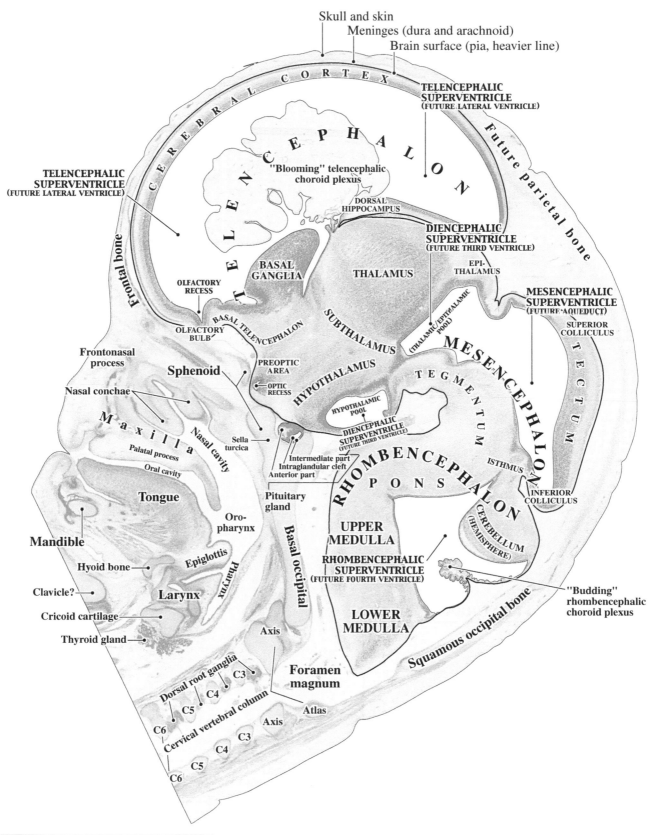

Skull and skin
Meninges (dura and arachnoid)
Brain surface (pia, heavier line)

C E R E B R A L   C O R T E X

**TELENCEPHALIC
SUPERVENTRICLE**
(FUTURE LATERAL VENTRICLE)

*Future parietal bone*

T E L E N C E P H A L O N

"Blooming" telencephalic
choroid plexus

**TELENCEPHALIC
SUPERVENTRICLE**
(FUTURE LATERAL VENTRICLE)

DORSAL
HIPPOCAMPUS

*Frontal bone*

**BASAL
GANGLIA**

**THALAMUS**

**DIENCEPHALIC
SUPERVENTRICLE**
(FUTURE THIRD VENTRICLE)

EPI-
THALAMUS

OLFACTORY
RECESS

**MESENCEPHALIC
SUPERVENTRICLE**
(FUTURE AQUEDUCT)

OLFACTORY
BULB

BASAL TELENCEPHALON

**SUBTHALAMUS**

(THALAMIC/EPITHALAMIC
POOL)

**MESENCEPHALON**

SUPERIOR
COLLICULUS

T E C T U M

**Frontonasal
process**

**Sphenoid**

PREOPTIC
AREA

**HYPOTHALAMUS**

T E G M E N T U M

Nasal conchae

OPTIC
RECESS

HYPOTHALAMIC
POOL

M a x i l l a

**DIENCEPHALIC
SUPERVENTRICLE**
(FUTURE THIRD VENTRICLE)

*Nasal cavity*

Palatal process

Sella
turcica

**RHOMBENCEPHALON**

ISTHMUS

INFERIOR
COLLICULUS

Oral cavity

Intermediate part
Intraglandular cleft
Anterior part

**PONS**

**CEREBELLUM**
(HEMISPHERE)

**Tongue**

Oro-
pharynx

Pituitary
gland

**Mandible**

*Pharynx*

Epiglottis

*Basal occipital*

**UPPER
MEDULLA**

Hyoid bone

Clavicle?

**Larynx**

**RHOMBENCEPHALIC
SUPERVENTRICLE**
(FUTURE FOURTH VENTRICLE)

Cricoid cartilage

"Budding"
rhombencephalic
choroid plexus

Thyroid gland

Axis

**LOWER
MEDULLA**

**Foramen
magnum**

Dorsal root ganglia

C3

C4

C5

*Squamous occipital bone*

C6

Atlas

Cervical vertebral column

Axis

C3

C4

C5

C6

FONT KEY:
**VENTRICULAR DIVISIONS – CAPITALS**
Major brain structure - Times **Bold CAPITALS**
All other structures - Times Roman or **Bold**

**PLATE 60C**

GW9 Sagittal, CR 40 mm, C6658
Level 4: Slide 71, Section 2
NEUROEPITHELIAL/
GLIOEPITHELIAL
DIVISIONS AND
DIFFERENTIATING
BRAIN
STRUCTURES

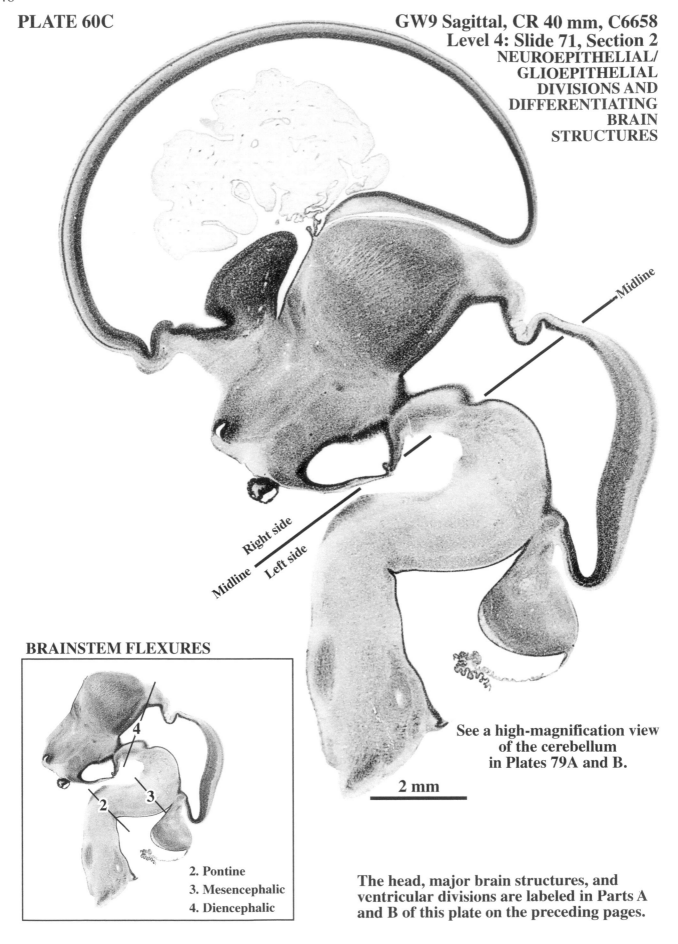

Midline

Right side

Midline    Left side

**BRAINSTEM FLEXURES**

2. Pontine
3. Mesencephalic
4. Diencephalic

See a high-magnification view
of the cerebellum
in Plates 79A and B.

2 mm

The head, major brain structures, and
ventricular divisions are labeled in Parts A
and B of this plate on the preceding pages.

**PLATE 60D**

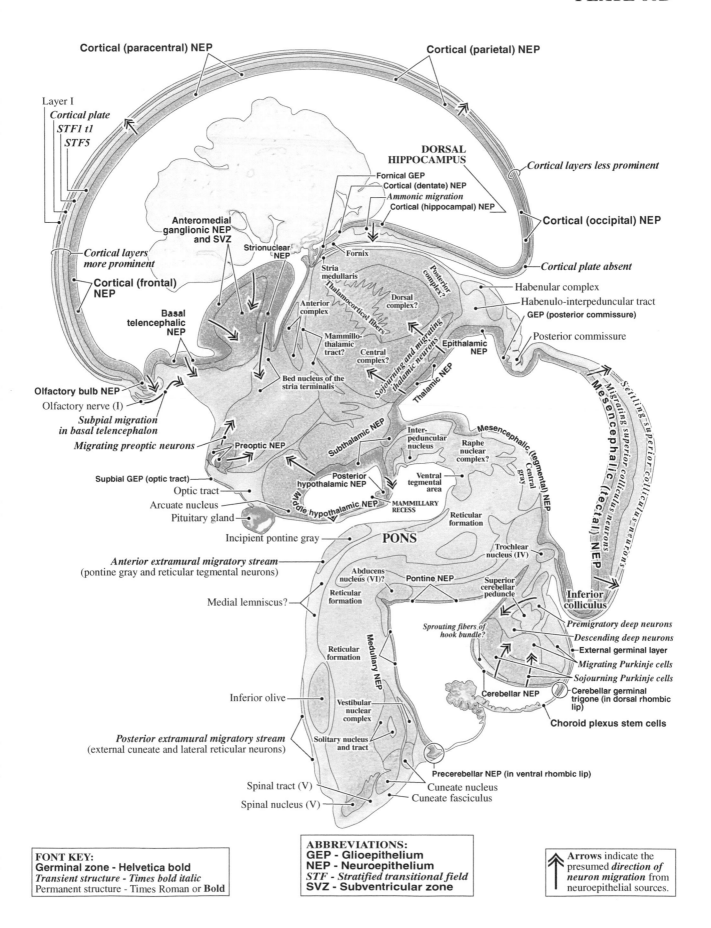

Cortical (paracentral) NEP

Cortical (parietal) NEP

Layer I
*Cortical plate*
*STF1 t1*
*STF5*

**DORSAL HIPPOCAMPUS**

Fornical GEP
Cortical (dentate) NEP
*Ammonic migration*
Cortical (hippocampal) NEP

*Cortical layers less prominent*

**Cortical (occipital) NEP**

**Anteromedial ganglionic NEP and SVZ**

Strionuclear NEP

Fornix

*Cortical layers more prominent*

**Cortical (frontal) NEP**

Stria medullaris

*Thalamocortical fibers*

*Posterior complex?*

Dorsal complex?

*Cortical plate absent*

Habenular complex
Habenulo-interpeduncular tract
**GEP (posterior commissure)**
Posterior commissure

Anterior complex

**Basal telencephalic NEP**

Mammillo-thalamic tract?

Central complex?

Epithalamic NEP

*Sojourning and migrating thalamic neurons*

Thalamic NEP

**Olfactory bulb NEP**
Olfactory nerve (I)

Bed nucleus of the stria terminalis

*Settling superior colliculus neurons*

*Migrating superior colliculus neurons*

**Mesencephalic (tectal) NEP**

*Subpial migration in basal telencephalon*

*Migrating preoptic neurons*

**Preoptic NEP**

**Subthalamic NEP**

Inter-peduncular nucleus

Raphe nuclear complex?

*Mesencephalic (tegmental) NEP*

Central gray

Subpial GEP (optic tract)
Optic tract
Arcuate nucleus
Pituitary gland

**Posterior hypothalamic NEP**

Ventral tegmental area

**Middle hypothalamic NEP**

MAMMILLARY RECESS

Reticular formation

Incipient pontine gray

**PONS**

Trochlear nucleus (IV)

**Inferior colliculus**

*Anterior extramural migratory stream*
(pontine gray and reticular tegmental neurons)

Abducens nucleus (VI)?

**Pontine NEP**

Superior cerebellar peduncle

Medial lemniscus?

Reticular formation

*Premigratory deep neurons*
*Descending deep neurons*
External germinal layer
*Migrating Purkinje cells*
*Sojourning Purkinje cells*
Cerebellar germinal trigone (in dorsal rhombic lip)

Reticular formation

**Medullary NEP**

*Sprouting fibers of hook bundle?*

**Cerebellar NEP**

Inferior olive

Vestibular nuclear complex

**Choroid plexus stem cells**

*Posterior extramural migratory stream*
(external cuneate and lateral reticular neurons)

Solitary nucleus and tract

**Precerebellar NEP (in ventral rhombic lip)**

Spinal tract (V)
Spinal nucleus (V)

Cuneate nucleus
Cuneate fasciculus

**FONT KEY:**
**Germinal zone - Helvetica bold**
*Transient structure - Times bold italic*
Permanent structure - Times Roman or **Bold**

**ABBREVIATIONS:**
**GEP** - Glioepithelium
**NEP** - Neuroepithelium
*STF* - Stratified transitional field
**SVZ** - Subventricular zone

**Arrows** indicate the presumed *direction of neuron migration* from neuroepithelial sources.

**GW9 Sagittal, CR 40 mm, C6658**
**Level 5: Slide 75, Section 2**
**HEAD STRUCTURES,**
**MAJOR BRAIN REGIONS,**
**AND VENTRICULAR**
**DIVISIONS**

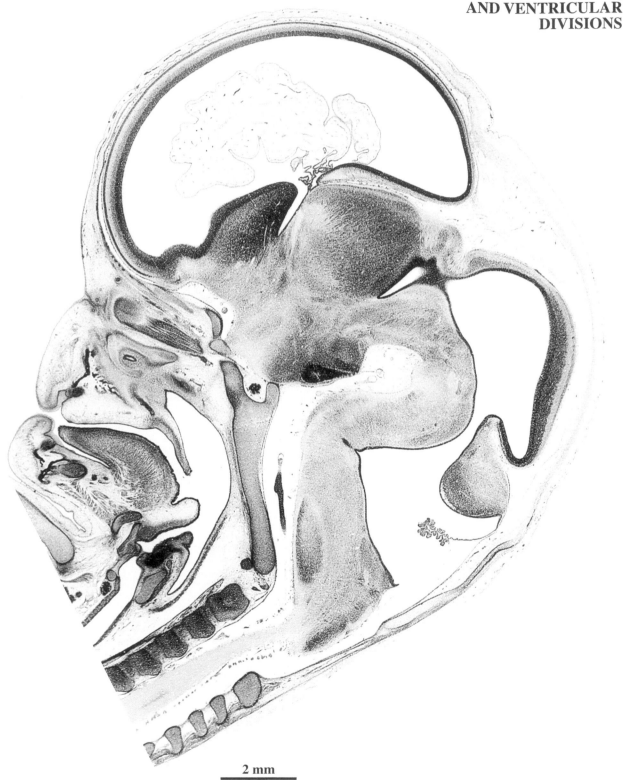

2 mm

**Neuroepithelial divisions, glioepithelial divisions, and differentiating structures are labeled in Parts C and D of this plate on the following pages.**

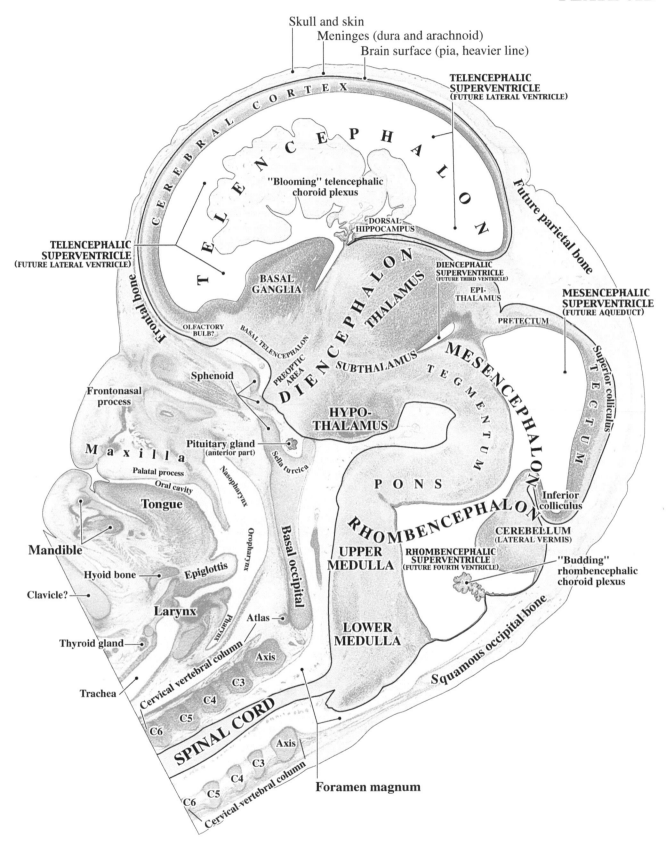

Skull and skin
Meninges (dura and arachnoid)
Brain surface (pia, heavier line)

**TELENCEPHALIC
SUPERVENTRICLE**
(FUTURE LATERAL VENTRICLE)

CEREBRAL CORTEX

TELENCEPHALON

"Blooming" telencephalic
choroid plexus

DORSAL
HIPPOCAMPUS

Future parietal bone

**TELENCEPHALIC
SUPERVENTRICLE**
(FUTURE LATERAL VENTRICLE)

DIENCEPHALON

DIENCEPHALIC
SUPERVENTRICLE
(FUTURE THIRD VENTRICLE)

EPI-
THALAMUS

**MESENCEPHALIC
SUPERVENTRICLE**
(FUTURE AQUEDUCT)

**BASAL
GANGLIA**

THALAMUS

PRETECTUM

Frontal bone

OLFACTORY
BULB?

BASAL TELENCEPHALON

SUBTHALAMUS

MESENCEPHALON

Superior colliculus

TECTUM

**Frontonasal
process**

PREOPTIC
AREA

**HYPO-
THALAMUS**

TEGMENTUM

**Sphenoid**

**Pituitary gland**
(anterior part)

Sella turcica

**M a x i l l a**

**PONS**

Inferior
colliculus

Nasopharynx

**Palatal process**

**Oral cavity**

**Tongue**

RHOMBENCEPHALON

**CEREBELLUM**
(LATERAL VERMIS)

**Mandible**

Oropharynx

Basal occipital

**UPPER
MEDULLA**

RHOMBENCEPHALIC
SUPERVENTRICLE
(FUTURE FOURTH VENTRICLE)

"Budding"
rhombencephalic
choroid plexus

**Hyoid bone**

**Epiglottis**

**Clavicle?**

**Larynx**

Atlas

**LOWER
MEDULLA**

**Thyroid gland**

Axis

Pharynx

Cervical vertebral column

**Trachea**

C3

C4

Squamous occipital bone

C5

C6

**SPINAL CORD**

Axis

C3

C4

C5

C6

Cervical vertebral column

**Foramen magnum**

---

**FONT KEY:**
**VENTRICULAR DIVISIONS – CAPITALS**
Major brain structure - Times **Bold CAPITALS**
All other structures - Times Roman or **Bold**

**PLATE 61C**

NEUROEPITHELIAL/
GLIOEPITHELIAL
DIVISIONS AND
DIFFERENTIATING
BRAIN
STRUCTURES

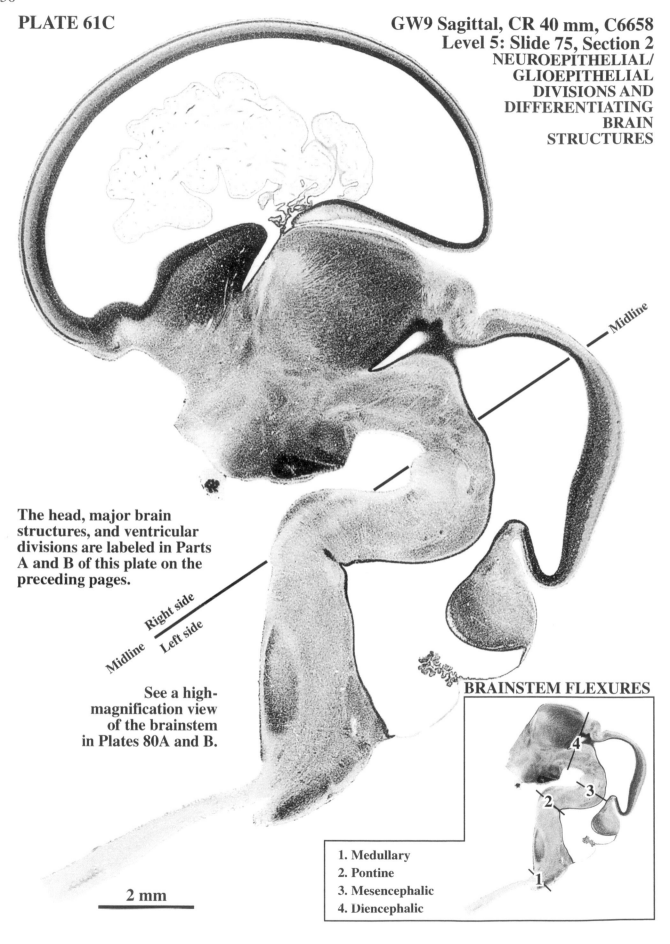

Midline

The head, major brain
structures, and ventricular
divisions are labeled in Parts
A and B of this plate on the
preceding pages.

Right side

Midline    Left side

See a high-
magnification view
of the brainstem
in Plates 80A and B.

**BRAINSTEM FLEXURES**

1. Medullary
2. Pontine
3. Mesencephalic
4. Diencephalic

2 mm

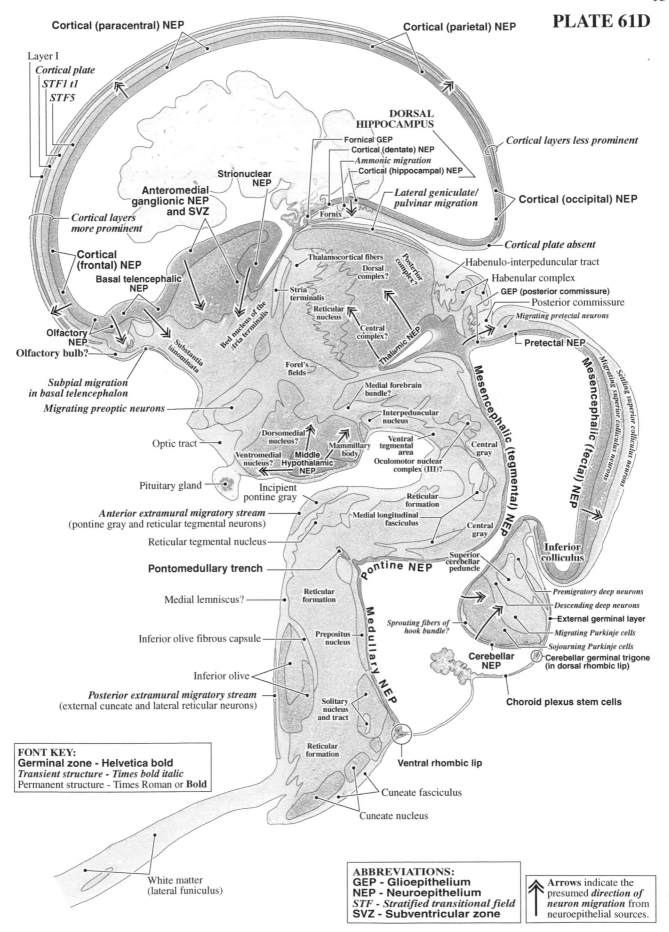

Cortical (paracentral) NEP

Cortical (parietal) NEP

Layer I
*Cortical plate*
*STF1 t1*
*STF5*

**DORSAL
HIPPOCAMPUS**

Fornical GEP
Cortical (dentata) NEP
*Ammonic migration*
Cortical (hippocampal) NEP

*Cortical layers less prominent*

Strionuclear
NEP

**Anteromedial
ganglionic NEP
and SVZ**

*Lateral geniculate/
pulvinar migration*

Fornix

**Cortical (occipital) NEP**

*Cortical layers
more prominent*

**Cortical
(frontal) NEP**

**Basal telencephalic
NEP**

Thalamocortical fibers

Dorsal
complex?

*Posterior
complex?*

*Cortical plate absent*

Habenulo-interpeduncular tract

Habenular complex

**GEP (posterior commissure)**

Posterior commissure

Stria
terminalis

Reticular
nucleus

*Central
complex?*

**Thalamic NEP**

*Migrating pretectal neurons*

**Pretectal NEP**

**Olfactory
NEP**

**Olfactory bulb?**

*Bed nucleus of the
stria terminalis*

Forel's
fields

*Substantia
innominata*

*Subpial migration
in basal telencephalon*

*Migrating preoptic neurons*

Medial forebrain
bundle?

Interpeduncular
nucleus

Optic tract

Dorsomedial
nucleus?

Ventromedial
nucleus?

**Middle
Hypothalamic
NEP**

*Mammillary
body*

Ventral
tegmental
area

Oculomotor nuclear
complex (III)?

Central
gray

**Mesencephalic (tegmental) NEP**

**Mesencephalic (tectal) NEP**

*Settling superior colliculus neurons*

*Migrating superior colliculus neurons*

Pituitary gland

Incipient
pontine gray

Reticular
formation

Medial longitudinal
fasciculus

Central
gray

**Inferior
colliculus**

*Anterior extramural migratory stream*
(pontine gray and reticular tegmental neurons)

Reticular tegmental nucleus

**Pontomedullary trench**

**Pontine NEP**

Superior
cerebellar
peduncle

*Premigratory deep neurons*

*Descending deep neurons*

Medial lemniscus?

Reticular
formation

**External germinal layer**

*Migrating Purkinje cells*

Inferior olive fibrous capsule

Prepositus
nucleus

**Medullary NEP**

*Sprouting fibers of
hook bundle?*

*Sojourning Purkinje cells*

**Cerebellar
NEP**

Cerebellar germinal trigone
(in dorsal rhombic lip)

Inferior olive

*Posterior extramural migratory stream*
(external cuneate and lateral reticular neurons)

Solitary
nucleus
and tract

**Choroid plexus stem cells**

Reticular
formation

**Ventral rhombic lip**

Cuneate fasciculus

Cuneate nucleus

**FONT KEY:**
**Germinal zone - Helvetica bold**
*Transient structure - Times bold italic*
Permanent structure - Times Roman or **Bold**

White matter
(lateral funiculus)

**ABBREVIATIONS:**
**GEP** - Glioepithelium
**NEP** - Neuroepithelium
*STF* - *Stratified transitional field*
**SVZ** - Subventricular zone

**Arrows** indicate the
presumed *direction of
neuron migration* from
neuroepithelial sources.

**PLATE 62A**

GW9 Sagittal, CR 40 mm, C6658
Level 6: Slide 79, Section 2
HEAD STRUCTURES,
MAJOR BRAIN REGIONS,
AND VENTRICULAR
DIVISIONS

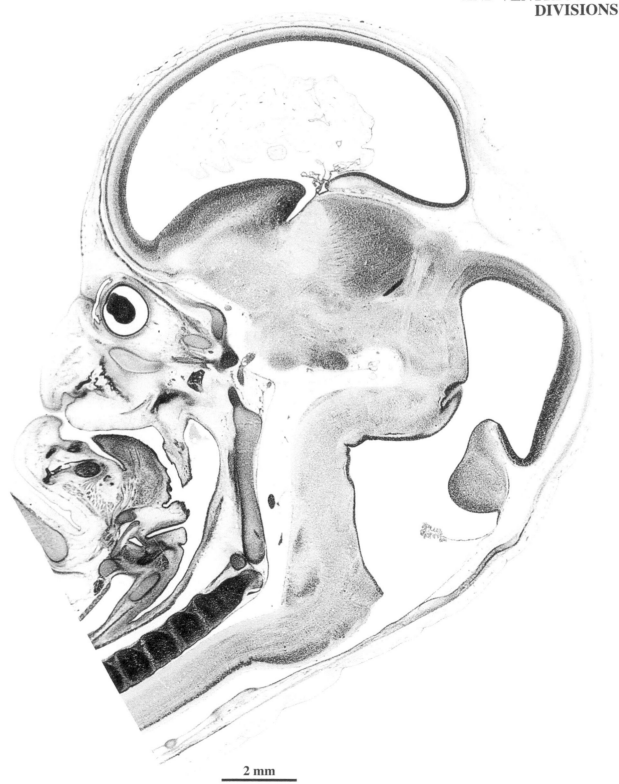

2 mm

Neuroepithelial divisions, glioepithelial divisions, and differentiating
structures are labeled in Parts C and D of this plate on the following pages.

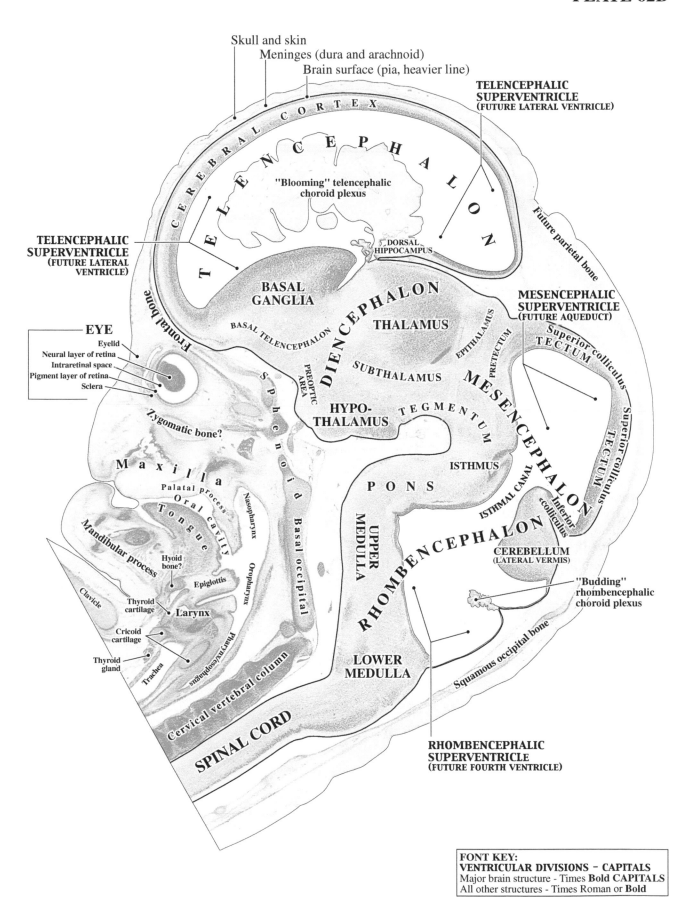

Skull and skin
Meninges (dura and arachnoid)
Brain surface (pia, heavier line)

**TELENCEPHALIC SUPERVENTRICLE**
**(FUTURE LATERAL VENTRICLE)**

CEREBRAL CORTEX

TELENCEPHALON

"Blooming" telencephalic choroid plexus

DORSAL HIPPOCAMPUS

Future parietal bone

**TELENCEPHALIC SUPERVENTRICLE**
**(FUTURE LATERAL VENTRICLE)**

Frontal bone

**BASAL GANGLIA**

BASAL TELENCEPHALON

DIENCEPHALON

**THALAMUS**

EPITHALAMUS

PRETECTUM

**MESENCEPHALIC SUPERVENTRICLE**
**(FUTURE AQUEDUCT)**

Superior colliculus TECTUM

**EYE**
Eyelid
Neural layer of retina
Intraretinal space
Pigment layer of retina
Sclera

PREOPTIC AREA

SUBTHALAMUS

**HYPO-THALAMUS**

TEGMENTUM

MESENCEPHALON

Superior colliculus TECTUM

Sphenoid

Zygomatic bone?

ISTHMUS

Inferior colliculus

Isthmal canal

M a x i l l a

Palatal process

Oral cavity

T o n g u e

Nasopharynx

Basal occipital

**PONS**

**UPPER MEDULLA**

**CEREBELLUM**
**(LATERAL VERMIS)**

Mandibular process

Hyoid bone?

Epiglottis

Oropharynx

**RHOMBENCEPHALON**

"Budding" rhombencephalic choroid plexus

Clavicle

Thyroid cartilage

**Larynx**

Cricoid cartilage

Thyroid gland

Trachea

Pharyngoesophagus

Cervical vertebral column

**LOWER MEDULLA**

Squamous occipital bone

**SPINAL CORD**

**RHOMBENCEPHALIC SUPERVENTRICLE**
**(FUTURE FOURTH VENTRICLE)**

FONT KEY:
**VENTRICULAR DIVISIONS – CAPITALS**
Major brain structure - Times **Bold CAPITALS**
All other structures - Times Roman or **Bold**

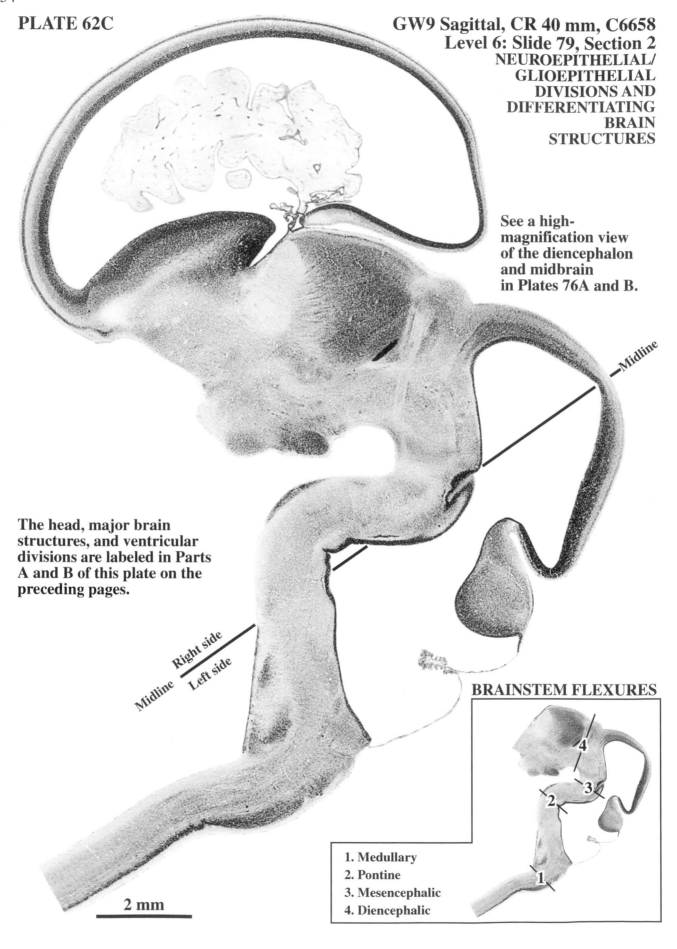

**PLATE 62C**

**GW9 Sagittal, CR 40 mm, C6658**
**Level 6: Slide 79, Section 2**
**NEUROEPITHELIAL/
GLIOEPITHELIAL
DIVISIONS AND
DIFFERENTIATING
BRAIN
STRUCTURES**

See a high-
magnification view
of the diencephalon
and midbrain
in Plates 76A and B.

Midline

The head, major brain
structures, and ventricular
divisions are labeled in Parts
A and B of this plate on the
preceding pages.

Right side
Left side
Midline

2 mm

**BRAINSTEM FLEXURES**

4
3
2
1

1. Medullary
2. Pontine
3. Mesencephalic
4. Diencephalic

**PLATE 62D**

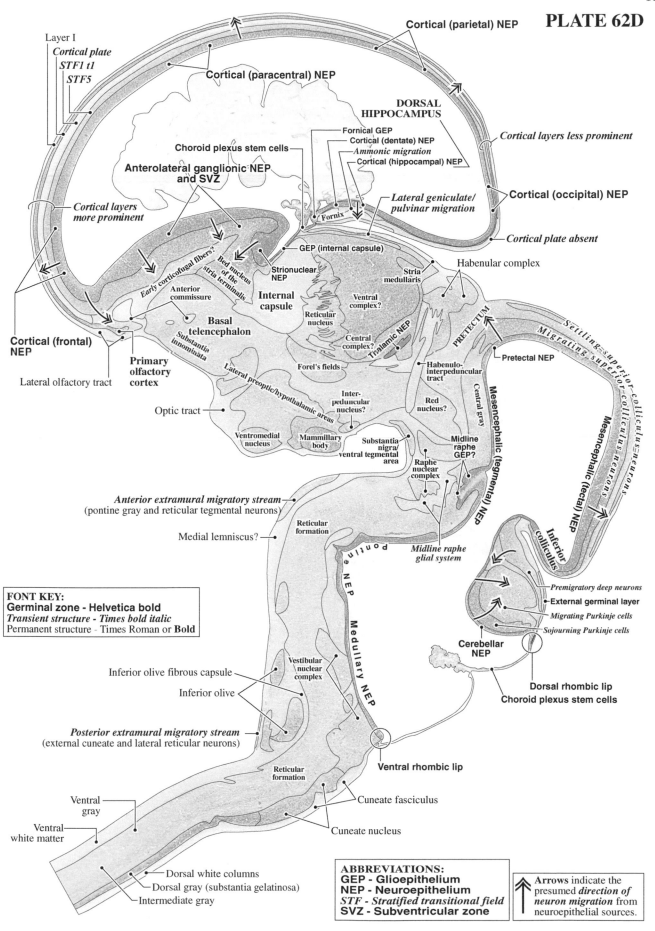

Layer I
*Cortical plate*
*STF1 t1*
*STF5*

Cortical (parietal) NEP

Cortical (paracentral) NEP

**DORSAL HIPPOCAMPUS**

Choroid plexus stem cells

**Anterolateral ganglionic NEP and SVZ**

Fornical GEP
Cortical (dentate) NEP
*Ammonic migration*
Cortical (hippocampal) NEP

*Cortical layers less prominent*

*Lateral geniculate/ pulvinar migration*

Cortical (occipital) NEP

*Cortical layers more prominent*

Fornix

GEP (internal capsule)

*Cortical plate absent*

Strionuclear NEP

Stria medullaris

Habenular complex

*Early corticofugal fibers?*

*Bed nucleus of the stria terminalis*

**Internal capsule**

Ventral complex?

**Cortical (frontal) NEP**

Anterior commissure

Reticular nucleus

**Basal telencephalon**

Central complex?

*Thalamic NEP*

PRETECTUM

*Settling superior colliculus neurons*

*Migrating superior colliculus neurons*

Primary olfactory cortex

*Substantia innominata*

Forel's fields

Habenulo-interpeduncular tract

Pretectal NEP

Lateral olfactory tract

*Lateral preoptic/hypothalamic areas*

Inter-peduncular nucleus?

Red nucleus?

Central gray

*Mesencephalic (tegmental) NEP*

*Mesencephalic (tectal) NEP*

Optic tract

Ventromedial nucleus

Mammillary body

Substantia nigra/ ventral tegmental area

Midline raphe GEP?

Raphe nuclear complex

**Inferior colliculus**

*Anterior extramural migratory stream*
(pontine gray and reticular tegmental neurons)

Reticular formation

*Midline raphe glial system*

Medial lemniscus?

*Pontine NEP*

*Premigratory deep neurons*
**External germinal layer**
*Migrating Purkinje cells*
*Sojourning Purkinje cells*

**Cerebellar NEP**

**FONT KEY:**
Germinal zone - Helvetica bold
*Transient structure - Times bold italic*
Permanent structure - Times Roman or **Bold**

*Medullary NEP*

Inferior olive fibrous capsule

Vestibular nuclear complex

**Dorsal rhombic lip**
**Choroid plexus stem cells**

Inferior olive

*Posterior extramural migratory stream*
(external cuneate and lateral reticular neurons)

Reticular formation

**Ventral rhombic lip**

Cuneate fasciculus

Ventral gray

Cuneate nucleus

Ventral white matter

Dorsal white columns
Dorsal gray (substantia gelatinosa)
Intermediate gray

**ABBREVIATIONS:**
**GEP** - Glioepithelium
**NEP** - Neuroepithelium
*STF* - *Stratified transitional field*
**SVZ** - Subventricular zone

**Arrows** indicate the presumed *direction of neuron migration* from neuroepithelial sources.

**PLATE 63A**

**GW9 Sagittal, CR 40 mm, C6658**
**Level 7: Slide 83, Section 1**
**HEAD STRUCTURES,**
**MAJOR BRAIN REGIONS,**
**AND VENTRICULAR**
**DIVISIONS**

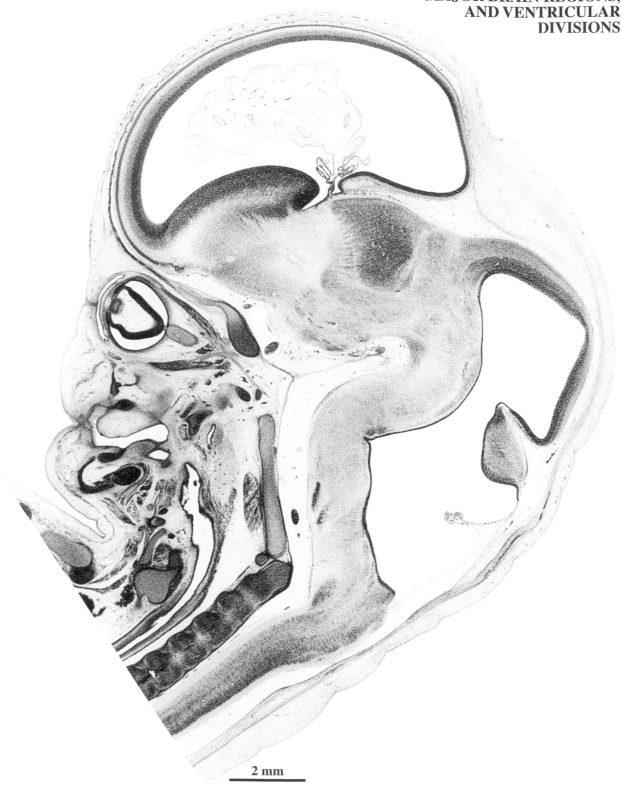

2 mm

Neuroepithelial divisions, glioepithelial divisions, and differentiating
structures are labeled in Parts C and D of this plate on the following pages.

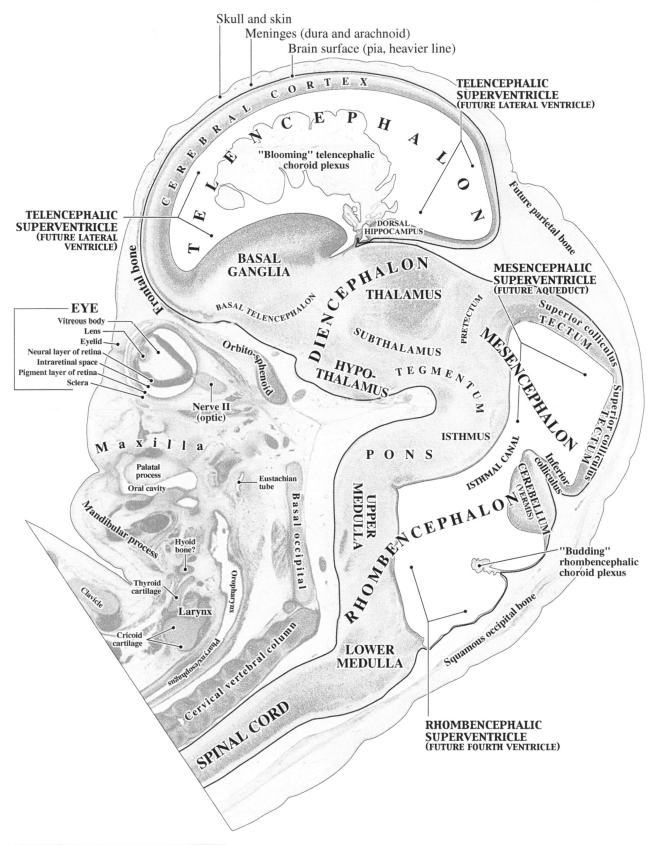

Skull and skin
Meninges (dura and arachnoid)
Brain surface (pia, heavier line)

**TELENCEPHALIC SUPERVENTRICLE**
**(FUTURE LATERAL VENTRICLE)**

CEREBRAL CORTEX

T E L E N C E P H A L O N

"Blooming" telencephalic choroid plexus

**DORSAL HIPPOCAMPUS**

Future parietal bone

**TELENCEPHALIC SUPERVENTRICLE (FUTURE LATERAL VENTRICLE)**

Frontal bone

**BASAL GANGLIA**

BASAL TELENCEPHALON

**DIENCEPHALON**

**THALAMUS**

**MESENCEPHALIC SUPERVENTRICLE (FUTURE AQUEDUCT)**

PRETECTUM

**MESENCEPHALON**

Superior colliculus **TECTUM**

Superior colliculus **TECTUM**

**EYE**
Vitreous body
Lens
Eyelid
Neural layer of retina
Intraretinal space
Pigment layer of retina
Sclera

Orbito-sphenoid

**SUBTHALAMUS**

**HYPO-THALAMUS**

T E G M E N T U M

Nerve II (optic)

ISTHMUS

**CEREBELLUM (VERMIS)**

Inferior colliculus

M a x i l l a

P O N S

ISTHMAL CANAL

Palatal process

Oral cavity

Eustachian tube

Basal occipital

**UPPER MEDULLA**

**RHOMBENCEPHALON**

"Budding" rhombencephalic choroid plexus

Mandibular process

Hyoid bone?

Clavicle

Thyroid cartilage

Larynx

Oropharynx

Cricoid cartilage

Pharynx/esophagus

Cervical vertebral column

**LOWER MEDULLA**

Squamous occipital bone

**SPINAL CORD**

**RHOMBENCEPHALIC SUPERVENTRICLE (FUTURE FOURTH VENTRICLE)**

FONT KEY:
**VENTRICULAR DIVISIONS - CAPITALS**
Major brain structure - Times **Bold CAPITALS**
All other structures - Times Roman or **Bold**

158

**GW9 Sagittal, CR 40 mm, C6658**
**Level 7: Slide 83, Section 1**
**NEUROEPITHELIAL/
GLIOEPITHELIAL
DIVISIONS AND
DIFFERENTIATING
BRAIN
STRUCTURES**

See a high-magnification
view of the lateral
forebrain in
Plates 73A and B.

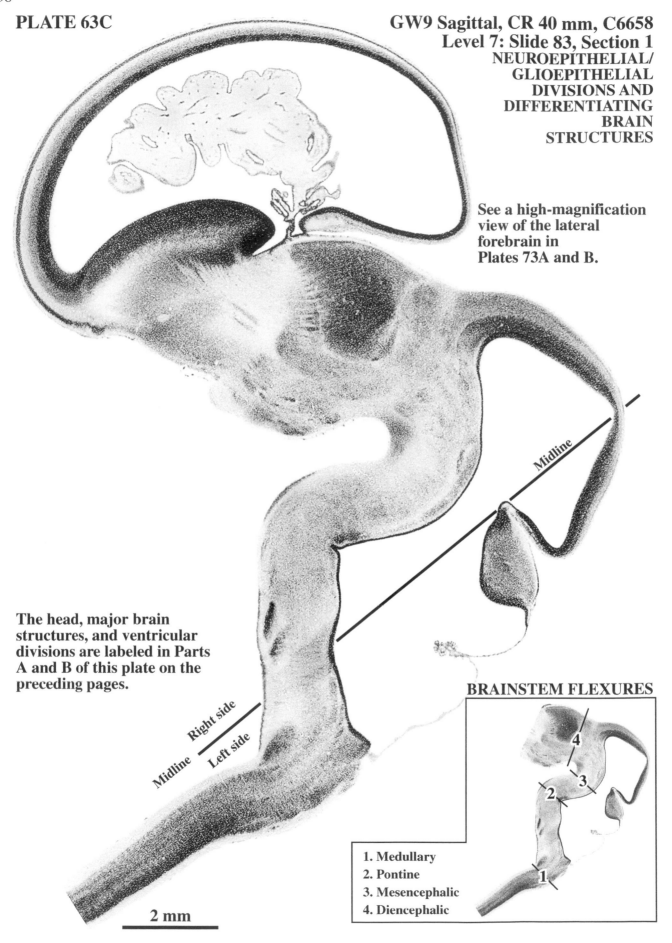

Midline

The head, major brain
structures, and ventricular
divisions are labeled in Parts
A and B of this plate on the
preceding pages.

Right side

Left side

Midline

**BRAINSTEM FLEXURES**

1. Medullary
2. Pontine
3. Mesencephalic
4. Diencephalic

2 mm

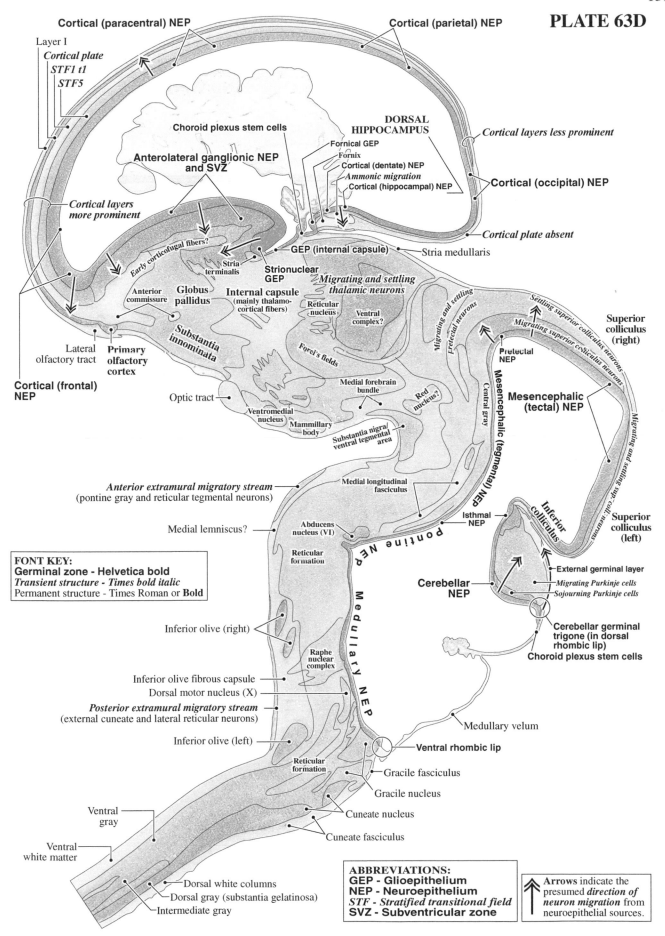

Cortical (paracentral) NEP

Cortical (parietal) NEP

Layer I
*Cortical plate*
*STF1 t1*
*STF5*

Choroid plexus stem cells

**DORSAL HIPPOCAMPUS**

*Cortical layers less prominent*

Fornical GEP
Fornix
Cortical (dentate) NEP
*Ammonic migration*
Cortical (hippocampal) NEP

**Anterolateral ganglionic NEP and SVZ**

**Cortical (occipital) NEP**

*Cortical layers more prominent*

*Cortical plate absent*

*Early corticofugal fibers?*

*Stria terminalis*

**GEP (internal capsule)**

Stria medullaris

**Strionuclear GEP**

*Migrating and settling thalamic neurons*

**Anterior commissure**

**Globus pallidus**

**Internal capsule**
(mainly thalamo-cortical fibers)

Reticular nucleus

Ventral complex?

*Migrating and settling*

*Prectal neurons*

*Settling superior colliculus neurons*

*Migrating superior colliculus neurons*

**Superior colliculus (right)**

*Substantia innominata*

*Forel's fields*

Prectal NEP

Lateral olfactory tract

**Primary olfactory cortex**

Medial forebrain bundle

*Red nucleus?*

**Mesencephalic (tectal) NEP**

**Cortical (frontal) NEP**

Optic tract

Ventromedial nucleus

Mammillary body

Substantia nigra/ventral tegmental area

*Central gray*

**Mesencephalic (tegmental) NEP**

*Migrating and settling sup.-coll.-neurons*

**Superior colliculus (left)**

*Anterior extramural migratory stream*
(pontine gray and reticular tegmental neurons)

Medial longitudinal fasciculus

**Inferior colliculus**

Medial lemniscus?

Abducens nucleus (VI)

Isthmal NEP

External germinal layer

**Reticular formation**

**Pontine NEP**

*Migrating Purkinje cells*
*Sojourning Purkinje cells*

**FONT KEY:**
Germinal zone - **Helvetica bold**
*Transient structure - Times bold italic*
Permanent structure - Times Roman or **Bold**

**Cerebellar NEP**

**Cerebellar germinal trigone (in dorsal rhombic lip)**
**Choroid plexus stem cells**

Inferior olive (right)

**Medullary NEP**

Raphe nuclear complex

Inferior olive fibrous capsule
Dorsal motor nucleus (X)

*Posterior extramural migratory stream*
(external cuneate and lateral reticular neurons)

Inferior olive (left)

Medullary velum

**Reticular formation**

**Ventral rhombic lip**

Gracile fasciculus

Gracile nucleus

Cuneate nucleus

Ventral gray

Cuneate fasciculus

Ventral white matter

Dorsal white columns
Dorsal gray (substantia gelatinosa)
Intermediate gray

**ABBREVIATIONS:**
**GEP** - Glioepithelium
**NEP** - Neuroepithelium
*STF* - Stratified transitional field
**SVZ** - Subventricular zone

**Arrows** indicate the presumed *direction of neuron migration* from neuroepithelial sources.

**PLATE 64A**

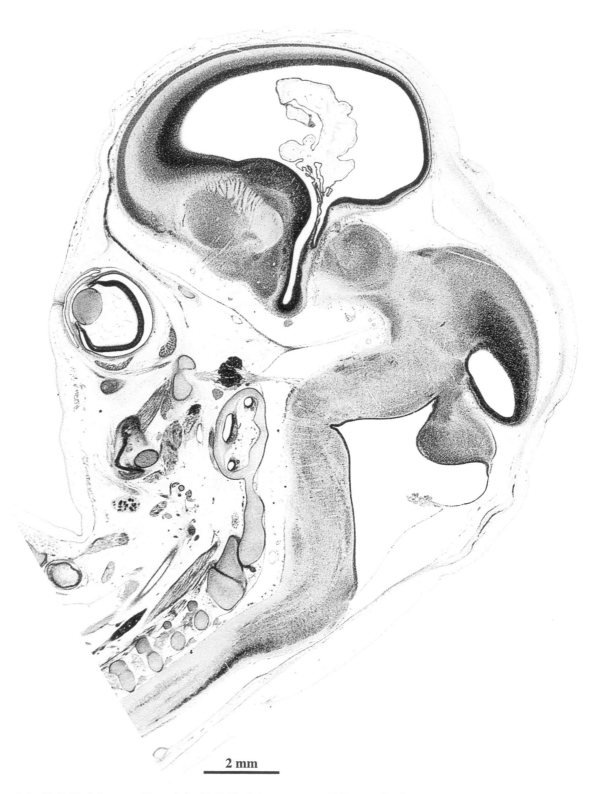

2 mm

Neuroepithelial divisions, glioepithelial divisions, and differentiating
structures are labeled in Parts C and D of this plate on the following pages.

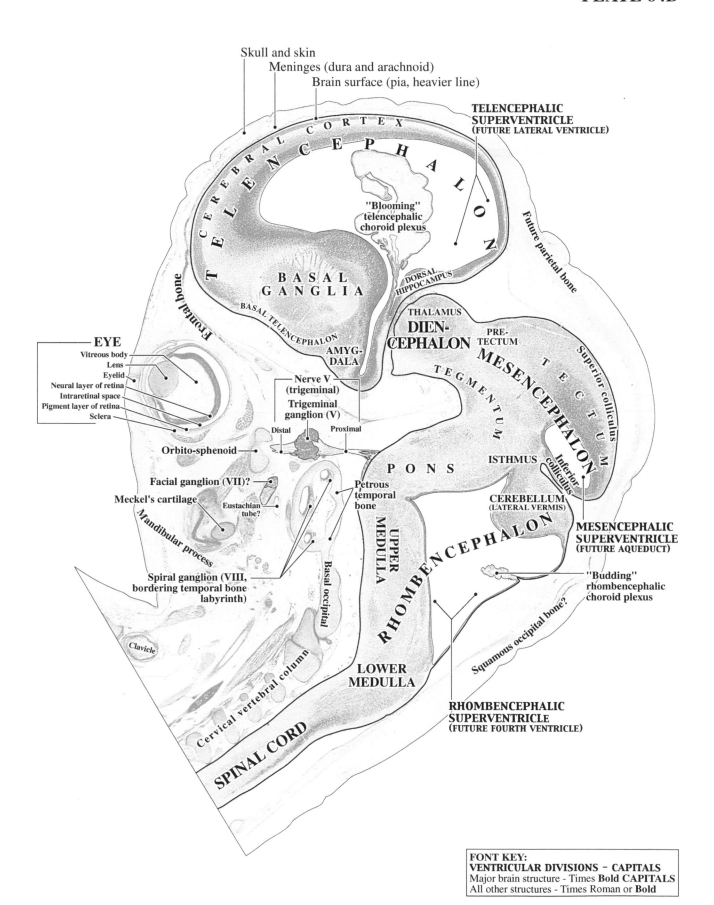

Skull and skin
Meninges (dura and arachnoid)
Brain surface (pia, heavier line)

**TELENCEPHALIC
SUPERVENTRICLE
(FUTURE LATERAL VENTRICLE)**

CEREBRAL CORTEX

TELENCEPHALON

"Blooming"
telencephalic
choroid plexus

Future parietal bone

**BASAL
GANGLIA**

DORSAL
HIPPOCAMPUS

BASAL TELENCEPHALON

THALAMUS

**DIEN-
CEPHALON**

PRE-
TECTUM

**MESENCEPHALON**

T E C T U M

Superior colliculus

**AMYG-
DALA**

T E G M E N T U M

**EYE**
Vitreous body
Lens
Eyelid
Neural layer of retina
Intraretinal space
Pigment layer of retina
Sclera

Nerve V
(trigeminal)

Trigeminal
ganglion (V)

ISTHMUS

Inferior
colliculus

Distal

Proximal

Orbito-sphenoid

**PONS**

**CEREBELLUM
(LATERAL VERMIS)**

Facial ganglion (VII)?

Petrous
temporal
bone

**MESENCEPHALIC
SUPERVENTRICLE
(FUTURE AQUEDUCT)**

Meckel's cartilage

Eustachian
tube?

**UPPER
MEDULLA**

**RHOMBENCEPHALON**

"Budding"
rhombencephalic
choroid plexus

Mandibular process

Spiral ganglion (VIII,
bordering temporal bone
labyrinth)

Basal occipital

Squamous occipital bone?

Clavicle

Cervical vertebral column

**LOWER
MEDULLA**

**RHOMBENCEPHALIC
SUPERVENTRICLE
(FUTURE FOURTH VENTRICLE)**

**SPINAL CORD**

**FONT KEY:**
**VENTRICULAR DIVISIONS – CAPITALS**
Major brain structure - Times **Bold CAPITALS**
All other structures - Times Roman or **Bold**

**PLATE 64C**

**NEUROEPITHELIAL/
GLIOEPITHELIAL
DIVISIONS AND
DIFFERENTIATING
BRAIN STRUCTURES**

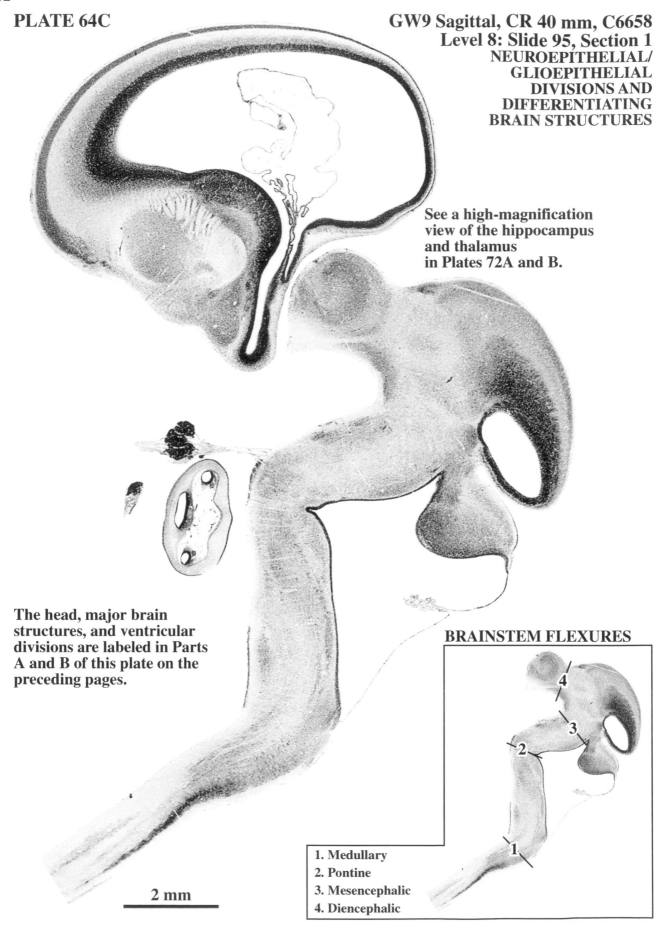

See a high-magnification
view of the hippocampus
and thalamus
in Plates 72A and B.

The head, major brain
structures, and ventricular
divisions are labeled in Parts
A and B of this plate on the
preceding pages.

**BRAINSTEM FLEXURES**

1. Medullary
2. Pontine
3. Mesencephalic
4. Diencephalic

**2 mm**

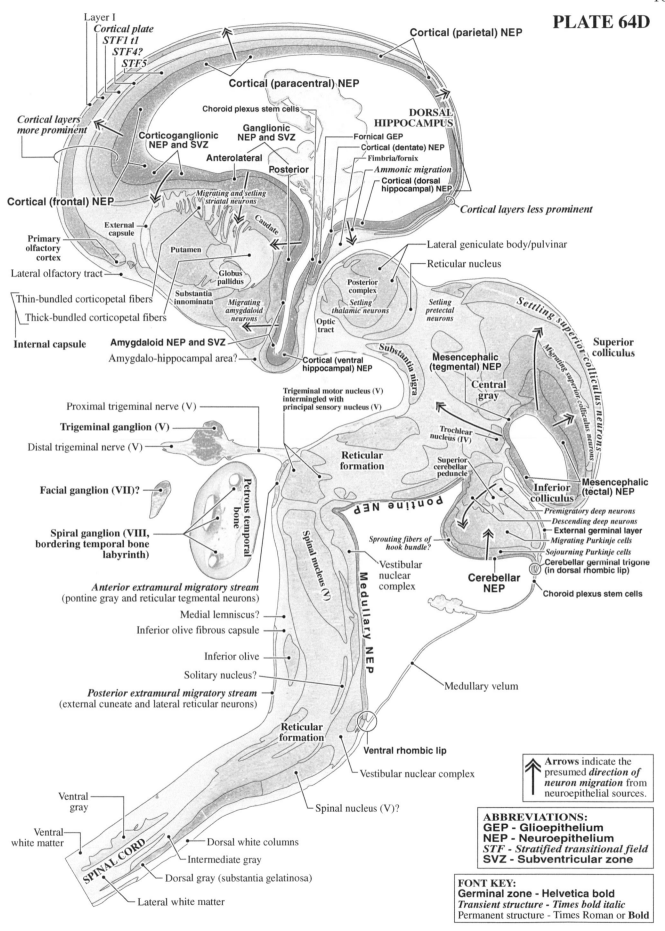

**PLATE 64D**

Layer I
*Cortical plate*
*STF1 t1*
*STF4?*
*STF5*

Cortical (parietal) NEP

Cortical (paracentral) NEP

Choroid plexus stem cells

**DORSAL HIPPOCAMPUS**

*Cortical layers more prominent*

Corticoganglionic NEP and SVZ

Ganglionic NEP and SVZ

Anterolateral

Posterior

Fornical GEP

Cortical (dentate) NEP

Fimbria/fornix

*Ammonic migration*

Cortical (dorsal hippocampal) NEP

*Cortical layers less prominent*

Cortical (frontal) NEP

*Migrating and setling striatal neurons*

External capsule

Caudate

Primary olfactory cortex

Putamen

Lateral olfactory tract

Globus pallidus

Substantia innominata

*Migrating amygdaloid neurons*

Thin-bundled corticopetal fibers

Thick-bundled corticopetal fibers

**Internal capsule**

**Amygdaloid NEP and SVZ**

Amygdalo-hippocampal area?

Cortical (ventral hippocampal) NEP

Lateral geniculate body/pulvinar

Reticular nucleus

Posterior complex

*Setling thalamic neurons*

*Setling pretectal neurons*

*Settling superior colliculus neurons*

Optic tract

*Substantia nigra*

Mesencephalic (tegmental) NEP

**Central gray**

*Migrating superior colliculus neurons*

**Superior colliculus**

Trigeminal motor nucleus (V) intermingled with principal sensory nucleus (V)

Proximal trigeminal nerve (V)

**Trigeminal ganglion (V)**

Distal trigeminal nerve (V)

Trochlear nucleus (IV)

**Reticular formation**

Superior cerebellar peduncle

**Inferior colliculus**

Mesencephalic (tectal) NEP

*Premigratory deep neurons*

*Descending deep neurons*

**External germinal layer**

*Migrating Purkinje cells*

*Sojourning Purkinje cells*

Cerebellar germinal trigone (in dorsal rhombic lip)

**Facial ganglion (VII)?**

Petrous temporal bone

*Sprouting fibers of hook bundle?*

Vestibular nuclear complex

**Cerebellar NEP**

**Spiral ganglion (VIII, bordering temporal bone labyrinth)**

**Pontine NEP**

**Medullary NEP**

Choroid plexus stem cells

*Anterior extramural migratory stream*
*(pontine gray and reticular tegmental neurons)*

Spinal nucleus (V)

Medial lemniscus?

Inferior olive fibrous capsule

Inferior olive

Solitary nucleus?

*Posterior extramural migratory stream*
*(external cuneate and lateral reticular neurons)*

Medullary velum

**Reticular formation**

**Ventral rhombic lip**

Vestibular nuclear complex

Ventral gray

Spinal nucleus (V)?

Ventral white matter

**SPINAL CORD**

Dorsal white columns

Intermediate gray

Dorsal gray (substantia gelatinosa)

Lateral white matter

**Arrows** indicate the presumed *direction of neuron migration* from neuroepithelial sources.

**ABBREVIATIONS:**
**GEP - Glioepithelium**
**NEP - Neuroepithelium**
*STF - Stratified transitional field*
**SVZ - Subventricular zone**

**FONT KEY:**
**Germinal zone - Helvetica bold**
*Transient structure - Times bold italic*
Permanent structure - Times Roman or **Bold**

**PLATE 65A**

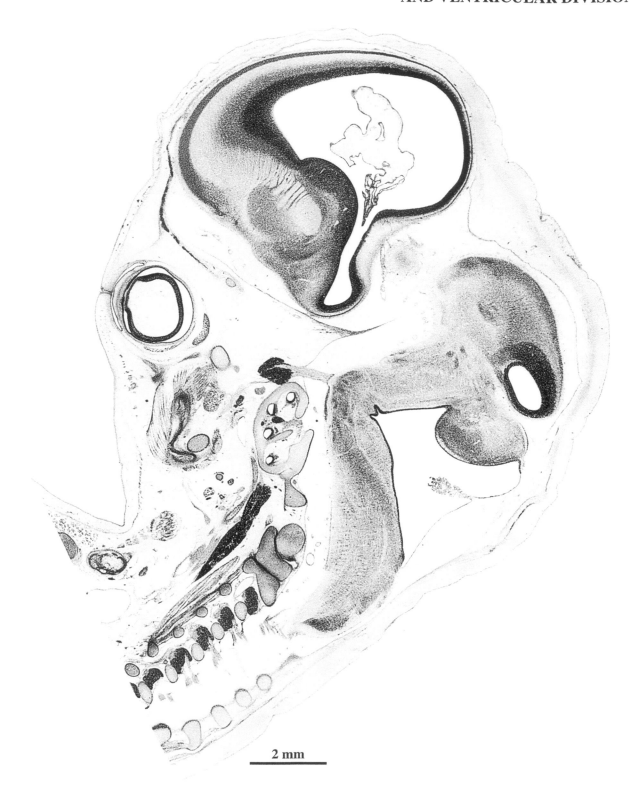

2 mm

**Neuroepithelial divisions, glioepithelial divisions, and differentiating
structures are labeled in Parts C and D of this plate on the following pages.**

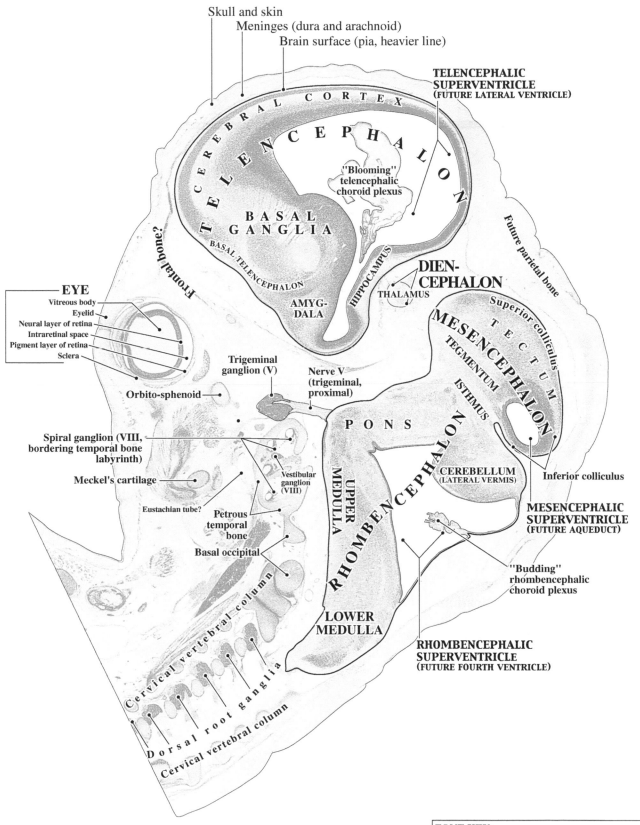

Skull and skin
Meninges (dura and arachnoid)
Brain surface (pia, heavier line)

**TELENCEPHALIC
SUPERVENTRICLE
(FUTURE LATERAL VENTRICLE)**

CEREBRAL CORTEX

TELENCEPHALON

"Blooming"
telencephalic
choroid plexus

BASAL
GANGLIA

*Future parietal bone*

BASAL TELENCEPHALON

HIPPOCAMPUS

**DIEN-
CEPHALON**

**THALAMUS**

*Superior colliculus*

*Frontal bone?*

**EYE**
Vitreous body
Eyelid
Neural layer of retina
Intraretinal space
Pigment layer of retina
Sclera

AMYG-
DALA

**MESENCEPHALON**

T E C T U M

TEGMENTUM

Trigeminal
ganglion (V)

Nerve V
(trigeminal,
proximal)

ISTHMUS

Orbito-sphenoid

P O N S

Inferior colliculus

Spiral ganglion (VIII,
bordering temporal bone
labyrinth)

Vestibular
ganglion
(VIII)

**CEREBELLUM
(LATERAL VERMIS)**

**MESENCEPHALIC
SUPERVENTRICLE
(FUTURE AQUEDUCT)**

Meckel's cartilage

UPPER
MEDULLA

RHOMBENCEPHALON

Eustachian tube?

Petrous
temporal
bone

"Budding"
rhombencephalic
choroid plexus

Basal occipital

**LOWER
MEDULLA**

**RHOMBENCEPHALIC
SUPERVENTRICLE
(FUTURE FOURTH VENTRICLE)**

*Cervical vertebral column*

*Dorsal root ganglia*

*Cervical vertebral column*

**FONT KEY:**
**VENTRICULAR DIVISIONS – CAPITALS**
Major brain structure - Times **Bold CAPITALS**
All other structures - Times Roman or **Bold**

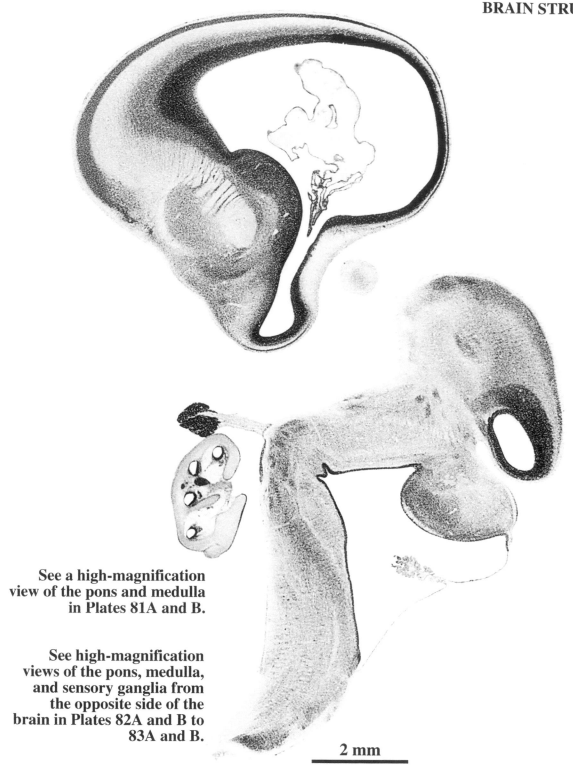

**See a high-magnification
view of the pons and medulla
in Plates 81A and B.**

**See high-magnification
views of the pons, medulla,
and sensory ganglia from
the opposite side of the
brain in Plates 82A and B to
83A and B.**

2 mm

The head, major brain structures, and ventricular divisions are
labeled in Parts A and B of this plate on the preceding pages.

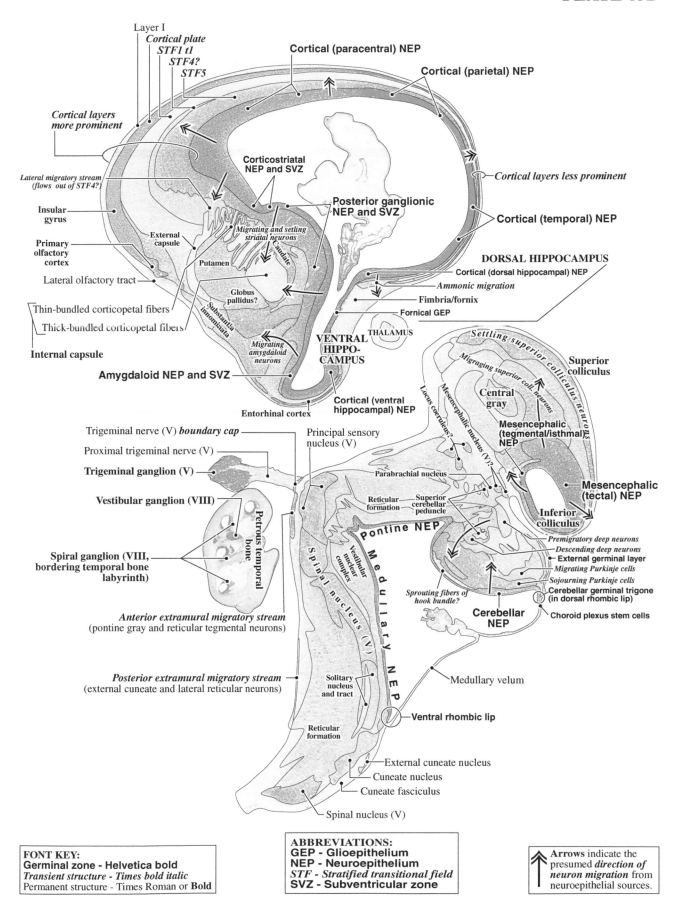

Layer I
*Cortical plate*
*STF1 t1*
*STF4?*
*STF5*

**Cortical (paracentral) NEP**

**Cortical (parietal) NEP**

*Cortical layers
more prominent*

**Corticostriatal
NEP and SVZ**

*Lateral migratory stream
(flows out of STF4?)*

*Cortical layers less prominent*

**Insular
gyrus**

**Primary
olfactory
cortex**

External
capsule

Putamen

*Migrating and settling
striatal neurons*

Caudate

**Posterior ganglionic
NEP and SVZ**

**Cortical (temporal) NEP**

Lateral olfactory tract

Thin-bundled corticopetal fibers

Thick-bundled corticopetal fibers

**Internal capsule**

Globus
pallidus?

Substantia
innominata

**DORSAL HIPPOCAMPUS**

Cortical (dorsal hippocampal) NEP

*Ammonic migration*

**Fimbria/fornix**

**Fornical GEP**

*Migrating
amygdaloid
neurons*

**VENTRAL
HIPPO-
CAMPUS**

THALAMUS

**Amygdaloid NEP and SVZ**

Entorhinal cortex

**Cortical (ventral
hippocampal) NEP**

*Settling superior colliculus neurons*

*Migrating superior coll. neurons*

**Superior
colliculus**

*Mesencephalic nucleus (V)?*

**Central
gray**

**Mesencephalic
(tegmental/isthmal)
NEP**

*Locus coeruleus?*

**Mesencephalic
(tectal) NEP**

Trigeminal nerve (V) *boundary cap*

Proximal trigeminal nerve (V)

**Trigeminal ganglion (V)**

**Vestibular ganglion (VIII)**

Principal sensory
nucleus (V)

Parabrachial nucleus

Reticular
formation

Superior
cerebellar
peduncle

**Inferior
colliculus**

*Premigratory deep neurons*
*Descending deep neurons*
**External germinal layer**
*Migrating Purkinje cells*
*Sojourning Purkinje cells*
**Cerebellar germinal trigone
(in dorsal rhombic lip)**

**Choroid plexus stem cells**

**Spiral ganglion (VIII,
bordering temporal bone
labyrinth)**

Petrous temporal
bone

**Pontine NEP**

Vestibular
nuclear
complex

**Spinal nucleus (V)**

*Medullary NEP*

*Sprouting fibers of
hook bundle?*

**Cerebellar
NEP**

*Anterior extramural migratory stream*
(pontine gray and reticular tegmental neurons)

*Posterior extramural migratory stream*
(external cuneate and lateral reticular neurons)

Solitary
nucleus
and tract

Medullary velum

**Ventral rhombic lip**

Reticular
formation

External cuneate nucleus
Cuneate nucleus
Cuneate fasciculus

Spinal nucleus (V)

FONT KEY:
**Germinal zone - Helvetica bold**
*Transient structure - Times bold italic*
Permanent structure - Times Roman or **Bold**

**ABBREVIATIONS:**
**GEP** - **Glioepithelium**
**NEP** - **Neuroepithelium**
*STF* - *Stratified transitional field*
**SVZ** - **Subventricular zone**

**Arrows** indicate the
presumed *direction of
neuron migration* from
neuroepithelial sources.

**PLATE 66A**

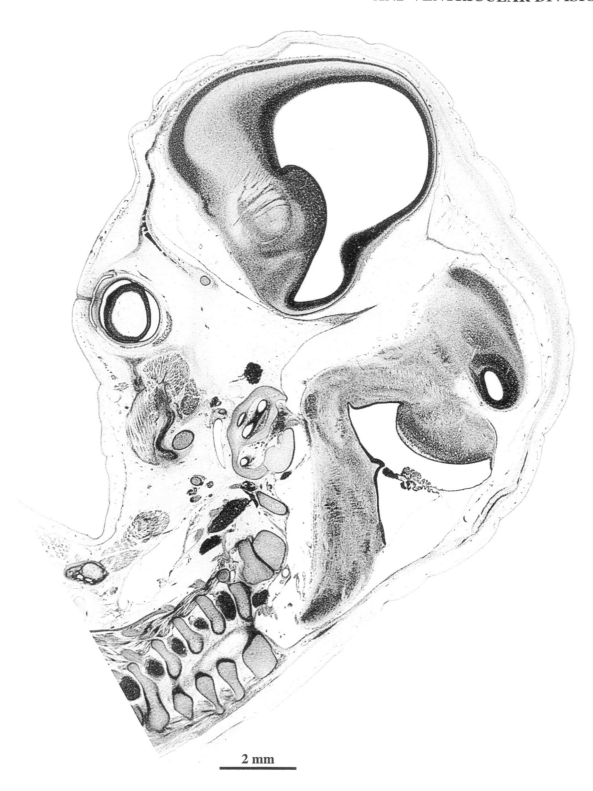

2 mm

Neuroepithelial divisions, glioepithelial divisions, and differentiating
structures are labeled in Parts C and D of this plate on the following pages.

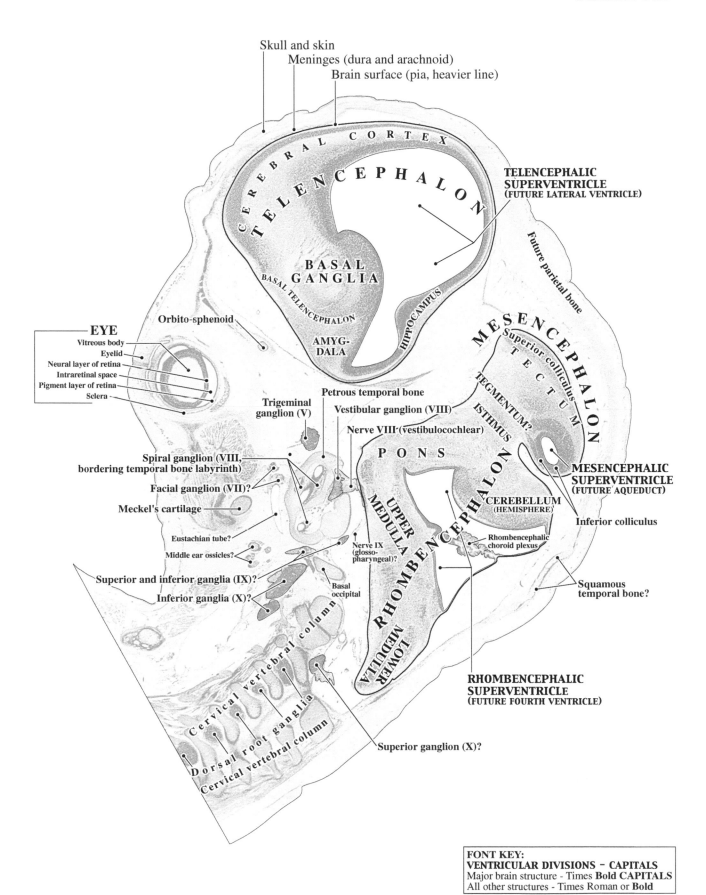

Skull and skin
Meninges (dura and arachnoid)
Brain surface (pia, heavier line)

CEREBRAL CORTEX

TELENCEPHALON

TELENCEPHALIC
SUPERVENTRICLE
(FUTURE LATERAL VENTRICLE)

Future parietal bone

BASAL
GANGLIA

BASAL TELENCEPHALON

HIPPOCAMPUS

MESENCEPHALON

AMYG-
DALA

Superior colliculus

T E C T U M

EYE
Vitreous body
Eyelid
Neural layer of retina
Intraretinal space
Pigment layer of retina
Sclera

Orbito-sphenoid

Trigeminal
ganglion (V)

Petrous temporal bone

Vestibular ganglion (VIII)

Nerve VIII (vestibulocochlear)

TEGMENTUM?

ISTHMUS

P O N S

MESENCEPHALIC
SUPERVENTRICLE
(FUTURE AQUEDUCT)

Spiral ganglion (VIII,
bordering temporal bone labyrinth)

Facial ganglion (VII)?

Meckel's cartilage

Eustachian tube?

Middle ear ossicles?

Superior and inferior ganglia (IX)?

Inferior ganglia (X)?

UPPER
MEDULLA

Nerve IX
(glosso-
pharyngeal)?

Basal
occipital

CEREBELLUM
(HEMISPHERE)

Rhombencephalic
choroid plexus

Inferior colliculus

RHOMBENCEPHALON

LOWER
MEDULLA

Squamous
temporal bone?

Cervical vertebral column

Dorsal root ganglia

Cervical vertebral column

RHOMBENCEPHALIC
SUPERVENTRICLE
(FUTURE FOURTH VENTRICLE)

Superior ganglion (X)?

FONT KEY:
**VENTRICULAR DIVISIONS – CAPITALS**
Major brain structure - Times **Bold CAPITALS**
All other structures - Times Roman or **Bold**

**PLATE 66C**

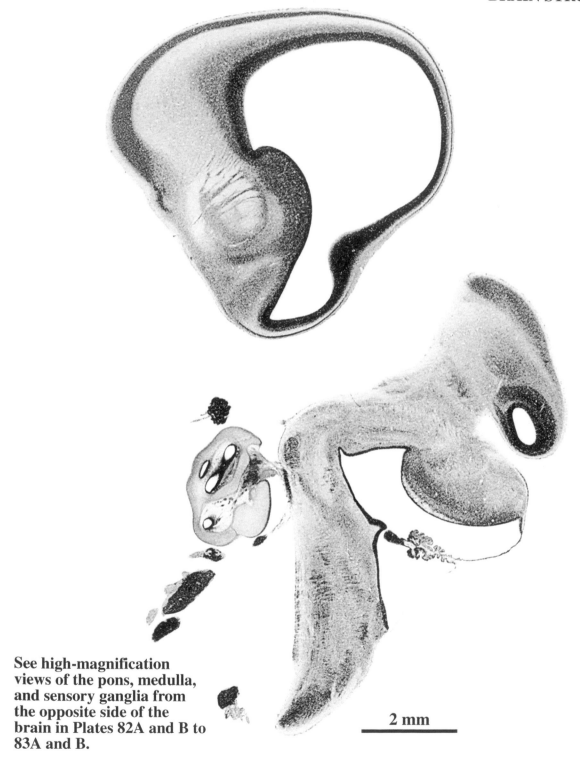

See high-magnification
views of the pons, medulla,
and sensory ganglia from
the opposite side of the
brain in Plates 82A and B to
83A and B.

2 mm

The head, major brain structures, and ventricular divisions are
labeled in Parts A and B of this plate on the preceding pages.

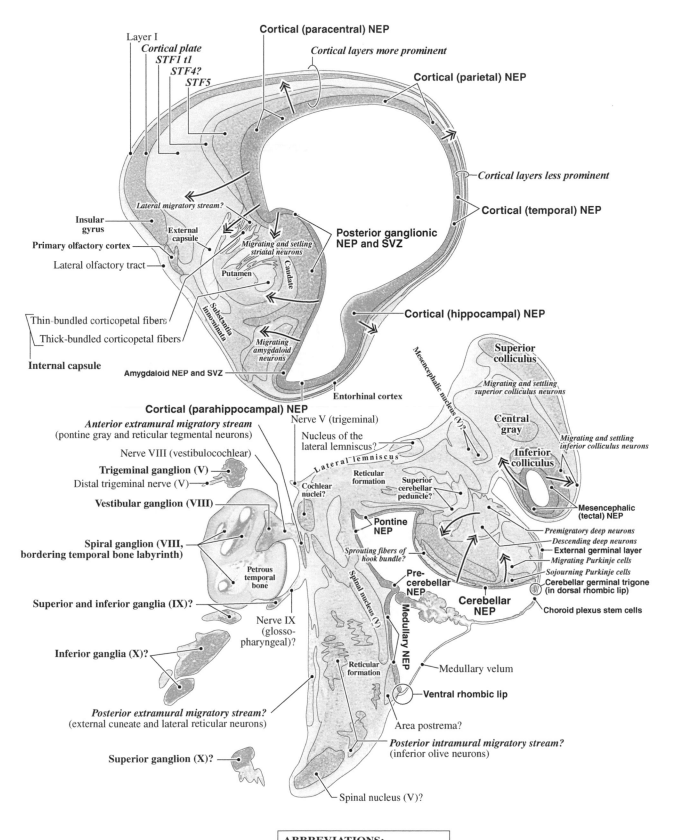

**PLATE 67A**

**GW9 Sagittal, CR 40 mm, C6658**
**Level 11: Slide 105, Section 2**
**HEAD STRUCTURES,**
**MAJOR BRAIN REGIONS,**
**AND VENTRICULAR DIVISIONS**

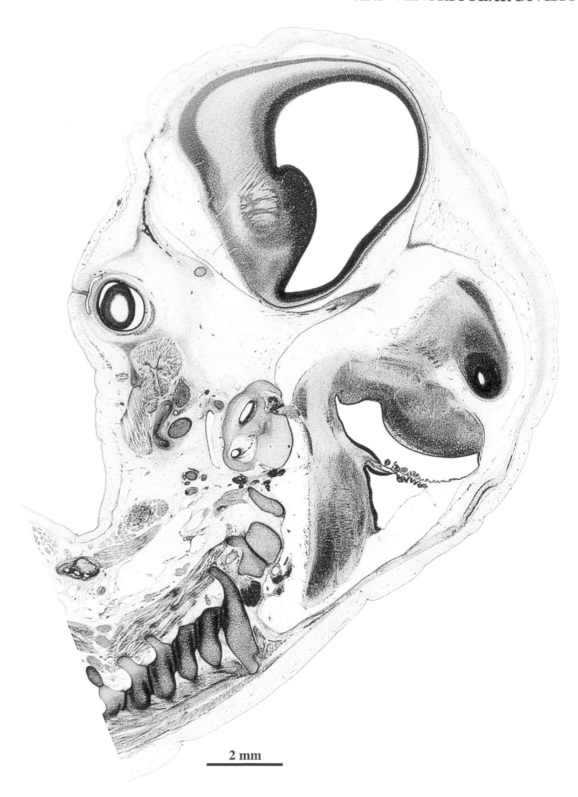

2 mm

**Neuroepithelial divisions, glioepithelial divisions, and differentiating structures are labeled in Parts C and D of this plate on the following pages.**

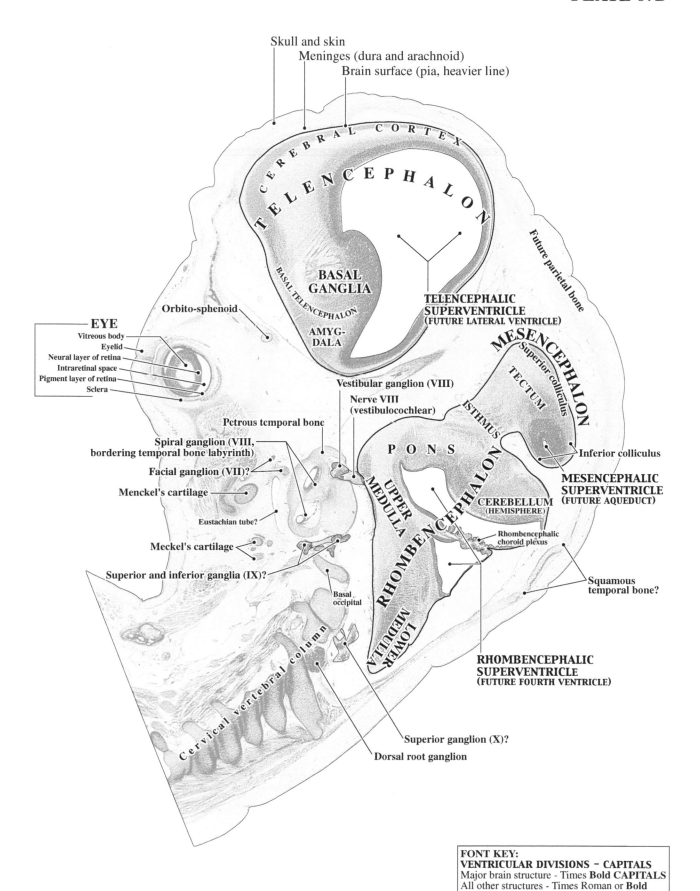

Skull and skin
Meninges (dura and arachnoid)
Brain surface (pia, heavier line)

C E R E B R A L   C O R T E X

T E L E N C E P H A L O N

BASAL
GANGLIA

BASAL TELENCEPHALON

AMYG-
DALA

Future parietal bone

TELENCEPHALIC
SUPERVENTRICLE
(FUTURE LATERAL VENTRICLE)

MESENCEPHALON

Superior colliculus

TECTUM

ISTHMUS

EYE
Vitreous body
Eyelid
Neural layer of retina
Intraretinal space
Pigment layer of retina
Sclera

Orbito-sphenoid

Vestibular ganglion (VIII)

Nerve VIII
(vestibulocochlear)

P O N S

Inferior colliculus

MESENCEPHALIC
SUPERVENTRICLE
(FUTURE AQUEDUCT)

Pctrous tcmporal bone

Spiral ganglion (VIII,
bordering temporal bone labyrinth)

Facial ganglion (VII)?

Menckel's cartilage

Eustachian tube?

Meckel's cartilage

Superior and inferior ganglia (IX)?

Basal
occipital

UPPER
MEDULLA

RHOMBENCEPHALON

CEREBELLUM
(HEMISPHERE)

Rhombencephalic
choroid plexus

LOWER
MEDULLA

Squamous
temporal bone?

RHOMBENCEPHALIC
SUPERVENTRICLE
(FUTURE FOURTH VENTRICLE)

Cervical vertebral column

Superior ganglion (X)?

Dorsal root ganglion

FONT KEY:
VENTRICULAR DIVISIONS − CAPITALS
Major brain structure - Times **Bold CAPITALS**
All other structures - Times Roman or **Bold**

174

**PLATE 67C**

**GW9 Sagittal, CR 40 mm, C6658**
**Level 11: Slide 105, Section 2**
**NEUROEPITHELIAL/GLIOEPITHELIAL**
**DIVISIONS AND DIFFERENTIATING**
**BRAIN STRUCTURES**

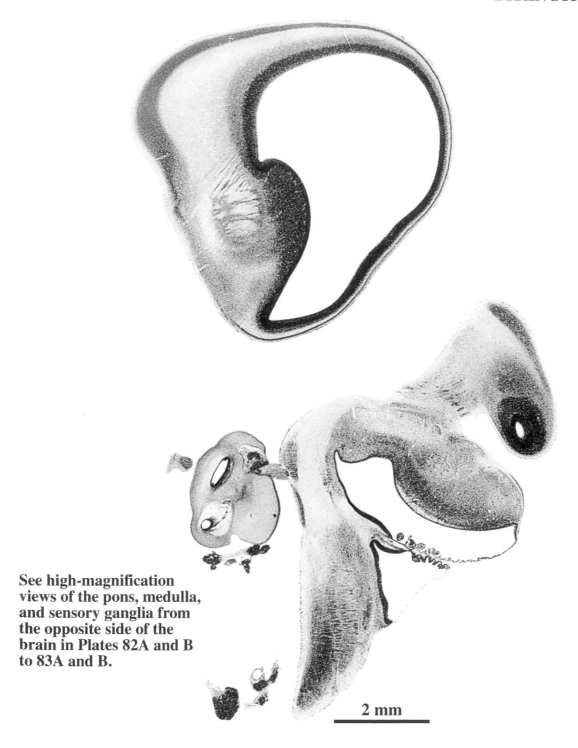

**See high-magnification views of the pons, medulla, and sensory ganglia from the opposite side of the brain in Plates 82A and B to 83A and B.**

2 mm

**The head, major brain structures, and ventricular divisions are labeled in Parts A and B of this plate on the preceding pages.**

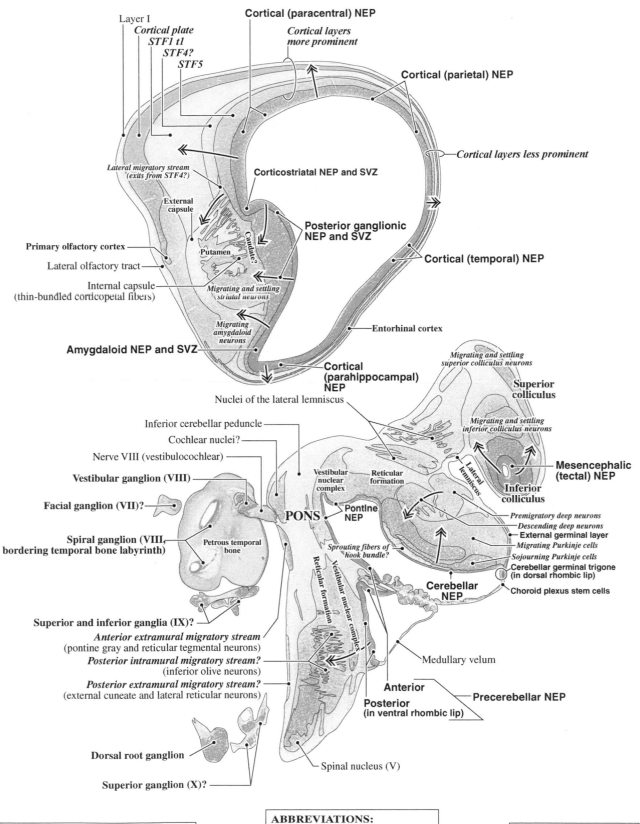

Layer I
*Cortical plate*
*STF1 t1*
*STF4?*
*STF5*

**Cortical (paracentral) NEP**

*Cortical layers more prominent*

**Cortical (parietal) NEP**

*Cortical layers less prominent*

*Lateral migratory stream (exits from STF4?)*

**Corticostriatal NEP and SVZ**

*External capsule*

**Posterior ganglionic NEP and SVZ**

Primary olfactory cortex

Lateral olfactory tract

Internal capsule
(thin-bundled corticopetal fibers)

*Caudate?*

*Putamen*

*Migrating and settling striatal neurons*

**Cortical (temporal) NEP**

*Migrating amygdaloid neurons*

Entorhinal cortex

**Amygdaloid NEP and SVZ**

**Cortical (parahippocampal) NEP**

*Migrating and settling superior colliculus neurons*

**Superior colliculus**

Nuclei of the lateral lemniscus

Inferior cerebellar peduncle

Cochlear nuclei?

Nerve VIII (vestibulocochlear)

**Vestibular ganglion (VIII)**

**Facial ganglion (VII)?**

**Spiral ganglion (VIII, bordering temporal bone labyrinth)**

*Petrous temporal bone*

*Vestibular nuclear complex*

*Reticular formation*

**PONS**

**Pontine NEP**

*Lateral lemniscus*

*Migrating and settling inferior colliculus neurons*

**Mesencephalic (tectal) NEP**

**Inferior colliculus**

*Premigratory deep neurons*
*Descending deep neurons*
External germinal layer
*Migrating Purkinje cells*
*Sojourning Purkinje cells*
Cerebellar germinal trigone
(in dorsal rhombic lip)
Choroid plexus stem cells

*Sprouting fibers of hook bundle?*

**Cerebellar NEP**

**Superior and inferior ganglia (IX)?**

*Anterior extramural migratory stream*
(pontine gray and reticular tegmental neurons)

*Posterior intramural migratory stream?*
(inferior olive neurons)

*Posterior extramural migratory stream?*
(external cuneate and lateral reticular neurons)

*Reticular formation*

*Vestibular nuclear complex*

Medullary velum

**Anterior**

**Posterior**
(in ventral rhombic lip)

**Precerebellar NEP**

**Dorsal root ganglion**

**Superior ganglion (X)?**

Spinal nucleus (V)

**FONT KEY:**
Germinal zone - **Helvetica bold**
*Transient structure - Times bold italic*
Permanent structure - Times Roman or **Bold**

**ABBREVIATIONS:**
**GEP** - Glioepithelium
**NEP** - Neuroepithelium
*STF - Stratified transitional field*
**SVZ** - Subventricular zone

**Arrows** indicate the presumed *direction of neuron migration* from neuroepithelial sources.

176

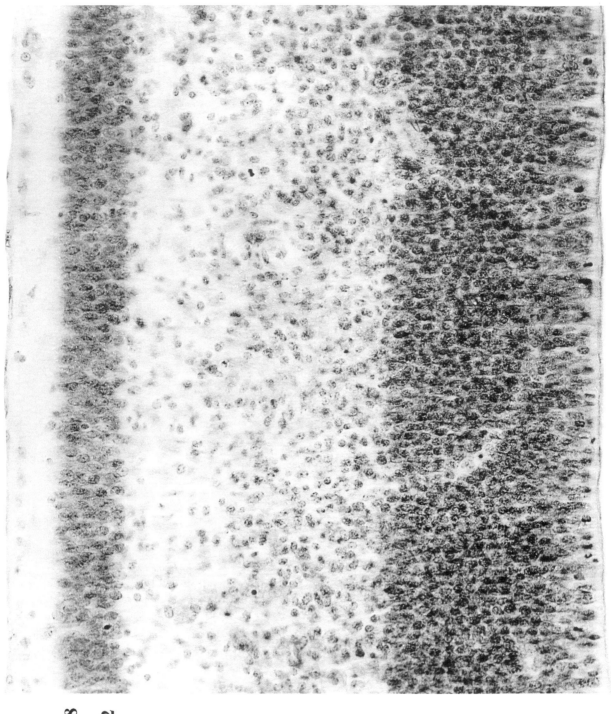

0.05 mm

GW9 Sagittal
CR 40 mm, C6658
Level 4:
Slide 71, Section 2

DORSAL
CORTEX

See the entire section in Plates 60A-D.

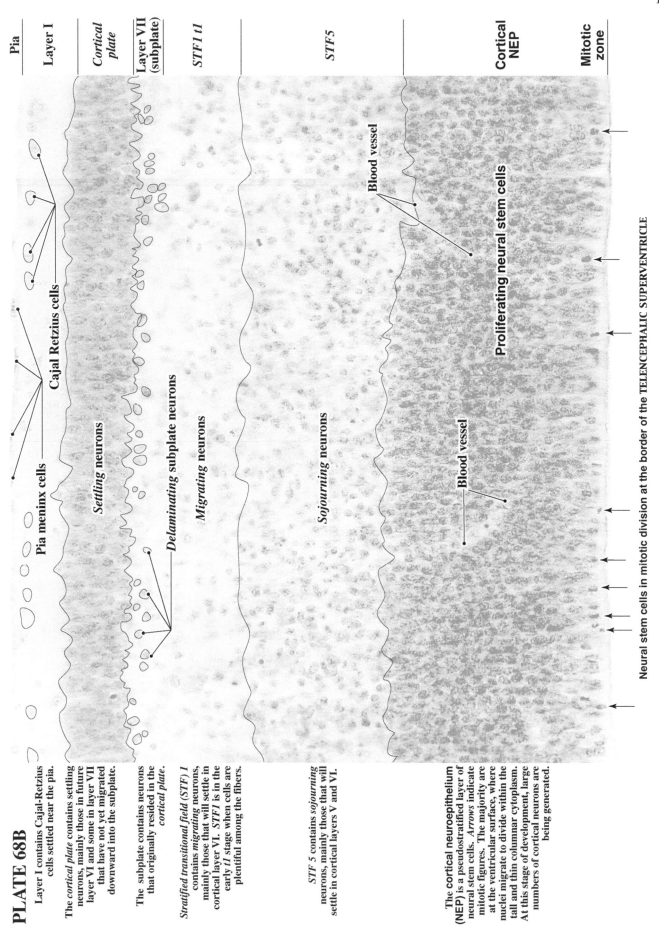

## PLATE 68B

Layer I contains Cajal-Retzius cells settled near the pia.

The *cortical plate* contains settling neurons, mainly those in future layer VI and some in layer VII that have not yet migrated downward into the subplate.

The subplate contains neurons that originally resided in the *cortical plate.*

*Stratified transitional field (STF) 1* contains *migrating* neurons, mainly those that will settle in cortical layer VI. *STF1* is in the early *t1* stage when cells are plentiful among the fibers.

*STF 5* contains *sojourning* neurons, mainly those that will settle in cortical layers V and VI.

The cortical neuroepithelium (NEP) is a pseudostratified layer of neural stem cells. *Arrows* indicate mitotic figures. The majority are at the ventricular surface, where nuclei migrate to divide within the tall and thin columnar cytoplasm. At this stage of development, large numbers of cortical neurons are being generated.

**Pia**

**Layer I**

*Cortical plate*

Layer VII (subplate)

*STF1 t1*

*STF5*

**Cortical NEP**

**Mitotic zone**

Cajal Retzius cells

Pia meninx cells

*Settling neurons*

*Delaminating subplate neurons*

*Migrating neurons*

*Sojourning neurons*

**Blood vessel**

**Proliferating neural stem cells**

**Blood vessel**

Neural stem cells in mitotic division at the border of the TELENCEPHALIC SUPERVENTRICLE

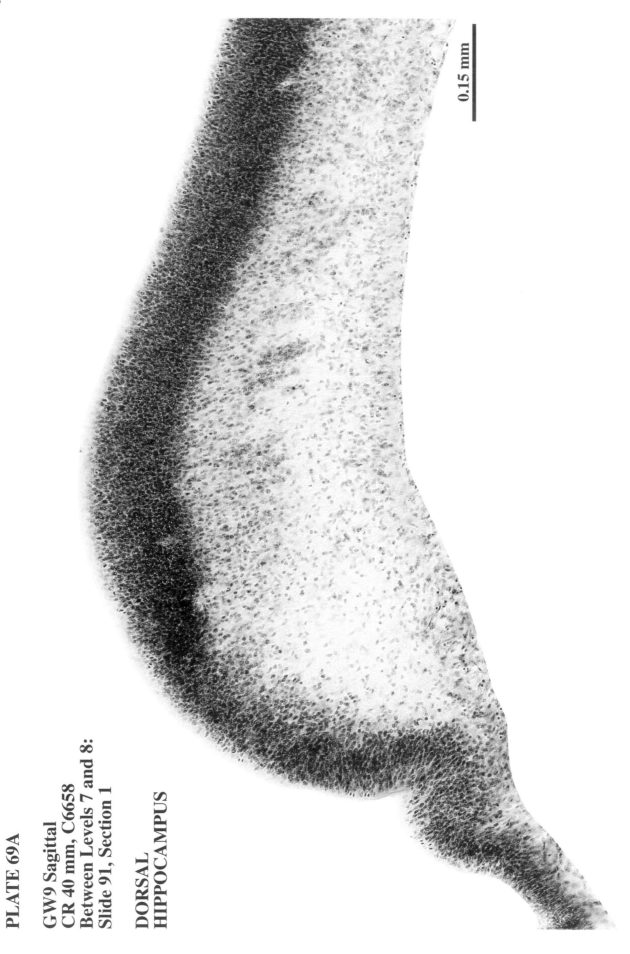

**PLATE 69A**

**GW9 Sagittal**
**CR 40 mm, C6658**
**Between Levels 7 and 8:**
**Slide 91, Section 1**

**DORSAL**
**HIPPOCAMPUS**

0.15 mm

See a low-magnification view of Level 7 in Plates 63A-D, Level 8 in Plates 64A-D.

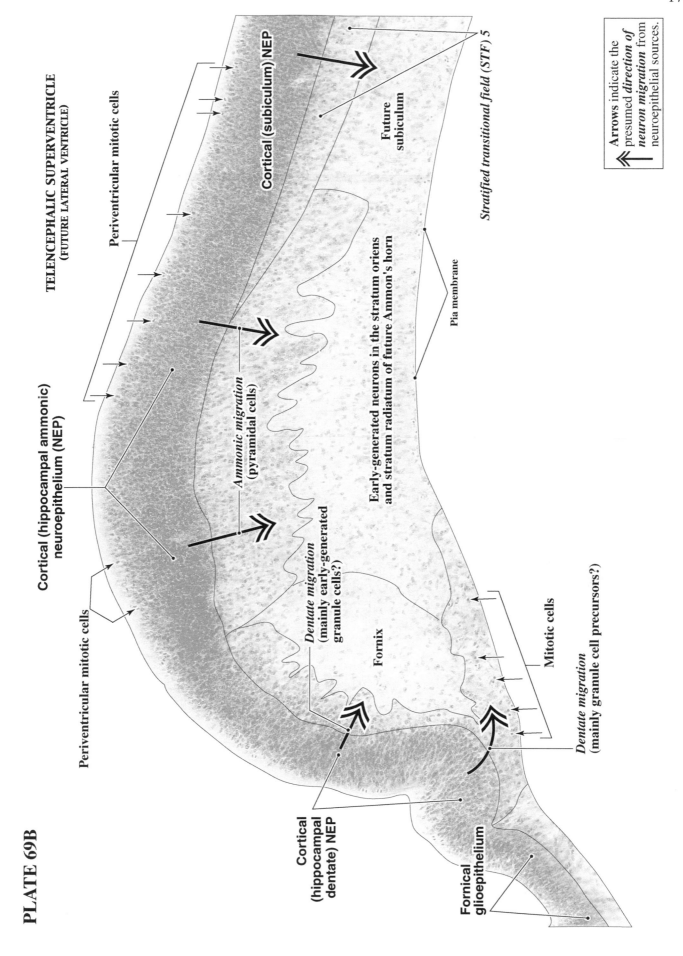

PLATE 69B

179

TELENCEPHALIC SUPERVENTRICLE
(FUTURE LATERAL VENTRICLE)

Periventricular mitotic cells

Cortical (subiculum) NEP

Stratified transitional field (STF) 5

Future subiculum

Pia membrane

Early-generated neurons in the stratum oriens and stratum radiatum of future Ammon's horn

Ammonic migration (pyramidal cells)

Cortical (hippocampal ammonic) neuroepithelium (NEP)

Periventricular mitotic cells

Dentate migration (mainly early-generated granule cells?)

Fornix

Mitotic cells

Dentate migration (mainly granule cell precursors?)

Cortical (hippocampal dentate) NEP

Fornical glioepithelium

Arrows indicate the presumed direction of neuron migration from neuroepithelial sources.

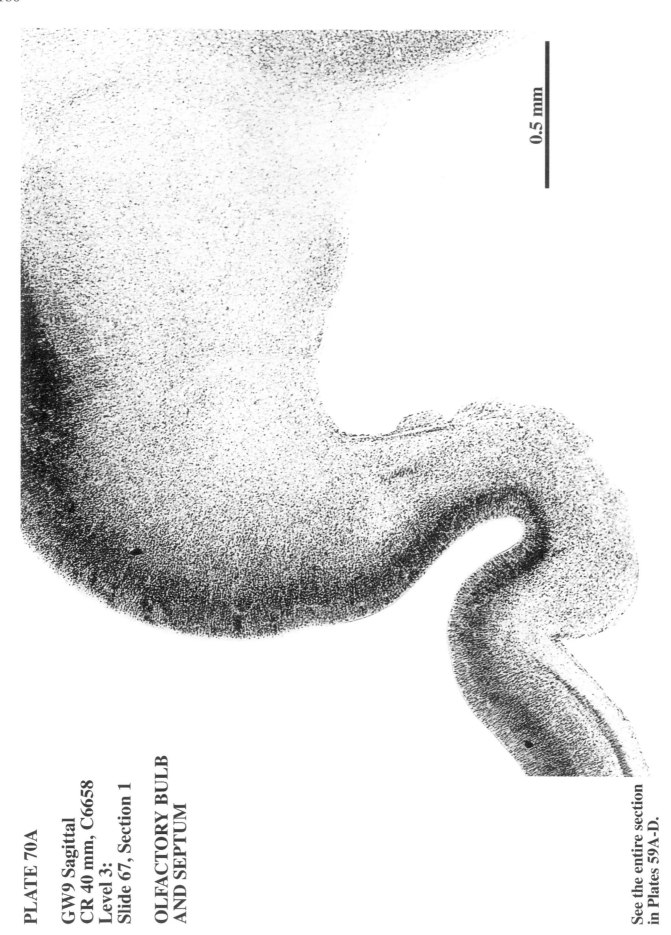

0.5 mm

PLATE 70A

GW9 Sagittal
CR 40 mm, C6658
Level 3:
Slide 67, Section 1

OLFACTORY BULB
AND SEPTUM

See the entire section
in Plates 59A-D.

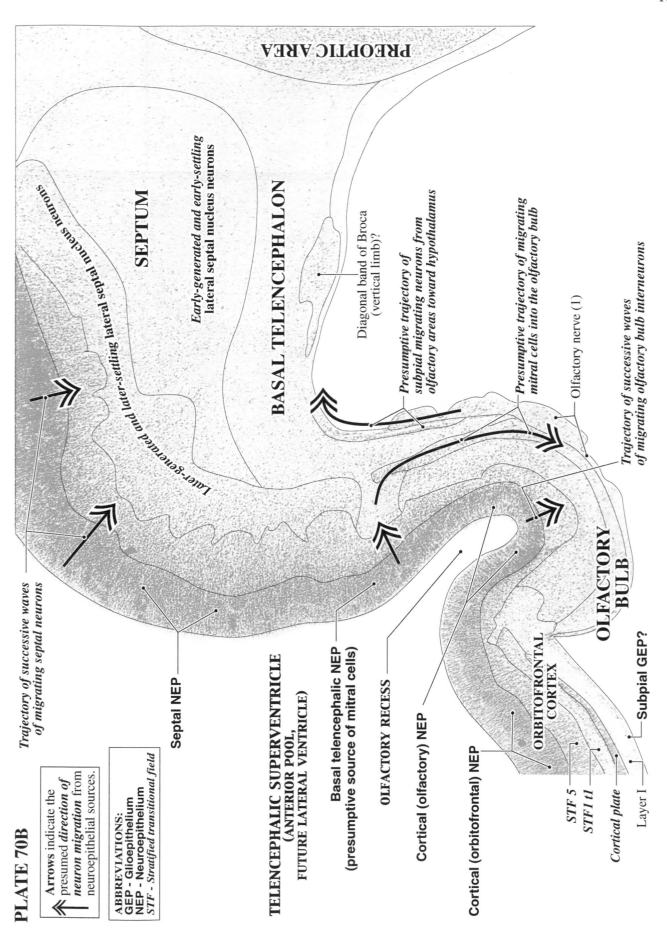

PLATE 70B

Arrows indicate the presumed *direction* of *neuron migration* from neuroepithelial sources.

ABBREVIATIONS:
GEP - Glioepithelium
NEP - Neuroepithelium
*STF - Stratified transitional field*

PREOPTIC AREA

SEPTUM

*Early-generated and early-settling lateral septal nucleus neurons*

BASAL TELENCEPHALON

Diagonal band of Broca (vertical limb)?

*Presumptive trajectory of subpial migrating neurons from olfactory areas toward hypothalamus*

*Presumptive trajectory of migrating mitral cells into the olfactory bulb*

Olfactory nerve (1)

*Trajectory of successive waves of migrating olfactory bulb interneurons*

*Later-generated and later-settling lateral septal nucleus neurons*

*Trajectory of successive waves of migrating septal neurons*

Septal NEP

TELENCEPHALIC SUPERVENTRICLE
(ANTERIOR POOL,
FUTURE LATERAL VENTRICLE)

Basal telencephalic NEP
(presumptive source of mitral cells)

OLFACTORY RECESS

Cortical (olfactory) NEP

Cortical (orbitofrontal) NEP

ORBITOFRONTAL CORTEX

OLFACTORY BULB

Subpial GEP?

*STF 5*

*STF1 t1*

*Cortical plate*

Layer 1

**PLATE 71A**

GW9 Sagittal
CR 40 mm, C6658
Level 3: Slide 67, Section 1

PREOPTIC AREA AND
ANTERIOR HYPOTHALAMUS

0.5 mm

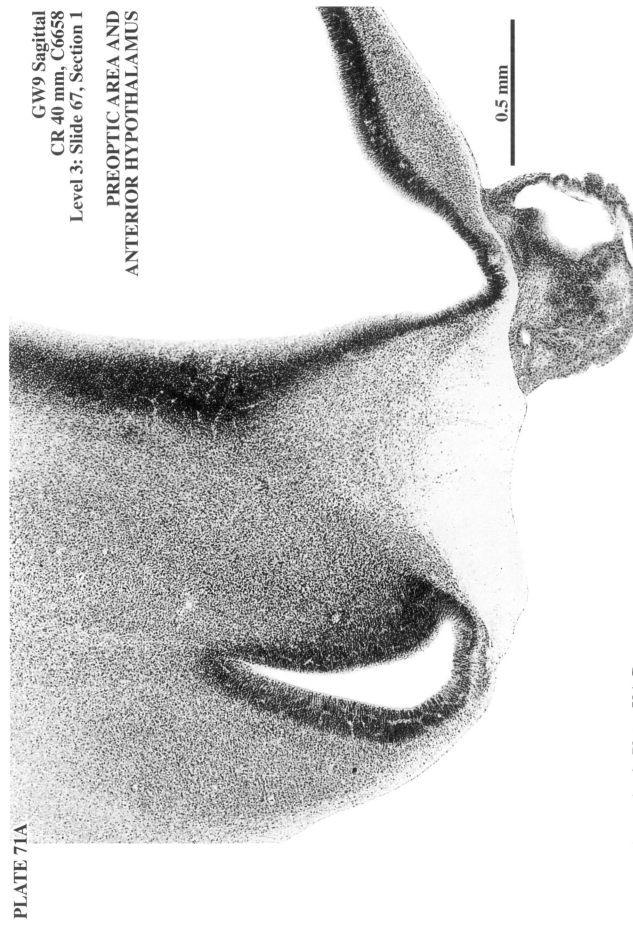

See the entire section in Plates 59A-D.

**PLATE 71B**

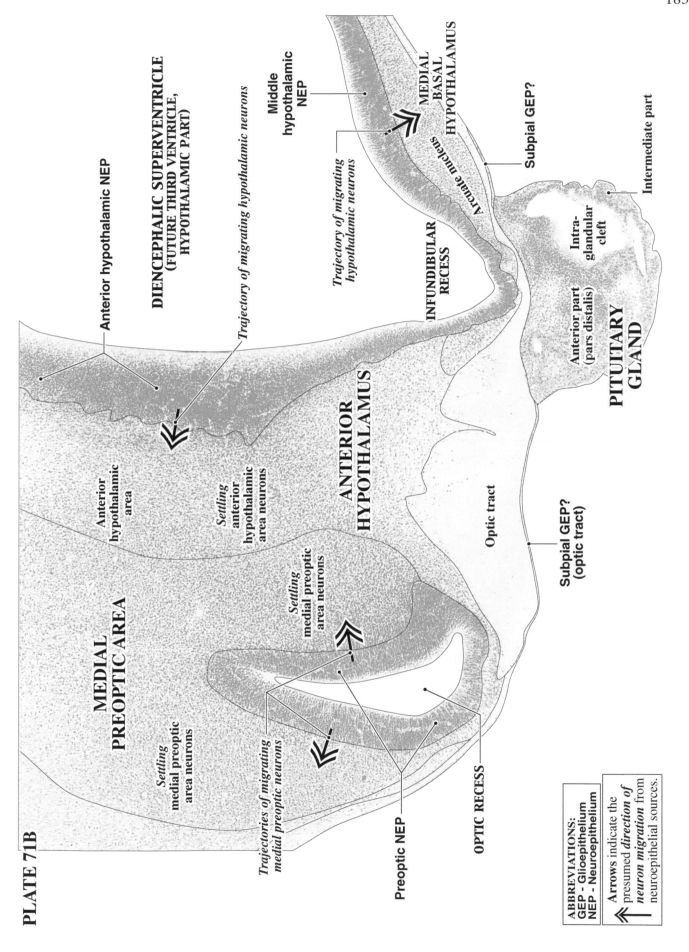

Anterior hypothalamic NEP

**DIENCEPHALIC SUPERVENTRICLE**
**(FUTURE THIRD VENTRICLE,**
**HYPOTHALAMIC PART)**

*Trajectory of migrating hypothalamic neurons*

Middle hypothalamic NEP

*Trajectory of migrating hypothalamic neurons*

Arcuate nucleus

**MEDIAL BASAL HYPOTHALAMUS**

**INFUNDIBULAR RECESS**

Subpial GEP?

Intermediate part

Intra-glandular cleft

Anterior part (pars distalis)

**PITUITARY GLAND**

Anterior hypothalamic area

*Settling anterior hypothalamic area neurons*

**ANTERIOR HYPOTHALAMUS**

Optic tract

**MEDIAL PREOPTIC AREA**

*Settling medial preoptic area neurons*

*Settling medial preoptic area neurons*

Subpial GEP? (optic tract)

*Trajectories of migrating medial preoptic neurons*

Preoptic NEP

**OPTIC RECESS**

**ABBREVIATIONS:**
GEP - Glioepithelium
NEP - Neuroepithelium

Arrows indicate the presumed *direction of neuron migration* from neuroepithelial sources.

**PLATE 72A**

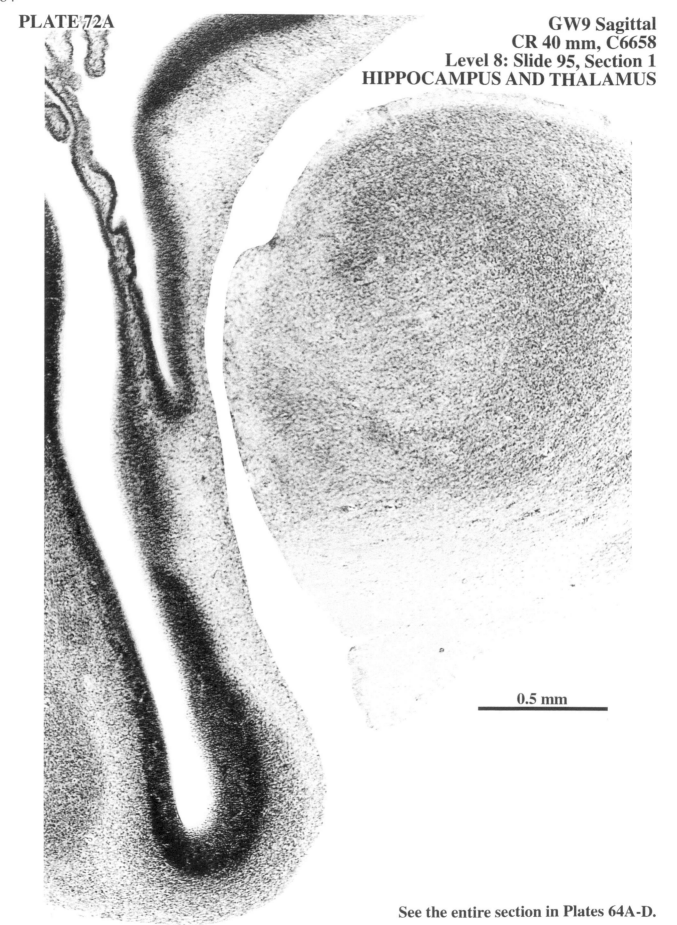

0.5 mm

See the entire section in Plates 64A-D.

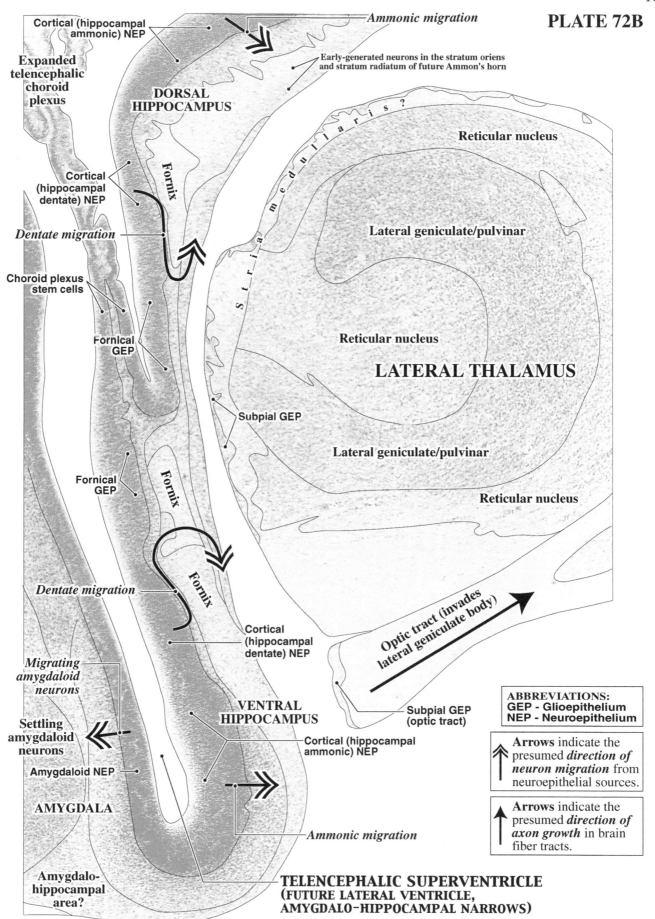

Cortical (hippocampal ammonic) NEP

*Ammonic migration*

Early-generated neurons in the stratum oriens and stratum radiatum of future Ammon's horn

**Expanded telencephalic choroid plexus**

**DORSAL HIPPOCAMPUS**

**Reticular nucleus**

Fornix

Cortical (hippocampal dentate) NEP

*Dentate migration*

**Lateral geniculate/pulvinar**

Choroid plexus stem cells

Fornical GEP

**Reticular nucleus**

**LATERAL THALAMUS**

Subpial GEP

**Lateral geniculate/pulvinar**

Fornical GEP

Fornix

**Reticular nucleus**

Fornix

*Dentate migration*

*Optic tract (invades lateral geniculate body)*

Cortical (hippocampal dentate) NEP

*Migrating amygdaloid neurons*

**Settling amygdaloid neurons**

Subpial GEP (optic tract)

**VENTRAL HIPPOCAMPUS**

Amygdaloid NEP

Cortical (hippocampal ammonic) NEP

**ABBREVIATIONS:**
**GEP** - Glioepithelium
**NEP** - Neuroepithelium

**AMYGDALA**

*Ammonic migration*

**Arrows** indicate the presumed *direction of neuron migration* from neuroepithelial sources.

**Amygdalo-hippocampal area?**

**Arrows** indicate the presumed *direction of axon growth* in brain fiber tracts.

**TELENCEPHALIC SUPERVENTRICLE**
**(FUTURE LATERAL VENTRICLE,**
**AMYGDALO-HIPPOCAMPAL NARROWS)**

*Stria medullaris ?*

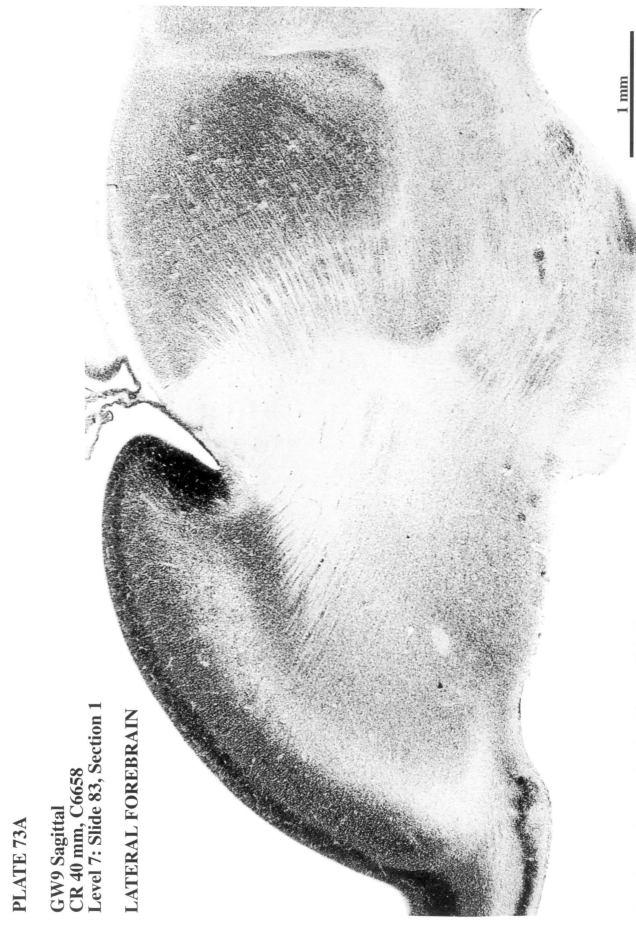

PLATE 73A

GW9 Sagittal
CR 40 mm, C6658
Level 7: Slide 83, Section 1

LATERAL FOREBRAIN

See the entire section in Plates 63A-D.

1 mm

# PLATE 73B

## TELENCEPHALIC SUPERVENTRICLE
### (FUTURE LATERAL VENTRICLE)

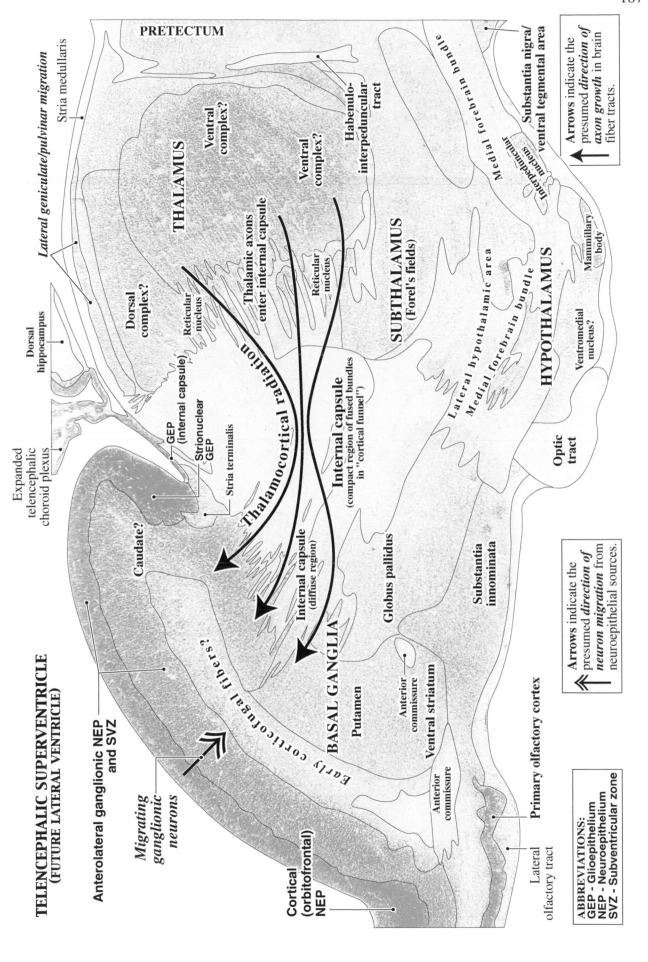

PRETECTUM

*Lateral geniculate/pulvinar migration*

Stria medullaris

Dorsal hippocampus

Expanded telencephalic choroid plexus

Anterolateral ganglionic NEP and SVZ

*Migrating ganglionic neurons*

Cortical (orbitofrontal) NEP

*Early corticofugal fibers?*

Caudate?

GEP (internal capsule)

Strionuclear GEP

Stria terminalis

Dorsal complex?

Reticular nucleus

THALAMUS

Ventral complex?

Thalamic axons enter internal capsule

Ventral complex?

Reticular nucleus

Habenulo-interpeduncular tract

*Thalamocortical radiation*

Internal capsule (diffuse region)

Internal capsule (compact region of fused bundles in "cortical funnel")

BASAL GANGLIA

Putamen

Globus pallidus

Anterior commissure

Ventral striatum

Anterior commissure

Substantia innominata

Primary olfactory cortex

Lateral olfactory tract

SUBTHALAMUS (Forel's fields)

Lateral hypothalamic area

Medial forebrain bundle

HYPOTHALAMUS

Ventromedial nucleus?

Optic tract

Medial forebrain bundle

Interpeduncular nucleus

Mammillary body

Substantia nigra/ventral tegmental area

Arrows indicate the presumed *direction of axon growth* in brain fiber tracts.

Arrows indicate the presumed *direction of neuron migration* from neuroepithelial sources.

ABBREVIATIONS:
GEP - Glioepithelium
NEP - Neuroepithelium
SVZ - Subventricular zone

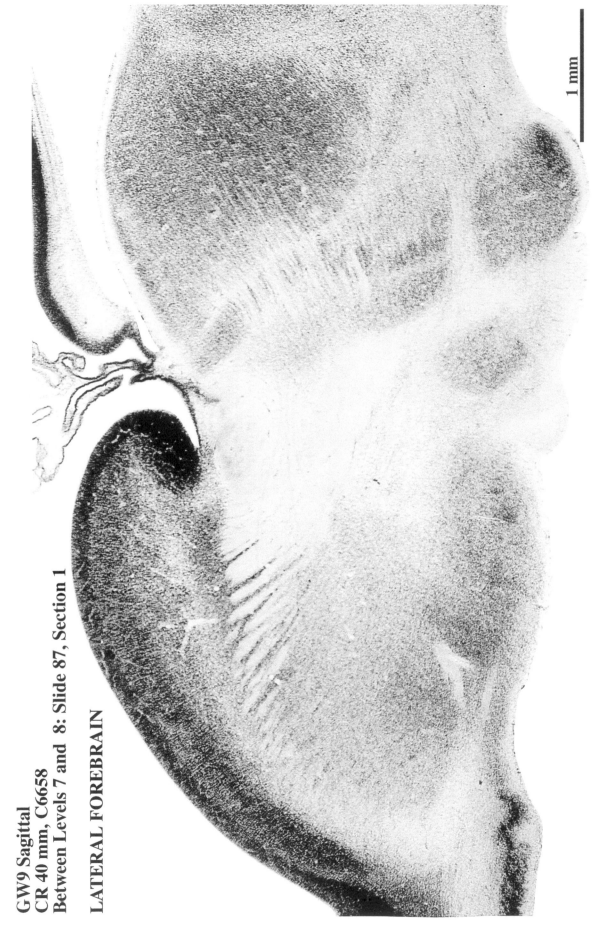

PLATE 74A

GW9 Sagittal
CR 40 mm, C6658
Between Levels 7 and 8: Slide 87, Section 1

LATERAL FOREBRAIN

See a low-magnification view of Level 7 in Plates 63A-D, Level 8 in Plates 64A-D.

1 mm

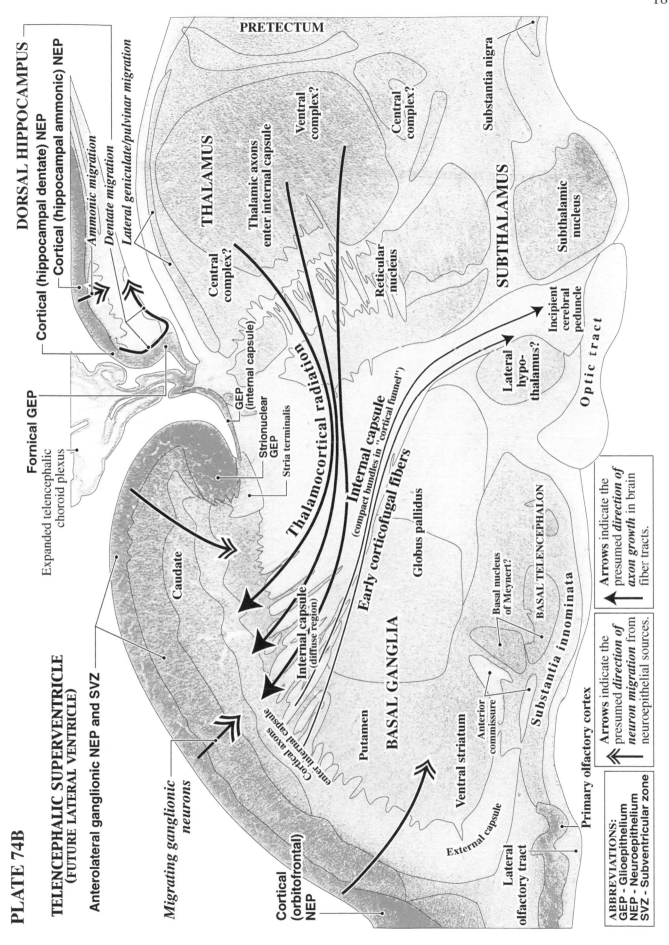

PLATE 74B

189

DORSAL HIPPOCAMPUS

TELENCEPHALIC SUPERVENTRICLE
(FUTURE LATERAL VENTRICLE)

Anterolateral ganglionic NEP and SVZ

Cortical (hippocampal dentate) NEP
Cortical (hippocampal ammonic) NEP

*Ammonic migration*
*Dentate migration*
*Lateral geniculate/pulvinar migration*

**PRETECTUM**

Substantia nigra

**THALAMUS**

Ventral
complex?

Central
complex?

Central
complex?

Thalamic axons
enter internal capsule

Reticular
nucleus

**SUBTHALAMUS**

Subthalamic
nucleus

Fornical GEP

Expanded telencephalic
choroid plexus

GEP
(internal capsule)

Strionuclear
GEP

Stria terminalis

**Thalamocortical radiation**

*Internal capsule*
(compact bundles in "cortical funnel")

*Early corticofugal fibers*

Lateral
hypo-
thalamus?

Incipient
cerebral
peduncle

Optic tract

Caudate

*Internal capsule*
(diffuse region)

Globus pallidus

*Cortical axons
enter internal capsule*

Putamen

**BASAL GANGLIA**

Ventral striatum

Anterior
commissure

Basal nucleus
of Meynert?

**BASAL TELENCEPHALON**

*Substantia innominata*

Primary olfactory cortex

*Migrating ganglionic
neurons*

External capsule

Cortical
(orbitofrontal)
NEP

Lateral
olfactory tract

Arrows indicate the
presumed *direction of
axon growth* in brain
fiber tracts.

Arrows indicate the
presumed *direction of
neuron migration* from
neuroepithelial sources.

ABBREVIATIONS:
GEP - Glioepithelium
NEP - Neuroepithelium
SVZ - Subventricular zone

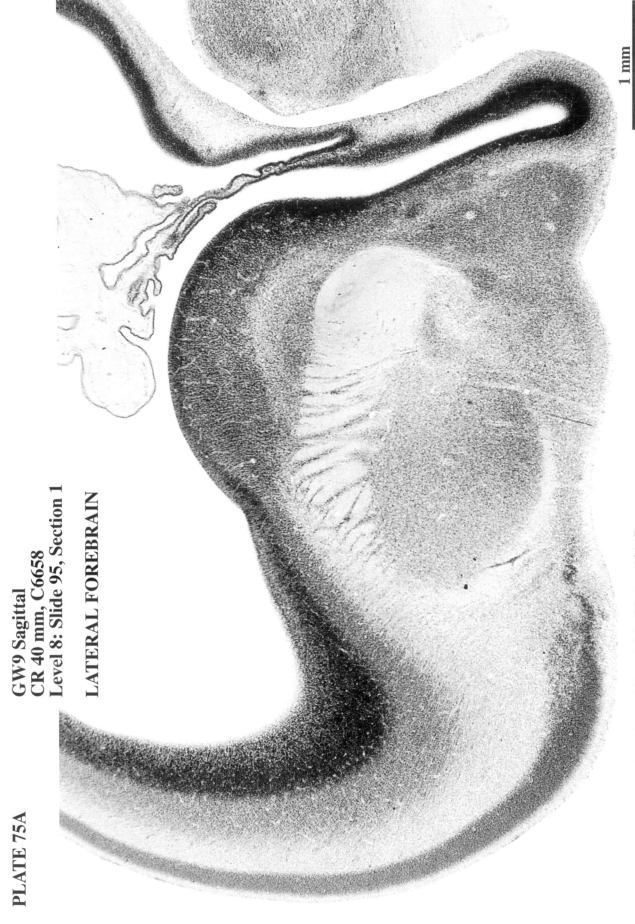

PLATE 75A

GW9 Sagittal
CR 40 mm, C6658
Level 8: Slide 95, Section 1

LATERAL FOREBRAIN

1 mm

See a low-magnification view of Level 8 in Plates 64A-D.

# PLATE 75B

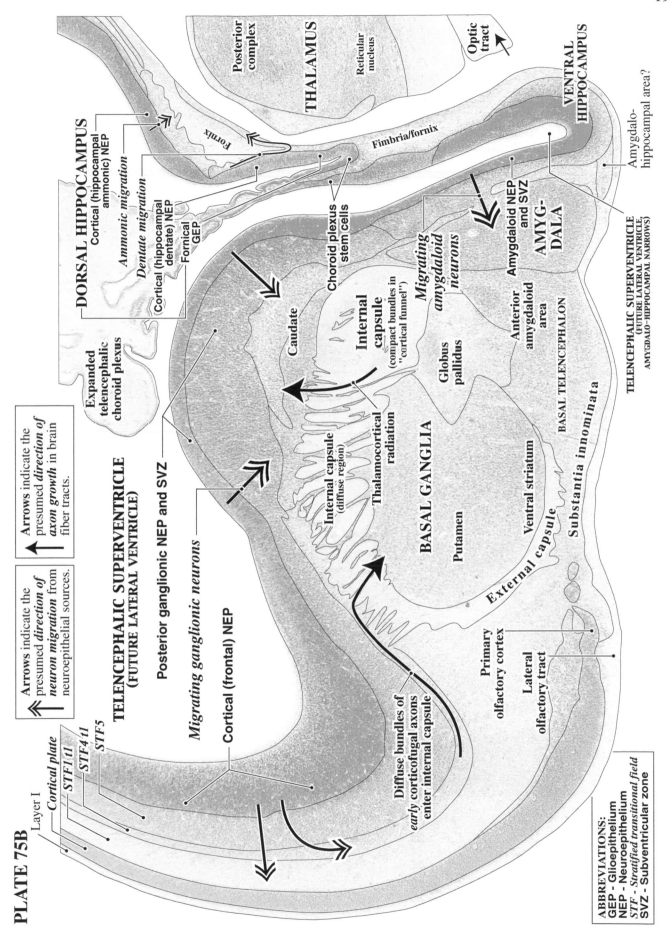

**Arrows** indicate the presumed *direction of neuron migration* from neuroepithelial sources.

**Arrows** indicate the presumed *direction of axon growth* in brain fiber tracts.

Layer I
Cortical plate
STF1-t1
STF4-t1
STF5

**TELENCEPHALIC SUPERVENTRICLE**
(FUTURE LATERAL VENTRICLE)

Posterior ganglionic NEP and SVZ

*Migrating ganglionic neurons*

Cortical (frontal) NEP

Expanded telencephalic choroid plexus

**DORSAL HIPPOCAMPUS**

Cortical (hippocampal ammonic) NEP

*Ammonic migration*

*Dentate migration*

Cortical (hippocampal dentate) NEP

Fornical GEP

Choroid plexus stem cells

Caudate

Internal capsule
(compact bundles in "cortical funnel")

*Migrating amygdaloid neurons*

Amygdaloid NEP and SVZ

**AMYG-DALA**

Anterior amygdaloid area

Globus pallidus

Internal capsule
(diffuse region)

Thalamocortical radiation

**BASAL GANGLIA**

Putamen

Ventral striatum

Substantia innominata

**BASAL TELENCEPHALON**

External capsule

Primary olfactory cortex

Lateral olfactory tract

Diffuse bundles of early corticofugal axons enter internal capsule

Fornix

Fimbria/fornix

Posterior complex

**THALAMUS**

Reticular nucleus

Optic tract

**VENTRAL HIPPOCAMPUS**

Amygdalo-hippocampal area?

**TELENCEPHALIC SUPERVENTRICLE**
(FUTURE LATERAL VENTRICLE, AMYGDALO-HIPPOCAMPAL NARROWS)

**ABBREVIATIONS:**
GEP - Glioepithelium
NEP - Neuroepithelium
STF - Stratified transitional field
SVZ - Subventricular zone

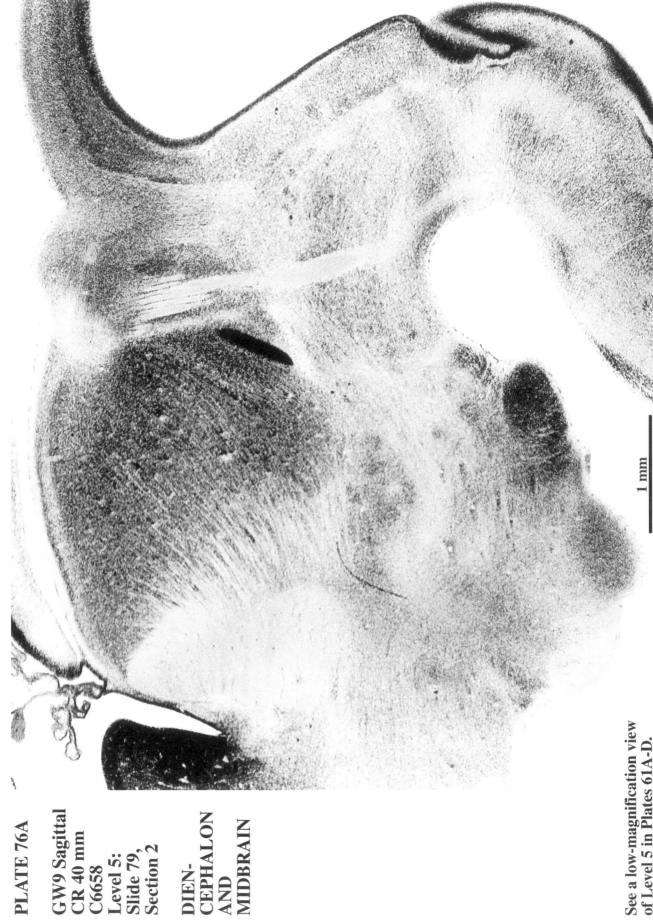

**PLATE 76A**

**GW9 Sagittal
CR 40 mm
C6658
Level 5:
Slide 79,
Section 2**

**DIEN-
CEPHALON
AND
MIDBRAIN**

1 mm

**See a low-magnification view
of Level 5 in Plates 61A-D.**

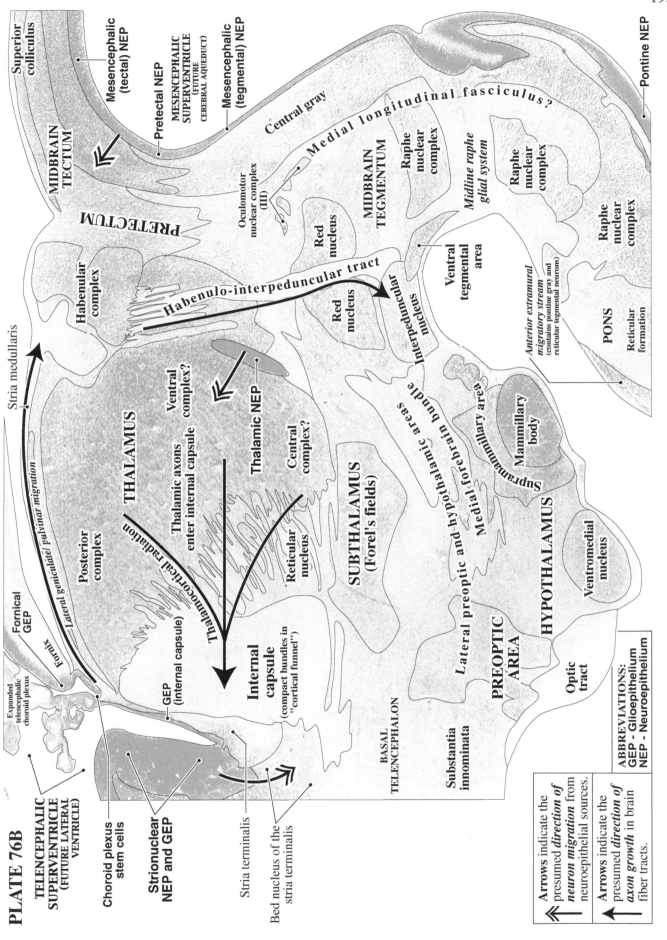

PLATE 76B

TELENCEPHALIC
SUPERVENTRICLE
(FUTURE LATERAL
VENTRICLE)

Choroid plexus stem cells

Strionuclear
NEP and GEP

Expanded telencephalic choroid plexus

Fornical GEP

Fornix

Lateral geniculate/ pulvinar migration

Stria medullaris

Posterior complex

GEP (internal capsule)

THALAMUS

Thalamic axons enter internal capsule

Thalamocortical radiation

Thalamic radiation (internal capsule)

Internal capsule (compact bundles in "cortical funnel")

Ventral complex?

Thalamic NEP

Central complex?

Reticular nucleus

Stria terminalis

Bed nucleus of the stria terminalis

BASAL TELENCEPHALON

Substantia innominata

PREOPTIC AREA

Lateral preoptic and hypothalamic areas

Optic tract

HYPOTHALAMUS

Ventromedial nucleus

SUBTHALAMUS (Forel's fields)

Medial forebrain bundle

Supramammillary area

Mammillary body

Interpeduncular nucleus

Habenulo-interpeduncular tract

Habenular complex

PRETECTUM

MIDBRAIN TECTUM

Superior colliculus

Mesencephalic (tectal) NEP

Pretectal NEP

MESENCEPHALIC SUPERVENTRICLE (FUTURE CEREBRAL AQUEDUCT)

Mesencephalic (tegmental) NEP

Central gray

Oculomotor nuclear complex (III)

Medial longitudinal fasciculus?

Red nucleus

Red nucleus

MIDBRAIN TEGMENTUM

Raphe nuclear complex

Midline raphe glial system

Ventral tegmental area

Anterior extramural migratory stream (contains pontine gray and reticular tegmental neurons)

Raphe nuclear complex

Raphe nuclear complex

PONS

Reticular formation

Pontine NEP

Arrows indicate the presumed *direction of neuron migration* from neuroepithelial sources.

Arrows indicate the presumed *direction of axon growth* in brain fiber tracts.

ABBREVIATIONS:
GEP - Glioepithelium
NEP - Neuroepithelium

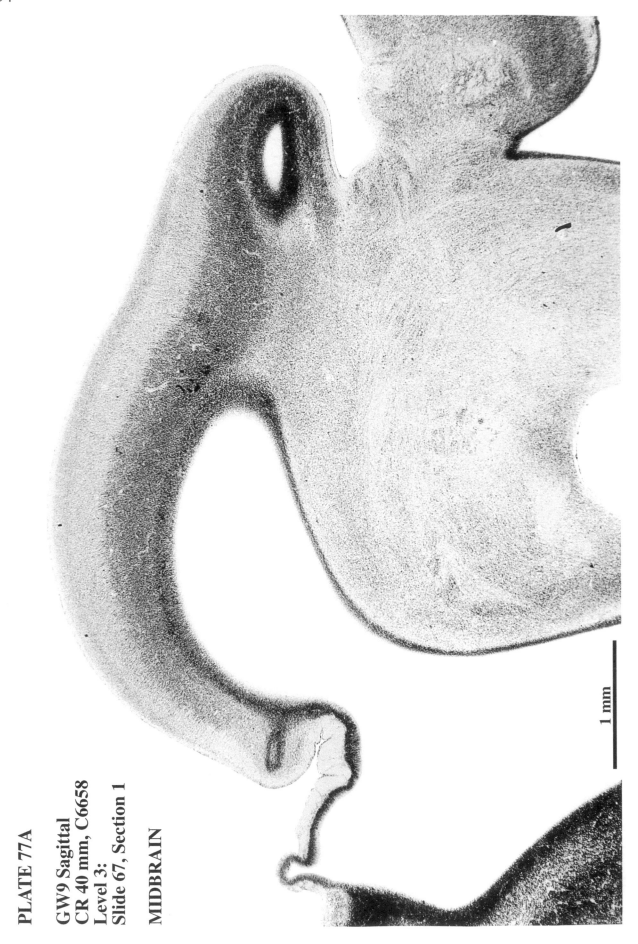

PLATE 77A

GW9 Sagittal
CR 40 mm, C6658
Level 3:
Slide 67, Section 1

MIDBRAIN

1 mm

See a low-magnification view of Level 3 in Plates 59A-D.

**PLATE 77B**

ABBREVIATIONS:
GEP - Glioepithelium
NEP - Neuroepithelium

Arrows indicate the presumed *direction of neuron migration* from neuroepithelial sources.

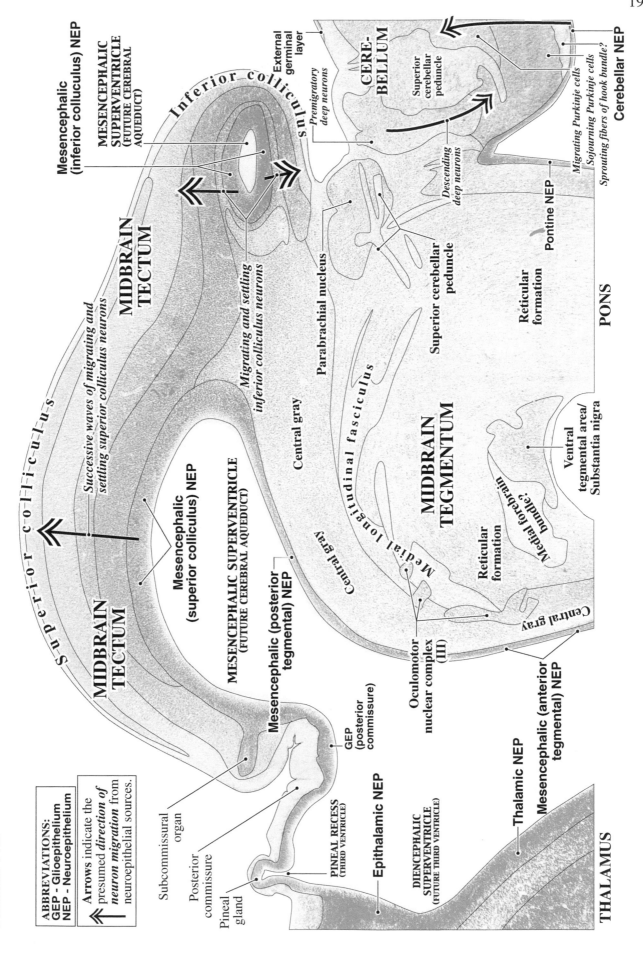

Mesencephalic (inferior colliculus) NEP

MESENCEPHALIC SUPERVENTRICLE (FUTURE CEREBRAL AQUEDUCT)

*Inferior colliculus*

External germinal layer

*Premigratory deep neurons*

CERE-BELLUM

Superior cerebellar peduncle

Cerebellar NEP

*Migrating Purkinje cells*
*Sojourning Purkinje cells*
*Sprouting fibers of hook bundle?*

**MIDBRAIN TECTUM**

*Successive waves of migrating and settling superior colliculus neurons*

*Superior colliculus*

*Migrating and settling inferior colliculus neurons*

Parabrachial nucleus

*Descending deep neurons*

Pontine NEP

**MIDBRAIN TECTUM**

Mesencephalic (superior colliculus) NEP

MESENCEPHALIC SUPERVENTRICLE (FUTURE CEREBRAL AQUEDUCT)

Mesencephalic (posterior tegmental) NEP

Central gray

*Central gray*

*Medial longitudinal fasciculus*

Reticular formation

**PONS**

**MIDBRAIN TEGMENTUM**

Reticular formation

*Medial forebrain bundle?*

Ventral tegmental area/Substantia nigra

Oculomotor nuclear complex (III)

*Central gray*

GEP (posterior commissure)

Subcommissural organ

Posterior commissure

Pineal gland

PINEAL RECESS (THIRD VENTRICLE)

Epithalamic NEP

DIENCEPHALIC SUPERVENTRICLE (FUTURE THIRD VENTRICLE)

Thalamic NEP

Mesencephalic (anterior tegmental) NEP

**THALAMUS**

196

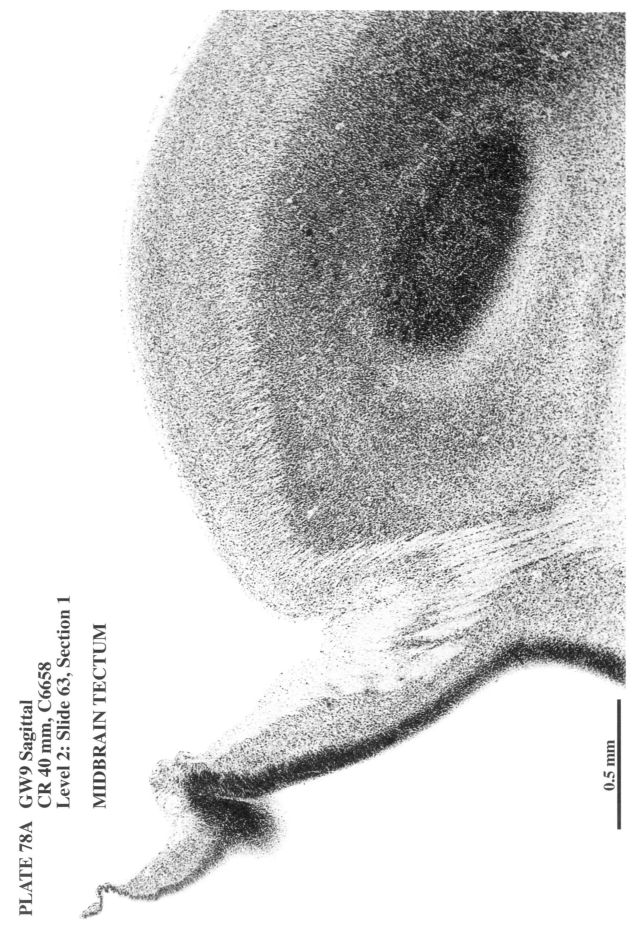

PLATE 78A   GW9 Sagittal
CR 40 mm, C6658
Level 2: Slide 63, Section 1

MIDBRAIN TECTUM

0.5 mm

See a low-magnification view of Level 2 in Plates 58A-D.

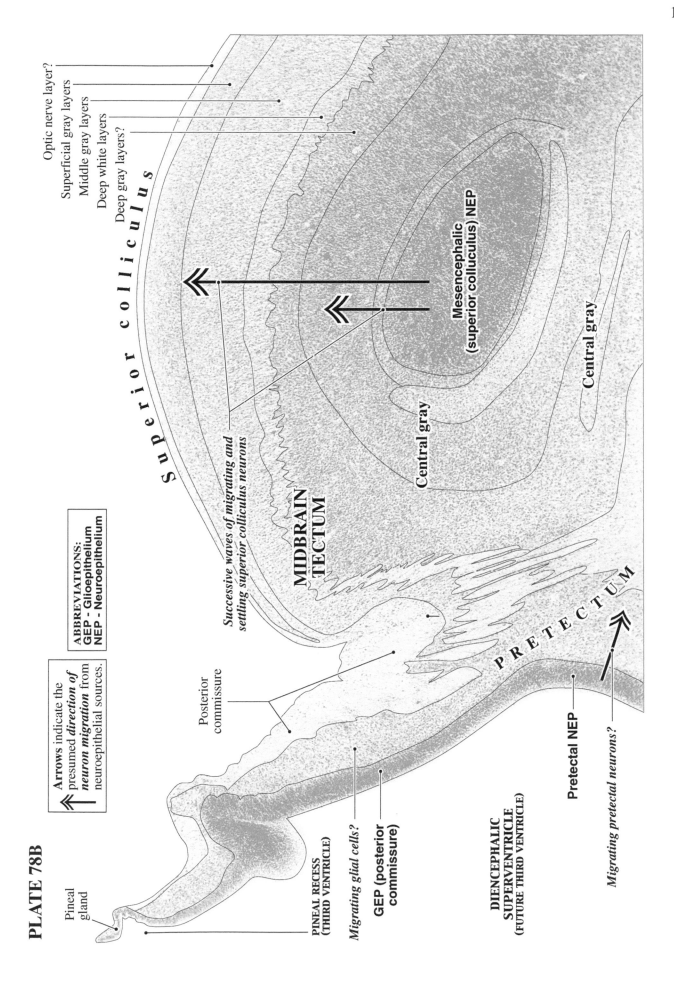

**PLATE 78B**

Arrows indicate the presumed *direction of neuron migration* from neuroepithelial sources.

**ABBREVIATIONS:**
**GEP** - Glioepithelium
**NEP** - Neuroepithelium

S u p e r i o r   c o l l i c u l u s

Optic nerve layer?
Superficial gray layers
Middle gray layers
Deep white layers
Deep gray layers?

*Successive waves of migrating and settling superior colliculus neurons*

**MIDBRAIN TECTUM**

Mesencephalic (superior colliculus) NEP

Central gray

Central gray

**PRETECTUM**

Posterior commissure

Pineal gland

**PINEAL RECESS (THIRD VENTRICLE)**

*Migrating glial cells?*

**GEP (posterior commissure)**

**DIENCEPHALIC SUPERVENTRICLE (FUTURE THIRD VENTRICLE)**

**Pretectal NEP**

*Migrating pretectal neurons?*

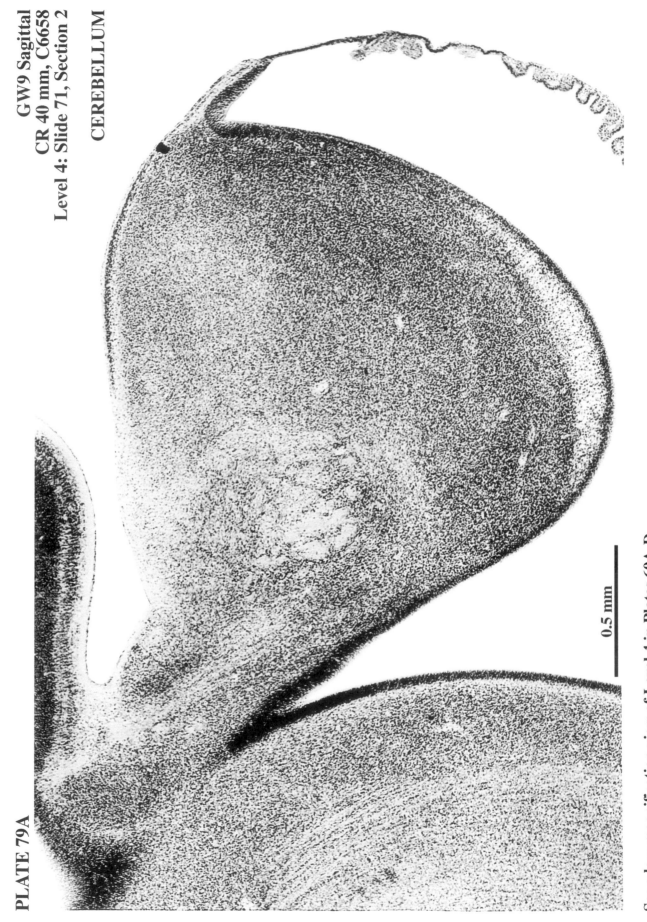

**PLATE 79A**

GW9 Sagittal
CR 40 mm, C6658
Level 4: Slide 71, Section 2
CEREBELLUM

0.5 mm

See a low-magnification view of Level 4 in Plates 60A-D.

198

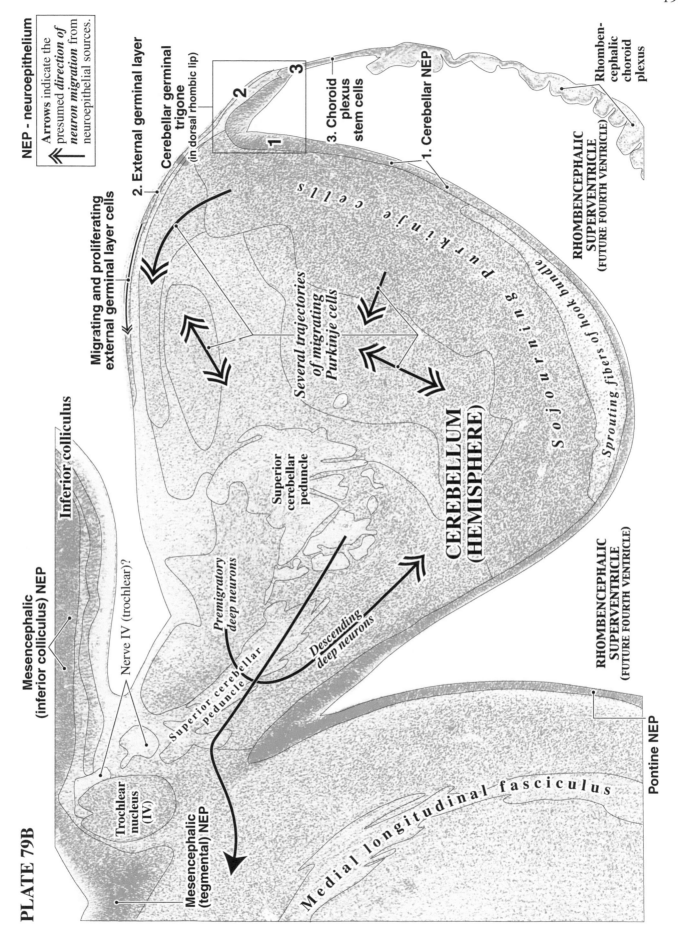

199

PLATE 79B

NEP - neuroepithelium

Arrows indicate the presumed *direction of neuron migration* from neuroepithelial sources.

2. External germinal layer

Cerebellar germinal trigone (in dorsal rhombic lip)

3. Choroid plexus stem cells

1. Cerebellar NEP

Rhomben-cephalic choroid plexus

RHOMBENCEPHALIC SUPERVENTRICLE (FUTURE FOURTH VENTRICLE)

*Sojourning purkinje cells*

*Sprouting fibers of hook bundle*

Migrating and proliferating external germinal layer cells

*Several trajectories of migrating Purkinje cells*

**CEREBELLUM (HEMISPHERE)**

Inferior colliculus

Superior cerebellar peduncle

Mesencephalic (inferior colliculus) NEP

Nerve IV (trochlear)?

*Premigratory deep neurons*

*Descending deep neurons*

RHOMBENCEPHALIC SUPERVENTRICLE (FUTURE FOURTH VENTRICLE)

Trochlear nucleus (IV)

Superior cerebellar peduncle

Mesencephalic (tegmental) NEP

*Medial longitudinal fasciculus*

Pontine NEP

200

PLATE 80A

GW9 Sagittal, CR 40 mm, C6658
Level 5: Slide 75, Section 2

PONS/MEDULLA

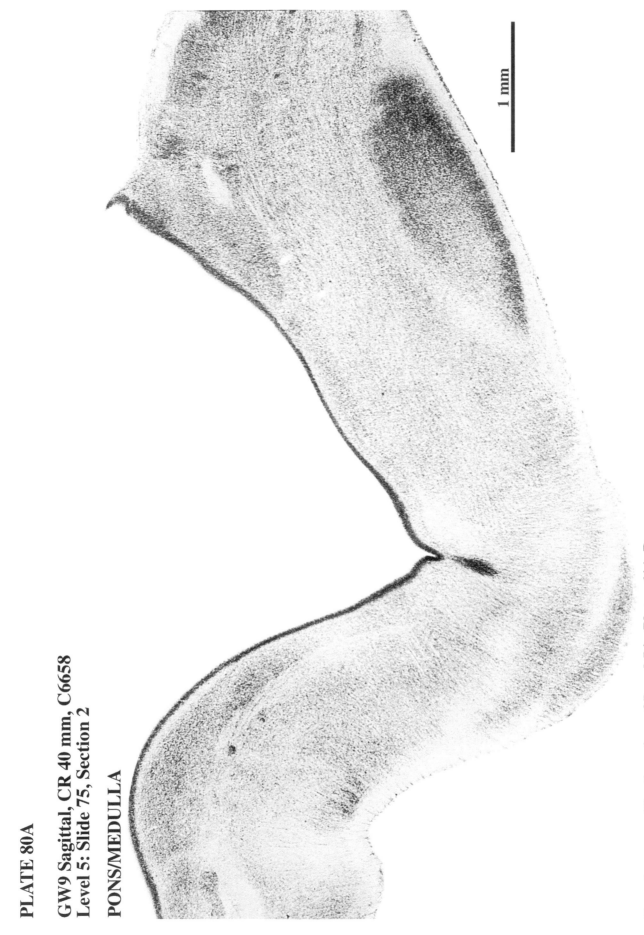

1 mm

See a low-magnification view of Level 5 in Plates 61A-D.

**PLATE 80B**

NEP - neuroepithelium

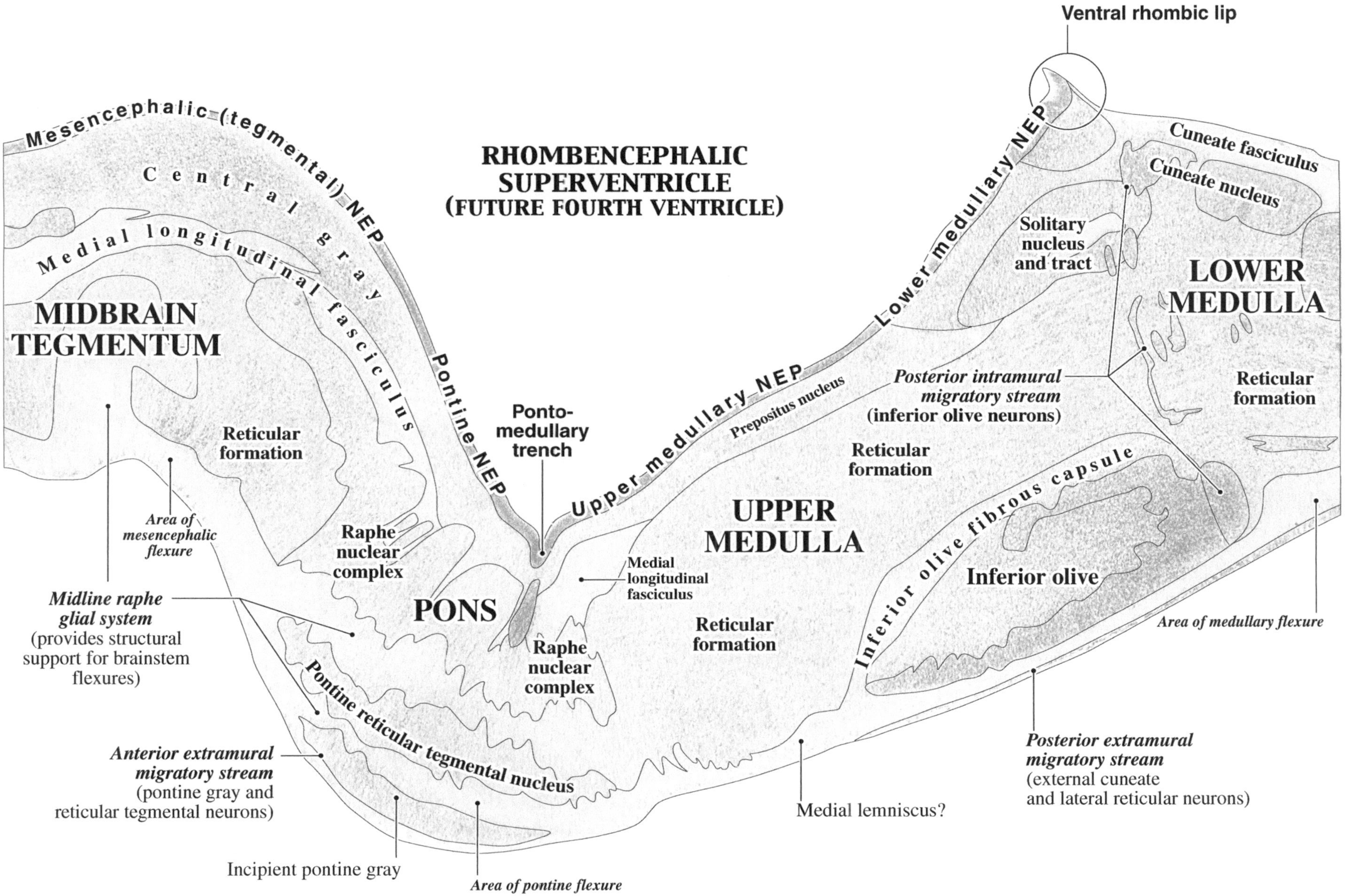

Ventral rhombic lip

**RHOMBENCEPHALIC SUPERVENTRICLE (FUTURE FOURTH VENTRICLE)**

Lower medullary NEP

Upper medullary NEP

Pontine NEP

Mesencephalic (tegmental) NEP

**LOWER MEDULLA**

Cuneate fasciculus

Cuneate nucleus

Reticular formation

Solitary nucleus and tract

Area of medullary flexure

Posterior intramural migratory stream (inferior olive neurons)

Posterior extramural migratory stream (external cuneate and lateral reticular neurons)

Prepositus nucleus

Inferior olive fibrous capsule

Reticular formation

Inferior olive

**UPPER MEDULLA**

Medial lemniscus?

Reticular formation

Medial longitudinal fasciculus

Raphe nuclear complex

Ponto-medullary trench

**PONS**

Raphe nuclear complex

Area of pontine flexure

Pontine reticular tegmental nucleus

Incipient pontine gray

Central gray

Medial longitudinal fasciculus

Reticular formation

**MIDBRAIN TEGMENTUM**

Area of mesencephalic flexure

Midline raphe glial system (provides structural support for brainstem flexures)

Anterior extramural migratory stream (pontine gray and reticular tegmental neurons)

**PLATE 81A**

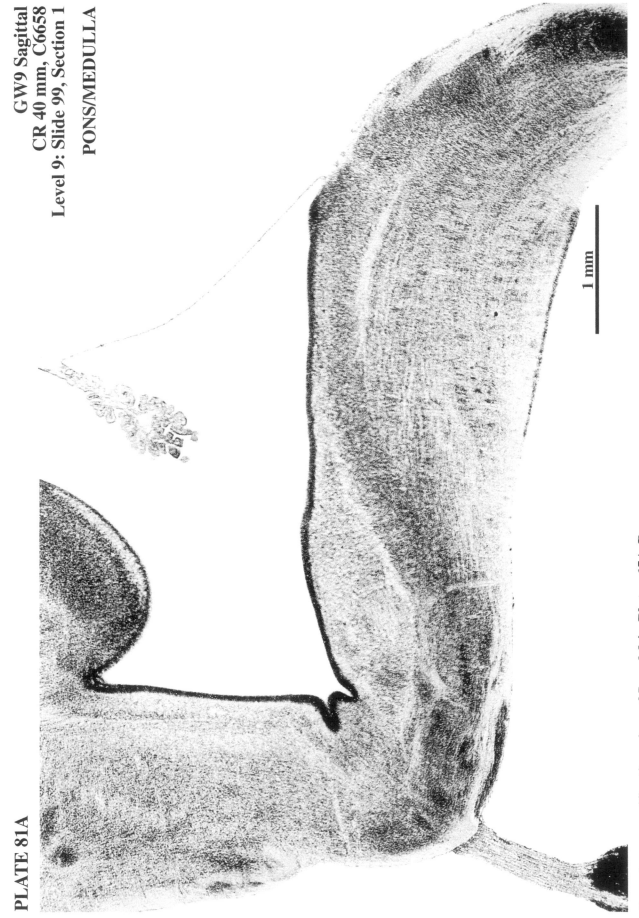

1 mm

See a low-magnification view of Level 9 in Plates 65A-D.

**NEP – neuroepithelium**

**PLATE 81B**

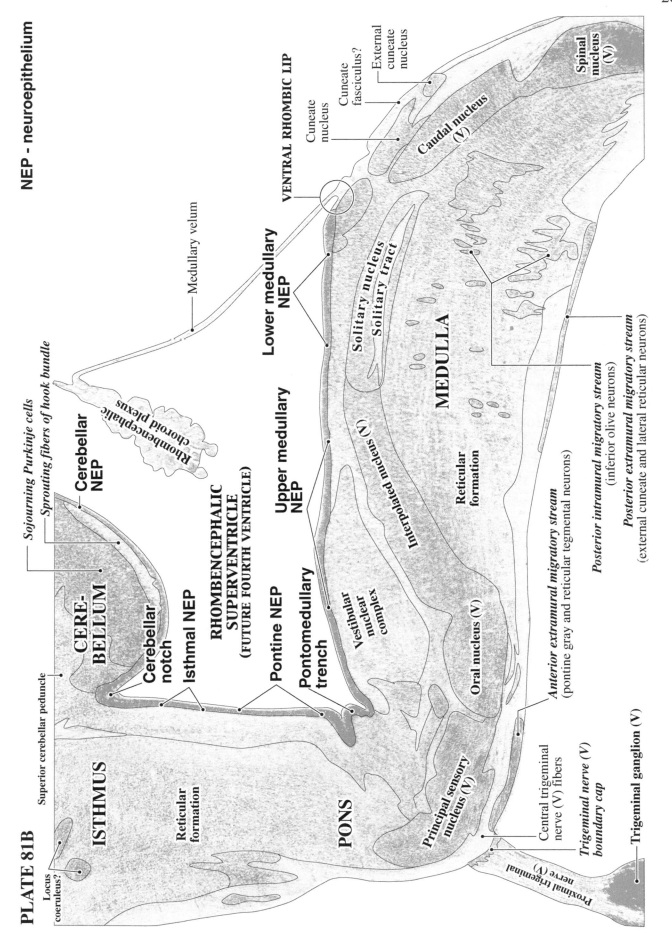

Locus coeruleus?

Superior cerebellar peduncle

Reticular formation

**ISTHMUS**

Sojourning Purkinje cells

Sprouting fibers of hook bundle

**CERE-BELLUM**

**Cerebellar NEP**

Cerebellar notch

Isthmal NEP

Medullary velum

*Rhombencephalic choroid plexus*

**RHOMBENCEPHALIC SUPERVENTRICLE**
**(FUTURE FOURTH VENTRICLE)**

Pontine NEP

Pontomedullary trench

**Upper medullary NEP**

**Lower medullary NEP**

**VENTRAL RHOMBIC LIP**

Cuneate nucleus

Cuneate fasciculus?

External cuneate nucleus

**Caudal nucleus (V)**

**Spinal nucleus (V)**

Solitary nucleus
Solitary tract

Interpolated nucleus (V)

**MEDULLA**

Reticular formation

Vestibular nuclear complex

Oral nucleus (V)

**PONS**

Principal sensory nucleus (V)

Anterior extramural migratory stream
(pontine gray and reticular tegmental neurons)

Posterior intramural migratory stream
(inferior olive neurons)

Posterior extramural migratory stream
(external cuneate and lateral reticular neurons)

Central trigeminal nerve (V) fibers

*Trigeminal nerve (V) boundary cap*

**Trigeminal ganglion (V)**

*Proximal trigeminal nerve (V)*

**PLATE 82A**

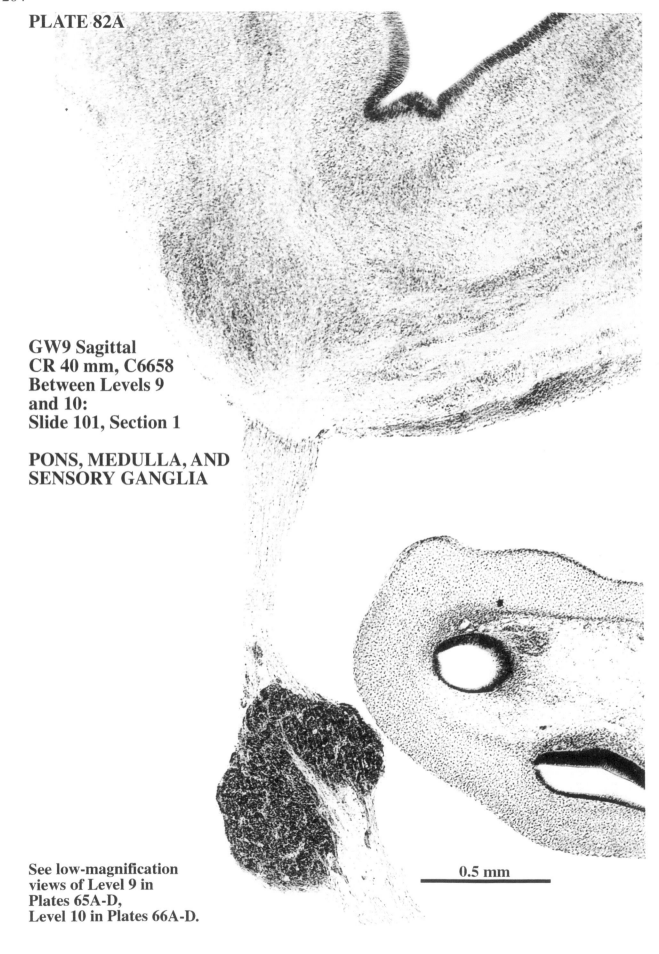

**GW9 Sagittal
CR 40 mm, C6658
Between Levels 9
and 10:
Slide 101, Section 1**

**PONS, MEDULLA, AND
SENSORY GANGLIA**

0.5 mm

**See low-magnification
views of Level 9 in
Plates 65A-D,
Level 10 in Plates 66A-D.**

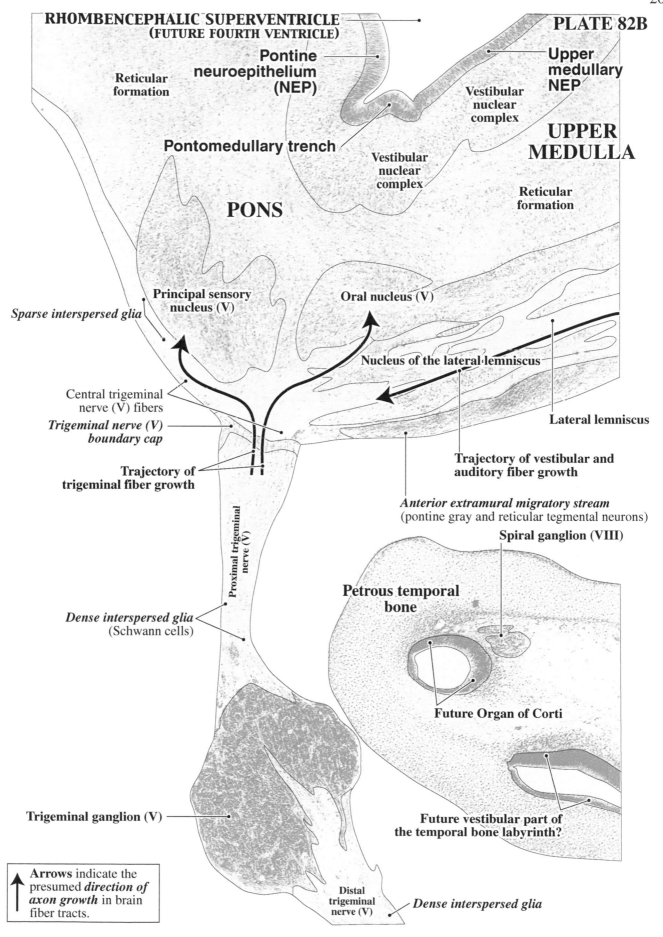

RHOMBENCEPHALIC SUPERVENTRICLE
(FUTURE FOURTH VENTRICLE)

Pontine
neuroepithelium
(NEP)

Upper
medullary
NEP

UPPER
MEDULLA

Reticular
formation

Vestibular
nuclear
complex

Pontomedullary trench

Vestibular
nuclear
complex

PONS

Reticular
formation

*Sparse interspersed glia*

Principal sensory
nucleus (V)

Oral nucleus (V)

Nucleus of the lateral lemniscus

Central trigeminal
nerve (V) fibers

*Trigeminal nerve (V)
boundary cap*

Lateral lemniscus

Trajectory of
trigeminal fiber growth

Trajectory of vestibular and
auditory fiber growth

*Anterior extramural migratory stream*
(pontine gray and reticular tegmental neurons)

Spiral ganglion (VIII)

Proximal trigeminal nerve (V)

*Dense interspersed glia*
(Schwann cells)

Petrous temporal
bone

Future Organ of Corti

Trigeminal ganglion (V)

Future vestibular part of
the temporal bone labyrinth?

**Arrows** indicate the
presumed *direction of
axon growth* in brain
fiber tracts.

Distal
trigeminal
nerve (V)

*Dense interspersed glia*

206

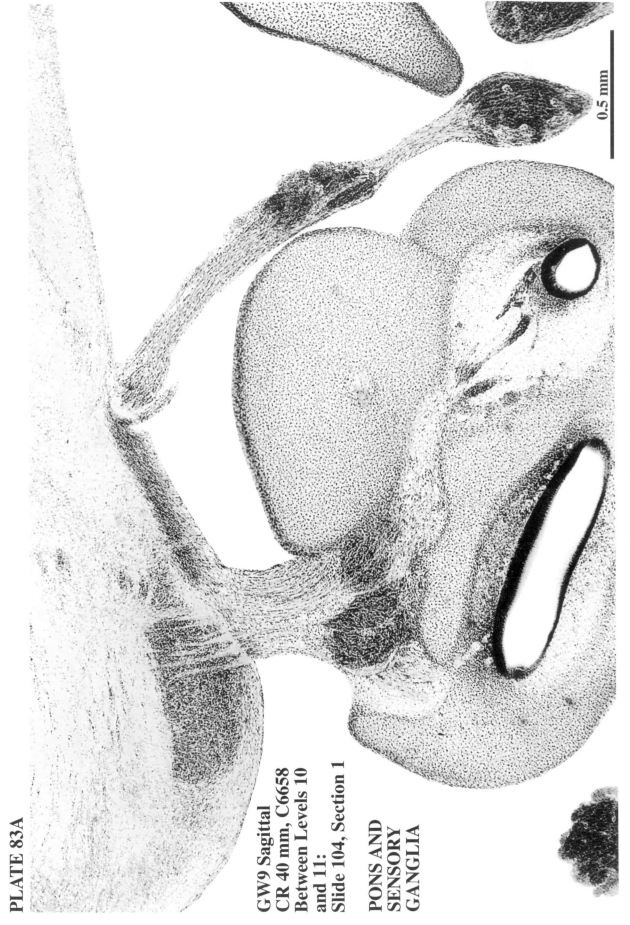

PLATE 83A

GW9 Sagittal
CR 40 mm, C6658
Between Levels 10
and 11:
Slide 104, Section 1

PONS AND
SENSORY
GANGLIA

0.5 mm

See low-magnification views of Level 10 in Plates 66A-D, Level 11 in Plates 67A-D.

PLATE 83B

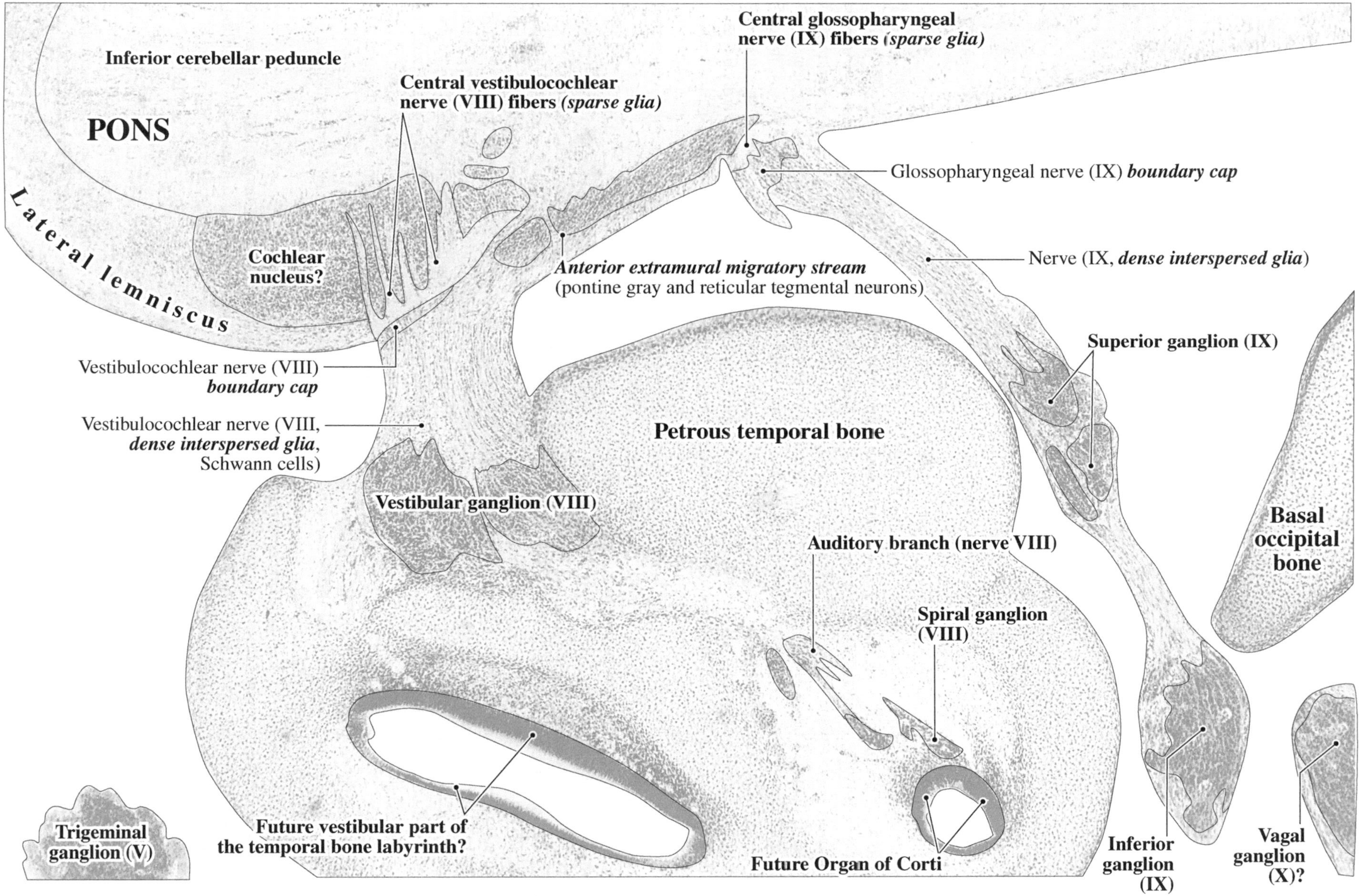

207

# PART V: GW9 HORIZONTAL

This is specimen number 886 in the Carnegie Collection, designated here as C886. A normal fetus with a crown-rump length (CR) of 43 mm was collected in 1914 after a hysterectomy due to pelvic inflammation. The fetus is estimated to be at gestational week (GW) 9. The entire fetus was fixed in Bouin's, embedded in celloidin, and 100 µm sections were cut in a plane midway between the horizontal and coronal planes (**Figure 4**). Because many of the sections do not contain the cerebral cortex and the brainstem is cut in a crosswise direction, it more closely resembles a horizontally sectioned brain. All sections were stained with hematoxylin and eosin. Since there is no photograph of C886's brain before it was embedded and cut, a specimen from Hochstetter (1919) is used to show the external features of the brain at GW9 (**Figure 4**). **Levels 1-11**, generally larger sections containing the cerebral hemispheres, are shown at low magnification in **Plates 84A/B-93A/B**. The core parts of the sections in **Levels 3-11** are also shown at high magnification in **Plates 85C/D-93C/D**. **Levels 12-19**, generally smaller sections containing only the brainstem, are shown at a high magnification in **Plates 94A/B-101A/B**. To more efficiently use page space, all plates are in landscape orientation (anterior/ventral: left side of photograph, bottom of page; posterior/dorsal: right side of photograph, top of page).

C886 is similar to C6658 in the level of brain maturation. The chief reason for including this specimen is to provide a different perspective for viewing brain structure at GW9. In the cerebral cortex, the **neuroepithelium** is prominent as the sole germinal matrix; the **stratified transitional field (STF)** consists of **STF1**, **STF5**, and **STF4** only in lateral areas. The anterolateral (thicker) to dorsomedial (thinner) maturation gradient in the cortical plate and STF layers are evident. In anterolateral parts of the cerebral cortex, streams of neurons and glia appear to leave **STF4** and enter the **lateral migratory stream**. The hippocampus contains **ammonic and dentate migrations**, but there is no evidence of a pyramidal in Ammon's horn or a dentate gyrus. A massive **neuroepithelium/subventricular zone** overlies the amygdala, nucleus accumbens, and stria-

tum (caudate and putamen) where neurons (and glia) are being generated.

The cerebellum is a thick, smooth plate overlying the posterior pons and medulla, and a definite **neuroepithelium** at the ventricular surface, indicating some Purkinje cells are still being generated. Many Purkinje cells are sojourning in a dense layer outside the neuroepithelium, and others are migrating upward. Many of the deep neurons are superficial in the cerebellum, but some are migrating downward to intermingle with upwardly migrating Purkinje cells. The cortical surface is partially covered by an **external germinal layer (egl)** that is actively producing neuronal stem cells, as it grows over the surface of the cerebellar cortex.

The third ventricle, aqueduct, and fourth ventricle are lined by thin **neuroepithelia**. The midbrain tegmentum, pons, and medulla have the thinnest neuroepithelia indicating that only the latest generated neurons are being produced at this time. The thick precerebellar neuroepithelium is an exception in the medulla. Thicker neuroepithelia are in the cerebellum (see above) and midbrain tectum, indicating many neurons are still being generated, although the majority of the neurons in these sites are already postmitotic. The neuroepithelium is still more thick in the hypothalamus and thalamus, in accordance with the later maturation of the diencephalon compared to the rest of the brainstem.

Neurons throughout the diencephalon, midbrain tegmentum, pons, and medulla are migrating and settling. This specimen shows a very prominent migration of subthalamic nucleus neurons from the posterior hypothalamic neuroepithelium. Except for the subthalamic nucleus, nuclear divisions are very indistinct throughout the diencephalon. More definition is seen in the midbrain tegmentum, pons, and medulla. The **anterior extramural** and **posterior extramural migratory streams** are dense subpial accumulations in the medulla and pons.

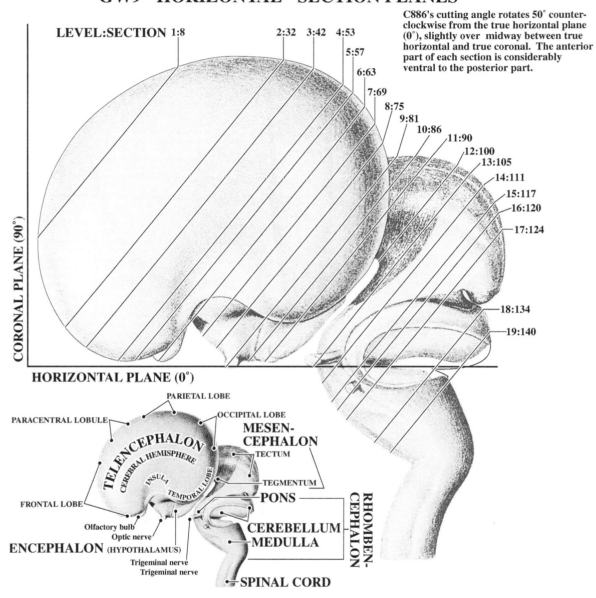

# GW9 "HORIZONTAL" SECTION PLANES

LEVEL:SECTION 1:8     2:32   3:42   4:53

5:57

6:63

7:69

8:75

9:81

10:86

11:90

12:100

13:105

14:111

15:117

16:120

17:124

18:134

19:140

C886's cutting angle rotates 50° counter-clockwise from the true horizontal plane (0°), slightly over midway between true horizontal and true coronal. The anterior part of each section is considerably ventral to the posterior part.

CORONAL PLANE (90°)

HORIZONTAL PLANE (0°)

PARIETAL LOBE

PARACENTRAL LOBULE

OCCIPITAL LOBE

MESEN-CEPHALON

TELENCEPHALON

CEREBRAL HEMISPHERE

TECTUM

INSULA

TEMPORAL LOBE

TEGMENTUM

FRONTAL LOBE

PONS

RHOMBEN-CEPHALON

Olfactory bulb

Optic nerve

CEREBELLUM

MEDULLA

ENCEPHALON (HYPOTHALAMUS)

Trigeminal nerve

Trigeminal nerve

SPINAL CORD

**Figure 4.** The lateral view of the brain and upper cervical spinal cord from a specimen with a crown-rump length of 38 mm (modified from Figure 43, Table VII, Hochstetter, 1919) serves to show the approximate locations and cutting angles of the illustrated sections of C886 in the following pages. The small inset identifies the major structural features. The line in the cerebellum and dorsal edges of the pons and medulla is the cut edge of the medullary velum.

210

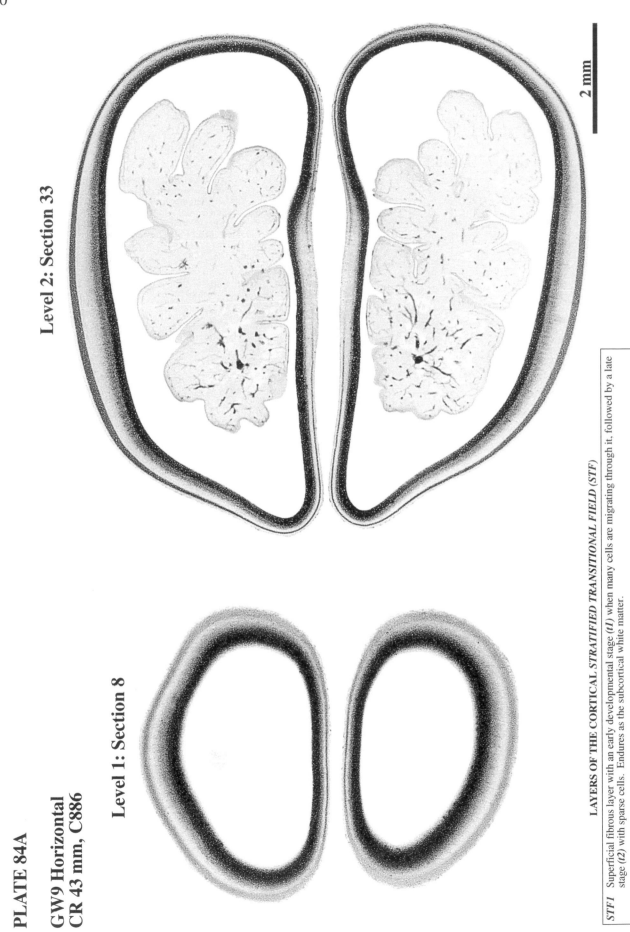

**PLATE 84A**

**GW9 Horizontal**
**CR 43 mm, C886**

**Level 1: Section 8**

**Level 2: Section 33**

2 mm

**LAYERS OF THE CORTICAL *STRATIFIED TRANSITIONAL FIELD (STF)***

*STF1*   Superficial fibrous layer with an early developmental stage (*t1*) when many cells are migrating through it, followed by a late stage (*t2*) with sparse cells. Endures as the subcortical white matter.

*STF5*   Deep cellular layer that is prominent during the first trimester, the first sojourn zone to appear outside the germinal matrix.

PLATE 84B

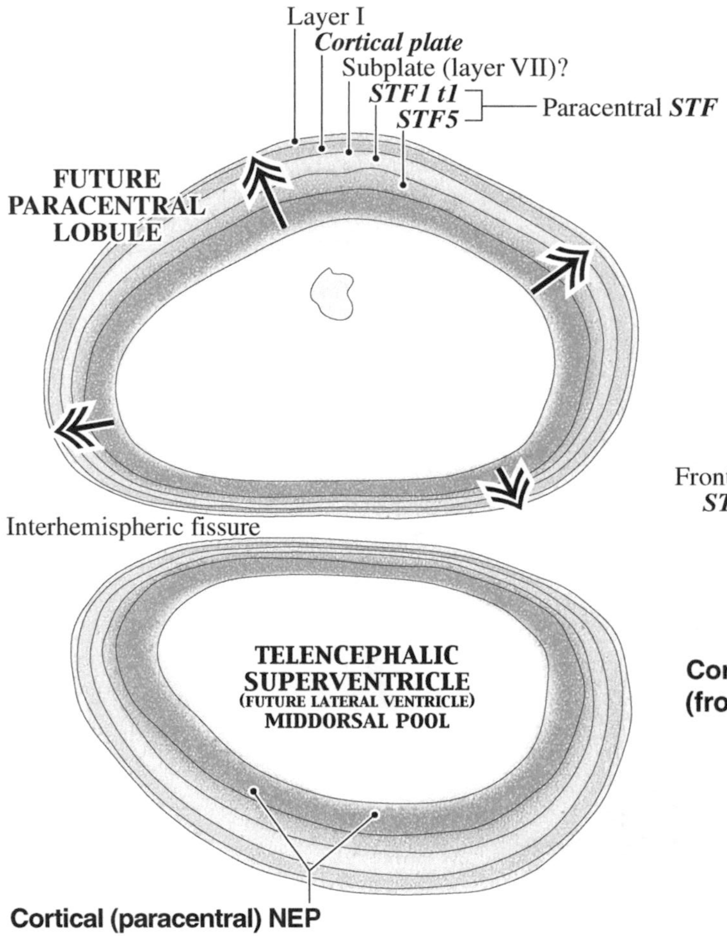

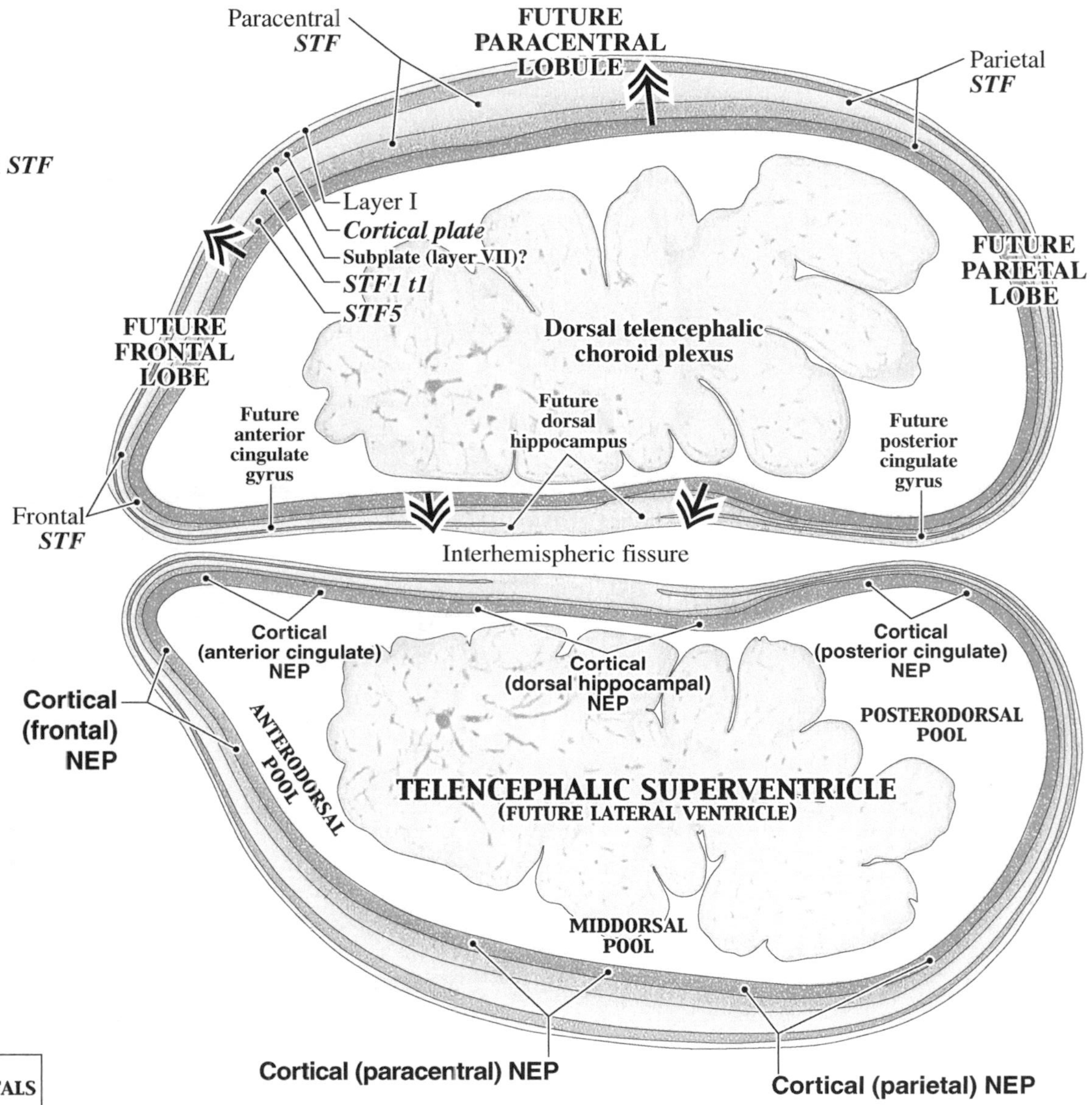

211

Parietal *STF*

**FUTURE PARIETAL LOBE**

Future posterior cingulate gyrus

Dorsal telencephalic choroid plexus

**FUTURE PARACENTRAL LOBULE**

Paracentral *STF*

Layer I
*Cortical plate*
Subplate (layer VII)?
*STF1 t1*
*STF5*

Future dorsal hippocampus

Future anterior cingulate gyrus

**FUTURE FRONTAL LOBE**

Frontal *STF*

Interhemispheric fissure

Cortical (posterior cingulate) NEP

**POSTERODORSAL POOL**

Cortical (dorsal hippocampal) NEP

**TELENCEPHALIC SUPERVENTRICLE**
**(FUTURE LATERAL VENTRICLE)**

**MIDDORSAL POOL**

Cortical (anterior cingulate) NEP

**ANTERODORSAL POOL**

**Cortical (frontal) NEP**

**Cortical (parietal) NEP**

**Cortical (paracentral) NEP**

Layer I
*Cortical plate*
Subplate (layer VII)?
*STF1 t1*
*STF5*

Paracentral *STF*

**FUTURE PARACENTRAL LOBULE**

Interhemispheric fissure

**TELENCEPHALIC SUPERVENTRICLE**
**(FUTURE LATERAL VENTRICLE)**
**MIDDORSAL POOL**

**Cortical (paracentral) NEP**

**Neuroepithelium - NEP**

Arrows indicate the presumed *direction of neuron migration* from neuroepithelial sources.

FONT KEY:
VENTRICULAR DIVISIONS – CAPITALS
Germinal zone - **Helvetica bold**
*Transient structure - Times bold italic*
Permanent structure - Times Roman or **Bold**

**PLATE 85A**

**GW9 Horizontal**
**CR 43 mm, C886**
**Level 3: Section 42**

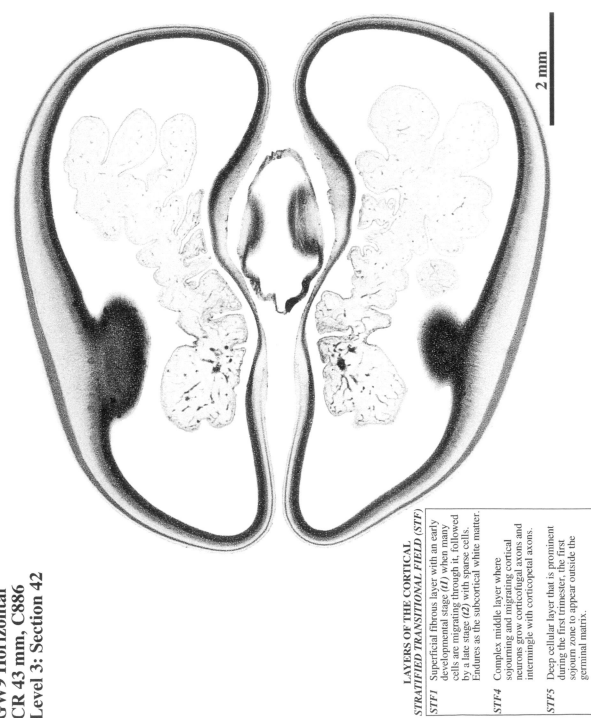

2 mm

**LAYERS OF THE CORTICAL**
**STRATIFIED TRANSITIONAL FIELD (STF)**

*STF1*  Superficial fibrous layer with an early
developmental stage (*t1*) when many
cells are migrating through it, followed
by a late stage (*t2*) with sparse cells.
Endures as the subcortical white matter.

*STF4*  Complex middle layer where
sojourning and migrating cortical
neurons grow corticofugal axons and
intermingle with corticopetal axons.

*STF5*  Deep cellular layer that is prominent
during the first trimester, the first
sojourn zone to appear outside the
germinal matrix.

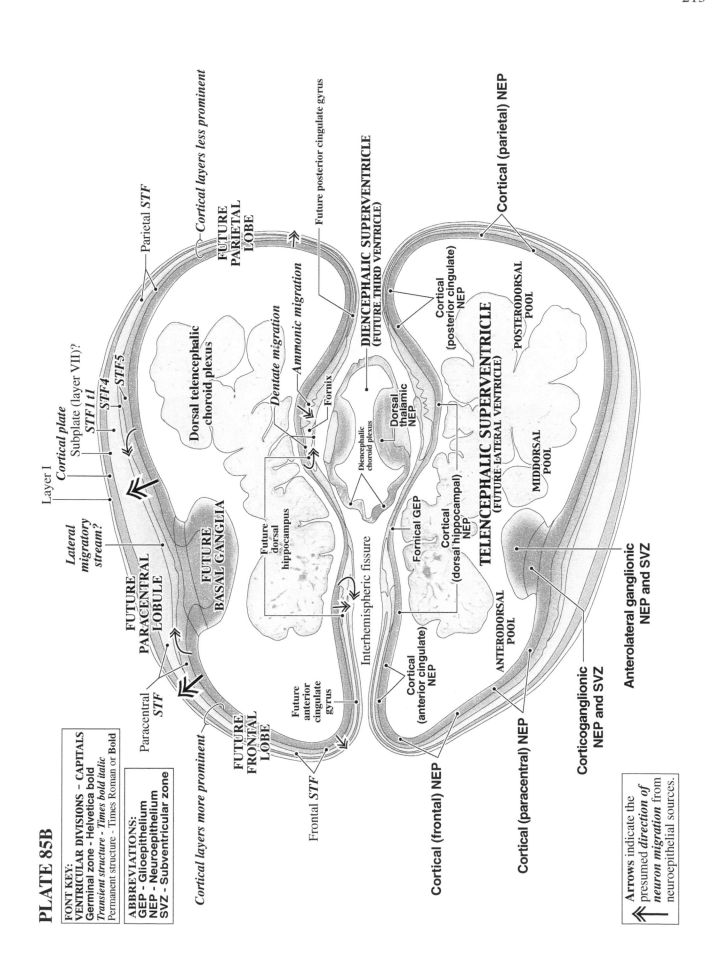

PLATE 85B

FONT KEY:
VENTRICULAR DIVISIONS – CAPITALS
Germinal zone - Helvetica bold
*Transient structure - Times bold italic*
Permanent structure - Times Roman or Bold

ABBREVIATIONS:
GEP - Glioepithelium
NEP - Neuroepithelium
SVZ - Subventricular zone

*Arrows* indicate the presumed *direction of neuron migration* from neuroepithelial sources.

Layer I
*Cortical plate*
Subplate (layer VII)?
*STF1 t1*
*STF4*
*STF5*
Parietal *STF*
*Cortical layers less prominent*

FUTURE PARIETAL LOBE

Dorsal telencephalic choroid plexus

*Dentate migration*

*Ammonic migration*

Future posterior cingulate gyrus

DIENCEPHALIC SUPERVENTRICLE
(FUTURE THIRD VENTRICLE)

Fornix

*Future dorsal hippocampus*

Diencephalic choroid plexus

Dorsal thalamic NEP

Cortical (posterior cingulate) NEP

Cortical (parietal) NEP

POSTERODORSAL POOL

MIDDORSAL POOL

TELENCEPHALIC SUPERVENTRICLE
(FUTURE LATERAL VENTRICLE)

Fornical GEP

Cortical (dorsal hippocampal) NEP

*Lateral migratory stream?*

FUTURE PARACENTRAL LOBULE

FUTURE BASAL GANGLIA

Interhemispheric fissure

Future anterior cingulate gyrus

Paracentral *STF*

FUTURE FRONTAL LOBE

Frontal *STF*

*Cortical layers more prominent*

ANTERODORSAL POOL

Cortical (anterior cingulate) NEP

Cortical (frontal) NEP

Cortical (paracentral) NEP

Corticoganglionic NEP and SVZ

Anterolateral ganglionic NEP and SVZ

213

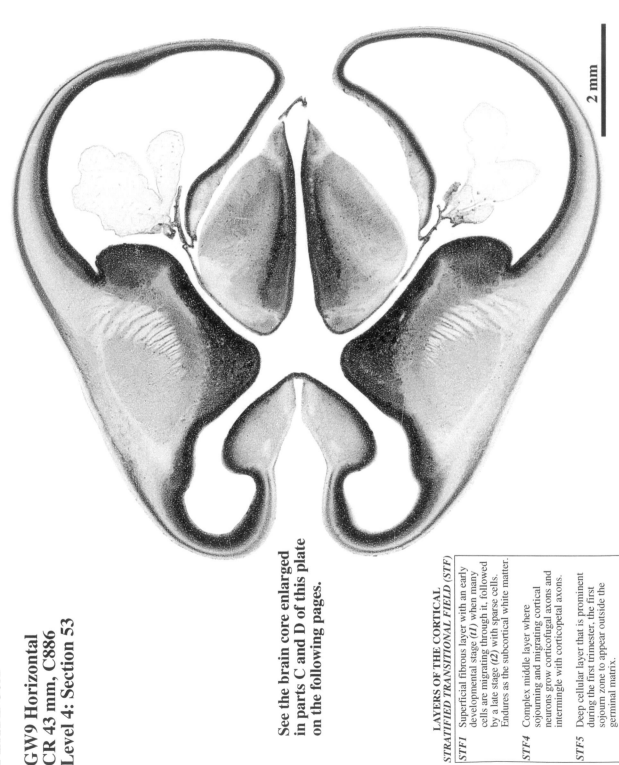

**PLATE 86A**

**GW9 Horizontal**
**CR 43 mm, C886**
**Level 4: Section 53**

**See the brain core enlarged**
**in parts C and D of this plate**
**on the following pages.**

**LAYERS OF THE CORTICAL**
*STRATIFIED TRANSITIONAL FIELD (STF)*

*STF1* Superficial fibrous layer with an early developmental stage (*t1*) when many cells are migrating through it, followed by a late stage (*t2*) with sparse cells. Endures as the subcortical white matter.

*STF4* Complex middle layer where sojourning and migrating cortical neurons grow corticofugal axons and intermingle with corticopetal axons.

*STF5* Deep cellular layer that is prominent during the first trimester, the first sojourn zone to appear outside the germinal matrix.

2 mm

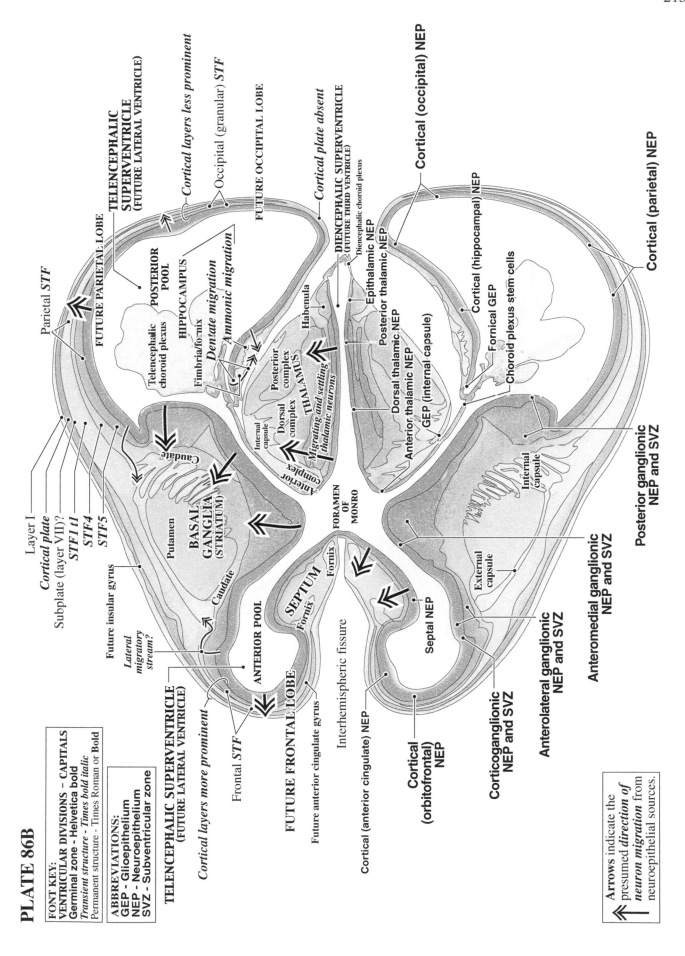

PLATE 86B

FONT KEY:
VENTRICULAR DIVISIONS – CAPITALS
Germinal zone - Helvetica bold
*Transient structure - Times bold italic*
Permanent structure - Times Roman or Bold

ABBREVIATIONS:
GEP - Glioepithelium
NEP - Neuroepithelium
SVZ - Subventricular zone

**TELENCEPHALIC SUPERVENTRICLE**
(FUTURE LATERAL VENTRICLE)

*Cortical layers more prominent*

*Frontal STF*

**FUTURE FRONTAL LOBE**

Future anterior cingulate gyrus

Interhemispheric fissure

Cortical (anterior cingulate) NEP

**Cortical (orbitofrontal) NEP**

**Corticoganglionic NEP and SVZ**

**Anterolateral ganglionic NEP and SVZ**

**Anteromedial ganglionic NEP and SVZ**

**Posterior ganglionic NEP and SVZ**

**Cortical (parietal) NEP**

**Cortical (occipital) NEP**

Cortical (parietal) NEP

Cortical (hippocampal) NEP

Cortical (hippocampal) GEP

Fornical GEP

Choroid plexus stem cells

Diencephalic choroid plexus

**DIENCEPHALIC SUPERVENTRICLE**
(FUTURE THIRD VENTRICLE)

*Cortical plate absent*

Epithalamic NEP

Posterior thalamic NEP

Dorsal thalamic NEP

Anterior thalamic NEP

GEP (internal capsule)

Habenula

Layer I

*Cortical plate*

Subplate (layer VII)?

*STF1 t1*

*STF4*

*STF5*

Future insular gyrus

*Lateral migratory stream?*

Putamen

Caudate

**BASAL GANGLIA (STRIATUM)**

Caudate

**SEPTUM**

*Fornix*

Fornix

Fornix

**ANTERIOR POOL**

*Frontal STF*

Septal NEP

External capsule

Internal capsule

Internal capsule

**FORAMEN OF MONRO**

*Anterior complex*

*Posterior complex*

**THALAMUS**

Dorsal complex

*Migrating (and settling) thalamic neurons*

Internal capsule

Fimbria/fornix

Telencephalic choroid plexus

**HIPPOCAMPUS**

*Dentate migration*

*Ammonic migration*

**POSTERIOR POOL**

*Parietal STF*

**FUTURE PARIETAL LOBE**

**TELENCEPHALIC SUPERVENTRICLE**
(FUTURE LATERAL VENTRICLE)

*Cortical layers less prominent*

Occipital (granular) *STF*

**FUTURE OCCIPITAL LOBE**

**Arrows** indicate the presumed *direction of neuron migration* from neuroepithelial sources.

215

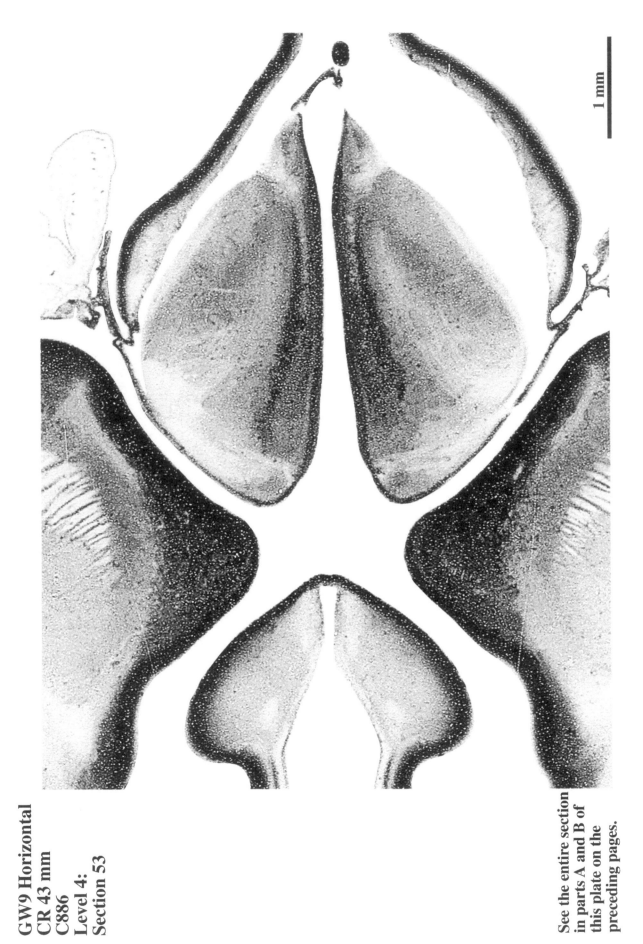

**PLATE 86C**

**GW9 Horizontal**
**CR 43 mm**
**C886**
**Level 4:**
**Section 53**

See the entire section in parts A and B of this plate on the preceding pages.

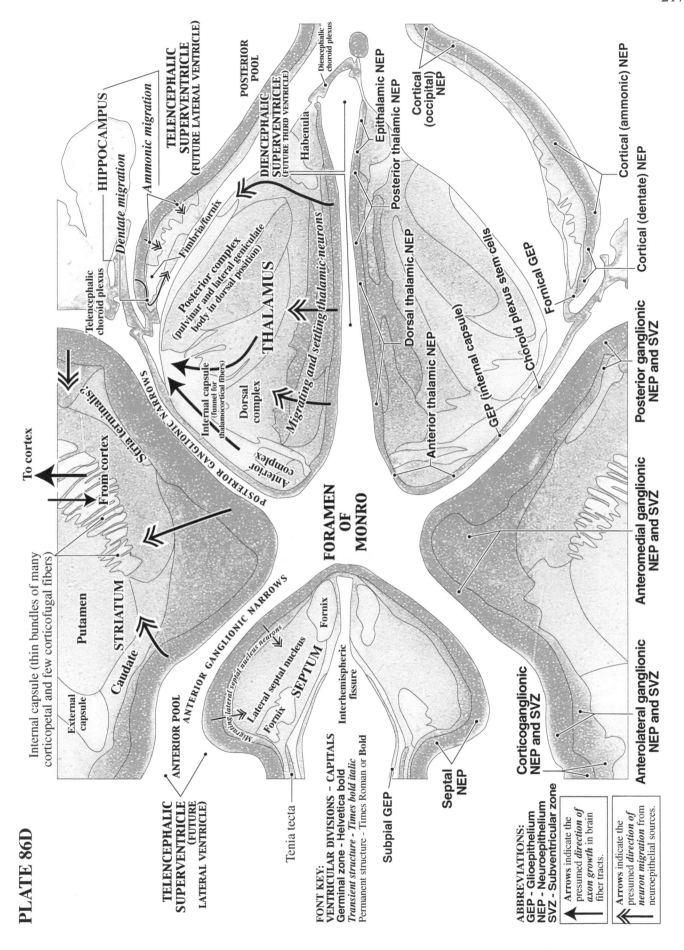

**PLATE 86D**

Internal capsule (thin bundles of many corticopetal and few corticofugal fibers)

To cortex

From cortex

HIPPOCAMPUS

*Ammonic migration*

TELENCEPHALIC SUPERVENTRICLE (FUTURE LATERAL VENTRICLE)

POSTERIOR POOL

*Dentate migration*

Telencephalic choroid plexus

Fimbria/fornix

DIENCEPHALIC SUPERVENTRICLE (FUTURE THIRD VENTRICLE)

Diencephalic choroid plexus

Habenula

Epithalamic NEP

Posterior thalamic NEP

Cortical (occipital) NEP

Posterior complex (pulvinar and lateral geniculate body in dorsal position)

THALAMUS

*Migrating and settling thalamic neurons*

Dorsal thalamic NEP

Cortical (ammonic) NEP

Cortical (dentate) NEP

Fornical GEP

Choroid plexus stem cells

GEP (internal capsule)

Anterior thalamic NEP

*Strial terminalis?*

POSTERIOR GANGLIONIC NARROWS

External capsule

Putamen

Internal capsule (tunnel for thalamocortical fibers)

Dorsal complex

Anterior complex

STRIATUM

Caudate

ANTERIOR POOL

ANTERIOR GANGLIONIC NARROWS

TELENCEPHALIC SUPERVENTRICLE (FUTURE LATERAL VENTRICLE)

Posterior ganglionic NEP and SVZ

Anteromedial ganglionic NEP and SVZ

FORAMEN OF MONRO

Corticoganglionic NEP and SVZ

Anterolateral ganglionic NEP and SVZ

*Migrating lateral septal nucleus neurons*

Lateral septal nucleus

Fornix

SEPTUM

Fornix

Interhemispheric fissure

Tenia tecta

Subpial GEP

Septal NEP

**FONT KEY:**
VENTRICULAR DIVISIONS – CAPITALS
Germinal zone – Helvetica bold
*Transient structure* – Times bold italic
Permanent structure – Times Roman or Bold

**ABBREVIATIONS:**
GEP - Glioepithelium
NEP - Neuroepithelium
SVZ - Subventricular zone

Arrows indicate the presumed *direction of axon growth* in brain fiber tracts.

Arrows indicate the presumed *direction of neuron migration* from neuroepithelial sources.

**PLATE 87A**

**GW9 Horizontal**
**CR 43 mm, C886**
**Level 5: Section 57**

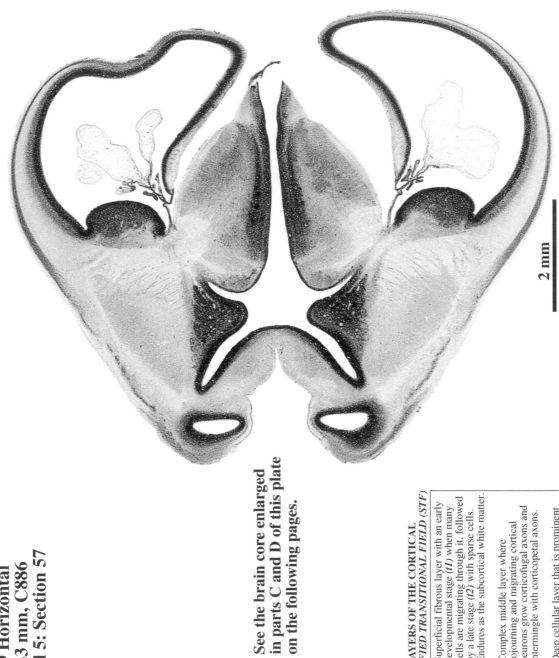

2 mm

See the brain core enlarged
in parts C and D of this plate
on the following pages.

**LAYERS OF THE CORTICAL**
*STRATIFIED TRANSITIONAL FIELD (STF)*

*STF1*   Superficial fibrous layer with an early
developmental stage *(t1)* when many
cells are migrating through it, followed
by a late stage *(t2)* with sparse cells.
Endures as the subcortical white matter.

*STF4*   Complex middle layer where
sojourning and migrating cortical
neurons grow corticofugal axons and
intermingle with corticopetal axons.

*STF5*   Deep cellular layer that is prominent
during the first trimester, the first
sojourn zone to appear outside the
germinal matrix.

219

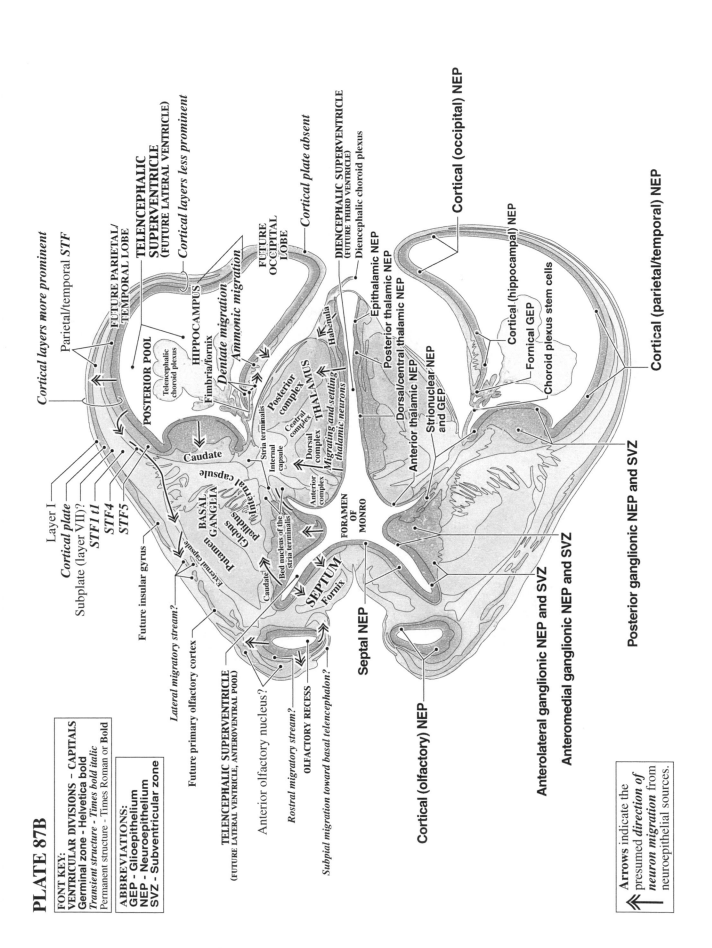

PLATE 87B

FONT KEY:
VENTRICULAR DIVISIONS – CAPITALS
Germinal zone - Helvetica bold
Transient structure - Times bold italic
Permanent structure - Times Roman or Bold

ABBREVIATIONS:
GEP - Glioepithelium
NEP - Neuroepithelium
SVZ - Subventricular zone

Arrows indicate the presumed direction of neuron migration from neuroepithelial sources.

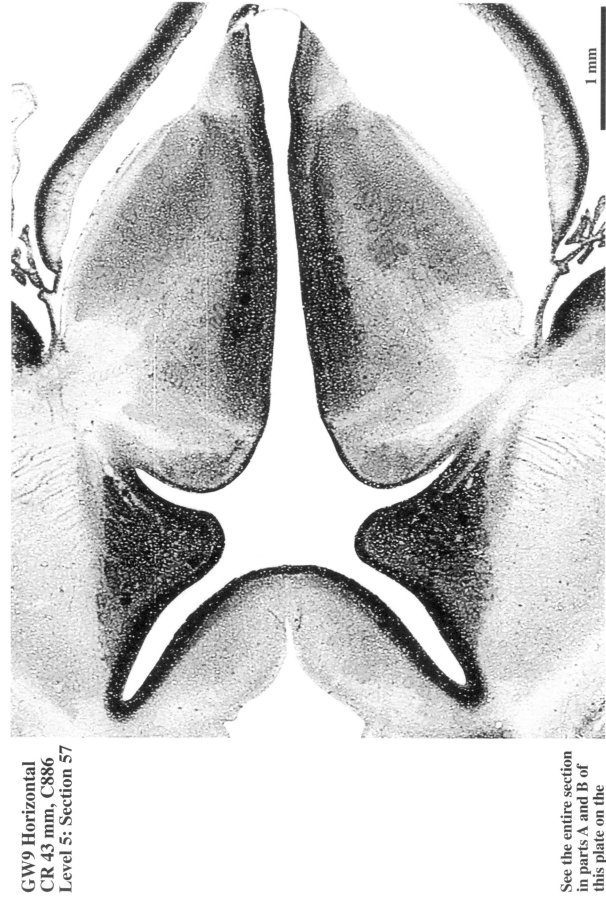

**PLATE 87C**

**GW9 Horizontal
CR 43 mm, C886
Level 5: Section 57**

See the entire section
in parts A and B of
this plate on the
preceding pages.

1 mm

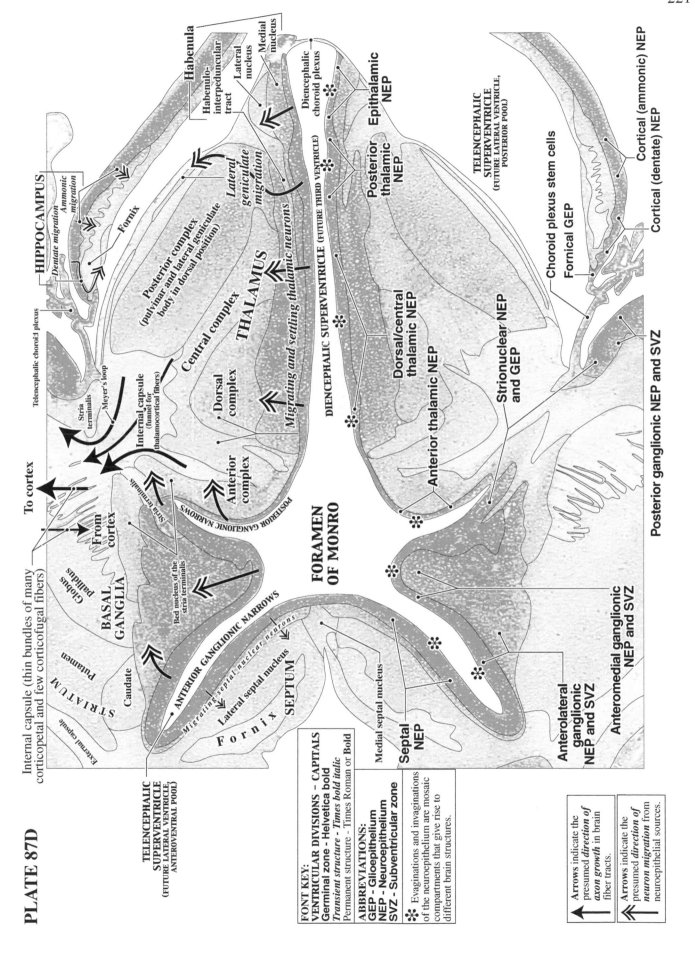

**PLATE 87D**

Internal capsule (thin bundles of many corticopetal and few corticofugal fibers)

To cortex

From cortex

HIPPOCAMPUS

Dentate migration
Ammonic migration

Fornix

Telencephalic choroid plexus

Stria terminalis

Meyer's loop

Internal capsule (tunnel for thalamocortical fibers)

Posterior complex (pulvinar and lateral geniculate body in dorsal position)

Lateral geniculate migration

Lateral nucleus

Medial nucleus

Habenula

Habenulo-interpeduncular tract

Diencephalic choroid plexus

Epithalamic NEP

Posterior thalamic NEP

Central complex

THALAMUS

Dorsal complex

Migrating and settling thalamic neurons

DIENCEPHALIC SUPERVENTRICLE (FUTURE THIRD VENTRICLE)

TELENCEPHALIC SUPERVENTRICLE (FUTURE LATERAL VENTRICLE, POSTERIOR POOL)

Choroid plexus stem cells

Fornical GEP

Cortical (ammonic) NEP

Cortical (dentate) NEP

Dorsal/central thalamic NEP

Strionuclear NEP and GEP

Anterior thalamic NEP

Anterior complex

POSTERIOR GANGLIONIC NARROWS

Stria terminalis

Bed nucleus of the stria terminalis

BASAL GANGLIA

Globus pallidus

STRIATUM

Putamen

Caudate

External capsule

ANTERIOR GANGLIONIC NARROWS

Migrating septal nuclear neurons

Lateral septal nucleus

SEPTUM

Fornix

Medial septal nucleus

Septal NEP

FORAMEN OF MONRO

Anterolateral ganglionic NEP and SVZ

Anteromedial ganglionic NEP and SVZ

Posterior ganglionic NEP and SVZ

**TELENCEPHALIC SUPERVENTRICLE** (FUTURE LATERAL VENTRICLE, ANTEROVENTRAL POOL)

FONT KEY:
VENTRICULAR DIVISIONS – CAPITALS
Germinal zone - Helvetica bold
*Transient structure - Times bold italic*
Permanent structure - Times Roman or Bold

ABBREVIATIONS:
GEP - Glioepithelium
NEP - Neuroepithelium
SVZ - Subventricular zone

✱ Evaginations and invaginations of the neuroepithelium are mosaic compartments that give rise to different brain structures.

Arrows indicate the presumed *direction of axon growth* in brain fiber tracts.

Arrows indicate the presumed *direction of neuron migration* from neuroepithelial sources.

**PLATE 88A**

**GW9 Horizontal**
**CR 43 mm, C886**
**Level 6: Section 63**

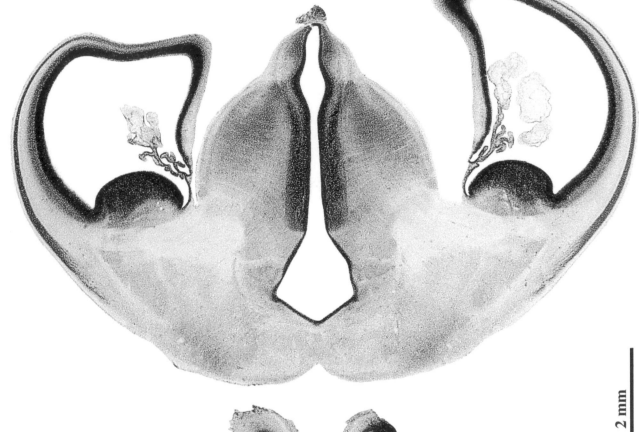

2 mm

See the brain core enlarged in parts C and D of this plate on the following pages.

LAYERS OF THE CORTICAL
*STRATIFIED TRANSITIONAL FIELD (STF)*

*STF1*  Superficial fibrous layer with an early developmental stage (*t1*) when many cells are migrating through it, followed by a late stage (*t2*) with sparse cells. Endures as the subcortical white matter.

*STF4*  Complex middle layer where sojourning and migrating cortical neurons grow corticofugal axons and intermingle with corticopetal axons.

*STF5*  Deep cellular layer that is prominent during the first trimester, the first sojourn zone to appear outside the germinal matrix.

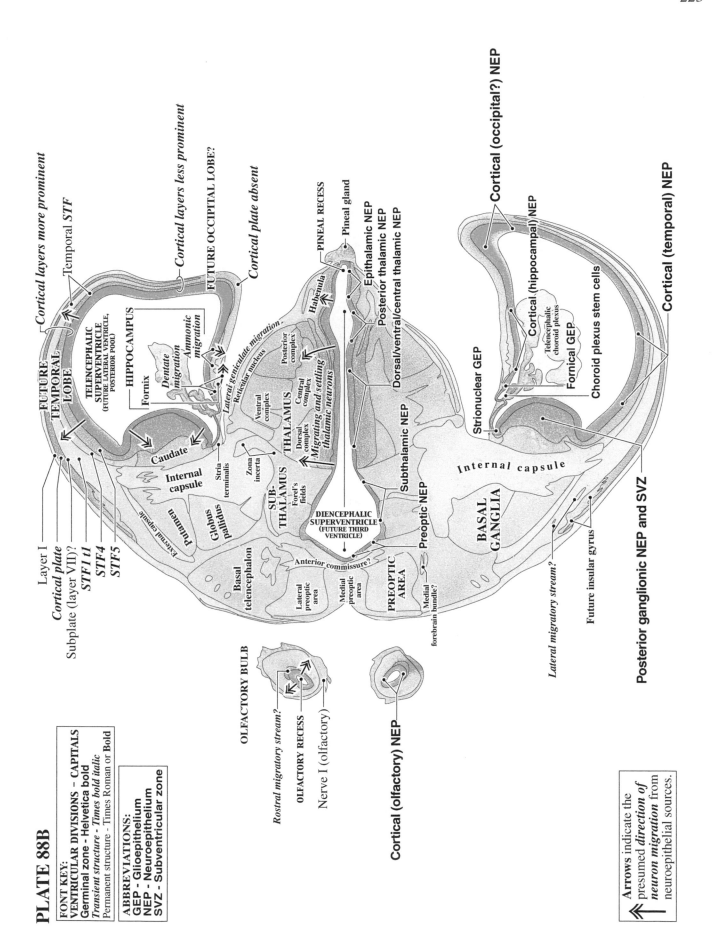

PLATE 88B

FONT KEY:
VENTRICULAR DIVISIONS – CAPITALS
Germinal zone - Helvetica bold
*Transient structure - Times bold italic*
Permanent structure - Times Roman or **Bold**

ABBREVIATIONS:
GEP - Glioepithelium
NEP - Neuroepithelium
SVZ - Subventricular zone

Layer I
*Cortical plate*
Subplate (layer VII)?
*STF1 t1*
*STF4*
*STF5*

*Cortical layers more prominent*
Temporal *STF*

*Cortical layers less prominent*
FUTURE OCCIPITAL LOBE?
**Cortical plate absent**

FUTURE
TEMPORAL
LOBE

TELENCEPHALIC
SUPERVENTRICLE
(FUTURE LATERAL VENTRICLE,
POSTERIOR POOL)

HIPPOCAMPUS
Fornix
*Dentate migration*
*Ammonic migration*
*Lateral geniculate migration*
Reticular nucleus

Caudate
Internal
capsule
Putamen
External capsule
Globus
pallidus
SUB-
THALAMUS
Forel's
fields
Stria
terminalis
Zona
incerta

Ventral
complex
Posterior
complex
Dorsal
complex
Central
complex
THALAMUS
*Migrating and settling
thalamic neurons*

PINEAL RECESS
Pineal gland
Epithalamic NEP
Posterior thalamic NEP
Dorsal/ventral/central thalamic NEP
Habenula

Cortical (occipital?) NEP

Cortical (hippocampal) NEP

Strionuclear GEP
Telencephalic
choroid plexus
Fornical GEP
Choroid plexus stem cells

Internal capsule

DIENCEPHALIC
SUPERVENTRICLE
(FUTURE THIRD
VENTRICLE)

Subthalamic NEP
Preoptic NEP

BASAL
GANGLIA

**Cortical (temporal) NEP**

Basal
telencephalon
Lateral
preoptic
area
Medial
preoptic
area
PREOPTIC
AREA
Medial
forebrain bundle?
*Anterior commissure?*

Future insular gyrus
*Lateral migratory stream?*

**Posterior ganglionic NEP and SVZ**

OLFACTORY BULB
*Rostral migratory stream?*
OLFACTORY RECESS
Nerve I (olfactory)

Cortical (olfactory) NEP

**Cortical (olfactory) NEP**

Arrows indicate the
presumed *direction of
neuron migration* from
neuroepithelial sources.

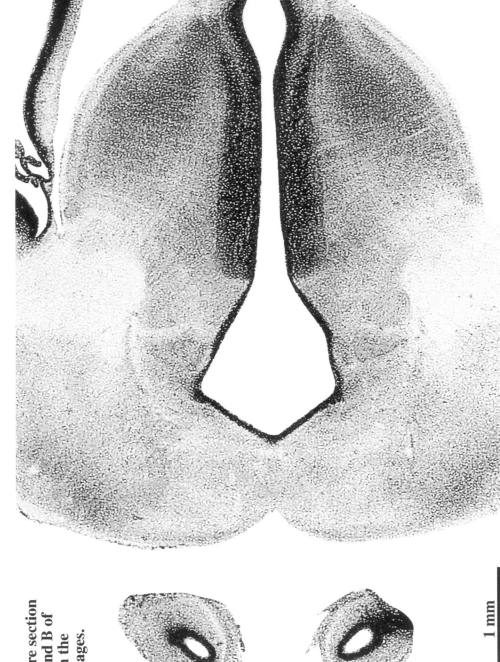

**PLATE 88C**

**GW9 Horizontal
CR 43 mm, C886
Level 6: Section 63**

**See the entire section
in parts A and B of
this plate on the
preceding pages.**

1 mm

PLATE 88D

225

**FONT KEY:**
VENTRICULAR DIVISIONS - CAPITALS
Germinal zone - Helvetica bold
*Transient structure - Times bold italic*
Permanent structure - Times Roman or Bold

**ABBREVIATIONS:**
GEP - Glioepithelium
NEP - Neuroepithelium
SVZ - Subventricular zone

✱ Evaginations and invaginations of the neuroepithelium are mosaic compartments that give rise to different brain structures.

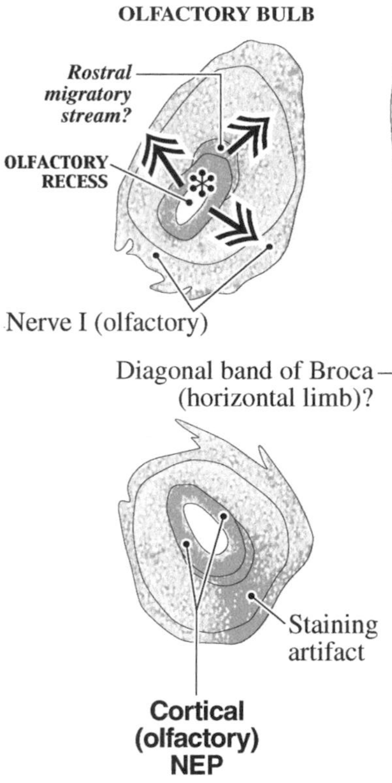

Arrows indicate the presumed *direction of axon growth* in brain fiber tracts.

Arrows indicate the presumed *direction of neuron migration* from neuroepithelial sources.

HIPPOCAMPUS

TELENCEPHALIC SUPERVENTRICLE (FUTURE LATERAL VENTRICLE, POSTERIOR POOL)

Cortical (occipital?) NEP

Cortical (ammonic) NEP

**Habenula**

Medial nucleus
Lateral nucleus
Habenulo-interpeduncular tract

Pineal gland

PINEAL RECESS

Epithalamic NEP

Posterior thalamic NEP

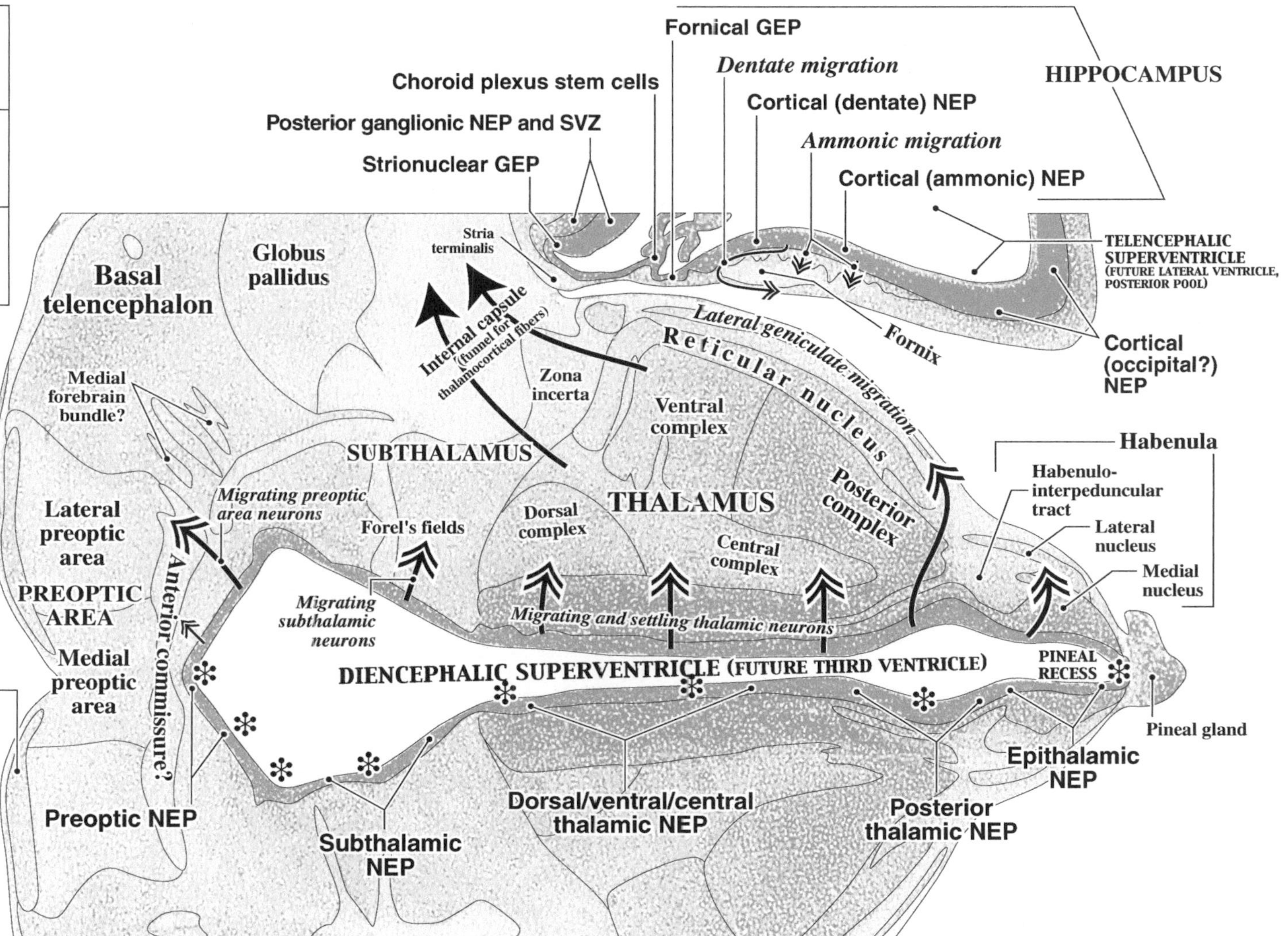

Fornical GEP
*Dentate migration*
Cortical (dentate) NEP
*Ammonic migration*

*Fornix*

*Lateral geniculate migration*

Reticular nucleus

Posterior complex

Ventral complex

THALAMUS

Central complex

*Migrating and settling thalamic neurons*

DIENCEPHALIC SUPERVENTRICLE (FUTURE THIRD VENTRICLE)

Dorsal/ventral/central thalamic NEP

Choroid plexus stem cells
Posterior ganglionic NEP and SVZ
Strionuclear GEP

*Stria terminalis*

*Zona incerta*

Dorsal complex

SUBTHALAMUS

*Internal capsule (thalamocortical fibers)*

Forel's fields

*Migrating subthalamic neurons*

Subthalamic NEP

**Basal telencephalon**

**Globus pallidus**

*Medial forebrain bundle?*

*Migrating preoptic area neurons*

Lateral preoptic area
PREOPTIC AREA
Medial preoptic area

*Anterior commissure?*

Preoptic NEP

BASAL GANGLIA

Internal capsule

OLFACTORY BULB

*Rostral migratory stream?*
OLFACTORY RECESS

Nerve I (olfactory)

Diagonal band of Broca (horizontal limb)?

Cortical (olfactory) NEP

Staining artifact

226

**PLATE 89A**

**GW9 Horizontal**
**CR 43 mm, C886**
**Level 7: Section 69**

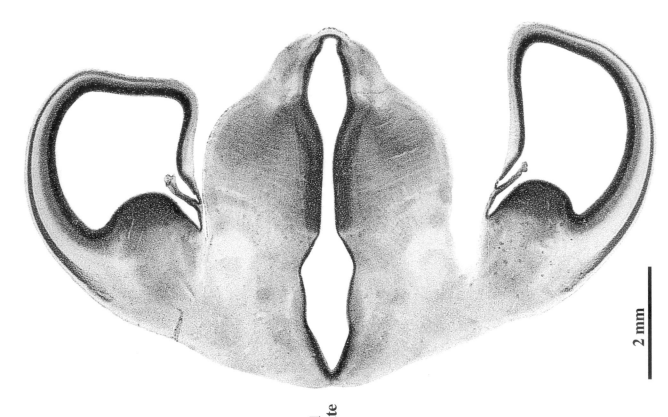

2 mm

**See the brain core enlarged**
**in parts C and D of this plate**
**on the following pages.**

**LAYERS OF THE CORTICAL**
***STRATIFIED TRANSITIONAL FIELD (STF)***

*STF1* Superficial fibrous layer with an early
developmental stage (*t1*) when many
cells are migrating through it, followed
by a late stage (*t2*) with sparse cells.
Endures as the subcortical white matter.

*STF4* Complex middle layer where
sojourning and migrating cortical
neurons grow corticofugal axons and
intermingle with corticopetal axons.

*STF5* Deep cellular layer that is prominent
during the first trimester, the first
sojourn zone to appear outside the
germinal matrix.

# PLATE 89B

**FONT KEY:**
VENTRICULAR DIVISIONS – CAPITALS
Germinal zone - **Helvetica bold**
*Transient structure - Times bold italic*
Permanent structure - Times Roman or **Bold**

**ABBREVIATIONS:**
GEP - Glioepithelium
NEP - Neuroepithelium
SVZ - Subventricular zone

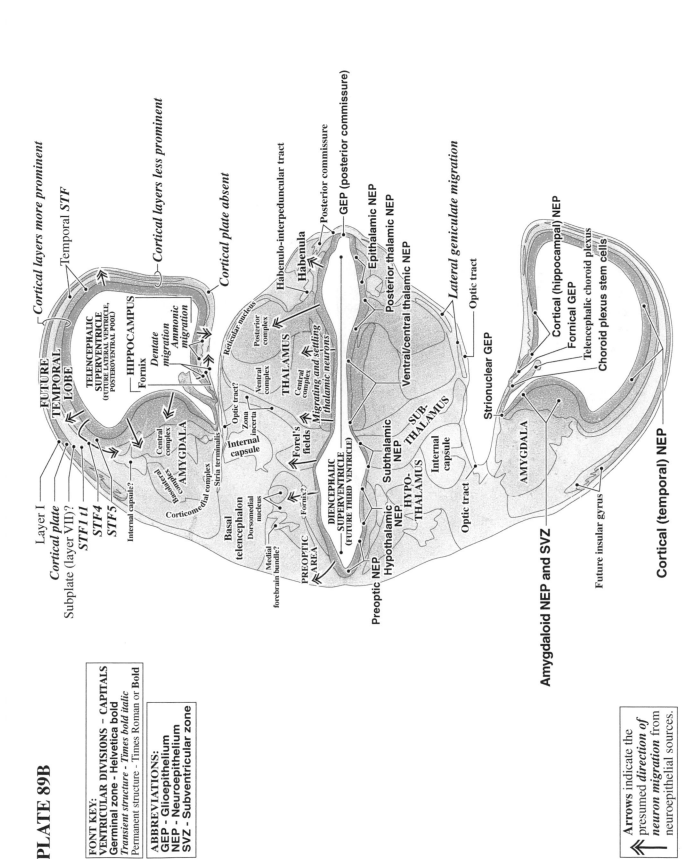

Layer I
*Cortical plate*
Subplate (layer VII)?
*STF1 t1*
*STF4*
*STF5*

Internal capsule?

*Cortical layers more prominent*

Temporal *STF*

**FUTURE**
**TEMPORAL**
**LOBE**

*Cortical layers less prominent*

*Cortical plate absent*

**TELENCEPHALIC**
**SUPERVENTRICLE**
(FUTURE LATERAL VENTRICLE,
POSTEROVENTRAL POOL)

**HIPPOCAMPUS**

*Fornix*

*Dentate*
*migration*

*Ammonic*
*migration*

Basolateral
complex

Central
complex

**AMYGDALA**

Corticomedial complex

Stria terminalis

*Internal*
*capsule*

Optic tract?

*Zona*
*incerta*

*Ventral*
*complex*

Reticular nucleus

Posterior
complex

**THALAMUS**

Central
complex

*Migrating and settling*
*thalamic neurons*

*Forel's*
*fields*

Posterior commissure

Habenulo-interpeduncular tract

**Habenula**

**GEP (posterior commissure)**

**Epithalamic NEP**

Posterior-thalamic NEP

**Ventral/central thalamic NEP**

*Lateral geniculate migration*

Optic tract

**Strionuclear GEP**

**AMYGDALA**

Future insular gyrus

Optic tract

*Internal*
*capsule*

**SUB-**
**THALAMUS**

**Subthalamic**
**NEP**

**HYPO-**
**THALAMUS**

**Hypothalamic**
**NEP**

Basal
telencephalon

Dorsomedial
nucleus

Medial
forebrain bundle?

*Fornix?*

**PREOPTIC**
**AREA**

**DIENCEPHALIC**
**SUPERVENTRICLE**
(FUTURE THIRD VENTRICLE)

**Preoptic NEP**

**Amygdaloid NEP and SVZ**

**Cortical (temporal) NEP**

Cortical (hippocampal) NEP

**Fornical GEP**

Telencephalic choroid plexus

Choroid plexus stem cells

Arrows indicate the
presumed *direction of*
*neuron migration* from
neuroepithelial sources.

228

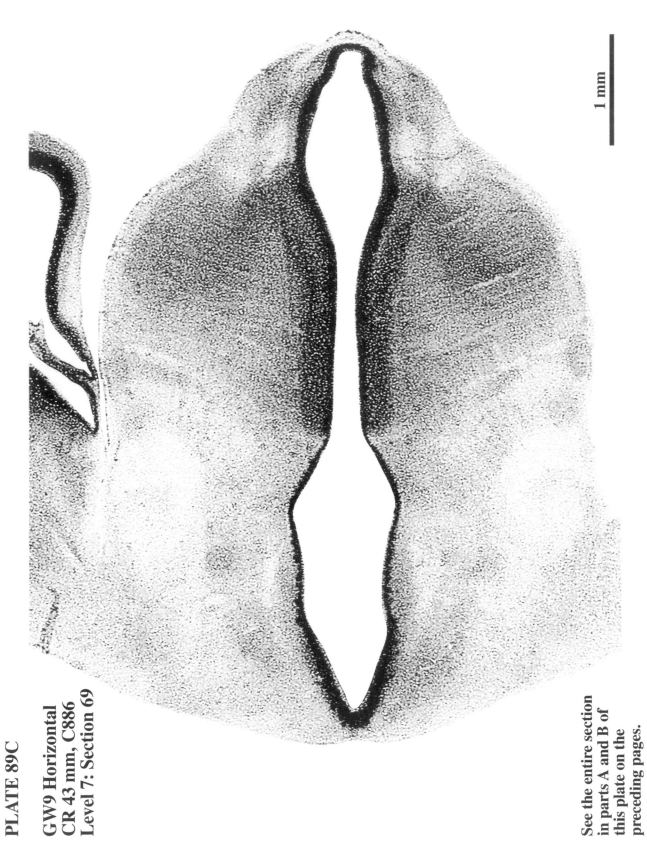

1 mm

**PLATE 89C**

**GW9 Horizontal**
**CR 43 mm, C886**
**Level 7: Section 69**

See the entire section
in parts A and B of
this plate on the
preceding pages.

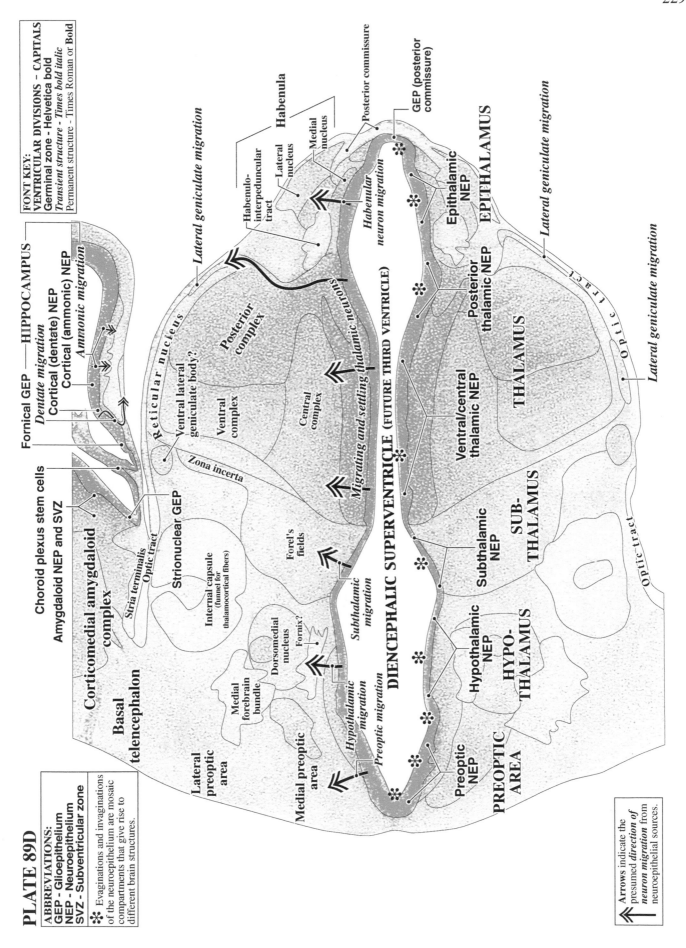

PLATE 89D

**ABBREVIATIONS:**
GEP - Glioepithelium
NEP - Neuroepithelium
SVZ - Subventricular zone

✳ Evaginations and invaginations of the neuroepithelium are mosaic compartments that give rise to different brain structures.

FONT KEY:
VENTRICULAR DIVISIONS - CAPITALS
Germinal zone - Helvetica bold
*Transient structure - Times bold italic*
Permanent structure - Times Roman or Bold

Choroid plexus stem cells

Amygdaloid NEP and SVZ

Corticomedial amygdaloid complex

Basal telencephalon

Fornical GEP —— HIPPOCAMPUS

*Dentate migration*
Cortical (dentate) NEP
Cortical (ammonic) NEP
*Ammonic migration*

*Lateral geniculate migration*

Habenula

Habenulo-interpeduncular tract

Lateral nucleus

Medial nucleus

Posterior commissure

GEP (posterior commissure)

EPITHALAMUS

Epithalamic NEP

Posterior thalamic NEP

*Lateral geniculate migration*

*Habenular neuron migration*

Reticular nucleus

Posterior complex

Ventral lateral geniculate body?

Ventral complex

Central complex

*Migrating and settling thalamic neurons*

Zona incerta

Stria terminalis

Optic tract

Strionuclear GEP

Internal capsule (funnel for thalamocortical fibers)

Forel's fields

Medial forebrain bundle

Dorsomedial nucleus

Fornix?

Lateral preoptic area

Medial preoptic area

*Subthalamic migration*

*Hypothalamic migration*

*Preoptic migration*

**DIENCEPHALIC SUPERVENTRICLE** (FUTURE THIRD VENTRICLE)

Ventral/central thalamic NEP

THALAMUS

Subthalamic NEP

SUB-THALAMUS

Hypothalamic NEP

HYPO-THALAMUS

Preoptic NEP

PREOPTIC AREA

*Lateral geniculate migration*

Optic tract

*Lateral geniculate migration*

Optic tract

⋘ Arrows indicate the presumed *direction of neuron migration* from neuroepithelial sources.

**PLATE 90A**

**GW9 Horizontal**
**CR 43 mm, C886**
**Level 8: Section 75**

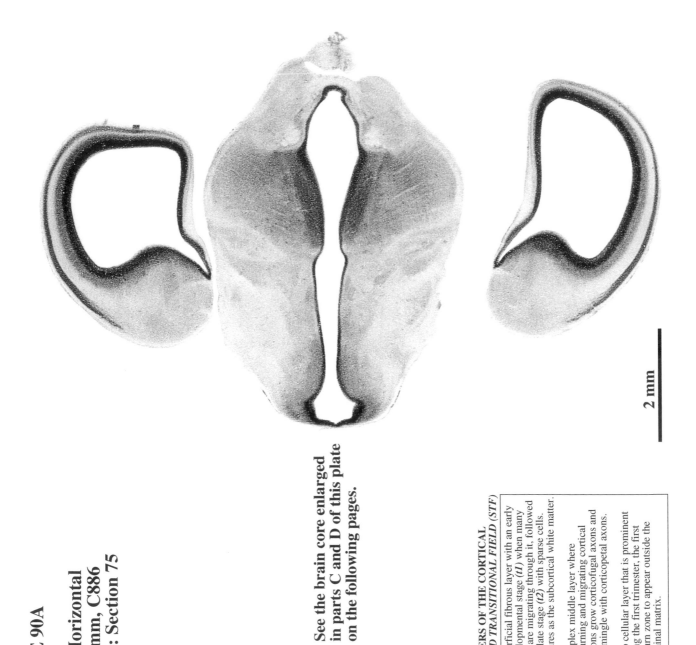

See the brain core enlarged
in parts C and D of this plate
on the following pages.

2 mm

**LAYERS OF THE CORTICAL**
*STRATIFIED TRANSITIONAL FIELD (STF)*

*STF1* Superficial fibrous layer with an early
developmental stage (*t1*) when many
cells are migrating through it, followed
by a late stage (*t2*) with sparse cells.
Endures as the subcortical white matter.

*STF4* Complex middle layer where
sojourning and migrating cortical
neurons grow corticofugal axons and
intermingle with corticopetal axons.

*STF5* Deep cellular layer that is prominent
during the first trimester, the first
sojourn zone to appear outside the
germinal matrix.

# PLATE 90B

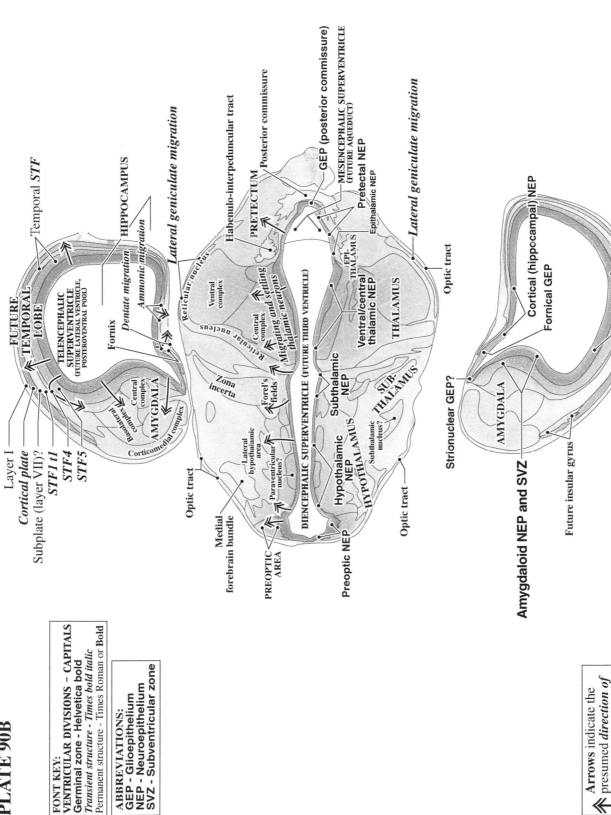

Layer I
*Cortical plate*
Subplate (layer VII)?
*STF1 t1*
*STF4*
*STF5*

**FUTURE TEMPORAL LOBE**

Temporal *STF*

**HIPPOCAMPUS**

*Lateral geniculate migration*

*Dentate migration*

*Ammonic migration*

**TELENCEPHALIC SUPERVENTRICLE**
(FUTURE LATERAL VENTRICLE, POSTEROVENTRAL POOL)

Fornix

Commissural complex
Central complex
**AMYGDALA**

Corticomedial complex

Optic tract

Medial forebrain bundle

Lateral hypothalamic area

Paraventricular nucleus?

**PREOPTIC AREA**

Preoptic NEP

**DIENCEPHALIC SUPERVENTRICLE (FUTURE THIRD VENTRICLE)**

Zona incerta

Reticular nucleus

Forel's fields

Central complex

Ventral complex

Reticular nucleus

Migrating and settling thalamic neurons

Habenulo-interpeduncular tract

Posterior commissure

**PRETECTUM**

**GEP (posterior commissure)**

**MESENCEPHALIC SUPERVENTRICLE**
(FUTURE AQUEDUCT)

Pretectal NEP

Epithalamic NEP

EPI-THALAMUS

Ventral/central thalamic NEP

**THALAMUS**

*Lateral geniculate migration*

Optic tract

Subthalamic NEP

**HYPOTHALAMUS**
Hypothalamic NEP

Subthalamic nucleus?

**SUB-THALAMUS**

Optic tract

Strionuclear GEP?

Cortical (hippocampal) NEP

Fornical GEP

**AMYGDALA**

Future insular gyrus

**Amygdaloid NEP and SVZ**

**Cortical (temporal) NEP**

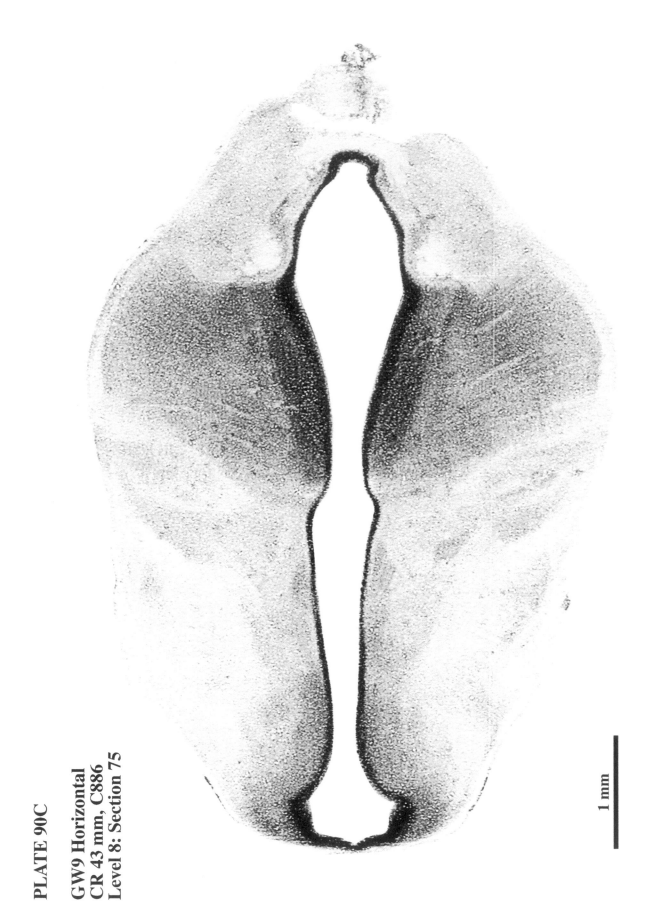

232

**PLATE 90C**

**GW9 Horizontal**
**CR 43 mm, C886**
**Level 8: Section 75**

See the entire section in parts A and B of this plate on the preceding pages.

1 mm

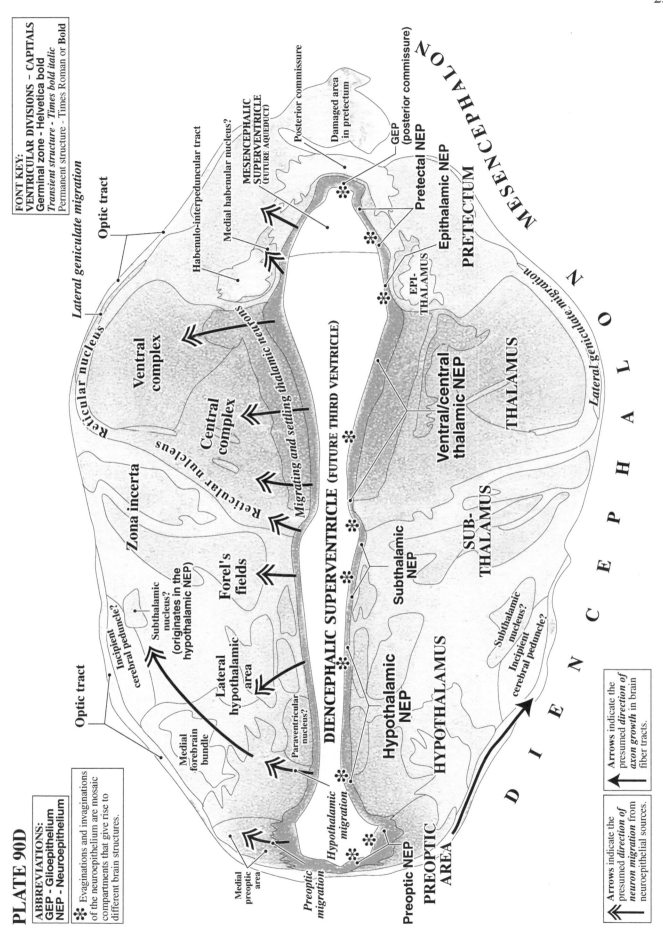

# PLATE 90D

**ABBREVIATIONS:**
GEP - Glioepithelium
NEP - Neuroepithelium

✻ Evaginations and invaginations of the neuroepithelium are mosaic compartments that give rise to different brain structures.

**FONT KEY:**
VENTRICULAR DIVISIONS – CAPITALS
Germinal zone - **Helvetica bold**
*Transient structure - Times bold italic*
Permanent structure - Times Roman or **Bold**

Arrows indicate the presumed *direction of neuron migration* from neuroepithelial sources.

Arrows indicate the presumed *direction of axon growth* in brain fiber tracts.

234

**PLATE 91A**

**GW9 Horizontal**
**CR 43 mm, C886**
**Level 9: Section 81**

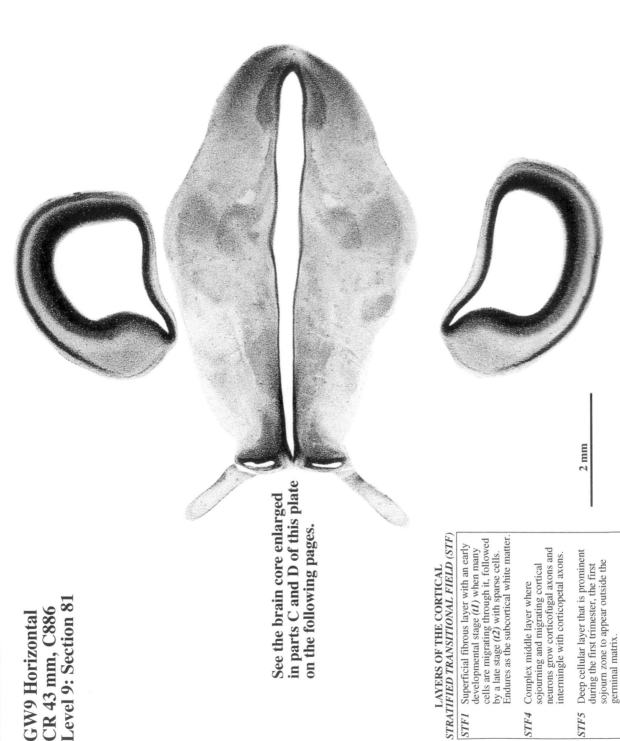

See the brain core enlarged
in parts C and D of this plate
on the following pages.

2 mm

**LAYERS OF THE CORTICAL**
*STRATIFIED TRANSITIONAL FIELD (STF)*

*STF1* Superficial fibrous layer with an early
developmental stage (*t1*) when many
cells are migrating through it, followed
by a late stage (*t2*) with sparse cells.
Endures as the subcortical white matter.

*STF4* Complex middle layer where
sojourning and migrating cortical
neurons grow corticofugal axons and
intermingle with corticopetal axons.

*STF5* Deep cellular layer that is prominent
during the first trimester, the first
sojourn zone to appear outside the
germinal matrix.

# PLATE 91B

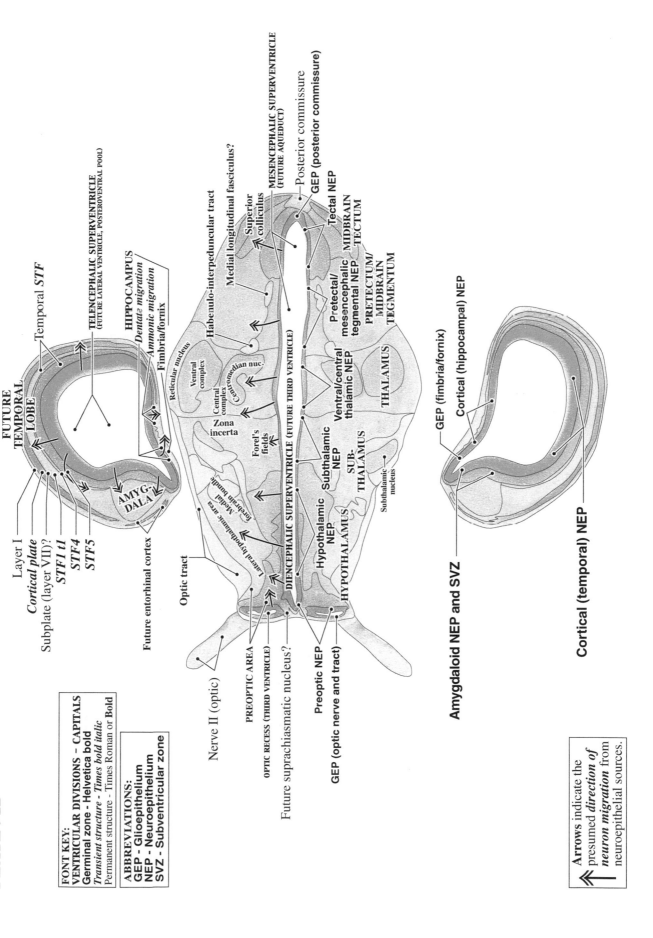

FUTURE TEMPORAL LOBE

Temporal *STF*

Layer I
*Cortical plate*
Subplate (layer VII)?
*STF1 t1*
*STF4*
*STF5*

TELENCEPHALIC SUPERVENTRICLE
(FUTURE LATERAL VENTRICLE, POSTEROVENTRAL POOL)

HIPPOCAMPUS
*Dentate migration*
*Ammonic migration*
*Fimbria/fornix*

AMYG-DALA

Future entorhinal cortex

Optic tract

Nerve II (optic)

PREOPTIC AREA

OPTIC RECESS (THIRD VENTRICLE)

Future suprachiasmatic nucleus?

GEP (optic nerve and tract)

Preoptic NEP

Reticular nucleus

Ventral complex

Central complex

Centromedian nuc.

Zona incerta

Forel's fields

*Medial forebrain bundle*

*Lateral hypothalamic area*

DIENCEPHALIC SUPERVENTRICLE (FUTURE THIRD VENTRICLE)

Habenulo-interpeduncular tract

Medial longitudinal fasciculus?

Superior colliculus

MESENCEPHALIC SUPERVENTRICLE
(FUTURE AQUEDUCT)

Posterior commissure

GEP (posterior commissure)

Tectal NEP

MIDBRAIN TECTUM

Pretectal/mesencephalic tegmental NEP

PRETECTUM/MIDBRAIN TEGMENTUM

THALAMUS

Ventral/central thalamic NEP

Subthalamic NEP

SUB-THALAMUS

Subthalamic nucleus

Hypothalamic NEP

HYPOTHALAMUS

**Amygdaloid NEP and SVZ**

GEP (fimbria/fornix)

Cortical (hippocampal) NEP

**Cortical (temporal) NEP**

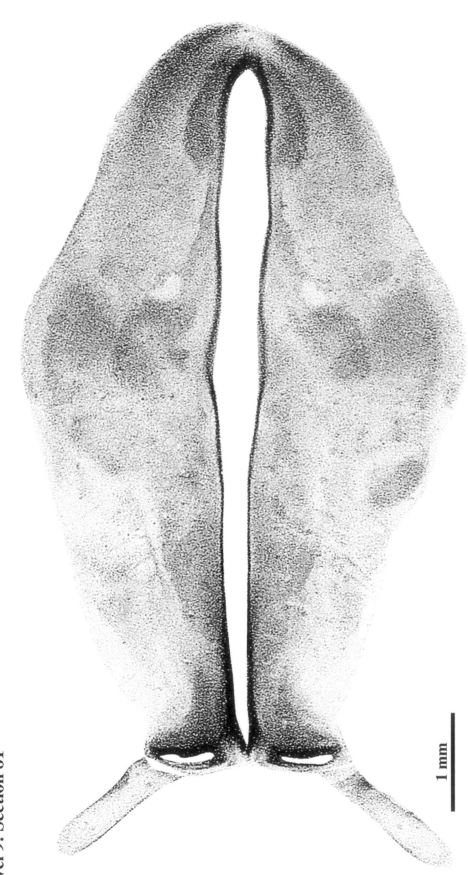

1 mm

See the entire section in parts A and B of this plate on the preceding pages.

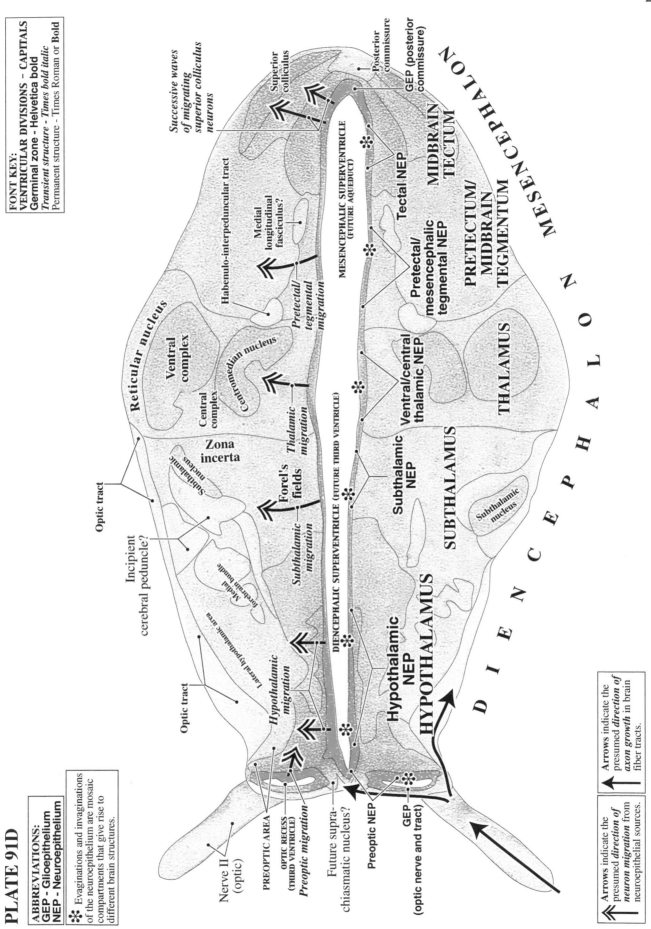

PLATE 91D

**ABBREVIATIONS:**
**GEP** - Glioepithelium
**NEP** - Neuroepithelium

✱ Evaginations and invaginations of the neuroepithelium are mosaic compartments that give rise to different brain structures.

FONT KEY:
VENTRICULAR DIVISIONS – CAPITALS
**Germinal zone** - **Helvetica bold**
*Transient structure* - *Times bold italic*
Permanent structure - Times Roman or Bold

Nerve II (optic)

PREOPTIC AREA

OPTIC RECESS (THIRD VENTRICLE)
*Preoptic migration*

Future supra-chiasmatic nucleus?

Preoptic NEP

GEP (optic nerve and tract)

Optic tract

*Hypothalamic migration*

*Lateral hypothalamic area*

*Medial forebrain bundle*

**Hypothalamic NEP**

**HYPOTHALAMUS**

DIENCEPHALIC SUPERVENTRICLE (FUTURE THIRD VENTRICLE)

Incipient cerebral peduncle?

Optic tract

*Subthalamic nucleus*

Zona incerta

*Forel's fields*

*Subthalamic migration*

**Subthalamic NEP**

**SUBTHALAMUS**

*Subthalamic nucleus*

Reticular nucleus

**Ventral complex**

Central complex

Centromedian nucleus

*Thalamic migration*

Habenulo-interpeduncular tract

*Medial longitudinal fasciculus?*

*Pretectal/ tegmental migration*

MESENCEPHALIC SUPERVENTRICLE (FUTURE AQUEDUCT)

**Ventral/central thalamic NEP**

**Pretectal/ mesencephalic tegmental NEP**

**THALAMUS**

**PRETECTUM/ MIDBRAIN TEGMENTUM**

*Successive waves of migrating superior colliculus neurons*

Superior colliculus

Posterior commissure

GEP (posterior commissure)

**Tectal NEP**

**MIDBRAIN TECTUM**

D I E N C E P H A L O N

M E S E N C E P H A L O N

Arrows indicate the presumed *direction of neuron migration* from neuroepithelial sources.

Arrows indicate the presumed *direction of axon growth* in brain fiber tracts.

**PLATE 92A**

**GW9 Horizontal**
**CR 43 mm, C886**
**Level 10: Section 86**

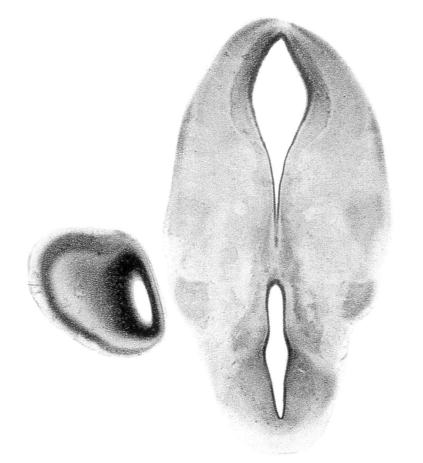

2 mm

**See the brain core enlarged**
**in parts C and D of this plate**
**on the following pages.**

**LAYERS OF THE CORTICAL**
*STRATIFIED TRANSITIONAL FIELD (STF)*

*STF1*  Superficial fibrous layer with an early
developmental stage (*t1*) when many
cells are migrating through it, followed
by a late stage (*t2*) with sparse cells.
Endures as the subcortical white matter.

*STF4*  Complex middle layer where
sojourning and migrating cortical
neurons grow corticofugal axons and
intermingle with corticopetal axons.

*STF5*  Deep cellular layer that is prominent
during the first trimester, the first
sojourn zone to appear outside the
germinal matrix.

PLATE 92B

FUTURE
TEMPORAL
LOBE

Temporal *STF*

Layer I
*Cortical plate*
Subplate (layer VII)?
*STF1 t1*
*STF4*
*STF5*

**TELENCEPHALIC SUPERVENTRICLE** (FUTURE LATERAL VENTRICLE, VENTRAL POOL)

Nucleus of the optic tract

Auditory radiation?

FONT KEY:
VENTRICULAR DIVISIONS – CAPITALS
Germinal zone - Helvetica bold
*Transient structure - Times bold italic*
Permanent structure - Times Roman or Bold

**NEP - Neuroepithelium**

Optic tract

**DIENCEPHALIC SUPERVENTRICLE**
(FUTURE THIRD VENTRICLE)

Optic chiasm

Medial
geniculate
body

Mammillo
thalamic
tract?

**Habenulo-interpeduncular tract**
**Medial longitudinal fasciculus?**

Superior
colliculus

**MESENCEPHALIC SUPERVENTRICLE**
(FUTURE AQUEDUCT)

Posterior commissure

Mesencephalic tectal NEP

**MIDBRAIN
TECTUM**

Mesencephalic
tegmental NEP

Oculomotor
nuclear complex
(III)?

**MIDBRAIN
TEGMENTUM**

**THALAMUS**

Subthalamic
nucleus

Medial
forebrain
bundle

Supramammillary
area

Lateral
hypothalamic area

Ventromedial
nucleus?

**Hypothalamic
NEP**

**HYPOTHALAMUS**

SUB-
THALAMUS

Cortical (temporal) NEP

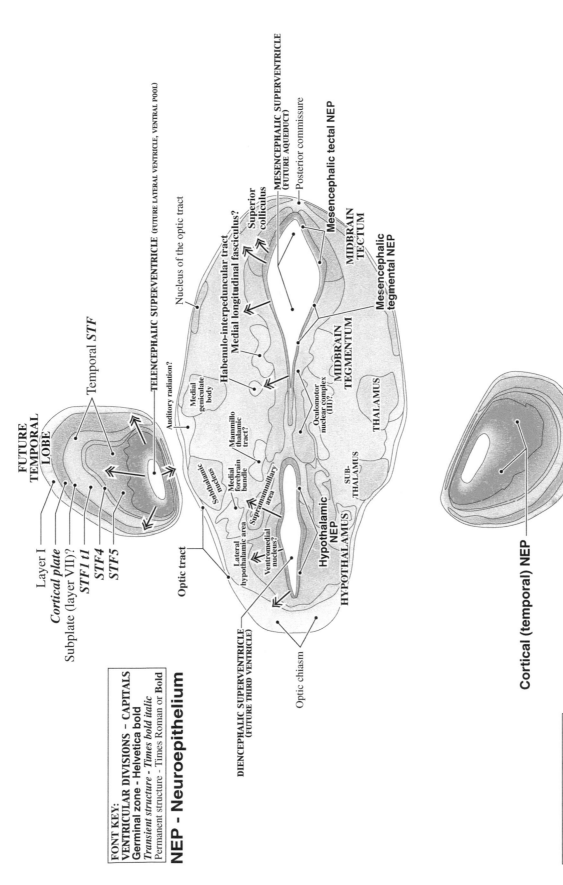

**Arrows** indicate the
presumed *direction of*
*neuron migration* from
neuroepithelial sources.

240

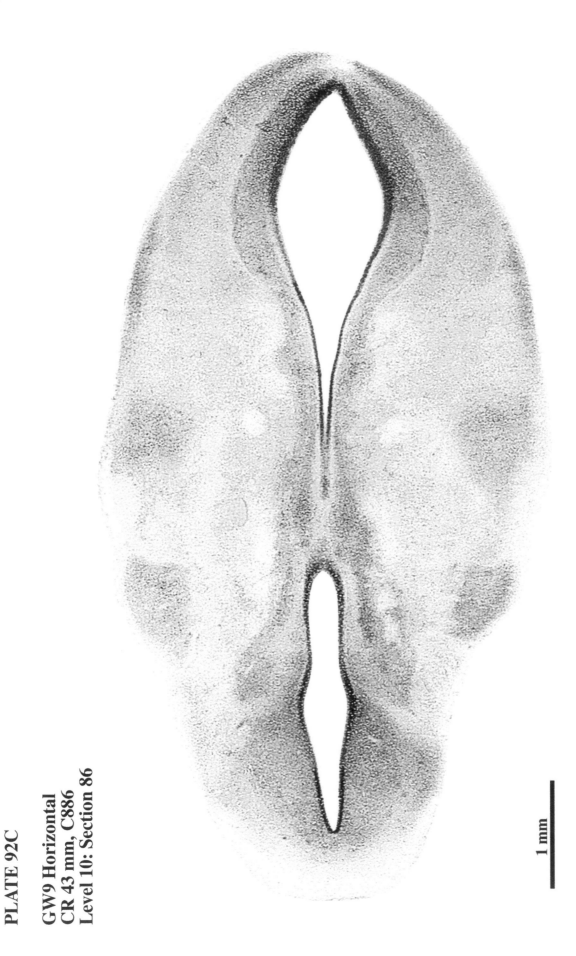

See the entire section in parts A and B of this plate on the preceding pages.

1 mm

PLATE 92C

GW9 Horizontal
CR 43 mm, C886
Level 10: Section 86

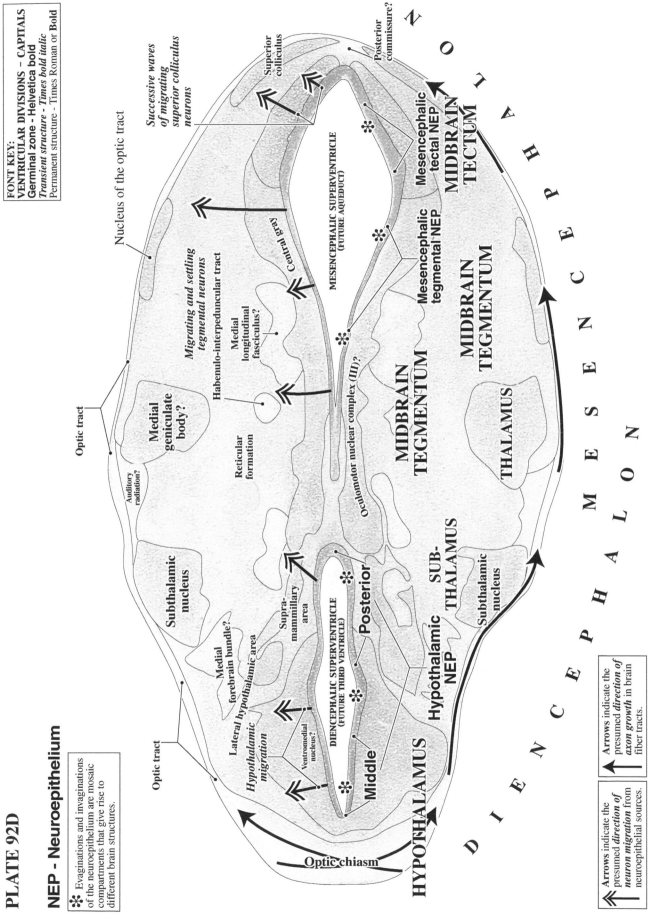

PLATE 92D

## NEP - Neuroepithelium

✳ Evaginations and invaginations of the neuroepithelium are mosaic compartments that give rise to different brain structures.

FONT KEY:
VENTRICULAR DIVISIONS – CAPITALS
Germinal zone - **Helvetica bold**
Transient structure - *Times bold italic*
Permanent structure - Times Roman or Bold

Successive waves of migrating superior colliculus neurons

Superior colliculus

Posterior commissure?

Nucleus of the optic tract

Mesencephalic tectal NEP

**MIDBRAIN TECTUM**

✳ MESENCEPHALIC SUPERVENTRICLE (FUTURE AQUEDUCT)

Central gray

Optic tract

Migrating and settling tegmental neurons

Habenulo-interpeduncular tract

Medial geniculate body?

Medial longitudinal fasciculus?

Mesencephalic tegmental NEP

✳ Oculomotor nuclear complex (III)?

**MIDBRAIN TEGMENTUM**

**MIDBRAIN TEGMENTUM**

Auditory radiation?

Reticular formation

**THALAMUS**

Subthalamic nucleus

Supra-mammillary area

**SUB-THALAMUS**

Subthalamic nucleus

Medial forebrain bundle?

✳ Posterior

**Hypothalamic NEP**

Optic tract

Lateral hypothalamic area

✳ DIENCEPHALIC SUPERVENTRICLE (FUTURE THIRD VENTRICLE)

Hypothalamic migration

Ventromedial nucleus?

✳ Middle

✳ **HYPOTHALAMUS**

Optic chiasm

**M E S E N C E P H A L O N**

**D I E N C E P H A L O N**

Arrows indicate the presumed *direction of neuron migration* from neuroepithelial sources.

Arrows indicate the presumed *direction of axon growth* in brain fiber tracts.

242

PLATE 93A

GW9 Horizontal
CR 43 mm, C886
Level 11: Section 90

See the brain core enlarged
in parts C and D of this plate
on the following pages.

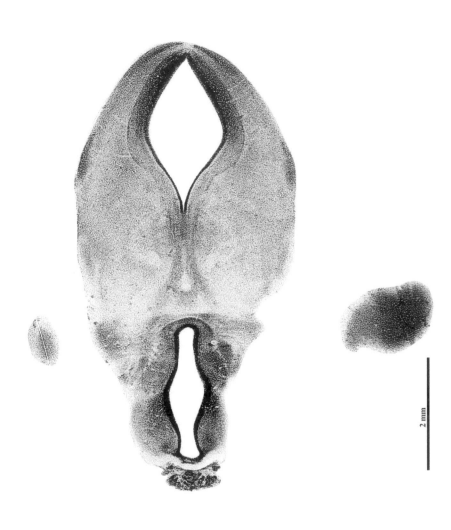

2 mm

# PLATE 93B

**FONT KEY:**
VENTRICULAR DIVISIONS - CAPITALS
Germinal zone - Helvetica bold
*Transient structure - Times bold italic*
Permanent structure - Times Roman or Bold

# NEP - Neuroepithelium

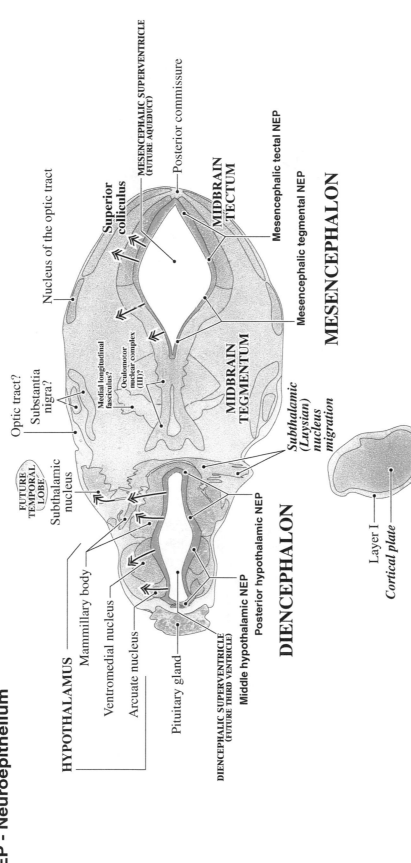

Arrows indicate the presumed *direction of neuron migration* from neuroepithelial sources.

244

**PLATE 93C**

**GW9 Horizontal**
**CR 43 mm, C886**
**Level 11: Section 90**

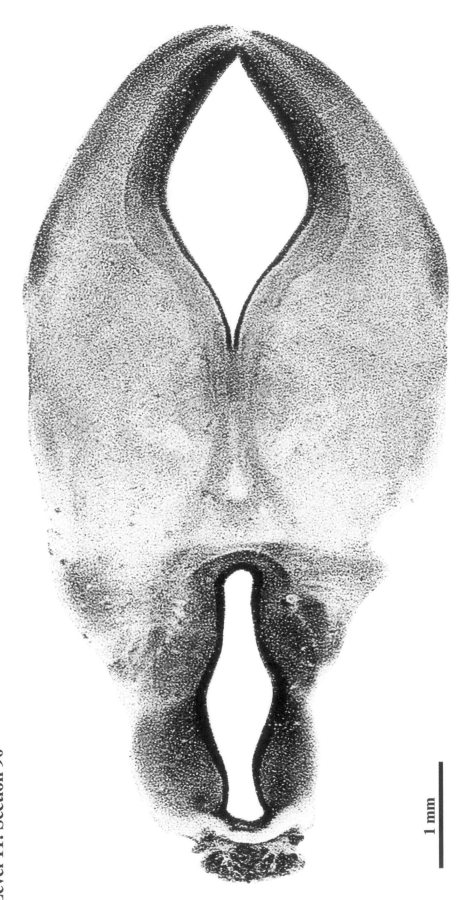

1 mm

See the entire section in parts A and B of this plate on the preceding pages.

# PLATE 93D

## NEP - Neuroepithelium

✱ Evaginations and invaginations of the neuroepithelium are mosaic compartments that give rise to different brain structures.

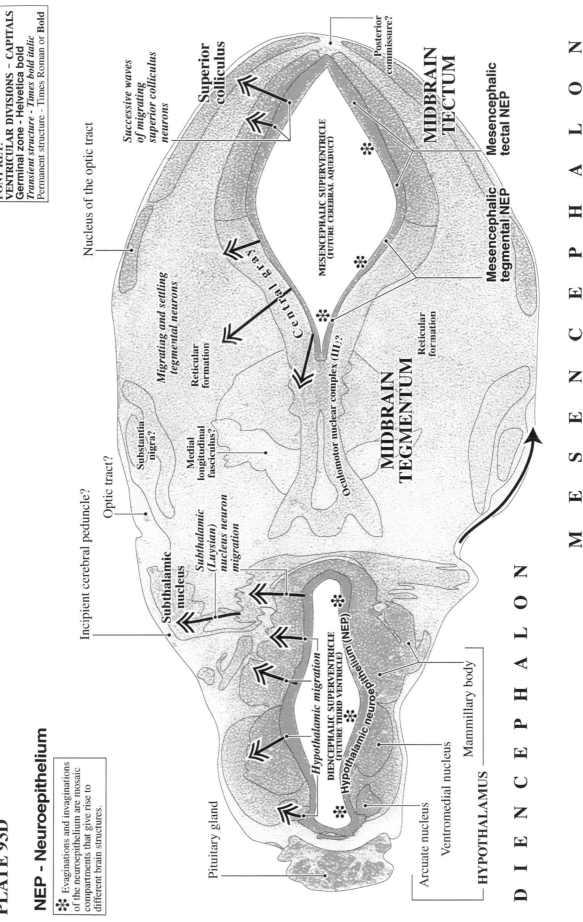

Successive waves of migrating superior colliculus neurons

**Superior colliculus**

Posterior commissure?

**MIDBRAIN TECTUM**

Mesencephalic tectal NEP

Nucleus of the optic tract

MESENCEPHALIC SUPERVENTRICLE (FUTURE CEREBRAL AQUEDUCT)

Mesencephalic tegmental NEP

Migrating and settling tegmental neurons

Central gray

Reticular formation

**MIDBRAIN TEGMENTUM**

Reticular formation

Optic tract?

Substantia nigra?

Medial longitudinal fasciculus?

Oculomotor nuclear complex (III,?)

Incipient cerebral peduncle?

**Subthalamic nucleus**

Subthalamic (Luysian) nucleus neuron migration

Hypothalamic migration

DIENCEPHALIC SUPERVENTRICLE (FUTURE THIRD VENTRICLE)

Hypothalamic neuroepithelium (NEP)

Mammillary body

Pituitary gland

Arcuate nucleus

Ventromedial nucleus

**HYPOTHALAMUS**

D I E N C E P H A L O N

M E S E N C E P H A L O N

Arrows indicate the presumed *direction of neuron migration* from neuroepithelial sources.

Arrows indicate the presumed *direction of axon growth* in brain fiber tracts.

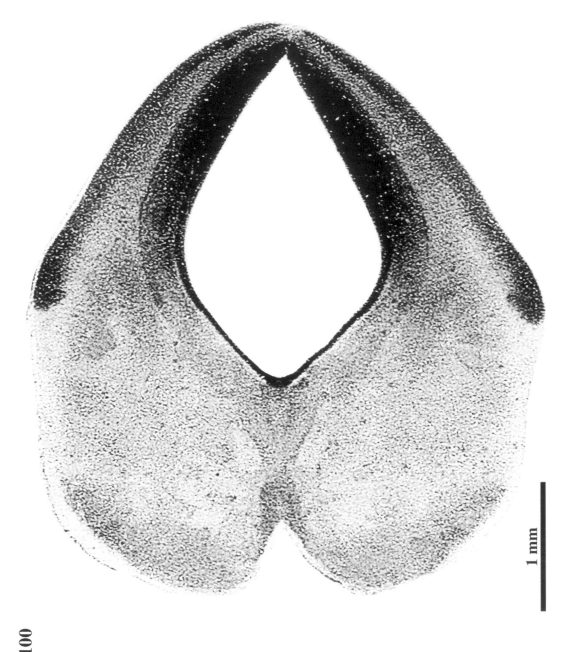

**PLATE 94A**

**GW9 Horizontal**
**CR 43 mm, C886**
**Level 12: Section 100**

1 mm

**Levels 12 to 19 are shown only at higher magnification.**

# PLATE 94B

**ABBREVIATIONS:**
**GEP - Glioepithelium**
**NEP - Neuroepithelium**

✳ Evaginations and invaginations of the neuroepithelium are mosaic compartments that give rise to different brain structures.

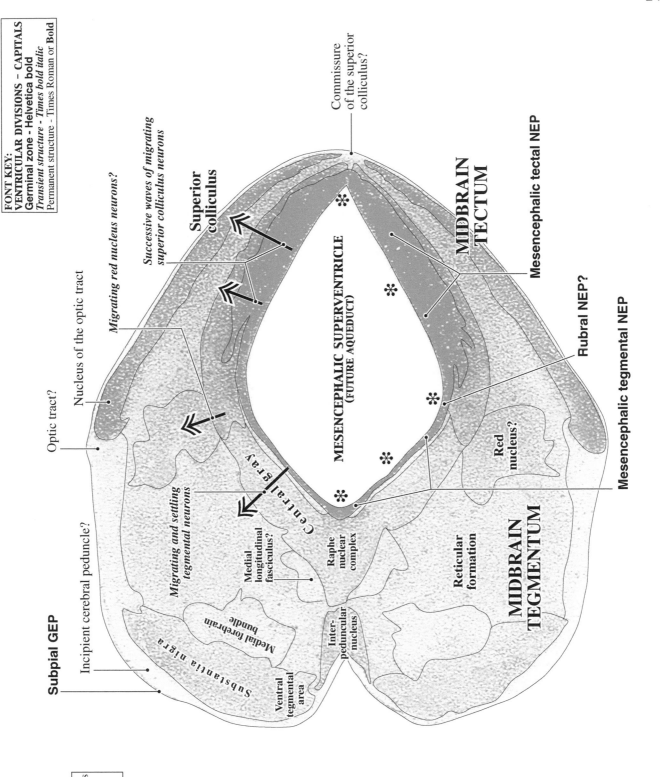

⫷ Arrows indicate the presumed *direction of neuron migration* from neuroepithelial sources.

Commissure of the superior colliculus?

Successive waves of migrating superior colliculus neurons

*Migrating red nucleus neurons?*

**Superior colliculus**

Nucleus of the optic tract

Optic tract?

**Subpial GEP**

Incipient cerebral peduncle?

*Substantia nigra*

**Medial forebrain bundle**

Ventral tegmental area

*Migrating and settling tegmental neurons*

Medial longitudinal fasciculus?

Central gray

Raphe nuclear complex

Inter-peduncular nucleus

Reticular formation

**MIDBRAIN TEGMENTUM**

Red nucleus?

**MESENCEPHALIC SUPERVENTRICLE (FUTURE AQUEDUCT)**

**MIDBRAIN TECTUM**

Mesencephalic tectal NEP

Rubral NEP?

Mesencephalic tegmental NEP

248

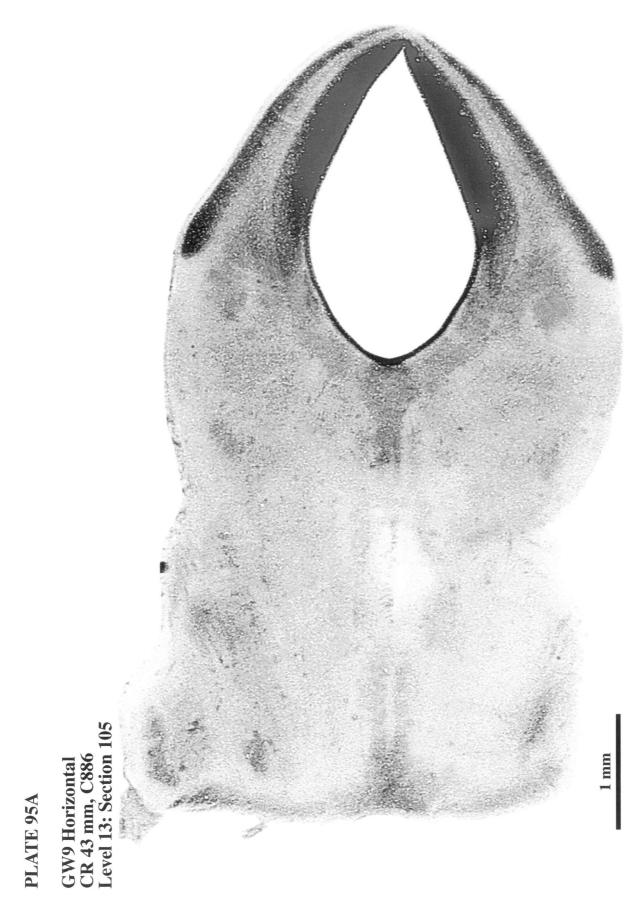

PLATE 95A

GW9 Horizontal
CR 43 mm, C886
Level 13: Section 105

Levels 12 to 19 are shown only at higher magnification.

1 mm

# PLATE 95B

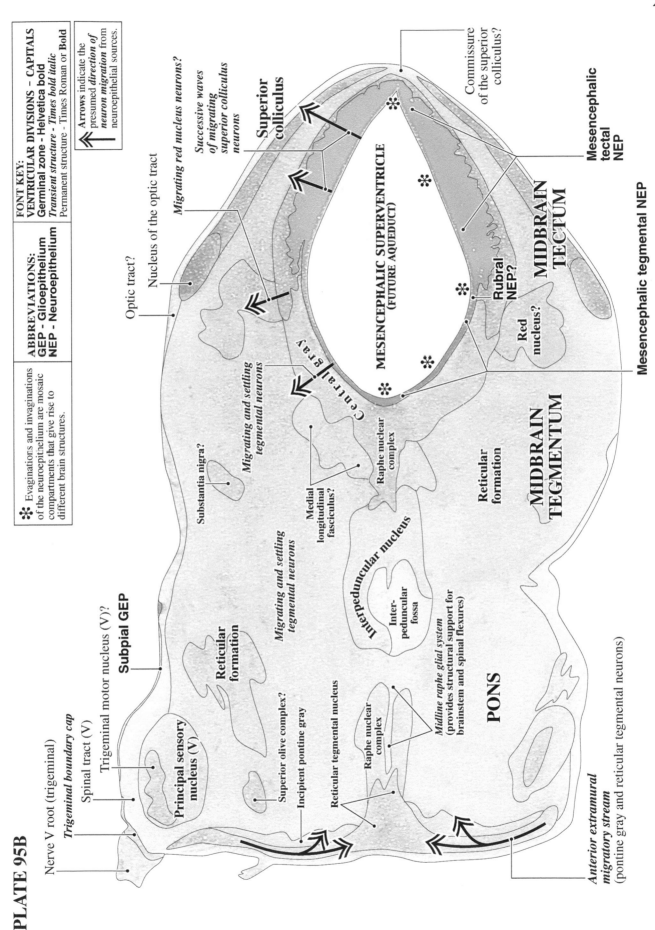

**FONT KEY:**
VENTRICULAR DIVISIONS – CAPITALS
Germinal zone - **Helvetica bold**
*Transient structure - Times bold italic*
Permanent structure - Times Roman or **Bold**

⇐ Arrows indicate the presumed *direction of neuron migration* from neuroepithelial sources.

**ABBREVIATIONS:**
GEP - Glioepithelium
NEP - Neuroepithelium

✻ Evaginations and invaginations of the neuroepithelium are mosaic compartments that give rise to different brain structures.

Nerve V root (trigeminal)
*Trigeminal boundary cap*
Spinal tract (V)
Trigeminal motor nucleus (V)?
**Subpial GEP**

**Principal sensory nucleus (V)**
*Reticular formation*
Superior olive complex?
Incipient pontine gray
Reticular tegmental nucleus
Raphe nuclear complex

*Migrating and settling tegmental neurons*

Substantia nigra?

*Migrating and settling tegmental neurons*

Medial longitudinal fasciculus?

Raphe nuclear complex

**Interpeduncular nucleus**

Inter-peduncular fossa

*Midline raphe glial system* (provides structural support for brainstem and spinal flexures)

Reticular formation

**MIDBRAIN TEGMENTUM**

**PONS**

*Anterior extramural migratory stream* (pontine gray and reticular tegmental neurons)

*Central gray*

**MESENCEPHALIC SUPERVENTRICLE** (FUTURE AQUEDUCT)

Optic tract?
Nucleus of the optic tract

*Migrating red nucleus neurons?*

*Successive waves of migrating superior colliculus neurons*

**Superior colliculus**

Commissure of the superior colliculus?

**Mesencephalic tectal NEP**

Rubral NEP?

**MIDBRAIN TECTUM**

Red nucleus?

**Mesencephalic tegmental NEP**

250

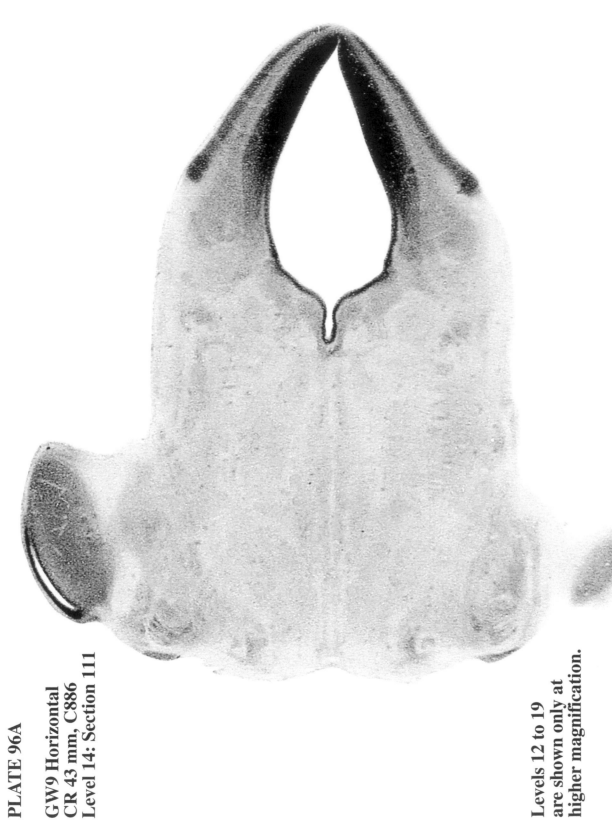

PLATE 96A

GW9 Horizontal
CR 43 mm, C886
Level 14: Section 111

Levels 12 to 19
are shown only at
higher magnification.

1 mm

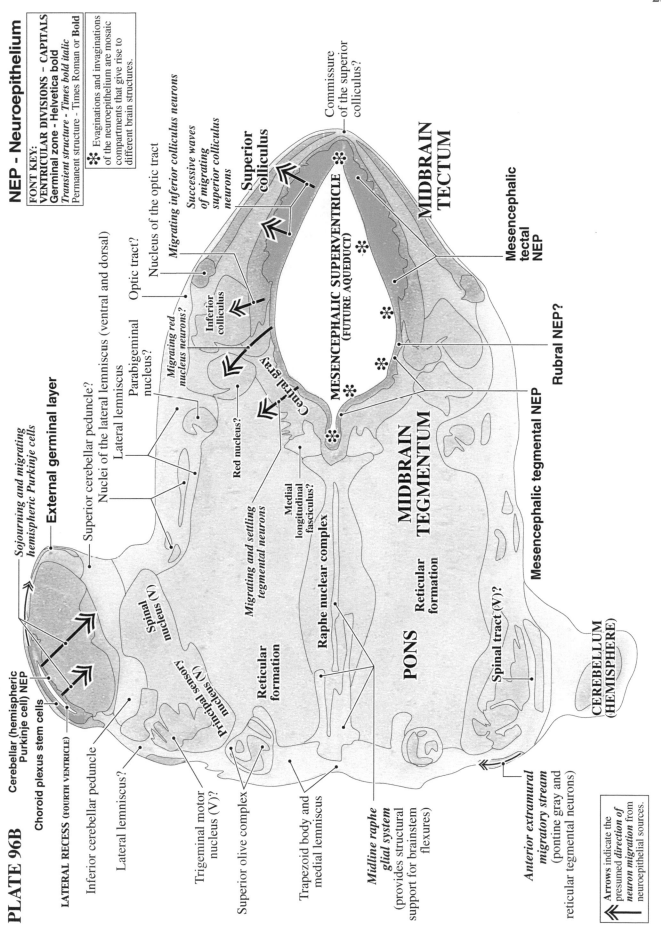

**PLATE 96B**

**NEP - Neuroepithelium**

FONT KEY:
VENTRICULAR DIVISIONS – CAPITALS
Germinal zone - Helvetica bold
Transient structure - Times bold italic
Permanent structure - Times Roman or **Bold**

✳ Evaginations and invaginations of the neuroepithelium are mosaic compartments that give rise to different brain structures.

Arrows indicate the presumed direction of neuron migration from neuroepithelial sources.

252

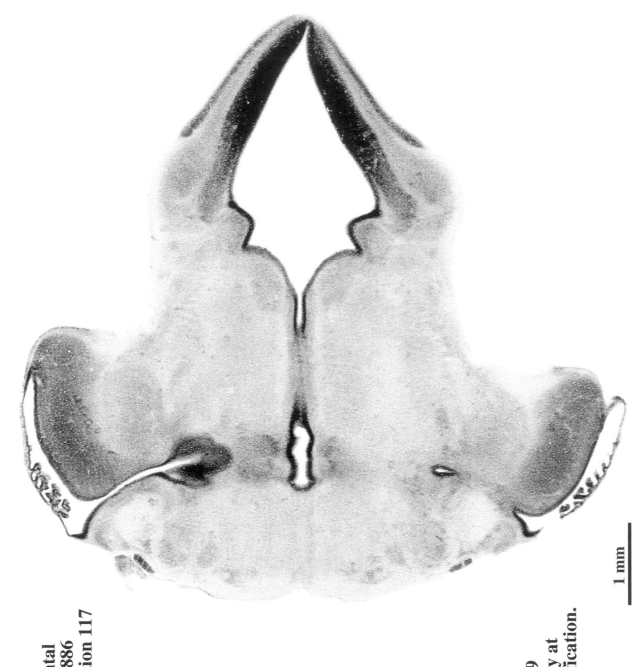

PLATE 97A

GW9 Horizontal
CR 43 mm, C886
Level 15: Section 117

Levels 12 to 19
are shown only at
higher magnification.

1 mm

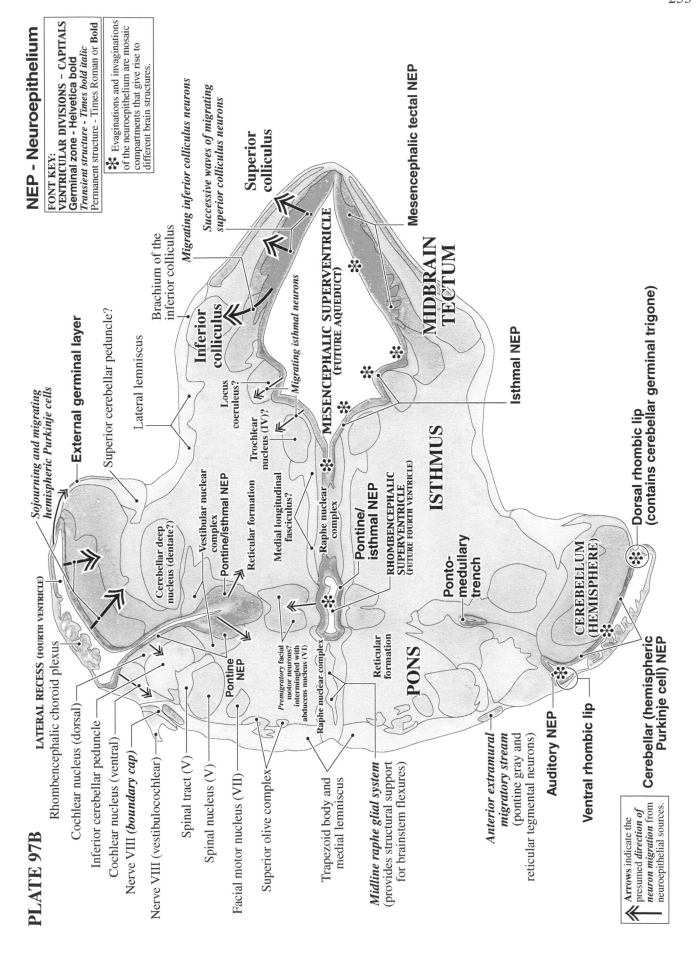

PLATE 97B

253

## NEP - Neuroepithelium

FONT KEY:
VENTRICULAR DIVISIONS – CAPITALS
Germinal zone - Helvetica bold
Transient structure - Times bold italic
Permanent structure - Times Roman or Bold

✱ Evaginations and invaginations of the neuroepithelium are mosaic compartments that give rise to different brain structures.

LATERAL RECESS (FOURTH VENTRICLE)
Rhombencephalic choroid plexus
Cochlear nucleus (dorsal)
Inferior cerebellar peduncle
Cochlear nucleus (ventral)
Nerve VIII (boundary cap)
Nerve VIII (vestibulocochlear)
Spinal tract (V)
Spinal nucleus (V)
Facial motor nucleus (VII)
Superior olive complex
Trapezoid body and medial lemniscus
Midline raphe glial system (provides structural support for brainstem flexures)
Anterior extramural migratory stream (pontine gray and reticular tegmental neurons)
Auditory NEP
Ventral rhombic lip
Cerebellar (hemispheric Purkinje cell) NEP

Sojourning and migrating hemispheric Purkinje cells
External germinal layer
Superior cerebellar peduncle?
Lateral lemniscus
Brachium of the inferior colliculus
Migrating inferior colliculus neurons
Successive waves of migrating superior colliculus neurons
Superior colliculus
Mesencephalic tectal NEP
MIDBRAIN TECTUM
Isthmal NEP
Dorsal rhombic lip (contains cerebellar germinal trigone)

Inferior colliculus
Locus coeruleus?
Trochlear nucleus (IV)?
Migrating isthmal neurons
MESENCEPHALIC SUPERVENTRICLE (FUTURE AQUEDUCT)
Cerebellar deep nucleus (dentate?)
Vestibular nuclear complex
Pontine/isthmal NEP
Reticular formation
Medial longitudinal fasciculus?
Raphe nuclear complex
Pontine/ isthmal NEP
RHOMBENCEPHALIC SUPERVENTRICLE (FUTURE FOURTH VENTRICLE)
ISTHMUS
Pontine NEP
Premigratory facial motor neurons? intermingled with abducens nucleus (VI)
Reticular formation
Raphe nuclear complex
PONS
Ponto-medullary trench
CEREBELLUM (HEMISPHERE)

Arrows indicate the presumed direction of neuron migration from neuroepithelial sources.

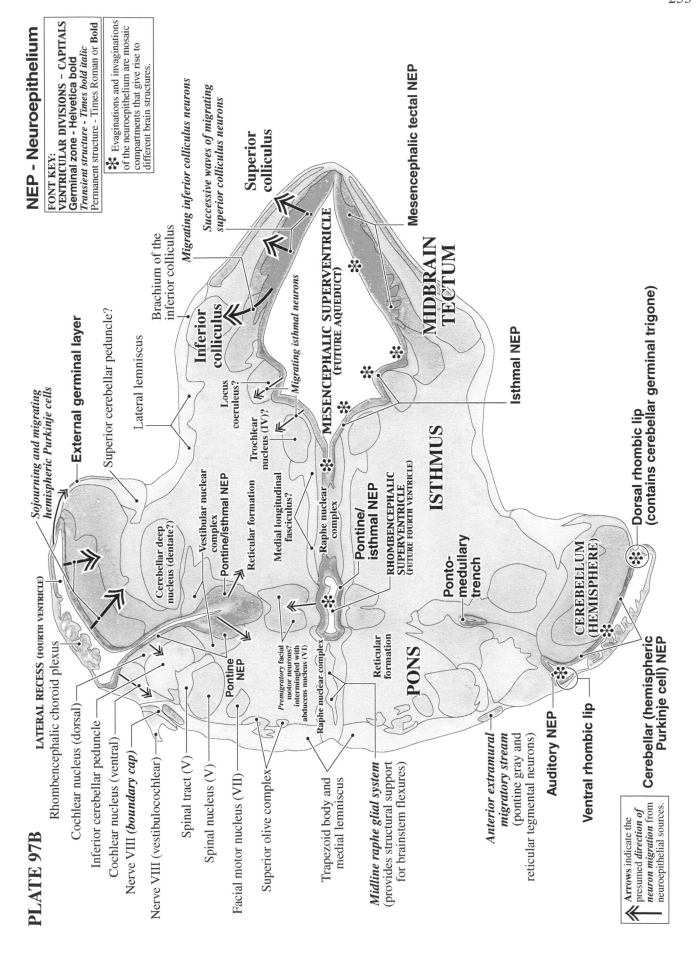

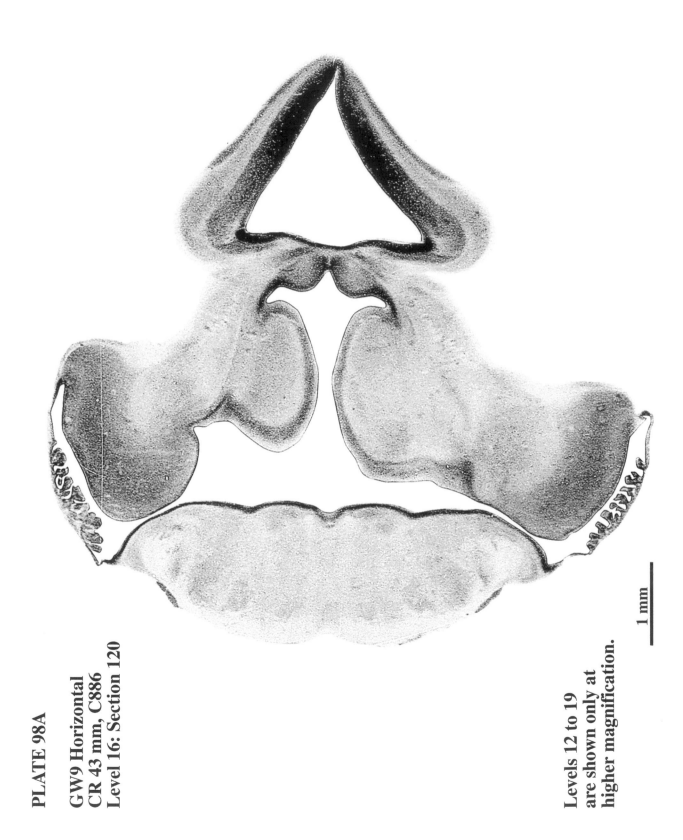

PLATE 98A

GW9 Horizontal
CR 43 mm, C886
Level 16: Section 120

Levels 12 to 19
are shown only at
higher magnification.

1 mm

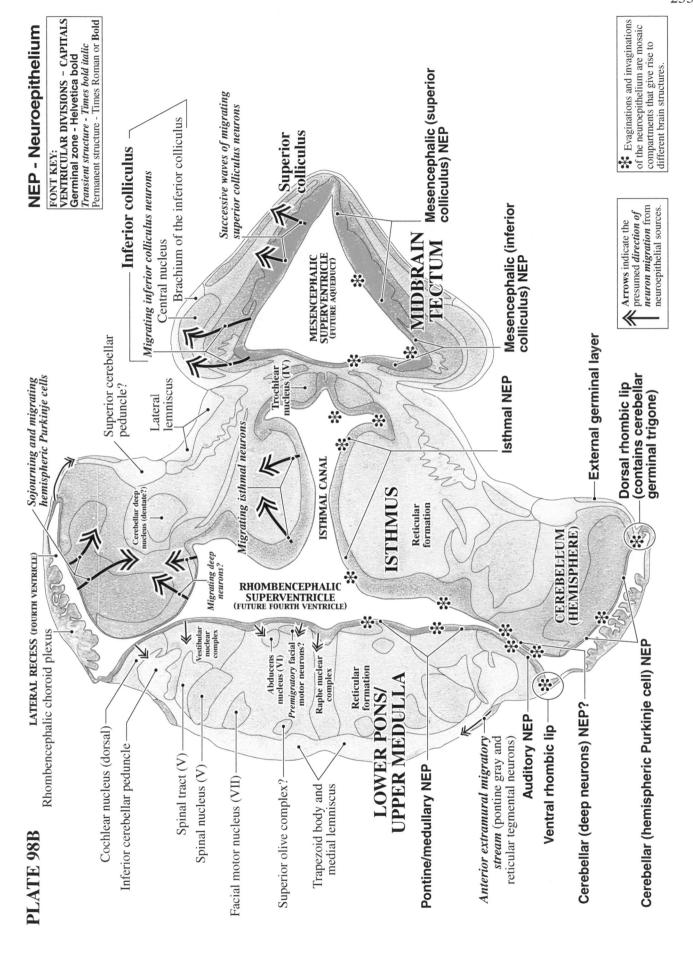

255

PLATE 98B

**NEP - Neuroepithelium**

FONT KEY:
VENTRICULAR DIVISIONS – CAPITALS
Germinal zone – Helvetica bold
*Transient structure - Times bold italic*
Permanent structure - Times Roman or **Bold**

LATERAL RECESS (FOURTH VENTRICLE)
Rhombencephalic choroid plexus

**Inferior colliculus**

Brachium of the inferior colliculus

Central nucleus

*Migrating inferior colliculus neurons*

*Successive waves of migrating superior colliculus neurons*

**Superior colliculus**

MESENCEPHALIC SUPERVENTRICLE (FUTURE AQUEDUCT)

Mesencephalic (superior colliculus) NEP

**MIDBRAIN TECTUM**

Mesencephalic (inferior colliculus) NEP

Isthmal NEP

External germinal layer

Dorsal rhombic lip (contains cerebellar germinal trigone)

Cerebellar (deep neurons) NEP?

Cerebellar (hemispheric Purkinje cell) NEP

Ventral rhombic lip

Auditory NEP

**Pontine/medullary NEP**

*Anterior extramural migratory stream (pontine gray and reticular tegmental neurons)*

Trapezoid body and medial lemniscus

Superior olive complex?

Facial motor nucleus (VII)

Spinal nucleus (V)

Spinal tract (V)

Inferior cerebellar peduncle

Cochlear nucleus (dorsal)

**LOWER PONS/ UPPER MEDULLA**

Reticular formation

Raphe nuclear complex

*Premigratory facial motor neurons?*

Abducens nucleus (VI)

Vestibular nuclear complex

RHOMBENCEPHALIC SUPERVENTRICLE (FUTURE FOURTH VENTRICLE)

*Migrating deep neurons?*

**ISTHMUS**

Reticular formation

**CEREBELLUM (HEMISPHERE)**

ISTHMAL CANAL

*Migrating isthmal neurons*

Trochlear nucleus (IV)

Lateral lemniscus

Cerebellar deep nucleus (dentate?)

Superior cerebellar peduncle?

*Sojourning and migrating hemispheric Purkinje cells*

Evaginations and invaginations of the neuroepithelium are mosaic compartments that give rise to different brain structures.

Arrows indicate the presumed *direction of neuron migration* from neuroepithelial sources.

256

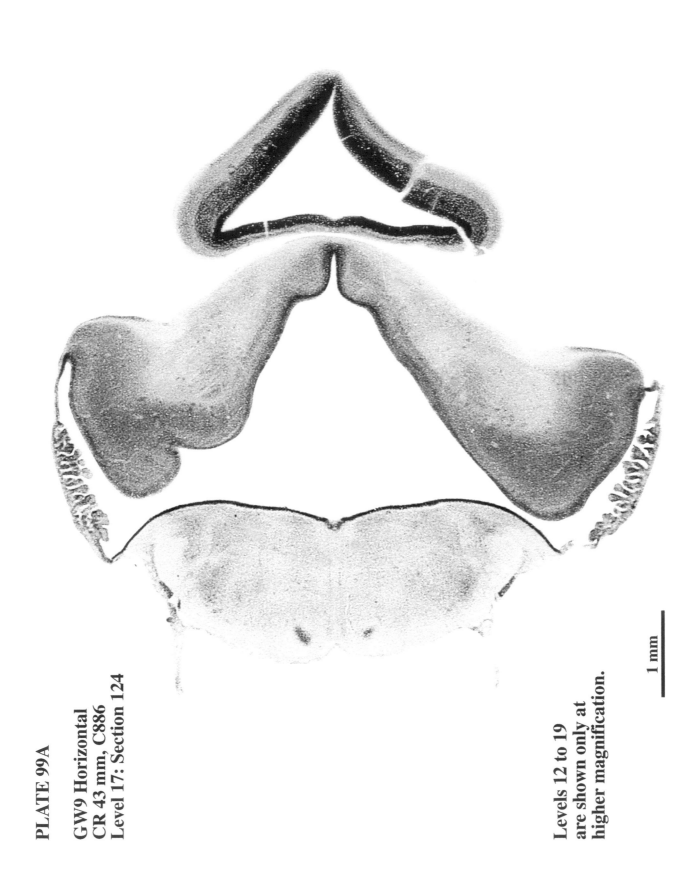

PLATE 99A

GW9 Horizontal
CR 43 mm, C886
Level 17: Section 124

Levels 12 to 19
are shown only at
higher magnification.

1 mm

# PLATE 99B

**NEP - Neuroepithelium**

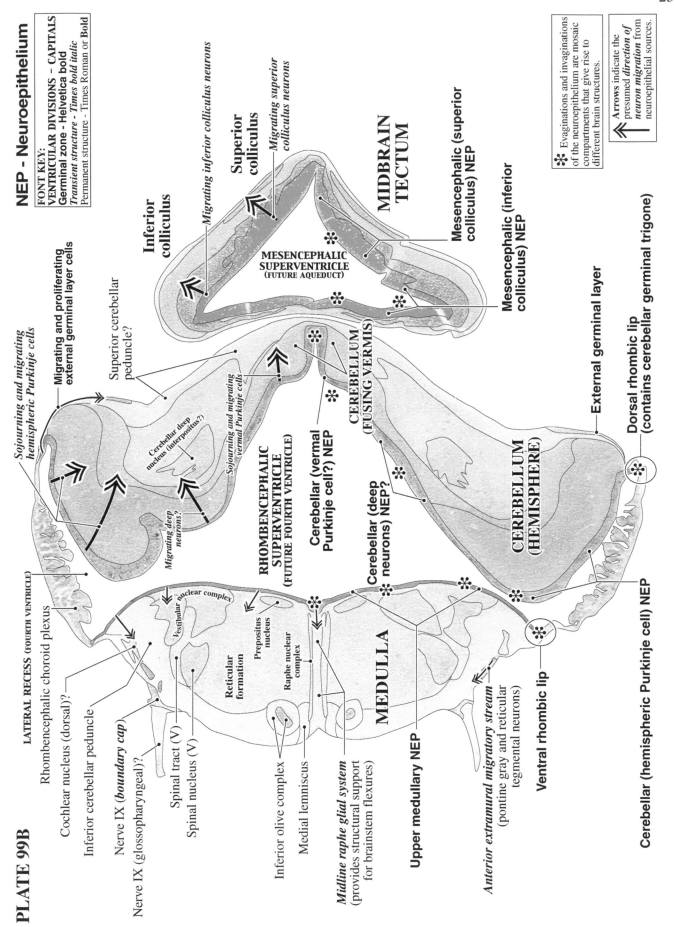

LATERAL RECESS (FOURTH VENTRICLE)

Rhombencephalic choroid plexus

Cochlear nucleus (dorsal)?

Inferior cerebellar peduncle

Nerve IX (*boundary cap*)

Nerve IX (glossopharyngeal)?

Spinal tract (V)

Spinal nucleus (V)

Inferior olive complex

Medial lemniscus

*Midline raphe glial system*
*(provides structural support*
*for brainstem flexures)*

**Upper medullary NEP**

*Anterior extramural migratory stream*
*(pontine gray and reticular*
*tegmental neurons)*

**Ventral rhombic lip**

**Cerebellar (hemispheric Purkinje cell) NEP**

*Sojourning and migrating*
*hemispheric Purkinje cells*

**Migrating and proliferating**
**external germinal layer cells**

Superior cerebellar
peduncle?

*Cerebellar deep*
*nuclei (interpositus?)*

*Sojourning and migrating*
*vermal Purkinje cells*

*Migrating deep*
*neurons?*

RHOMBENCEPHALIC
SUPERVENTRICLE
(FUTURE FOURTH VENTRICLE)

**Cerebellar (vermal Purkinje cell?) NEP**

CEREBELLUM
(FUSING VERMIS)

**Cerebellar (deep**
**neurons) NEP?**

**CEREBELLUM**
**(HEMISPHERE)**

**External germinal layer**

**Dorsal rhombic lip**
**(contains cerebellar germinal trigone)**

Vestibular nuclear complex

Prepositus
nucleus

**Reticular**
**formation**

Raphe nuclear
complex

**MEDULLA**

*Migrating inferior colliculus neurons*

*Migrating superior*
*colliculus neurons*

**Inferior**
**colliculus**

**Superior**
**colliculus**

MESENCEPHALIC
SUPERVENTRICLE
(FUTURE AQUEDUCT)

MIDBRAIN
TECTUM

**Mesencephalic (superior**
**colliculus) NEP**

**Mesencephalic (inferior**
**colliculus) NEP**

✳ Evaginations and invaginations
of the neuroepithelium are mosaic
compartments that give rise to
different brain structures.

⇐ *Arrows* indicate the
presumed *direction of*
*neuron migration* from
neuroepithelial sources.

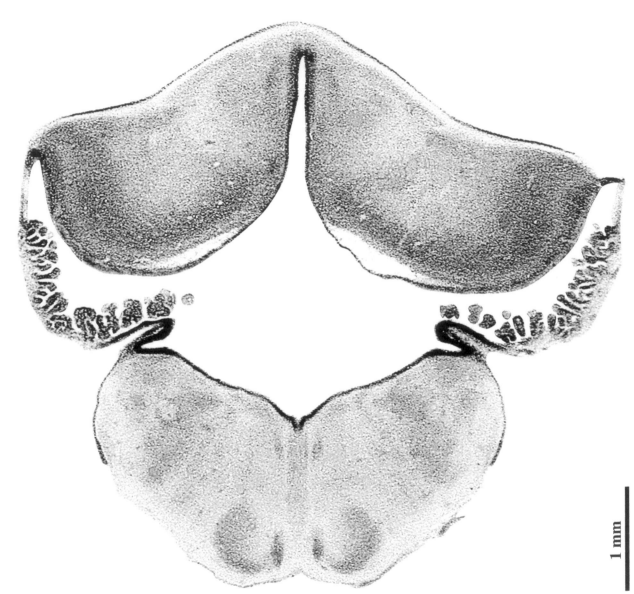

**PLATE 100A**

**GW9 Horizontal**
**CR 43 mm, C886**
**Level 18: Section 134**

1 mm

**Levels 12 to 19**
**are shown only at**
**higher magnification.**

# PLATE 100B

## NEP - Neuroepithelium

FONT KEY:
VENTRICULAR DIVISIONS – CAPITALS
Germinal zone - Helvetica bold
*Transient structure - Times bold italic*
Permanent structure - Times Roman or Bold

✱ For a detailed description of the development of the hook bundle, see Altman and Bayer (1996) pp. 71-73, 202-204.

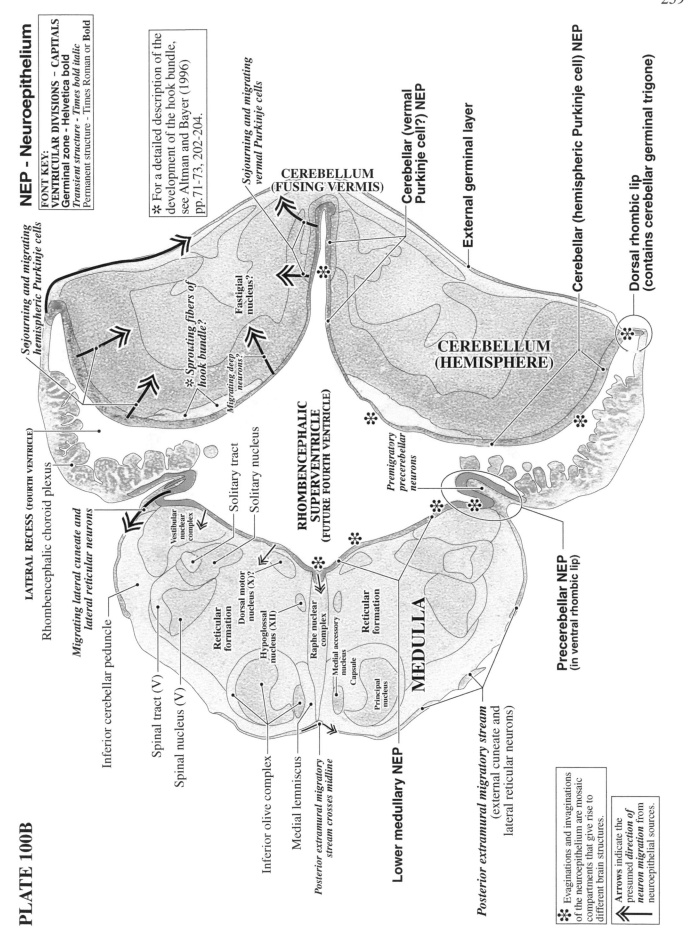

✱ Evaginations and invaginations of the neuroepithelium are mosaic compartments that give rise to different brain structures.

⟵ Arrows indicate the presumed *direction of neuron migration* from neuroepithelial sources.

260

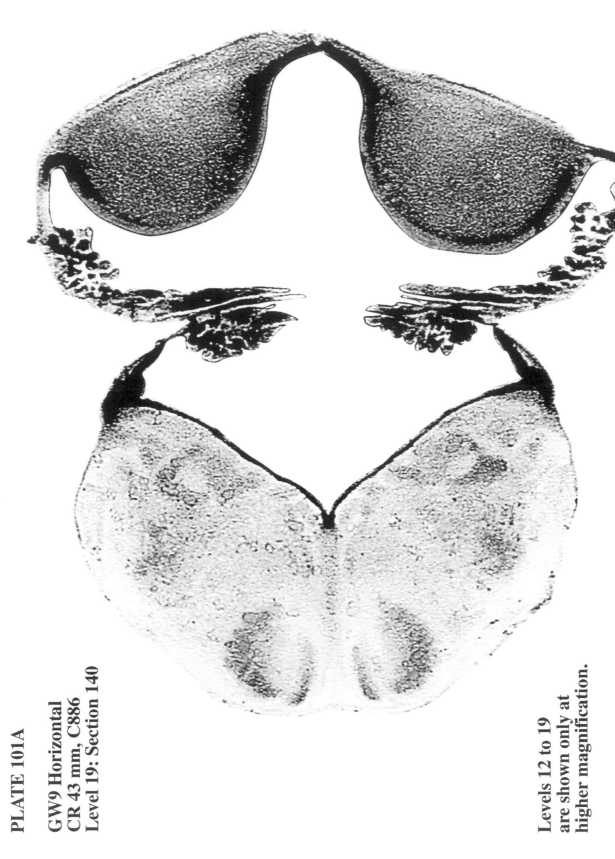

**PLATE 101A**

**GW9 Horizontal**
**CR 43 mm, C886**
**Level 19: Section 140**

**Levels 12 to 19**
**are shown only at**
**higher magnification.**

1 mm

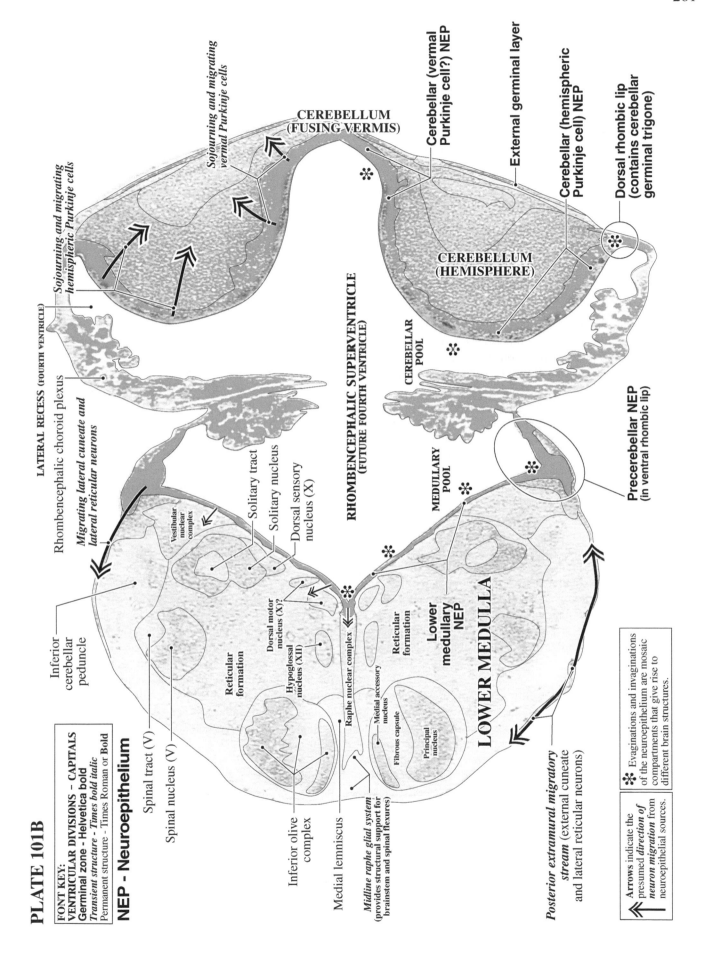

PLATE 101B

FONT KEY:
VENTRICULAR DIVISIONS - CAPITALS
Germinal zone - Helvetica bold
Transient structure - Times bold italic
Permanent structure - Times Roman or Bold

NEP - Neuroepithelium

Arrows indicate the presumed direction of neuron migration from neuroepithelial sources.

✱ Evaginations and invaginations of the neuroepithelium are mosaic compartments that give rise to different brain structures.

CEREBELLUM (FUSING VERMIS)

Sojourning and migrating vermal Purkinje cells

Sojourning and migrating hemispheric Purkinje cells

Cerebellar (vermal Purkinje cell?) NEP

External germinal layer

Cerebellar (hemispheric Purkinje cell) NEP

Cerebellar (hemispheric Purkinje cell) NEP

Dorsal rhombic lip (contains cerebellar germinal trigone)

CEREBELLUM (HEMISPHERE)

LATERAL RECESS (FOURTH VENTRICLE)

Rhombencephalic choroid plexus

Migrating lateral cuneate and lateral reticular neurons

RHOMBENCEPHALIC SUPERVENTRICLE (FUTURE FOURTH VENTRICLE)

CEREBELLAR POOL

MEDULLARY POOL

Precerebellar NEP (in ventral rhombic lip)

Inferior cerebellar peduncle

Vestibular nuclear complex

Solitary tract

Solitary nucleus

Dorsal sensory nucleus (X)

Dorsal motor nucleus (X)?

Hypoglossal nucleus (XII)

Raphe nuclear complex

Medial accessory nucleus

Fibrous capsule

Principal nucleus

Reticular formation

Lower medullary NEP

Reticular formation

LOWER MEDULLA

Spinal tract (V)

Spinal nucleus (V)

Inferior olive complex

Medial lemniscus

Midline raphe glial system (provides structural support for brainstem and spinal flexures)

Posterior extramural migratory stream (external cuneate and lateral reticular neurons)

# PART VI: GW9 CORONAL

This specimen is human fetus number 282 with a crown-rump length (CR) of 42 mm estimated to be at gestational week (GW) 9 (Minot Collection histological record number 841, referred to here as M841). The fetus was embedded in paraffin, cut in 10-μm thick sections, and stained with borax carmine and Lyon's blue. No information is available on date of collection (sometime between 1900 and 1910) and the kind of fixative. Since there is no photograph of this brain before it was embedded and cut, a specimen from Hochstetter (1919) that is comparable to M841 has been modified to show the approximate section plane and external features of the brain at GW9 (inset, **Figure 5**). Like most of the specimens in this Volume, the sections are not cut exactly in one plane; M841 is midway between coronal and horizontal. Since the cerebral cortex is in every section and the brainstem is cut in a more horizontal orientation, the brain more closely resembles a coronally sectioned brain. Photographs of 22 sections (**Levels 1-22**) are illustrated at low magnification in **Plates 102-121**. Excellent tissue detail is preserved at low magnification, but there is a fine granular precipitate visible at high magnification.

M841 is similar to the other GW9 specimens in the level of brain maturation. The chief reason for including this specimen is to provide a third perspective for viewing brain structure at GW9. In the cerebral cortex, the *neuroepithelium* is prominent as the sole germinal matrix; the *stratified transitional field (STF)* consists of *STF1*, *STF5*, and *STF4* only in lateral areas. The anterolateral (thicker) to dorsomedial (thinner) maturation gradient in the cortical plate and *STF* layers are evident. In this specimen, the olfactory evagination is most evident coming from the basal telencephalic rather than from the cerebral cortical neuroepithelium. In anterolateral parts of the cerebral cortex, streams of neurons and glia appear to leave *STF4* and enter the *lateral migratory stream*. A massive *neuroepithelium/subventricular zone* overlies the amyg-

dala, nucleus accumbens, and striatum (caudate and putamen) where neurons (and glia) are being generated.

The cerebellum has a definite *neuroepithelium* at the ventricular surface. Most of the Purkinje cells are sojourning in a thick dense layer outside the neuroepithelium, and others are migrating upward. Many of the deep neurons are superficial in the cerebellum, but some are migrating downward to intermingle with upwardly migrating Purkinje cells. The cortical surface is partially covered by an *external germinal layer (egl)* that is actively producing neuronal stem cells, as it grows over the surface of the cerebellar cortex.

The third ventricle, aqueduct, and fourth ventricle are lined by thin *neuroepithelia*. The midbrain tegmentum, pons, and medulla have the thinnest neuroepithelia indicating that only the latest generated neurons are being produced at this time. The thick precerebellar neuroepithelium is an exception in the medulla. Thicker neuroepithelia are in the cerebellum (see above) and midbrain tectum, indicating many neurons are still being generated, although the majority of the neurons in these sites are already postmitotic. The neuroepithelium is still thicker in the hypothalamus and thalamus, in accordance with the later maturation of the diencephalon compared to the rest of the brainstem.

Neurons throughout the diencephalon, midbrain tegmentum, pons, and medulla are migrating and settling. Nuclear divisions are very indistinct throughout the diencephalon because more neurons are migrating and have not yet settled permanently. More definition is seen in the midbrain tegmentum, pons, and medulla where cell migration is waning. As with the other GW9 speciments, the *anterior extramural* and *posterior extramural migratory streams* are dense subpial accumulations in the medulla and pons.

# GW9 "CORONAL" SECTION PLANES

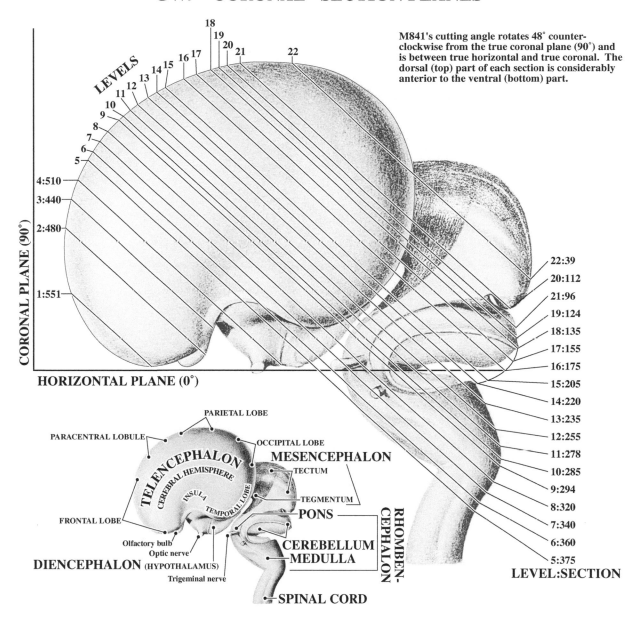

M841's cutting angle rotates 48° counter-clockwise from the true coronal plane (90°) and is between true horizontal and true coronal. The dorsal (top) part of each section is considerably anterior to the ventral (bottom) part.

LEVELS

18
19
20 21
16 17
15
14
13
12
11
10
9
8
7
6
5
22

CORONAL PLANE (90°)

4:510
3:440
2:480
1:551

HORIZONTAL PLANE (0°)

22:39
20:112
21:96
19:124
18:135
17:155
16:175
15:205
14:220
13:235
12:255
11:278
10:285
9:294
8:320
7:340
6:360
5:375

LEVEL:SECTION

PARIETAL LOBE
PARACENTRAL LOBULE
OCCIPITAL LOBE
TELENCEPHALON
CEREBRAL HEMISPHERE
MESENCEPHALON
TECTUM
INSULA
TEGMENTUM
TEMPORAL LOBE
PONS
FRONTAL LOBE
RHOMBEN-CEPHALON
Olfactory bulb
Optic nerve
CEREBELLUM
MEDULLA
DIENCEPHALON (HYPOTHALAMUS)
Trigeminal nerve
SPINAL CORD

**Figure 5.** The lateral view of the brain and upper cervical spinal cord from a specimen with a crown-rump length of 38 mm (modified from Figure 43, Table VII, Hochstetter, 1919) serves to show the approximate locations and cutting angles of the illustrated sections of M841 in the following pages. The small inset identifies the major structural features. The line in the cerebellum and dorsal edges of the pons and medulla is the cut edge of the medullary velum.

# PLATE 102A

**GW9 Coronal
CR 42 mm
M841**

## Level 1: Section 551

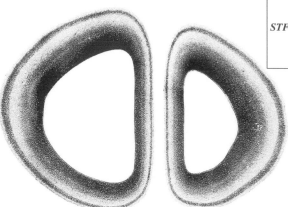

## Level 2: Section 480

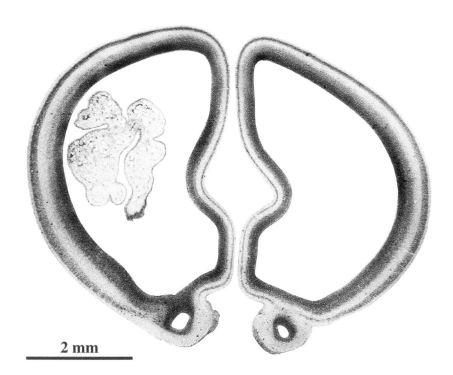

2 mm

## Level 1: Section 551

Interhemispheric fissure

Layer I
*Cortical plate*
Subplate (layer VII)?
*STF1 t1*
*STF5*
— Frontal *STF*

**FUTURE
FRONTAL
LOBE**

**Cortical (frontal) NEP**

**TELENCEPHALIC
SUPERVENTRICLE
(FUTURE LATERAL
VENTRICLE)
ANTERIOR POOL**

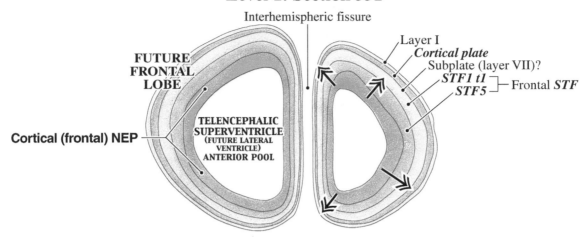

## Level 2: Section 480

Interhemispheric fissure

Layer I
*Cortical plate*
Subplate (layer VII)?
*STF1 t1*
*STF5*
— Frontal *STF*

**Cortical (frontal) NEP**

Cingulate *STF*

**FUTURE
FRONTAL
LOBE**

**Telencephalic
choroid plexus**

**TELENCEPHALIC
SUPERVENTRICLE
(FUTURE LATERAL
VENTRICLE)
ANTERIOR POOL**

**Cortical
(cingulate)
NEP**

Tenia tecta

**Cortical (orbito-
frontal) NEP**

**Cortical (olfactory) NEP**

Orbitofrontal *STF*

Anterior olfactory nucleus?

**OLFACTORY RECESS**

Nerve I (olfactory)

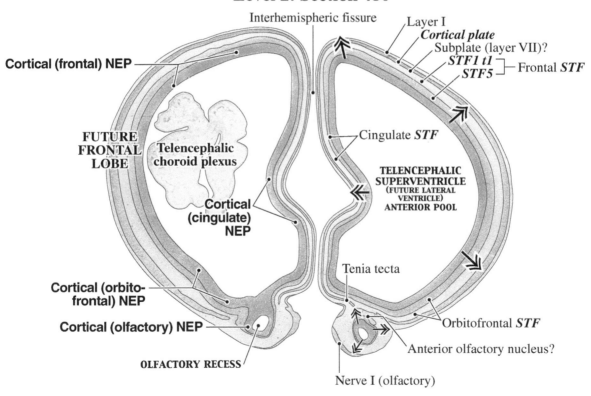

Cortical folding in the midline
is a shrinkage artifact.

**Arrows** indicate the
presumed *direction of
neuron migration* from
neuroepithelial sources.

**PLATE 103A**

**GW9 Coronal
CR 42 mm
M841**

### Level 3: Section 440

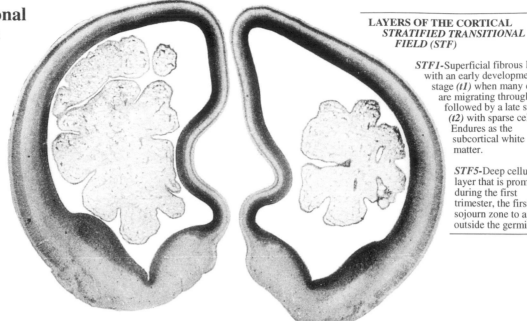

### Level 4: Section 410

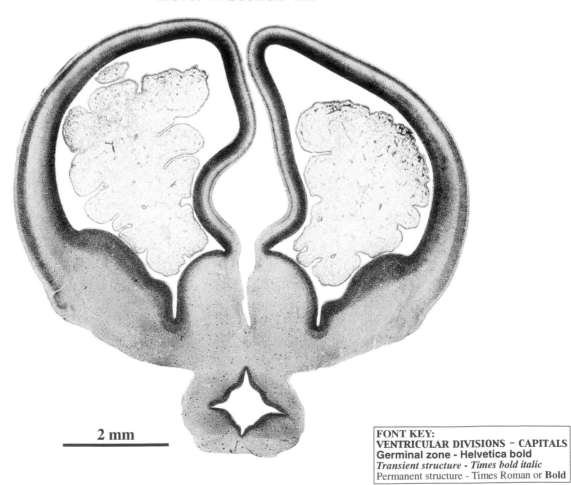

2 mm

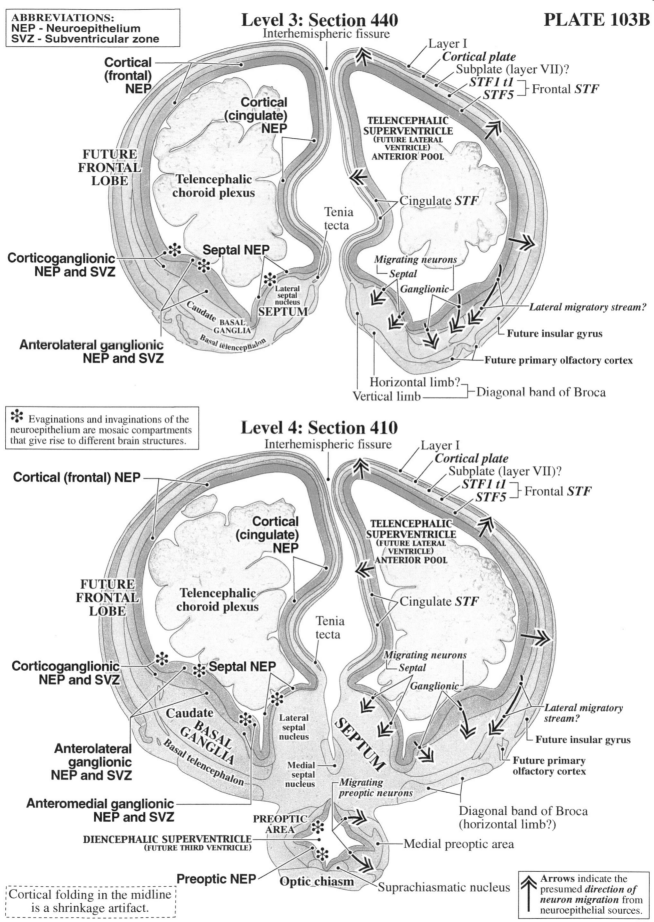

**ABBREVIATIONS:**
NEP - Neuroepithelium
SVZ - Subventricular zone

## Level 3: Section 440

Interhemispheric fissure

Cortical
(frontal)
NEP

Cortical
(cingulate)
NEP

Telencephalic
choroid plexus

FUTURE
FRONTAL
LOBE

Tenia
tecta

Layer I
*Cortical plate*
Subplate (layer VII)?
*STF1 t1*
*STF5* ⎤ Frontal *STF*

TELENCEPHALIC
SUPERVENTRICLE
(FUTURE LATERAL
VENTRICLE)
ANTERIOR POOL

Cingulate *STF*

Corticoganglionic
NEP and SVZ

Septal NEP

Lateral
septal
nucleus

Caudate

BASAL
GANGLIA

SEPTUM

*Basal telencephalon*

Anterolateral ganglionic
NEP and SVZ

*Migrating neurons*
└ *Septal*
*Ganglionic*

*Lateral migratory stream?*

Future insular gyrus

Future primary olfactory cortex

Horizontal limb?
Vertical limb ⎦ Diagonal band of Broca

✱ Evaginations and invaginations of the neuroepithelium are mosaic compartments that give rise to different brain structures.

## Level 4: Section 410

Interhemispheric fissure

Cortical (frontal) NEP

Cortical
(cingulate)
NEP

Telencephalic
choroid plexus

FUTURE
FRONTAL
LOBE

Tenia
tecta

Layer I
*Cortical plate*
Subplate (layer VII)?
*STF1 t1*
*STF5* ⎤ Frontal *STF*

TELENCEPHALIC
SUPERVENTRICLE
(FUTURE LATERAL
VENTRICLE)
ANTERIOR POOL

Cingulate *STF*

Corticoganglionic
NEP and SVZ

Septal NEP

Caudate

BASAL
GANGLIA

*Basal telencephalon*

Lateral
septal
nucleus

SEPTUM

*Migrating neurons*
└ *Septal*
*Ganglionic*

*Lateral migratory
stream?*

Future insular gyrus

Future primary
olfactory cortex

Anterolateral
ganglionic
NEP and SVZ

Anteromedial ganglionic
NEP and SVZ

Medial
septal
nucleus

*Migrating
preoptic neurons*

Diagonal band of Broca
(horizontal limb?)

PREOPTIC
AREA

Medial preoptic area

DIENCEPHALIC SUPERVENTRICLE
(FUTURE THIRD VENTRICLE)

Preoptic NEP

Optic chiasm

Suprachiasmatic nucleus

Cortical folding in the midline
is a shrinkage artifact.

**Arrows** indicate the
presumed *direction of
neuron migration* from
neuroepithelial sources.

**PLATE 104A**

**GW9 Coronal
CR 42 mm
M841
Level 5:
Section 375**

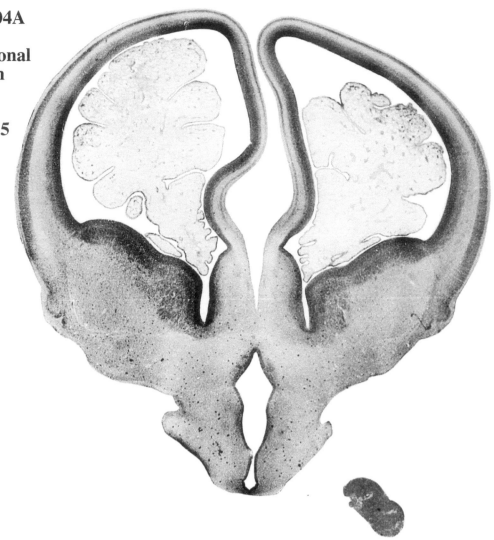

2 mm

**LAYERS OF THE CORTICAL**
*STRATIFIED TRANSITIONAL FIELD (STF)*

| | |
|---|---|
| *STF1* | Superficial fibrous layer with an early developmental stage *(t1)* when many cells are migrating through it, followed by a late stage *(t2)* with sparse cells. Endures as the subcortical white matter. |
| *STF5* | Deep cellular layer that is prominent during the first trimester, the first sojourn zone to appear outside the germinal matrix. |

**FONT KEY:**
**VENTRICULAR DIVISIONS – CAPITALS**
**Germinal zone - Helvetica bold**
*Transient structure - Times bold italic*
Permanent structure - Times Roman or **Bold**

**PLATE 104B**

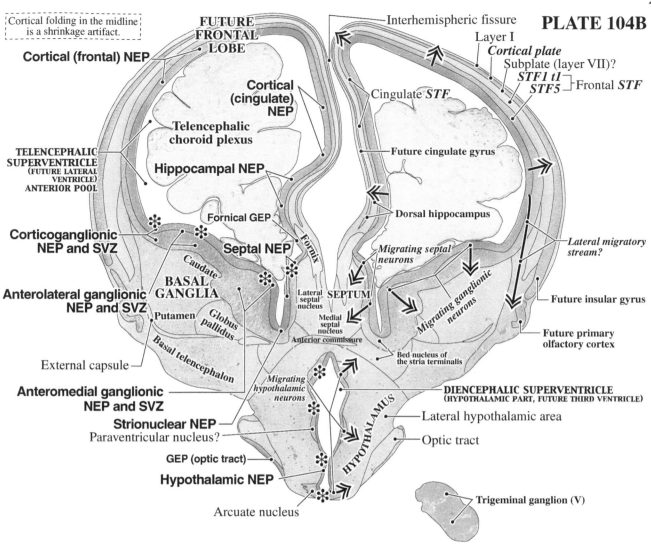

Cortical folding in the midline is a shrinkage artifact.

**FUTURE FRONTAL LOBE**

**Cortical (frontal) NEP**

Interhemispheric fissure

Layer I
*Cortical plate*
Subplate (layer VII)?
*STF1 t1*
*STF5* ⎤ Frontal *STF*

**Cortical (cingulate) NEP**

Cingulate *STF*

**Telencephalic choroid plexus**

Future cingulate gyrus

**TELENCEPHALIC SUPERVENTRICLE (FUTURE LATERAL VENTRICLE) ANTERIOR POOL**

**Hippocampal NEP**

Fornical GEP

Dorsal hippocampus

*Lateral migratory stream?*

**Corticoganglionic NEP and SVZ**

*Migrating septal neurons*

**Septal NEP**

Caudate

Fornix

**BASAL GANGLIA**

**Anterolateral ganglionic NEP and SVZ**

Lateral septal nucleus

**SEPTUM**

*Migrating ganglionic neurons*

Future insular gyrus

Putamen

Globus pallidus

Medial septal nucleus

Future primary olfactory cortex

Anterior commissure

External capsule

Basal telencephalon

Bed nucleus of the stria terminalis

**Anteromedial ganglionic NEP and SVZ**

*Migrating hypothalamic neurons*

**DIENCEPHALIC SUPERVENTRICLE (HYPOTHALAMIC PART, FUTURE THIRD VENTRICLE)**

**Strionuclear NEP**
Paraventricular nucleus?

Lateral hypothalamic area

Optic tract

**GEP (optic tract)**

HYPOTHALAMUS

**Hypothalamic NEP**

Arcuate nucleus

Trigeminal ganglion (V)

✳ Evaginations and invaginations of the neuroepithelium are mosaic compartments that give rise to different brain structures.

**ABBREVIATIONS:**
**GEP** - Glioepithelium
**G/EP** - Glioepithelium/ependyma
**NEP** - Neuroepithelium
**SVZ** - Subventricular zone

**Arrows** indicate the presumed *direction of neuron migration* from neuroepithelial sources.

**SPINAL CORD**

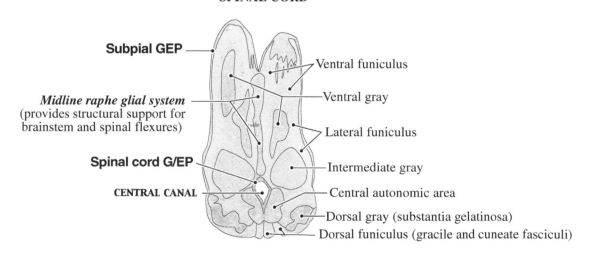

**Subpial GEP**

Ventral funiculus

*Midline raphe glial system* (provides structural support for brainstem and spinal flexures)

Ventral gray

Lateral funiculus

**Spinal cord G/EP**

Intermediate gray

**CENTRAL CANAL**

Central autonomic area

Dorsal gray (substantia gelatinosa)

Dorsal funiculus (gracile and cuneate fasciculi)

**PLATE 105A**

**GW9 Coronal**
**CR 42 mm**
**M841**
**Level 6:**
**Section 360**

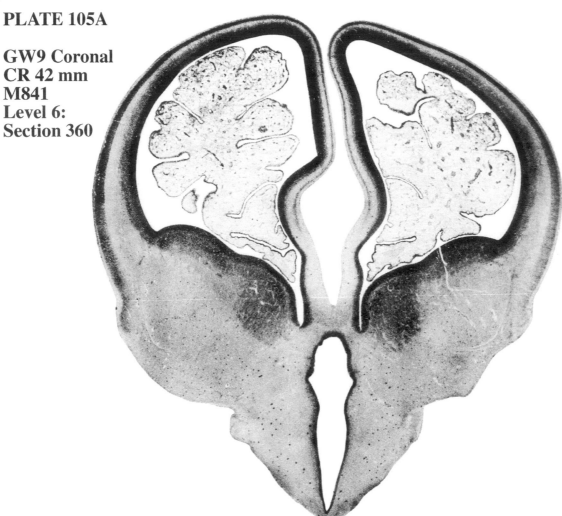

2 mm

**LAYERS OF THE CORTICAL**
*STRATIFIED TRANSITIONAL FIELD (STF)*

| | |
|---|---|
| ***STF1*** | Superficial fibrous layer with an early developmental stage *(t1)* when many cells are migrating through it, followed by a late stage *(t2)* with sparse cells. Endures as the subcortical white matter. |
| ***STF5*** | Deep cellular layer that is prominent during the first trimester, the first sojourn zone to appear outside the germinal matrix. |

FONT KEY:
**VENTRICULAR DIVISIONS – CAPITALS**
**Germinal zone - Helvetica bold**
***Transient structure - Times bold italic***
Permanent structure - Times Roman or **Bold**

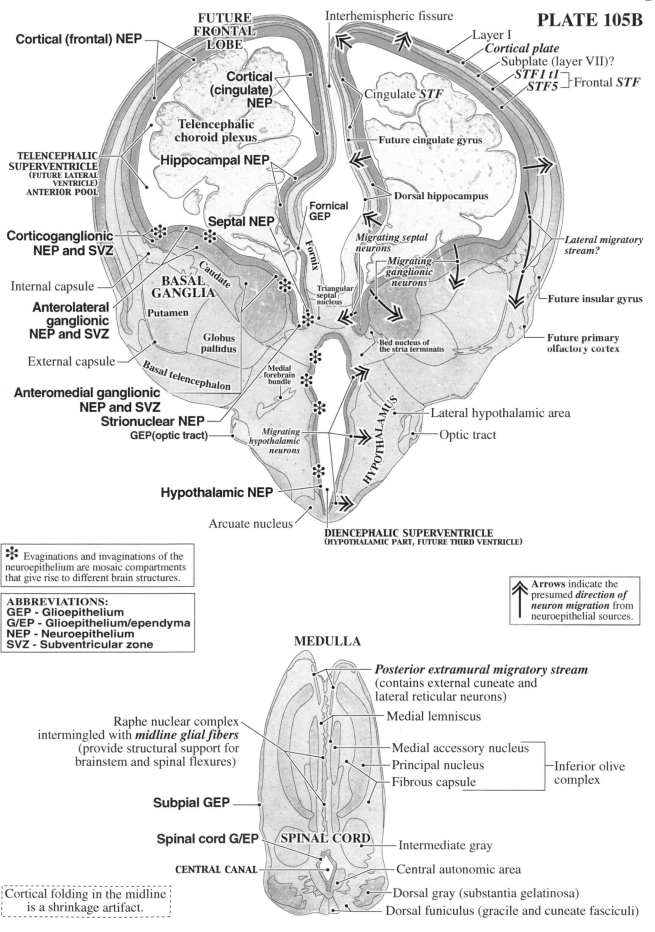

271

PLATE 105B

Interhemispheric fissure

FUTURE FRONTAL LOBE

Cortical (frontal) NEP

Cortical (cingulate) NEP

Telencephalic choroid plexus

Hippocampal NEP

TELENCEPHALIC SUPERVENTRICLE (FUTURE LATERAL VENTRICLE) ANTERIOR POOL

Corticoganglionic NEP and SVZ

Internal capsule

BASAL GANGLIA

Caudate

Anterolateral ganglionic NEP and SVZ

Putamen

External capsule

Globus pallidus

Basal telencephalon

Anteromedial ganglionic NEP and SVZ

Strionuclear NEP

GEP(optic tract)

Medial forebrain bundle

Migrating hypothalamic neurons

Hypothalamic NEP

Arcuate nucleus

Septal NEP

Fornical GEP

Fornix

Triangular septal nucleus

Layer I
Cortical plate
Subplate (layer VII)?
STF1 t1
STF5
Frontal STF

Cingulate STF

Future cingulate gyrus

Dorsal hippocampus

Migrating septal neurons

Migrating ganglionic neurons

Lateral migratory stream?

Future insular gyrus

Bed nucleus of the stria terminalis

Future primary olfactory cortex

HYPOTHALAMUS

Lateral hypothalamic area

Optic tract

DIENCEPHALIC SUPERVENTRICLE
(HYPOTHALAMIC PART, FUTURE THIRD VENTRICLE)

✳ Evaginations and invaginations of the neuroepithelium are mosaic compartments that give rise to different brain structures.

ABBREVIATIONS:
GEP - Glioepithelium
G/EP - Glioepithelium/ependyma
NEP - Neuroepithelium
SVZ - Subventricular zone

Arrows indicate the presumed direction of neuron migration from neuroepithelial sources.

MEDULLA

Posterior extramural migratory stream (contains external cuneate and lateral reticular neurons)

Medial lemniscus

Raphe nuclear complex intermingled with midline glial fibers (provide structural support for brainstem and spinal flexures)

Medial accessory nucleus

Principal nucleus

Inferior olive complex

Fibrous capsule

Subpial GEP

Spinal cord G/EP

SPINAL CORD

Intermediate gray

CENTRAL CANAL

Central autonomic area

Dorsal gray (substantia gelatinosa)

Cortical folding in the midline is a shrinkage artifact.

Dorsal funiculus (gracile and cuneate fasciculi)

**PLATE 106A**

**GW9 Coronal**
**CR 42 mm**
**M841**
**Level 7:**
**Section**
**340**

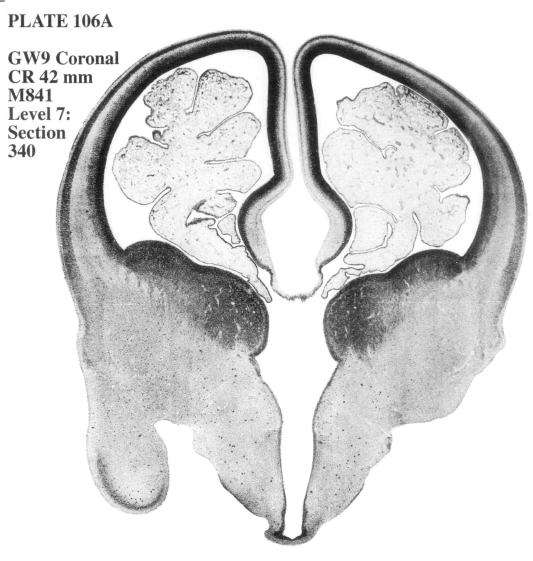

**LAYERS OF THE CORTICAL**
*STRATIFIED TRANSITIONAL FIELD (STF)*

*STF1*  Superficial fibrous layer with an early
developmental stage *(t1)* when many cells are
migrating through it, followed by a late stage *(t2)*
with sparse cells.  Endures as the subcortical white
matter.

*STF4*  Complex middle layer where sojourning and
migrating cortical neurons grow corticofugal axons
and intermingle with corticopetal axons.

*STF5*  Deep cellular layer that is prominent during the
first trimester, the first sojourn zone to appear
outside the germinal matrix.

2 mm

**FONT KEY:**
**VENTRICULAR DIVISIONS − CAPITALS**
**Germinal zone - Helvetica bold**
*Transient structure - Times bold italic*
Permanent structure - Times Roman or **Bold**

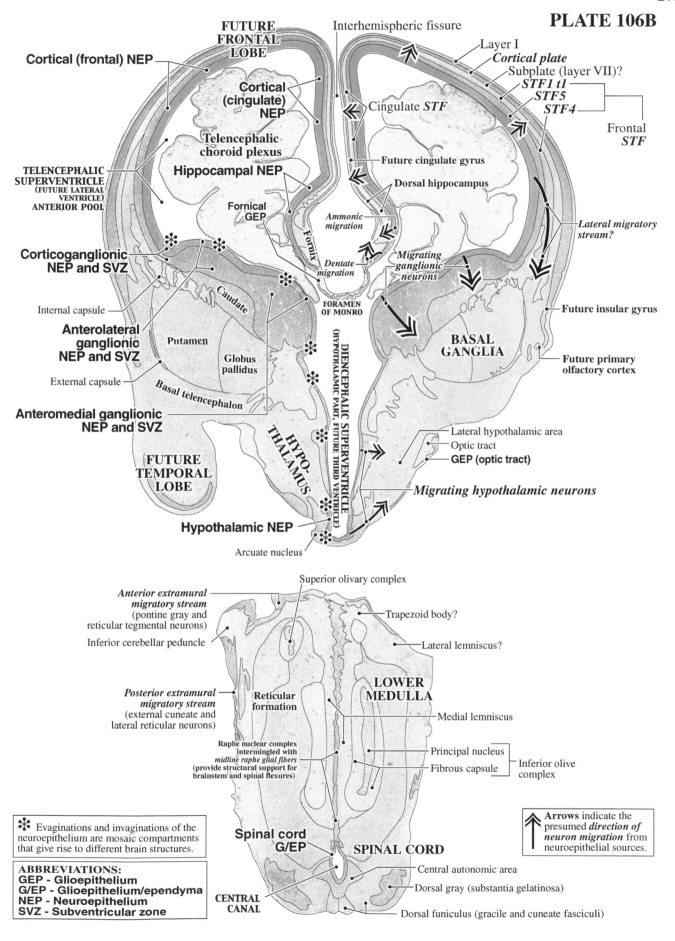

**FUTURE FRONTAL LOBE**

Interhemispheric fissure

Layer I
*Cortical plate*
Subplate (layer VII)?
*STF1 t1*
*STF5*
*STF4*

**Cortical (frontal) NEP**

**Cortical (cingulate) NEP**

Frontal *STF*

Cingulate *STF*

**Telencephalic choroid plexus**

Future cingulate gyrus

**Hippocampal NEP**

Dorsal hippocampus

*Lateral migratory stream?*

**TELENCEPHALIC SUPERVENTRICLE (FUTURE LATERAL VENTRICLE) ANTERIOR POOL**

Fornical GEP

*Ammonic migration*

Fornix

*Migrating ganglionic neurons*

**Corticoganglionic NEP and SVZ**

Caudate

*Dentate migration*

**FORAMEN OF MONRO**

Future insular gyrus

Internal capsule

**Anterolateral ganglionic NEP and SVZ**

Putamen

Globus pallidus

**BASAL GANGLIA**

External capsule

Basal telencephalon

Future primary olfactory cortex

**Anteromedial ganglionic NEP and SVZ**

**DIENCEPHALIC SUPERVENTRICLE (HYPOTHALAMIC PART, FUTURE THIRD VENTRICLE)**

**FUTURE TEMPORAL LOBE**

**HYPO-THALAMUS**

Lateral hypothalamic area
Optic tract
**GEP (optic tract)**

*Migrating hypothalamic neurons*

**Hypothalamic NEP**

Arcuate nucleus

Superior olivary complex

*Anterior extramural migratory stream* (pontine gray and reticular tegmental neurons)

Trapezoid body?

Inferior cerebellar peduncle

Lateral lemniscus?

*Posterior extramural migratory stream* (external cuneate and lateral reticular neurons)

Reticular formation

**LOWER MEDULLA**

Medial lemniscus

Raphe nuclear complex intermingled with *midline raphe glial fibers* (provide structural support for brainstem and spinal flexures)

Principal nucleus
Fibrous capsule

Inferior olive complex

**Spinal cord G/EP**

**SPINAL CORD**

Central autonomic area

**CENTRAL CANAL**

Dorsal gray (substantia gelatinosa)

Dorsal funiculus (gracile and cuneate fasciculi)

✱ Evaginations and invaginations of the neuroepithelium are mosaic compartments that give rise to different brain structures.

**ABBREVIATIONS:**
**GEP** - Glioepithelium
**G/EP** - Glioepithelium/ependyma
**NEP** - Neuroepithelium
**SVZ** - Subventricular zone

**Arrows** indicate the presumed *direction of neuron migration* from neuroepithelial sources.

**PLATE 107A**

**GW9 Coronal**
**CR 42 mm**
**M841**
**Level 8:**
**Section**
**320**

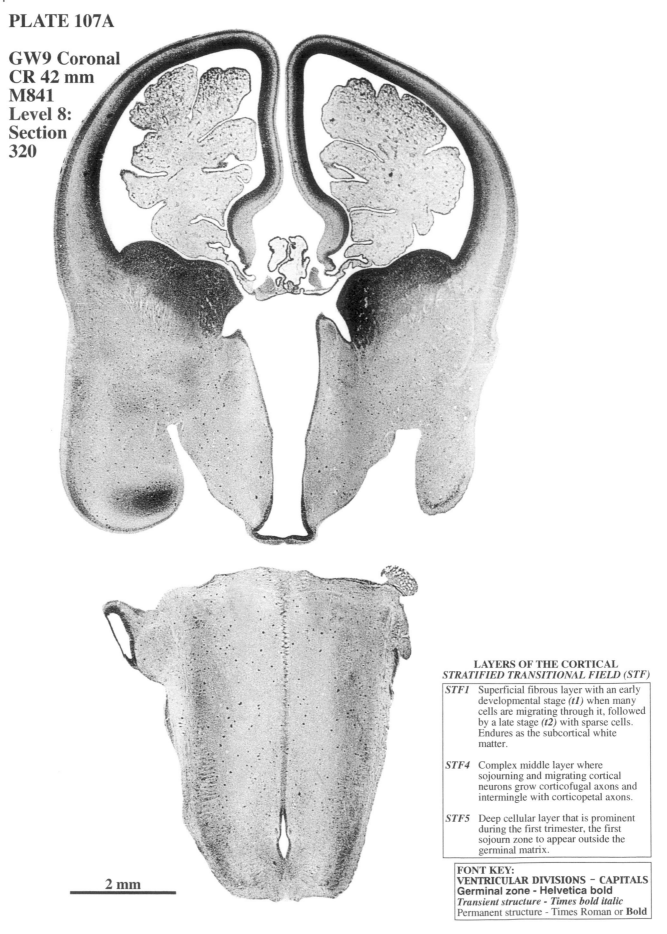

2 mm

**LAYERS OF THE CORTICAL**
*STRATIFIED TRANSITIONAL FIELD (STF)*

| | |
|---|---|
| *STF1* | Superficial fibrous layer with an early developmental stage *(t1)* when many cells are migrating through it, followed by a late stage *(t2)* with sparse cells. Endures as the subcortical white matter. |
| *STF4* | Complex middle layer where sojourning and migrating cortical neurons grow corticofugal axons and intermingle with corticopetal axons. |
| *STF5* | Deep cellular layer that is prominent during the first trimester, the first sojourn zone to appear outside the germinal matrix. |

**FONT KEY:**
**VENTRICULAR DIVISIONS – CAPITALS**
**Germinal zone - Helvetica bold**
*Transient structure - Times bold italic*
Permanent structure - Times Roman or **Bold**

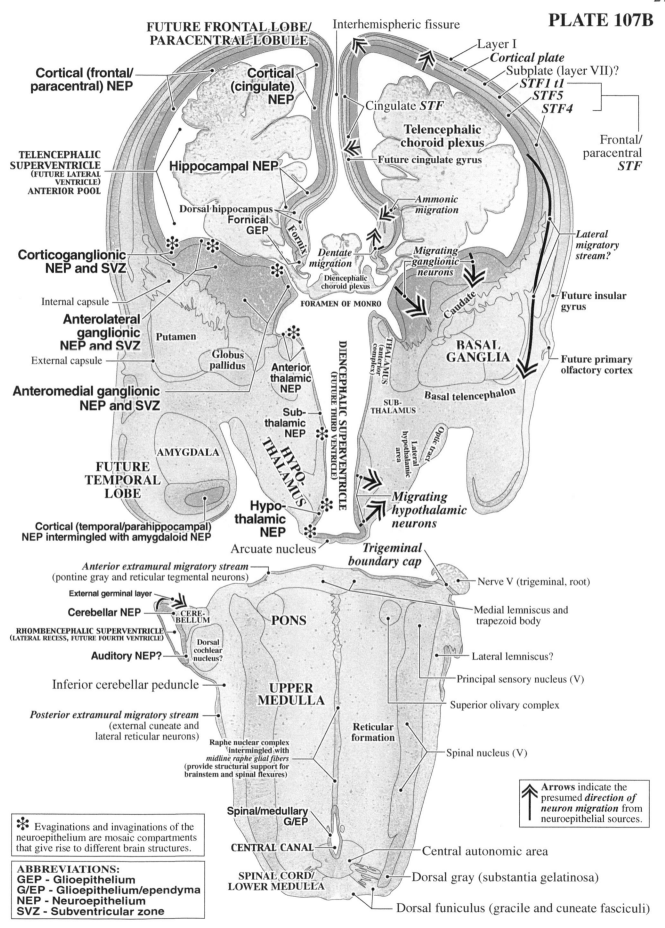

FUTURE FRONTAL LOBE/
PARACENTRAL LOBULE

Interhemispheric fissure

Layer I
*Cortical plate*
Subplate (layer VII)?
*STF1 t1*
*STF5*
*STF4*

Cortical (frontal/
paracentral) NEP

Cortical
(cingulate)
NEP

Cingulate *STF*

**Telencephalic
choroid plexus**

Frontal/
paracentral
*STF*

TELENCEPHALIC
SUPERVENTRICLE
(FUTURE LATERAL
VENTRICLE)
ANTERIOR POOL

**Hippocampal NEP**

Future cingulate gyrus

Dorsal hippocampus

Fornical
GEP

*Ammonic
migration*

*Lateral
migratory
stream?*

Fornix

*Dentate
migration*

*Migrating
ganglionic
neurons*

**Corticoganglionic
NEP and SVZ**

Diencephalic
choroid plexus

**FORAMEN OF MONRO**

Caudate

Future insular
gyrus

Internal capsule

THALAMUS
(anterior
complex)

**BASAL
GANGLIA**

**Anterolateral
ganglionic
NEP and SVZ**

Putamen

Globus
pallidus

Anterior
thalamic
NEP

DIENCEPHALIC SUPERVENTRICLE
(FUTURE THIRD VENTRICLE)

SUB-
THALAMUS

Basal telencephalon

Future primary
olfactory cortex

External capsule

**Anteromedial ganglionic
NEP and SVZ**

Sub-
thalamic
NEP

HYPO-
THALAMUS

Optic tract

Lateral
hypothalamic
area

**AMYGDALA**

**FUTURE
TEMPORAL
LOBE**

Hypo-
thalamic
NEP

*Migrating
hypothalamic
neurons*

Cortical (temporal/parahippocampal)
NEP intermingled with amygdaloid NEP

Arcuate nucleus

*Trigeminal
boundary cap*

*Anterior extramural migratory stream*
(pontine gray and reticular tegmental neurons)

Nerve V (trigeminal, root)

External germinal layer

**Cerebellar NEP**

CERE-
BELLUM

**PONS**

Medial lemniscus and
trapezoid body

RHOMBENCEPHALIC SUPERVENTRICLE
(LATERAL RECESS, FUTURE FOURTH VENTRICLE)

Dorsal
cochlear
nucleus?

Lateral lemniscus?

**Auditory NEP?**

Principal sensory nucleus (V)

Inferior cerebellar peduncle

**UPPER
MEDULLA**

Superior olivary complex

*Posterior extramural migratory stream*
(external cuneate and
lateral reticular neurons)

Reticular
formation

Spinal nucleus (V)

Raphe nuclear complex
intermingled with
*midline raphe glial fibers*
(provide structural support for
brainstem and spinal flexures)

**Spinal/medullary
G/EP**

**CENTRAL CANAL**

Central autonomic area

**SPINAL CORD/
LOWER MEDULLA**

Dorsal gray (substantia gelatinosa)

Dorsal funiculus (gracile and cuneate fasciculi)

**Arrows** indicate the
presumed *direction of
neuron migration* from
neuroepithelial sources.

✳ Evaginations and invaginations of the
neuroepithelium are mosaic compartments
that give rise to different brain structures.

**ABBREVIATIONS:**
GEP - Glioepithelium
G/EP - Glioepithelium/ependyma
NEP - Neuroepithelium
SVZ - Subventricular zone

**PLATE 108A**

**GW9 Coronal**
**CR 42 mm**
**M841**
**Level 9:**
**Section**
**294**

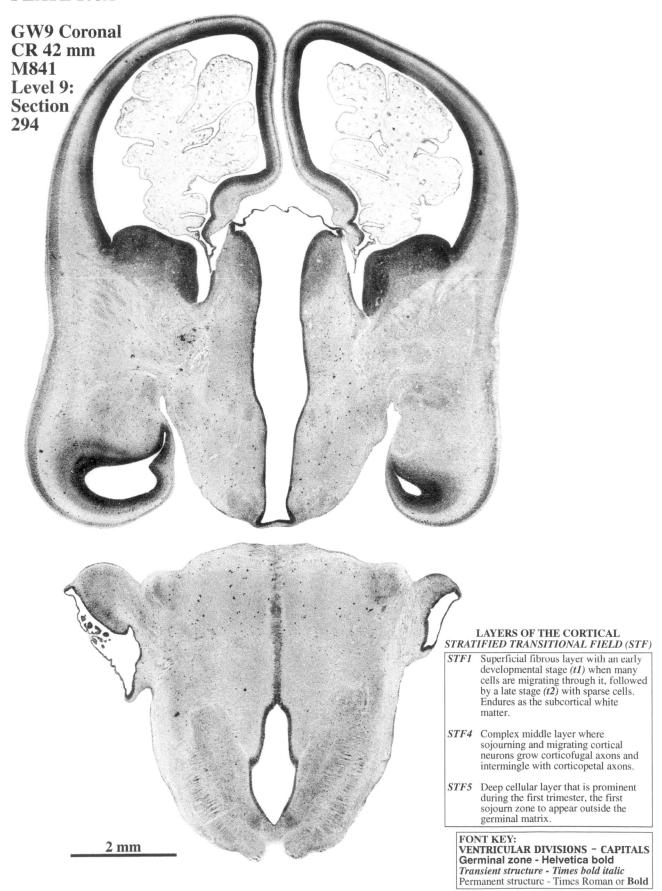

2 mm

**LAYERS OF THE CORTICAL**
*STRATIFIED TRANSITIONAL FIELD (STF)*

| | |
|---|---|
| *STF1* | Superficial fibrous layer with an early developmental stage *(t1)* when many cells are migrating through it, followed by a late stage *(t2)* with sparse cells. Endures as the subcortical white matter. |
| *STF4* | Complex middle layer where sojourning and migrating cortical neurons grow corticofugal axons and intermingle with corticopetal axons. |
| *STF5* | Deep cellular layer that is prominent during the first trimester, the first sojourn zone to appear outside the germinal matrix. |

**FONT KEY:**
**VENTRICULAR DIVISIONS – CAPITALS**
**Germinal zone - Helvetica bold**
*Transient structure - Times bold italic*
Permanent structure - Times Roman or **Bold**

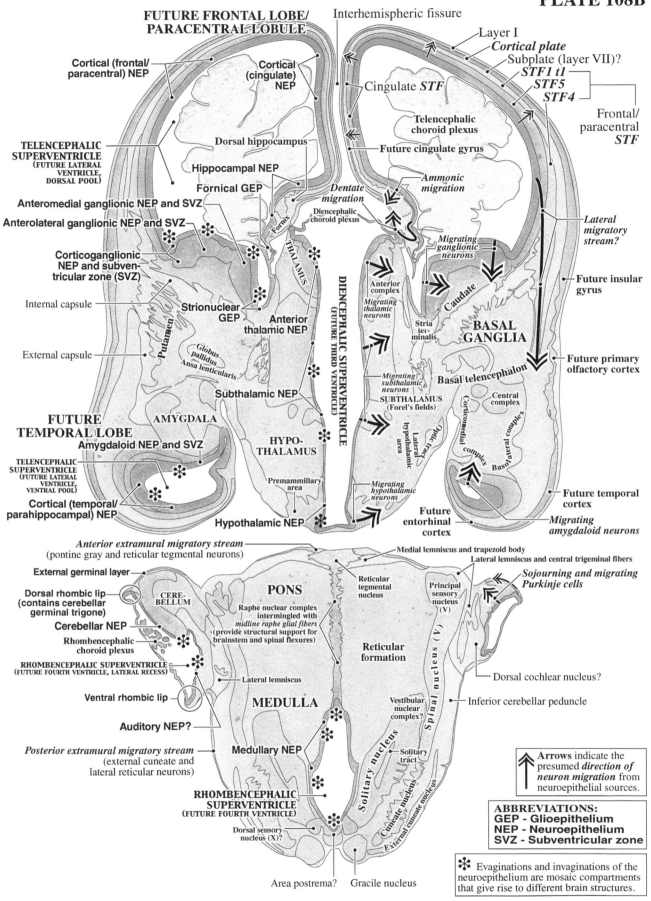

**PLATE 109A**

**GW9 Coronal
CR 42 mm
M841
Level 10:
Section
285**

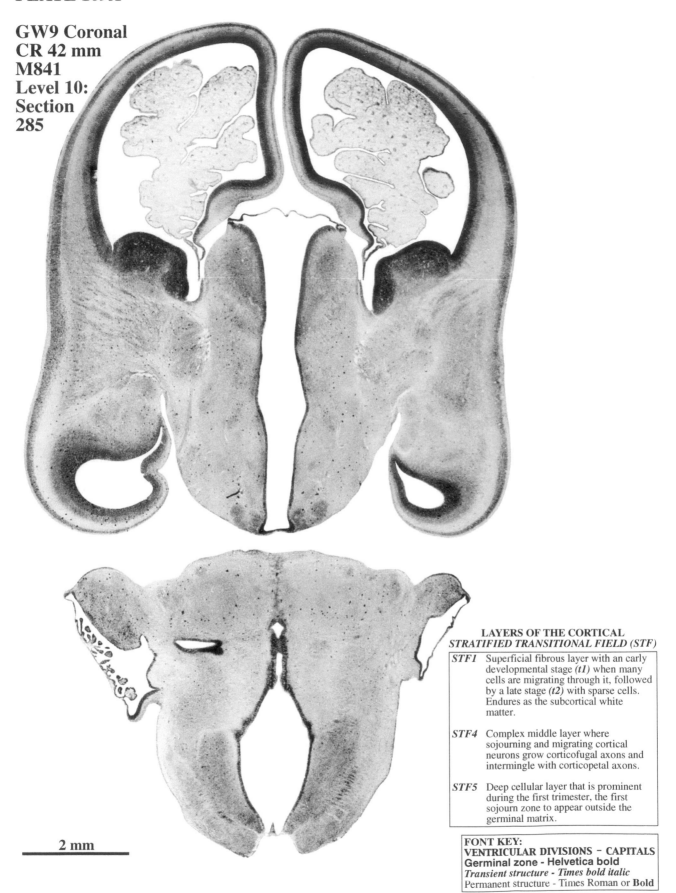

**2 mm**

**LAYERS OF THE CORTICAL
*STRATIFIED TRANSITIONAL FIELD (STF)***

| | |
|---|---|
| ***STF1*** | Superficial fibrous layer with an early developmental stage *(t1)* when many cells are migrating through it, followed by a late stage *(t2)* with sparse cells. Endures as the subcortical white matter. |
| ***STF4*** | Complex middle layer where sojourning and migrating cortical neurons grow corticofugal axons and intermingle with corticopetal axons. |
| ***STF5*** | Deep cellular layer that is prominent during the first trimester, the first sojourn zone to appear outside the germinal matrix. |

**FONT KEY:
VENTRICULAR DIVISIONS – CAPITALS
Germinal zone - Helvetica bold
*Transient structure - Times bold italic*
Permanent structure - Times Roman or Bold**

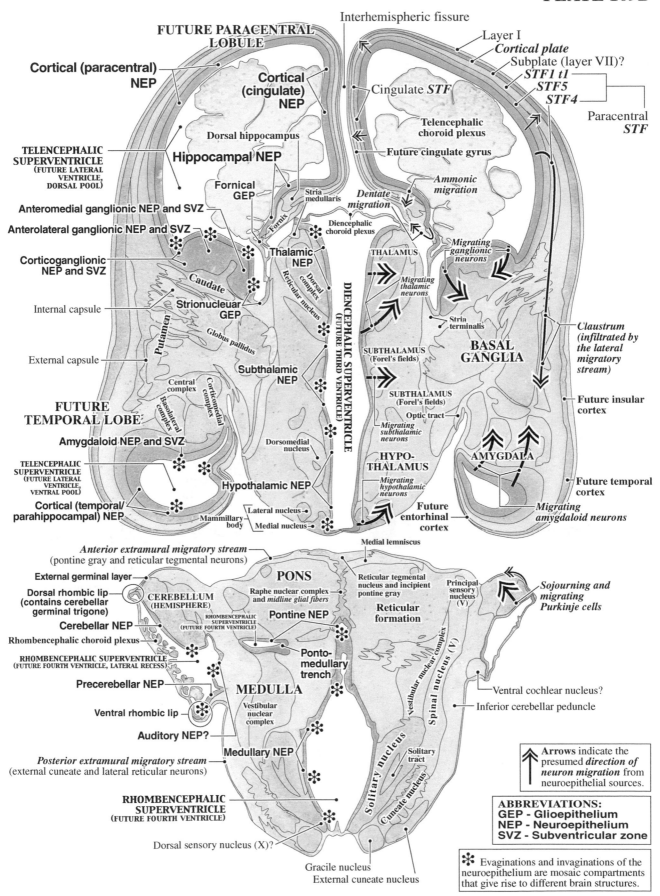

Interhemispheric fissure

**FUTURE PARACENTRAL LOBULE**

Layer I
*Cortical plate*
Subplate (layer VII)?
*STF1 t1*
*STF5*
*STF4*
Paracentral *STF*

**Cortical (paracentral) NEP**

**Cortical (cingulate) NEP**

Cingulate *STF*

Dorsal hippocampus

**TELENCEPHALIC SUPERVENTRICLE**
(FUTURE LATERAL VENTRICLE, DORSAL POOL)

**Hippocampal NEP**

Telencephalic choroid plexus

Future cingulate gyrus

*Ammonic migration*

Fornical GEP

Stria medullaris

*Dentate migration*

**Anteromedial ganglionic NEP and SVZ**

Diencephalic choroid plexus

*Migrating ganglionic neurons*

**Anterolateral ganglionic NEP and SVZ**

Fornix

**Thalamic NEP**

THALAMUS

**Corticoganglionic NEP and SVZ**

Caudate

Dorsal complex

Reticular nucleus

*Migrating thalamic neurons*

Internal capsule

**Strionucleuar GEP**

Stria terminalis

*Claustrum (infiltrated by the lateral migratory stream)*

Putamen

Globus pallidus

SUBTHALAMUS (Forel's fields)

**BASAL GANGLIA**

External capsule

**Subthalamic NEP**

Central complex

Corticomedial complex

SUBTHALAMUS (Forel's fields)

Optic tract

**Future insular cortex**

**FUTURE TEMPORAL LOBE**

Basolateral complex

DIENCEPHALIC SUPERVENTRICLE (FUTURE THIRD VENTRICLE)

*Migrating subthalamic neurons*

**Amygdaloid NEP and SVZ**

Dorsomedial nucleus

**HYPO-THALAMUS**

**AMYGDALA**

**TELENCEPHALIC SUPERVENTRICLE**
(FUTURE LATERAL VENTRICLE, VENTRAL POOL)

**Hypothalamic NEP**

*Migrating hypothalamic neurons*

**Future temporal cortex**

**Cortical (temporal/ parahippocampal) NEP**

Lateral nucleus

Mammillary body

Medial nucleus

**Future entorhinal cortex**

*Migrating amygdaloid neurons*

Medial lemniscus

*Anterior extramural migratory stream*
(pontine gray and reticular tegmental neurons)

Reticular tegmental nucleus and incipient pontine gray

Principal sensory nucleus (V)

*Sojourning and migrating Purkinje cells*

**External germinal layer**

**PONS**

**Dorsal rhombic lip**
(contains cerebellar germinal trigone)

**CEREBELLUM** (HEMISPHERE)

Raphe nuclear complex and *midline glial fibers*

**Reticular formation**

**Cerebellar NEP**

RHOMBENCEPHALIC SUPERVENTRICLE (FUTURE FOURTH VENTRICLE)

**Pontine NEP**

Vestibular nuclear complex

Spinal nucleus (V)

**Rhombencephalic choroid plexus**

**Ponto-medullary trench**

**RHOMBENCEPHALIC SUPERVENTRICLE**
(FUTURE FOURTH VENTRICLE, LATERAL RECESS)

**MEDULLA**

**Ventral cochlear nucleus?**

**Precerebellar NEP**

Vestibular nuclear complex

**Inferior cerebellar peduncle**

**Ventral rhombic lip**

**Auditory NEP?**

**Medullary NEP**

Solitary nucleus

Solitary tract

*Posterior extramural migratory stream*
(external cuneate and lateral reticular neurons)

Cuneate nucleus

**RHOMBENCEPHALIC SUPERVENTRICLE**
(FUTURE FOURTH VENTRICLE)

**Arrows** indicate the presumed *direction of neuron migration* from neuroepithelial sources.

Dorsal sensory nucleus (X)?

**ABBREVIATIONS:**
**GEP** - Glioepithelium
**NEP** - Neuroepithelium
**SVZ** - Subventricular zone

Gracile nucleus
External cuneate nucleus

✱ Evaginations and invaginations of the neuroepithelium are mosaic compartments that give rise to different brain structures.

**PLATE 110A**

**GW9 Coronal
CR 42 mm
M841
Level 11:
Section
270**

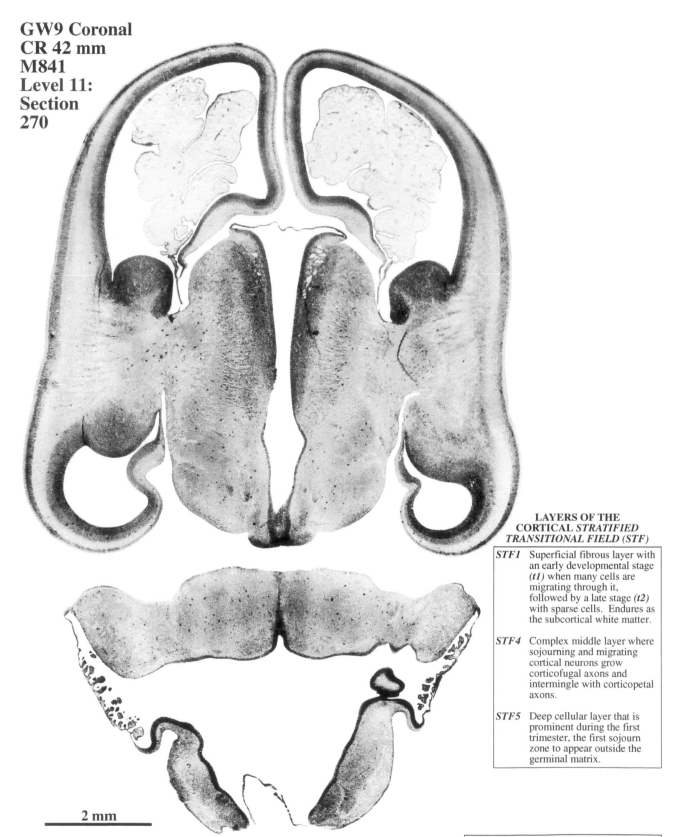

2 mm

**LAYERS OF THE
CORTICAL *STRATIFIED
TRANSITIONAL FIELD (STF)***

*STF1* Superficial fibrous layer with an early developmental stage *(t1)* when many cells are migrating through it, followed by a late stage *(t2)* with sparse cells. Endures as the subcortical white matter.

*STF4* Complex middle layer where sojourning and migrating cortical neurons grow corticofugal axons and intermingle with corticopetal axons.

*STF5* Deep cellular layer that is prominent during the first trimester, the first sojourn zone to appear outside the germinal matrix.

**FONT KEY:**
**VENTRICULAR DIVISIONS – CAPITALS**
**Germinal zone - Helvetica bold**
*Transient structure - Times bold italic*
Permanent structure - Times Roman or **Bold**

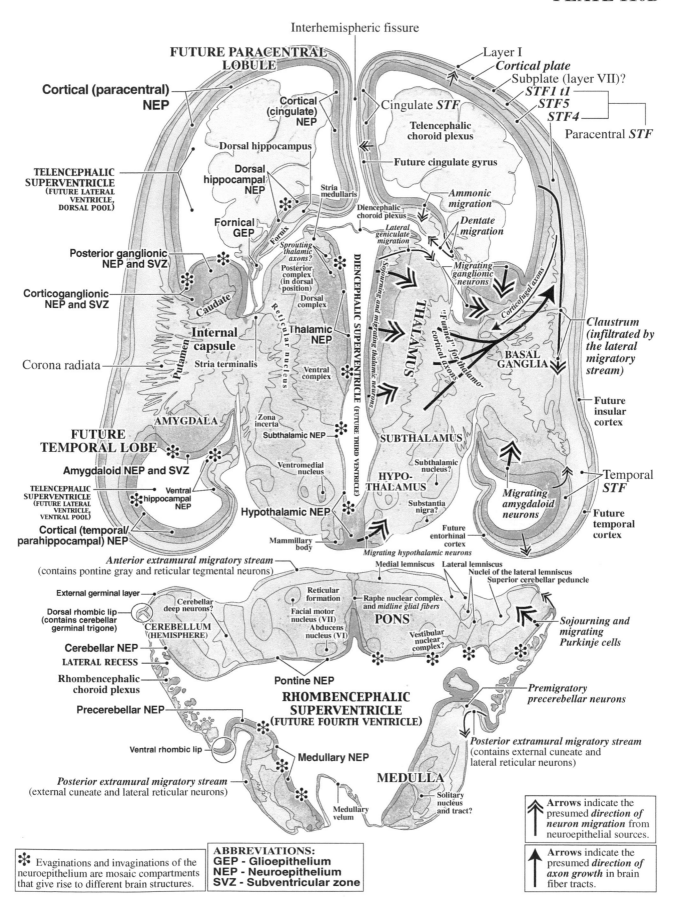

Interhemispheric fissure

**FUTURE PARACENTRAL LOBULE**

Layer I
*Cortical plate*
Subplate (layer VII)?
*STF1 t1*
*STF5*
*STF4*
Paracentral *STF*

**Cortical (paracentral) NEP**

Cortical (cingulate) NEP

Cingulate *STF*

Telencephalic choroid plexus

Dorsal hippocampus

Dorsal hippocampal NEP

Stria medullaris

Future cingulate gyrus

**TELENCEPHALIC SUPERVENTRICLE**
(FUTURE LATERAL VENTRICLE, DORSAL POOL)

Diencephalic choroid plexus

*Ammonic migration*

*Dentate migration*

Fornical GEP

Fornix

*Sprouting thalamic axons?*

*Lateral geniculate migration*

**Posterior ganglionic NEP and SVZ**

Posterior complex (in dorsal position)

*Migrating ganglionic neurons*

**Corticoganglionic NEP and SVZ**

Caudate

Dorsal complex

*Sojourning and migrating thalamic neurons*

DIENCEPHALIC SUPERVENTRICLE (FUTURE THIRD VENTRICLE)

*"Funnel" for thalamo-cortical axons*

*Corticofugal axons*

*Claustrum (infiltrated by the lateral migratory stream)*

**Internal capsule**

Putamen

Reticular nucleus

Thalamic NEP

THALAMUS

**BASAL GANGLIA**

Stria terminalis

Corona radiata

Ventral complex

**Future insular cortex**

**AMYGDALA**

Zona incerta

Subthalamic NEP

**SUBTHALAMUS**

Subthalamic nucleus?

**FUTURE TEMPORAL LOBE**

Ventromedial nucleus

**HYPO-THALAMUS**

Substantia nigra?

*Migrating amygdaloid neurons*

Temporal *STF*

**Amygdaloid NEP and SVZ**

**TELENCEPHALIC SUPERVENTRICLE**
(FUTURE LATERAL VENTRICLE, VENTRAL POOL)

Ventral hippocampal NEP

Hypothalamic NEP

**Future temporal cortex**

**Cortical (temporal/ parahippocampal) NEP**

Mammillary body

Future entorhinal cortex

*Migrating hypothalamic neurons*

*Anterior extramural migratory stream*
(contains pontine gray and reticular tegmental neurons)

Medial lemniscus

Lateral lemniscus

Nuclei of the lateral lemniscus
Superior cerebellar peduncle

External germinal layer

Reticular formation

Dorsal rhombic lip
(contains cerebellar germinal trigone)

Cerebellar deep neurons?

Raphe nuclear complex and *midline glial fibers*

**CEREBELLUM (HEMISPHERE)**

Facial motor nucleus (VII)

Abducens nucleus (VI)

**PONS**

Vestibular nuclear complex?

*Sojourning and migrating Purkinje cells*

**Cerebellar NEP**

**LATERAL RECESS**

**Rhombencephalic choroid plexus**

**Precerebellar NEP**

*Premigratory precerebellar neurons*

Pontine NEP

**RHOMBENCEPHALIC SUPERVENTRICLE**
(FUTURE FOURTH VENTRICLE)

Ventral rhombic lip

Medullary NEP

*Posterior extramural migratory stream*
(contains external cuneate and lateral reticular neurons)

**MEDULLA**

*Posterior extramural migratory stream*
(external cuneate and lateral reticular neurons)

Medullary velum

Solitary nucleus and tract?

✳ Evaginations and invaginations of the neuroepithelium are mosaic compartments that give rise to different brain structures.

**ABBREVIATIONS:**
**GEP** - Glioepithelium
**NEP** - Neuroepithelium
**SVZ** - Subventricular zone

**Arrows** indicate the presumed *direction of neuron migration* from neuroepithelial sources.

**Arrows** indicate the presumed *direction of axon growth* in brain fiber tracts.

**PLATE 111A**

**GW9 Coronal
CR 42 mm
M841
Level 12:
Section 255**

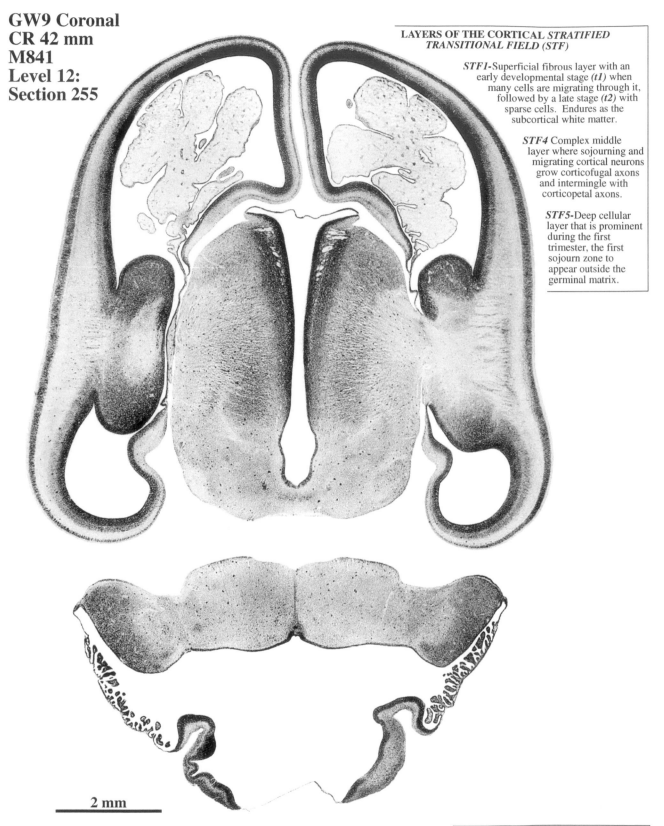

2 mm

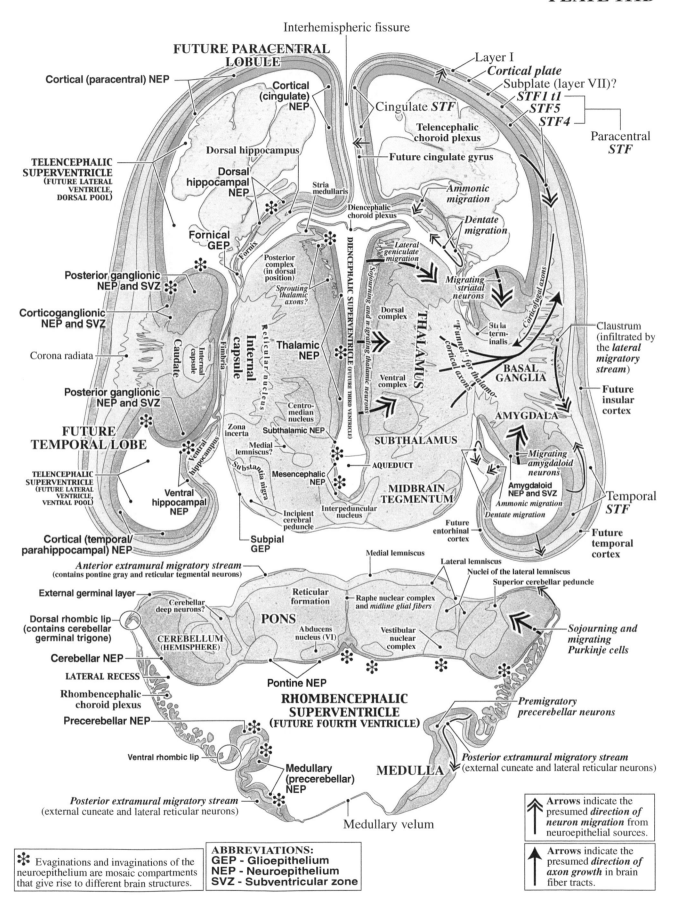

Interhemispheric fissure

FUTURE PARACENTRAL LOBULE

Cortical (paracentral) NEP

Cortical (cingulate) NEP

Cingulate *STF*

Layer I
*Cortical plate*
Subplate (layer VII)?
*STF1 t1*
*STF5*
*STF4*

Paracentral *STF*

Telencephalic choroid plexus

Future cingulate gyrus

Dorsal hippocampus

*Ammonic migration*

Dorsal hippocampal NEP

Stria medullaris

Diencephalic choroid plexus

*Dentate migration*

TELENCEPHALIC SUPERVENTRICLE (FUTURE LATERAL VENTRICLE, DORSAL POOL)

Fornical GEP

Fornix

Posterior complex (in dorsal position)

*Lateral geniculate migration*

*Sprouting thalamic axons?*

*Migrating striatal neurons*

*Corticofugal axons*

Claustrum (infiltrated by the *lateral migratory stream*)

Posterior ganglionic NEP and SVZ

Corticoganglionic NEP and SVZ

Corona radiata

Caudate

Internal capsule

Fimbria

Internal capsule

*Reticular nucleus*

Thalamic NEP

Dorsal complex

THALAMUS

Stria terminalis

"Funnel" for thalamo-cortical axons

BASAL GANGLIA

Future insular cortex

Posterior ganglionic NEP and SVZ

FUTURE TEMPORAL LOBE

Zona incerta

Centro-median nucleus

Ventral complex

AMYGDALA

TELENCEPHALIC SUPERVENTRICLE (FUTURE LATERAL VENTRICLE, VENTRAL POOL)

*Ventral hippocampus*

Medial lemniscus?

Subthalamic NEP

SUBTHALAMUS

*Migrating amygdaloid neurons*

Ventral hippocampal NEP

Substantia nigra

Mesencephalic NEP

AQUEDUCT

Amygdaloid NEP and SVZ

*Ammonic migration*

Temporal *STF*

Cortical (temporal/parahippocampal) NEP

Incipient cerebral peduncle

Interpeduncular nucleus

MIDBRAIN TEGMENTUM

*Dentate migration*

Future entorhinal cortex

Future temporal cortex

Subpial GEP

Medial lemniscus

Lateral lemniscus

Nuclei of the lateral lemniscus

*Anterior extramural migratory stream* (contains pontine gray and reticular tegmental neurons)

External germinal layer

Reticular formation

Superior cerebellar peduncle

Cerebellar deep neurons?

PONS

Raphe nuclear complex and *midline glial fibers*

Dorsal rhombic lip (contains cerebellar germinal trigone)

CEREBELLUM (HEMISPHERE)

Abducens nucleus (VI)

Vestibular nuclear complex

*Sojourning and migrating Purkinje cells*

Cerebellar NEP

LATERAL RECESS

Pontine NEP

Rhombencephalic choroid plexus

Precerebellar NEP

RHOMBENCEPHALIC SUPERVENTRICLE (FUTURE FOURTH VENTRICLE)

*Premigratory precerebellar neurons*

Ventral rhombic lip

Medullary (precerebellar) NEP

MEDULLA

*Posterior extramural migratory stream* (external cuneate and lateral reticular neurons)

*Posterior extramural migratory stream* (external cuneate and lateral reticular neurons)

Medullary velum

**Arrows** indicate the presumed *direction of neuron migration* from neuroepithelial sources.

**Arrows** indicate the presumed *direction of axon growth* in brain fiber tracts.

✳ Evaginations and invaginations of the neuroepithelium are mosaic compartments that give rise to different brain structures.

ABBREVIATIONS:
**GEP** - Glioepithelium
**NEP** - Neuroepithelium
**SVZ** - Subventricular zone

**PLATE 112A**

**GW9 Coronal**
**CR 42 mm**
**M841**
**Level 13:**
**Section 235**

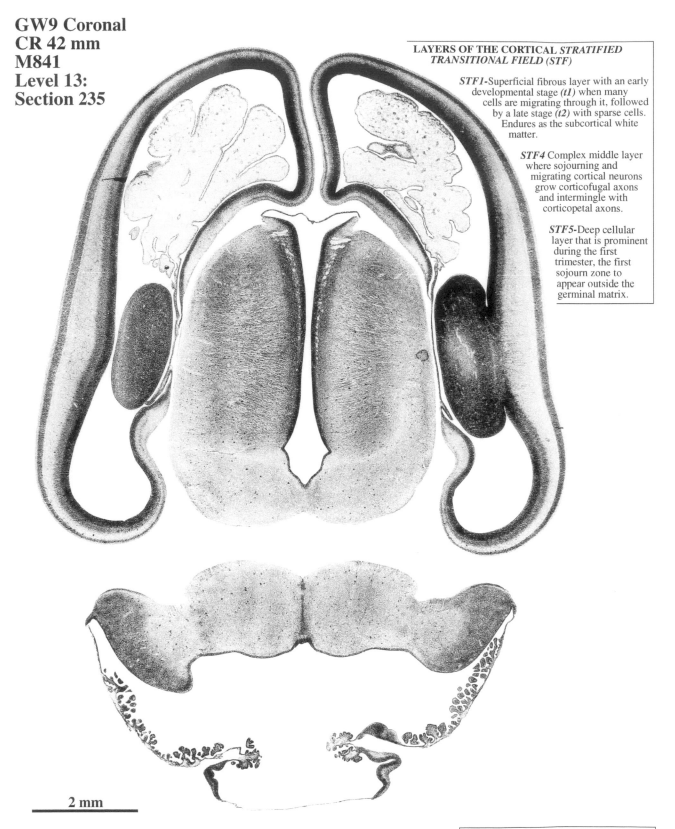

2 mm

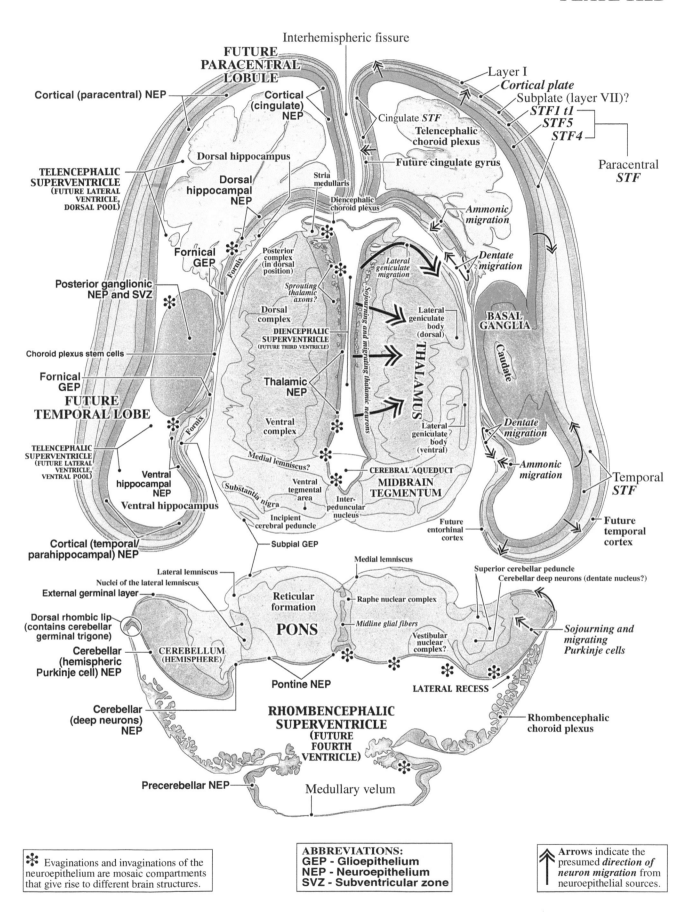

Interhemispheric fissure

FUTURE
PARACENTRAL
LOBULE

Cortical (paracentral) NEP

Cortical
(cingulate)
NEP

Layer I
*Cortical plate*
Subplate (layer VII)?
*STF1 t1*
*STF5*
*STF4*

Cingulate *STF*

Telencephalic
choroid plexus

Paracentral
*STF*

Dorsal hippocampus

Future cingulate gyrus

TELENCEPHALIC
SUPERVENTRICLE
(FUTURE LATERAL
VENTRICLE,
DORSAL POOL)

Dorsal
hippocampal
NEP

Stria
medullaris

*Ammonic
migration*

Diencephalic
choroid plexus

Fornical
GEP

Posterior
complex
(in dorsal
position)

*Lateral
geniculate
migration*

*Dentate
migration*

*Sprouting
thalamic
axons?*

Posterior ganglionic
NEP and SVZ

Dorsal
complex

Lateral
geniculate
body
(dorsal)

BASAL
GANGLIA

DIENCEPHALIC
SUPERVENTRICLE
(FUTURE THIRD VENTRICLE)

Choroid plexus stem cells

THALAMUS

Caudate

Fornical
GEP
FUTURE
TEMPORAL LOBE

Thalamic
NEP

TELENCEPHALIC
SUPERVENTRICLE
(FUTURE LATERAL
VENTRICLE,
VENTRAL POOL)

Ventral
complex

Lateral
geniculate
body
(ventral)

*Dentate
migration*

Ventral
hippocampal
NEP

*Medial lemniscus?*

CEREBRAL AQUEDUCT

*Ammonic
migration*

Temporal
*STF*

Ventral hippocampus

*Substantia nigra*

Ventral
tegmental
area

MIDBRAIN
TEGMENTUM

Cortical (temporal/
parahippocampal) NEP

Incipient
cerebral peduncle

Inter-
peduncular
nucleus

Future
entorhinal
cortex

Future
temporal
cortex

Subpial GEP

Lateral lemniscus

Medial lemniscus

Superior cerebellar peduncle
Cerebellar deep neurons (dentate nucleus?)

Nuclei of the lateral lemniscus

Reticular
formation

Raphe nuclear complex

External germinal layer

Dorsal rhombic lip
(contains cerebellar
germinal trigone)

PONS

*Midline glial fibers*

Cerebellar
(hemispheric Purkinje cell) NEP

CEREBELLUM
(HEMISPHERE)

*Vestibular
nuclear
complex?*

*Sojourning and
migrating
Purkinje cells*

Pontine NEP

LATERAL RECESS

Cerebellar
(deep neurons)
NEP

RHOMBENCEPHALIC
SUPERVENTRICLE
(FUTURE
FOURTH
VENTRICLE)

Rhombencephalic
choroid plexus

Precerebellar NEP

Medullary velum

✳ Evaginations and invaginations of the
neuroepithelium are mosaic compartments
that give rise to different brain structures.

**ABBREVIATIONS:**
**GEP** - Glioepithelium
**NEP** - Neuroepithelium
**SVZ** - Subventricular zone

**Arrows** indicate the
presumed *direction of
neuron migration* from
neuroepithelial sources.

**PLATE 113A**

**GW9 Coronal**
**CR 42 mm**
**M841**
**Level 14:**
**Section 220**

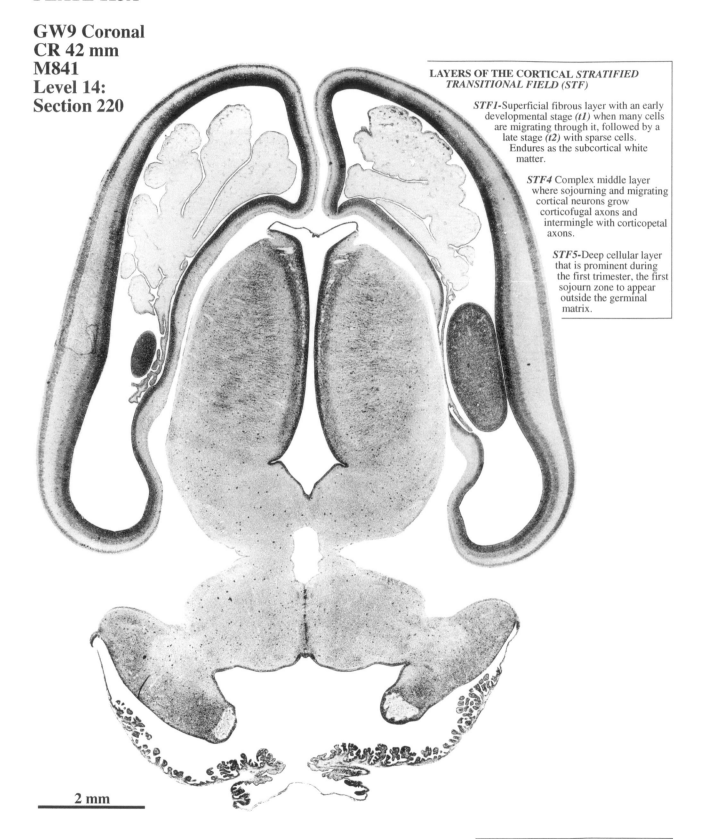

2 mm

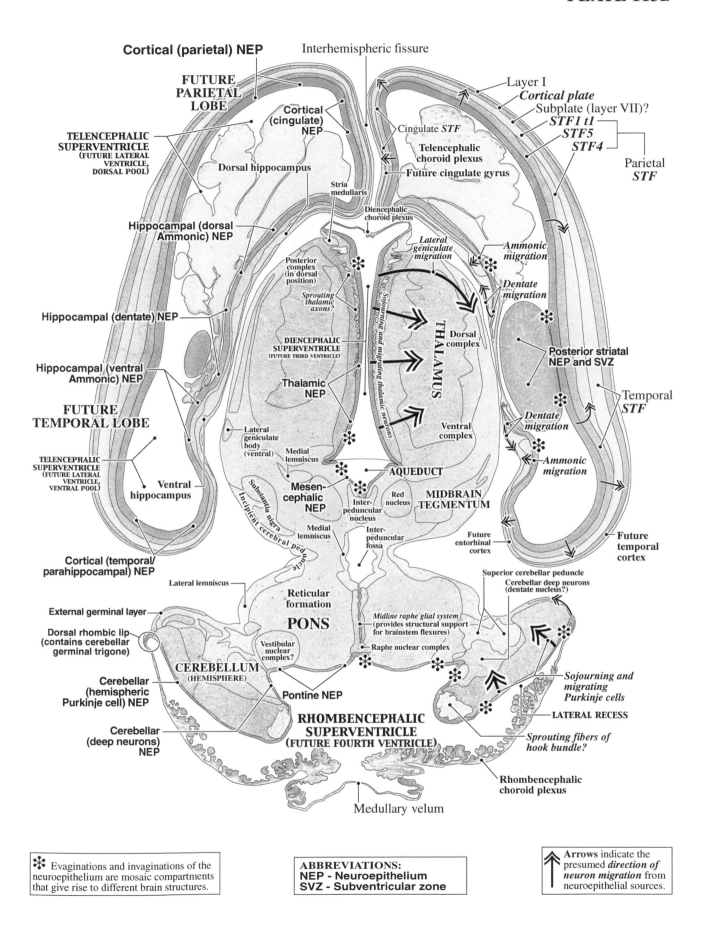

**Cortical (parietal) NEP**

Interhemispheric fissure

**FUTURE PARIETAL LOBE**

Layer I
*Cortical plate*
Subplate (layer VII)?
*STF1 t1*
*STF5*
*STF4*
Parietal *STF*

Cortical (cingulate) NEP

Cingulate *STF*

**TELENCEPHALIC SUPERVENTRICLE (FUTURE LATERAL VENTRICLE, DORSAL POOL)**

Telencephalic choroid plexus

Dorsal hippocampus

Stria medullaris

Future cingulate gyrus

Diencephalic choroid plexus

**Hippocampal (dorsal Ammonic) NEP**

Posterior complex (in dorsal position)

*Lateral geniculate migration*

*Ammonic migration*

*Sprouting thalamic axons?*

*Dentate migration*

**Hippocampal (dentate) NEP**

THALAMUS

Dorsal complex

**DIENCEPHALIC SUPERVENTRICLE (FUTURE THIRD VENTRICLE)**

*Sojourning and migrating thalamic neurons*

**Posterior striatal NEP and SVZ**

**Hippocampal (ventral Ammonic) NEP**

**Thalamic NEP**

**FUTURE TEMPORAL LOBE**

Lateral geniculate body (ventral)

Ventral complex

*Dentate migration*

Temporal *STF*

**TELENCEPHALIC SUPERVENTRICLE (FUTURE LATERAL VENTRICLE, VENTRAL POOL)**

Medial lemniscus

**AQUEDUCT**

*Ammonic migration*

**Ventral hippocampus**

*Substantia nigra*

*Incipient cerebral peduncle*

**Mesencephalic NEP**

Interpeduncular nucleus

Red nucleus

**MIDBRAIN TEGMENTUM**

**Cortical (temporal/ parahippocampal) NEP**

Medial lemniscus

Interpeduncular fossa

Future entorhinal cortex

Future temporal cortex

Lateral lemniscus

**Reticular formation**

**PONS**

Superior cerebellar peduncle
Cerebellar deep neurons (dentate nucleus?)

**External germinal layer**

*Midline raphe glial system (provides structural support for brainstem flexures)*

**Dorsal rhombic lip (contains cerebellar germinal trigone)**

Vestibular nuclear complex?

Raphe nuclear complex

**CEREBELLUM (HEMISPHERE)**

*Sojourning and migrating Purkinje cells*

**Cerebellar (hemispheric Purkinje cell) NEP**

Pontine NEP

**LATERAL RECESS**

**Cerebellar (deep neurons) NEP**

**RHOMBENCEPHALIC SUPERVENTRICLE (FUTURE FOURTH VENTRICLE)**

*Sprouting fibers of hook bundle?*

Rhombencephalic choroid plexus

Medullary velum

✳ Evaginations and invaginations of the neuroepithelium are mosaic compartments that give rise to different brain structures.

**ABBREVIATIONS:**
**NEP** - Neuroepithelium
**SVZ** - Subventricular zone

**Arrows** indicate the presumed *direction of neuron migration* from neuroepithelial sources.

# PLATE 114A

**GW9 Coronal
CR 42 mm
M841
Level 15:
Section 205**

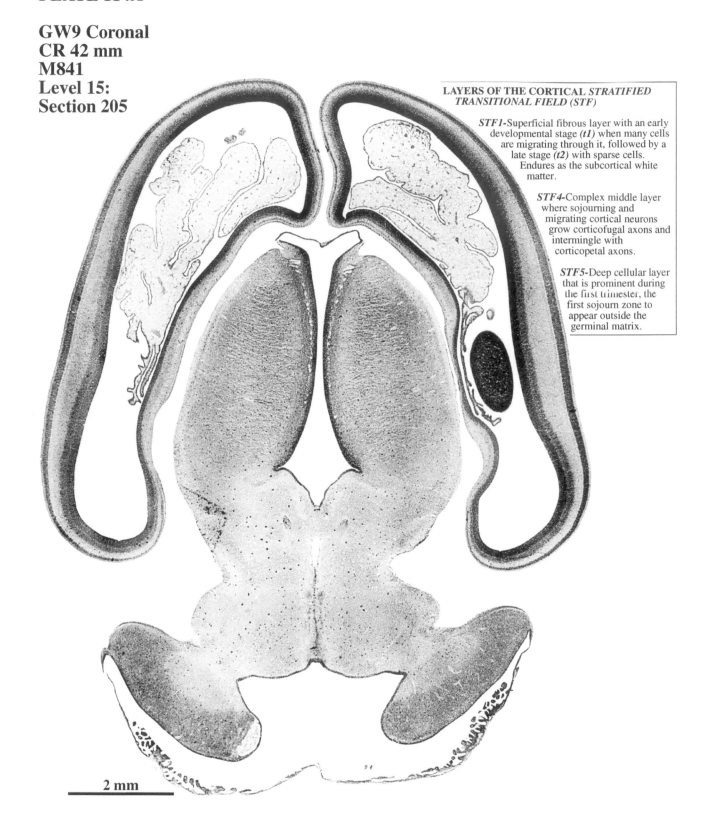

**LAYERS OF THE CORTICAL *STRATIFIED
TRANSITIONAL FIELD (STF)***

*STF1-*Superficial fibrous layer with an early
developmental stage *(t1)* when many cells
are migrating through it, followed by a
late stage *(t2)* with sparse cells.
Endures as the subcortical white
matter.

*STF4-*Complex middle layer
where sojourning and
migrating cortical neurons
grow corticofugal axons and
intermingle with
corticopetal axons.

*STF5-*Deep cellular layer
that is prominent during
the first trimester, the
first sojourn zone to
appear outside the
germinal matrix.

2 mm

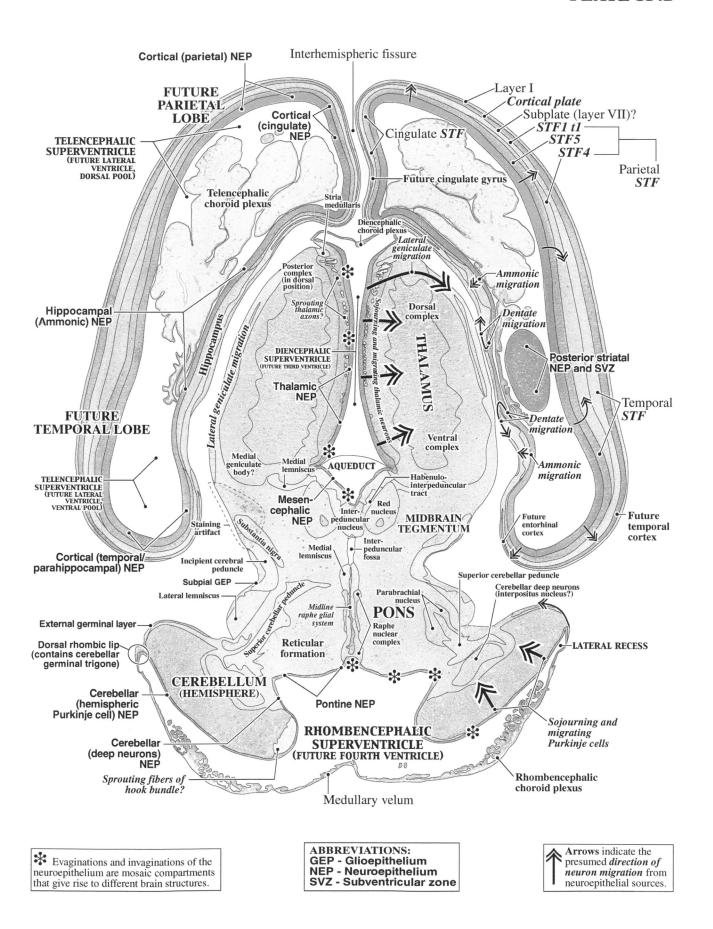

Cortical (parietal) NEP

Interhemispheric fissure

Layer I
*Cortical plate*
Subplate (layer VII)?
*STF1 t1*
*STF5*
*STF4*

Parietal *STF*

**FUTURE PARIETAL LOBE**

Cortical (cingulate) NEP

Cingulate *STF*

**TELENCEPHALIC SUPERVENTRICLE**
(FUTURE LATERAL VENTRICLE, DORSAL POOL)

Telencephalic choroid plexus

Stria medullaris

Diencephalic choroid plexus

Future cingulate gyrus

*Lateral geniculate migration*

*Ammonic migration*

Posterior complex (in dorsal position)

*Sprouting thalamic axons?*

Dorsal complex

*Dentate migration*

Hippocampal (Ammonic) NEP

Hippocampus

*Sojourning and migrating thalamic neurons*

**THALAMUS**

Posterior striatal NEP and SVZ

**DIENCEPHALIC SUPERVENTRICLE**
(FUTURE THIRD VENTRICLE)

*Lateral geniculate migration*

Thalamic NEP

*Dentate migration*

Temporal *STF*

**FUTURE TEMPORAL LOBE**

Ventral complex

Medial geniculate body?

Medial lemniscus

**AQUEDUCT**

Habenulo-interpeduncular tract

*Ammonic migration*

**TELENCEPHALIC SUPERVENTRICLE**
(FUTURE LATERAL VENTRICLE, VENTRAL POOL)

Mesen-cephalic NEP

Inter-peduncular nucleus

Red nucleus

**MIDBRAIN TEGMENTUM**

Future entorhinal cortex

**Future temporal cortex**

Staining artifact

*Substantia nigra*

Inter-peduncular fossa

Cortical (temporal/ parahippocampal) NEP

Incipient cerebral peduncle

Medial lemniscus

Subpial GEP

Lateral lemniscus

*Midline raphe glial system*

Parabrachial nucleus

**PONS**

Superior cerebellar peduncle

Cerebellar deep neurons (interpositus nucleus?)

External germinal layer

Dorsal rhombic lip (contains cerebellar germinal trigone)

*Superior cerebellar peduncle*

Reticular formation

Raphe nuclear complex

**LATERAL RECESS**

Cerebellar (hemispheric Purkinje cell) NEP

**CEREBELLUM (HEMISPHERE)**

Pontine NEP

Cerebellar (deep neurons) NEP

**RHOMBENCEPHALIC SUPERVENTRICLE**
(FUTURE FOURTH VENTRICLE)

*Sojourning and migrating Purkinje cells*

*Sprouting fibers of hook bundle?*

Rhombencephalic choroid plexus

Medullary velum

✱ Evaginations and invaginations of the neuroepithelium are mosaic compartments that give rise to different brain structures.

**ABBREVIATIONS:**
GEP - Glioepithelium
NEP - Neuroepithelium
SVZ - Subventricular zone

**Arrows** indicate the presumed *direction of neuron migration* from neuroepithelial sources.

**PLATE 115A**

**GW9 Coronal**
**CR 42 mm**
**M841**
**Level 16:**
**Section 175**

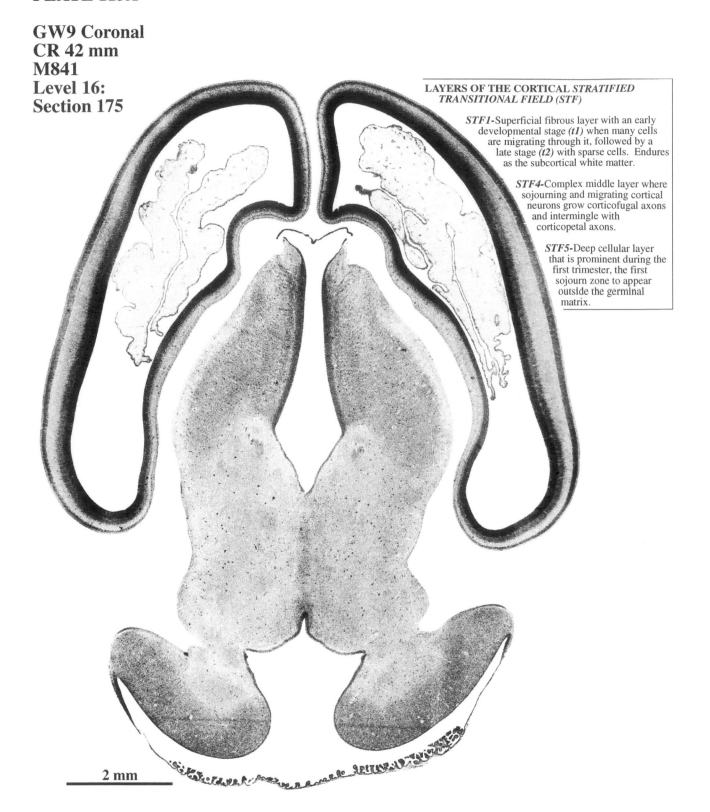

**LAYERS OF THE CORTICAL *STRATIFIED TRANSITIONAL FIELD (STF)***

***STF1-***Superficial fibrous layer with an early developmental stage *(t1)* when many cells are migrating through it, followed by a late stage *(t2)* with sparse cells. Endures as the subcortical white matter.

***STF4-***Complex middle layer where sojourning and migrating cortical neurons grow corticofugal axons and intermingle with corticopetal axons.

***STF5-***Deep cellular layer that is prominent during the first trimester, the first sojourn zone to appear outside the germinal matrix.

2 mm

FONT KEY:
**VENTRICULAR DIVISIONS – CAPITALS**
**Germinal zone - Helvetica bold**
***Transient structure - Times bold italic***
Permanent structure - Times Roman or **Bold**

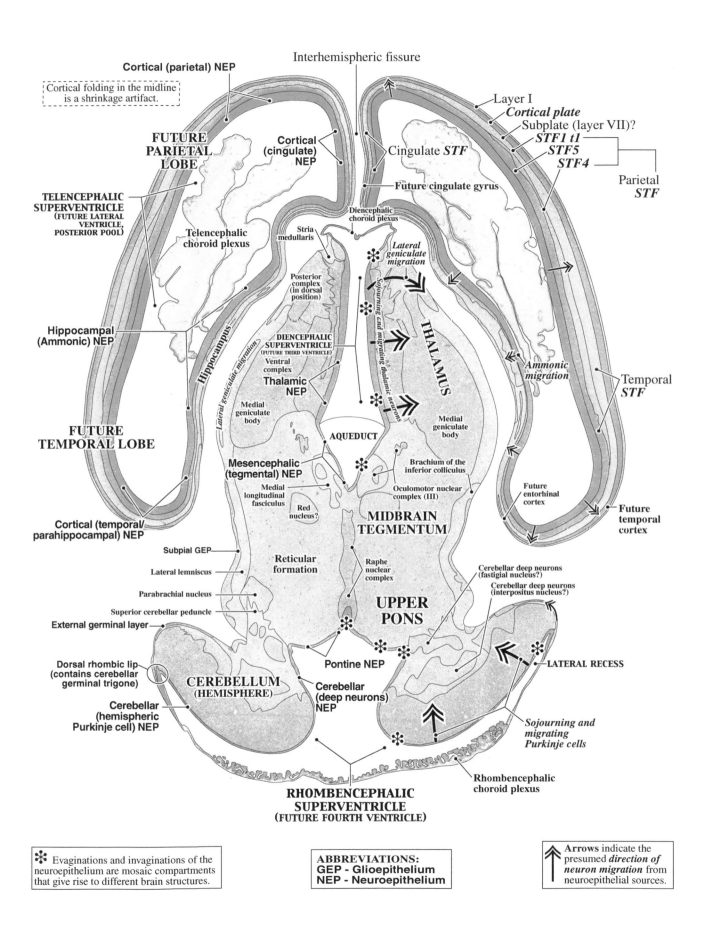

Interhemispheric fissure

Cortical (parietal) NEP

Cortical folding in the midline
is a shrinkage artifact.

**FUTURE PARIETAL LOBE**

Cortical (cingulate) NEP

Cingulate *STF*

Future cingulate gyrus

Layer I
*Cortical plate*
Subplate (layer VII)?
*STF1 t1*
*STF5*
*STF4*

Parietal
*STF*

**TELENCEPHALIC SUPERVENTRICLE**
(FUTURE LATERAL VENTRICLE, POSTERIOR POOL)

Telencephalic choroid plexus

Stria medullaris

Diencephalic choroid plexus

*Lateral geniculate migration*

Posterior complex (in dorsal position)

**THALAMUS**

Hippocampal (Ammonic) NEP

*Lateral geniculate migration*

*Hippocampus*

**DIENCEPHALIC SUPERVENTRICLE**
(FUTURE THIRD VENTRICLE)

Ventral complex

Thalamic NEP

*Sojourning and migrating thalamic neurons*

*Ammonic migration*

Temporal
*STF*

Medial geniculate body

Medial geniculate body

**FUTURE TEMPORAL LOBE**

**AQUEDUCT**

Mesencephalic (tegmental) NEP

Brachium of the inferior colliculus

Oculomotor nuclear complex (III)

Future entorhinal cortex

Medial longitudinal fasciculus

Red nucleus?

**MIDBRAIN TEGMENTUM**

**Future temporal cortex**

Cortical (temporal/parahippocampal) NEP

Subpial GEP

Lateral lemniscus

Parabrachial nucleus

Superior cerebellar peduncle

External germinal layer

Reticular formation

Raphe nuclear complex

**UPPER PONS**

Cerebellar deep neurons (fastigial nucleus?)

Cerebellar deep neurons (interpositus nucleus?)

Dorsal rhombic lip (contains cerebellar germinal trigone)

**CEREBELLUM (HEMISPHERE)**

Cerebellar (deep neurons) NEP

Pontine NEP

**LATERAL RECESS**

Cerebellar (hemispheric Purkinje cell) NEP

*Sojourning and migrating Purkinje cells*

Rhombencephalic choroid plexus

**RHOMBENCEPHALIC SUPERVENTRICLE**
(FUTURE FOURTH VENTRICLE)

✣ Evaginations and invaginations of the neuroepithelium are mosaic compartments that give rise to different brain structures.

**ABBREVIATIONS:**
**GEP** - Glioepithelium
**NEP** - Neuroepithelium

**Arrows** indicate the presumed *direction of neuron migration* from neuroepithelial sources.

**PLATE 116A**

**GW9 Coronal**
**CR 42 mm**
**M841**
**Level 17: Section 155**

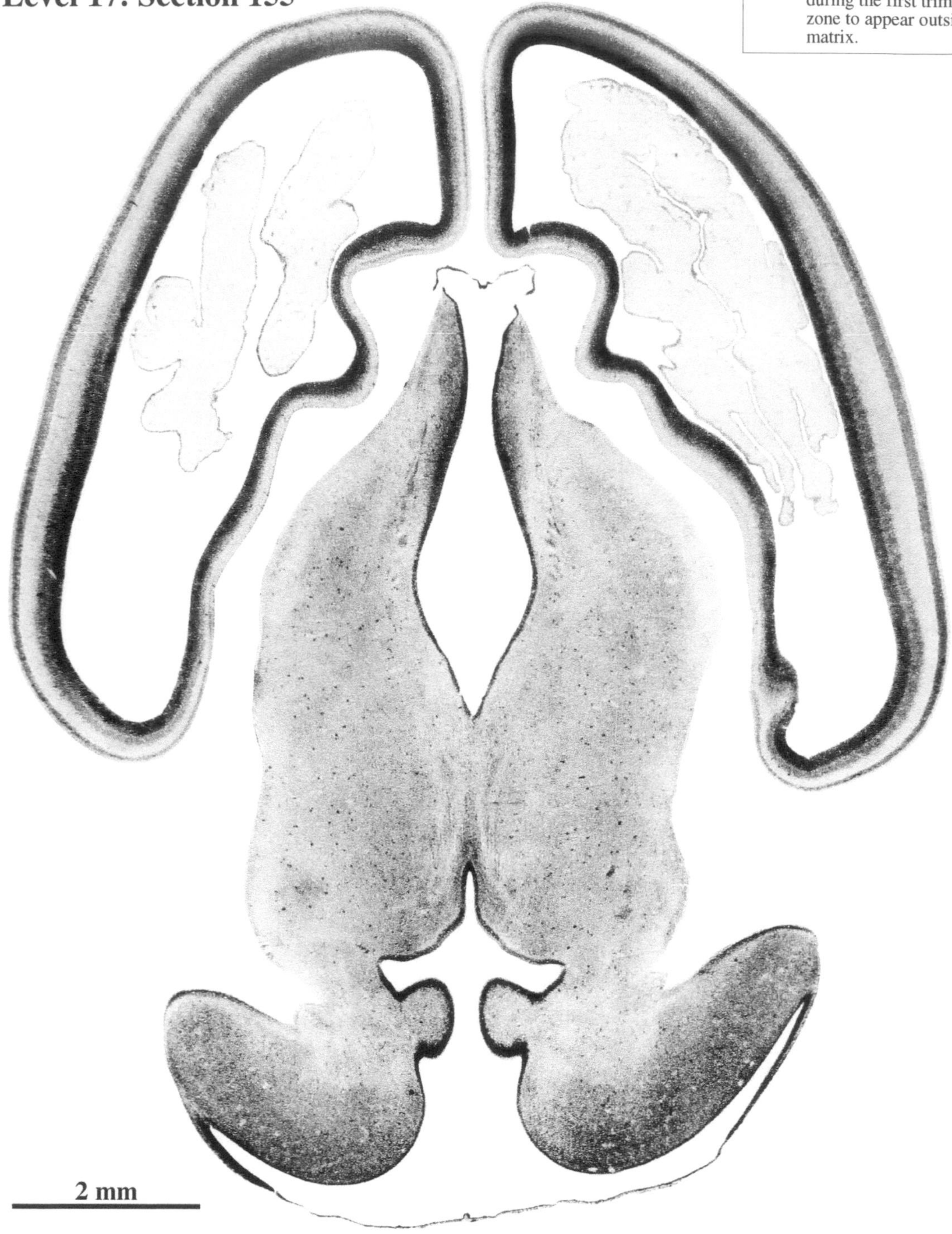

2 mm

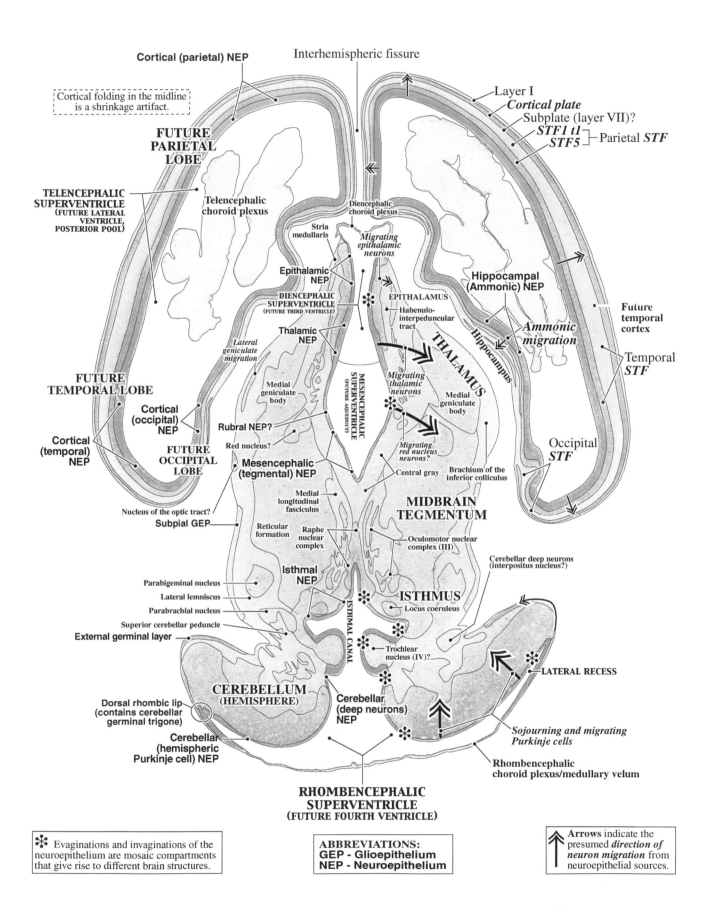

Cortical (parietal) NEP

Interhemispheric fissure

Layer I
*Cortical plate*
Subplate (layer VII)?
*STF1 t1*
*STF5* — Parietal *STF*

Cortical folding in the midline
is a shrinkage artifact.

**FUTURE
PARIETAL
LOBE**

**TELENCEPHALIC
SUPERVENTRICLE**
(FUTURE LATERAL
VENTRICLE,
POSTERIOR POOL)

Telencephalic
choroid plexus

Diencephalic
choroid plexus

Stria
medullaris

*Migrating
epithalamic
neurons*

Epithalamic
NEP

**DIENCEPHALIC
SUPERVENTRICLE**
(FUTURE THIRD VENTRICLE)

EPITHALAMUS

Habenulo-
interpeduncular
tract

Hippocampal
(Ammonic) NEP

Future
temporal
cortex

Thalamic
NEP

THALAMUS

Hippocampus

*Ammonic
migration*

**FUTURE
TEMPORAL LOBE**

*Lateral
geniculate
migration*

Medial
geniculate
body

MESENCEPHALIC SUPERVENTRICLE
(FUTURE AQUEDUCT)

*Migrating
thalamic
neurons*

Medial
geniculate
body

Temporal
*STF*

Cortical
(occipital)
NEP

Rubral NEP?

Red nucleus?

*Migrating
red nucleus
neurons?*

Central gray

Brachium of the
inferior colliculus

Occipital
*STF*

Cortical
(temporal)
NEP

**FUTURE
OCCIPITAL
LOBE**

**Mesencephalic
(tegmental) NEP**

Nucleus of the optic tract?

Subpial GEP

Medial
longitudinal
fasciculus

**MIDBRAIN
TEGMENTUM**

Reticular
formation

Raphe
nuclear
complex

Oculomotor nuclear
complex (III)

Parabigeminal nucleus

Lateral lemniscus

Parabrachial nucleus

Superior cerebellar peduncle

External germinal layer

**Isthmal
NEP**

ISTHMAL CANAL

**ISTHMUS**

Locus coeruleus

Cerebellar deep neurons
(interpositus nucleus?)

Trochlear
nucleus (IV)?

**LATERAL RECESS**

**CEREBELLUM
(HEMISPHERE)**

Dorsal rhombic lip
(contains cerebellar
germinal trigone)

Cerebellar
(hemispheric
Purkinje cell) NEP

Cerebellar
(deep neurons)
NEP

*Sojourning and migrating
Purkinje cells*

Rhombencephalic
choroid plexus/medullary velum

**RHOMBENCEPHALIC
SUPERVENTRICLE**
(FUTURE FOURTH VENTRICLE)

✳ Evaginations and invaginations of the
neuroepithelium are mosaic compartments
that give rise to different brain structures.

**ABBREVIATIONS:
GEP** - Glioepithelium
**NEP** - Neuroepithelium

**Arrows** indicate the
presumed *direction of
neuron migration* from
neuroepithelial sources.

**PLATE 117A**

**GW9 Coronal**
**CR 42 mm**
**M841**
**Level 18: Section 135**

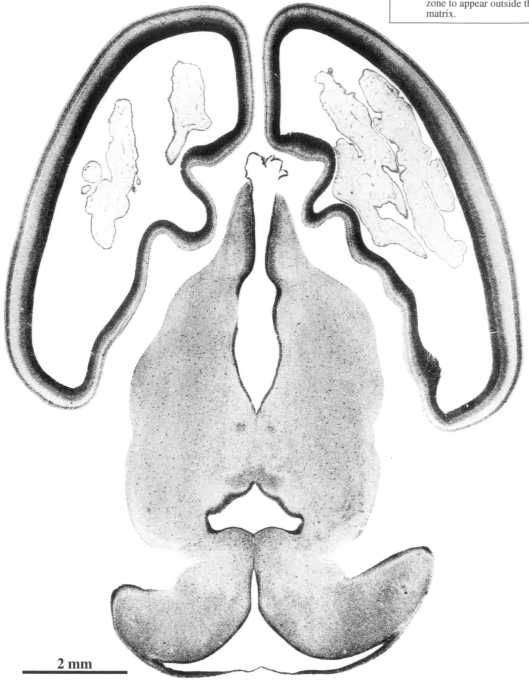

2 mm

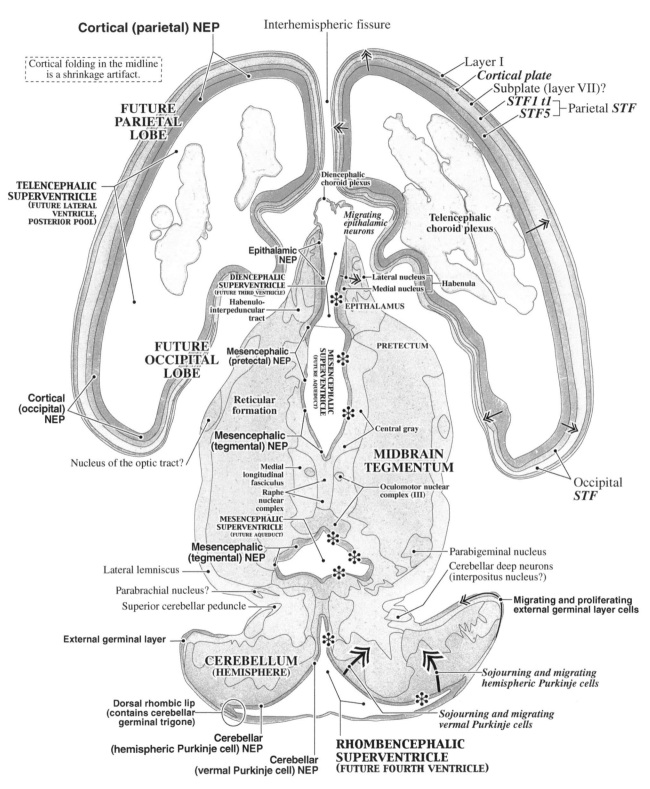

**Cortical (parietal) NEP**

Interhemispheric fissure

Cortical folding in the midline
is a shrinkage artifact.

Layer I
*Cortical plate*
Subplate (layer VII)?
*STF1 t1*
*STF5* Parietal *STF*

**FUTURE
PARIETAL
LOBE**

**TELENCEPHALIC
SUPERVENTRICLE**
(FUTURE LATERAL
VENTRICLE,
POSTERIOR POOL)

Diencephalic
choroid plexus

Telencephalic
choroid plexus

*Migrating
epithalamic
neurons*

Epithalamic
NEP

**DIENCEPHALIC
SUPERVENTRICLE**
(FUTURE THIRD VENTRICLE)

Lateral nucleus
Medial nucleus
Habenula

**EPITHALAMUS**

Habenulo-
interpeduncular
tract

**FUTURE
OCCIPITAL
LOBE**

Mesencephalic
(pretectal) NEP

**PRETECTUM**

**MESENCEPHALIC
SUPERVENTRICLE**
(FUTURE AQUEDUCT)

Cortical
(occipital)
NEP

Reticular
formation

Mesencephalic
(tegmental) NEP

Central gray

**MIDBRAIN
TEGMENTUM**

Nucleus of the optic tract?

Medial
longitudinal
fasciculus

Occipital
*STF*

Raphe
nuclear
complex

Oculomotor nuclear
complex (III)

**MESENCEPHALIC
SUPERVENTRICLE**
(FUTURE AQUEDUCT)

Mesencephalic
(tegmental) NEP

Parabigeminal nucleus

Lateral lemniscus

Cerebellar deep neurons
(interpositus nucleus?)

Parabrachial nucleus?

**Migrating and proliferating
external germinal layer cells**

Superior cerebellar peduncle

External germinal layer

*Sojourning and migrating
hemispheric Purkinje cells*

**CEREBELLUM
(HEMISPHERE)**

Dorsal rhombic lip
(contains cerebellar
germinal trigone)

*Sojourning and migrating
vermal Purkinje cells*

Cerebellar
(hemispheric Purkinje cell) NEP

Cerebellar
(vermal Purkinje cell) NEP

**RHOMBENCEPHALIC
SUPERVENTRICLE**
(FUTURE FOURTH VENTRICLE)

✳ Evaginations and invaginations of the
neuroepithelium are mosaic compartments
that give rise to different brain structures.

**NEP - Neuroepithelium**

**Arrows** indicate the
presumed *direction of
neuron migration* from
neuroepithelial sources.

# PLATE 118A

**GW9 Coronal**
**CR 42 mm**
**M841**
**Level 19: Section 124**

**LAYERS OF THE CORTICAL**
*STRATIFIED TRANSITIONAL FIELD (STF)*

| | |
|---|---|
| *STF1* | Superficial fibrous layer with an early developmental stage *(t1)* when many cells are migrating through it, followed by a late stage *(t2)* with sparse cells. Endures as the subcortical white matter. |
| *STF5* | Deep cellular layer that is prominent during the first trimester, the first sojourn zone to appear outside the germinal matrix. |

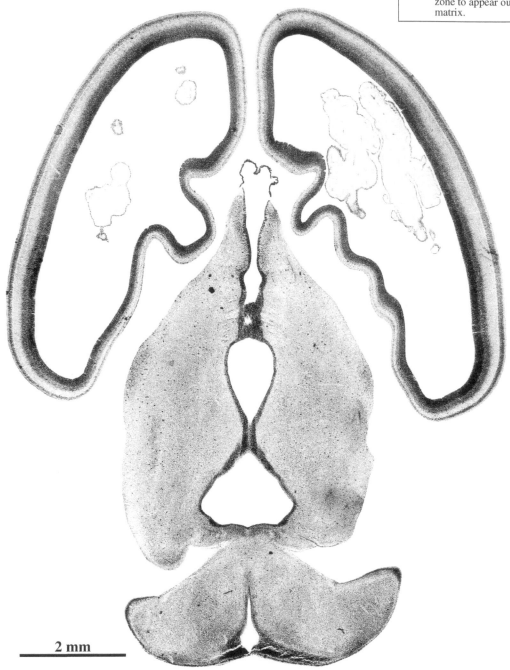

2 mm

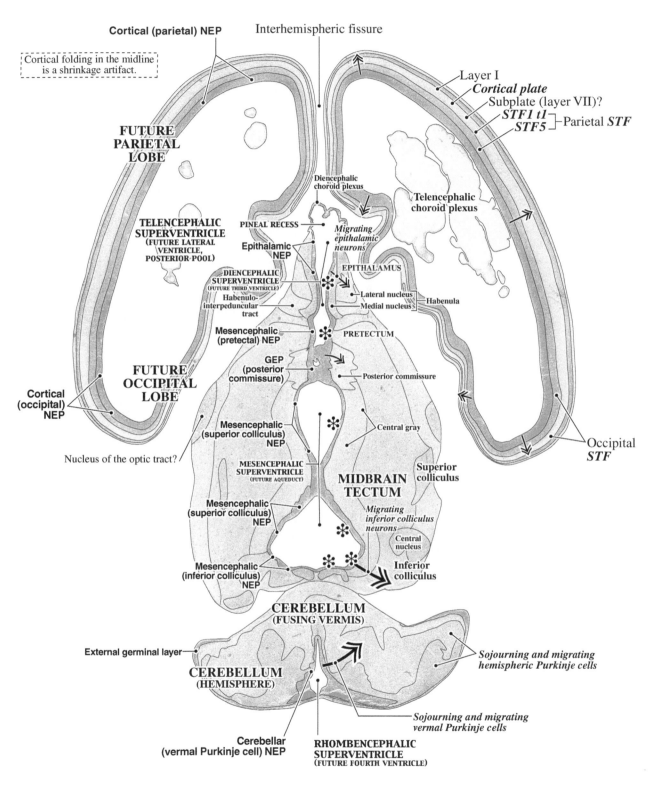

Cortical (parietal) NEP

Interhemispheric fissure

Cortical folding in the midline is a shrinkage artifact.

Layer I
*Cortical plate*
Subplate (layer VII)?
*STF1 t1*
*STF5* — Parietal *STF*

**FUTURE PARIETAL LOBE**

Diencephalic choroid plexus

Telencephalic choroid plexus

**TELENCEPHALIC SUPERVENTRICLE**
(FUTURE LATERAL VENTRICLE, POSTERIOR POOL)

PINEAL RECESS

*Migrating epithalamic neurons*

Epithalamic NEP

EPITHALAMUS

**DIENCEPHALIC SUPERVENTRICLE**
(FUTURE THIRD VENTRICLE)

Lateral nucleus
Medial nucleus — Habenula

Habenulo-interpeduncular tract

Mesencephalic (pretectal) NEP

PRETECTUM

**FUTURE OCCIPITAL LOBE**

GEP (posterior commissure)

Posterior commissure

Cortical (occipital) NEP

Nucleus of the optic tract?

Mesencephalic (superior colliculus) NEP

Central gray

Occipital *STF*

**MESENCEPHALIC SUPERVENTRICLE**
(FUTURE AQUEDUCT)

Superior colliculus

**MIDBRAIN TECTUM**

Mesencephalic (superior colliculus) NEP

*Migrating inferior colliculus neurons*

Central nucleus

Mesencephalic (inferior colliculus) NEP

Inferior colliculus

**CEREBELLUM (FUSING VERMIS)**

External germinal layer

*Sojourning and migrating hemispheric Purkinje cells*

**CEREBELLUM (HEMISPHERE)**

*Sojourning and migrating vermal Purkinje cells*

Cerebellar (vermal Purkinje cell) NEP

**RHOMBENCEPHALIC SUPERVENTRICLE**
(FUTURE FOURTH VENTRICLE)

✳ Evaginations and invaginations of the neuroepithelium are mosaic compartments that give rise to different brain structures.

**ABBREVIATIONS:**
**GEP** - Glioepithelium
**NEP** - Neuroepithelium

**Arrows** indicate the presumed *direction of neuron migration* from neuroepithelial sources.

# PLATE 119A

**GW9 Coronal**
**CR 42 mm**
**M841**
**Level 20: Section 112**

**LAYERS OF THE CORTICAL**
*STRATIFIED TRANSITIONAL FIELD (STF)*

| | |
|---|---|
| *STF1* | Superficial fibrous layer with an early developmental stage *(t1)* when many cells are migrating through it, followed by a late stage *(t2)* with sparse cells. Endures as the subcortical white matter. |
| *STF5* | Deep cellular layer that is prominent during the first trimester, the first sojourn zone to appear outside the germinal matrix. |

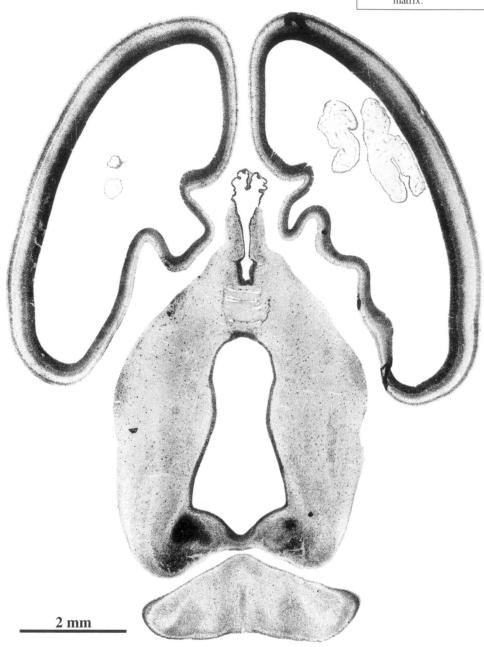

2 mm

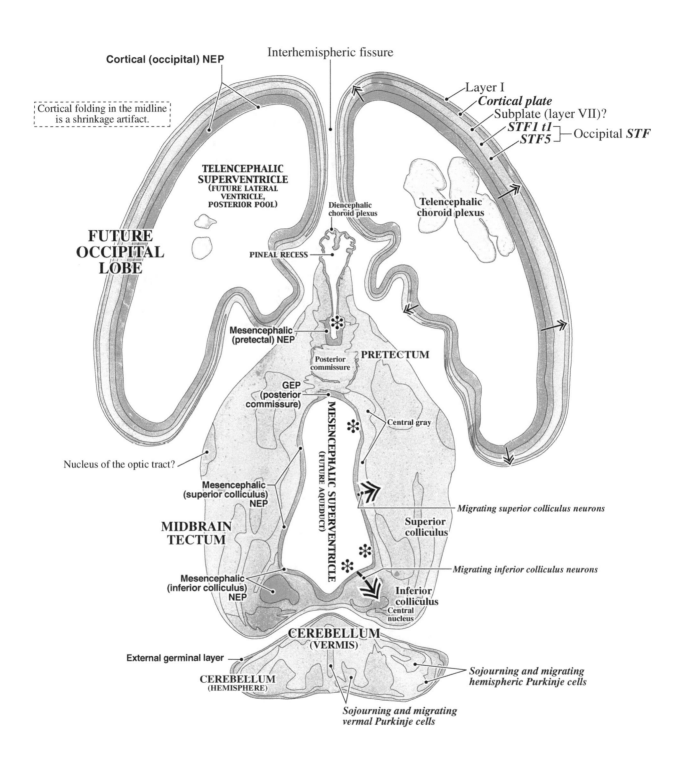

Cortical (occipital) NEP

Interhemispheric fissure

Layer I
Cortical plate
Subplate (layer VII)?
STF1 t1
STF5 ⎤ Occipital STF

Cortical folding in the midline
is a shrinkage artifact.

TELENCEPHALIC
SUPERVENTRICLE
(FUTURE LATERAL
VENTRICLE,
POSTERIOR POOL)

Diencephalic
choroid plexus

Telencephalic
choroid plexus

FUTURE
OCCIPITAL
LOBE

PINEAL RECESS

Mesencephalic
(pretectal) NEP

PRETECTUM

Posterior
commissure

GEP
(posterior
commissure)

Central gray

Nucleus of the optic tract?

MESENCEPHALIC SUPERVENTRICLE
(FUTURE AQUEDUCT)

Mesencephalic
(superior colliculus)
NEP

Migrating superior colliculus neurons

Superior
colliculus

MIDBRAIN
TECTUM

Mesencephalic
(inferior colliculus)
NEP

Migrating inferior colliculus neurons

Inferior
colliculus
Central
nucleus

CEREBELLUM
(VERMIS)

External germinal layer

Sojourning and migrating
hemispheric Purkinje cells

CEREBELLUM
(HEMISPHERE)

Sojourning and migrating
vermal Purkinje cells

✻ Evaginations and invaginations of the
neuroepithelium are mosaic compartments
that give rise to different brain structures.

ABBREVIATIONS:
GEP - Glioepithelium
NEP - Neuroepithelium

Arrows indicate the
presumed direction of
neuron migration from
neuroepithelial sources.

**PLATE 120A**

**GW9 Coronal**
**CR 42 mm**
**M841**
**Level 21: Section 96**

**LAYERS OF THE CORTICAL**
*STRATIFIED TRANSITIONAL FIELD (STF)*

*STF1*  Superficial fibrous layer with an early
developmental stage *(t1)* when many
cells are migrating through it, followed
by a late stage *(t2)* with sparse cells.
Endures as the subcortical white matter.

*STF5*  Deep cellular layer that is prominent
during the first trimester, the first sojourn
zone to appear outside the germinal
matrix.

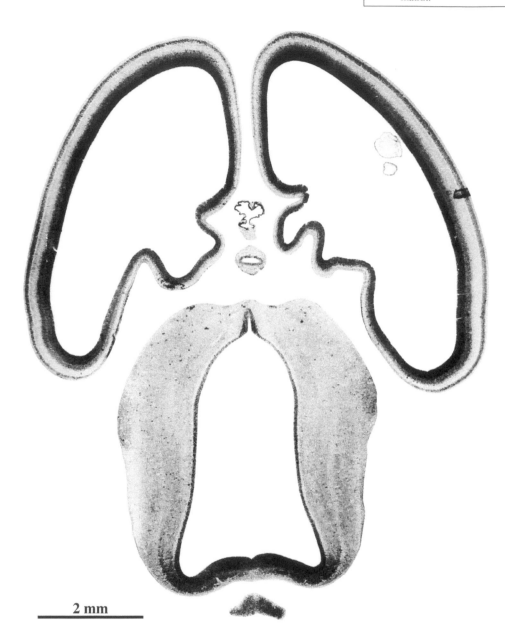

2 mm

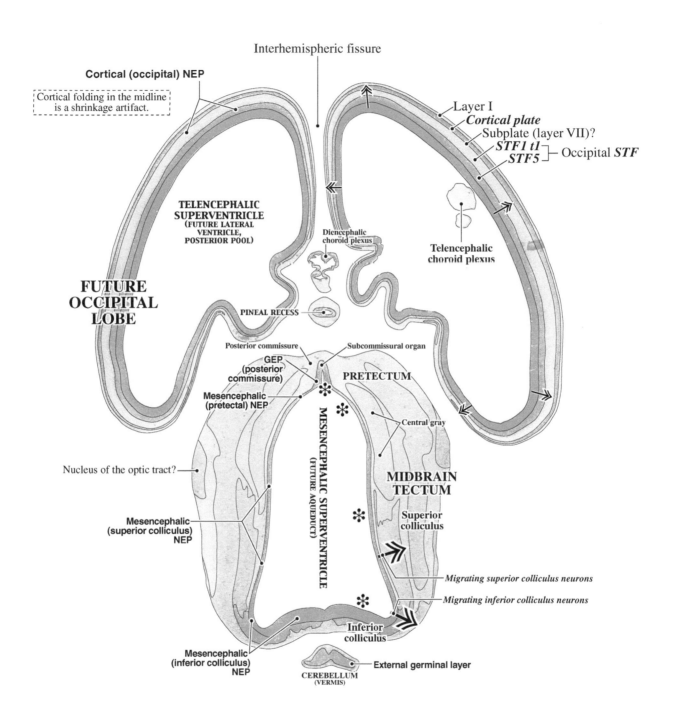

Interhemispheric fissure

**Cortical (occipital) NEP**

Cortical folding in the midline
is a shrinkage artifact.

Layer I
*Cortical plate*
Subplate (layer VII)?
*STF1 t1*
*STF5* ⎤— Occipital *STF*

**TELENCEPHALIC
SUPERVENTRICLE**
(FUTURE LATERAL
VENTRICLE,
POSTERIOR POOL)

Diencephalic
choroid plexus

Telencephalic
choroid plexus

**FUTURE
OCCIPITAL
LOBE**

**PINEAL RECESS**

Posterior commissure

Subcommissural organ

GEP
(posterior
commissure)

**PRETECTUM**

Mesencephalic
(pretectal) NEP

Central gray

Nucleus of the optic tract?

**MESENCEPHALIC SUPERVENTRICLE**
(FUTURE AQUEDUCT)

**MIDBRAIN
TECTUM**

Mesencephalic
(superior colliculus)
NEP

Superior
colliculus

*Migrating superior colliculus neurons*

*Migrating inferior colliculus neurons*

Inferior
colliculus

Mesencephalic
(inferior colliculus)
NEP

External germinal layer

**CEREBELLUM
(VERMIS)**

✳ Evaginations and invaginations of the
neuroepithelium are mosaic compartments
that give rise to different brain structures.

**ABBREVIATIONS:
GEP - Glioepithelium
NEP - Neuroepithelium**

**Arrows** indicate the
presumed *direction of
neuron migration* from
neuroepithelial sources.

# PLATE 121A

**GW9 Coronal**
**CR 42 mm**
**M841**
**Level 22: Section 39**

**LAYERS OF THE CORTICAL**
*STRATIFIED TRANSITIONAL FIELD (STF)*

| | |
|---|---|
| *STF1* | Superficial fibrous layer with an early developmental stage *(t1)* when many cells are migrating through it, followed by a late stage *(t2)* with sparse cells. Endures as the subcortical white matter. |
| *STF5* | Deep cellular layer that is prominent during the first trimester, the first sojourn zone to appear outside the germinal matrix. |

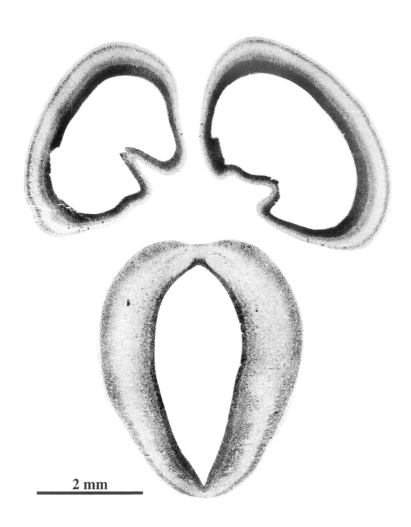

2 mm

**FONT KEY:**
**VENTRICULAR DIVISIONS – CAPITALS**
**Germinal zone - Helvetica bold**
*Transient structure - Times bold italic*
Permanent structure - Times Roman or **Bold**

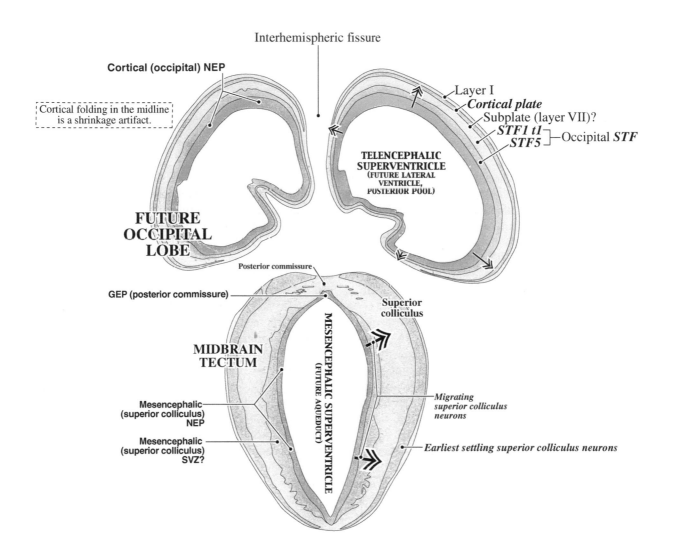

Interhemispheric fissure

Cortical (occipital) NEP

Cortical folding in the midline is a shrinkage artifact.

Layer I
*Cortical plate*
Subplate (layer VII)?
*STF1 t1*
*STF5* ⎤ Occipital *STF*

**TELENCEPHALIC SUPERVENTRICLE**
(FUTURE LATERAL VENTRICLE, POSTERIOR POOL)

**FUTURE OCCIPITAL LOBE**

Posterior commissure

GEP (posterior commissure)

Superior colliculus

**MIDBRAIN TECTUM**

**MESENCEPHALIC SUPERVENTRICLE**
(FUTURE AQUEDUCT)

Mesencephalic (superior colliculus) NEP

*Migrating superior colliculus neurons*

Mesencephalic (superior colliculus) SVZ?

*Earliest settling superior colliculus neurons*

✳ Evaginations and invaginations of the neuroepithelium are mosaic compartments that give rise to different brain structures.

**ABBREVIATIONS:**
GEP - Glioepithelium
NEP - Neuroepithelium
SVZ - Subventricular zone

**Arrows** indicate the presumed *direction of neuron migration* from neuroepithelial sources.

# PART VII: GW8 SAGITTAL

This is specimen number145 in the Carnegie Collection, designated here as C145, a male with a crown-rump length (CR) of 33 mm estimated to be at gestational week (GW) 8. The entire fetus was cut in the sagittal plane. Information on the date of specimen collection, fixative, section thickness, and embedding medium was not available to us. The sections are thick (probably between 50 to 100 μm) and appear to be embedded in celloidin. Since there is no photograph of C145's brain before histological processing, a specimen from Hochstetter (1919) that is comparable in age to C145 is used to show external brain features at GW8 (**A, Figure 6**). C145's brain structures are easier to understand because sections are closely parallel to the midline; **Figure 6** shows the approximate slight rotations in horizontal (**B**) and vertical (**C**) dimensions. Photographs of 10 sections (**Levels 1-10**) are illustrated at low magnification in four parts (**Plates 122A-D** through **131A-D**). The **A/B** parts show the brain in place in the skull; the **C/D** parts show only the brain (and some peripheral ganglia) at slightly higher magnification. **Plates 132-147** show high magnification views of various parts of the brain. All of the high-magnification plates are rotated 90° (landscape orientation) to show photographs at higer magnification in the available page space.

C145 is considerably less mature than the GW9 specimens. One of the most notable features of this specimen is the larger volume of the brain ventricles when compared to the brain parenchyma (areas where neurons migrate, settle, and differentiate). The largest structure in each of the brain's major subdivisions is the *superventricles* in their cores. For example, the telencephalon is largely occupied by the telencephalic superventricle. The thickness of the parenchyma is a key to the degree of maturation of the various brain structures. It is thickest in the medulla, pons, and midbrain tegmentum where most of the neurons have been generated and thinner in the cerebellum, midbrain tectum, and diencephalon. Within the telencephalon, the cerebral cortex has a very thin parenchyma, while the basal telen-

cephalon and parts of the basal ganglia have thick parenchymal components.

Throughout the cerebral cortex, the *neuroepithelium* is prominent as the sole germinal matrix. The *stratified transitional field (STF)* contains *STF1* and *STF5* only in lateral areas. The pronounced anterolateral (thicker) to dorsomedial (thinner) maturation gradient is evident in both the *cortical plate* and the *STF* layers. The olfactory bulb is just beginning to evaginate in front of the basal telencephalic neuroepithelium. Neurons are just beginning to migrate in the hippocampus. A massive *neuroepithelium/subventricular zone* overlies the amygdala, nucleus accumbens, and striatum (caudate and putamen) where neurons (and glia) are being generated.

The cerebellum has a thicker *neuroepithelium*, indicating many Purkinje cells are being generated. Earlier-generated Purkinje cells are sojourning in a dense layer outside the neuroepithelium. Most of the deep neurons are superficial in the cerebellum. The *external germinal layer (egl)* is barely visible eminating from the germinal trigone in the dorsal rhombic lip.

In sections near the midline, the brainstem *neuroepithelium* varies in thickness. It is thinner in the midbrain tegmentum, pons, and medulla in accordance with an earlier maturation gradient. Most neurons have been generated in these structures and are settling. In the cerebellum (see above), midbrain tectum, and diencephalon, the neuroepithelium is thicker indicating that substantial neurogenesis is happening in these structures, but many neurons have already been produced. However, more lateral sections have a thick *precerebellar neuroepithelium* in the medulla. Since the pontine gray is totally absent, nearly all of those neurons have yet to be generated in the precerebellar neuroepithelium; some of the earliest-generated pontine gray neurons are migrating in the *anterior extramural migratory stream*.

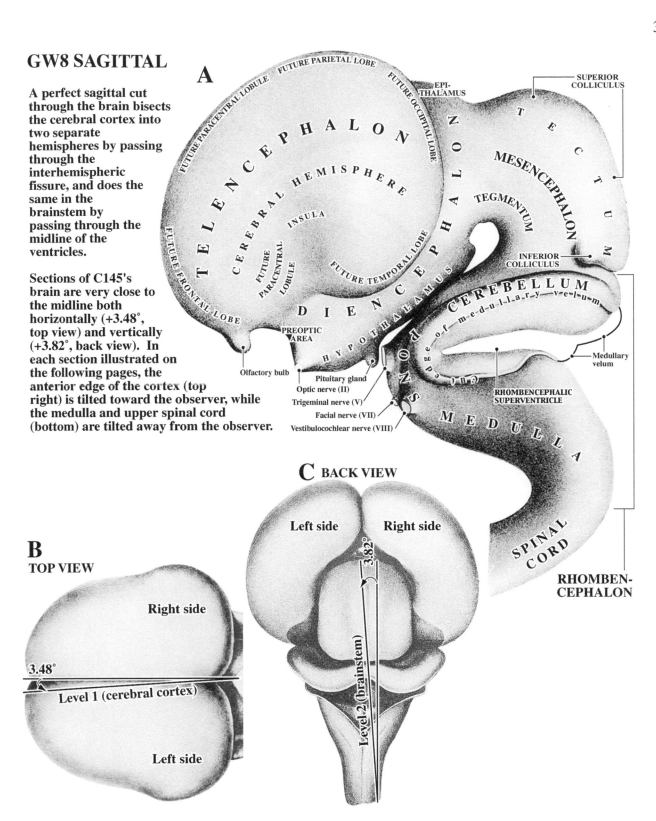

# GW8 SAGITTAL

**A**

A perfect sagittal cut through the brain bisects the cerebral cortex into two separate hemispheres by passing through the interhemispheric fissure, and does the same in the brainstem by passing through the midline of the ventricles.

Sections of C145's brain are very close to the midline both horizontally (+3.48°, top view) and vertically (+3.82°, back view). In each section illustrated on the following pages, the anterior edge of the cortex (top right) is tilted toward the observer, while the medulla and upper spinal cord (bottom) are tilted away from the observer.

FUTURE PARACENTRAL LOBULE
FUTURE PARIETAL LOBE
FUTURE OCCIPITAL LOBE
EPI-THALAMUS
SUPERIOR COLLICULUS

TELENCEPHALON
CEREBRAL HEMISPHERE
INSULA
FUTURE PARACENTRAL LOBULE
FUTURE FRONTAL LOBE
FUTURE TEMPORAL LOBE
DIENCEPHALON
HYPOTHALAMUS
PREOPTIC AREA

MESENCEPHALON
TECTUM
TEGMENTUM
INFERIOR COLLICULUS
CEREBELLUM
caudal edge of medullary velum
PONS
Medullary velum
RHOMBENCEPHALIC SUPERVENTRICLE
MEDULLA

Olfactory bulb
Pituitary gland
Optic nerve (II)
Trigeminal nerve (V)
Facial nerve (VII)
Vestibulocochlear nerve (VIII)

SPINAL CORD

RHOMBEN-CEPHALON

**C** BACK VIEW

Left side    Right side

3.82°

Level 2 (brainstem)

**B**

**TOP VIEW**

Right side

3.48°

Level 1 (cerebral cortex)

Left side

**Figure 6. A**, The lateral view of the brain and upper cervical spinal cord from a specimen with a crown-rump length of 27 mm (modified from Figure 39, Table VII, Hochstetter, 1919) identifies external features of a brain similar to C145 (CR 33 mm). **B**, Top view of the brain with a crown-rump length of 38 mm (modified from Figure 45, Table VIII, Hochstetter, 1919) shows how C145's sections rotate from a line parallel to the horizontal midline in the interhemispheric fissure. **C**, Back view of the brain in **B** (modified from Figure 44, Table VIII, Hochstetter, 1919) shows how C145's sections rotate from a line parallel to the vertical midline in the brainstem and upper cervical spinal cord.

**PLATE 122A**

GW8 Sagittal, CR 33 mm, C145
Level 1: Slide 23, Section 2
HEAD STRUCTURES,
MAJOR BRAIN REGIONS,
AND VENTRICULAR
DIVISIONS

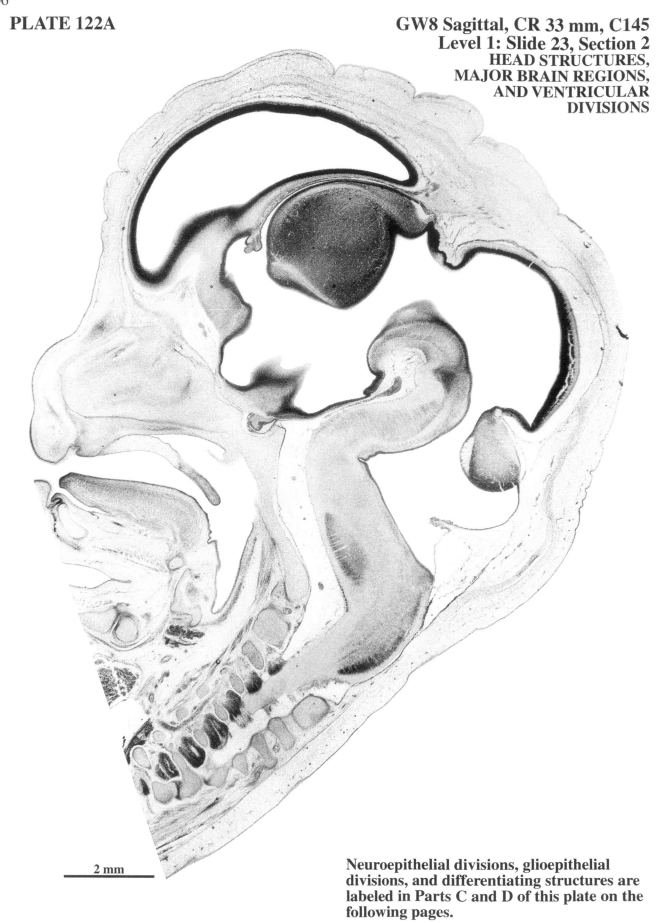

2 mm

Neuroepithelial divisions, glioepithelial
divisions, and differentiating structures are
labeled in Parts C and D of this plate on the
following pages.

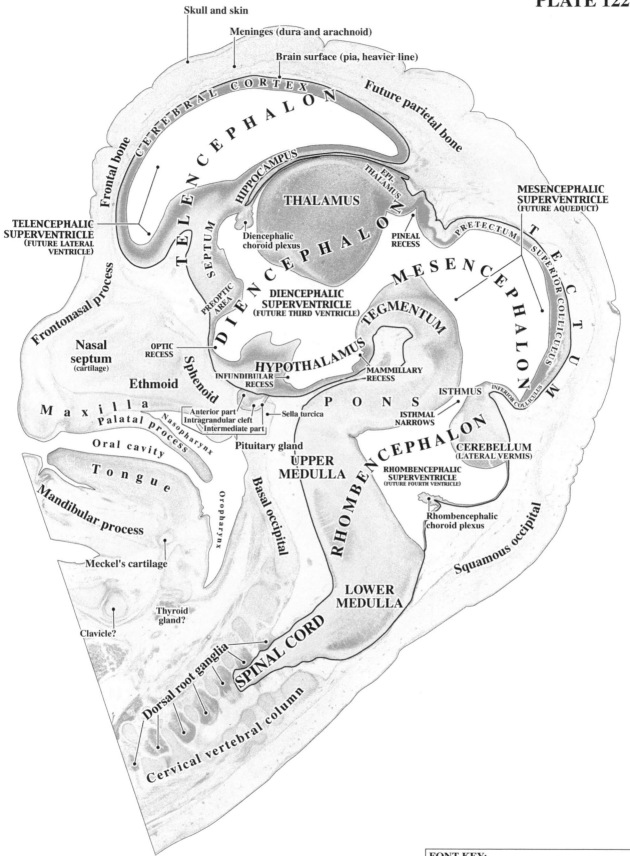

Skull and skin

Meninges (dura and arachnoid)

Brain surface (pia, heavier line)

CEREBRAL CORTEX

Future parietal bone

**TELENCEPHALON**

Frontal bone

HIPPOCAMPUS

**THALAMUS**

EPI.
THALAMUS

**MESENCEPHALIC
SUPERVENTRICLE
(FUTURE AQUEDUCT)**

**TELENCEPHALIC
SUPERVENTRICLE
(FUTURE LATERAL
VENTRICLE)**

SEPTUM

Diencephalic
choroid plexus

PINEAL
RECESS

PRETECTUM

**MESENCEPHALON**

SUPERIOR COLLICULUS

**TECTUM**

Frontonasal process

PREOPTIC
AREA

**DIENCEPHALON**

**DIENCEPHALIC
SUPERVENTRICLE
(FUTURE THIRD VENTRICLE)**

**TEGMENTUM**

**Nasal
septum**
(cartilage)

OPTIC
RECESS

Sphenoid

**HYPOTHALAMUS**

MAMMILLARY
RECESS

**Ethmoid**

INFUNDIBULAR
RECESS

ISTHMUS

INFERIOR COLLICULUS

**Maxilla**

Anterior part
Intragrandular cleft
Intermediate part

Sella turcica

**P O N S**

ISTHMAL
NARROWS

Palatal process

Nasopharynx

Pituitary gland

**CEREBELLUM**
(LATERAL VERMIS)

Oral cavity

**UPPER
MEDULLA**

**RHOMBENCEPHALIC
SUPERVENTRICLE**
(FUTURE FOURTH VENTRICLE)

**T o n g u e**

Oropharynx

**RHOMBENCEPHALON**

Rhombencephalic
choroid plexus

Mandibular process

Basal occipital

Squamous occipital

Meckel's cartilage

**LOWER
MEDULLA**

Thyroid
gland?

Clavicle?

Dorsal root ganglia

**SPINAL CORD**

Cervical vertebral column

**FONT KEY:**
**VENTRICULAR DIVISIONS – CAPITALS**
Major brain structure - Times **Bold CAPITALS**
All other structures - Times Roman or **Bold**

**PLATE 122C**

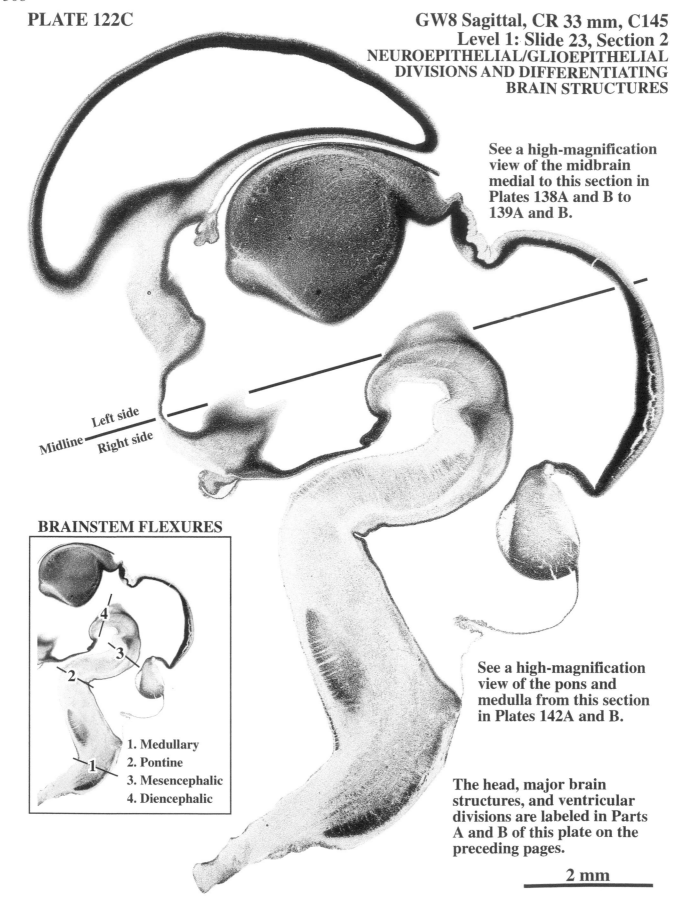

See a high-magnification
view of the midbrain
medial to this section in
Plates 138A and B to
139A and B.

Left side

Midline — Right side

**BRAINSTEM FLEXURES**

1. Medullary
2. Pontine
3. Mesencephalic
4. Diencephalic

See a high-magnification
view of the pons and
medulla from this section
in Plates 142A and B.

The head, major brain
structures, and ventricular
divisions are labeled in Parts
A and B of this plate on the
preceding pages.

2 mm

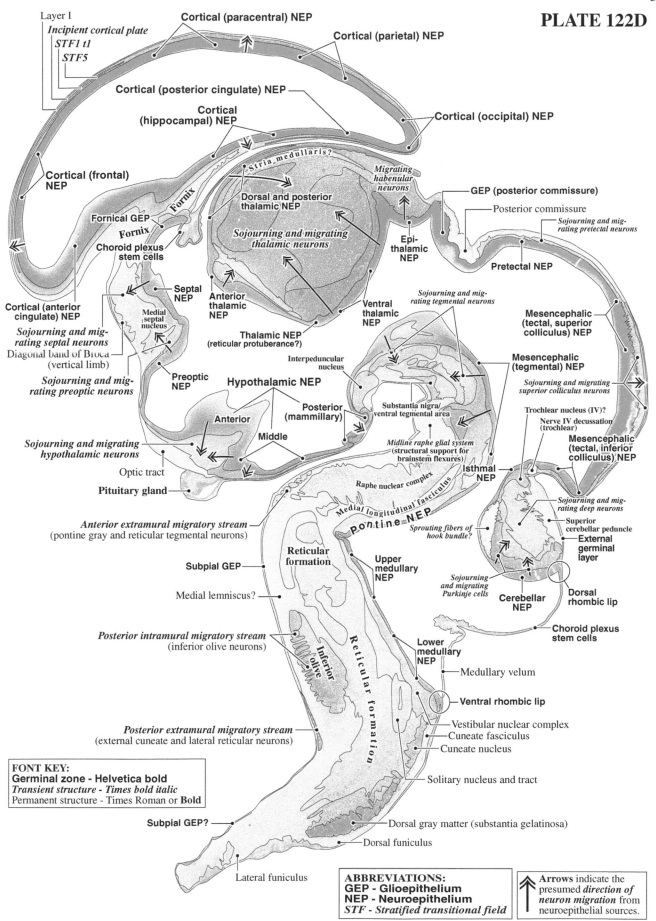

Layer I
*Incipient cortical plate*
*STF1 t1*
*STF5*

Cortical (paracentral) NEP

Cortical (parietal) NEP

Cortical (posterior cingulate) NEP

Cortical (hippocampal) NEP

Cortical (occipital) NEP

Cortical (frontal) NEP

*Stria medullaris?*

*Migrating habenular neurons*

GEP (posterior commissure)

Posterior commissure

**Dorsal and posterior thalamic NEP**

*Sojourning and migrating thalamic neurons*

Epi-thalamic NEP

*Sojourning and migrating pretectal neurons*

**Fornical GEP**
*Fornix*
*Fornix*

**Choroid plexus stem cells**

**Pretectal NEP**

Cortical (anterior cingulate) NEP

*Sojourning and migrating septal neurons*

Diagonal band of Broca (vertical limb)

**Septal NEP**

**Medial septal nucleus**

**Anterior thalamic NEP**

**Ventral thalamic NEP**

*Sojourning and migrating tegmental neurons*

**Mesencephalic (tectal, superior colliculus) NEP**

**Mesencephalic (tegmental) NEP**

**Thalamic NEP (reticular protuberance?)**

Interpeduncular nucleus

*Sojourning and migrating superior colliculus neurons*

*Sojourning and migrating preoptic neurons*

**Preoptic NEP**

**Hypothalamic NEP**

Substantia nigra/ventral tegmental area

Trochlear nucleus (IV)?
Nerve IV decussation (trochlear)

**Mesencephalic (tectal, inferior colliculus) NEP**

**Anterior**

**Posterior (mammillary)**

**Middle**

*Sojourning and migrating hypothalamic neurons*

Optic tract

*Midline raphe glial system (structural support for brainstem flexures)*

**Isthmal NEP**

*Sojourning and migrating deep neurons*

Superior cerebellar peduncle

**Pituitary gland**

Raphe nuclear complex

*Medial longitudinal fasciculus*

**Pontine NEP**

*Sprouting fibers of hook bundle?*

**External germinal layer**

*Anterior extramural migratory stream (pontine gray and reticular tegmental neurons)*

**Reticular formation**

**Upper medullary NEP**

*Sojourning and migrating Purkinje cells*

**Cerebellar NEP**

**Dorsal rhombic lip**

**Subpial GEP**

Medial lemniscus?

**Choroid plexus stem cells**

*Posterior intramural migratory stream (inferior olive neurons)*

*Inferior olive*

**Lower medullary NEP**

Medullary velum

**Reticular formation**

**Ventral rhombic lip**

Vestibular nuclear complex
Cuneate fasciculus
Cuneate nucleus

*Posterior extramural migratory stream (external cuneate and lateral reticular neurons)*

Solitary nucleus and tract

**FONT KEY:**
**Germinal zone - Helvetica bold**
*Transient structure - Times bold italic*
Permanent structure - Times Roman or **Bold**

**Subpial GEP?**

Dorsal gray matter (substantia gelatinosa)

Dorsal funiculus

Lateral funiculus

**ABBREVIATIONS:**
**GEP - Glioepithelium**
**NEP - Neuroepithelium**
*STF - Stratified transitional field*

**Arrows** indicate the presumed *direction of neuron migration* from neuroepithelial sources.

**PLATE 123A**

**GW8 Sagittal, CR 33 mm, C145**
**Level 2: Slide 22, Section 2**
**HEAD STRUCTURES,**
**MAJOR BRAIN REGIONS,**
**AND VENTRICULAR**
**DIVISIONS**

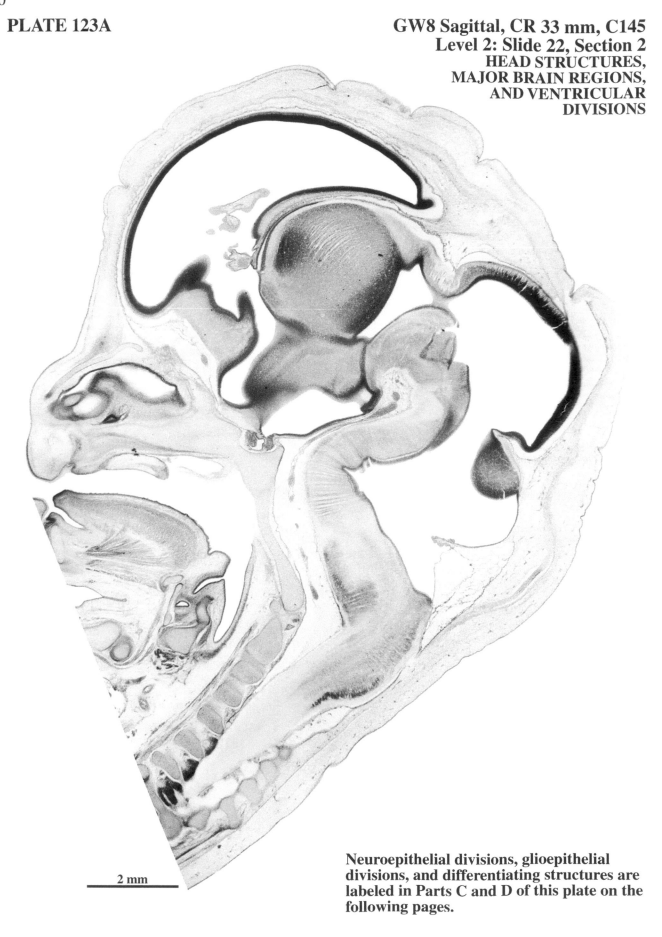

2 mm

Neuroepithelial divisions, glioepithelial
divisions, and differentiating structures are
labeled in Parts C and D of this plate on the
following pages.

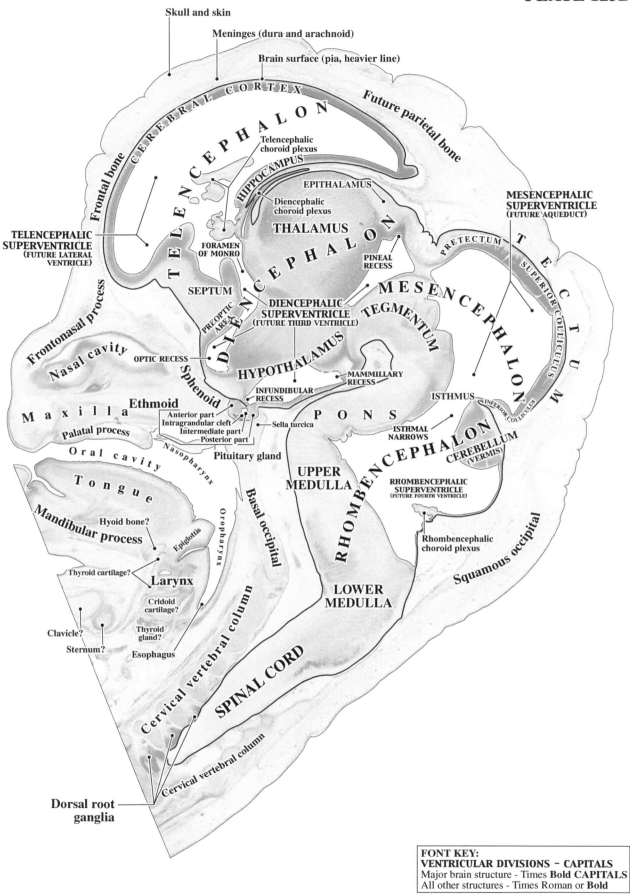

Skull and skin

Meninges (dura and arachnoid)

Brain surface (pia, heavier line)

C E R E B R A L   C O R T E X

Future parietal bone

Frontal bone

TELENCEPHALON

Telencephalic
choroid plexus

HIPPOCAMPUS

EPITHALAMUS

Diencephalic
choroid plexus

THALAMUS

MESENCEPHALIC
SUPERVENTRICLE
(FUTURE AQUEDUCT)

TELENCEPHALIC
SUPERVENTRICLE
(FUTURE LATERAL
VENTRICLE)

FORAMEN
OF MONRO

DIENCEPHALON

PRETECTUM

T
E
C
T
U
M

SUPERIOR COLLICULUS

SEPTUM

PINEAL
RECESS

MESENCEPHALON

Frontonasal process

PREOPTIC
AREA

DIENCEPHALIC
SUPERVENTRICLE
(FUTURE THIRD VENTRICLE)

TEGMENTUM

Nasal cavity

OPTIC RECESS

HYPOTHALAMUS

MAMMILLARY
RECESS

Sphenoid

INFUNDIBULAR
RECESS

ISTHMUS

Maxilla

Ethmoid

Anterior part
Intragrandular cleft
Intermediate part
Posterior part

Sella turcica

P O N S

ISTHMAL
NARROWS

INFERIOR
COLLICULUS

RHOMBENCEPHALON

CEREBELLUM
(VERMIS)

Palatal process

Nasopharynx

Pituitary gland

O r a l   c a v i t y

UPPER
MEDULLA

RHOMBENCEPHALIC
SUPERVENTRICLE
(FUTURE FOURTH VENTRICLE)

T o n g u e

Oropharynx

Mandibular process

Hyoid bone?

Basal occipital

Squamous occipital

Epiglottis

Rhombencephalic
choroid plexus

Thyroid cartilage?

Larynx

Cridoid
cartilage?

LOWER
MEDULLA

Clavicle?

Thyroid
gland?

Sternum?

Esophagus

Cervical vertebral column

SPINAL CORD

Cervical vertebral column

Dorsal root
ganglia

FONT KEY:
VENTRICULAR DIVISIONS – CAPITALS
Major brain structure - Times **Bold CAPITALS**
All other structures - Times Roman or **Bold**

**PLATE 123C**

GW8 Sagittal, CR 33 mm, C145
Level 2: Slide 22, Section 2
NEUROEPITHELIAL/GLIOEPI-
THELIAL DIVISIONS AND
DIFFERENTIATING BRAIN
STRUCTURES

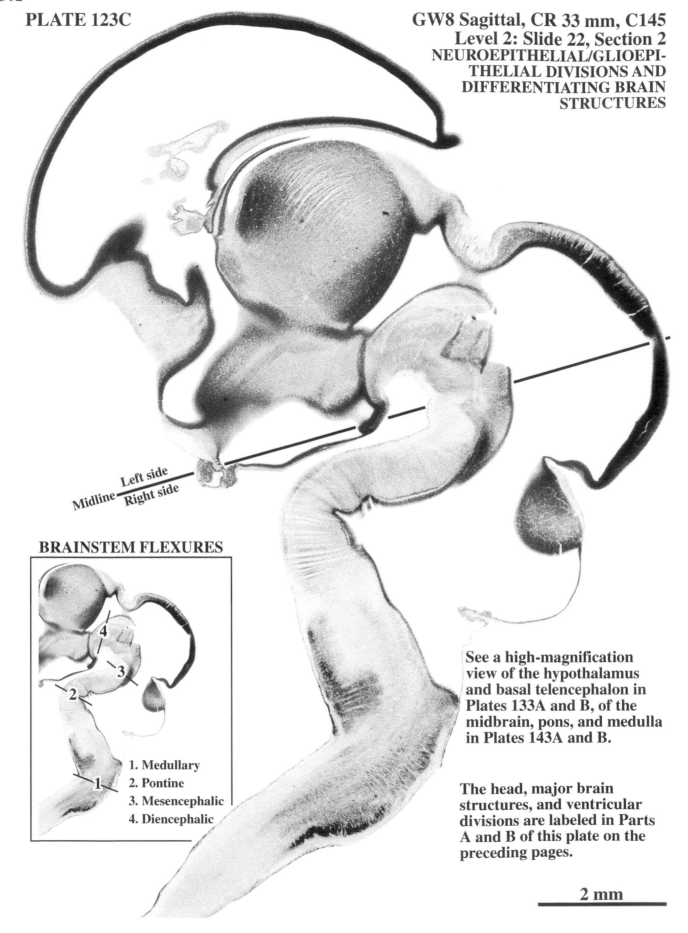

Left side
Midline
Right side

**BRAINSTEM FLEXURES**

1. Medullary
2. Pontine
3. Mesencephalic
4. Diencephalic

See a high-magnification view of the hypothalamus and basal telencephalon in Plates 133A and B, of the midbrain, pons, and medulla in Plates 143A and B.

The head, major brain structures, and ventricular divisions are labeled in Parts A and B of this plate on the preceding pages.

2 mm

**PLATE 123D**

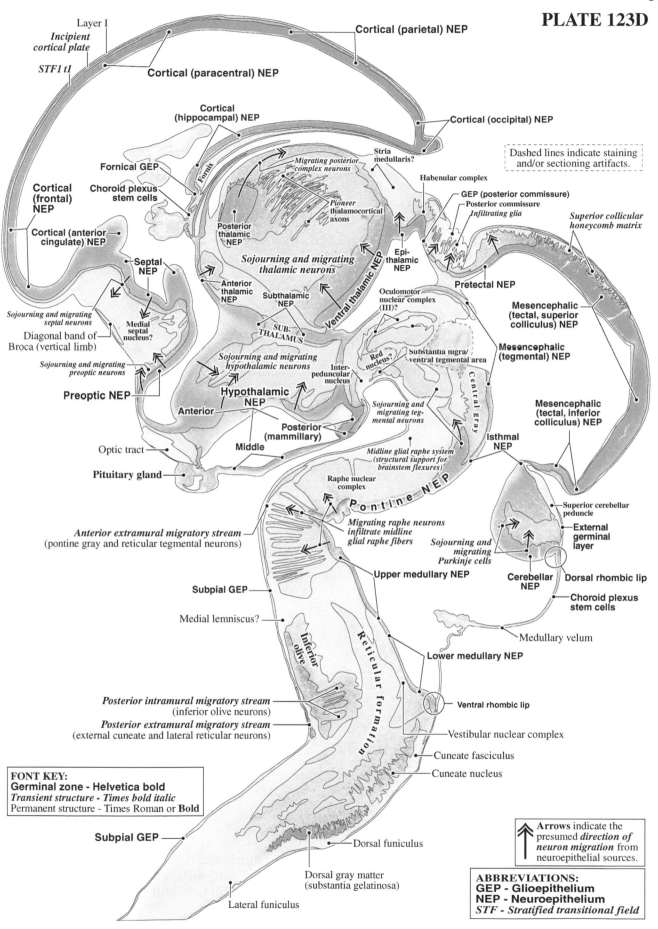

Layer I
*Incipient cortical plate*
*STF1 tI*

**Cortical (parietal) NEP**

**Cortical (paracentral) NEP**

Cortical (hippocampal) NEP

**Cortical (frontal) NEP**

**Fornical GEP**
**Choroid plexus stem cells**

*Fornix*

*Migrating posterior complex neurons*

Stria medullaris?

**Cortical (occipital) NEP**

Dashed lines indicate staining and/or sectioning artifacts.

*Pioneer thalamocortical axons*

Posterior thalamic NEP

Habenular complex

**GEP (posterior commissure)**
Posterior commissure
*Infiltrating glia*

*Superior collicular honeycomb matrix*

**Cortical (anterior cingulate) NEP**

**Septal NEP**

*Sojourning and migrating thalamic neurons*

Anterior thalamic NEP

Subthalamic NEP

**SUB-THALAMUS**

**Ventral thalamic NEP**

Oculomotor nuclear complex (III)?

Epi-thalamic NEP

**Pretectal NEP**

**Mesencephalic (tectal, superior colliculus) NEP**

**Mesencephalic (tegmental) NEP**

*Sojourning and migrating septal neurons*

**Diagonal band of Broca (vertical limb)**

Medial septal nucleus?

Red nucleus?

Substantia nigra/ ventral tegmental area

*Central gray*

*Sojourning and migrating preoptic neurons*

**Preoptic NEP**

*Sojourning and migrating hypothalamic neurons*

**Hypothalamic NEP**

Anterior

Inter-peduncular nucleus

*Sojourning and migrating tegmental neurons*

**Mesencephalic (tectal, inferior colliculus) NEP**

Posterior (mammillary)

**Middle**

**Isthmal NEP**

**Optic tract**

**Pituitary gland**

*Midline glial raphe system (structural support for brainstem flexures)*

Raphe nuclear complex

**Superior cerebellar peduncle**

**External germinal layer**

*Sojourning and migrating Purkinje cells*

**Pontine NEP**

*Migrating raphe neurons infiltrate midline glial raphe fibers*

**Cerebellar NEP**

**Dorsal rhombic lip**

*Anterior extramural migratory stream (pontine gray and reticular tegmental neurons)*

**Choroid plexus stem cells**

**Subpial GEP**

**Upper medullary NEP**

Medullary velum

Medial lemniscus?

*Inferior olive*

**Lower medullary NEP**

*Reticular formation*

**Ventral rhombic lip**

*Posterior intramural migratory stream (inferior olive neurons)*

*Posterior extramural migratory stream (external cuneate and lateral reticular neurons)*

Vestibular nuclear complex

Cuneate fasciculus

Cuneate nucleus

**FONT KEY:**
Germinal zone - Helvetica bold
*Transient structure - Times bold italic*
Permanent structure - Times Roman or **Bold**

Dorsal funiculus

**Subpial GEP**

Dorsal gray matter (substantia gelatinosa)

Lateral funiculus

**Arrows** indicate the presumed *direction of neuron migration* from neuroepithelial sources.

**ABBREVIATIONS:**
**GEP - Glioepithelium**
**NEP - Neuroepithelium**
*STF - Stratified transitional field*

**PLATE 124A**

GW8 Sagittal, CR 33 mm, C145
Level 3: Slide 20, Section 2
HEAD STRUCTURES,
MAJOR BRAIN REGIONS,
AND VENTRICULAR
DIVISIONS

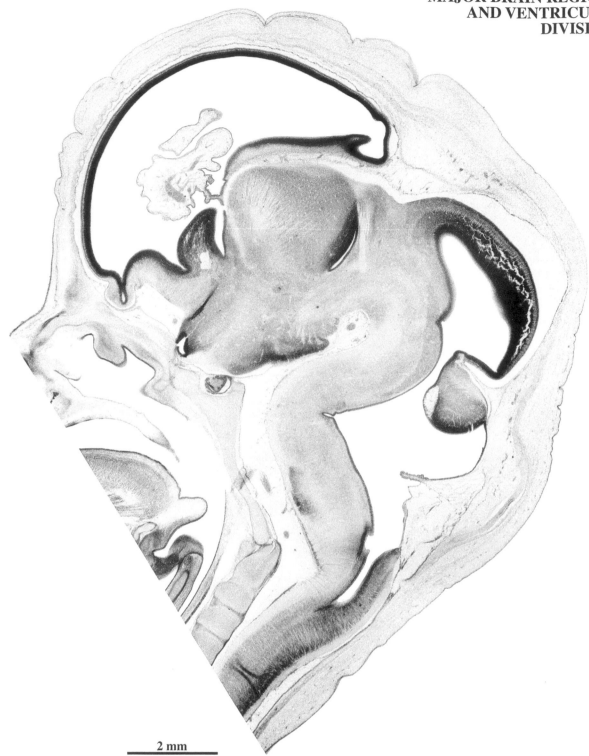

2 mm

Neuroepithelial divisions, glioepithelial
divisions, and differentiating structures are
labeled in Parts C and D of this plate on the
following pages.

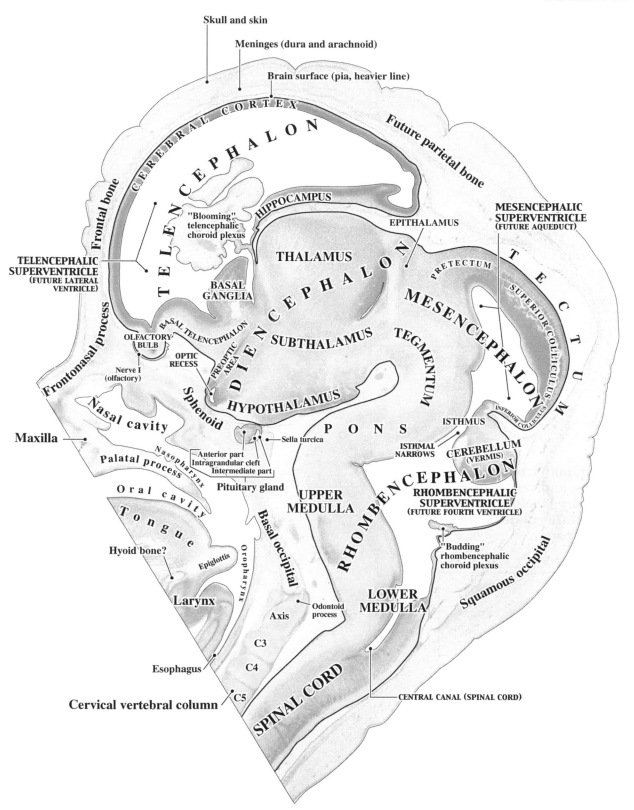

Skull and skin

Meninges (dura and arachnoid)

Brain surface (pia, heavier line)

CEREBRAL CORTEX

TELENCEPHALON

Future parietal bone

Frontal bone

HIPPOCAMPUS

EPITHALAMUS

MESENCEPHALIC
SUPERVENTRICLE
(FUTURE AQUEDUCT)

"Blooming"
telencephalic
choroid plexus

THALAMUS

PRETECTUM

TELENCEPHALIC
SUPERVENTRICLE
(FUTURE LATERAL
VENTRICLE)

BASAL
GANGLIA

DIENCEPHALON

MESENCEPHALON

T E C T U M

SUPERIOR COLLICULUS

Frontonasal process

OLFACTORY
BULB

BASAL TELENCEPHALON

SUBTHALAMUS

TEGMENTUM

OPTIC
RECESS

PREOPTIC
AREA

Nerve I
(olfactory)

Sphenoid

HYPOTHALAMUS

INFERIOR COLLICULUS

Nasal cavity

ISTHMUS

Maxilla

P O N S

ISTHMAL
NARROWS

CEREBELLUM
(VERMIS)

Sella turcica

Anterior part
Intragrandular cleft
Intermediate part

Nasopharynx

Palatal process

Pituitary gland

UPPER
MEDULLA

RHOMBENCEPHALON

RHOMBENCEPHALIC
SUPERVENTRICLE
(FUTURE FOURTH VENTRICLE)

Oral cavity

Tongue

"Budding"
rhombencephalic
choroid plexus

Hyoid bone?

Epiglottis

Oropharynx

Basal occipital

Larynx

LOWER
MEDULLA

Squamous occipital

Axis

Odontoid
process

C3

Esophagus

C4

SPINAL CORD

Cervical vertebral column

C5

CENTRAL CANAL (SPINAL CORD)

FONT KEY:
**VENTRICULAR DIVISIONS – CAPITALS**
Major brain structure - Times **Bold CAPITALS**
All other structures - Times Roman or **Bold**

**PLATE 124C**

GW8 Sagittal, CR 33 mm, C145
Level 3: Slide 20, Section 2
NEUROEPITHELIAL/GLIOEPI-
THELIAL DIVISIONS AND
DIFFERENTIATING BRAIN
STRUCTURES

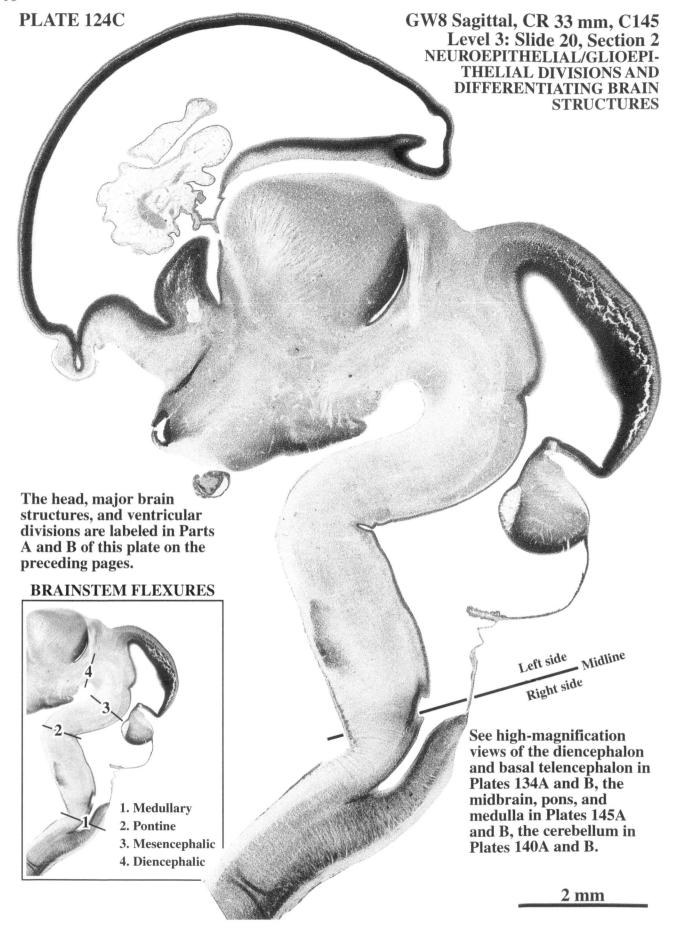

The head, major brain
structures, and ventricular
divisions are labeled in Parts
A and B of this plate on the
preceding pages.

**BRAINSTEM FLEXURES**

1. Medullary
2. Pontine
3. Mesencephalic
4. Diencephalic

Left side — Midline
Right side

See high-magnification
views of the diencephalon
and basal telencephalon in
Plates 134A and B, the
midbrain, pons, and
medulla in Plates 145A
and B, the cerebellum in
Plates 140A and B.

2 mm

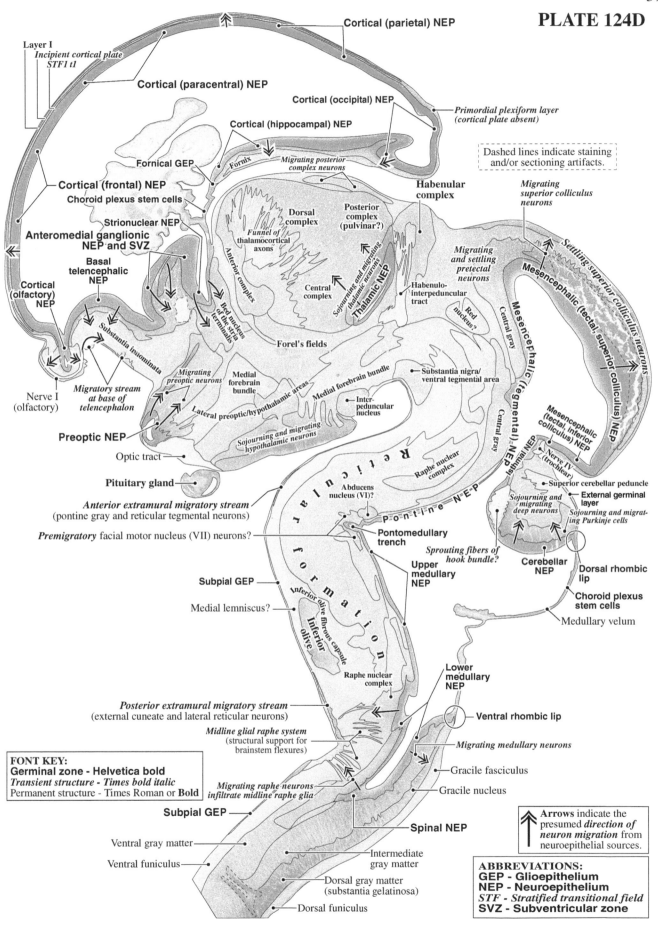

317

PLATE 124D

Cortical (parietal) NEP

Layer I
*Incipient cortical plate*
*STF1 t1*

Cortical (paracentral) NEP

Cortical (occipital) NEP

Cortical (hippocampal) NEP

*Primordial plexiform layer*
*(cortical plate absent)*

Dashed lines indicate staining
and/or sectioning artifacts.

Fornical GEP

Fornix

*Migrating posterior*
*complex neurons*

Habenular
complex

*Migrating*
*superior colliculus*
*neurons*

Cortical (frontal) NEP

Choroid plexus stem cells

Strionuclear NEP

**Anteromedial ganglionic**
**NEP and SVZ**

Basal
telencephalic
NEP

Dorsal
complex

Posterior
complex
(pulvinar?)

*Funnel of*
*thalamocortical*
*axons*

*Anterior complex*

Central
complex

*Sojourning and migrating*
*thalamic neurons*

**Thalamic NEP**

*Migrating*
*and settling*
*pretectal*
*neurons*

Habenulo-
interpeduncular
tract

**Mesencephalic (tectal, superior colliculus) NEP**

*Settling superior colliculus neurons*

Cortical
(olfactory)
NEP

*Red nucleus*
*of the stria*
*terminalis*

Forel's fields

Red
nucleus?

*Substantia nigra/*
*ventral tegmental area*

**Mesencephalic (tegmental) NEP**

Central
gray

Central
gray

Nerve I
(olfactory)

*Substantia innominata*

*Migratory stream*
*at base of*
*telencephalon*

*Migrating*
*preoptic neurons*

Medial
forebrain
bundle

*Medial forebrain bundle*

Inter-
peduncular
nucleus

**Isthmial NEP**

Nerve IV
(trochlear)

**Mesencephalic**
**(tectal, inferior**
**colliculus) NEP**

**Preoptic NEP**

*Lateral preoptic/hypothalamic areas*

*Sojourning and migrating*
*hypothalamic neurons*

**R e t i c u l a r**

Raphe nuclear
complex

Superior cerebellar peduncle

External germinal
layer

Optic tract

Pituitary gland

**Anterior extramural migratory stream**
*(pontine gray and reticular tegmental neurons)*

Abducens
nucleus (VI)?

**P o n t i n e  N E P**

*Sojourning and*
*migrating*
*deep neurons*

*Sojourning and migrat-*
*ing Purkinje cells*

*Premigratory* facial motor nucleus (VII) neurons?

**f o r m a t i o n**

Pontomedullary
trench

*Sprouting fibers of*
*hook bundle?*

**Cerebellar**
**NEP**

**Dorsal rhombic**
**lip**

Upper
medullary
**NEP**

**Subpial GEP**

*Inferior olive fibrous capsule*

Medial lemniscus?

**Choroid plexus**
**stem cells**

Medullary velum

*Inferior*
*olive*

Raphe nuclear
complex

Lower
medullary
**NEP**

**Posterior extramural migratory stream**
*(external cuneate and lateral reticular neurons)*

**Ventral rhombic lip**

*Migrating medullary neurons*

*Midline glial raphe system*
*(structural support for*
*brainstem flexures)*

Gracile fasciculus

Gracile nucleus

**FONT KEY:**
Germinal zone - **Helvetica bold**
*Transient structure - Times bold italic*
Permanent structure - Times Roman or **Bold**

*Migrating raphe neurons*
*infiltrate midline raphe glia*

**Subpial GEP**

**Spinal NEP**

Ventral gray matter

Ventral funiculus

Intermediate
gray matter

**Arrows** indicate the
presumed *direction of*
*neuron migration* from
neuroepithelial sources.

Dorsal gray matter
(substantia gelatinosa)

Dorsal funiculus

**ABBREVIATIONS:**
**GEP** - **Glioepithelium**
**NEP** - **Neuroepithelium**
*STF* - *Stratified transitional field*
**SVZ** - **Subventricular zone**

GW8 Sagittal, CR 33 mm, C145
Level 4: Slide 19, Section 2
HEAD STRUCTURES,
MAJOR BRAIN REGIONS,
AND VENTRICULAR
DIVISIONS

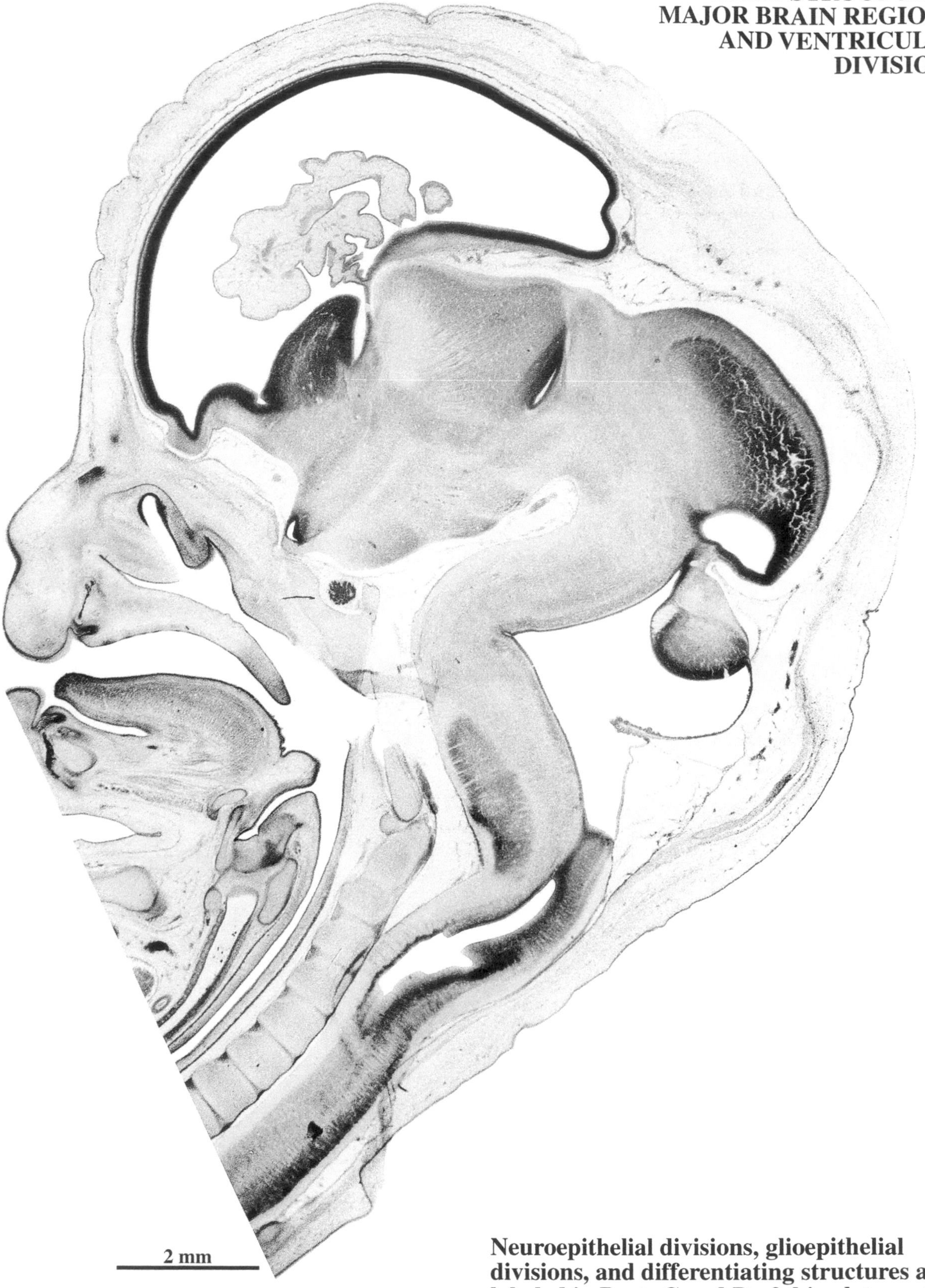

2 mm

Neuroepithelial divisions, glioepithelial
divisions, and differentiating structures are
labeled in Parts C and D of this plate on the
following pages.

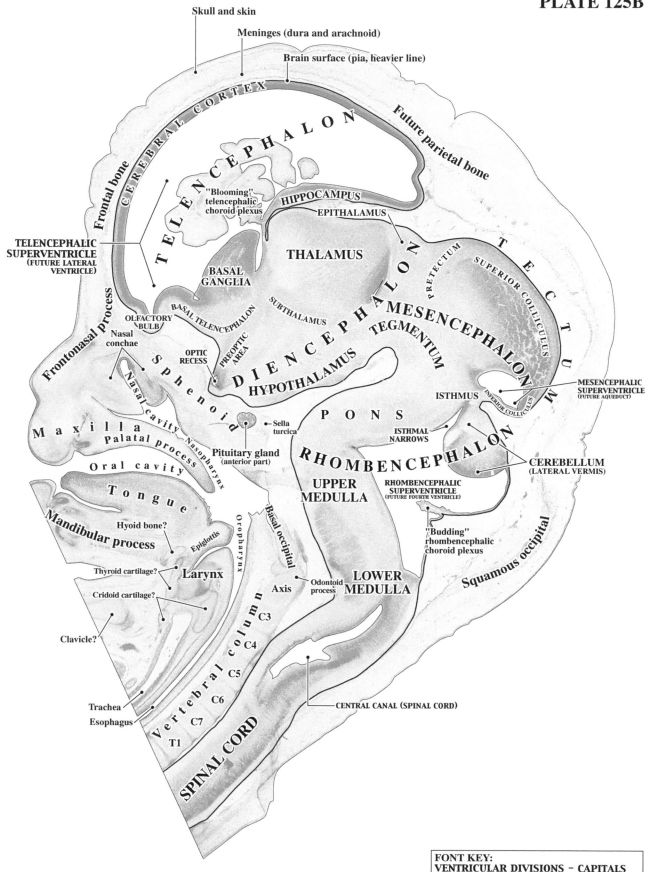

Skull and skin

Meninges (dura and arachnoid)

Brain surface (pia, heavier line)

CEREBRAL CORTEX

Future parietal bone

TELENCEPHALON

Frontal bone

"Blooming" telencephalic choroid plexus

HIPPOCAMPUS

EPITHALAMUS

THALAMUS

PRETECTUM

SUPERIOR COLLICULUS

T E C T U M

TELENCEPHALIC SUPERVENTRICLE (FUTURE LATERAL VENTRICLE)

BASAL GANGLIA

MESENCEPHALON

Frontonasal process

OLFACTORY BULB

BASAL TELENCEPHALON

SUBTHALAMUS

Olfactory

Nasal conchae

PREOPTIC AREA

DIENCEPHALON

TEGMENTUM

Sphenoid

OPTIC RECESS

HYPOTHALAMUS

ISTHMUS

MESENCEPHALIC SUPERVENTRICLE (FUTURE AQUEDUCT)

Nasal cavity

Nasopharynx

Sella turcica

INFERIOR COLLICULUS

Maxilla

Palatal process

Pituitary gland (anterior part)

P O N S

ISTHMAL NARROWS

CEREBELLUM (LATERAL VERMIS)

Oral cavity

RHOMBENCEPHALON

T o n g u e

UPPER MEDULLA

RHOMBENCEPHALIC SUPERVENTRICLE (FUTURE FOURTH VENTRICLE)

Mandibular process

Hyoid bone?

Oropharynx

Basal occipital

"Budding" rhombencephalic choroid plexus

Epiglottis

Thyroid cartilage?

Larynx

LOWER MEDULLA

Squamous occipital

Cridoid cartilage?

Axis

Odontoid process

Clavicle?

Vertebral column

C3

C4

Trachea

C5

Esophagus

C6

CENTRAL CANAL (SPINAL CORD)

C7

T1

SPINAL CORD

FONT KEY:
**VENTRICULAR DIVISIONS – CAPITALS**
Major brain structure - Times **Bold CAPITALS**
All other structures - Times Roman or **Bold**

**PLATE 125C**

NEUROEPITHELIAL/GLIOEPI-
THELIAL DIVISIONS AND
DIFFERENTIATING
BRAIN STRUCTURES

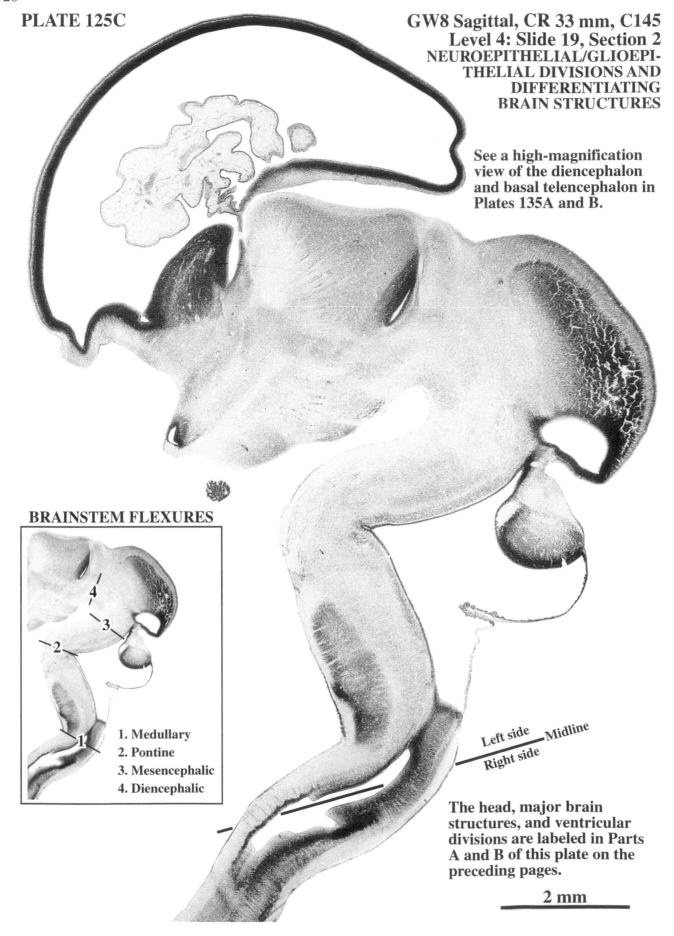

See a high-magnification
view of the diencephalon
and basal telencephalon in
Plates 135A and B.

**BRAINSTEM FLEXURES**

1. Medullary
2. Pontine
3. Mesencephalic
4. Diencephalic

Left side — Midline
Right side

The head, major brain
structures, and ventricular
divisions are labeled in Parts
A and B of this plate on the
preceding pages.

2 mm

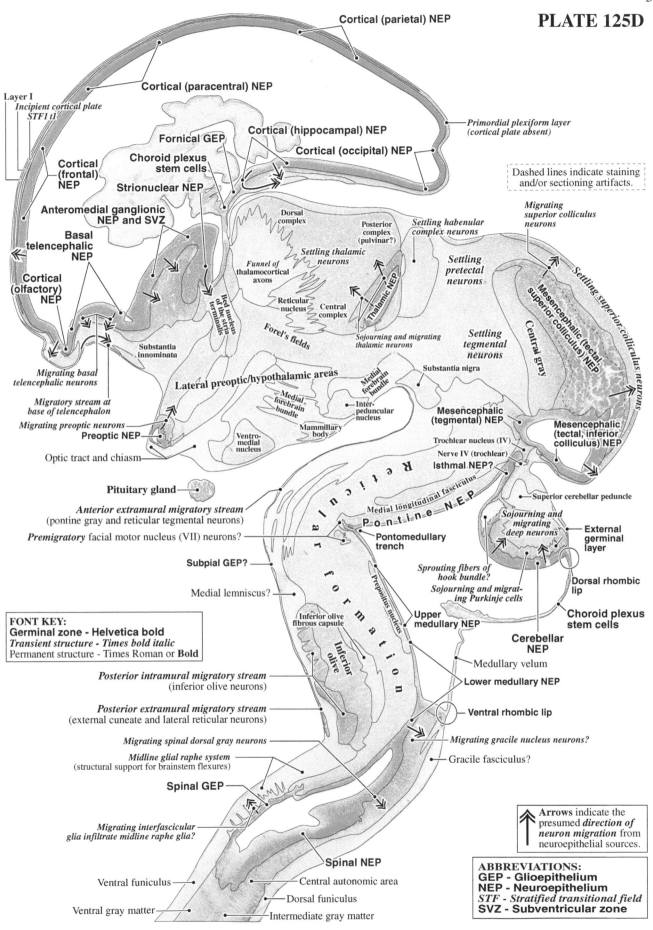

Cortical (parietal) NEP

Cortical (paracentral) NEP

Layer I
*Incipient cortical plate STF1 t1*

Fornical GEP

Choroid plexus stem cells

Cortical (hippocampal) NEP

Cortical (occipital) NEP

*Primordial plexiform layer (cortical plate absent)*

Dashed lines indicate staining and/or sectioning artifacts.

**Cortical (frontal) NEP**

**Strionuclear NEP**

**Anteromedial ganglionic NEP and SVZ**

**Basal telencephalic NEP**

**Cortical (olfactory) NEP**

Dorsal complex

Posterior complex (pulvinar?)

*Settling habenular complex neurons*

*Migrating superior colliculus neurons*

*Settling thalamic neurons*

*Settling pretectal neurons*

*Funnel of thalamocortical axons*

Reticular nucleus

Central complex

*Thalamic NEP*

*Settling superior colliculus neurons*

*Mesencephalic (tectal, superior colliculus) NEP*

Central gray

*Settling tegmental neurons*

*Forel's fields*

*Sojourning and migrating thalamic neurons*

Bed nucleus of the stria terminalis

*Migrating basal telencephalic neurons*

Substantia innominata

*Migratory stream at base of telencephalon*

Lateral preoptic/hypothalamic areas

Medial forebrain bundle

Substantia nigra

*Medial forebrain bundle*

Inter-peduncular nucleus

**Mesencephalic (tegmental) NEP**

**Mesencephalic (tectal, inferior colliculus) NEP**

*Migrating preoptic neurons*

**Preoptic NEP**

Mammillary body

Ventro-medial nucleus

Trochlear nucleus (IV)

Nerve IV (trochlear)

**Isthmal NEP?**

Optic tract and chiasm

Superior cerebellar peduncle

**Pituitary gland**

*Reticular*

*Medial longitudinal fasciculus*

**Pontine NEP**

*Sojourning and migrating deep neurons*

**External germinal layer**

*Anterior extramural migratory stream* (pontine gray and reticular tegmental neurons)

**Dorsal rhombic lip**

*Premigratory facial motor nucleus (VII) neurons?*

Pontomedullary trench

*Sprouting fibers of hook bundle?*

*Sojourning and migrating Purkinje cells*

**Choroid plexus stem cells**

**Subpial GEP?**

*formation*

Prepositus nucleus

**Upper medullary NEP**

**Cerebellar NEP**

Medial lemniscus?

Inferior olive fibrous capsule

Inferior olive

Medullary velum

**Lower medullary NEP**

**FONT KEY:**
Germinal zone - Helvetica bold
*Transient structure - Times bold italic*
Permanent structure - Times Roman or **Bold**

*Posterior intramural migratory stream* (inferior olive neurons)

*Posterior extramural migratory stream* (external cuneate and lateral reticular neurons)

**Ventral rhombic lip**

*Migrating gracile nucleus neurons?*

*Migrating spinal dorsal gray neurons*

Gracile fasciculus?

*Midline glial raphe system* (structural support for brainstem flexures)

**Spinal GEP**

*Migrating interfascicular glia infiltrate midline raphe glia?*

**Spinal NEP**

Ventral funiculus

Central autonomic area

Ventral gray matter

Dorsal funiculus

Intermediate gray matter

**Arrows** indicate the presumed *direction of neuron migration* from neuroepithelial sources.

**ABBREVIATIONS:**
**GEP** - Glioepithelium
**NEP** - Neuroepithelium
*STF* - Stratified transitional field
**SVZ** - Subventricular zone

**PLATE 126A**

GW8 Sagittal, CR 33 mm, C145
Level 5: Slide 18, Section 2
HEAD STRUCTURES,
MAJOR BRAIN REGIONS,
AND VENTRICULAR
DIVISIONS

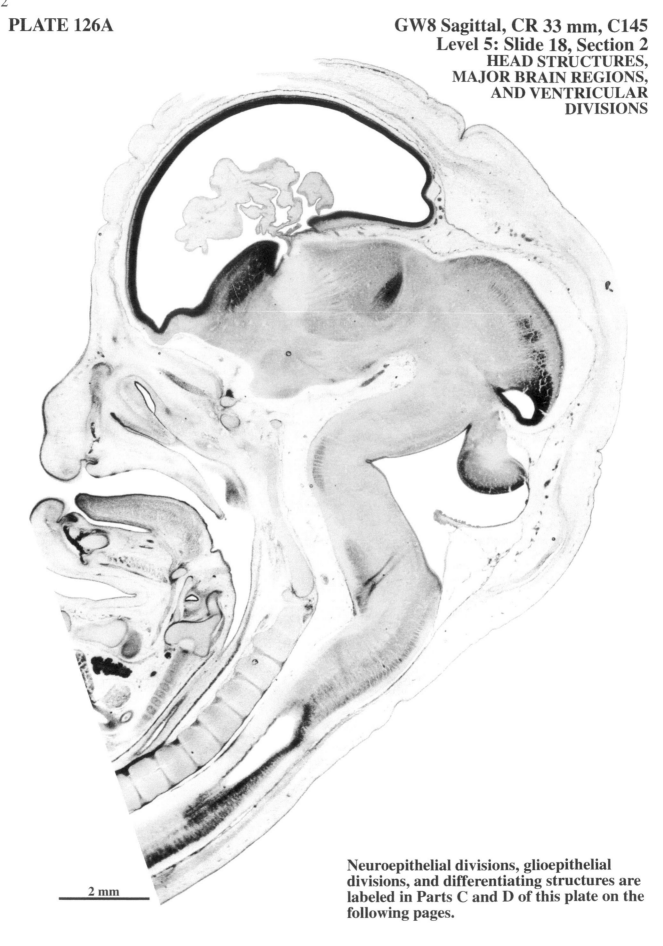

2 mm

Neuroepithelial divisions, glioepithelial divisions, and differentiating structures are labeled in Parts C and D of this plate on the following pages.

Dashed lines indicate staining and/or sectioning artifacts.

Skull and skin

Meninges (dura and arachnoid)

Brain surface (pia, heavier line)

CEREBRAL CORTEX

Frontal bone

TELENCEPHALON

Future parietal bone

"Blooming" telencephalic choroid plexus

HIPPOCAMPUS

EPITHALAMUS

THALAMUS

TELENCEPHALIC SUPERVENTRICLE (FUTURE LATERAL VENTRICLE)

BASAL GANGLIA

PRETECTUM

SUPERIOR COLLICULUS

T E C T U M

OLFACTORY BULB

BASAL TELENCEPHALON

DIENCEPHALON

MESENCEPHALON

Frontonasal process

Nasal conchae

SUBTHALAMUS

TEGMENTUM

Nasal cavity

HYPOTHALAMUS

ISTHMUS

INFERIOR COLLICULUS

MESENCEPHALIC SUPERVENTRICLE (FUTURE AQUEDUCT)

Sphenoid

Sella turcica

PONS

Maxilla

ISTHMAL NARROWS

CEREBELLUM (HEMISPHERE)

Palatal process

Nasopharynx

Basal occipital

RHOMBENCEPHALON

Oral cavity

UPPER MEDULLA

RHOMBENCEPHALIC SUPERVENTRICLE (FUTURE FOURTH VENTRICLE)

Tongue

Mandibular process

Hyoid bone?

Oropharynx

"Budding" rhombencephalic choroid plexus

Epiglottis

Thyroid cartilage?

Larynx

Atlas?

LOWER MEDULLA

Squamous occipital

Cridoid cartilage?

Odontoid process

Clavicle?

Axis

Sternum?

C3

Tracheal cartilagenous rings

C4

C5

Vertebral column

C6

Esophagus

C7

T1

T2

CENTRAL CANAL (SPINAL CORD)

T3

SPINAL CORD

FONT KEY:
VENTRICULAR DIVISIONS – CAPITALS
Major brain structure - Times **Bold CAPITALS**
All other structures - Times Roman or **Bold**

**PLATE 126C**

**NEUROEPITHELIAL/GLIOEPI-
THELIAL DIVISIONS AND
DIFFERENTIATING
BRAIN STRUCTURES**

See a high-magnification
view of the diencephalon
and basal telencephalon in
Plates 136A and B.

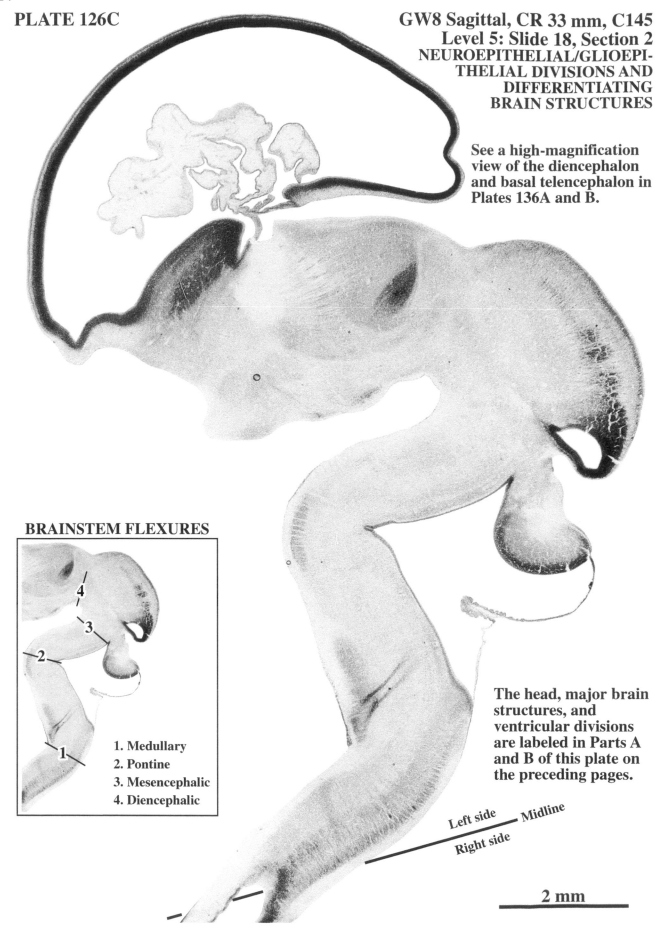

**BRAINSTEM FLEXURES**

1. Medullary
2. Pontine
3. Mesencephalic
4. Diencephalic

The head, major brain
structures, and
ventricular divisions
are labeled in Parts A
and B of this plate on
the preceding pages.

Left side — Midline
Right side

**2 mm**

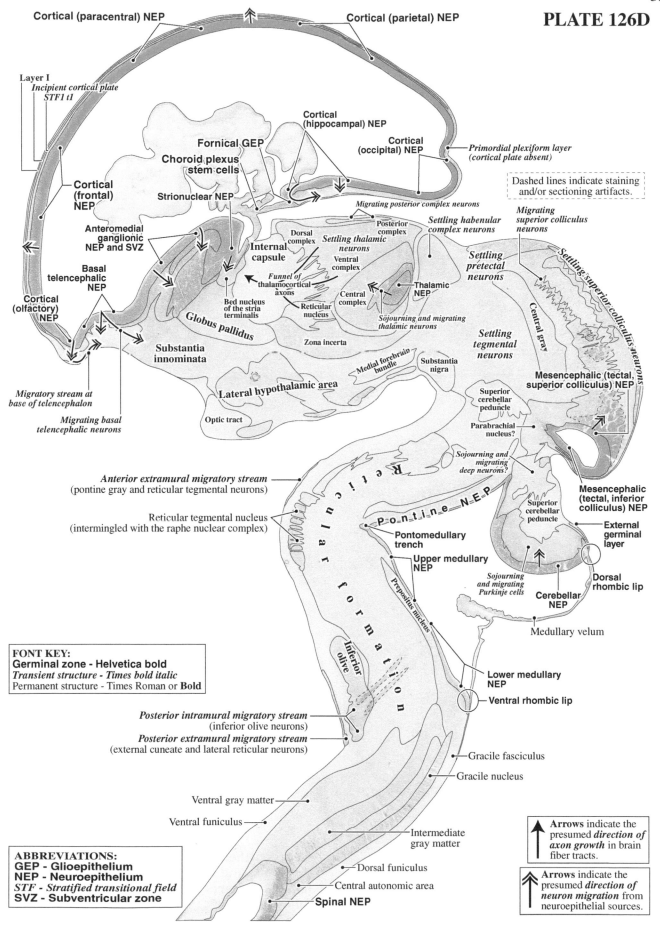

Cortical (paracentral) NEP

Cortical (parietal) NEP

Layer I
*Incipient cortical plate
STF1 t1*

Cortical
(hippocampal) NEP

**Fornical GEP**

**Choroid plexus
stem cells**

Cortical
(occipital) NEP

*Primordial plexiform layer
(cortical plate absent)*

**Cortical
(frontal)
NEP**

**Strionuclear NEP**

Dashed lines indicate staining
and/or sectioning artifacts.

*Migrating posterior complex neurons*

**Anteromedial
ganglionic
NEP and SVZ**

*Dorsal
complex*

Posterior
complex

*Settling thalamic
neurons*

*Settling habenular
complex neurons*

*Migrating
superior colliculus
neurons*

**Internal
capsule**

Ventral
complex

*Settling
pretectal
neurons*

**Basal
telencephalic
NEP**

*Funnel of
thalamocortical
axons*

Central
complex

**Thalamic
NEP**

Settling superior colliculus neurons

**Cortical
(olfactory)
NEP**

*Bed nucleus
of the stria
terminalis*

Reticular
nucleus

*Sojourning and migrating
thalamic neurons*

*Central gray*

*Globus pallidus*

Zona incerta

*Settling
tegmental
neurons*

*Migratory stream at
base of telencephalon*

**Substantia
innominata**

*Medial forebrain
bundle*

Substantia
nigra

**Mesencephalic (tectal,
superior colliculus) NEP**

*Migrating basal
telencephalic neurons*

*Lateral hypothalamic area*

Superior
cerebellar
peduncle

Optic tract

Parabrachial
nucleus?

*Sojourning and
migrating
deep neurons?*

**Mesencephalic
(tectal, inferior
colliculus) NEP**

*Anterior extramural migratory stream*
(pontine gray and reticular tegmental neurons)

*Reticular formation*

Superior
cerebellar
peduncle

**External
germinal
layer**

Reticular tegmental nucleus
(intermingled with the raphe nuclear complex)

**Pontine NEP**

**Pontomedullary
trench**

*Sojourning
and migrating
Purkinje cells*

**Dorsal
rhombic lip**

**Upper medullary
NEP**

**Cerebellar
NEP**

*Prepositus nucleus*

Medullary velum

*Inferior
olive*

**Lower medullary
NEP**

*Posterior intramural migratory stream*
(inferior olive neurons)

**Ventral rhombic lip**

*Posterior extramural migratory stream*
(external cuneate and lateral reticular neurons)

Gracile fasciculus

Gracile nucleus

Ventral gray matter

Ventral funiculus

Intermediate
gray matter

**FONT KEY:**
Germinal zone - Helvetica bold
*Transient structure - Times bold italic*
Permanent structure - Times Roman or **Bold**

Dorsal funiculus

Central autonomic area

**ABBREVIATIONS:**
**GEP** - Glioepithelium
**NEP** - Neuroepithelium
*STF* - Stratified transitional field
**SVZ** - Subventricular zone

**Spinal NEP**

**Arrows** indicate the
presumed *direction of
axon growth* in brain
fiber tracts.

**Arrows** indicate the
presumed *direction of
neuron migration* from
neuroepithelial sources.

**GW8 Sagittal, CR 33 mm, C145**
**Level 6: Slide 16, Section 2**
**Left side of brain**
**HEAD STRUCTURES,**
**MAJOR BRAIN REGIONS,**
**AND VENTRICULAR**
**DIVISIONS**

2 mm

Neuroepithelial divisions, glioepithelial divisions, and differentiating structures are labeled in Parts C and D of this plate on the following pages.

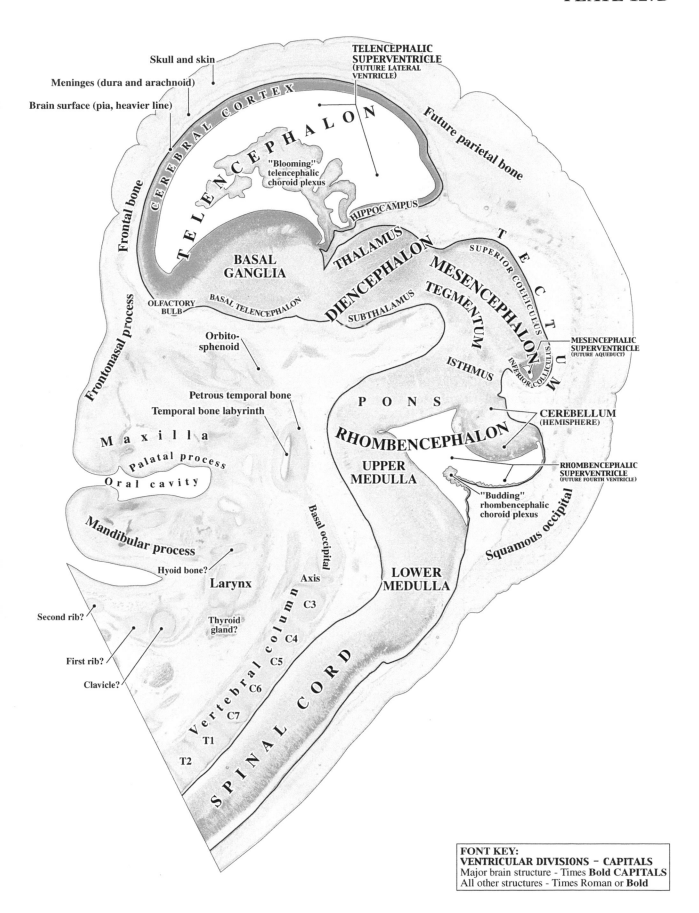

Skull and skin

Meninges (dura and arachnoid)

Brain surface (pia, heavier line)

**TELENCEPHALIC SUPERVENTRICLE**
**(FUTURE LATERAL VENTRICLE)**

*CEREBRAL CORTEX*

**TELENCEPHALON**

Future parietal bone

Frontal bone

"Blooming" telencephalic choroid plexus

HIPPOCAMPUS

**BASAL GANGLIA**

**THALAMUS**

**DIENCEPHALON**

*T E C T U M*

SUPERIOR COLLICULUS

**MESENCEPHALON**

OLFACTORY BULB

BASAL TELENCEPHALON

SUBTHALAMUS

**TEGMENTUM**

"INFERIOR COLLICULUS"

**MESENCEPHALIC SUPERVENTRICLE**
**(FUTURE AQUEDUCT)**

Frontonasal process

Orbito-sphenoid

ISTHMUS

Petrous temporal bone

Temporal bone labyrinth

**P O N S**

**CEREBELLUM**
**(HEMISPHERE)**

M a x i l l a

**RHOMBENCEPHALON**

Palatal process

**UPPER MEDULLA**

**RHOMBENCEPHALIC SUPERVENTRICLE**
**(FUTURE FOURTH VENTRICLE)**

O r a l   c a v i t y

"Budding" rhombencephalic choroid plexus

Mandibular process

Squamous occipital

Hyoid bone?

Basal occipital

Larynx

Axis

C3

Second rib?

Thyroid gland?

C4

**LOWER MEDULLA**

First rib?

C5

Clavicle?

C6

*V e r t e b r a l   c o l u m n*

C7

T1

*S P I N A L   C O R D*

T2

FONT KEY:
**VENTRICULAR DIVISIONS – CAPITALS**
Major brain structure - Times **Bold CAPITALS**
All other structures - Times Roman or **Bold**

**PLATE 127C**

GW8 Sagittal, CR 33 mm, C145
Level 6: Slide 16, Section 2
Left side of brain
NEUROEPITHELIAL/
GLIOEPITHELIAL
DIVISIONS AND
DIFFERENTIATING
BRAIN STRUCTURES

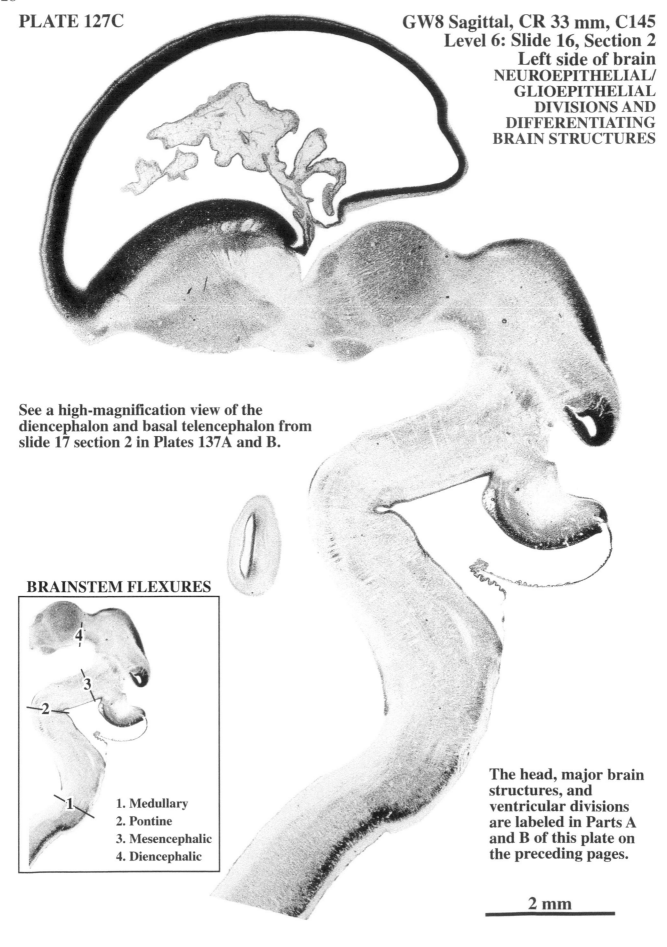

See a high-magnification view of the diencephalon and basal telencephalon from slide 17 section 2 in Plates 137A and B.

**BRAINSTEM FLEXURES**

1. Medullary
2. Pontine
3. Mesencephalic
4. Diencephalic

The head, major brain structures, and ventricular divisions are labeled in Parts A and B of this plate on the preceding pages.

**2 mm**

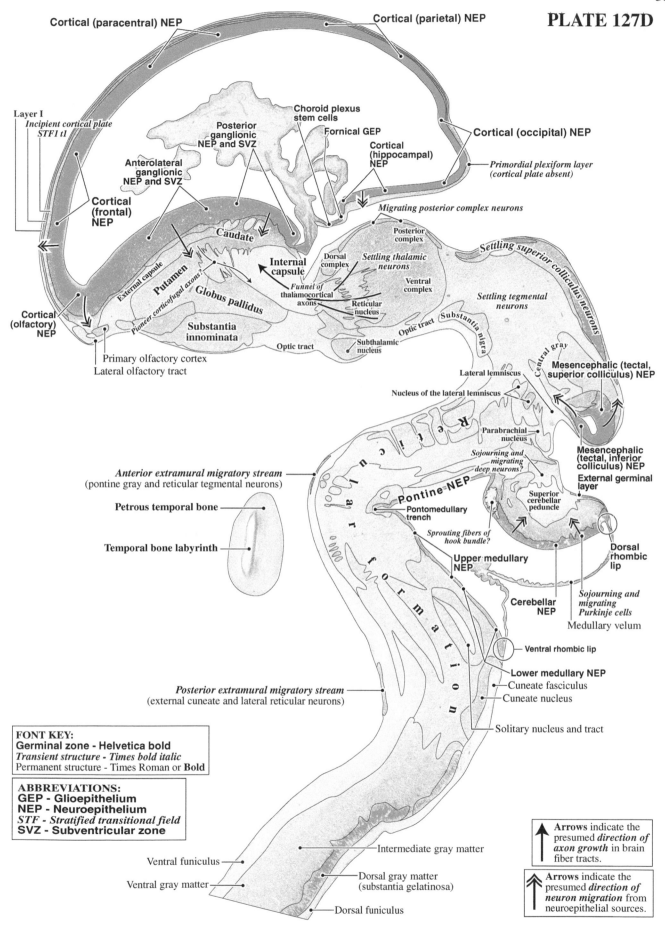

Cortical (paracentral) NEP

Cortical (parietal) NEP

Layer I
*Incipient cortical plate*
*STF1 t1*

Posterior
ganglionic
NEP and SVZ

Choroid plexus
stem cells

Fornical GEP

Cortical (occipital) NEP

Anterolateral
ganglionic
NEP and SVZ

Cortical
(hippocampal)
NEP

*Primordial plexiform layer*
*(cortical plate absent)*

Cortical
(frontal)
NEP

Caudate

*Migrating posterior complex neurons*

Posterior
complex

*Settling superior colliculus neurons*

Internal
capsule

Dorsal
complex

*Settling thalamic*
*neurons*

Ventral
complex

*Settling tegmental*
*neurons*

External capsule

Putamen

Globus pallidus

*Funnel of*
*thalamocortical*
*axons*

Reticular
nucleus

Cortical
(olfactory)
NEP

*Pioneer corticofugal axons?*

Substantia
innominata

Optic tract

Subthalamic
nucleus

Optic tract

Substantia nigra

*Central gray*

Mesencephalic (tectal,
superior colliculus) NEP

Primary olfactory cortex
Lateral olfactory tract

Lateral lemniscus

Nucleus of the lateral lemniscus

Parabrachial
nucleus

Mesencephalic
(tectal, inferior
colliculus) NEP

*Anterior extramural migratory stream*
(pontine gray and reticular tegmental neurons)

*R e t i c u l a r*

*Sojourning and*
*migrating*
*deep neurons?*

External germinal
layer

Petrous temporal bone

Pontine NEP

Pontomedullary
trench

Superior
cerebellar
peduncle

Temporal bone labyrinth

*f o r m a t i o n*

*Sprouting fibers of*
*hook bundle?*

Upper medullary
NEP

Dorsal
rhombic
lip

*Sojourning and*
*migrating*
*Purkinje cells*

Cerebellar
NEP

Medullary velum

Ventral rhombic lip

*Posterior extramural migratory stream*
(external cuneate and lateral reticular neurons)

Lower medullary NEP
Cuneate fasciculus
Cuneate nucleus

Solitary nucleus and tract

**FONT KEY:**
Germinal zone - Helvetica bold
*Transient structure - Times bold italic*
Permanent structure - Times Roman or **Bold**

**ABBREVIATIONS:**
**GEP** - Glioepithelium
**NEP** - Neuroepithelium
*STF* - *Stratified transitional field*
**SVZ** - Subventricular zone

Intermediate gray matter

Ventral funiculus

Dorsal gray matter
(substantia gelatinosa)

Ventral gray matter

**Arrows** indicate the
presumed *direction of*
*axon growth* in brain
fiber tracts.

Dorsal funiculus

**Arrows** indicate the
presumed *direction of*
*neuron migration* from
neuroepithelial sources.

**PLATE 128A**

GW8 Sagittal, CR 33 mm, C145
Level 7: Slide 15, Section 1
Left side of brain
HEAD STRUCTURES,
MAJOR BRAIN REGIONS,
AND VENTRICULAR
DIVISIONS

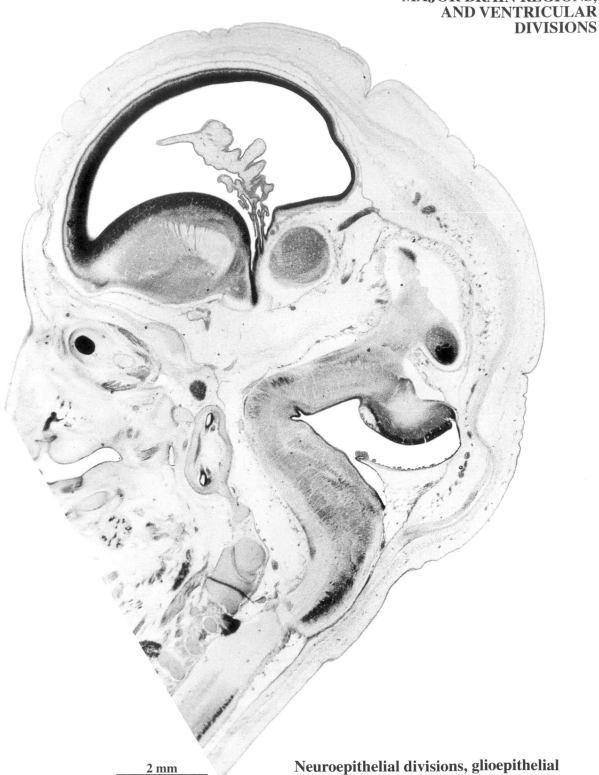

2 mm

Neuroepithelial divisions, glioepithelial divisions, and differentiating structures are labeled in Parts C and D of this plate on the following pages.

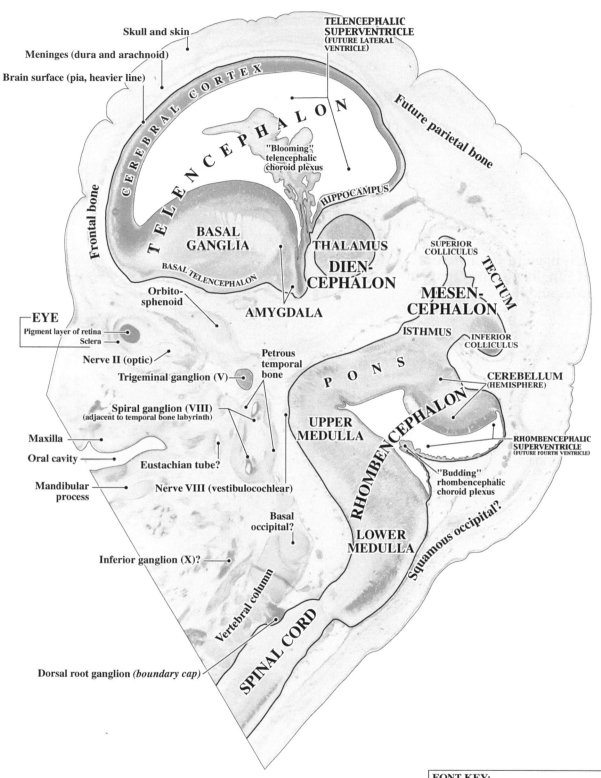

Skull and skin

Meninges (dura and arachnoid)

Brain surface (pia, heavier line)

TELENCEPHALIC
SUPERVENTRICLE
(FUTURE LATERAL
VENTRICLE)

CEREBRAL CORTEX

Future parietal bone

TELENCEPHALON

"Blooming"
telencephalic
choroid plexus

HIPPOCAMPUS

Frontal bone

BASAL
GANGLIA

THALAMUS
DIEN-
CEPHALON

SUPERIOR
COLLICULUS

TECTUM

BASAL TELENCEPHALON

MESEN-
CEPHALON

Orbito-
sphenoid

AMYGDALA

ISTHMUS

INFERIOR
COLLICULUS

EYE

Pigment layer of retina

Sclera

Nerve II (optic)

Trigeminal ganglion (V)

Petrous
temporal
bone

P O N S

CEREBELLUM
(HEMISPHERE)

Spiral ganglion (VIII)
(adjacent to temporal bone labyrinth)

UPPER
MEDULLA

RHOMBENCEPHALON

RHOMBENCEPHALIC
SUPERVENTRICLE
(FUTURE FOURTH VENTRICLE)

Maxilla

Oral cavity

Mandibular
process

Eustachian tube?

Nerve VIII (vestibulocochlear)

"Budding"
rhombencephalic
choroid plexus

Basal
occipital?

LOWER
MEDULLA

Squamous occipital?

Inferior ganglion (X)?

Vertebral column

Dorsal root ganglion (*boundary cap*)

SPINAL CORD

FONT KEY:
**VENTRICULAR DIVISIONS – CAPITALS**
Major brain structure - Times **Bold CAPITALS**
All other structures - Times Roman or **Bold**

**PLATE 128C**

NEUROEPITHELIAL/
GLIOEPITHELIAL
DIVISIONS AND
DIFFERENTIATING
BRAIN STRUCTURES

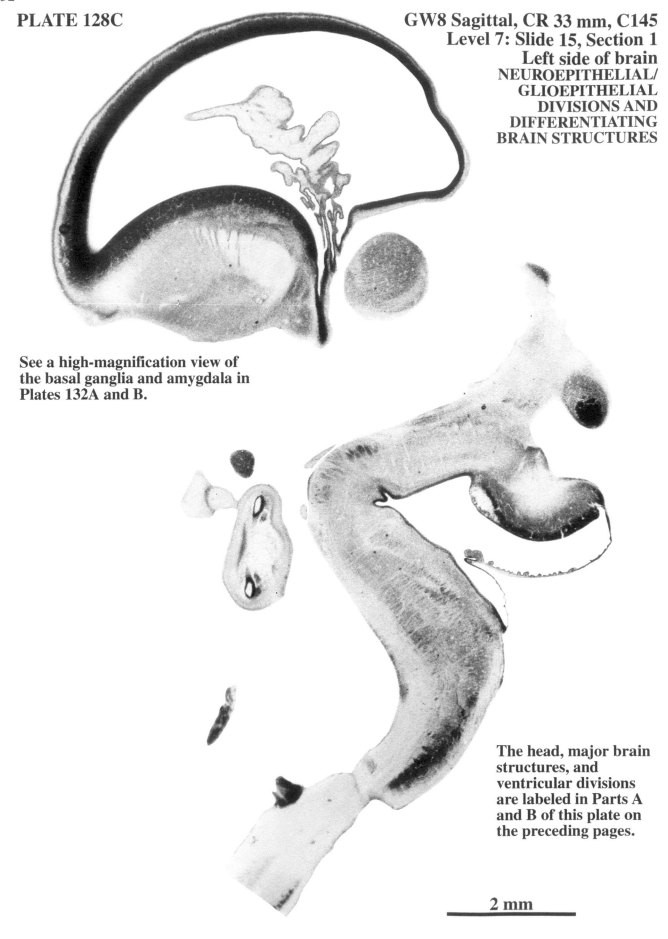

See a high-magnification view of
the basal ganglia and amygdala in
Plates 132A and B.

The head, major brain
structures, and
ventricular divisions
are labeled in Parts A
and B of this plate on
the preceding pages.

2 mm

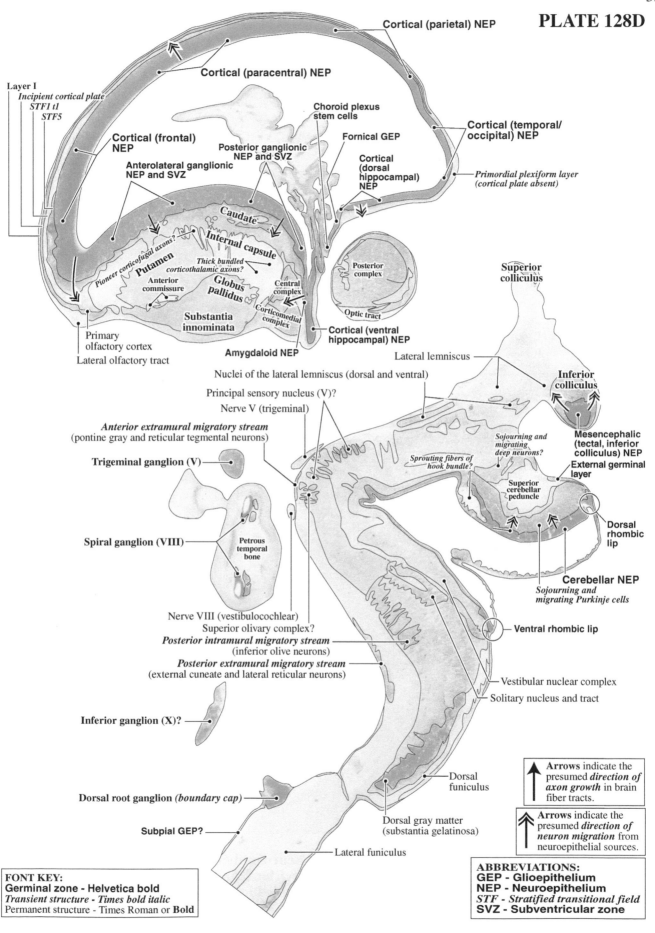

Cortical (parietal) NEP

Cortical (paracentral) NEP

Layer I
*Incipient cortical plate*
*STF1 t1*
*STF5*

Choroid plexus stem cells

Fornical GEP

Cortical (temporal/occipital) NEP

Cortical (frontal) NEP

Posterior ganglionic NEP and SVZ

Cortical (dorsal hippocampal) NEP

Anterolateral ganglionic NEP and SVZ

*Primordial plexiform layer (cortical plate absent)*

Caudate

Internal capsule

*Pioneer corticofugal axons?*

Putamen

*Thick bundled corticothalamic axons?*

Anterior commissure

Globus pallidus

Central complex

Posterior complex

Substantia innominata

Corticomedial complex

Superior colliculus

Primary olfactory cortex

Lateral olfactory tract

Cortical (ventral hippocampal) NEP

Optic tract

Inferior colliculus

Amygdaloid NEP

Lateral lemniscus

Mesencephalic (tectal, inferior colliculus) NEP

Nuclei of the lateral lemniscus (dorsal and ventral)

Principal sensory nucleus (V)?

Nerve V (trigeminal)

*Sojourning and migrating deep neurons?*

*Sprouting fibers of hook bundle?*

External germinal layer

*Anterior extramural migratory stream*
(pontine gray and reticular tegmental neurons)

Superior cerebellar peduncle

Trigeminal ganglion (V)

Dorsal rhombic lip

Spiral ganglion (VIII)

Petrous temporal bone

Cerebellar NEP
*Sojourning and migrating Purkinje cells*

Nerve VIII (vestibulocochlear)

Superior olivary complex?

Ventral rhombic lip

*Posterior intramural migratory stream*
(inferior olive neurons)

*Posterior extramural migratory stream*
(external cuneate and lateral reticular neurons)

Vestibular nuclear complex

Solitary nucleus and tract

Inferior ganglion (X)?

Dorsal funiculus

Dorsal root ganglion *(boundary cap)*

Dorsal gray matter (substantia gelatinosa)

Subpial GEP?

Lateral funiculus

Arrows indicate the presumed *direction of axon growth* in brain fiber tracts.

Arrows indicate the presumed *direction of neuron migration* from neuroepithelial sources.

FONT KEY:
Germinal zone - Helvetica bold
*Transient structure - Times bold italic*
Permanent structure - Times Roman or **Bold**

ABBREVIATIONS:
GEP - Glioepithelium
NEP - Neuroepithelium
*STF - Stratified transitional field*
SVZ - Subventricular zone

**PLATE 129A**

GW8 Sagittal, CR 33 mm, C145
Level 8: Slide 13, Section 2
Left side of brain
HEAD STRUCTURES,
MAJOR BRAIN REGIONS,
AND VENTRICULAR
DIVISIONS

2 mm

Neuroepithelial divisions, glioepithelial divisions, and differentiating structures are labeled in Parts C and D of this plate on the following pages.

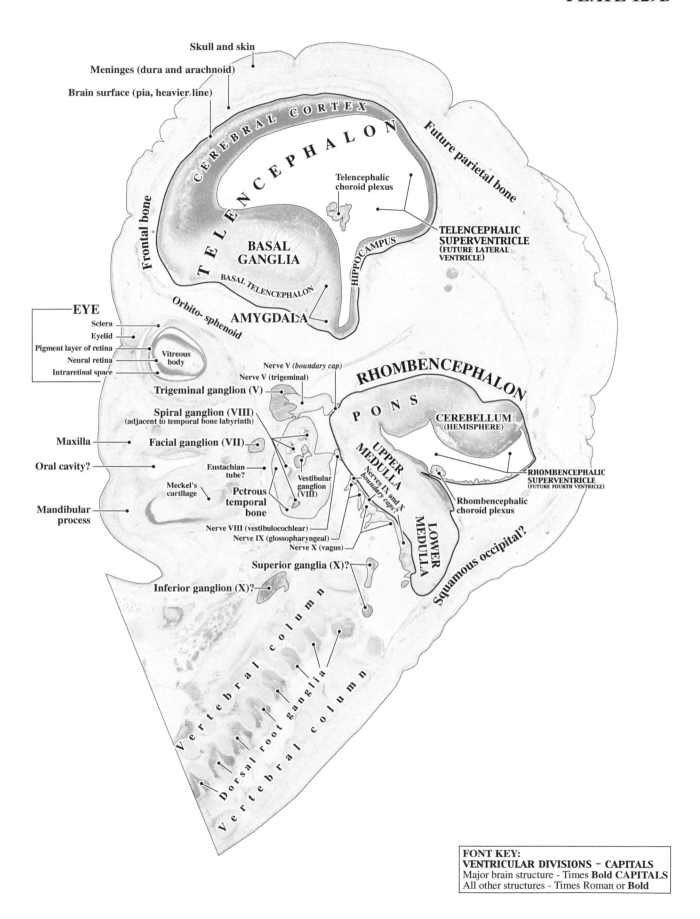

Skull and skin

Meninges (dura and arachnoid)

Brain surface (pia, heavier line)

CEREBRAL CORTEX

TELENCEPHALON

Future parietal bone

Telencephalic
choroid plexus

Frontal bone

BASAL
GANGLIA

TELENCEPHALIC
SUPERVENTRICLE
(FUTURE LATERAL
VENTRICLE)

BASAL TELENCEPHALON

HIPPOCAMPUS

EYE

Orbito- sphenoid

AMYGDALA

Sclera

Eyelid

Pigment layer of retina

Neural retina

Intraretinal space

Vitreous
body

Nerve V (boundary cap)

Nerve V (trigeminal)

RHOMBENCEPHALON

Trigeminal ganglion (V)

PONS

CEREBELLUM
(HEMISPHERE)

Spiral ganglion (VIII)
(adjacent to temporal bone labyrinth)

UPPER
MEDULLA

Maxilla

Facial ganglion (VII)

RHOMBENCEPHALIC
SUPERVENTRICLE
(FUTURE FOURTH VENTRICLE)

Oral cavity?

Eustachian
tube?

Nerves IX and X
boundary caps?

Rhombencephalic
choroid plexus

Meckel's
cartilage

Vestibular
ganglion
(VIII)

Mandibular
process

Petrous
temporal
bone

LOWER
MEDULLA

Nerve VIII (vestibulocochlear)

Squamous occipital?

Nerve IX (glossopharyngeal)

Nerve X (vagus)

Superior ganglia (X)?

Inferior ganglion (X)?

Vertebral column

Vertebral column

Dorsal root ganglia

FONT KEY:
VENTRICULAR DIVISIONS – CAPITALS
Major brain structure - Times Bold CAPITALS
All other structures - Times Roman or Bold

**PLATE 129C**

GW8 Sagittal, CR 33 mm, C145
Level 8: Slide 13, Section 2
Left side of brain
NEUROEPITHELIAL/GLIOEPITHELIAL
DIVISIONS AND DIFFERENTIATING
BRAIN STRUCTURES

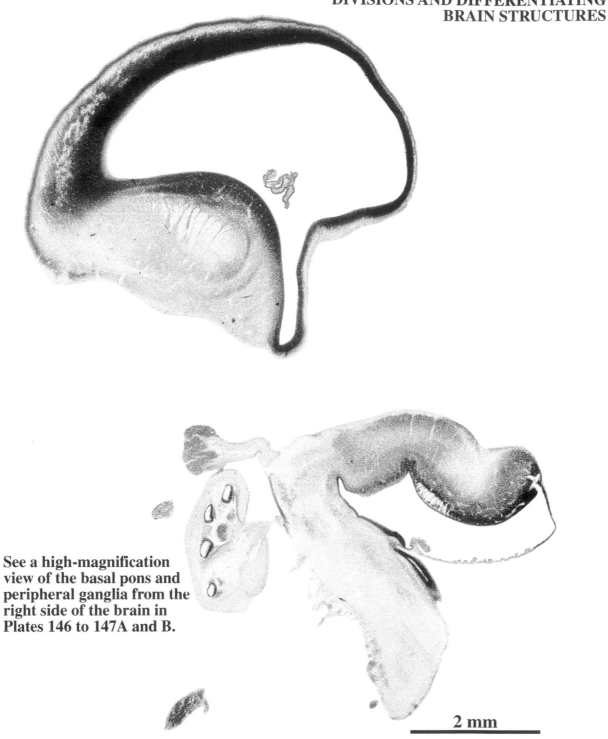

See a high-magnification
view of the basal pons and
peripheral ganglia from the
right side of the brain in
Plates 146 to 147A and B.

2 mm

The head, major brain structures, and ventricular
divisions are labeled in Parts A and B of this plate
on the preceding pages.

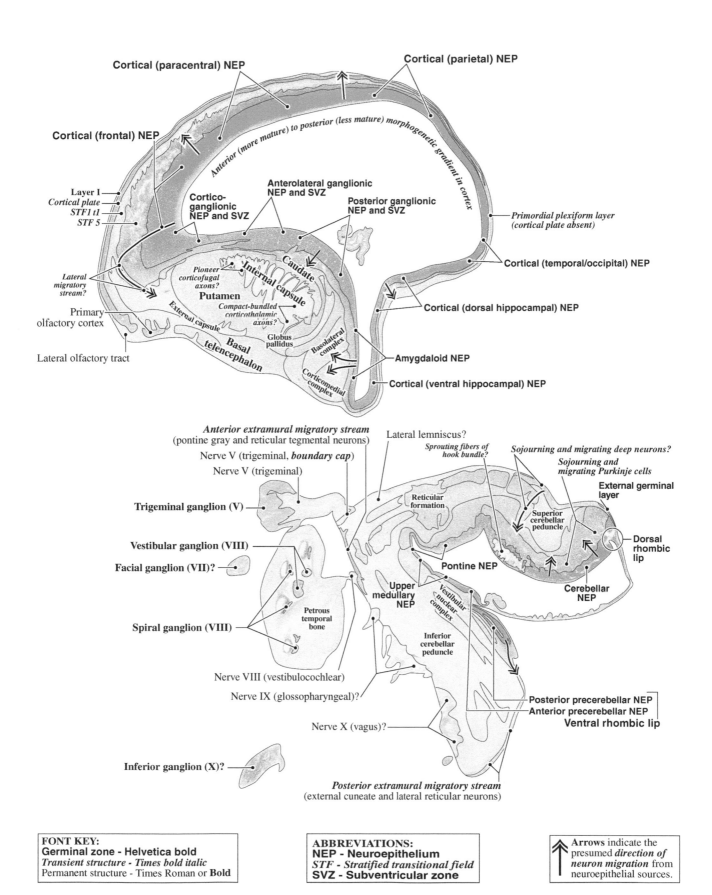

Cortical (paracentral) NEP

Cortical (parietal) NEP

Cortical (frontal) NEP

*Anterior (more mature) to posterior (less mature) morphogenetic gradient in cortex*

Layer I
*Cortical plate*
*STF1 t1*
*STF 5*

**Cortico-
ganglionic
NEP and SVZ**

**Anterolateral ganglionic
NEP and SVZ**

**Posterior ganglionic
NEP and SVZ**

*Primordial plexiform layer
(cortical plate absent)*

*Lateral
migratory
stream?*

*Pioneer
corticofugal
axons?*

**Caudate**

*Internal capsule*

Cortical (temporal/occipital) NEP

**Putamen**

*Compact-bundled
corticothalamic
axons?*

Cortical (dorsal hippocampal) NEP

Primary
olfactory cortex

*External capsule*

Globus
pallidus

*Basolateral
complex*

Lateral olfactory tract

**Basal
telencephalon**

*Corticomedial
complex*

**Amygdaloid NEP**

Cortical (ventral hippocampal) NEP

*Anterior extramural migratory stream*
(pontine gray and reticular tegmental neurons)

Lateral lemniscus?

*Sprouting fibers of
hook bundle?*

*Sojourning and migrating deep neurons?*

*Sojourning and
migrating Purkinje cells*

Nerve V (trigeminal, *boundary cap*)

Nerve V (trigeminal)

Reticular
formation

**External germinal
layer**

**Trigeminal ganglion (V)**

*Superior
cerebellar
peduncle*

Dorsal
rhombic
lip

**Vestibular ganglion (VIII)**

**Facial ganglion (VII)?**

**Pontine NEP**

Upper
medullary
NEP

*Vestibular
nuclear
complex*

**Cerebellar
NEP**

Petrous
temporal
bone

**Spiral ganglion (VIII)**

Inferior
cerebellar
peduncle

Nerve VIII (vestibulocochlear)

Nerve IX (glossopharyngeal)?

**Posterior precerebellar NEP**
**Anterior precerebellar NEP**
**Ventral rhombic lip**

Nerve X (vagus)?

**Inferior ganglion (X)?**

*Posterior extramural migratory stream*
(external cuneate and lateral reticular neurons)

---

**FONT KEY:**
Germinal zone - **Helvetica bold**
*Transient structure - **Times bold italic***
Permanent structure - Times Roman or **Bold**

**ABBREVIATIONS:**
**NEP** - Neuroepithelium
*STF* - *Stratified transitional field*
**SVZ** - **Subventricular zone**

**Arrows** indicate the
presumed *direction of
neuron migration* from
neuroepithelial sources.

**PLATE 130A**

GW8 Sagittal, CR 33 mm, C145
Level 9: Slide 12, Section 4
**Left side of brain**
HEAD STRUCTURES,
MAJOR BRAIN REGIONS,
AND VENTRICULAR
DIVISIONS

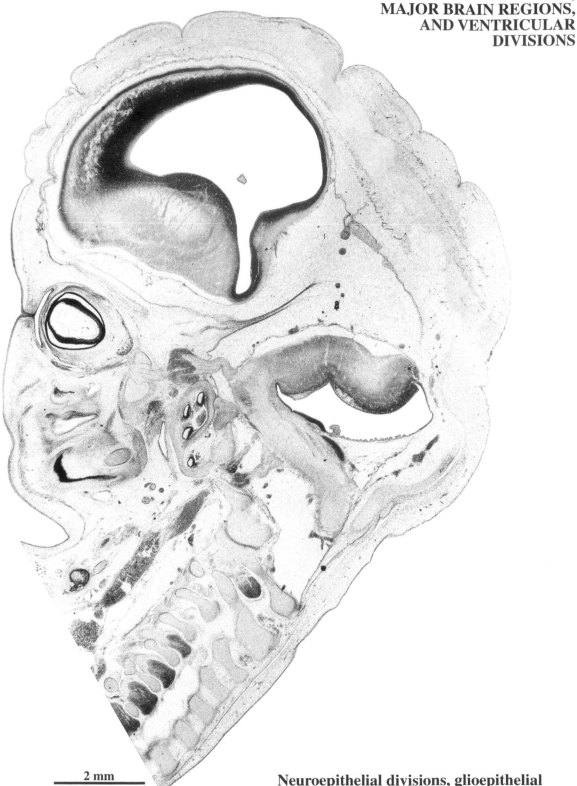

2 mm

**Neuroepithelial divisions, glioepithelial divisions, and differentiating structures are labeled in Parts C and D of this plate on the following pages.**

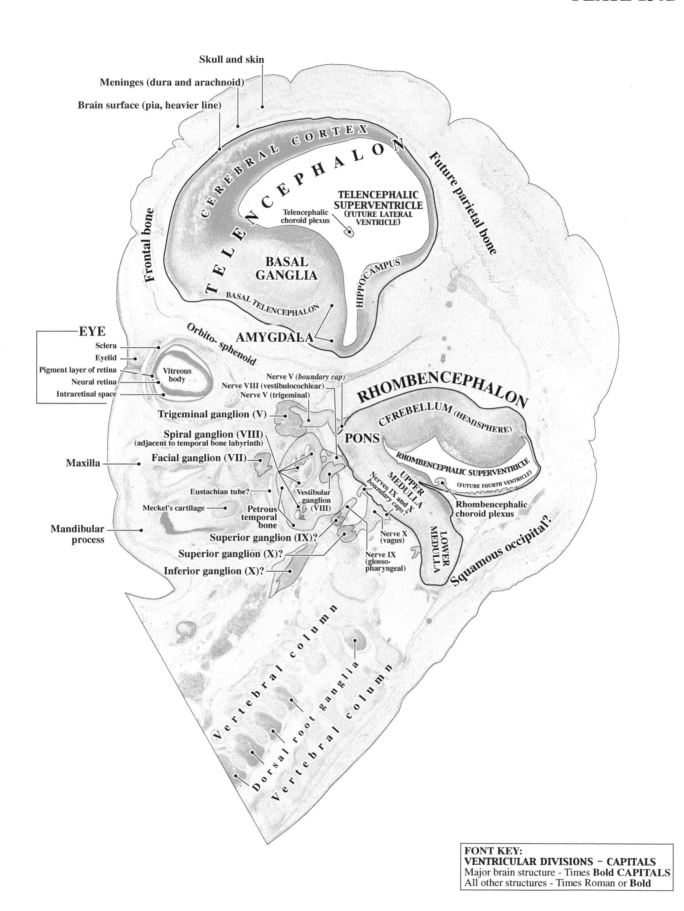

Skull and skin

Meninges (dura and arachnoid)

Brain surface (pia, heavier line)

CEREBRAL CORTEX

TELENCEPHALON

Future parietal bone

Frontal bone

**TELENCEPHALIC SUPERVENTRICLE (FUTURE LATERAL VENTRICLE)**

Telencephalic choroid plexus

**BASAL GANGLIA**

HIPPOCAMPUS

BASAL TELENCEPHALON

**AMYGDALA**

EYE
- Sclera
- Eyelid
- Pigment layer of retina
- Neural retina
- Intraretinal space

Vitreous body

Orbito-sphenoid

Nerve V (*boundary cap*)

Nerve VIII (vestibulocochlear)

Nerve V (trigeminal)

**RHOMBENCEPHALON**

CEREBELLUM (HEMISPHERE)

**PONS**

Trigeminal ganglion (V)

Spiral ganglion (VIII)
(adjacent to temporal bone labyrinth)

Facial ganglion (VII)

RHOMBENCEPHALIC SUPERVENTRICLE

(FUTURE FOURTH VENTRICLE)

UPPER MEDULLA

Nerves IX and X *boundary caps?*

Maxilla

Eustachian tube?

Meckel's cartilage

Vestibular ganglion (VIII)

Petrous temporal bone

Rhombencephalic choroid plexus

LOWER MEDULLA

Mandibular process

Superior ganglion (IX)?

Superior ganglion (X)?

Nerve X (vagus)

Inferior ganglion (X)?

Nerve IX (glosso-pharyngeal)

Squamous occipital?

Vertebral column

Dorsal root ganglia

Vertebral column

**FONT KEY:**
**VENTRICULAR DIVISIONS – CAPITALS**
Major brain structure - Times **Bold CAPITALS**
All other structures - Times Roman or **Bold**

**PLATE 130C**

GW8 Sagittal, CR 33 mm, C145
Level 9: Slide 12, Section 4
**Left side of brain**
NEUROEPITHELIAL/GLIOEPITHELIAL
DIVISIONS AND DIFFERENTIATING
BRAIN STRUCTURES

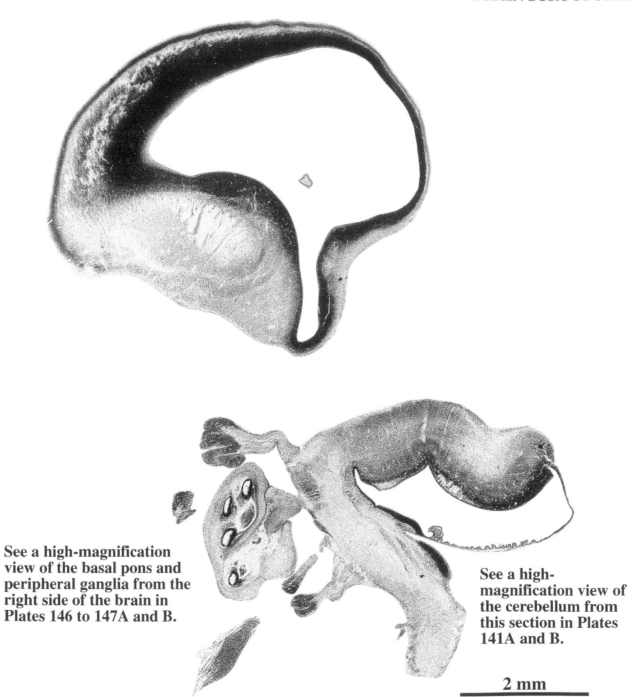

See a high-magnification
view of the basal pons and
peripheral ganglia from the
right side of the brain in
Plates 146 to 147A and B.

See a high-
magnification view of
the cerebellum from
this section in Plates
141A and B.

**2 mm**

The head, major brain structures, and ventricular
divisions are labeled in Parts A and B of this plate
on the preceding pages.

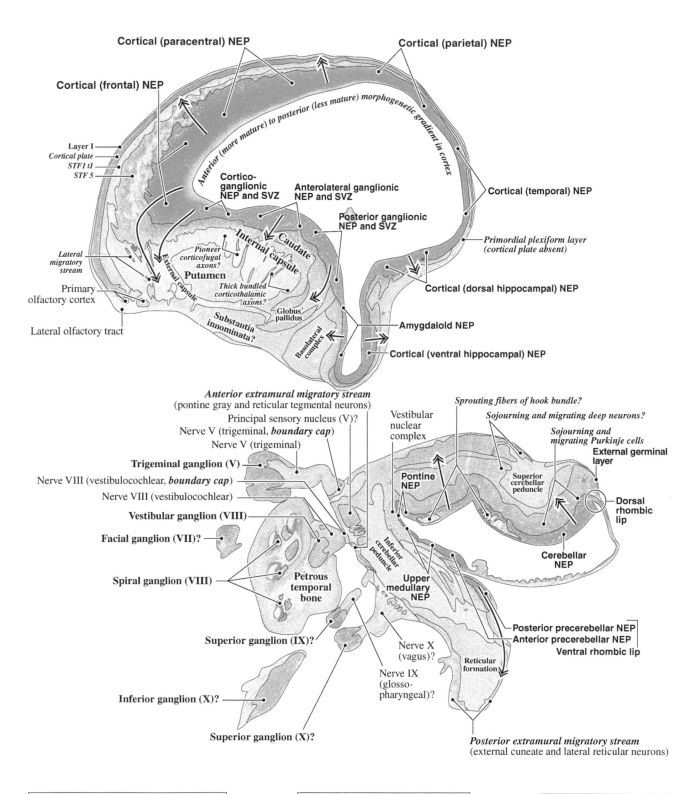

Cortical (paracentral) NEP

Cortical (parietal) NEP

Cortical (frontal) NEP

*Anterior (more mature) to posterior (less mature) morphogenetic gradient in cortex*

Layer I
*Cortical plate*
*STF1 tI*
*STF 5*

Cortico-
ganglionic
NEP and SVZ

Anterolateral ganglionic
NEP and SVZ

Posterior ganglionic
NEP and SVZ

Cortical (temporal) NEP

*Pioneer
corticofugal
axons?*

*Internal capsule*

*Caudate*

*Primordial plexiform layer
(cortical plate absent)*

*Lateral
migratory
stream*

*External capsule*

**Putamen**

*Thick bundled
corticothalamic
axons?*

Primary
olfactory cortex

*Globus
pallidus*

Cortical (dorsal hippocampal) NEP

*Substantia
innominata?*

Lateral olfactory tract

*Basolateral
complex*

Amygdalold NEP

Cortical (ventral hippocampal) NEP

*Anterior extramural migratory stream*
(pontine gray and reticular tegmental neurons)

Principal sensory nucleus (V)?

Nerve V (trigeminal, *boundary cap*)

Nerve V (trigeminal)

**Trigeminal ganglion (V)**

Nerve VIII (vestibulocochlear, *boundary cap*)

Nerve VIII (vestibulocochlear)

**Vestibular ganglion (VIII)**

**Facial ganglion (VII)?**

**Spiral ganglion (VIII)**

**Petrous
temporal
bone**

**Superior ganglion (IX)?**

Nerve X
(vagus)?

Nerve IX
(glosso-
pharyngeal)?

**Inferior ganglion (X)?**

**Superior ganglion (X)?**

Vestibular
nuclear
complex

*Sprouting fibers of hook bundle?*

*Sojourning and migrating deep neurons?*

*Sojourning and
migrating Purkinje cells*

**External germinal
layer**

**Pontine
NEP**

*Superior
cerebellar
peduncle*

**Dorsal
rhombic
lip**

*Inferior
cerebellar
peduncle*

**Cerebellar
NEP**

**Upper
medullary
NEP**

Posterior precerebellar NEP
Anterior precerebellar NEP
Ventral rhombic lip

*Reticular
formation*

*Posterior extramural migratory stream*
(external cuneate and lateral reticular neurons)

FONT KEY:
Germinal zone - **Helvetica bold**
Transient structure - *Times bold italic*
Permanent structure - Times Roman or **Bold**

**ABBREVIATIONS:**
**NEP** - Neuroepithelium
*STF* - Stratified transitional field
**SVZ** - Subventricular zone

**Arrows** indicate the
presumed *direction of
neuron migration* from
neuroepithelial sources.

**GW8 Sagittal, CR 33 mm, C145**
**Level 10: Slide 11, Section 4**
**Left side of brain**
**HEAD STRUCTURES,**
**MAJOR BRAIN REGIONS,**
**AND VENTRICULAR**
**DIVISIONS**

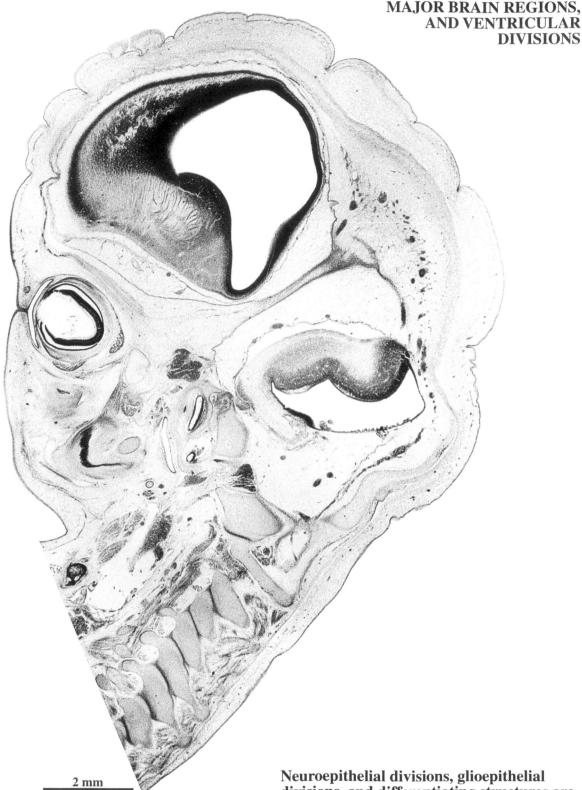

2 mm

**Neuroepithelial divisions, glioepithelial divisions, and differentiating structures are labeled in Parts C and D of this plate on the following pages.**

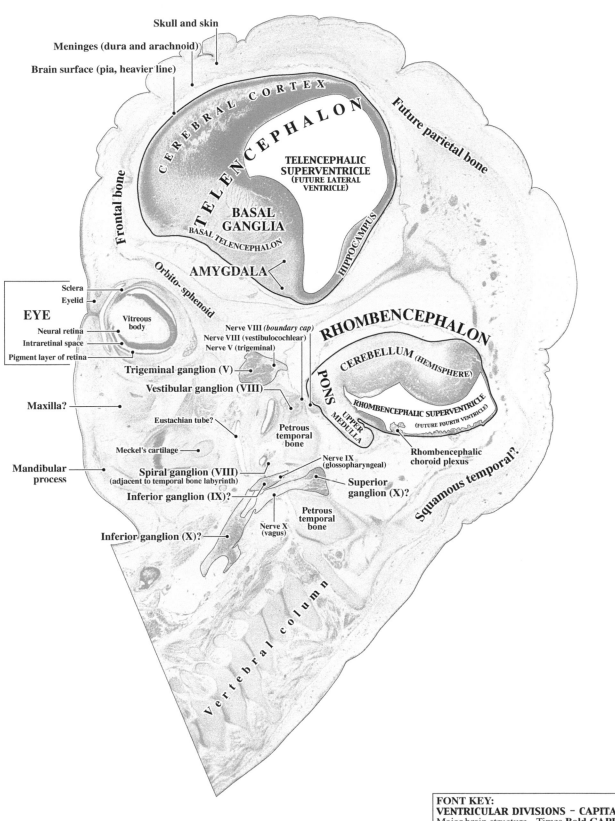

GW8 Sagittal, CR 33 mm, C145
Level 10: Slide 11, Section 4
**Left side of brain**
NEUROEPITHELIAL/GLIOEPITHELIAL
DIVISIONS AND DIFFERENTIATING
BRAIN STRUCTURES

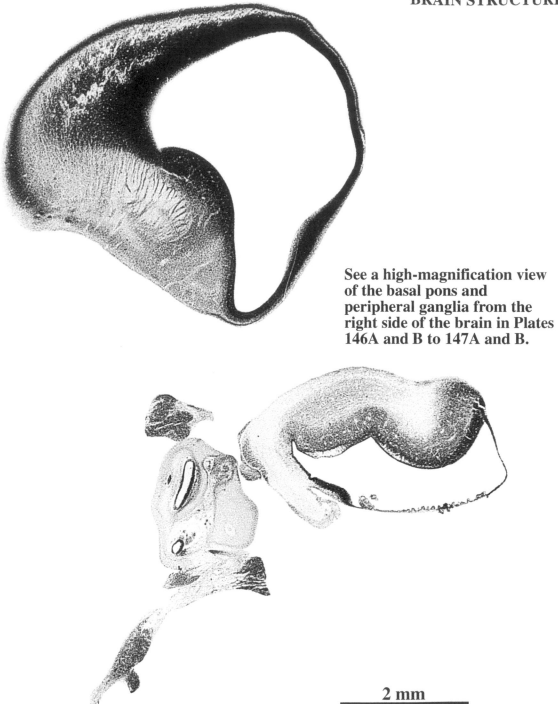

See a high-magnification view
of the basal pons and
peripheral ganglia from the
right side of the brain in Plates
146A and B to 147A and B.

2 mm

The head, major brain structures, and ventricular
divisions are labeled in Parts A and B of this plate
on the preceding pages.

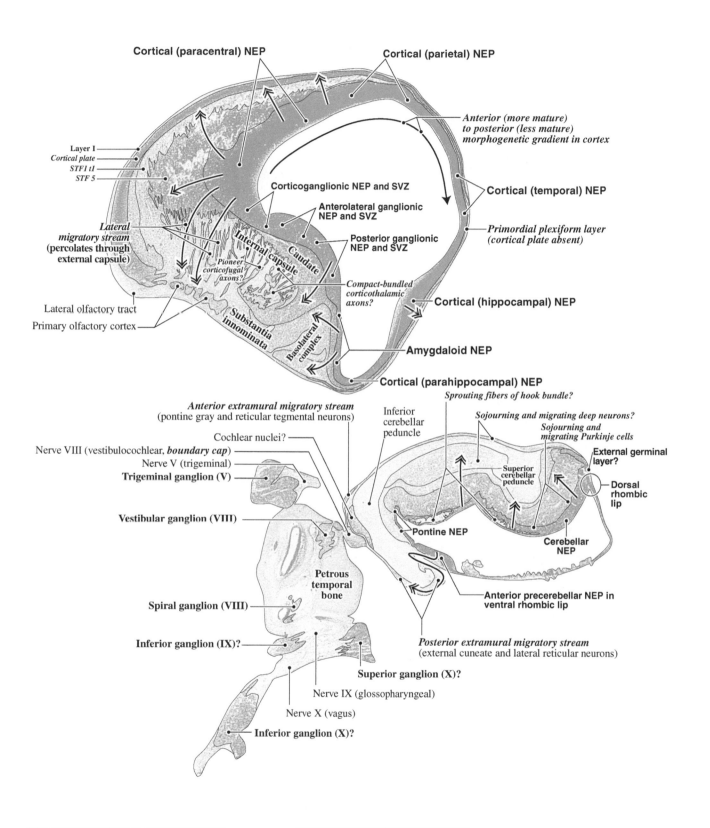

Cortical (paracentral) NEP

Cortical (parietal) NEP

*Anterior (more mature)
to posterior (less mature)
morphogenetic gradient in cortex*

Layer I

*Cortical plate*

*STF1 t1*

*STF 5*

Corticoganglionic NEP and SVZ

Anterolateral ganglionic
NEP and SVZ

Posterior ganglionic
NEP and SVZ

Cortical (temporal) NEP

*Primordial plexiform layer
(cortical plate absent)*

*Lateral
migratory stream
(percolates through
external capsule)*

*Internal capsule*

*Caudate*

*Pioneer
corticofugal
axons?*

*Compact-bundled
corticothalamic
axons?*

Cortical (hippocampal) NEP

Lateral olfactory tract

Primary olfactory cortex

*Substantia
innominata*

*Basolateral
complex*

Amygdaloid NEP

Cortical (parahippocampal) NEP

*Anterior extramural migratory stream*
*(pontine gray and reticular tegmental neurons)*

*Sprouting fibers of hook bundle?*

Inferior
cerebellar
peduncle

*Sojourning and migrating deep neurons?*

*Sojourning and
migrating Purkinje cells*

Cochlear nuclei?

Nerve VIII (vestibulocochlear, *boundary cap*)

Nerve V (trigeminal)

**Trigeminal ganglion (V)**

Superior
cerebellar
peduncle

**External germinal
layer?**

**Dorsal
rhombic
lip**

**Vestibular ganglion (VIII)**

Pontine NEP

**Cerebellar
NEP**

**Petrous
temporal
bone**

**Spiral ganglion (VIII)**

**Inferior ganglion (IX)?**

**Anterior precerebellar NEP in
ventral rhombic lip**

*Posterior extramural migratory stream*
*(external cuneate and lateral reticular neurons)*

**Superior ganglion (X)?**

Nerve IX (glossopharyngeal)

Nerve X (vagus)

**Inferior ganglion (X)?**

FONT KEY:
**Germinal zone - Helvetica bold**
*Transient structure - Times bold italic*
Permanent structure - Times Roman or **Bold**

**ABBREVIATIONS:**
**NEP - Neuroepithelium**
*STF - Stratified transitional field*
**SVZ - Subventricular zone**

**Arrows** indicate the
presumed *direction of
neuron migration* from
neuroepithelial sources.

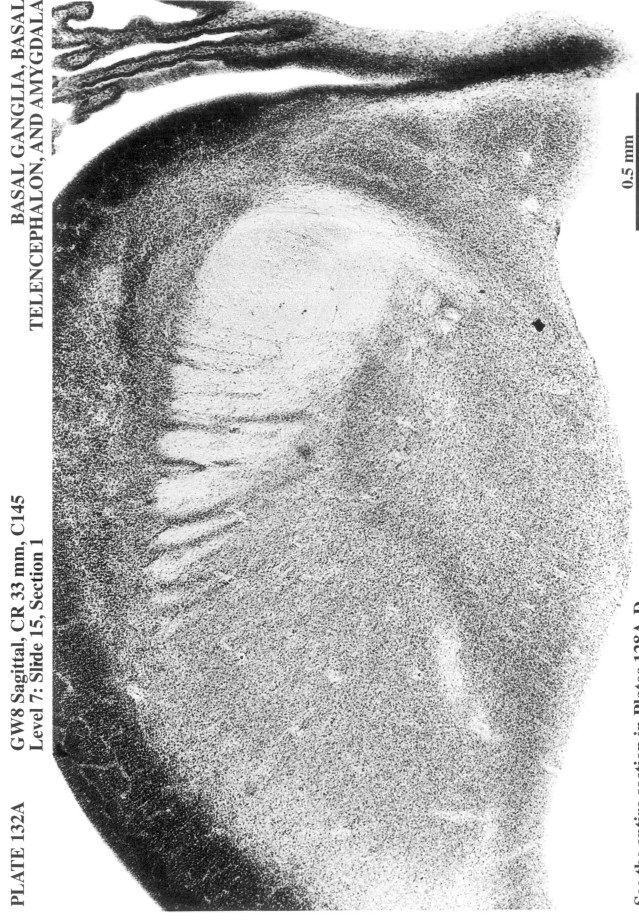

PLATE 132A

GW8 Sagittal, CR 33 mm, C145
Level 7: Slide 15, Section 1

BASAL GANGLIA, BASAL
TELENCEPHALON, AND AMYGDALA

0.5 mm

See the entire section in Plates 128A-D.

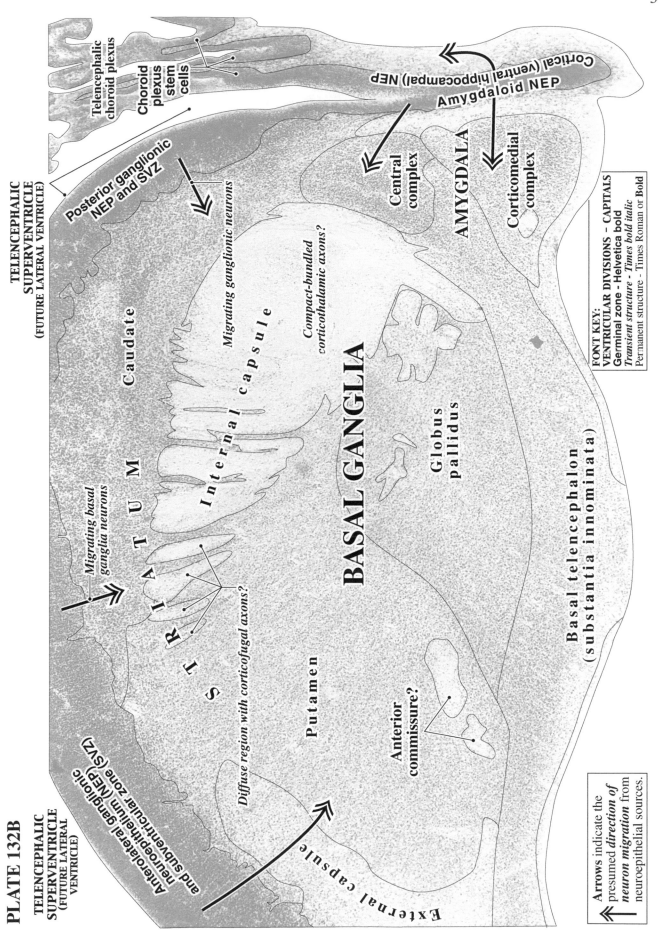

**PLATE 132B**

**TELENCEPHALIC SUPERVENTRICLE (FUTURE LATERAL VENTRICLE)**

**TELENCEPHALIC SUPERVENTRICLE (FUTURE LATERAL VENTRICLE)**

Telencephalic choroid plexus

Choroid plexus stem cells

*Cortical (ventral hippocampal) NEP*

*Amygdaloid NEP*

Posterior ganglionic NEP and SVZ

**Central complex**

**AMYGDALA**

**Corticomedial complex**

*Migrating ganglionic neurons*

*Compact-bundled corticothalamic axons?*

**Caudate**

Internal capsule

**S T R I A T U M**

*Migrating basal ganglia neurons*

**BASAL GANGLIA**

**Globus pallidus**

*Diffuse region with corticofugal axons?*

**Putamen**

**Basal telencephalon (substantia innominata)**

*Anterior commissure?*

**Anterolateral ganglionic neuroepithelium (NEP) and subventricular zone (SVZ)**

*External capsule*

FONT KEY:
VENTRICULAR DIVISIONS – CAPITALS
Germinal zone - Helvetica bold
*Transient structure - Times bold italic*
Permanent structure - Times Roman or Bold

Arrows indicate the presumed *direction of neuron migration* from neuroepithelial sources.

348

PLATE 133A

GW8 Sagittal, CR 33 mm, C145
Level 2: Slide 22, Section 2

HYPOTHALAMUS AND PREOPTIC AREA

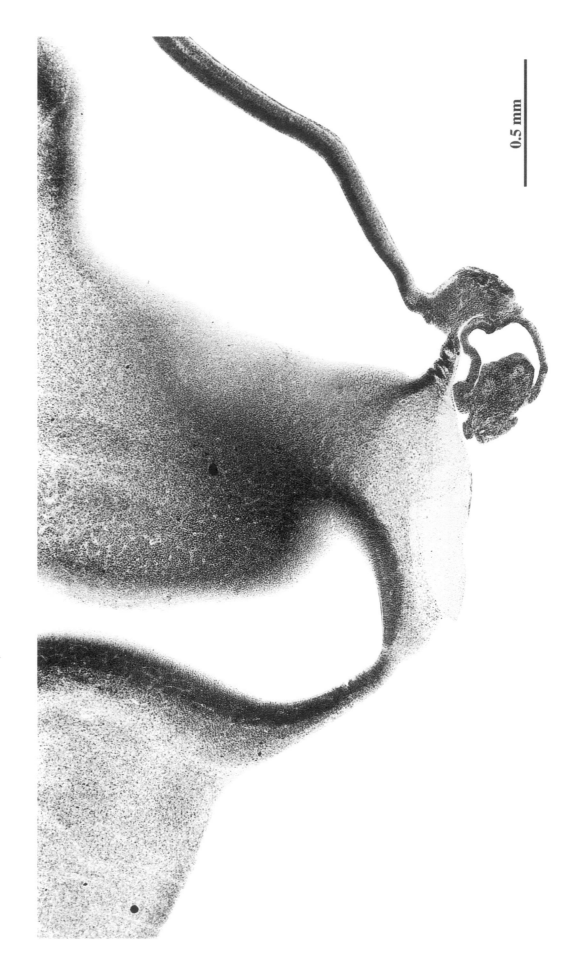

See the entire section in Plates 123A-D.

**PLATE 133B**

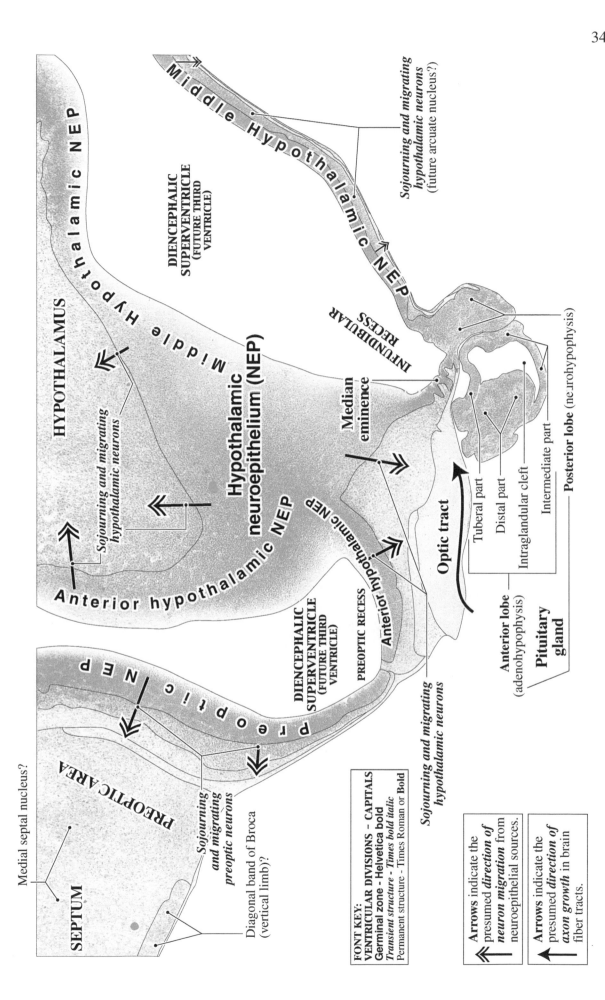

Medial septal nucleus?

SEPTUM

PREOPTIC AREA

Diagonal band of Broca (vertical limb)?

Preoptic NEP

*Sojourning and migrating preoptic neurons*

DIENCEPHALIC SUPERVENTRICLE (FUTURE THIRD VENTRICLE)

PREOPTIC RECESS

Anterior hypothalamic NEP

*Sojourning and migrating hypothalamic neurons*

Anterior hypothalamic NEP

**Hypothalamic neuroepithelium (NEP)**

*Sojourning and migrating hypothalamic neurons*

HYPOTHALAMUS

Middle hypothalamic NEP

Hypothalamic NEP

Middle Hypothalamic NEP

DIENCEPHALIC SUPERVENTRICLE (FUTURE THIRD VENTRICLE)

*Sojourning and migrating hypothalamic neurons (future arcuate nucleus?)*

INFUNDIBULAR RECESS

**Median eminence**

**Optic tract**

Tuberal part

Distal part

Intraglandular cleft

Intermediate part

**Posterior lobe** (neurohypophysis)

**Anterior lobe** (adenohypophysis)

**Pituitary gland**

FONT KEY:
VENTRICULAR DIVISIONS – CAPITALS
Germinal zone - **Helvetica bold**
Transient structure - ***Times bold italic***
Permanent structure - Times Roman or **Bold**

**Arrows** indicate the presumed ***direction of neuron migration*** from neuroepithelial sources.

**Arrows** indicate the presumed ***direction of axon growth*** in brain fiber tracts.

DIENCEPHALON AND BASAL TELENCEPHALON

PLATE 134A

GW8 Sagittal,
CR 33 mm, C145
Level 3:
Slide 20, Section 2

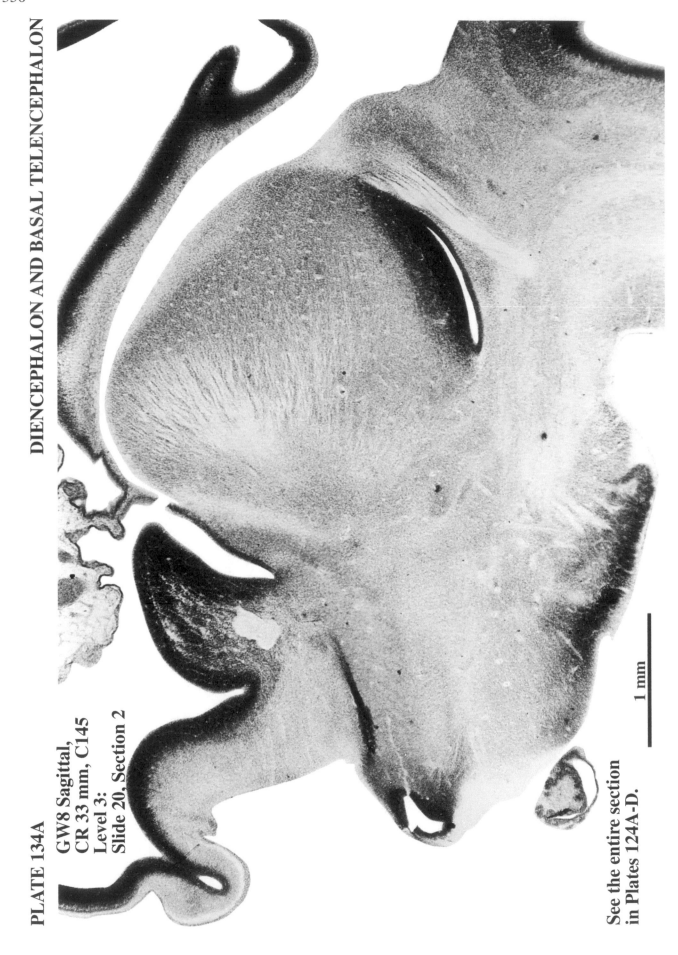

1 mm

See the entire section
in Plates 124A-D.

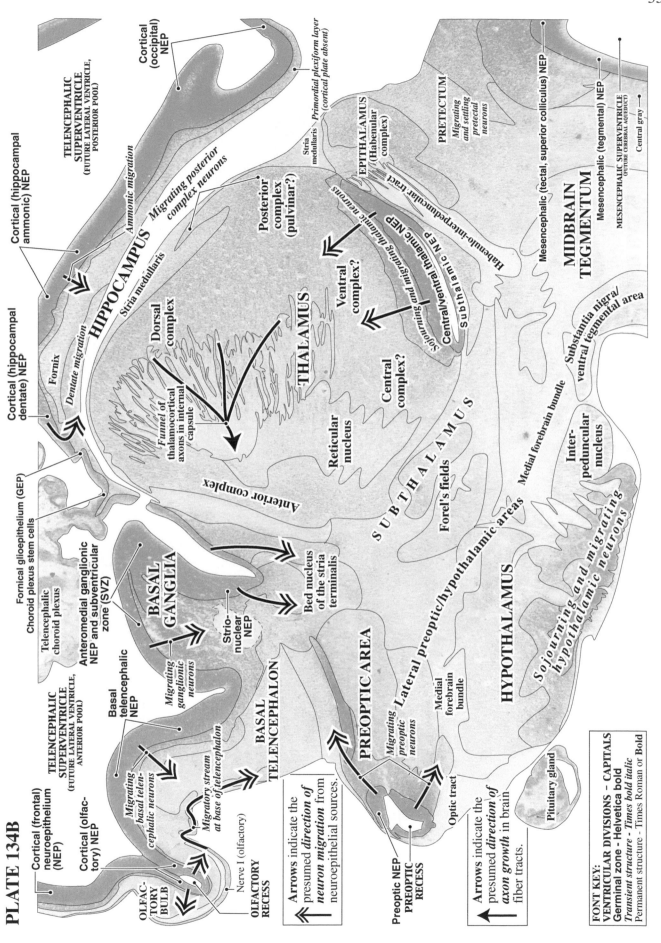

PLATE 134B

351

FONT KEY:
VENTRICULAR DIVISIONS – CAPITALS
Germinal zone - Helvetica bold
*Transient structure - Times bold italic*
Permanent structure - Times Roman or Bold

Cortical (frontal) neuroepithelium (NEP)

Cortical (olfactory) NEP

Fornical glioepithelium (GEP)
Choroid plexus stem cells

Telencephalic choroid plexus

Anteromedial ganglionic NEP and subventricular zone (SVZ)

**TELENCEPHALIC SUPERVENTRICLE (FUTURE LATERAL VENTRICLE, ANTERIOR POOL)**

Cortical (hippocampal dentate) NEP

Cortical (hippocampal ammonic) NEP

**TELENCEPHALIC SUPERVENTRICLE (FUTURE LATERAL VENTRICLE, POSTERIOR POOL)**

Cortical (occipital) NEP

*Primordial plexiform layer (cortical plate absent)*

Stria medullaris

**EPITHALAMUS (Habenular complex)**

**PRETECTUM**
*Migrating and settling pretectal neurons*

*Habenulo-interpeduncular tract*

*Central/ventral thalamic NEP*

*Sojourning and migrating thalamic neurons*

Subthalamic

Mesencephalic (tectal, superior colliculus) NEP

**MIDBRAIN TEGMENTUM**

Mesencephalic (tegmental) NEP

**MESENCEPHALIC SUPERVENTRICLE (FUTURE CEREBRAL AQUEDUCT)**

Central gray

*Migrating posterior complex neurons*

**HIPPOCAMPUS**

*Ammonic migration*

Stria medullaris

*Dentate migration*

Fornix

Dorsal complex

Funnel of thalamocortical axons in internal capsule

Posterior complex (pulvinar?)

Ventral complex?

**THALAMUS**

*Anterior complex*

Reticular nucleus

Central complex?

**S U B T H A L A M U S**

Forel's fields

Substantia nigra/ ventral tegmental area

Inter-peduncular nucleus

Medial forebrain bundle

*Sojourning and migrating hypothalamic neurons*

**BASAL GANGLIA**

*Basal telencephalic NEP*

*Migrating ganglionic neurons*

Strio-nuclear NEP

Bed nucleus of the stria terminalis

**BASAL TELENCEPHALON**

*Migrating basal telen-cephalic neurons*

*Migratory stream at base of telencephalon*

Nerve I (olfactory)

**OLFAC-TORY BULB**

**OLFACTORY RECESS**

**PREOPTIC AREA**

*Lateral preoptic/hypothalamic areas*

Medial forebrain bundle

**HYPOTHALAMUS**

*Migrating preoptic neurons*

Optic tract

Preoptic NEP
**PREOPTIC RECESS**

Pituitary gland

Arrows indicate the presumed *direction of neuron migration* from neuroepithelial sources.

Arrows indicate the presumed *direction of axon growth* in brain fiber tracts.

DIENCEPHALON AND BASAL TELENCEPHALON

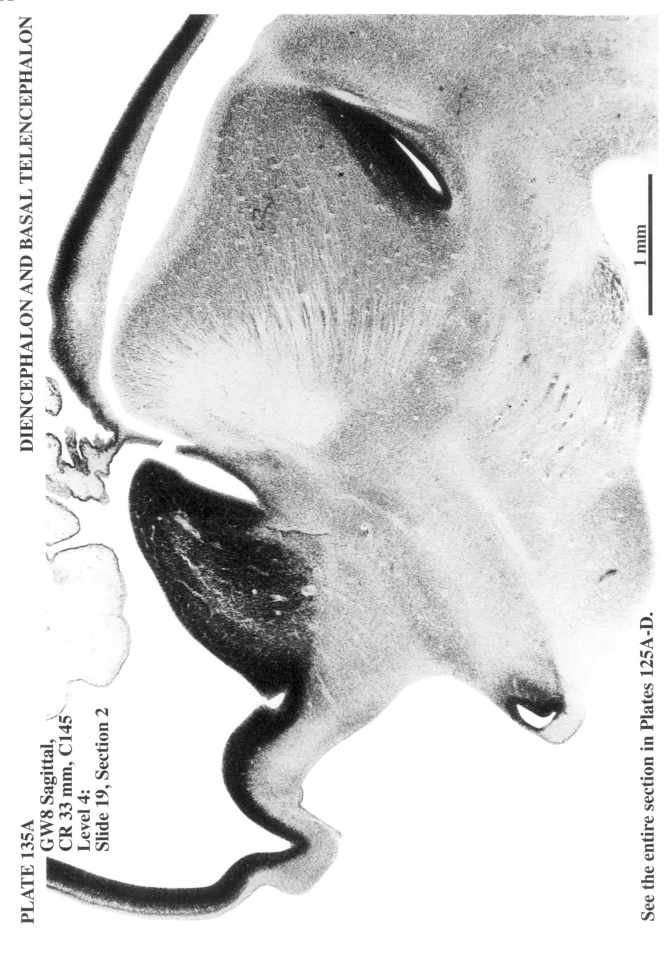

PLATE 135A

GW8 Sagittal,
CR 33 mm, C145
Level 4:
Slide 19, Section 2

1 mm

See the entire section in Plates 125A-D.

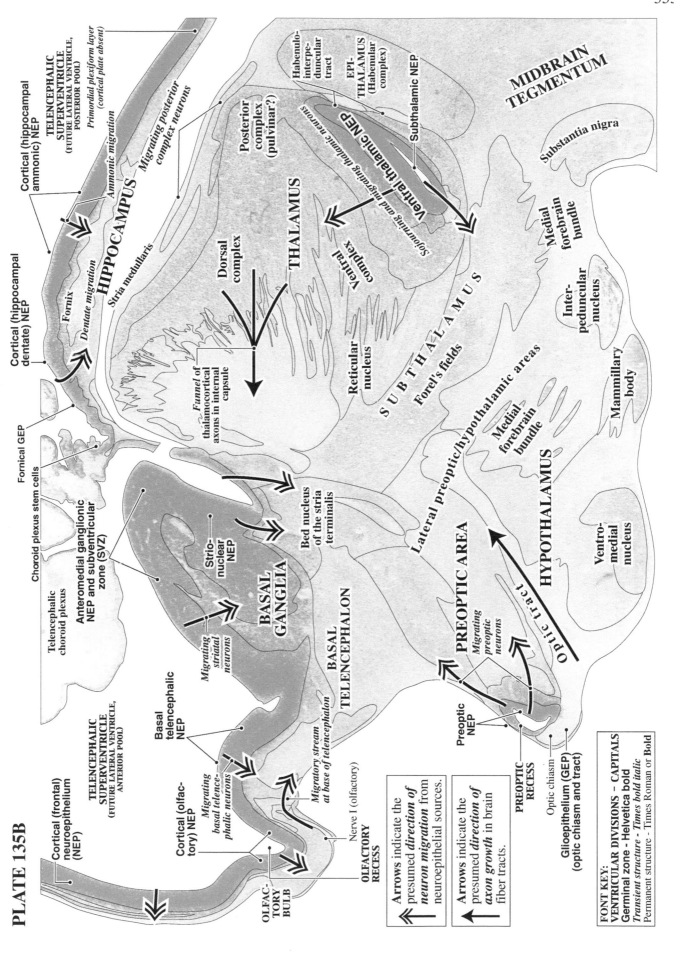

**PLATE 135B**

353

MIDBRAIN TEGMENTUM

FONT KEY:
VENTRICULAR DIVISIONS – CAPITALS
Germinal zone - Helvetica bold
*Transient structure - Times bold italic*
Permanent structure - Times Roman or **Bold**

Arrows indicate the presumed *direction of neuron migration* from neuroepithelial sources.

Arrows indicate the presumed *direction of axon growth* in brain fiber tracts.

354

PLATE 136A

BASAL TELENCEPHALON AND DIENCEPHALON

GW8 Sagittal, CR 33 mm, C145
Level 5: Slide 18, Section 2

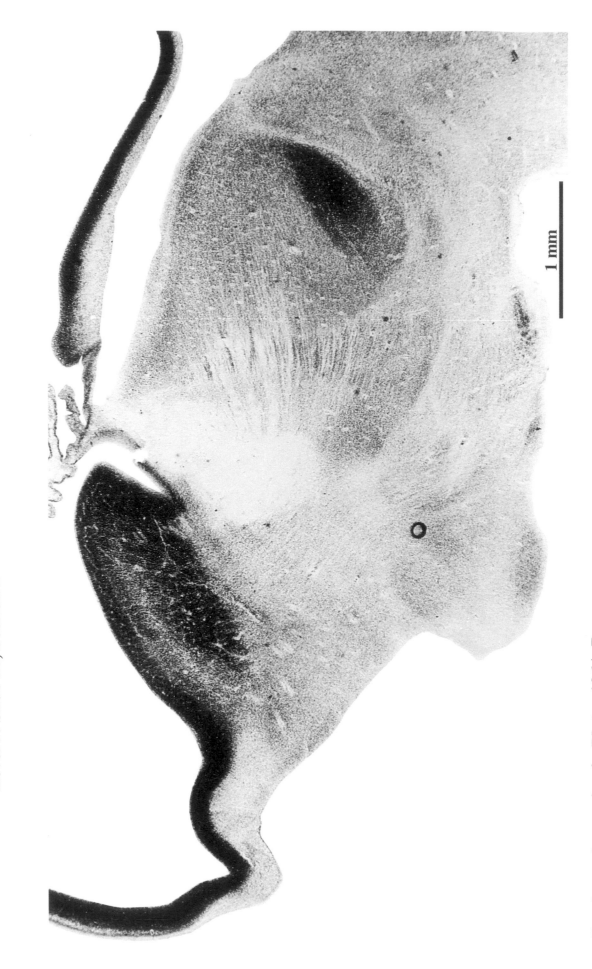

1 mm

See the entire section in Plates 126A-D.

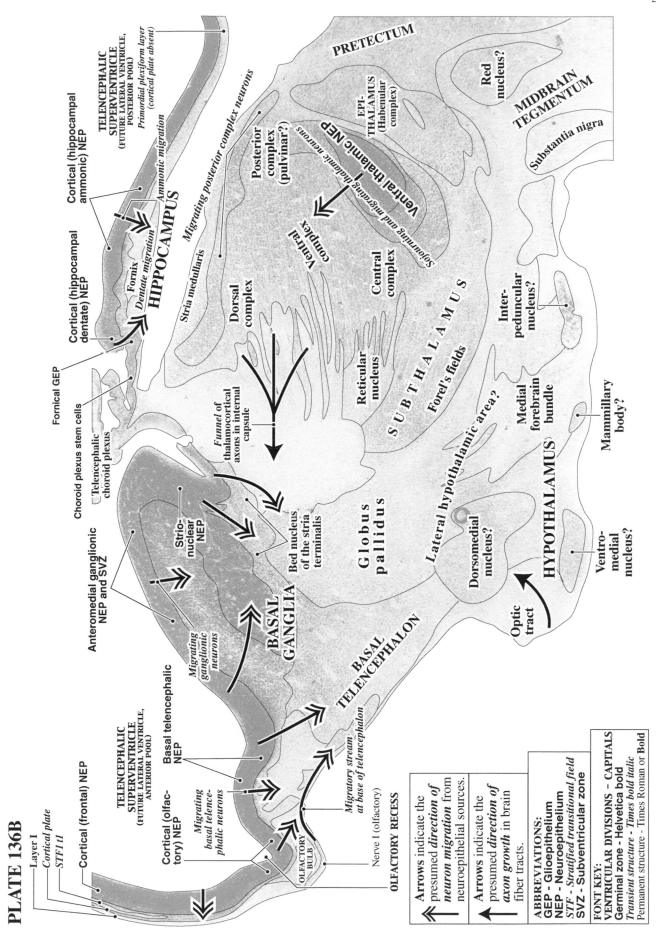

## PLATE 136B

Layer I
Cortical plate
STF1 t1
Cortical (frontal) NEP

**TELENCEPHALIC SUPERVENTRICLE**
(FUTURE LATERAL VENTRICLE, ANTERIOR POOL)
Basal telencephalic NEP

Cortical (olfactory) NEP

*Migrating basal telencephalic neurons*

**OLFACTORY BULB**

**OLFACTORY RECESS**

Nerve I (olfactory)

*Migratory stream at base of telencephalon*

Anteromedial ganglionic NEP and SVZ

Strio-nuclear NEP

*Migrating ganglionic neurons*

**BASAL GANGLIA**

**BASAL TELENCEPHALON**

Bed nucleus of the stria terminalis

**G l o b u s p a l l i d u s**

Optic tract

**HYPOTHALAMUS**

Dorsomedial nucleus?

Lateral hypothalamic area?

Ventro-medial nucleus?

Mammillary body?

Medial forebrain bundle

Inter-peduncular nucleus?

**S U B T H A L A M U S**

Forel's fields

Reticular nucleus

Central complex

*Sojourning and migrating thalamic neurons*

**Ventral thalamic NEP**

Ventral complex

Dorsal complex

*Funnel of thalamocortical axons in internal capsule*

Stria medullaris

*Migrating posterior complex neurons*

Posterior complex (pulvinar?)

**EPI-THALAMUS** (Habenular complex)

**PRETECTUM**

Red nucleus?

**MIDBRAIN TEGMENTUM**

Substantia nigra

Fornical GEP

Choroid plexus stem cells

Telencephalic choroid plexus

Cortical (hippocampal dentate) NEP

Cortical (hippocampal ammonic) NEP

**TELENCEPHALIC SUPERVENTRICLE**
(FUTURE LATERAL VENTRICLE, POSTERIOR POOL)
*Primordial plexiform layer (cortical plate absent)*

*Ammonic migration*

**HIPPOCAMPUS**

*Dentate migration*

Fornix

Arrows indicate the presumed *direction of neuron migration* from neuroepithelial sources.

Arrows indicate the presumed *direction of axon growth* in brain fiber tracts.

**ABBREVIATIONS:**
**GEP** - Glioepithelium
**NEP** - Neuroepithelium
*STF - Stratified transitional field*
**SVZ** - Subventricular zone

**FONT KEY:**
VENTRICULAR DIVISIONS – CAPITALS
Germinal zone - Helvetica bold
*Transient structure - Times bold italic*
Permanent structure - Times Roman or **Bold**

BASAL TELENCEPHALON AND DIENCEPHALON

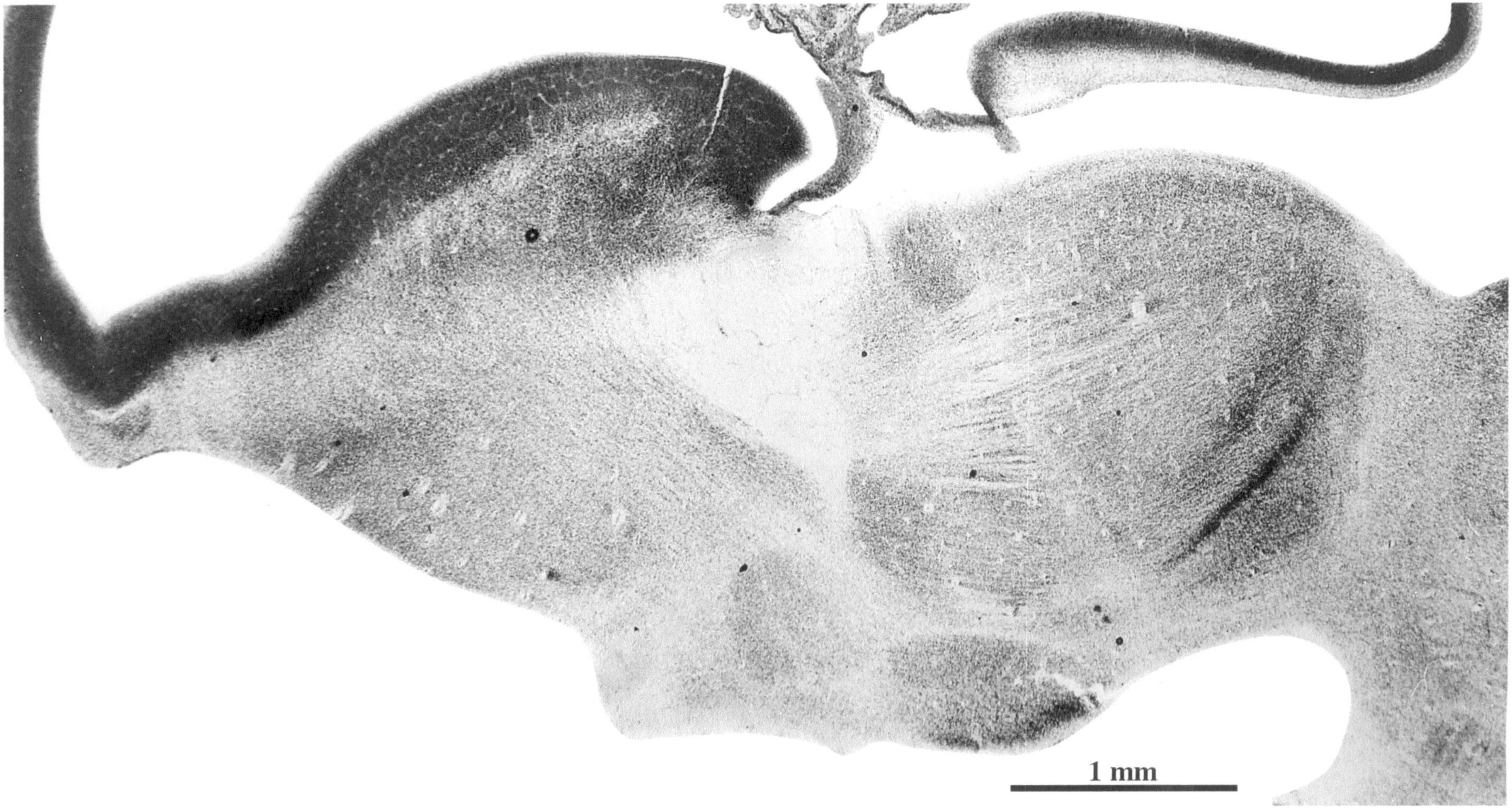

PLATE 137A

GW8 Sagittal, CR 33 mm, C145
Between levels 5 and 6:
Slide 17, Section 2

1 mm

See level 5 in Plates 126A-D, level 6 in Plates 127A-D.

# PLATE 137B

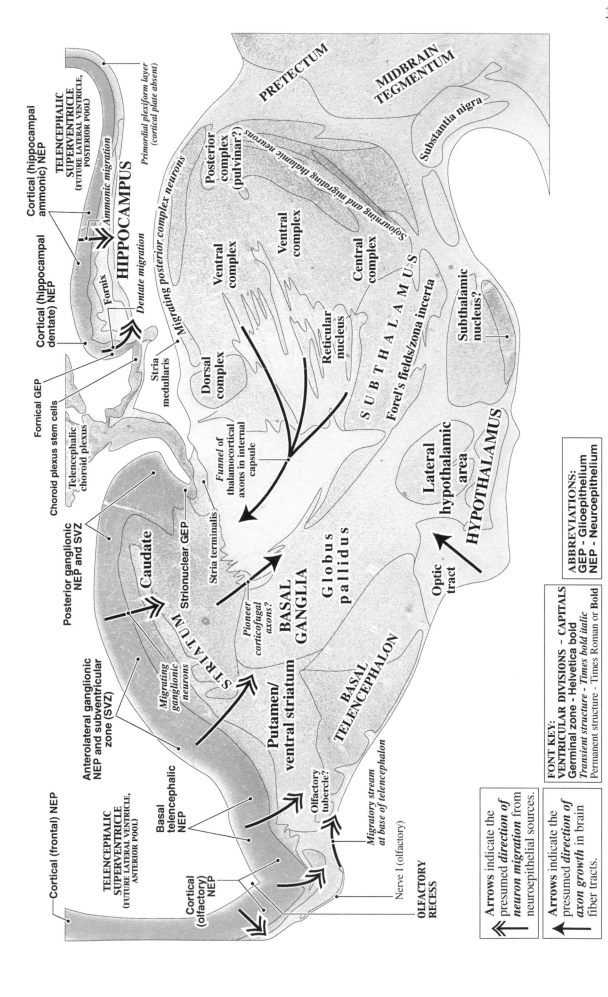

357

**Labels (top to bottom, left region):**

Cortical (frontal) NEP

**TELENCEPHALIC SUPERVENTRICLE** (FUTURE LATERAL VENTRICLE, ANTERIOR POOL)

Basal telencephalic NEP

Cortical (olfactory) NEP

**OLFACTORY RECESS**

Nerve I (olfactory)

Migratory stream at base of telencephalon

Olfactory tubercle?

Anterolateral ganglionic NEP and subventricular zone (SVZ)

Posterior ganglionic NEP and SVZ

Telencephalic choroid plexus

Choroid plexus stem cells

Fornical GEP

Cortical (hippocampal dentate) NEP

Cortical (hippocampal ammonic) NEP

**TELENCEPHALIC SUPERVENTRICLE** (FUTURE LATERAL VENTRICLE, POSTERIOR POOL)

Primordial plexiform layer (cortical plate absent)

Ammonic migration

Dentate migration

Fornix

Stria medullaris

**HIPPOCAMPUS**

Migrating posterior complex neurons

Posterior complex (pulvinar?)

Ventral complex

Ventral complex

Dorsal complex

Somatomotor and migrating thalamic neurons

Central complex

Reticular nucleus

Funnel of thalamocortical axons in internal capsule

Stria terminalis

Strionuclear GEP

**Caudate**

Migrating ganglionic neurons

**STRIATUM**

Pioneer corticofugal axons?

**BASAL GANGLIA**

Putamen/ ventral striatum

**Globus pallidus**

**BASAL TELENCEPHALON**

Optic tract

**HYPOTHALAMUS**

Lateral hypothalamic area

Subthalamic nucleus?

**S U B T H A L A M U S**

Forel's fields/zona incerta

Substantia nigra

**PRETECTUM**

**MIDBRAIN TEGMENTUM**

**ABBREVIATIONS:**
GEP - Glioepithelium
NEP - Neuroepithelium

**FONT KEY:**
VENTRICULAR DIVISIONS - CAPITALS
Germinal zone - Helvetica bold
Transient structure - Times bold italic
Permanent structure - Times Roman or Bold

⇐ Arrows indicate the presumed *direction of neuron migration* from neuroepithelial sources.

← Arrows indicate the presumed *direction of axon growth* in brain fiber tracts.

MIDBRAIN TECTUM AND TEGMENTUM

PLATE 138A

GW8 Sagittal, CR 33 mm, C145
Medial to level 1:
Slide 24, Section 2

See level 1 in
Plates 122A-D.

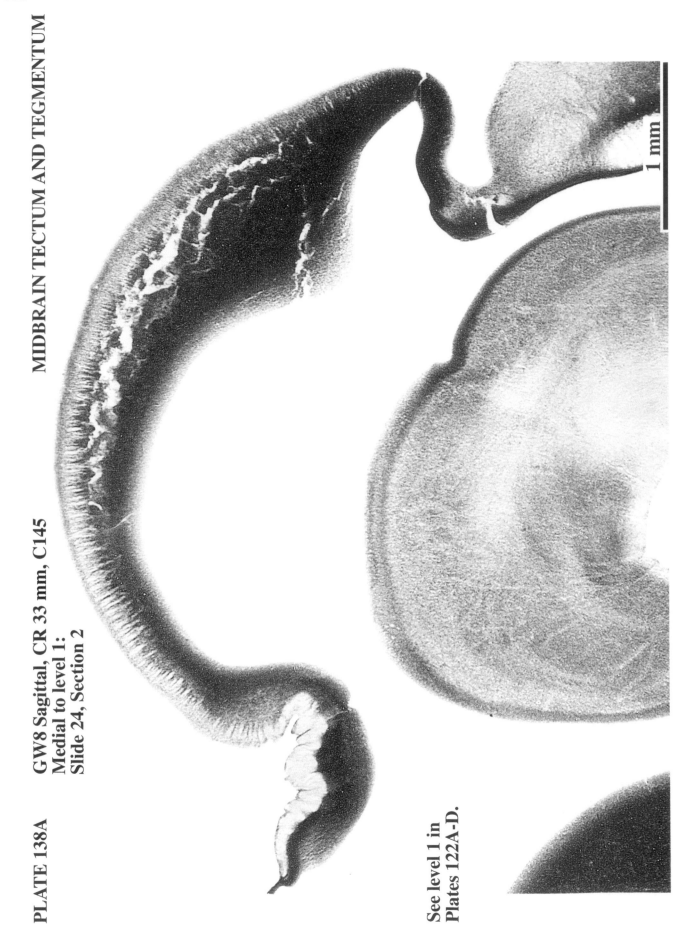

1 mm

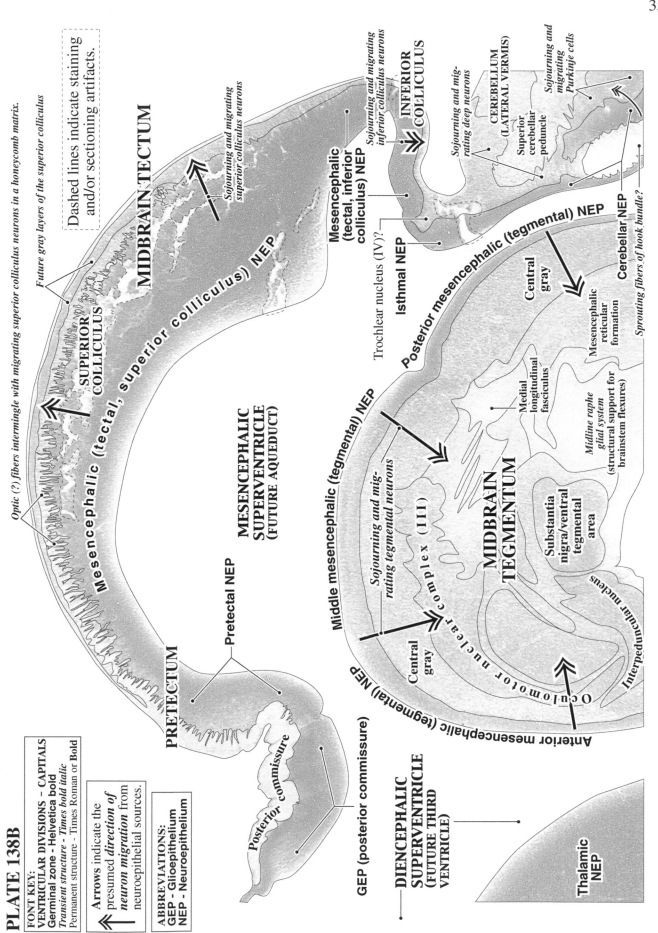

359

PLATE 138B

FONT KEY:
VENTRICULAR DIVISIONS – CAPITALS
Germinal zone - Helvetica bold
Transient structure - *Times bold italic*
Permanent structure - Times Roman or **Bold**

Arrows indicate the presumed *direction of neuron migration* from neuroepithelial sources.

ABBREVIATIONS:
GEP - Glioepithelium
NEP - Neuroepithelium

Dashed lines indicate staining and/or sectioning artifacts.

MIDBRAIN TECTUM

*Sojourning and migrating superior colliculus neurons*

*Future gray layers of the superior colliculus*

*Optic (?) fibers intermingle with migrating superior colliculus neurons in a honeycomb matrix.*

SUPERIOR COLLICULUS

Mesencephalic (tectal, superior colliculus) NEP

PRETECTUM

Pretectal NEP

MESENCEPHALIC SUPERVENTRICLE (FUTURE AQUEDUCT)

*Posterior commissure*

GEP (posterior commissure)

DIENCEPHALIC SUPERVENTRICLE (FUTURE THIRD VENTRICLE)

Thalamic NEP

Middle mesencephalic (tegmental) NEP

Anterior mesencephalic (tegmental) NEP

*Sojourning and migrating tegmental neurons*

*Oculomotor complex (III)*

Central gray

Oculomotor nuclear complex

MIDBRAIN TEGMENTUM

Substantia nigra/ventral tegmental area

Interpeduncular nucleus

*Medial longitudinal fasciculus*

*Midline raphe glial system (structural support for brainstem flexures)*

*Mesencephalic reticular formation*

Central gray

Posterior mesencephalic (tegmental) NEP

Mesencephalic (tectal, inferior colliculus) NEP

*Sojourning and migrating inferior colliculus neurons*

INFERIOR COLLICULUS

Trochlear nucleus (IV)?

Isthmal NEP

CEREBELLUM (LATERAL VERMIS)

*Sojourning and migrating deep neurons*

Superior cerebellar peduncle

*Sojourning and migrating Purkinje cells*

Cerebellar NEP

*Sprouting fibers of hook bundle?*

MIDBRAIN TEGMENTUM

GW8 Sagittal, CR 33 mm, C145
Level 1: Slide 23, Section 2

PLATE 139A

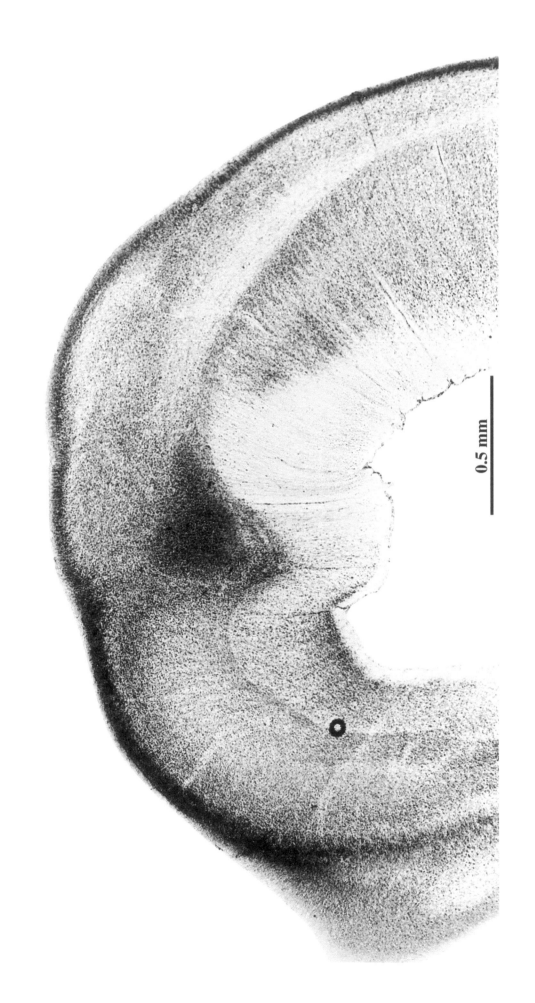

0.5 mm

See the entire section in Plates 122A-D.

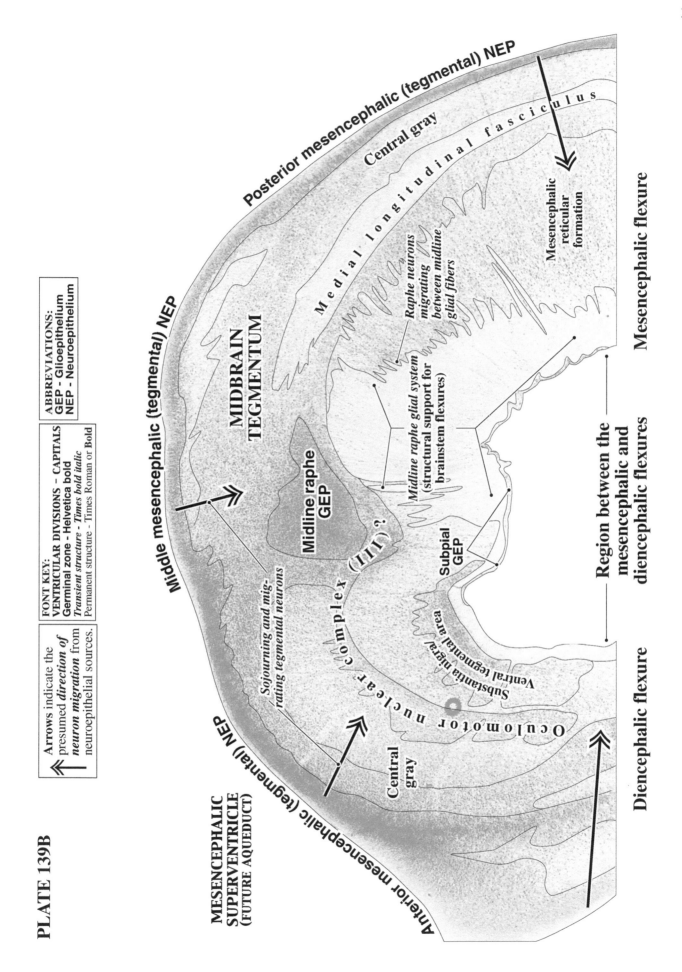

PLATE 139B

361

**ABBREVIATIONS:**
GEP - Glioepithelium
NEP - Neuroepithelium

**FONT KEY:**
VENTRICULAR DIVISIONS – CAPITALS
Germinal zone - **Helvetica bold**
*Transient structure - Times bold italic*
Permanent structure - Times Roman or Bold

Arrows indicate the presumed *direction of neuron migration* from neuroepithelial sources.

Posterior mesencephalic (tegmental) NEP

Central gray

*Medial longitudinal fasciculus*

*Raphe neurons migrating between midline glial fibers*

*Mesencephalic reticular formation*

Mesencephalic flexure

Middle mesencephalic (tegmental) NEP

**MIDBRAIN TEGMENTUM**

**Midline raphe GEP**

*Midline raphe glial system (structural support for brainstem flexures)*

**Subpial GEP**

Region between the mesencephalic and diencephalic flexures

*Sojourning and migrating tegmental neurons*

*Oculomotor nuclear complex (IIIv)?*

*Substantia nigra/ Ventral tegmental area*

**MESENCEPHALIC SUPERVENTRICLE (FUTURE AQUEDUCT)**

**Anterior mesencephalic (tegmental) NEP**

Central gray

Diencephalic flexure

CEREBELLUM: LATERAL VERMIS

PLATE 140A

GW8 Sagittal
CR 33 mm, C145
Level 3:
Slide 20,
Section 2

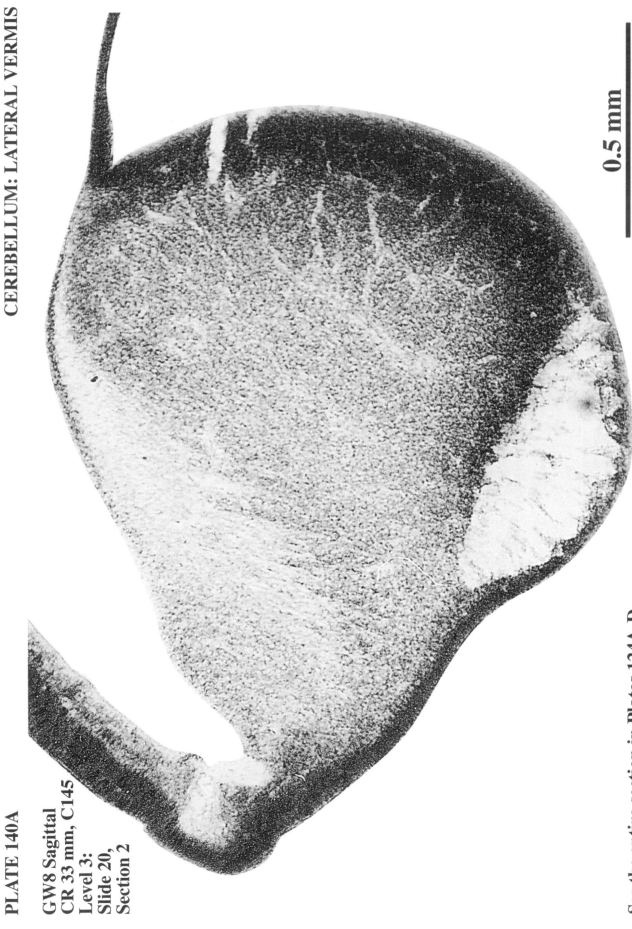

0.5 mm

See the entire section in Plates 124A-D.

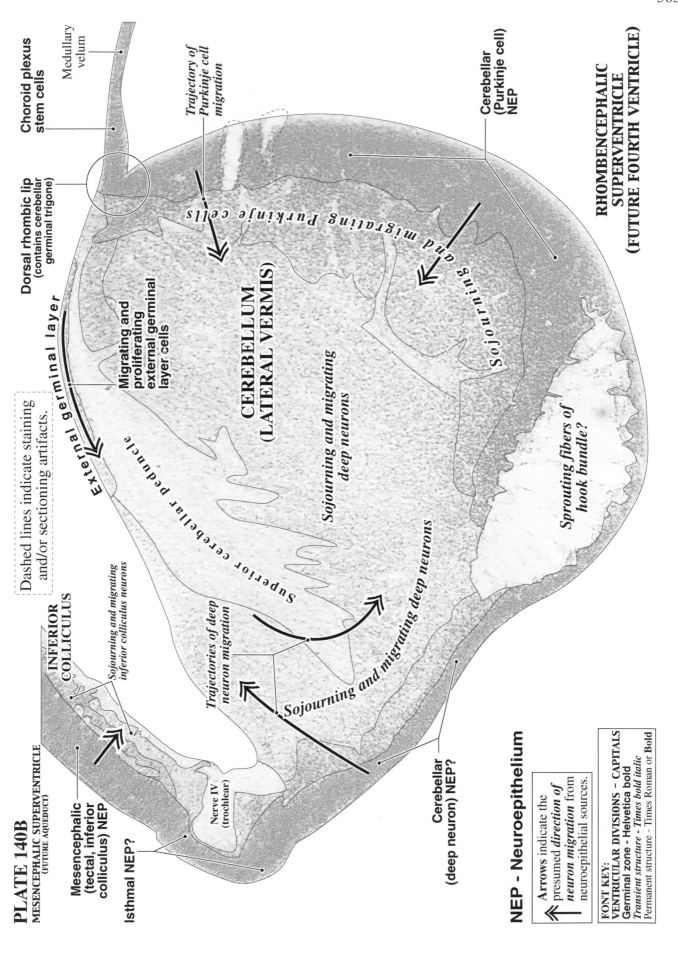

363

**PLATE 140B**
MESENCEPHALIC SUPERVENTRICLE
(FUTURE AQUEDUCT)

Choroid plexus
stem cells

Medullary
velum

Trajectory of
Purkinje cell
migration

Cerebellar
(Purkinje cell)
NEP

**RHOMBENCEPHALIC
SUPERVENTRICLE
(FUTURE FOURTH VENTRICLE)**

Dorsal rhombic lip
(contains cerebellar
germinal trigone)

*Sojourning and migrating Purkinje cells*

Dashed lines indicate staining
and/or sectioning artifacts.

**Migrating and
proliferating
external germinal
layer cells**

**External germinal layer**

**CEREBELLUM
(LATERAL VERMIS)**

*Superior cerebellar peduncle*

*Sojourning and migrating
deep neurons*

*Sojourning
and migrating deep neurons*

*Sprouting fibers of
hook bundle?*

**INFERIOR
COLLICULUS**

*Sojourning and migrating
inferior colliculus neurons*

*Trajectories of deep
neuron migration*

*Sojourning and migrating deep neurons*

**Mesencephalic
(tectal, inferior
colliculus) NEP**

**Isthmal NEP?**

*Nerve IV
(trochlear)*

**Cerebellar
(deep neuron) NEP?**

**NEP - Neuroepithelium**

Arrows indicate the
presumed *direction of
neuron migration* from
neuroepithelial sources.

FONT KEY:
VENTRICULAR DIVISIONS – CAPITALS
Germinal zone – Helvetica bold
*Transient structure - Times bold italic*
Permanent structure - Times Roman or Bold

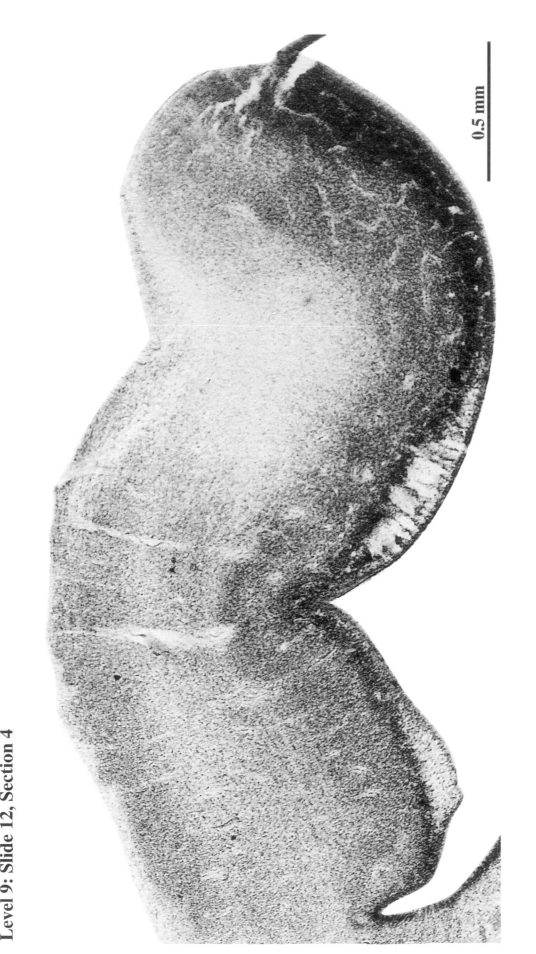

CEREBELLUM: HEMISPHERE

PLATE 141A

GW8 Sagittal
CR 33 mm, C145
Level 9: Slide 12, Section 4

0.5 mm

See the entire section in Plates 130A-D.

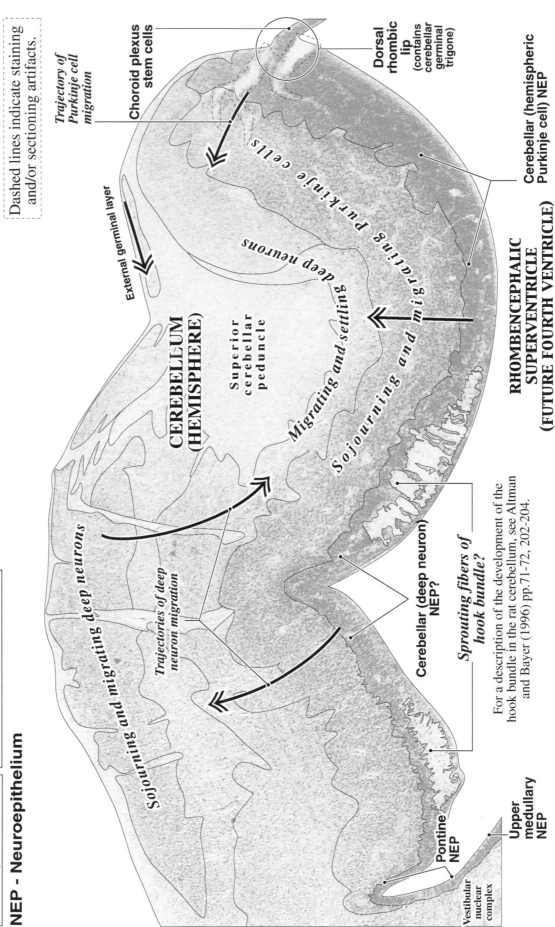

365

PLATE 141B

Arrows indicate the presumed *direction of neuron migration* from neuroepithelial sources.

**NEP - Neuroepithelium**

FONT KEY:
VENTRICULAR DIVISIONS – CAPITALS
**Germinal zone - Helvetica bold**
*Transient structure - Times bold italic*
Permanent structure - Times Roman or **Bold**

Dashed lines indicate staining and/or sectioning artifacts.

*Trajectory of Purkinje cell migration*

**Choroid plexus stem cells**

**Dorsal rhombic lip** (contains cerebellar germinal trigone)

**Cerebellar (hemispheric Purkinje cell) NEP**

*External germinal layer*

*Migrating Purkinje cells*

*deep neurons*

**CEREBELLUM (HEMISPHERE)**

*Superior cerebellar peduncle*

*Migrating and settling deep neurons*

*Sojourning and migrating*

**RHOMBENCEPHALIC SUPERVENTRICLE (FUTURE FOURTH VENTRICLE)**

*Sojourning and migrating deep neurons*

*Trajectories of deep neuron migration*

**Cerebellar (deep neuron) NEP?**

*Sprouting fibers of hook bundle?*

For a description of the development of the hook bundle in the rat cerebellum, see Altman and Bayer (1996) pp.71-72, 202-204.

**Pontine NEP**

**Upper medullary NEP**

Vestibular nuclear complex

PONS AND MEDULLA

PLATE 142A

GW8 Sagittal
CR 33 mm, C145
Level 1: Slide 23, Section 2

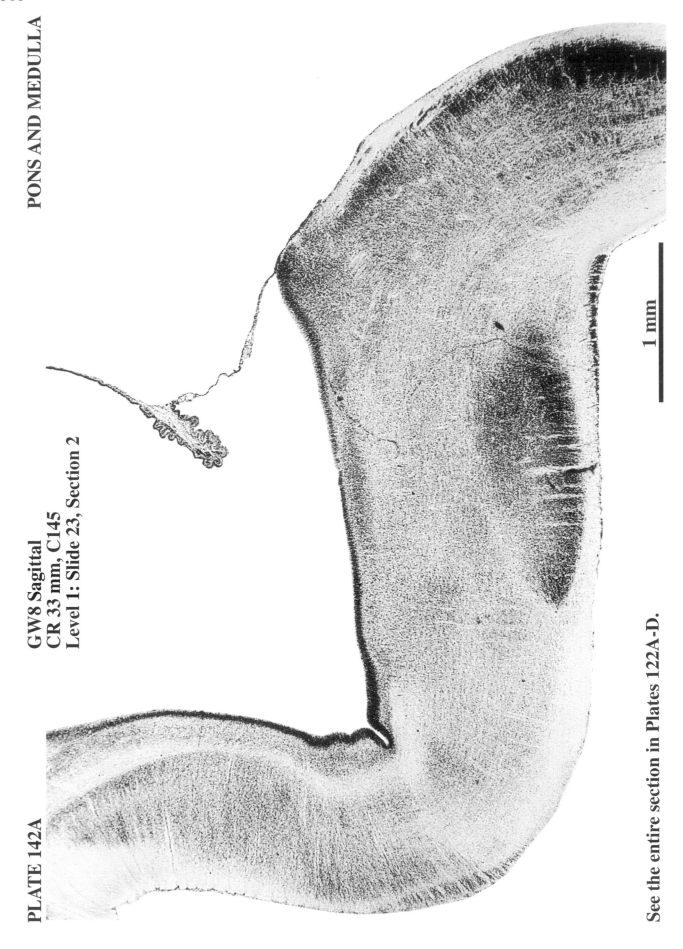

1 mm

See the entire section in Plates 122A-D.

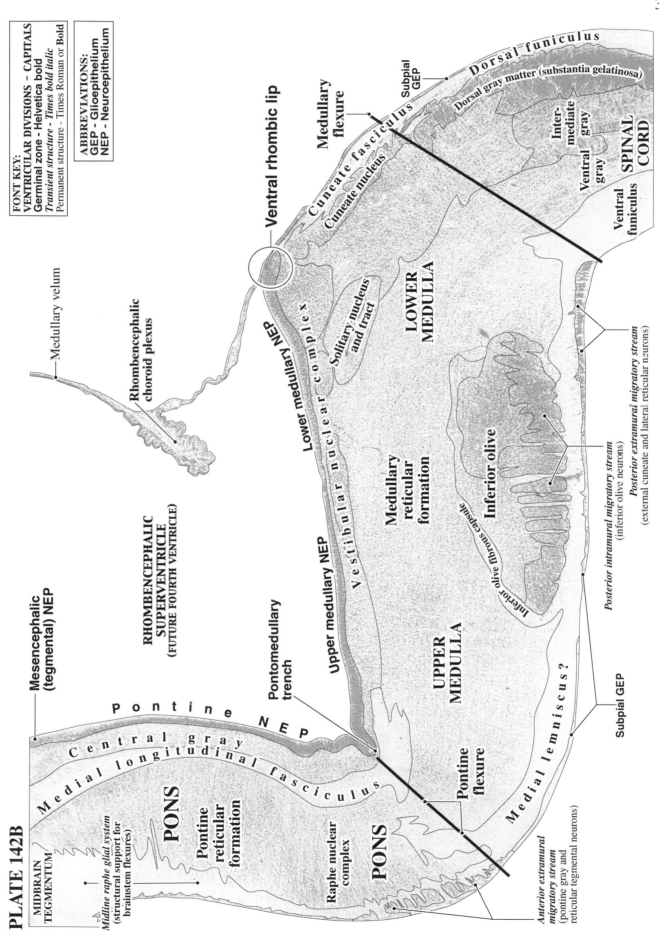

PLATE 142B

FONT KEY:
VENTRICULAR DIVISIONS – CAPITALS
Germinal zone - Helvetica bold
Transient structure - Times bold italic
Permanent structure - Times Roman or Bold

ABBREVIATIONS:
GEP - Glioepithelium
NEP - Neuroepithelium

MIDBRAIN TEGMENTUM

Mesencephalic (tegmental) NEP

Midline raphe glial system (structural support for brainstem flexures)

Medullary velum

Rhombencephalic choroid plexus

RHOMBENCEPHALIC SUPERVENTRICLE (FUTURE FOURTH VENTRICLE)

Pontine NEP

Central gray

Medial longitudinal fasciculus

PONS

Pontine reticular formation

Raphe nuclear complex

PONS

Pontomedullary trench

Upper medullary NEP

Lower medullary NEP

Vestibular nuclear complex

Ventral rhombic lip

Medullary flexure

Cuneate fasciculus

Cuneate nucleus

Dorsal funiculus

Subpial GEP

Dorsal gray matter (substantia gelatinosa)

Dorsal gray matter

Inter-mediate gray

Ventral gray

Ventral gray

SPINAL CORD

Ventral funiculus

Solitary nucleus and tract

LOWER MEDULLA

Medullary reticular formation

Inferior olive

Inferior olive fibrous capsule

UPPER MEDULLA

Pontine flexure

Medial lemniscus

Medial lemniscus ?

Subpial GEP

Posterior intramural migratory stream (inferior olive neurons)

Posterior extramural migratory stream (external cuneate and lateral reticular neurons)

Anterior extramural migratory stream (pontine gray and reticular tegmental neurons)

PONS AND MEDULLA

PLATE 143A

GW8 Sagittal
CR 33 mm, C145
Level 2: Slide 22, Section 2

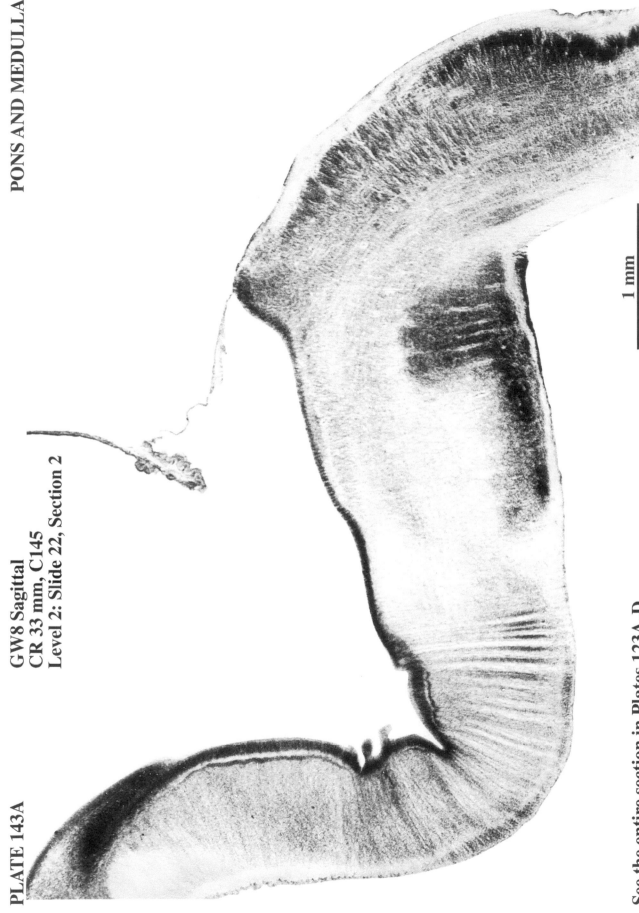

1 mm

See the entire section in Plates 123A-D.

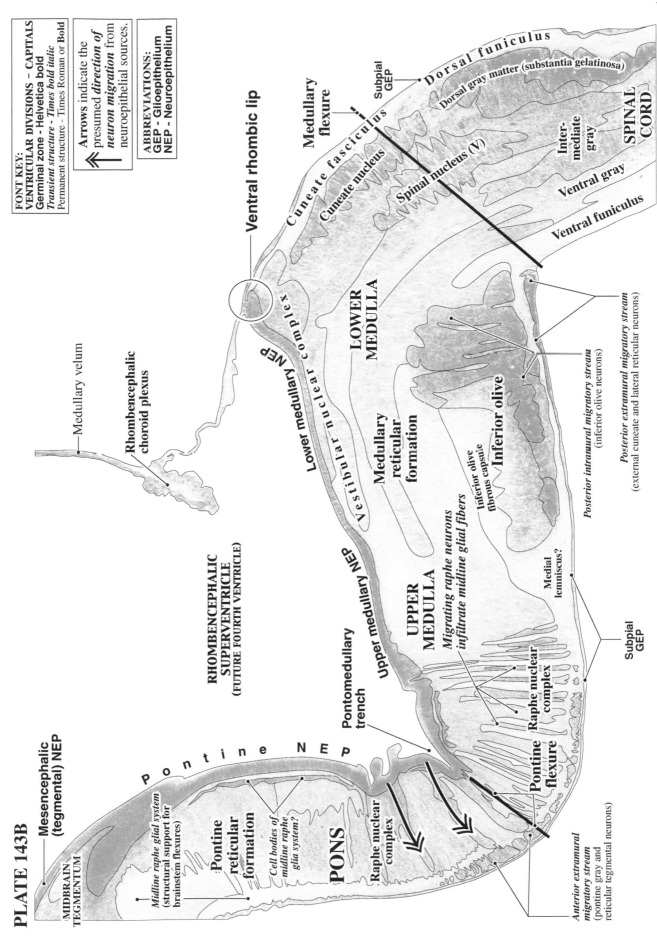

## PLATE 143B

**FONT KEY:**
VENTRICULAR DIVISIONS – CAPITALS
Germinal zone - Helvetica bold
*Transient structure - Times bold italic*
Permanent structure - Times Roman or Bold

⟸ Arrows indicate the presumed *direction of neuron migration* from neuroepithelial sources.

**ABBREVIATIONS:**
GEP - Glioepithelium
NEP - Neuroepithelium

Medullary velum

Rhombencephalic choroid plexus

Ventral rhombic lip

Medullary flexure

Dorsal funiculus

Subpial GEP

Dorsal gray matter (substantia gelatinosa)

Cuneate fasciculus

Cuneate nucleus

Spinal nucleus (V)

Inter-mediate gray

**SPINAL CORD**

Ventral gray

Ventral funiculus

Lower medullary nuclear complex

Vestibular nuclear complex

NEP

**LOWER MEDULLA**

Medullary reticular formation

Inferior olive

Inferior olive fibrous capsule

*Migrating raphe neurons infiltrate midline glial fibers*

Posterior intramural migratory stream (inferior olive neurons)

Posterior extramural migratory stream (external cuneate and lateral reticular neurons)

Upper medullary NEP

**RHOMBENCEPHALIC SUPERVENTRICLE (FUTURE FOURTH VENTRICLE)**

Pontomedullary trench

**UPPER MEDULLA**

Medial lemniscus?

Subpial GEP

*Raphe nuclear complex*

Pontine flexure

**P o n t i n e   N E P**

*Midline raphe glial system (structural support for brainstem flexures)*

*Cell bodies of midline raphe glia system?*

Pontine reticular formation

**PONS**

*Raphe nuclear complex*

Mesencephalic (tegmental) NEP

MIDBRAIN TEGMENTUM

*Anterior extramural migratory stream (pontine gray and reticular tegmental neurons)*

BRAINSTEM

GW8 Sagittal
CR 33 mm, C145
Between levels 2 and 3:
Slide 21, Section 2

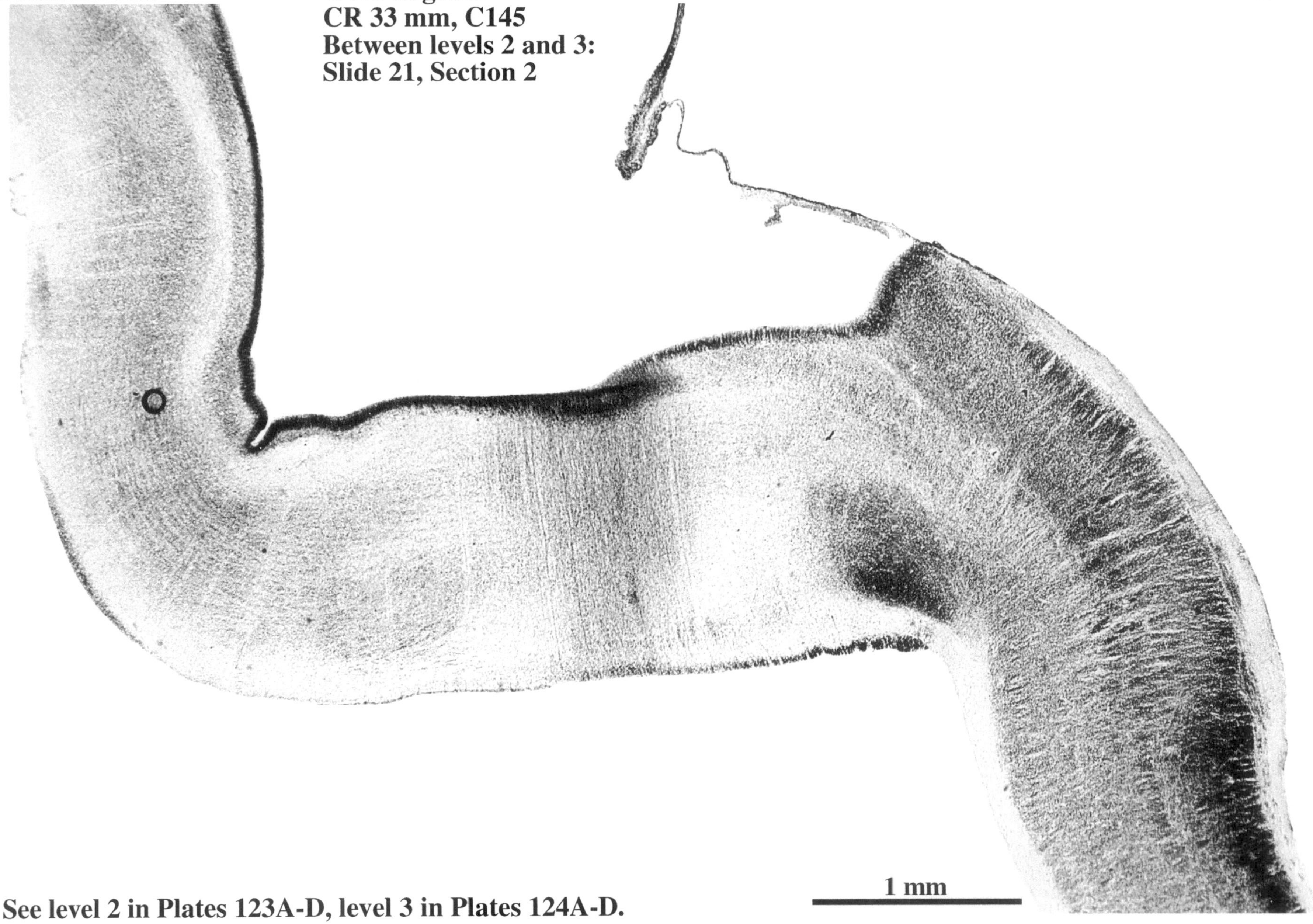

1 mm

See level 2 in Plates 123A-D, level 3 in Plates 124A-D.

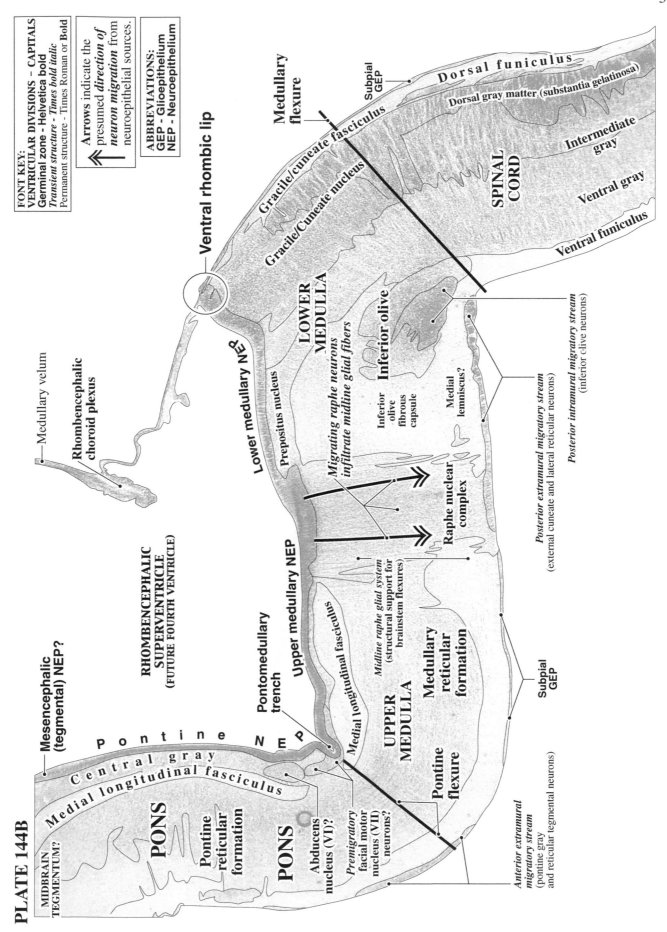

371

PLATE 144B

FONT KEY:
VENTRICULAR DIVISIONS – CAPITALS
Germinal zone – Helvetica bold
*Transient structure - Times bold italic*
Permanent structure - Times Roman or Bold

⬅ Arrows indicate the presumed *direction of neuron migration* from neuroepithelial sources.

ABBREVIATIONS:
GEP - Glioepithelium
NEP - Neuroepithelium

Medullary velum

Rhombencephalic choroid plexus

Ventral rhombic lip

Medullary flexure

Gracile/cuneate fasciculus

Gracile/Cuneate nucleus

Subpial GEP

Dorsal funiculus

Dorsal gray matter (substantia gelatinosa)

Intermediate gray

SPINAL CORD

Ventral gray

Ventral funiculus

LOWER MEDULLA

Inferior olive

Lower medullary NEP

Prepositus nucleus

*Migrating raphe neurons infiltrate midline glial fibers*

Inferior olive fibrous capsule

Medial lemniscus?

Posterior extramural migratory stream
(external cuneate and lateral reticular neurons)

*Posterior intramural migratory stream
(inferior olive neurons)*

Raphe nuclear complex

Upper medullary NEP

RHOMBENCEPHALIC SUPERVENTRICLE
(FUTURE FOURTH VENTRICLE)

Pontomedullary trench

Medial longitudinal fasciculus

*Midline raphe glial system
(structural support for brainstem flexures)*

UPPER MEDULLA

Medullary reticular formation

Subpial GEP

Pontine flexure

Mesencephalic (tegmental) NEP?

MIDBRAIN TEGMENTUM?

P o n t i n e    N E P

C e n t r a l   g r a y

M e d i a l   l o n g i t u d i n a l   f a s c i c u l u s

PONS

Pontine reticular formation

PONS

Abducens nucleus (VI)?

*Premigratory facial motor nucleus (VII) neurons?*

*Anterior extramural migratory stream
(pontine gray and reticular tegmental neurons)*

**PLATE 145A**

**GW8 Sagittal**
**CR 33 mm, C145**
**Level 3: Slide 20, Section 2**

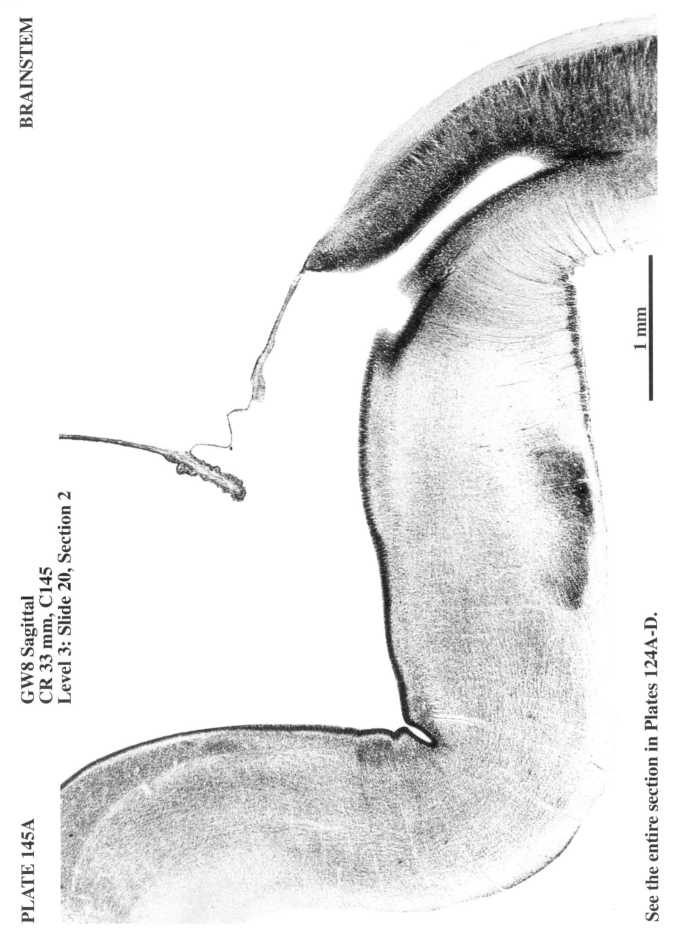

See the entire section in Plates 124A-D.

1 mm

**PLATE 145B**

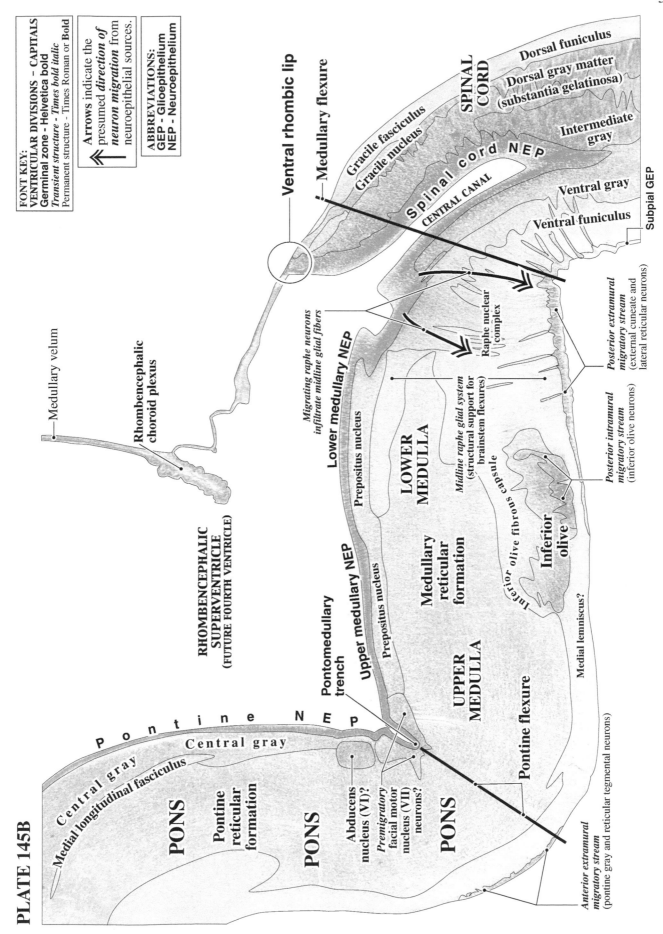

Medullary velum

Rhombencephalic
choroid plexus

**RHOMBENCEPHALIC
SUPERVENTRICLE**
(FUTURE FOURTH VENTRICLE)

Ventral rhombic lip

Medullary flexure

Gracile fasciculus
Gracile nucleus

**SPINAL
CORD**

Dorsal funiculus
Dorsal gray matter
(substantia gelatinosa)

Spinal cord NEP

Intermediate
gray

CENTRAL CANAL

Ventral gray

Ventral funiculus

Subpial GEP

*Migrating raphe neurons
infiltrate midline glial fibers*

*Lower medullary NEP*

Prepositus nucleus

**LOWER
MEDULLA**

Raphe nuclear
complex

*Posterior extramural
migratory stream*
(external cuneate and
lateral reticular neurons)

*Midline raphe glial system*
(structural support for
brainstem flexures)

*Posterior intramural
migratory stream*
(inferior olive neurons)

*Inferior olive fibrous capsule*

**Inferior
olive**

*Upper medullary NEP*

Prepositus nucleus

**Medullary
reticular
formation**

**UPPER
MEDULLA**

Medial lemniscus?

Pontomedullary
trench

**P o n t i n e  N E P**

Central gray

*Central gray*

*Medial longitudinal fasciculus*

**PONS**

Pontine
reticular
formation

**PONS**

Abducens
nucleus (VI)?

Premigratory
facial motor
nucleus (VII) neurons?

**PONS**

Pontine flexure

*Anterior extramural
migratory stream*
(pontine gray and reticular tegmental neurons)

## LATERAL PONS, MEDULLA, AND SENSORY GANGLIA

**PLATE 146A**

**GW8 Sagittal**
**CR 33 mm, C145**
**Similar to Levels 8, 9, and 10:**
**Slide 27, Section 3**

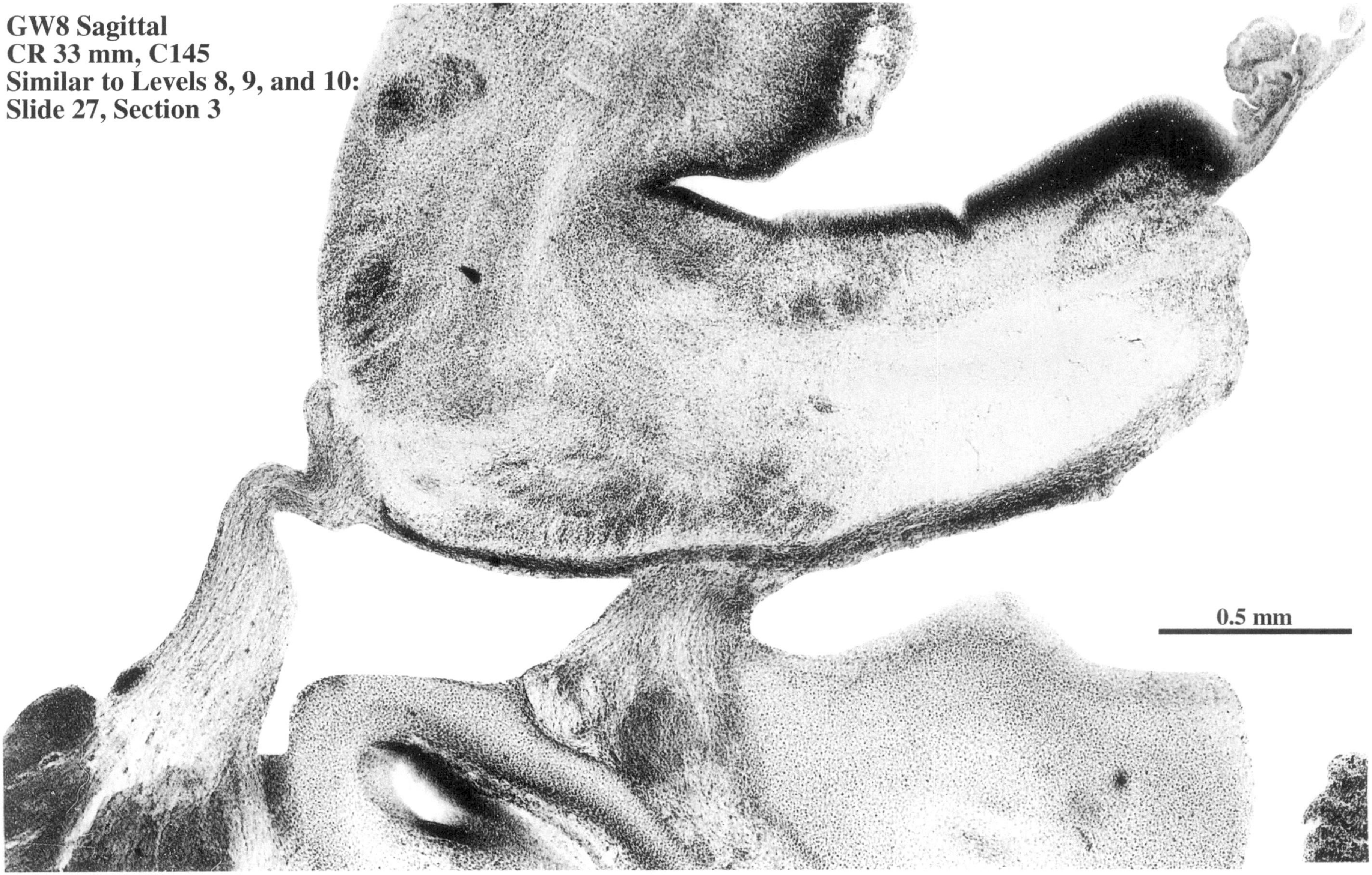

0.5 mm

See similar areas from the left side of the brain in Plates 129A-D to 132A-D.

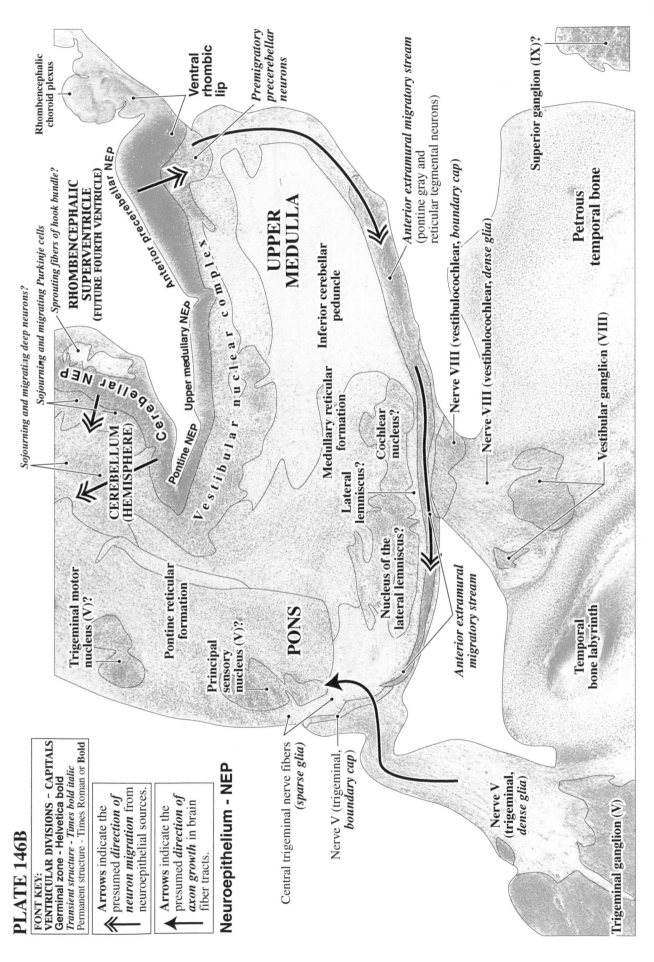

375

PLATE 146B

FONT KEY:
VENTRICULAR DIVISIONS – CAPITALS
Germinal zone - **Helvetica bold**
Transient structure - *Times bold italic*
Permanent structure - Times Roman or **Bold**

⇐ Arrows indicate the presumed *direction of neuron migration* from neuroepithelial sources.

← Arrows indicate the presumed *direction of axon growth* in brain fiber tracts.

**Neuroepithelium - NEP**

*Sojourning and migrating deep neurons?*

*Sojourning and migrating Purkinje cells*

*Sprouting fibers of hook bundle?*

Rhombencephalic choroid plexus

**Ventral rhombic lip**

*Premigratory precerebellar neurons*

*Anterior extramural migratory stream*
*(pontine gray and reticular tegmental neurons)*

Superior ganglion (IX)?

**RHOMBENCEPHALIC SUPERVENTRICLE**
(FUTURE FOURTH VENTRICLE)

*Anterior precerebellar NEP*

**Cerebellar NEP**

**CEREBELLUM (HEMISPHERE)**

*Upper medullary NEP*

*Pontine NEP*

V e s t i b u l a r   n u c l e a r   c o m p l e x

**UPPER MEDULLA**

**Inferior cerebellar peduncle**

Petrous temporal bone

Nerve VIII (vestibulocochlear, *boundary cap*)

Nerve VIII (vestibulocochlear, *dense glia*)

Vestibular ganglion (VIII)

**Trigeminal motor nucleus (V)?**

**Pontine reticular formation**

**Principal sensory nucleus (V)?**

**Medullary reticular formation**

Cochlear nucleus?

*Lateral lemniscus?*

**PONS**

Nucleus of the lateral lemniscus?

*Anterior extramural migratory stream*

Temporal bone labyrinth

*Central trigeminal nerve fibers (sparse glia)*

Nerve V (trigeminal, *boundary cap*)

**Nerve V (trigeminal, *dense glia*)**

Trigeminal ganglion (V)

376

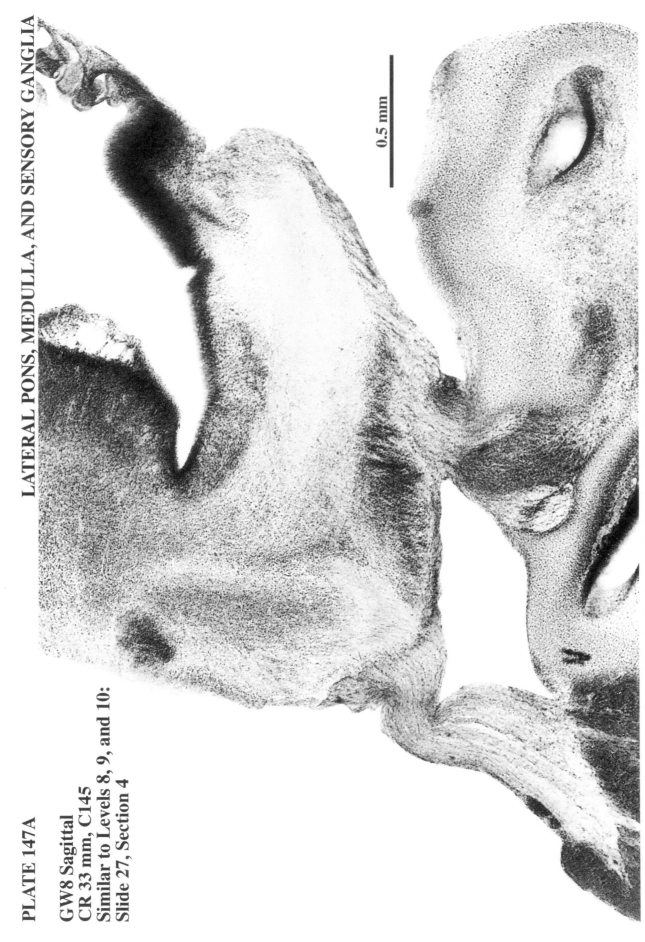

0.5 mm

PLATE 147A

GW8 Sagittal
CR 33 mm, C145
Similar to Levels 8, 9, and 10:
Slide 27, Section 4

See similar areas from the left side of the brain in Plates 129A-D to 132A-D.

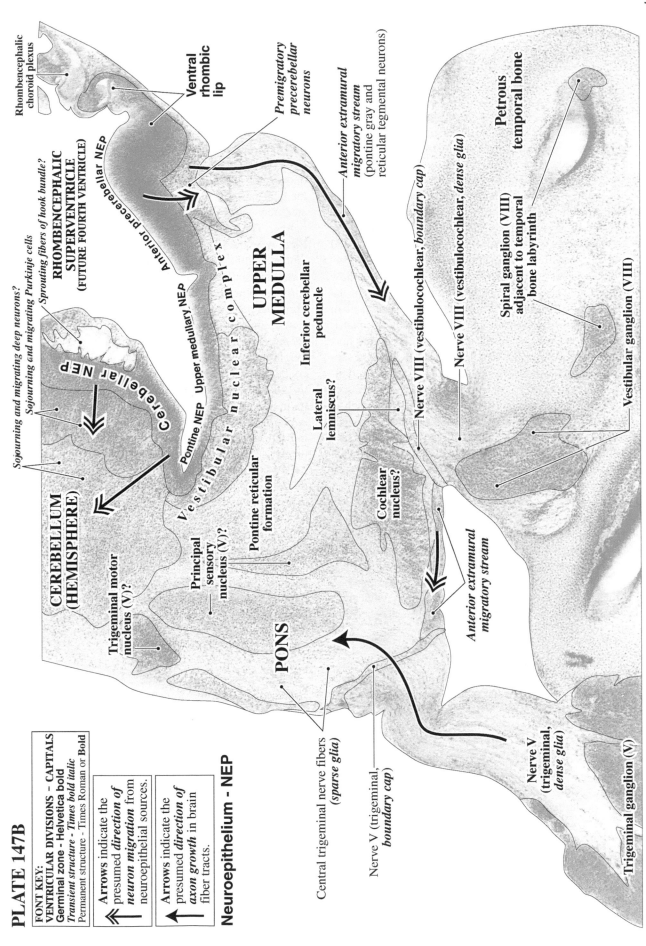

## PLATE 147B

Rhombencephalic choroid plexus

**Ventral rhombic lip**

*Premigratory precerebellar neurons*

*Anterior extramural migratory stream (pontine gray and reticular tegmental neurons)*

**Petrous temporal bone**

*Sojourning and migrating deep neurons?*
*Sojourning and migrating Purkinje cells*
Sprouting fibers of hook bundle?

Anterior precerebellar NEP

**RHOMBENCEPHALIC SUPERVENTRICLE (FUTURE FOURTH VENTRICLE)**

Pontine NEP  Upper medullary NEP

Cerebellar NEP

V e s t i b u l a r  n u c l e a r  c o m p l e x

**UPPER MEDULLA**

Inferior cerebellar peduncle

Nerve VIII (vestibulocochlear, *dense glia*)

Nerve VIII (vestibulocochlear, *boundary cap*)

Spiral ganglion (VIII) adjacent to temporal bone labyrinth

**Vestibular ganglion (VIII)**

**CEREBELLUM (HEMISPHERE)**

Trigeminal motor nucleus (V)?

Principal sensory nucleus (V)?

Pontine reticular formation

Lateral lemniscus?

Cochlear nucleus?

*Anterior extramural migratory stream*

**PONS**

Central trigeminal nerve fibers (*sparse glia*)

Nerve V (trigeminal, *boundary cap*)

Nerve V (trigeminal, *dense glia*)

**Trigeminal ganglion (V)**

# PART VIII: GW8 CORONAL

This is specimen number 9226 in the Carnegie Collection, designated here as C9226. A normal male fetus with a crown-rump length (CR) of 31 mm was collected in 1954, and is estimated to be in gestational week (GW) 8. The entire fetus was embedded in paraffin mixed with 8% celloidin, cut transversely in 10 μm thick sections, and stained with azan. The histology of this specimen is remarkable, and the sections are nearly perfectly bilateral. Since there is no photograph of this brain before it was embedded and cut, a specimen from Hochstetter (1919) that is only partially comparable to C9226 has been modified to show the approximate section plane and external features of the brain at GW8 (**Figure 7**). Like most of the specimens in this Volume, the sections are not cut exactly in one plane; C9226's cortex is cut midway between coronal and horizontal planes. Since the cerebral cortex is in every section and the brainstem is cut in a more horizontal orientation, the brain more closely resembles a coronally sectioned brain. Unfortunately, the Hochstetter specimen is less mature (CR27 mm) and we could not find a drawing of a brain specimen that would fit C9226. The C9226 sections through the cortex and brainstem are not in the same plane when transferred to Hochstetter's CR27 mm specimen. Instead, brainstem planes of section appear to fan upward and downward from sections in the cortex. We interpret this to indicate that the brain flexures are more loosely folded in the Hochstetter specimen than in C9226. If one "squeezes" the brainstem to make the folds tighter, the cortex and brainstem planes would line up. Photographs of 23 sections (**Levels 1-23**) are illustrated at low magnification in **Plates 148-167**. High-magnification views of different areas of the cerebral cortex are shown in **Plates 168-169**.

C9226 is similar to the other GW8 specimens and shows brain maturation in a different perspective. Each of the brain's major subdivisions has a large *superventricle* in the cores, especially the telencephalon and the rhombencephalon. Midline sagittal sections have large diencephalic and mesencephalic superventricles because the cuts are parallel to their dorsoventral and anteroposterior axes (*see* C145 in **Part VII**). C9226's coronal sections show the slit-like shapes of the diencephalic and mesencephalic superventricles in the midline.

The parenchyma, the area between the superficial border of the *neuroepithelium (NEP) / subventricular zone (SVZ)* and the pial membrane, is the region where neurons migrate, settle, and differentiate. The thicknesses of the neuroepithelium and the parenchyma are clues to the level of maturation of a developing brain structure.

The parenchyma is thick and bordered by a thin NEP in the medulla, pons, and midbrain tegmentum, indicating that most neurons have been generated in these structures. Furthermore, the lack of dense accumulations of cells just outside the NEP in the midbrain tegmentum, pons, and medulla indicate that very few neurons are being generated, few are migrating, and most are settled and differentiating. There are two exceptions in the medulla and pons. First, near the pontomedullary trench, presumptive facial motor neurons are migrating toward their ventral pontine/medullary settling sites. Second, the *precerebellar neuroepithelium* in the medulla is thicker and generating pontine gray (and possibly other neurons); many precerebellar neurons are migrating in the *anterior and posterior extramural migratory streams*. The cerebellar NEP is thicker than that in the pons and medulla. The cerebellar parenchyma contains a very dense Purkinje cell sojourn zone outside the NEP and presumptive earlier-generated deep neurons lie in a superficial position. Like C145, the *external germinal layer (egl)* is barely visible emanating from the germinal trigone in the dorsal rhombic lip. The mesencephalic tectal NEP is thicker than the tegmental NEP and its very thin parenchyma contains dense sojourning and migrating tectal neurons adjacent to the NEP. The tectum is one of the most immature brain structures.

The diencephalic NEP is thicker indicating that many neurons are still being generated even though there is also a thick parenchyma, especially in the thalamus. That is because the thalamus is very large in the mature human brain. There are dense accumulations of young neurons in sojourn zones outside the hypothalamic and thalamic NEPs, indicating that cell migration is more active than final settling and differentiation.

Within the telencephalon, the cerebral cortex has a thick NEP and a very thin parenchyma, indicating that it is the most immature brain structure. The cerebral cortical NEP is the sole germinal matrix. The *stratified transitional field (STF)* contains *STF1* and *STF5* only in lateral areas. The pronounced anterolateral (thicker) to dorsomedial (thinner) maturation gradient is evident in both the *cortical plate* and the *STF* layers. In contrast, both the basal telencephalic NEP/SVZ and parenchyma are thick. That is because the basal telencephalon contains many early-generated neuronal populations (for example, globus pallidus and substantia innominata) and massive late-generated populations (striatal neurons in the caudate and putamen). Most of the neurons settling in the basal telencephalon at GW8 are those of the early-generated populations.

379

# GW8 "CORONAL" SECTION PLANES

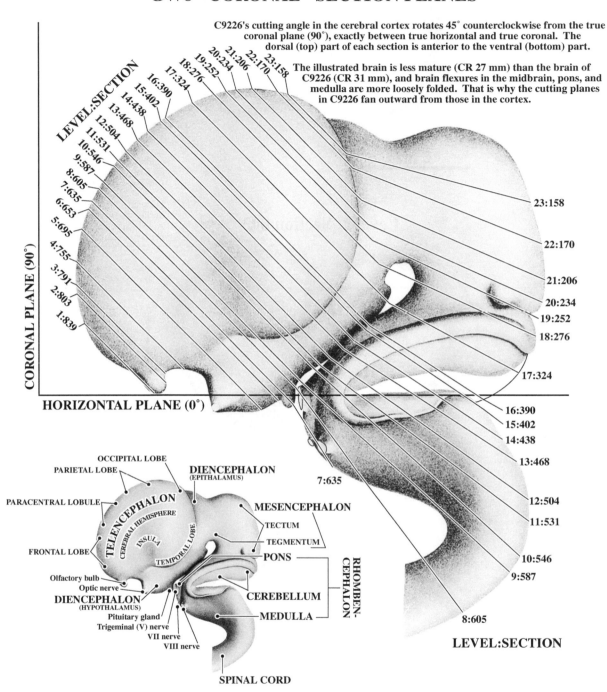

C9226's cutting angle in the cerebral cortex rotates 45° counterclockwise from the true coronal plane (90°), exactly between true horizontal and true coronal. The dorsal (top) part of each section is anterior to the ventral (bottom) part.

The illustrated brain is less mature (CR 27 mm) than the brain of C9226 (CR 31 mm), and brain flexures in the midbrain, pons, and medulla are more loosely folded. That is why the cutting planes in C9226 fan outward from those in the cortex.

LEVEL:SECTION

17:324
16:390
15:402
14:438
13:468
12:504
11:531
10:546
9:587
8:605
7:635
6:653
5:695
4:755
3:791
2:803
1:839

18:276
19:252
20:234
21:206
22:170
23:158

CORONAL PLANE (90°)

HORIZONTAL PLANE (0°)

23:158
22:170
21:206
20:234
19:252
18:276
17:324
16:390
15:402
14:438
13:468
12:504
11:531
10:546
9:587
8:605
7:635

LEVEL:SECTION

OCCIPITAL LOBE
PARIETAL LOBE
PARACENTRAL LOBULE
FRONTAL LOBE
Olfactory bulb
Optic nerve
DIENCEPHALON (HYPOTHALAMUS)
Pituitary gland
Trigeminal (V) nerve
VII nerve
VIII nerve
SPINAL CORD

TELENCEPHALON
CEREBRAL HEMISPHERE
INSULA
TEMPORAL LOBE

DIENCEPHALON (EPITHALAMUS)
MESENCEPHALON
TECTUM
TEGMENTUM
PONS
CEREBELLUM
MEDULLA
RHOMBEN-CEPHALON

**Figure 7.** The lateral view of the brain and upper cervical spinal cord from a specimen with a crown-rump length of 27 mm (modified from Figure 37, Table VII, Hochstetter, 1919) serves to show the approximate locations and cutting angles of the illustrated sections of C9226 in the following pages. The small inset identifies the major structural features. The line in the cerebellum and dorsal edges of the pons and medulla is the cut edge of the medullary velum.

**PLATE 148A**

**GW8 Coronal
CR 31 mm
C9226**

## Level 1: Section 820

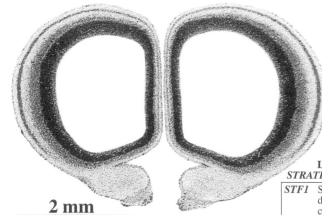

2 mm

**LAYERS OF THE CORTICAL
*STRATIFIED TRANSITIONAL FIELD (STF)***

*STF1*  Superficial fibrous layer with an early
developmental stage *(t1)* when many
cells are migrating through it, followed
by a late stage *(t2)* with sparse cells.
Endures as the subcortical white matter.

*STF5*  Deep cellular layer that is prominent
during the first trimester, the first sojourn
zone to appear outside the germinal
matrix.

## Level 2: Section 803

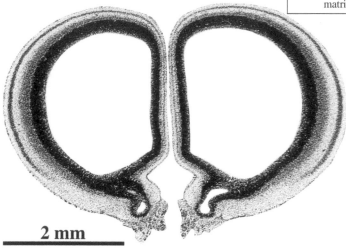

2 mm

## Level 3: Section 791

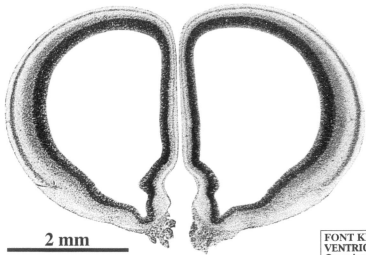

2 mm

**FONT KEY:
VENTRICULAR DIVISIONS – CAPITALS
Germinal zone - Helvetica bold
*Transient structure - Times bold italic*
Permanent structure - Times Roman or Bold**

381

**PLATE 148B**

## Level 1: Section 820

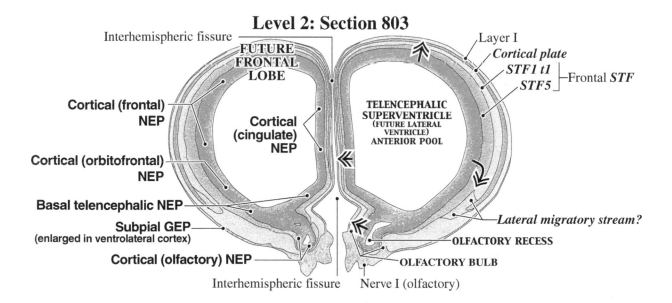

## Level 2: Section 803

## Level 3: Section 791

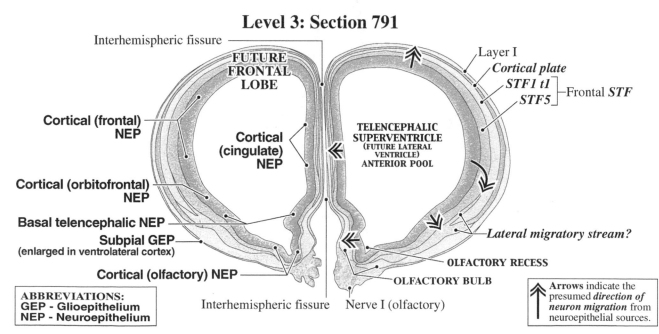

Arrows indicate the presumed *direction of neuron migration* from neuroepithelial sources.

**PLATE 149A**

**GW8 Coronal**
**CR 31 mm**
**C9226**

## Level 4: Section 755

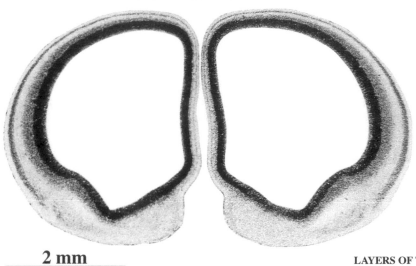

_____ 2 mm _____

## Level 5: Section 695

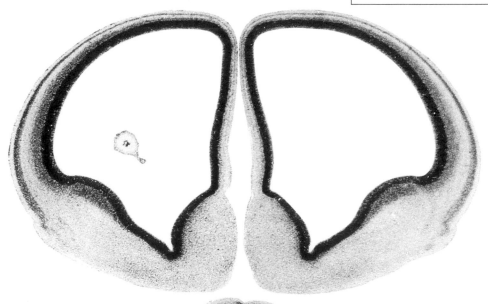

_____ 2 mm _____

### Level 4: Section 755

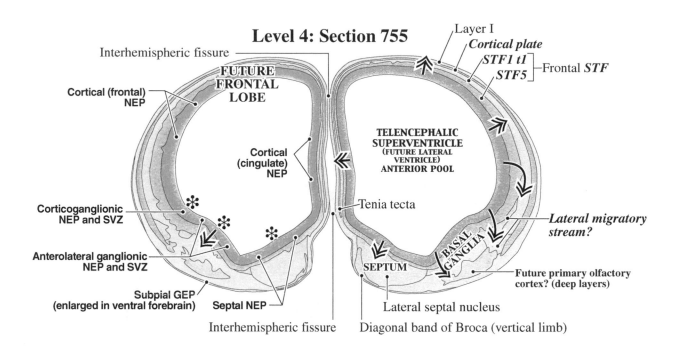

Interhemispheric fissure

Layer I
*Cortical plate*
*STF1 t1*
*STF5*
Frontal *STF*

**FUTURE FRONTAL LOBE**

Cortical (frontal) NEP

**TELENCEPHALIC SUPERVENTRICLE (FUTURE LATERAL VENTRICLE) ANTERIOR POOL**

Cortical (cingulate) NEP

*Lateral migratory stream?*

Corticoganglionic NEP and SVZ

Tenia tecta

Anterolateral ganglionic NEP and SVZ

**BASAL GANGLIA**

**SEPTUM**

Future primary olfactory cortex? (deep layers)

Subpial GEP (enlarged in ventral forebrain)

Septal NEP

Lateral septal nucleus

Interhemispheric fissure

Diagonal band of Broca (vertical limb)

### Level 5: Section 695

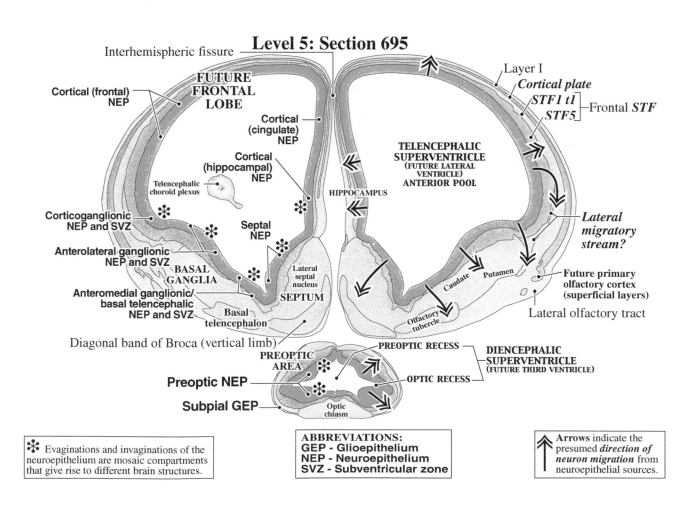

Interhemispheric fissure

**FUTURE FRONTAL LOBE**

Layer I
*Cortical plate*
*STF1 t1*
*STF5*
Frontal *STF*

Cortical (frontal) NEP

Cortical (cingulate) NEP

Cortical (hippocampal) NEP

**TELENCEPHALIC SUPERVENTRICLE (FUTURE LATERAL VENTRICLE) ANTERIOR POOL**

Telencephalic choroid plexus

HIPPOCAMPUS

Corticoganglionic NEP and SVZ

*Lateral migratory stream?*

Septal NEP

Anterolateral ganglionic NEP and SVZ

**BASAL GANGLIA**

Lateral septal nucleus

Anteromedial ganglionic/ basal telencephalic NEP and SVZ

**SEPTUM**

Caudate   Putamen

Future primary olfactory cortex (superficial layers)

Basal telencephalon

Olfactory tubercle

Lateral olfactory tract

Diagonal band of Broca (vertical limb)

**PREOPTIC RECESS**

**PREOPTIC AREA**

**DIENCEPHALIC SUPERVENTRICLE (FUTURE THIRD VENTRICLE)**

**Preoptic NEP**

**OPTIC RECESS**

**Subpial GEP**

Optic chiasm

✻ Evaginations and invaginations of the neuroepithelium are mosaic compartments that give rise to different brain structures.

**ABBREVIATIONS:**
GEP - Glioepithelium
NEP - Neuroepithelium
SVZ - Subventricular zone

**Arrows** indicate the presumed *direction of neuron migration* from neuroepithelial sources.

**PLATE 150A**

**GW8 Coronal**
**CR 31 mm**
**C9226**
**Level 6: Section 653**

**LAYERS OF THE CORTICAL**
*STRATIFIED TRANSITIONAL FIELD (STF)*

*STF1*  Superficial fibrous layer with an early
developmental stage *(t1)* when many
cells are migrating through it, followed
by a late stage *(t2)* with sparse cells.
Endures as the subcortical white matter.

*STF5*  Deep cellular layer that is prominent
during the first trimester, the first sojourn
zone to appear outside the germinal
matrix.

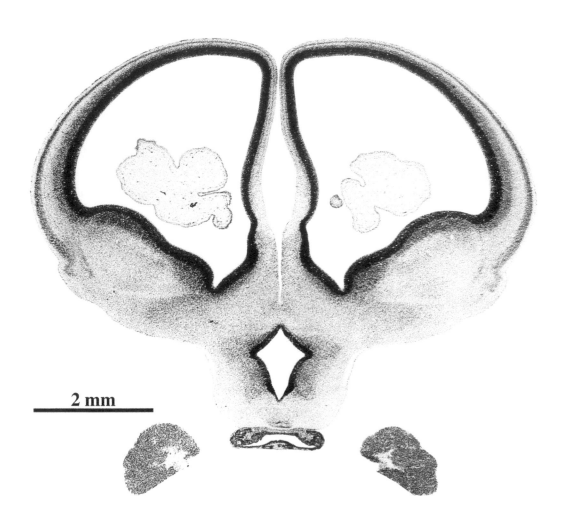

2 mm

**FONT KEY:**
**VENTRICULAR DIVISIONS − CAPITALS**
**Germinal zone - Helvetica bold**
*Transient structure - Times bold italic*
Permanent structure - Times Roman or **Bold**

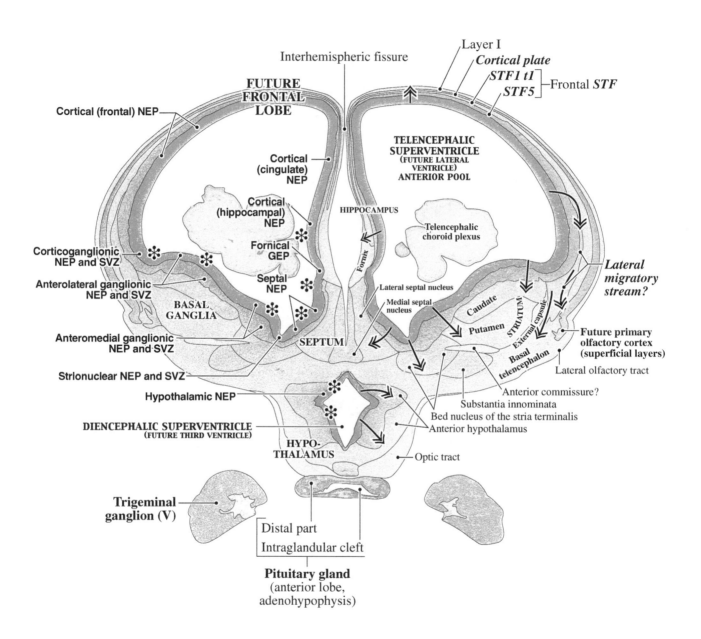

Interhemispheric fissure

Layer I
*Cortical plate*
*STF1 t1*
*STF5*
Frontal *STF*

**FUTURE FRONTAL LOBE**

Cortical (frontal) NEP

**TELENCEPHALIC SUPERVENTRICLE (FUTURE LATERAL VENTRICLE) ANTERIOR POOL**

Cortical (cingulate) NEP

HIPPOCAMPUS

Cortical (hippocampal) NEP

Telencephalic choroid plexus

Fornix

Fornical GEP

Septal NEP

Corticoganglionic NEP and SVZ

*Lateral migratory stream?*

Lateral septal nucleus

Anterolateral ganglionic NEP and SVZ

Medial septal nucleus

Caudate

**BASAL GANGLIA**

Putamen

STRIATUM

External capsule

**SEPTUM**

Anteromedial ganglionic NEP and SVZ

Basal telencephalon

Future primary olfactory cortex (superficial layers)

Strionuclear NEP and SVZ

Lateral olfactory tract

Hypothalamic NEP

Anterior commissure?

Substantia innominata

Bed nucleus of the stria terminalis

**DIENCEPHALIC SUPERVENTRICLE (FUTURE THIRD VENTRICLE)**

Anterior hypothalamus

**HYPO-THALAMUS**

Optic tract

**Trigeminal ganglion (V)**

Distal part

Intraglandular cleft

**Pituitary gland** (anterior lobe, adenohypophysis)

✻ Evaginations and invaginations of the neuroepithelium are mosaic compartments that give rise to different brain structures.

**ABBREVIATIONS:**
**GEP** - Glioepithelium
**NEP** - Neuroepithelium
**SVZ** - Subventricular zone

**Arrows** indicate the presumed *direction of neuron migration* from neuroepithelial sources.

# PLATE 151A

**GW8 Coronal
CR 31 mm
C9226
Level 7: Section 635**

**LAYERS OF THE CORTICAL
*STRATIFIED TRANSITIONAL FIELD (STF)*

*STF1*  Superficial fibrous layer with an early developmental stage *(t1)* when many cells are migrating through it, followed by a late stage *(t2)* with sparse cells. Endures as the subcortical white matter.

*STF5*  Deep cellular layer that is prominent during the first trimester, the first sojourn zone to appear outside the germinal matrix.

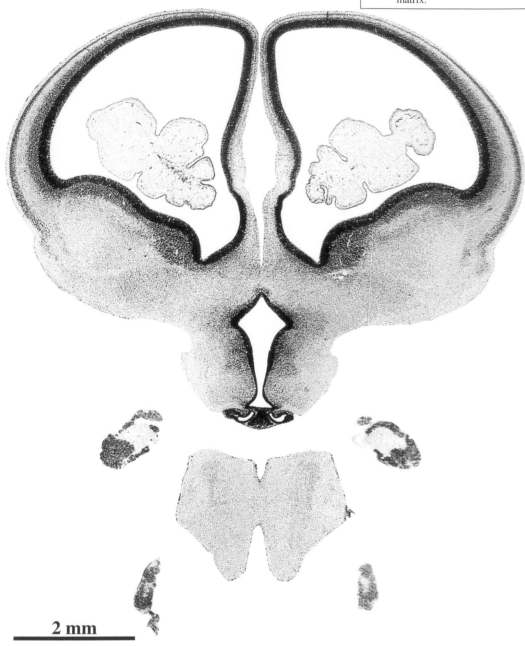

2 mm

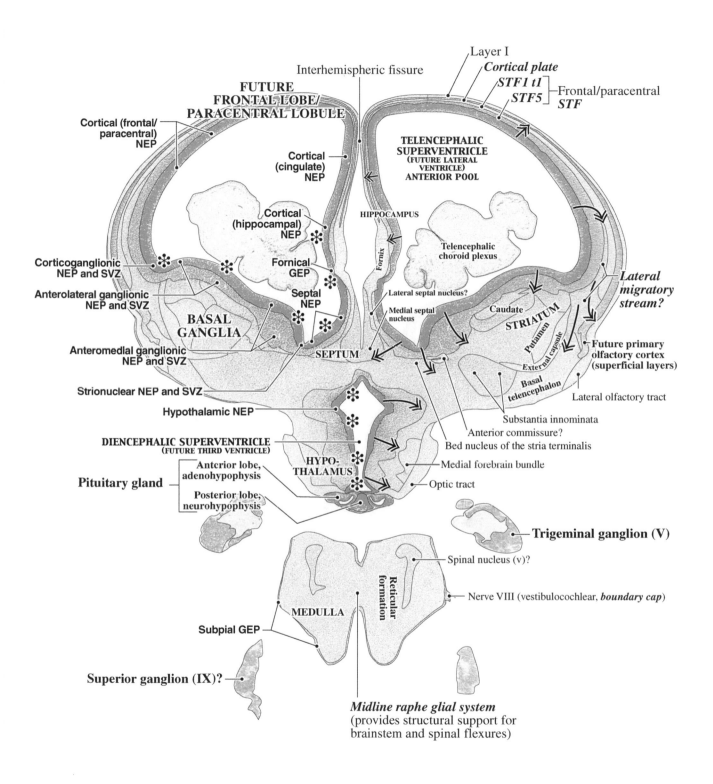

Layer I
*Cortical plate*
*STF1 t1*
*STF5*
Frontal/paracentral
*STF*

Interhemispheric fissure

**FUTURE FRONTAL LOBE/ PARACENTRAL LOBULE**

Cortical (frontal/ paracentral) NEP

Cortical (cingulate) NEP

**TELENCEPHALIC SUPERVENTRICLE** *(FUTURE LATERAL VENTRICLE)* **ANTERIOR POOL**

HIPPOCAMPUS

Cortical (hippocampal) NEP

Fornical GEP

Septal NEP

Telencephalic choroid plexus

Fornix

Corticoganglionic NEP and SVZ

Anterolateral ganglionic NEP and SVZ

**BASAL GANGLIA**

Lateral septal nucleus?

Medial septal nucleus

Caudate

**STRIATUM**

Putamen

*Lateral migratory stream?*

Anteromedial ganglionic NEP and SVZ

SEPTUM

External capsule

Future primary olfactory cortex (superficial layers)

Strionuclear NEP and SVZ

Hypothalamic NEP

Basal telencephalon

Lateral olfactory tract

**DIENCEPHALIC SUPERVENTRICLE** *(FUTURE THIRD VENTRICLE)*

**HYPO-THALAMUS**

Substantia innominata

Anterior commissure?

Bed nucleus of the stria terminalis

**Pituitary gland**

Anterior lobe, adenohypophysis

Posterior lobe; neurohypophysis

Medial forebrain bundle

Optic tract

**Trigeminal ganglion (V)**

Spinal nucleus (v)?

Reticular formation

Nerve VIII (vestibulocochlear, *boundary cap*)

**MEDULLA**

Subpial GEP

**Superior ganglion (IX)?**

*Midline raphe glial system*
(provides structural support for brainstem and spinal flexures)

✳ Evaginations and invaginations of the neuroepithelium are mosaic compartments that give rise to different brain structures.

**ABBREVIATIONS:**
GEP - Glioepithelium
NEP - Neuroepithelium
SVZ - Subventricular zone

**Arrows** indicate the presumed *direction of neuron migration* from neuroepithelial sources.

**PLATE 152A**

**GW8 Coronal**
**CR 31 mm**
**C9226**
**Level 8: Section 605**

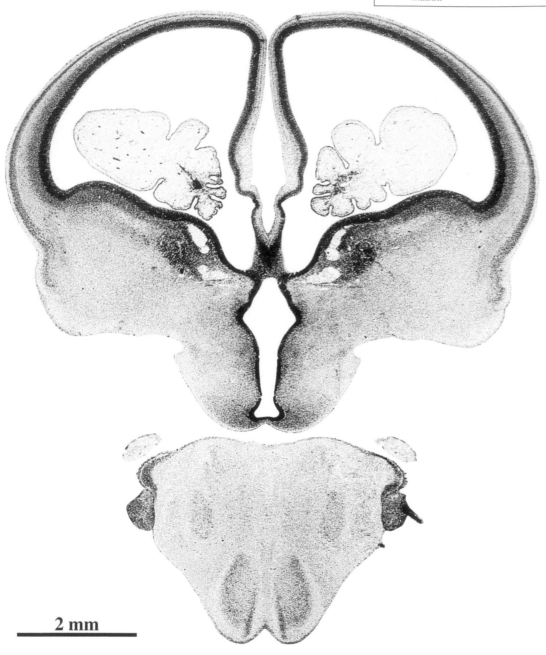

2 mm

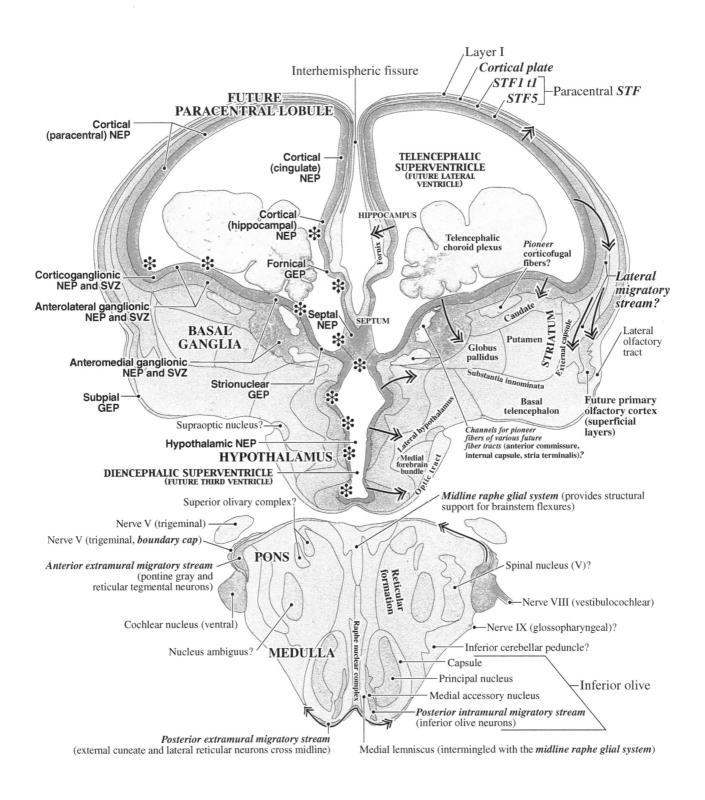

Interhemispheric fissure

Layer I
*Cortical plate*
*STF1 t1*
*STF5* — Paracentral *STF*

**FUTURE PARACENTRAL LOBULE**

Cortical (paracentral) NEP

Cortical (cingulate) NEP

**TELENCEPHALIC SUPERVENTRICLE** (FUTURE LATERAL VENTRICLE)

HIPPOCAMPUS

Cortical (hippocampal) NEP

Telencephalic choroid plexus

*Pioneer corticofugal fibers?*

Corticoganglionic NEP and SVZ

Fornical GEP

Fornix

SEPTUM

*Lateral migratory stream?*

Anterolateral ganglionic NEP and SVZ

Septal NEP

**BASAL GANGLIA**

Caudate

Putamen

**STRIATUM**

External capsule

Lateral olfactory tract

Anteromedial ganglionic NEP and SVZ

Globus pallidus

Strionuclear GEP

Substantia innominata

Subpial GEP

Basal telencephalon

Future primary olfactory cortex (superficial layers)

Supraoptic nucleus?

Hypothalamic NEP

Lateral hypothalamus

*Channels for pioneer fibers of various future fiber tracts (anterior commissure, internal capsule, stria terminalis)?*

**HYPOTHALAMUS**

**DIENCEPHALIC SUPERVENTRICLE** (FUTURE THIRD VENTRICLE)

Medial forebrain bundle

Optic tract

*Midline raphe glial system* (provides structural support for brainstem flexures)

Superior olivary complex?

Nerve V (trigeminal)

Nerve V (trigeminal, *boundary cap*)

*Anterior extramural migratory stream* (pontine gray and reticular tegmental neurons)

**PONS**

Reticular formation

Spinal nucleus (V)?

Nerve VIII (vestibulocochlear)

Cochlear nucleus (ventral)

Raphe nuclear complex

Nerve IX (glossopharyngeal)?

Nucleus ambiguus?

**MEDULLA**

Inferior cerebellar peduncle?

Capsule

Principal nucleus

Medial accessory nucleus

Inferior olive

*Posterior intramural migratory stream* (inferior olive neurons)

*Posterior extramural migratory stream* (external cuneate and lateral reticular neurons cross midline)

Medial lemniscus (intermingled with the *midline raphe glial system*)

✴ Evaginations and invaginations of the neuroepithelium are mosaic compartments that give rise to different brain structures.

**ABBREVIATIONS:**
**GEP** - Glioepithelium
**NEP** - Neuroepithelium
**SVZ** - Subventricular zone

**Arrows** indicate the presumed *direction of neuron migration* from neuroepithelial sources.

**PLATE 153A**

**GW8 Coronal**
**CR 31 mm**
**C9226**
**Level 9:**
**Section**
**587**

**See high-magnification views of the cerebral cortex from nearby sections in Plates 168A and B to 169A and B.**

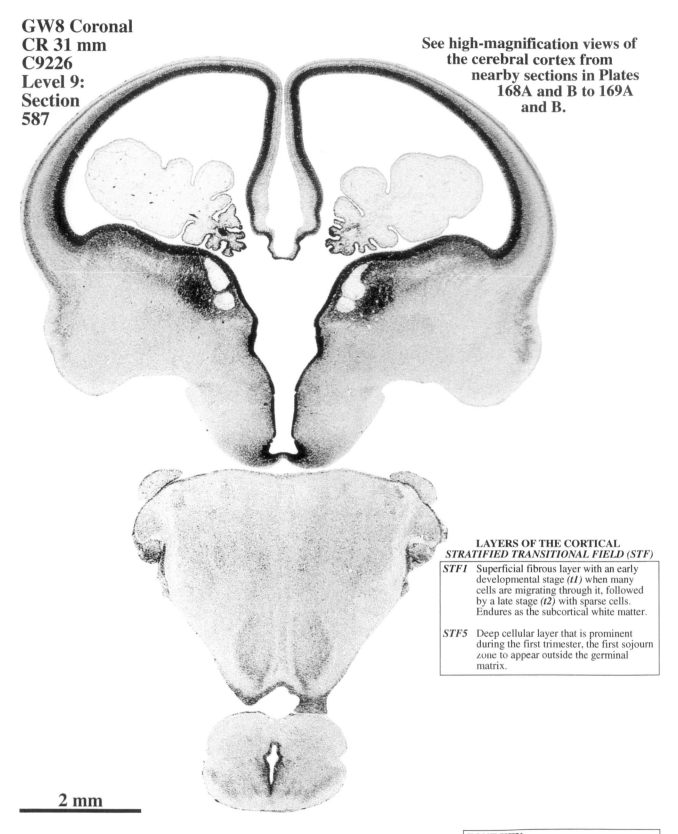

**LAYERS OF THE CORTICAL**
*STRATIFIED TRANSITIONAL FIELD (STF)*

*STF1*   Superficial fibrous layer with an early developmental stage *(t1)* when many cells are migrating through it, followed by a late stage *(t2)* with sparse cells. Endures as the subcortical white matter.

*STF5*   Deep cellular layer that is prominent during the first trimester, the first sojourn zone to appear outside the germinal matrix.

2 mm

FONT KEY:
VENTRICULAR DIVISIONS – CAPITALS
**Germinal zone - Helvetica bold**
*Transient structure - Times bold italic*
Permanent structure - Times Roman or **Bold**

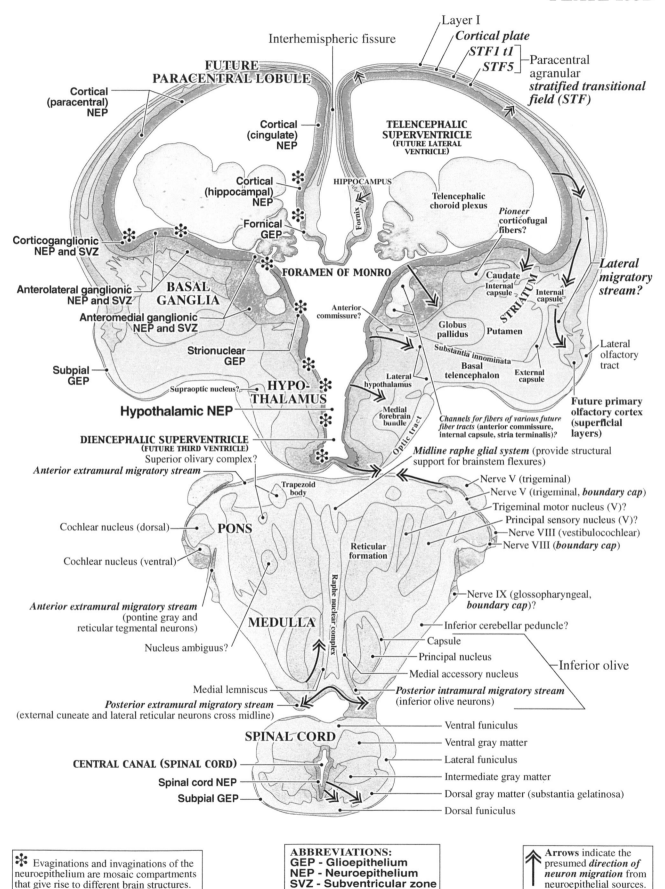

Layer I
Cortical plate
*STF1 t1*
*STF5*

Paracentral
agranular
*stratified transitional
field (STF)*

Interhemispheric fissure

FUTURE
PARACENTRAL LOBULE

Cortical
(paracentral)
NEP

Cortical
(cingulate)
NEP

TELENCEPHALIC
SUPERVENTRICLE
(FUTURE LATERAL
VENTRICLE)

Cortical
(hippocampal)
NEP

HIPPOCAMPUS

Telencephalic
choroid plexus

*Pioneer
corticofugal
fibers?*

*Lateral
migratory
stream?*

Fornical
GEP

Fornix

FORAMEN OF MONRO

Caudate
Internal
capsule

Corticoganglionic
NEP and SVZ

Internal
capsule?

BASAL
GANGLIA

Anterolateral ganglionic
NEP and SVZ

Anteromedial ganglionic
NEP and SVZ

Anterior
commissure?

Globus
pallidus

Putamen

STRIATUM

Lateral
olfactory
tract

Strionuclear
GEP

Substantia innominata

Basal
telencephalon

External
capsule

Subpial
GEP

Supraoptic nucleus?

HYPO-
THALAMUS

Lateral
hypothalamus

Future primary
olfactory cortex
(superficial
layers)

Hypothalamic NEP

Medial
forebrain
bundle

*Channels for fibers of various future
fiber tracts (anterior commissure,
internal capsule, stria terminalis)?*

DIENCEPHALIC SUPERVENTRICLE
(FUTURE THIRD VENTRICLE)

Superior olivary complex?

*Anterior extramural migratory stream*

Optic tract

*Midline raphe glial system* (provide structural
support for brainstem flexures)

Nerve V (trigeminal)

Trapezoid
body

Nerve V (trigeminal, *boundary cap*)

Trigeminal motor nucleus (V)?

Cochlear nucleus (dorsal)

PONS

Principal sensory nucleus (V)?

Nerve VIII (vestibulocochlear)

Nerve VIII (*boundary cap*)

Cochlear nucleus (ventral)

Reticular
formation

*Anterior extramural migratory stream*
(pontine gray and
reticular tegmental neurons)

Nerve IX (glossopharyngeal,
*boundary cap*)?

Raphe nuclear complex

MEDULLA

Inferior cerebellar peduncle?

Nucleus ambiguus?

Capsule

Principal nucleus

Medial accessory nucleus

Inferior olive

Medial lemniscus

*Posterior intramural migratory stream*
(inferior olive neurons)

*Posterior extramural migratory stream*
(external cuneate and lateral reticular neurons cross midline)

Ventral funiculus

Ventral gray matter

SPINAL CORD

Lateral funiculus

Intermediate gray matter

CENTRAL CANAL (SPINAL CORD)

Spinal cord NEP

Dorsal gray matter (substantia gelatinosa)

Subpial GEP

Dorsal funiculus

✳ Evaginations and invaginations of the
neuroepithelium are mosaic compartments
that give rise to different brain structures.

ABBREVIATIONS:
GEP - Glioepithelium
NEP - Neuroepithelium
SVZ - Subventricular zone

**Arrows** indicate the
presumed *direction of
neuron migration* from
neuroepithelial sources.

**PLATE 154A**

**GW8 Coronal
CR 31 mm
C9226
Level 10:
Section
546**

See high-magnification views of
the cerebral cortex from
nearby sections in Plates
168A and B to 169A
and B.

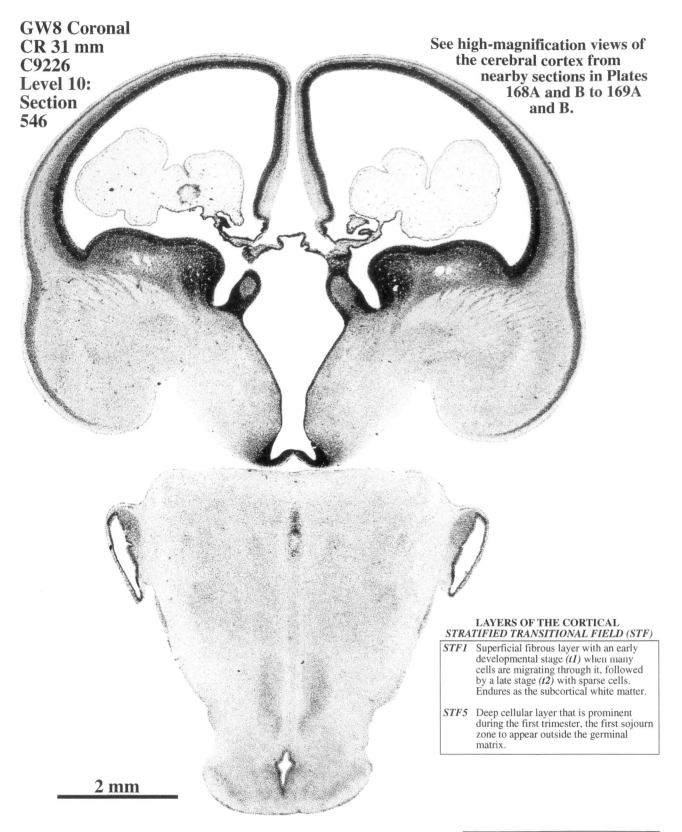

**LAYERS OF THE CORTICAL
*STRATIFIED TRANSITIONAL FIELD (STF)***

*STF1*  Superficial fibrous layer with an early
developmental stage *(t1)* when many
cells are migrating through it, followed
by a late stage *(t2)* with sparse cells.
Endures as the subcortical white matter.

*STF5*  Deep cellular layer that is prominent
during the first trimester, the first sojourn
zone to appear outside the germinal
matrix.

2 mm

**FONT KEY:
VENTRICULAR DIVISIONS − CAPITALS
Germinal zone - Helvetica bold
*Transient structure - Times bold italic*
Permanent structure - Times Roman or Bold**

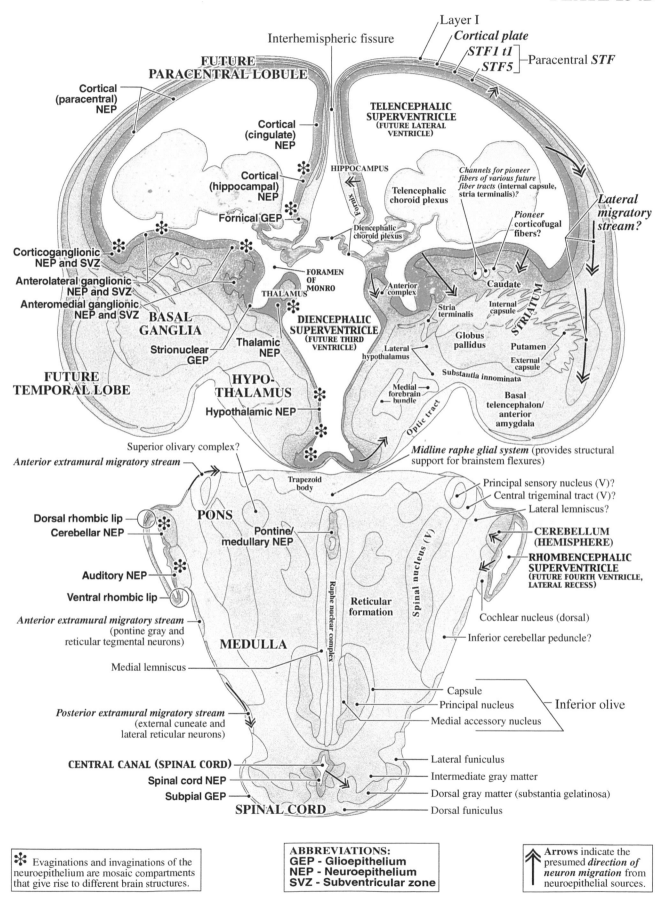

Interhemispheric fissure

Layer I
Cortical plate
STF1 t1
STF5
Paracentral STF

FUTURE
PARACENTRAL LOBULE

Cortical
(paracentral)
NEP

Cortical
(cingulate)
NEP

Cortical
(hippocampal)
NEP

Fornical GEP

Corticoganglionic
NEP and SVZ

Anterolateral ganglionic
NEP and SVZ

Anteromedial ganglionic
NEP and SVZ

BASAL
GANGLIA

Strionuclear
GEP

Thalamic
NEP

FUTURE
TEMPORAL LOBE

HYPO-
THALAMUS

Hypothalamic NEP

Superior olivary complex?

Anterior extramural migratory stream

Dorsal rhombic lip

Cerebellar NEP

Auditory NEP

Ventral rhombic lip

Anterior extramural migratory stream
(pontine gray and
reticular tegmental neurons)

Medial lemniscus

Posterior extramural migratory stream
(external cuneate and
lateral reticular neurons)

CENTRAL CANAL (SPINAL CORD)

Spinal cord NEP

Subpial GEP

SPINAL CORD

HIPPOCAMPUS

Telencephalic
choroid plexus

TELENCEPHALIC
SUPERVENTRICLE
(FUTURE LATERAL
VENTRICLE)

Channels for pioneer
fibers of various future
fiber tracts (internal capsule,
stria terminalis)?

Pioneer
corticofugal
fibers?

Diencephalic
choroid plexus

FORAMEN
OF
MONRO

THALAMUS

DIENCEPHALIC
SUPERVENTRICLE
(FUTURE THIRD
VENTRICLE)

Anterior
complex

Stria
terminalis

Caudate

Internal
capsule

STRIATUM

Lateral
hypothalamus

Globus
pallidus

Putamen

External
capsule

Medial
forebrain
bundle

Substantia innominata

Optic tract

Basal
telencephalon/
anterior
amygdala

Lateral
migratory
stream?

Midline raphe glial system (provides structural
support for brainstem flexures)

PONS

Trapezoid
body

Pontine/
medullary NEP

Principal sensory nucleus (V)?
Central trigeminal tract (V)?
Lateral lemniscus?

CEREBELLUM
(HEMISPHERE)

RHOMBENCEPHALIC
SUPERVENTRICLE
(FUTURE FOURTH VENTRICLE,
LATERAL RECESS)

Raphe nuclear complex

Reticular
formation

Spinal nucleus (V)

Cochlear nucleus (dorsal)

Inferior cerebellar peduncle?

MEDULLA

Capsule

Principal nucleus

Medial accessory nucleus

Inferior olive

Lateral funiculus

Intermediate gray matter

Dorsal gray matter (substantia gelatinosa)

Dorsal funiculus

✳ Evaginations and invaginations of the
neuroepithelium are mosaic compartments
that give rise to different brain structures.

ABBREVIATIONS:
GEP - Glioepithelium
NEP - Neuroepithelium
SVZ - Subventricular zone

Arrows indicate the
presumed direction of
neuron migration from
neuroepithelial sources.

**PLATE 155A**

**GW8 Coronal**
**CR 31 mm**
**C9226**
**Level 11:**
**Section**
**531**

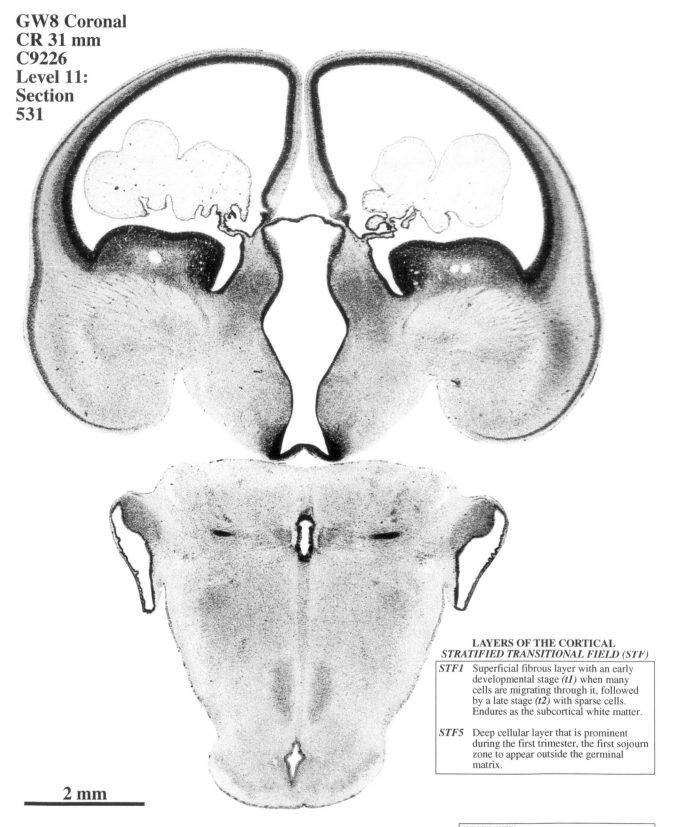

**2 mm**

**LAYERS OF THE CORTICAL**
*STRATIFIED TRANSITIONAL FIELD (STF)*

*STF1*  Superficial fibrous layer with an early
developmental stage *(t1)* when many
cells are migrating through it, followed
by a late stage *(t2)* with sparse cells.
Endures as the subcortical white matter.

*STF5*  Deep cellular layer that is prominent
during the first trimester, the first sojourn
zone to appear outside the germinal
matrix.

FONT KEY:
VENTRICULAR DIVISIONS – CAPITALS
Germinal zone - Helvetica bold
*Transient structure - Times bold italic*
Permanent structure - Times Roman or **Bold**

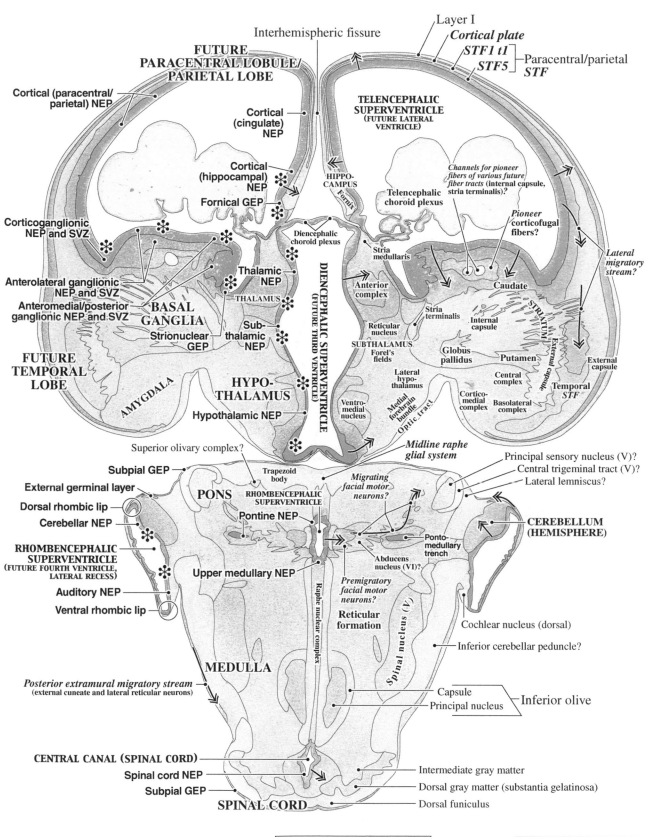

Interhemispheric fissure

Layer I
Cortical plate
*STF1 t1*
*STF5*
Paracentral/parietal
*STF*

FUTURE
PARACENTRAL LOBULE/
PARIETAL LOBE

Cortical (paracentral/
parietal) NEP

Cortical
(cingulate)
NEP

TELENCEPHALIC
SUPERVENTRICLE
(FUTURE LATERAL
VENTRICLE)

*Channels for pioneer
fibers of various future
fiber tracts (internal capsule,
stria terminalis)?*

Cortical
(hippocampal)
NEP

Fornical GEP

HIPPO-
CAMPUS
Fornix

Telencephalic
choroid plexus

*Pioneer
corticofugal
fibers?*

Corticoganglionic
NEP and SVZ

Diencephalic
choroid plexus

Stria
medullaris

*Lateral
migratory
stream?*

Anterolateral ganglionic
NEP and SVZ

Thalamic
NEP

Anterior
complex

Caudate

STRIATUM

Anteromedial/posterior
ganglionic NEP and SVZ

THALAMUS

BASAL
GANGLIA

Stria
terminalis

Internal
capsule

External capsule

Strionuclear
GEP

Sub-
thalamic
NEP

Reticular
nucleus

SUBTHALAMUS
Forel's fields

Globus
pallidus

Putamen

External
capsule

FUTURE
TEMPORAL
LOBE

AMYGDALA

HYPO-
THALAMUS

DIENCEPHALIC SUPERVENTRICLE
(FUTURE THIRD VENTRICLE)

Lateral
hypo-
thalamus

Central
complex

Temporal
STF

Hypothalamic NEP

Ventro-
medial
nucleus

Medial forebrain bundle
Optic tract

Cortico-
medial
complex

Basolateral
complex

Superior olivary complex?

*Midline raphe
glial system*

Principal sensory nucleus (V)?

Subpial GEP

Trapezoid
body

*Migrating
facial motor
neurons?*

Central trigeminal tract (V)?

Lateral lemniscus?

External germinal layer

PONS

RHOMBENCEPHALIC
SUPERVENTRICLE

Dorsal rhombic lip

Pontine NEP

Cerebellar NEP

Ponto-
medullary
trench

CEREBELLUM
(HEMISPHERE)

RHOMBENCEPHALIC
SUPERVENTRICLE
(FUTURE FOURTH VENTRICLE,
LATERAL RECESS)

Abducens
nucleus (VI)?

Auditory NEP

Upper medullary NEP

*Premigratory
facial motor
neurons?*

Spinal nucleus (V)

Ventral rhombic lip

Reticular
formation

Cochlear nucleus (dorsal)

Raphe nuclear complex

Inferior cerebellar peduncle?

MEDULLA

*Posterior extramural migratory stream*
(external cuneate and lateral reticular neurons)

Capsule

Principal nucleus

Inferior olive

CENTRAL CANAL (SPINAL CORD)

Intermediate gray matter

Spinal cord NEP

Dorsal gray matter (substantia gelatinosa)

Subpial GEP

Dorsal funiculus

SPINAL CORD

✳ Evaginations and invaginations of the
neuroepithelium are mosaic compartments
that give rise to different brain structures.

ABBREVIATIONS:
GEP - Glioepithelium
NEP - Neuroepithelium
SVZ - Subventricular zone

**Arrows** indicate the
presumed *direction of
neuron migration* from
neuroepithelial sources.

**PLATE 156A**

**GW8 Coronal
CR 31 mm
C9226
Level 12:
Section 504**

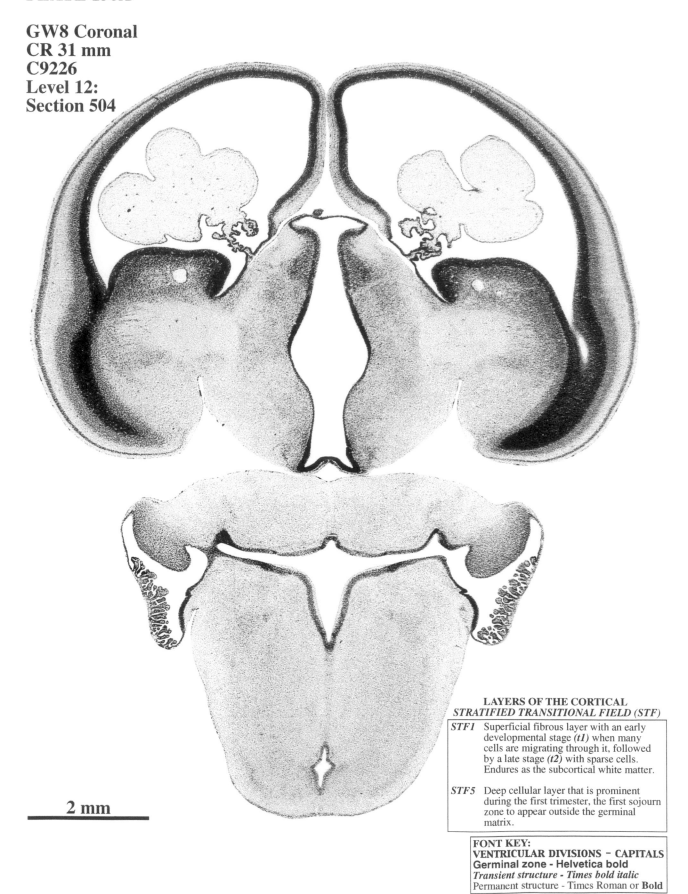

**2 mm**

**LAYERS OF THE CORTICAL**
*STRATIFIED TRANSITIONAL FIELD (STF)*

*STF1*  Superficial fibrous layer with an early developmental stage *(t1)* when many cells are migrating through it, followed by a late stage *(t2)* with sparse cells. Endures as the subcortical white matter.

*STF5*  Deep cellular layer that is prominent during the first trimester, the first sojourn zone to appear outside the germinal matrix.

**FONT KEY:**
**VENTRICULAR DIVISIONS – CAPITALS**
**Germinal zone - Helvetica bold**
***Transient structure - Times bold italic***
Permanent structure - Times Roman or **Bold**

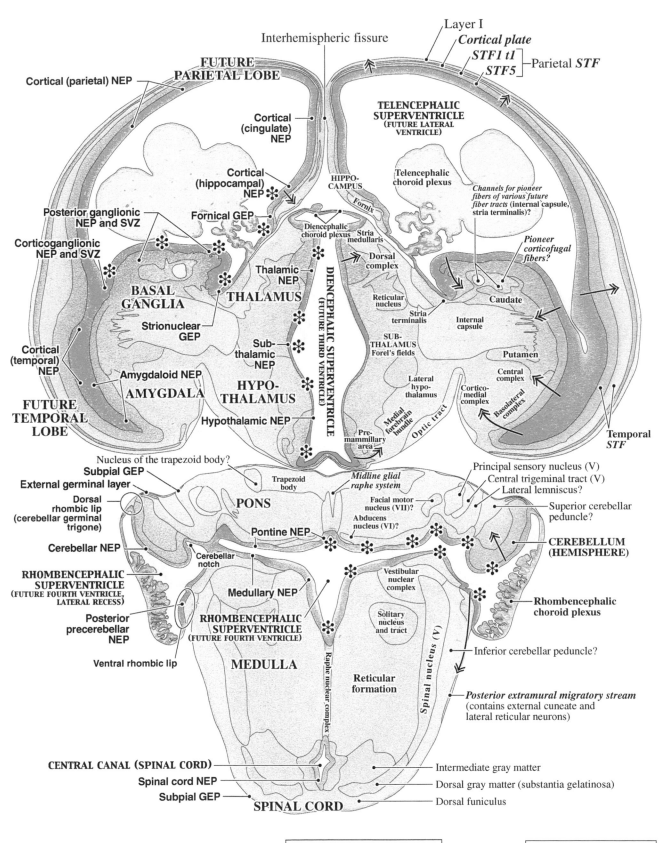

**FUTURE PARIETAL LOBE**

Cortical (parietal) NEP

Interhemispheric fissure

Layer I
*Cortical plate*
*STF1 t1*
*STF5* — Parietal *STF*

**TELENCEPHALIC SUPERVENTRICLE** (FUTURE LATERAL VENTRICLE)

Cortical (cingulate) NEP

Cortical (hippocampal) NEP ✱

Telencephalic choroid plexus

HIPPO-CAMPUS

*Channels for pioneer fibers of various future fiber tracts (internal capsule, stria terminalis)?*

Posterior ganglionic NEP and SVZ

Fornical GEP

Fornix

*Pioneer corticofugal fibers?*

Corticoganglionic NEP and SVZ

Diencephalic choroid plexus

Stria medullaris

Dorsal complex

Thalamic NEP

**BASAL GANGLIA**

**THALAMUS**

Reticular nucleus

Caudate

Strionuclear GEP

Stria terminalis

Internal capsule

Cortical (temporal) NEP

Sub-thalamic NEP

**SUB-THALAMUS** Forel's fields

Putamen

**DIENCEPHALIC SUPERVENTRICLE** (FUTURE THIRD VENTRICLE)

Amygdaloid NEP

**AMYGDALA**

**HYPO-THALAMUS**

Lateral hypo-thalamus

Central complex

**FUTURE TEMPORAL LOBE**

Hypothalamic NEP

Cortico-medial complex

Basolateral complex

Pre-mammillary area

*Medial forebrain bundle*

*Optic tract*

Temporal *STF*

Nucleus of the trapezoid body?

Subpial GEP

External germinal layer

Dorsal rhombic lip (cerebellar germinal trigone)

Trapezoid body

*Midline glial raphe system*

Principal sensory nucleus (V)
Central trigeminal tract (V)
Lateral lemniscus?

**PONS**

Facial motor nucleus (VII)?

Abducens nucleus (VI)?

Superior cerebellar peduncle?

Cerebellar NEP

Pontine NEP

**CEREBELLUM (HEMISPHERE)**

**RHOMBENCEPHALIC SUPERVENTRICLE** (FUTURE FOURTH VENTRICLE, LATERAL RECESS)

Cerebellar notch

Vestibular nuclear complex

Medullary NEP

Rhombencephalic choroid plexus

Posterior precerebellar NEP

**RHOMBENCEPHALIC SUPERVENTRICLE** (FUTURE FOURTH VENTRICLE)

Solitary nucleus and tract

Inferior cerebellar peduncle?

Ventral rhombic lip

**MEDULLA**

*Raphe nuclear complex*

*Spinal nucleus (V)*

Reticular formation

*Posterior extramural migratory stream* (contains external cuneate and lateral reticular neurons)

**CENTRAL CANAL (SPINAL CORD)**

Spinal cord NEP

Subpial GEP

**SPINAL CORD**

Intermediate gray matter

Dorsal gray matter (substantia gelatinosa)

Dorsal funiculus

✱ Evaginations and invaginations of the neuroepithelium are mosaic compartments that give rise to different brain structures.

**ABBREVIATIONS:**
GEP - Glioepithelium
NEP - Neuroepithelium
SVZ - Subventricular zone

**Arrows** indicate the presumed *direction of neuron migration* from neuroepithelial sources.

**PLATE 157A**

**GW8 Coronal**
**CR 31 mm**
**C9226**
**Level 13:**
**Section 468**

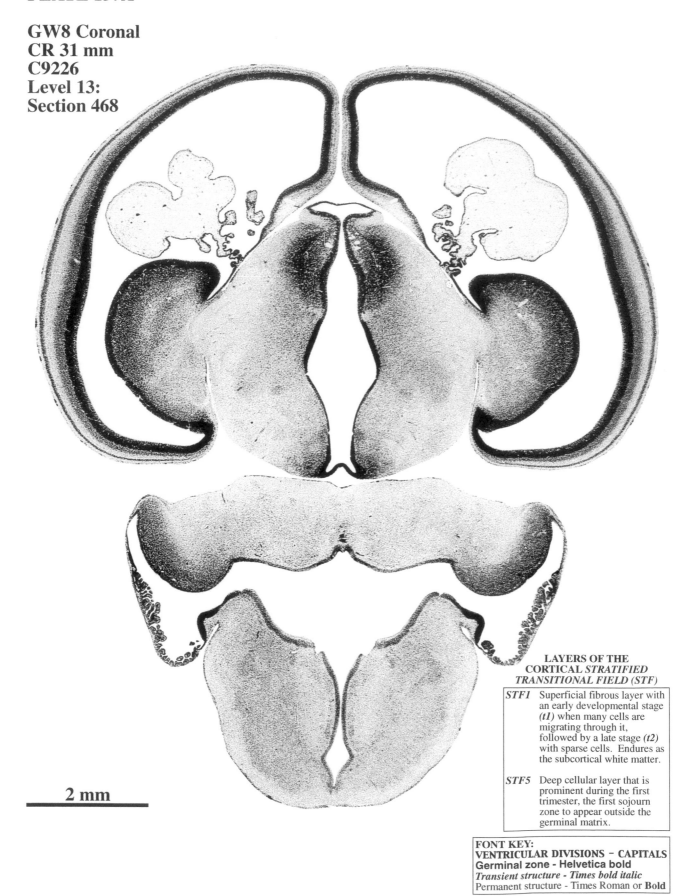

**2 mm**

**LAYERS OF THE**
**CORTICAL** *STRATIFIED*
*TRANSITIONAL FIELD (STF)*

*STF1* Superficial fibrous layer with
an early developmental stage
*(t1)* when many cells are
migrating through it,
followed by a late stage *(t2)*
with sparse cells. Endures as
the subcortical white matter.

*STF5* Deep cellular layer that is
prominent during the first
trimester, the first sojourn
zone to appear outside the
germinal matrix.

FONT KEY:
**VENTRICULAR DIVISIONS – CAPITALS**
**Germinal zone - Helvetica bold**
***Transient structure - Times bold italic***
Permanent structure - Times Roman or **Bold**

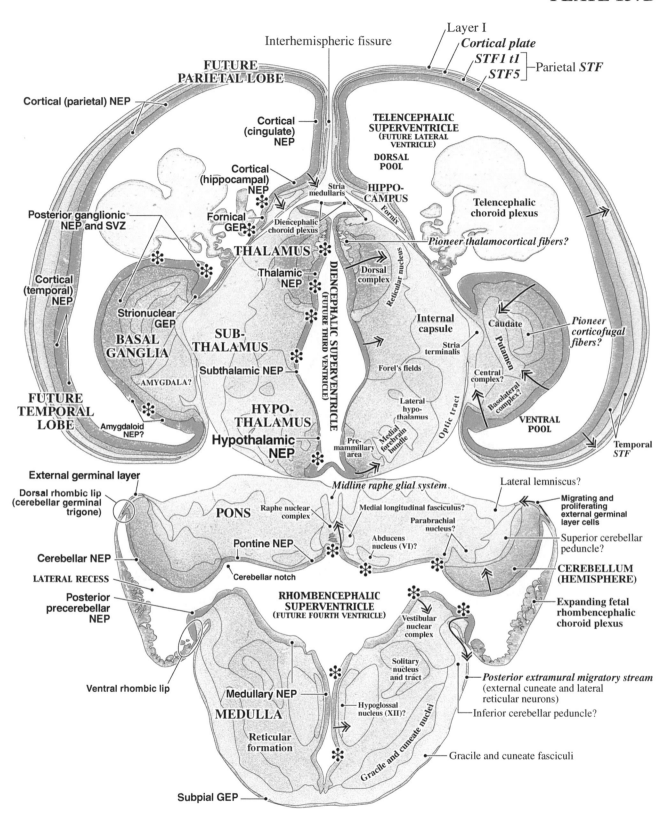

Interhemispheric fissure

Layer I
*Cortical plate*
*STF1 t1*
*STF5* — Parietal *STF*

**FUTURE PARIETAL LOBE**

Cortical (parietal) NEP

Cortical (cingulate) NEP

**TELENCEPHALIC SUPERVENTRICLE**
**(FUTURE LATERAL VENTRICLE)**

**DORSAL POOL**

Cortical (hippocampal) NEP

Stria medullaris

**HIPPO-CAMPUS**

Telencephalic choroid plexus

Posterior ganglionic NEP and SVZ

Fornical GEP

Diencephalic choroid plexus

Fornix

*Pioneer thalamocortical fibers?*

**THALAMUS**

Thalamic NEP

Dorsal complex

Reticular nucleus

Cortical (temporal) NEP

Strionuclear GEP

**DIENCEPHALIC SUPERVENTRICLE**
**(FUTURE THIRD VENTRICLE)**

**Internal capsule**

Caudate

*Pioneer corticofugal fibers?*

**BASAL GANGLIA**

**SUB-THALAMUS**

Subthalamic NEP

Stria terminalis

Putamen

—AMYGDALA?

Forel's fields

Central complex?

Basolateral complex?

**FUTURE TEMPORAL LOBE**

Amygdaloid NEP?

**HYPO-THALAMUS**

**Hypothalamic NEP**

Lateral hypo-thalamus

Optic tract

**VENTRAL POOL**

Pre-mammillary area

Medial forebrain bundle

Temporal *STF*

External germinal layer

Dorsal rhombic lip (cerebellar germinal trigone)

*Midline raphe glial system*

Lateral lemniscus?

Migrating and proliferating external germinal layer cells

**PONS**

Raphe nuclear complex

Medial longitudinal fasciculus?

Parabrachial nucleus?

Superior cerebellar peduncle?

Cerebellar NEP

Pontine NEP

Abducens nucleus (VI)?

**CEREBELLUM (HEMISPHERE)**

**LATERAL RECESS**

Cerebellar notch

Posterior precerebellar NEP

**RHOMBENCEPHALIC SUPERVENTRICLE**
**(FUTURE FOURTH VENTRICLE)**

Vestibular nuclear complex

**Expanding fetal rhombencephalic choroid plexus**

Ventral rhombic lip

Solitary nucleus and tract

*Posterior extramural migratory stream*
(external cuneate and lateral reticular neurons)

Medullary NEP

Hypoglossal nucleus (XII)?

Inferior cerebellar peduncle?

**MEDULLA**

**Reticular formation**

Gracile and cuneate nuclei

Gracile and cuneate fasciculi

Subpial GEP

✳ Evaginations and invaginations of the neuroepithelium are mosaic compartments that give rise to different brain structures.

**ABBREVIATIONS:**
**GEP** - Glioepithelium
**NEP** - Neuroepithelium
**SVZ** - Subventricular zone

⬆ **Arrows** indicate the presumed *direction of neuron migration* from neuroepithelial sources.

**PLATE 158A**

**GW8 Coronal**
**CR 31 mm**
**C9226**
**Level 14:**
**Section 438**

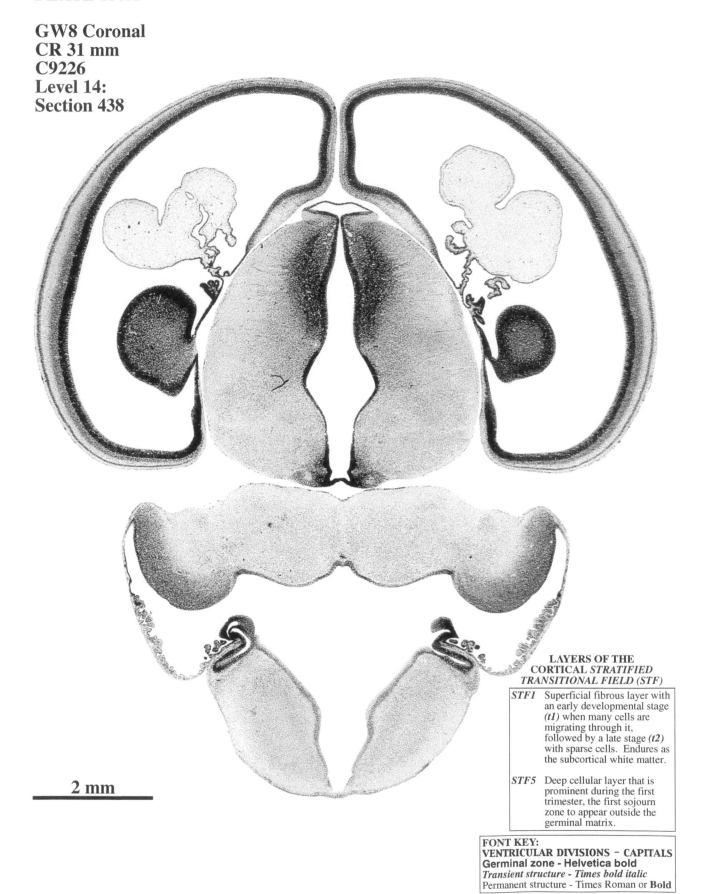

**2 mm**

**LAYERS OF THE**
**CORTICAL** *STRATIFIED*
*TRANSITIONAL FIELD (STF)*

*STF1* Superficial fibrous layer with
an early developmental stage
*(t1)* when many cells are
migrating through it,
followed by a late stage *(t2)*
with sparse cells. Endures as
the subcortical white matter.

*STF5* Deep cellular layer that is
prominent during the first
trimester, the first sojourn
zone to appear outside the
germinal matrix.

FONT KEY:
VENTRICULAR DIVISIONS – CAPITALS
Germinal zone - Helvetica bold
*Transient structure - Times bold italic*
Permanent structure - Times Roman or **Bold**

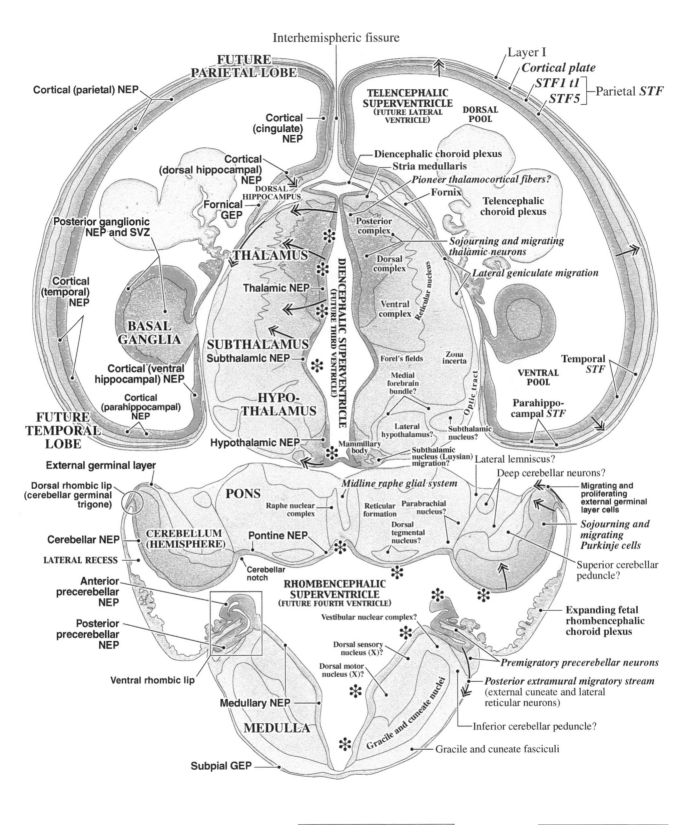

Interhemispheric fissure

FUTURE PARIETAL LOBE

Cortical (parietal) NEP

Cortical (cingulate) NEP

Cortical (dorsal hippocampal) NEP

DORSAL HIPPOCAMPUS

Fornical GEP

THALAMUS

Posterior ganglionic NEP and SVZ

Cortical (temporal) NEP

BASAL GANGLIA

Thalamic NEP

SUBTHALAMUS

Subthalamic NEP

Cortical (ventral hippocampal) NEP

HYPO-THALAMUS

Cortical (parahippocampal) NEP

FUTURE TEMPORAL LOBE

Hypothalamic NEP

External germinal layer

Dorsal rhombic lip (cerebellar germinal trigone)

Cerebellar NEP

LATERAL RECESS

Anterior precerebellar NEP

Posterior precerebellar NEP

Ventral rhombic lip

Medullary NEP

MEDULLA

Subpial GEP

Layer I
Cortical plate
STF1 t1
STF5 — Parietal STF

TELENCEPHALIC SUPERVENTRICLE (FUTURE LATERAL VENTRICLE)

DORSAL POOL

Diencephalic choroid plexus
Stria medullaris
Pioneer thalamocortical fibers?
Fornix

Telencephalic choroid plexus

Posterior complex

Sojourning and migrating thalamic neurons

Dorsal complex

Lateral geniculate migration

DIENCEPHALIC SUPERVENTRICLE (FUTURE THIRD VENTRICLE)

Ventral complex

Reticular nucleus

Forel's fields

Zona incerta

Medial forebrain bundle?

VENTRAL POOL

Temporal STF

Parahippo-campal STF

Optic tract

Lateral hypothalamus?

Subthalamic nucleus?

Mammillary body

Subthalamic nucleus (Luysian) migration?

Subthalamic nucleus?

Lateral lemniscus?

PONS

Raphe nuclear complex

Midline raphe glial system

Reticular formation

Parabrachial nucleus?

Deep cerebellar neurons?

Migrating and proliferating external germinal layer cells

CEREBELLUM (HEMISPHERE)

Pontine NEP

Dorsal tegmental nucleus?

Sojourning and migrating Purkinje cells

Cerebellar notch

RHOMBENCEPHALIC SUPERVENTRICLE (FUTURE FOURTH VENTRICLE)

Superior cerebellar peduncle?

Vestibular nuclear complex?

Expanding fetal rhombencephalic choroid plexus

Dorsal sensory nucleus (X)?

Dorsal motor nucleus (X)?

Premigratory precerebellar neurons

Posterior extramural migratory stream (external cuneate and lateral reticular neurons)

Gracile and cuneate nuclei

Inferior cerebellar peduncle?

Gracile and cuneate fasciculi

✳ Evaginations and invaginations of the neuroepithelium are mosaic compartments that give rise to different brain structures.

ABBREVIATIONS:
GEP - Glioepithelium
NEP - Neuroepithelium
SVZ - Subventricular zone

Arrows indicate the presumed direction of neuron migration from neuroepithelial sources.

**PLATE 159A**

**GW8 Coronal**
**CR 31 mm**
**C9226**
**Level 15:**
**Section**
**402**

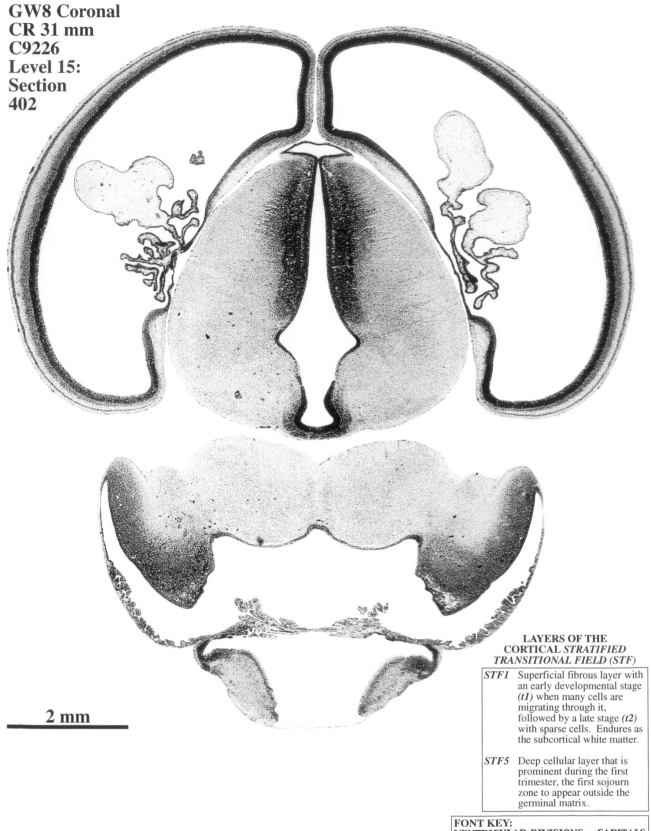

**2 mm**

**LAYERS OF THE**
**CORTICAL *STRATIFIED***
***TRANSITIONAL FIELD (STF)***

*STF1* Superficial fibrous layer with
an early developmental stage
*(t1)* when many cells are
migrating through it,
followed by a late stage *(t2)*
with sparse cells. Endures as
the subcortical white matter.

*STF5* Deep cellular layer that is
prominent during the first
trimester, the first sojourn
zone to appear outside the
germinal matrix.

**FONT KEY:**
**VENTRICULAR DIVISIONS – CAPITALS**
**Germinal zone - Helvetica bold**
***Transient structure - Times bold italic***
Permanent structure - Times Roman or **Bold**

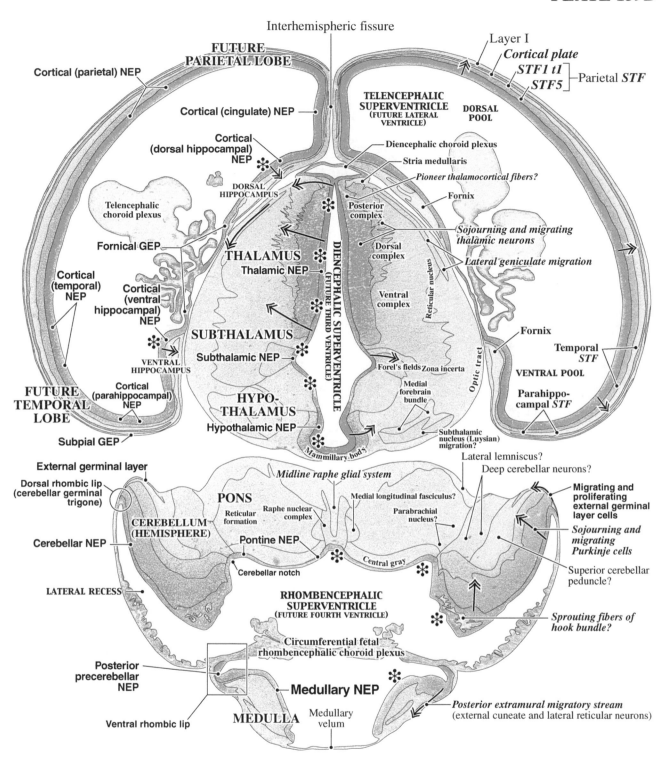

Interhemispheric fissure

**FUTURE PARIETAL LOBE**

Cortical (parietal) NEP

Cortical (cingulate) NEP

Cortical (dorsal hippocampal) NEP

DORSAL HIPPOCAMPUS

Telencephalic choroid plexus

Fornical GEP

**THALAMUS**

Thalamic NEP

Cortical (temporal) NEP

Cortical (ventral hippocampal) NEP

**SUBTHALAMUS**

Subthalamic NEP

VENTRAL HIPPOCAMPUS

**FUTURE TEMPORAL LOBE**

Cortical (parahippocampal) NEP

Subpial GEP

**HYPO-THALAMUS**

Hypothalamic NEP

Layer I
*Cortical plate*
*STF1 t1*
*STF5* ⎤ Parietal *STF*

**TELENCEPHALIC SUPERVENTRICLE (FUTURE LATERAL VENTRICLE)**

DORSAL POOL

Diencephalic choroid plexus

Stria medullaris

*Pioneer thalamocortical fibers?*

Fornix

Posterior complex

*Sojourning and migrating thalamic neurons*

Dorsal complex

*Lateral geniculate migration*

Reticular nucleus

Ventral complex

**DIENCEPHALIC SUPERVENTRICLE (FUTURE THIRD VENTRICLE)**

Fornix

Temporal *STF*

Optic tract

VENTRAL POOL

Parahippo-campal *STF*

Forel's fields

Zona incerta

Medial forebrain bundle

Subthalamic nucleus (Luysian) migration?

Lateral lemniscus?

Mammillary body

Deep cerebellar neurons?

**External germinal layer**

**Dorsal rhombic lip (cerebellar germinal trigone)**

**CEREBELLUM (HEMISPHERE)**

Cerebellar NEP

**LATERAL RECESS**

*Midline raphe glial system*

**PONS**

Reticular formation

Raphe nuclear complex

Pontine NEP

Cerebellar notch

*Medial longitudinal fasciculus?*

Parabrachial nucleus?

Central gray

**Migrating and proliferating external germinal layer cells**

*Sojourning and migrating Purkinje cells*

Superior cerebellar peduncle?

*Sprouting fibers of hook bundle?*

**Posterior precerebellar NEP**

**RHOMBENCEPHALIC SUPERVENTRICLE (FUTURE FOURTH VENTRICLE)**

Circumferential fetal rhombencephalic choroid plexus

Medullary NEP

Ventral rhombic lip

**MEDULLA**

Medullary velum

*Posterior extramural migratory stream (external cuneate and lateral reticular neurons)*

✳ Evaginations and invaginations of the neuroepithelium are mosaic compartments that give rise to different brain structures.

**ABBREVIATIONS:**
**GEP - Glioepithelium**
**NEP - Neuroepithelium**

**Arrows** indicate the presumed *direction of neuron migration* from neuroepithelial sources.

**PLATE 160A**

**GW8 Coronal**
**CR 31 mm**
**C9226**
**Level 16:**
**Section**
**390**

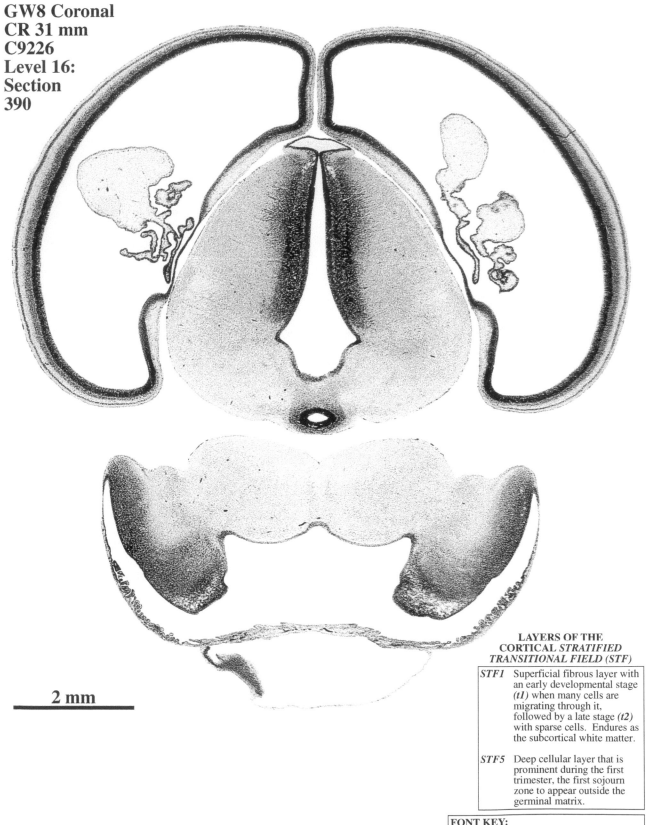

**2 mm**

**LAYERS OF THE**
**CORTICAL *STRATIFIED***
***TRANSITIONAL FIELD (STF)***

| | |
|---|---|
| *STF1* | Superficial fibrous layer with an early developmental stage *(t1)* when many cells are migrating through it, followed by a late stage *(t2)* with sparse cells.  Endures as the subcortical white matter. |
| *STF5* | Deep cellular layer that is prominent during the first trimester, the first sojourn zone to appear outside the germinal matrix. |

**FONT KEY:**
**VENTRICULAR DIVISIONS – CAPITALS**
**Germinal zone - Helvetica bold**
***Transient structure - Times bold italic***
Permanent structure - Times Roman or **Bold**

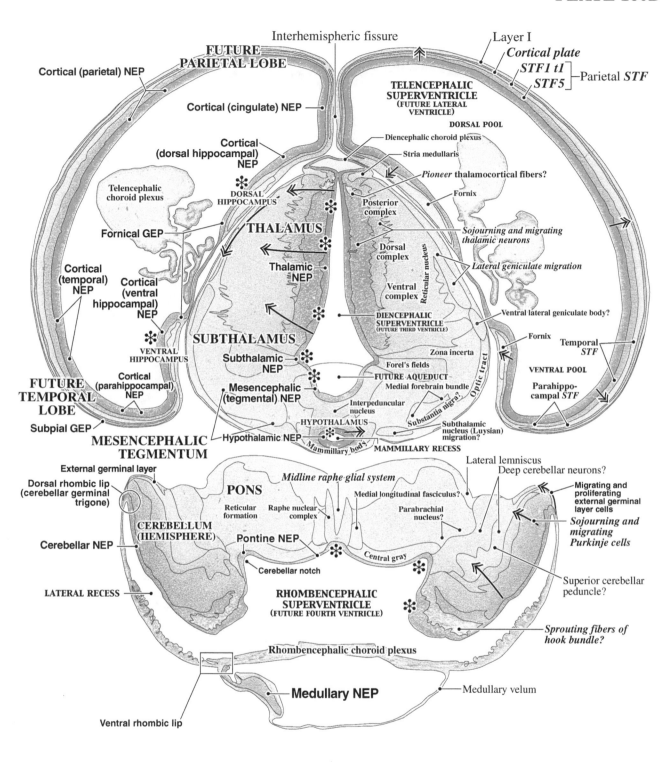

Interhemispheric fissure

Layer I

**FUTURE PARIETAL LOBE**

*Cortical plate*
*STF1 t1*
*STF5* — Parietal *STF*

Cortical (parietal) NEP

Cortical (cingulate) NEP

**TELENCEPHALIC SUPERVENTRICLE** (FUTURE LATERAL VENTRICLE)

**DORSAL POOL**

Diencephalic choroid plexus

Cortical (dorsal hippocampal) NEP

Stria medullaris

*Pioneer* thalamocortical fibers?

Telencephalic choroid plexus

**DORSAL HIPPOCAMPUS**

Posterior complex

Fornix

**THALAMUS**

*Sojourning and migrating thalamic neurons*

Fornical GEP

Thalamic NEP

Dorsal complex

*Lateral geniculate migration*

Cortical (temporal) NEP

Cortical (ventral hippocampal) NEP

Ventral complex

Reticular nucleus

Ventral lateral geniculate body?

**SUBTHALAMUS**

**DIENCEPHALIC SUPERVENTRICLE** (FUTURE THIRD VENTRICLE)

Fornix

Temporal *STF*

**VENTRAL HIPPOCAMPUS**

Subthalamic NEP

Zona incerta

**VENTRAL POOL**

Cortical (parahippocampal) NEP

Forel's fields

**FUTURE AQUEDUCT**

Parahippocampal *STF*

**FUTURE TEMPORAL LOBE**

Mesencephalic (tegmental) NEP

Medial forebrain bundle

Subpial GEP

Interpeduncular nucleus

Substantia nigra?

Subthalamic nucleus (Luysian) migration?

**MESENCEPHALIC TEGMENTUM**

Hypothalamic NEP

**HYPOTHALAMUS**

Mammillary bod's

**MAMMILLARY RECESS**

Optic tract

External germinal layer

Lateral lemniscus
Deep cerebellar neurons?

Dorsal rhombic lip (cerebellar germinal trigone)

*Midline raphe glial system*

**PONS**

Medial longitudinal fasciculus?

Migrating and proliferating external germinal layer cells

Cerebellar NEP

Reticular formation

Raphe nuclear complex

Parabrachial nucleus?

*Sojourning and migrating Purkinje cells*

**CEREBELLUM (HEMISPHERE)**

Pontine NEP

**LATERAL RECESS**

Cerebellar notch

Central gray

Superior cerebellar peduncle?

**RHOMBENCEPHALIC SUPERVENTRICLE** (FUTURE FOURTH VENTRICLE)

*Sprouting fibers of hook bundle?*

Rhombencephalic choroid plexus

**Medullary NEP**

Medullary velum

Ventral rhombic lip

✴ Evaginations and invaginations of the neuroepithelium are mosaic compartments that give rise to different brain structures.

**ABBREVIATIONS:**
**GEP** - Glioepithelium
**NEP** - Neuroepithelium

**Arrows** indicate the presumed *direction of neuron migration* from neuroepithelial sources.

**PLATE 161A**

**GW8 Coronal**
**CR 31 mm**
**C9226**
**Level 17:**
**Section 324**

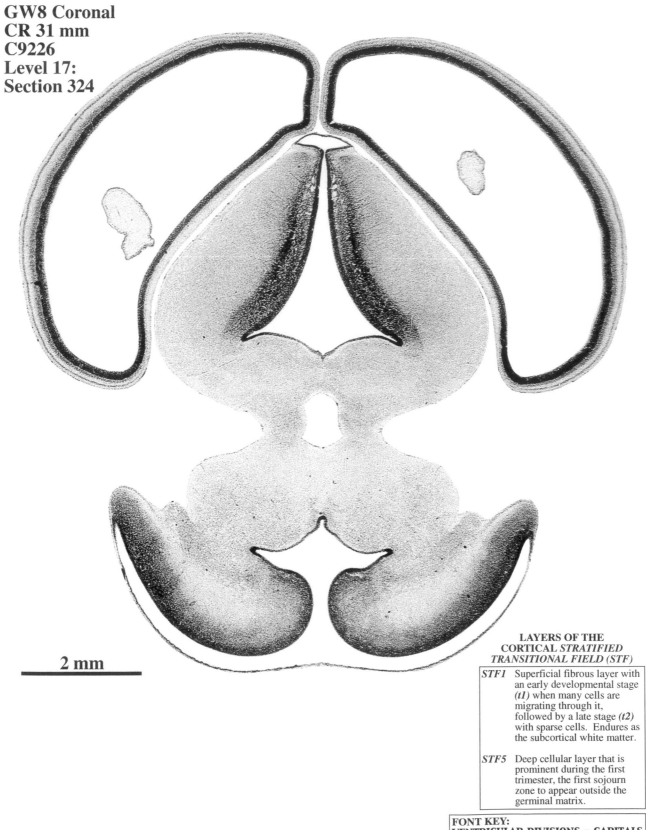

2 mm

**LAYERS OF THE**
**CORTICAL *STRATIFIED***
***TRANSITIONAL FIELD (STF)***

*STF1* Superficial fibrous layer with
an early developmental stage
*(t1)* when many cells are
migrating through it,
followed by a late stage *(t2)*
with sparse cells. Endures as
the subcortical white matter.

*STF5* Deep cellular layer that is
prominent during the first
trimester, the first sojourn
zone to appear outside the
germinal matrix.

FONT KEY:
**VENTRICULAR DIVISIONS – CAPITALS**
**Germinal zone - Helvetica bold**
***Transient structure - Times bold italic***
Permanent structure - Times Roman or **Bold**

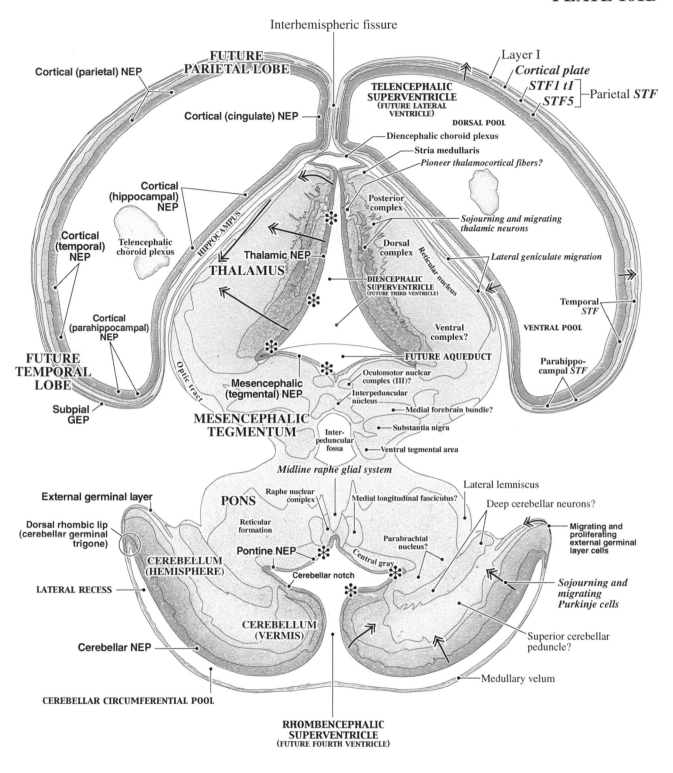

Interhemispheric fissure

FUTURE
PARIETAL-LOBE

Cortical (parietal) NEP

Cortical (cingulate) NEP

Cortical (hippocampal) NEP

Cortical (temporal) NEP

Telencephalic choroid plexus

HIPPOCAMPUS

Thalamic NEP

THALAMUS

Cortical (parahippocampal) NEP

FUTURE TEMPORAL LOBE

Subpial GEP

Optic tract

MESENCEPHALIC TEGMENTUM

Mesencephalic (tegmental) NEP

Layer I
*Cortical plate*
*STF1 t1*
*STF5*  }— Parietal *STF*

TELENCEPHALIC SUPERVENTRICLE
(FUTURE LATERAL VENTRICLE)

DORSAL POOL

Diencephalic choroid plexus

Stria medullaris

*Pioneer thalamocortical fibers?*

Posterior complex

*Sojourning and migrating thalamic neurons*

Dorsal complex

Reticular nucleus

*Lateral geniculate migration*

DIENCEPHALIC SUPERVENTRICLE
(FUTURE THIRD VENTRICLE)

Temporal *STF*

Ventral complex?

VENTRAL POOL

FUTURE AQUEDUCT

Parahippo-campal *STF*

Oculomotor nuclear complex (III)?

Interpeduncular nucleus

Medial forebrain bundle?

Substantia nigra

Inter-peduncular fossa

Ventral tegmental area

*Midline raphe glial system*

External germinal layer

Dorsal rhombic lip (cerebellar germinal trigone)

CEREBELLUM (HEMISPHERE)

LATERAL RECESS

Cerebellar NEP

Raphe nuclear complex

PONS

Reticular formation

Pontine NEP

Cerebellar notch

CEREBELLUM (VERMIS)

Medial longitudinal fasciculus?

Lateral lemniscus

Deep cerebellar neurons?

Migrating and proliferating external germinal layer cells

*Sojourning and migrating Purkinje cells*

Parabrachial nucleus?

Central gray

Superior cerebellar peduncle?

Medullary velum

CEREBELLAR CIRCUMFERENTIAL POOL

RHOMBENCEPHALIC SUPERVENTRICLE
(FUTURE FOURTH VENTRICLE)

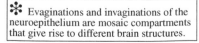

✳ Evaginations and invaginations of the neuroepithelium are mosaic compartments that give rise to different brain structures.

**ABBREVIATIONS:**
**GEP - Glioepithelium**
**NEP - Neuroepithelium**

**Arrows** indicate the presumed *direction of neuron migration* from neuroepithelial sources.

**PLATE 162A**

**GW8 Coronal**
**CR 31 mm**
**C9226**
**Level 18:**
**Section 276**

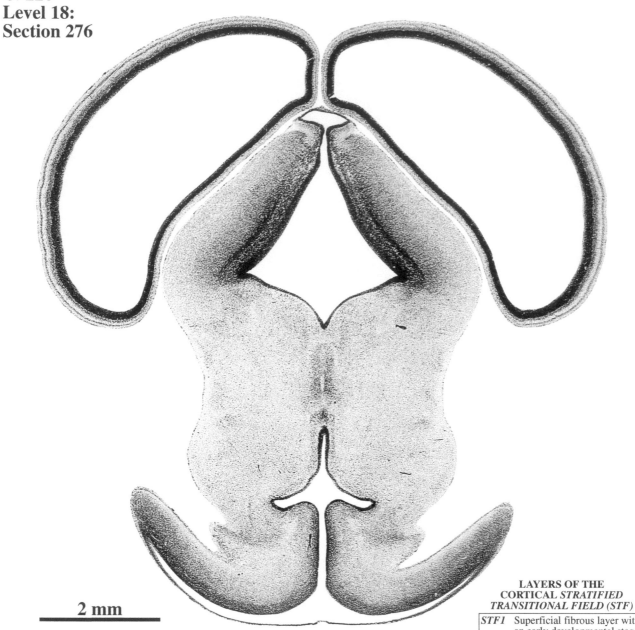

**2 mm**

**LAYERS OF THE**
**CORTICAL *STRATIFIED***
***TRANSITIONAL FIELD (STF)***

*STF1*  Superficial fibrous layer with
an early developmental stage
*(t1)* when many cells are
migrating through it,
followed by a late stage *(t2)*
with sparse cells. Endures as
the subcortical white matter.

*STF5*  Deep cellular layer that is
prominent during the first
trimester, the first sojourn
zone to appear outside the
germinal matrix.

**FONT KEY:**
**VENTRICULAR DIVISIONS − CAPITALS**
**Germinal zone - Helvetica bold**
***Transient structure - Times bold italic***
Permanent structure - Times Roman or **Bold**

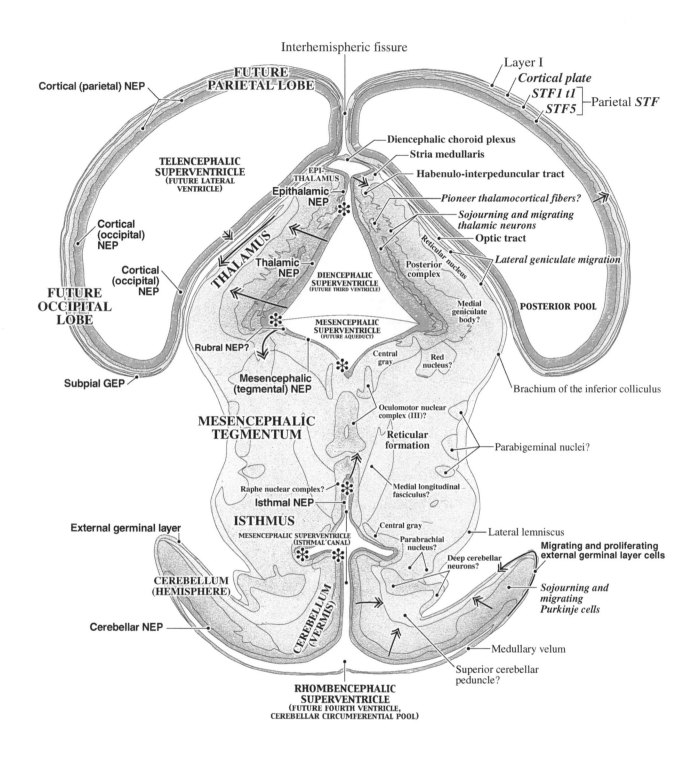

Interhemispheric fissure

**FUTURE PARIETAL LOBE**

Cortical (parietal) NEP

Layer I
*Cortical plate*
*STF1 t1*
*STF5* — Parietal *STF*

**TELENCEPHALIC SUPERVENTRICLE**
(FUTURE LATERAL VENTRICLE)

Diencephalic choroid plexus
Stria medullaris
Habenulo-interpeduncular tract

EPI-THALAMUS
Epithalamic NEP

*Pioneer thalamocortical fibers?*
*Sojourning and migrating thalamic neurons*
Optic tract

Cortical (occipital) NEP

THALAMUS
Thalamic NEP

Reticular nucleus

Posterior complex

*Lateral geniculate migration*

Cortical (occipital) NEP

**DIENCEPHALIC SUPERVENTRICLE**
(FUTURE THIRD VENTRICLE)

**FUTURE OCCIPITAL LOBE**

**MESENCEPHALIC SUPERVENTRICLE**
(FUTURE AQUEDUCT)

Medial geniculate body?

**POSTERIOR POOL**

Rubral NEP?

Central gray

Red nucleus?

Subpial GEP

Mesencephalic (tegmental) NEP

Brachium of the inferior colliculus

Oculomotor nuclear complex (III)?

**MESENCEPHALIC TEGMENTUM**

**Reticular formation**

Parabigeminal nuclei?

Raphe nuclear complex?

Medial longitudinal fasciculus?

Isthmal NEP

**ISTHMUS**

**MESENCEPHALIC SUPERVENTRICLE**
(ISTHMAL CANAL)

Central gray

Lateral lemniscus

External germinal layer

Parabrachial nucleus?

**Migrating and proliferating external germinal layer cells**

Deep cerebellar neurons?

**CEREBELLUM (HEMISPHERE)**

*Sojourning and migrating Purkinje cells*

Cerebellar NEP

**CEREBELLUM (VERMIS)**

Medullary velum

Superior cerebellar peduncle?

**RHOMBENCEPHALIC SUPERVENTRICLE**
(FUTURE FOURTH VENTRICLE, CEREBELLAR CIRCUMFERENTIAL POOL)

✱ Evaginations and invaginations of the neuroepithelium are mosaic compartments that give rise to different brain structures.

**ABBREVIATIONS:**
**GEP** - Glioepithelium
**NEP** - Neuroepithelium

**Arrows** indicate the presumed *direction of neuron migration* from neuroepithelial sources.

**PLATE 163A**

**GW8 Coronal**
**CR 31 mm**
**C9226**
**Level 19:**
**Section 252**

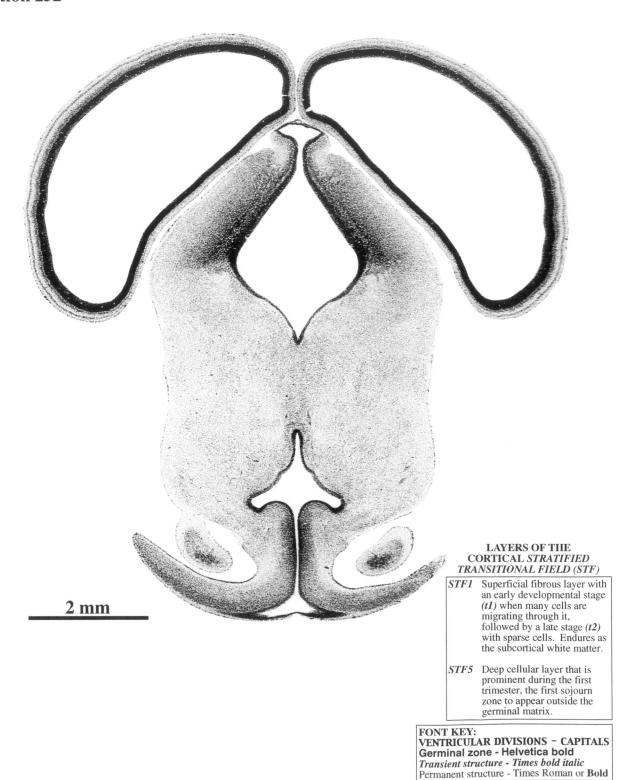

**2 mm**

**LAYERS OF THE**
**CORTICAL** *STRATIFIED*
*TRANSITIONAL FIELD (STF)*

*STF1*  Superficial fibrous layer with
an early developmental stage
*(t1)* when many cells are
migrating through it,
followed by a late stage *(t2)*
with sparse cells. Endures as
the subcortical white matter.

*STF5*  Deep cellular layer that is
prominent during the first
trimester, the first sojourn
zone to appear outside the
germinal matrix.

FONT KEY:
**VENTRICULAR DIVISIONS – CAPITALS**
**Germinal zone - Helvetica bold**
*Transient structure - Times bold italic*
Permanent structure - Times Roman or **Bold**

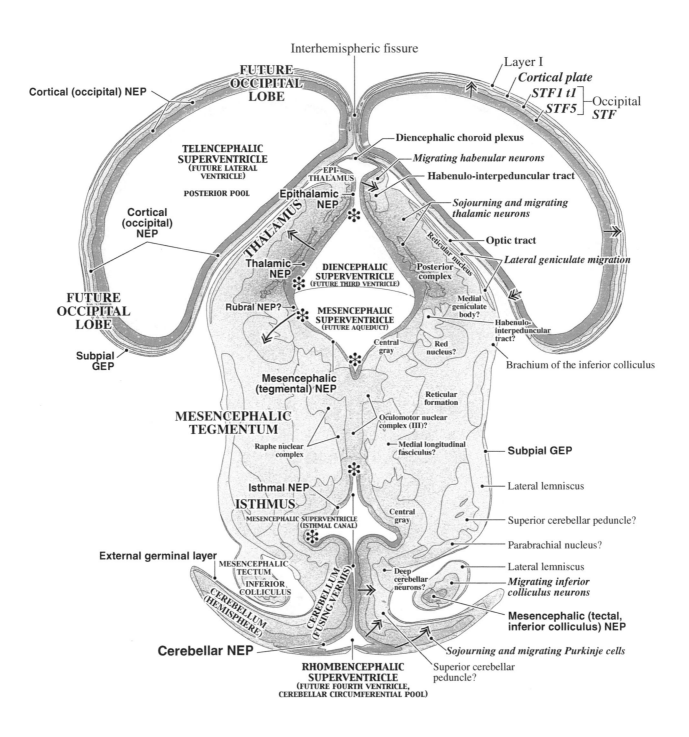

Interhemispheric fissure

**FUTURE OCCIPITAL LOBE**

Cortical (occipital) NEP

Layer I
*Cortical plate*
*STF1 t1*
*STF5* } Occipital *STF*

Diencephalic choroid plexus

**TELENCEPHALIC SUPERVENTRICLE** (FUTURE LATERAL VENTRICLE)

**POSTERIOR POOL**

EPI-THALAMUS

**Epithalamic NEP**

*Migrating habenular neurons*

Habenulo-interpeduncular tract

THALAMUS

*Sojourning and migrating thalamic neurons*

Cortical (occipital) NEP

**Thalamic NEP**

**DIENCEPHALIC SUPERVENTRICLE** (FUTURE THIRD VENTRICLE)

Reticular nucleus

Posterior complex

Optic tract

*Lateral geniculate migration*

**FUTURE OCCIPITAL LOBE**

Rubral NEP?

**MESENCEPHALIC SUPERVENTRICLE** (FUTURE AQUEDUCT)

Central gray

Medial geniculate body?

Red nucleus?

Habenulo-interpeduncular tract?

**Subpial GEP**

**Mesencephalic (tegmental) NEP**

Brachium of the inferior colliculus

**MESENCEPHALIC TEGMENTUM**

Reticular formation

Oculomotor nuclear complex (III)?

Raphe nuclear complex

Medial longitudinal fasciculus?

**Subpial GEP**

**Isthmal NEP**

**ISTHMUS**

**MESENCEPHALIC SUPERVENTRICLE** (ISTHMAL CANAL)

Central gray

Lateral lemniscus

Superior cerebellar peduncle?

Parabrachial nucleus?

**External germinal layer**

**MESENCEPHALIC TECTUM**

**INFERIOR COLLICULUS**

**CEREBELLUM (HEMISPHERE)**

**CEREBELLUM (FUSING VERMIS)**

Deep cerebellar neurons?

Lateral lemniscus

*Migrating inferior colliculus neurons*

**Mesencephalic (tectal, inferior colliculus) NEP**

**Cerebellar NEP**

*Sojourning and migrating Purkinje cells*

**RHOMBENCEPHALIC SUPERVENTRICLE** (FUTURE FOURTH VENTRICLE, CEREBELLAR CIRCUMFERENTIAL POOL)

Superior cerebellar peduncle?

✳ Evaginations and invaginations of the neuroepithelium are mosaic compartments that give rise to different brain structures.

**ABBREVIATIONS:**
**GEP** - Glioepithelium
**NEP** - Neuroepithelium

**Arrows** indicate the presumed *direction of neuron migration* from neuroepithelial sources.

**PLATE 164A**

**GW8 Coronal**
**CR 31 mm**
**C9226**
**Level 20:**
**Section 234**

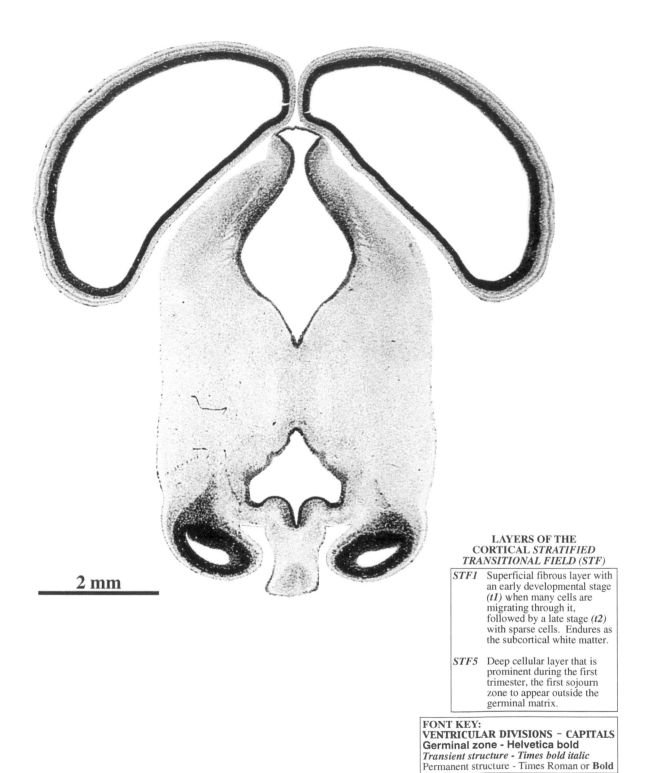

**2 mm**

**LAYERS OF THE**
**CORTICAL** *STRATIFIED*
*TRANSITIONAL FIELD (STF)*

| | |
|---|---|
| *STF1* | Superficial fibrous layer with an early developmental stage *(t1)* when many cells are migrating through it, followed by a late stage *(t2)* with sparse cells. Endures as the subcortical white matter. |
| *STF5* | Deep cellular layer that is prominent during the first trimester, the first sojourn zone to appear outside the germinal matrix. |

**FONT KEY:**
**VENTRICULAR DIVISIONS − CAPITALS**
**Germinal zone - Helvetica bold**
*Transient structure - Times bold italic*
Permanent structure - Times Roman or **Bold**

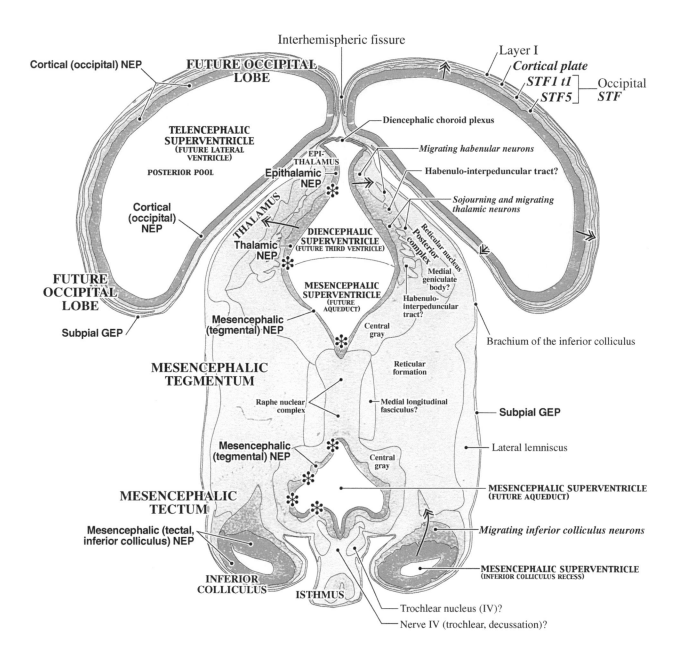

Interhemispheric fissure

Cortical (occipital) NEP

**FUTURE OCCIPITAL LOBE**

Layer I
*Cortical plate*
*STF1 t1*
*STF5*
Occipital *STF*

**TELENCEPHALIC SUPERVENTRICLE (FUTURE LATERAL VENTRICLE)**

**POSTERIOR POOL**

Diencephalic choroid plexus

*Migrating habenular neurons*

Habenulo-interpeduncular tract?

EPI-THALAMUS

Epithalamic NEP

*Sojourning and migrating thalamic neurons*

THALAMUS

Cortical (occipital) NEP

Thalamic NEP

**DIENCEPHALIC SUPERVENTRICLE (FUTURE THIRD VENTRICLE)**

Reticular nucleus Posterior complex

**FUTURE OCCIPITAL LOBE**

Medial geniculate body?

**MESENCEPHALIC SUPERVENTRICLE (FUTURE AQUEDUCT)**

Habenulo-interpeduncular tract?

Subpial GEP

Mesencephalic (tegmental) NEP

Central gray

Brachium of the inferior colliculus

**MESENCEPHALIC TEGMENTUM**

Reticular formation

Raphe nuclear complex

Medial longitudinal fasciculus?

**Subpial GEP**

Mesencephalic (tegmental) NEP

Central gray

Lateral lemniscus

**MESENCEPHALIC SUPERVENTRICLE (FUTURE AQUEDUCT)**

**MESENCEPHALIC TECTUM**

Mesencephalic (tectal, inferior colliculus) NEP

*Migrating inferior colliculus neurons*

**MESENCEPHALIC SUPERVENTRICLE (INFERIOR COLLICULUS RECESS)**

**INFERIOR COLLICULUS**

**ISTHMUS**

Trochlear nucleus (IV)?

Nerve IV (trochlear, decussation)?

✱ Evaginations and invaginations of the neuroepithelium are mosaic compartments that give rise to different brain structures.

**ABBREVIATIONS:**
**GEP - Glioepithelium**
**NEP - Neuroepithelium**

**Arrows** indicate the presumed *direction of neuron migration* from neuroepithelial sources.

**PLATE 165A**

**GW8 Coronal**
**CR 31 mm**
**C9226**
**Level 21:**
**Section 206**

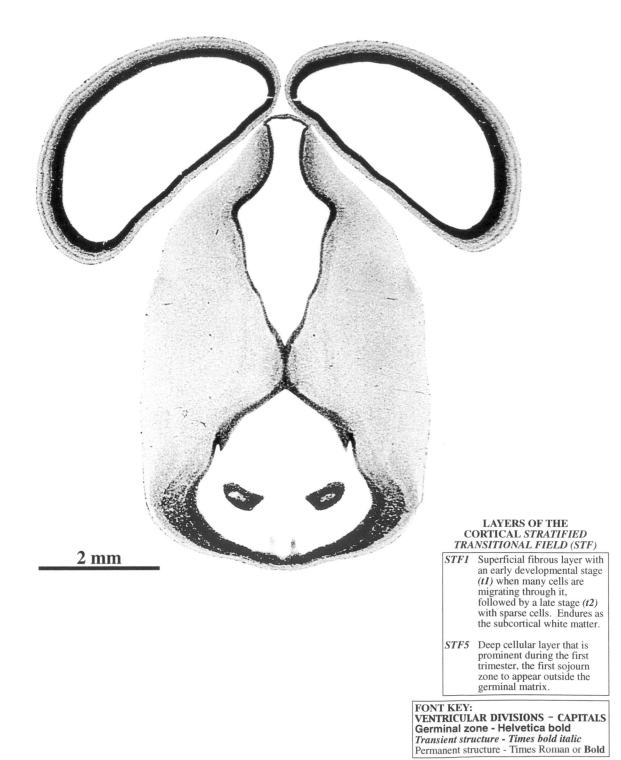

**2 mm**

**LAYERS OF THE**
**CORTICAL *STRATIFIED***
***TRANSITIONAL FIELD (STF)***

| | |
|---|---|
| *STF1* | Superficial fibrous layer with an early developmental stage *(t1)* when many cells are migrating through it, followed by a late stage *(t2)* with sparse cells. Endures as the subcortical white matter. |
| *STF5* | Deep cellular layer that is prominent during the first trimester, the first sojourn zone to appear outside the germinal matrix. |

**FONT KEY:**
**VENTRICULAR DIVISIONS − CAPITALS**
**Germinal zone - Helvetica bold**
***Transient structure - Times bold italic***
Permanent structure - Times Roman or **Bold**

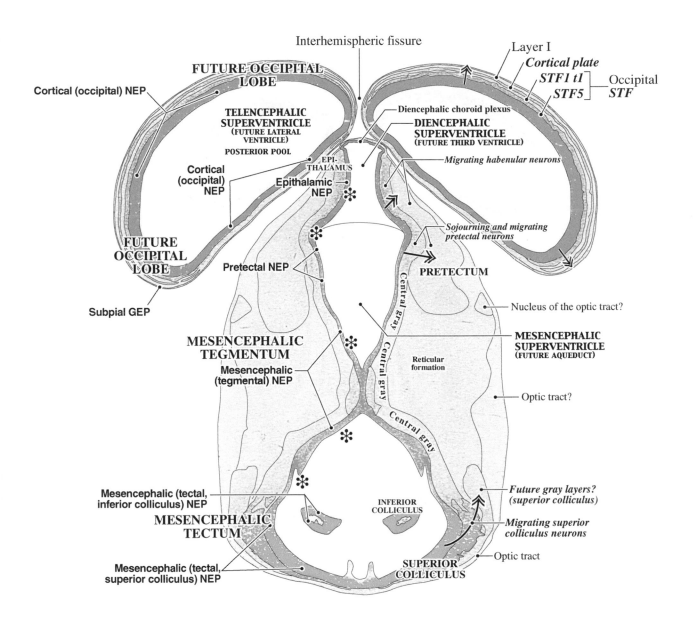

Interhemispheric fissure

FUTURE OCCIPITAL
LOBE

Cortical (occipital) NEP

TELENCEPHALIC
SUPERVENTRICLE
(FUTURE LATERAL
VENTRICLE)

POSTERIOR POOL

Cortical
(occipital)
NEP

EPI-
THALAMUS

Epithalamic
NEP

FUTURE
OCCIPITAL
LOBE

Pretectal NEP

Subpial GEP

MESENCEPHALIC
TEGMENTUM

Mesencephalic
(tegmental) NEP

Layer I
*Cortical plate*
*STF1 t1*
*STF5*

Occipital
*STF*

Diencephalic choroid plexus

DIENCEPHALIC
SUPERVENTRICLE
(FUTURE THIRD VENTRICLE)

*Migrating habenular neurons*

*Sojourning and migrating*
*pretectal neurons*

PRETECTUM

Central gray

Central gray

Central gray

Reticular
formation

Nucleus of the optic tract?

MESENCEPHALIC
SUPERVENTRICLE
(FUTURE AQUEDUCT)

Optic tract?

INFERIOR
COLLICULUS

Mesencephalic (tectal,
inferior colliculus) NEP

MESENCEPHALIC
TECTUM

Mesencephalic (tectal,
superior colliculus) NEP

*Future gray layers?*
*(superior colliculus)*

*Migrating superior*
*colliculus neurons*

Optic tract

SUPERIOR
COLLICULUS

✳ Evaginations and invaginations of the
neuroepithelium are mosaic compartments
that give rise to different brain structures.

ABBREVIATIONS:
**GEP - Glioepithelium**
**NEP - Neuroepithelium**

**Arrows** indicate the
presumed *direction of*
*neuron migration* from
neuroepithelial sources.

**PLATE 166A**

**GW8 Coronal**
**CR 31 mm**
**C9226**
**Level 22:**
**Section 170**

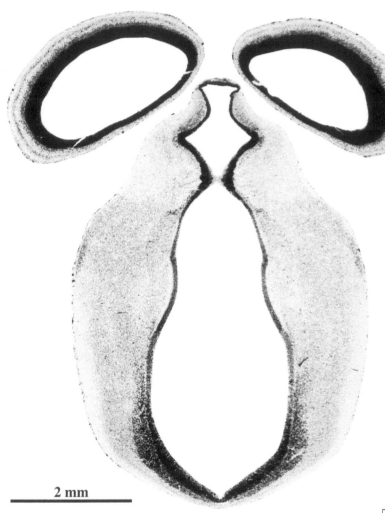

2 mm

**LAYERS OF THE**
**CORTICAL *STRATIFIED***
***TRANSITIONAL FIELD (STF)***

*STF1*  Superficial fibrous layer with
an early developmental stage
*(t1)* when many cells are
migrating through it,
followed by a late stage *(t2)*
with sparse cells. Endures as
the subcortical white matter.

*STF5*  Deep cellular layer that is
prominent during the first
trimester, the first sojourn
zone to appear outside the
germinal matrix.

**FONT KEY:**
**VENTRICULAR DIVISIONS – CAPITALS**
**Germinal zone - Helvetica bold**
***Transient structure - Times bold italic***
Permanent structure - Times Roman or **Bold**

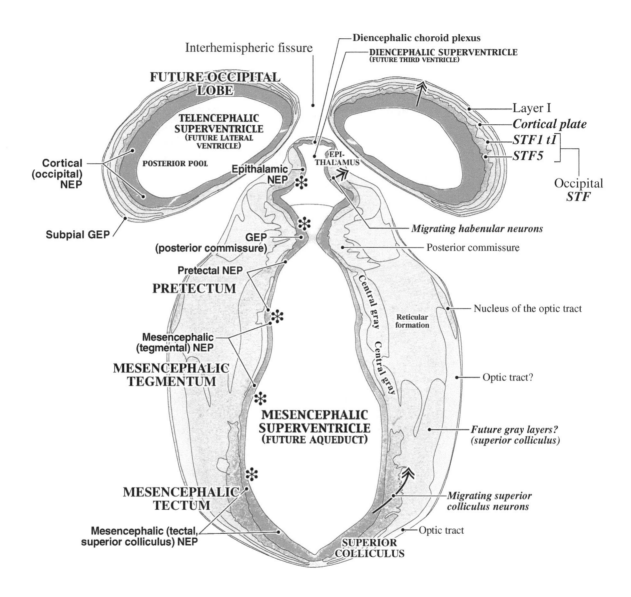

Interhemispheric fissure

Diencephalic choroid plexus

**DIENCEPHALIC SUPERVENTRICLE**
(FUTURE THIRD VENTRICLE)

**FUTURE OCCIPITAL LOBE**

Layer I

*Cortical plate*

**TELENCEPHALIC SUPERVENTRICLE**
(FUTURE LATERAL VENTRICLE)

*STF1 t1*

*STF5*

**POSTERIOR POOL**

Cortical (occipital) NEP

EPI-THALAMUS

Epithalamic NEP

Occipital *STF*

Subpial GEP

*Migrating habenular neurons*

GEP (posterior commissure)

Posterior commissure

Pretectal NEP

**PRETECTUM**

Nucleus of the optic tract

Central gray

Reticular formation

Mesencephalic (tegmental) NEP

Central gray

**MESENCEPHALIC TEGMENTUM**

Optic tract?

**MESENCEPHALIC SUPERVENTRICLE**
(FUTURE AQUEDUCT)

*Future gray layers?
(superior colliculus)*

**MESENCEPHALIC TECTUM**

*Migrating superior colliculus neurons*

Optic tract

Mesencephalic (tectal, superior colliculus) NEP

**SUPERIOR COLLICULUS**

✷ Evaginations and invaginations of the neuroepithelium are mosaic compartments that give rise to different brain structures.

**ABBREVIATIONS:**
**GEP - Glioepithelium**
**NEP - Neuroepithelium**

**Arrows** indicate the presumed *direction of neuron migration* from neuroepithelial sources.

**PLATE 167A**

**GW8 Coronal
CR 31 mm
C9226
Level 23:
Section 158**

2 mm

**LAYERS OF THE
CORTICAL *STRATIFIED
TRANSITIONAL FIELD (STF)***

*STF1*   Superficial fibrous layer with
an early developmental stage
*(t1)* when many cells are
migrating through it,
followed by a late stage *(t2)*
with sparse cells. Endures as
the subcortical white matter.

*STF5*   Deep cellular layer that is
prominent during the first
trimester, the first sojourn
zone to appear outside the
germinal matrix.

FONT KEY:
**VENTRICULAR DIVISIONS – CAPITALS**
**Germinal zone - Helvetica bold**
*Transient structure - Times bold italic*
Permanent structure - Times Roman or **Bold**

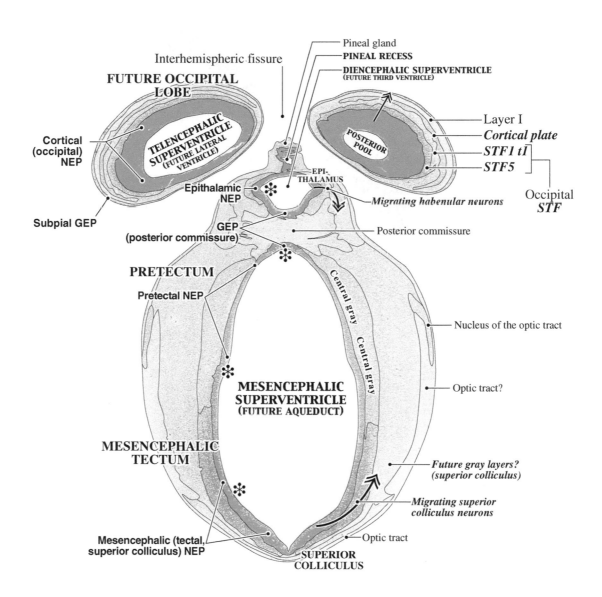

Pineal gland
PINEAL RECESS
DIENCEPHALIC SUPERVENTRICLE
(FUTURE THIRD VENTRICLE)

Interhemispheric fissure

**FUTURE OCCIPITAL
LOBE**

Cortical
(occipital)
NEP

TELENCEPHALIC
SUPERVENTRICLE
(FUTURE LATERAL
VENTRICLE)

POSTERIOR
POOL

Layer I
*Cortical plate*
*STF1 tI*
*STF5*

Occipital
*STF*

Subpial GEP

Epithalamic
NEP

EPI-
THALAMUS

*Migrating habenular neurons*

GEP
(posterior commissure)

Posterior commissure

**PRETECTUM**

Pretectal NEP

Central gray

Central gray

Nucleus of the optic tract

**MESENCEPHALIC
SUPERVENTRICLE
(FUTURE AQUEDUCT)**

Optic tract?

**MESENCEPHALIC
TECTUM**

*Future gray layers?
(superior colliculus)*

*Migrating superior
colliculus neurons*

Mesencephalic (tectal,
superior colliculus) NEP

Optic tract

**SUPERIOR
COLLICULUS**

✳ Evaginations and invaginations of the
neuroepithelium are mosaic compartments
that give rise to different brain structures.

**ABBREVIATIONS:
GEP - Glioepithelium
NEP - Neuroepithelium**

**Arrows** indicate the
presumed *direction of
neuron migration* from
neuroepithelial sources.

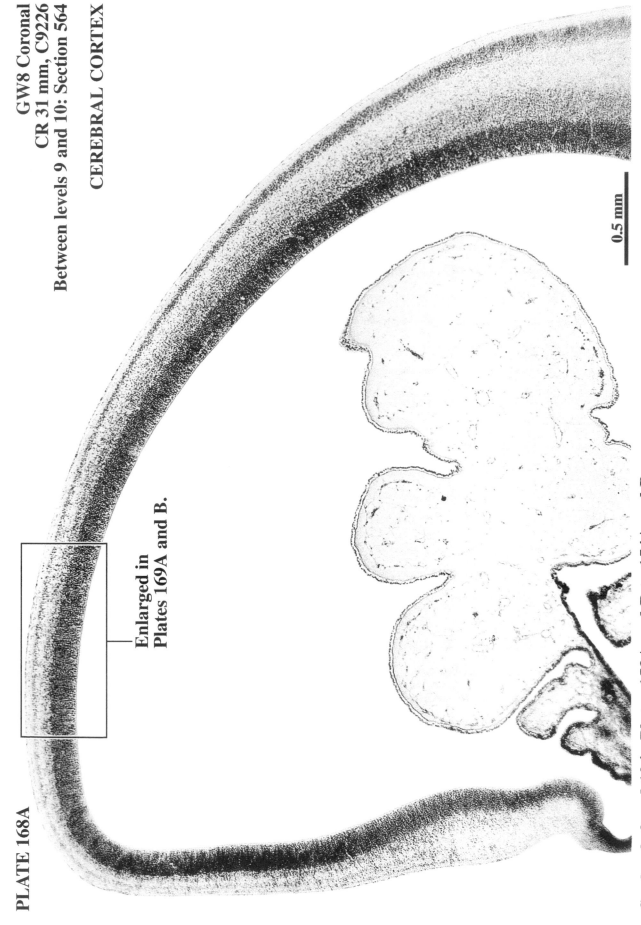

**PLATE 168A**

GW8 Coronal
CR 31 mm, C9226
Between levels 9 and 10: Section 564
CEREBRAL CORTEX

0.5 mm

Enlarged in
Plates 169A and B.

See levels 9 and 10 in Plates 153A and B to 154A and B.

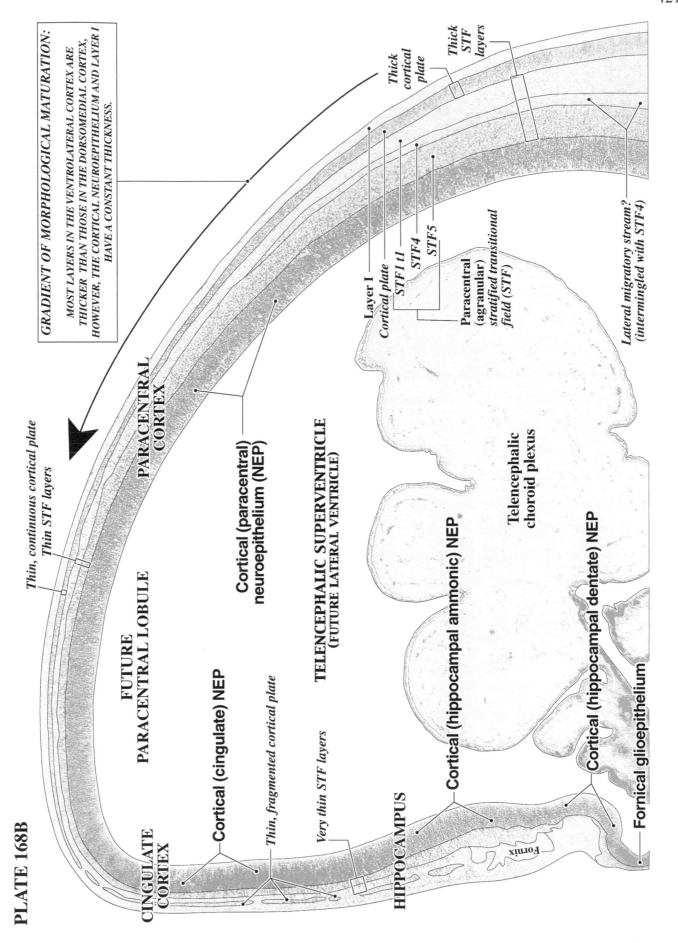

PLATE 168B

421

**GRADIENT OF MORPHOLOGICAL MATURATION:**

*MOST LAYERS IN THE VENTROLATERAL CORTEX ARE THICKER THAN THOSE IN THE DORSOMEDIAL CORTEX, HOWEVER, THE CORTICAL NEUROEPITHELIUM AND LAYER I HAVE A CONSTANT THICKNESS.*

Thick cortical plate

*Thick STF layers*

Layer I
Cortical plate
*STF1 t1*
*STF4*
*STF5*

Paracentral (agranular) stratified transitional field (STF)

Lateral migratory stream? (intermingled with STF4)

*Thin, continuous cortical plate*
*Thin STF layers*

FUTURE
PARACENTRAL LOBULE

PARACENTRAL
CORTEX

Cortical (paracentral) neuroepithelium (NEP)

TELENCEPHALIC SUPERVENTRICLE
(FUTURE LATERAL VENTRICLE)

Telencephalic choroid plexus

Cortical (cingulate) NEP

CINGULATE
CORTEX

*Thin, fragmented cortical plate*

*Very thin STF layers*

HIPPOCAMPUS

Cortical (hippocampal ammonic) NEP

Cortical (hippocampal dentate) NEP

Fornix

Fornical glioepithelium

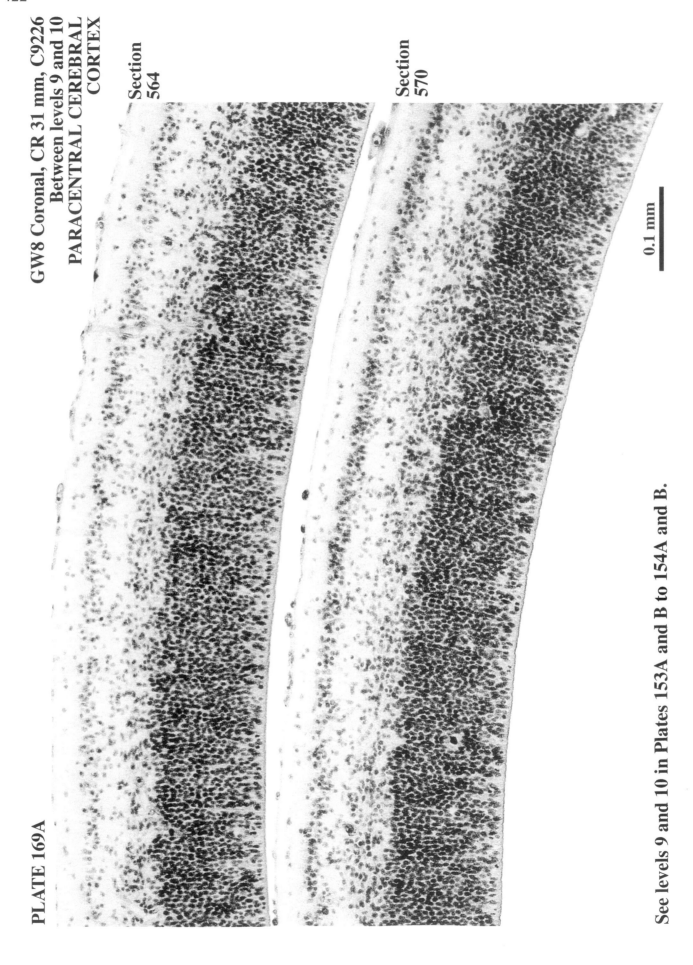

422

PLATE 169A

GW8 Coronal, CR 31 mm, C9226
Between levels 9 and 10
PARACENTRAL CEREBRAL
CORTEX

Section
564

Section
570

0.1 mm

See levels 9 and 10 in Plates 153A and B to 154A and B.

# PLATE 169B

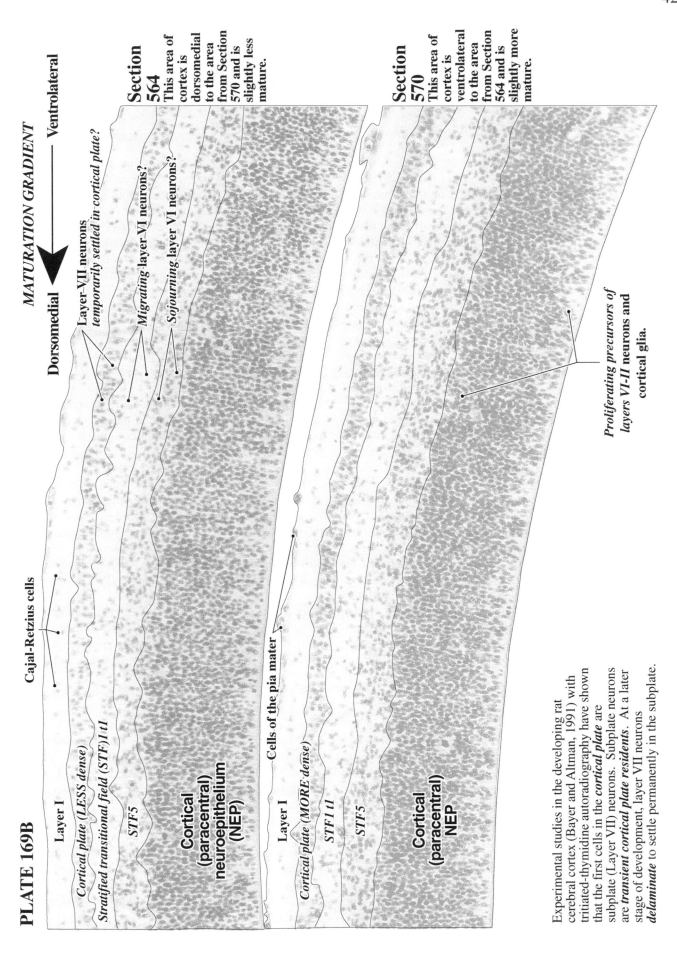

*MATURATION GRADIENT*

Dorsomedial ← → Ventrolateral

Cajal-Retzius cells

Layer I

Cortical plate (*LESS dense*)

Stratified transitional field (*STF*)1t1

*STF5*

**Cortical (paracentral) neuroepithelium (NEP)**

Layer-VII neurons *temporarily settled in cortical plate?*

*Migrating* layer-VI neurons?

*Sojourning* layer VI neurons?

**Section 564**

This area of cortex is dorsomedial to the area from Section 570 and is slightly less mature.

Cells of the pia mater

Layer I

Cortical plate (*MORE dense*)

*STF1t1*

*STF5*

**Cortical (paracentral) NEP**

**Section 570**

This area of cortex is ventrolateral to the area from Section 564 and is slightly more mature.

*Proliferating precursors of layers VI-II neurons and cortical glia.*

Experimental studies in the developing rat cerebral cortex (Bayer and Altman, 1991) with tritiated-thymidine autoradiography have shown that the first cells in the *cortical plate* are subplate (Layer VII) neurons. Subplate neurons are *transient cortical plate residents*. At a later stage of development, layer VII neurons *delaminate* to settle permanently in the subplate.

# PART IX: GW8 HORIZONTAL

This is specimen number 609 in the Carnegie Collection, designated here as C609. A normal female fetus with a crown-rump length (CR) of 32 mm was collected in 1916, and is estimated to be in gestational week (GW) 8. The entire fetus was embedded in paraffin, cut transversely in 50 μm thick sections, and stained with aluminum cochineal. Since there is no photograph of this brain before it was embedded and cut, a specimen from Hochstetter (1919) that is only partially comparable to C609 has been modified to show the approximate section plane and external features of the brain at GW8 (**Figure 8**). Like most of the specimens in this Volume, the sections are not cut exactly in one plane; C609's cortex is cut midway between coronal and horizontal planes, and is presented as a "horizontal" brain. The C609 section planes through the cortex and brainstem are not at the same angle when transferred to Hochstetter's CR27 mm specimen. Instead, brainstem planes of section appear to fan upward and downward from sections in the cortex apparently around a fulcrum centering in the invagination of the medullary velum overlying the rhombencephalic superventricle. We interpret this to indicate that the brain flexures are more loosely folded in the Hochstetter specimen than in C609. But it is difficult to determine how the brainstem is folded in C609 to make the section planes line up with those in the cortex. Photographs of 23 sections (**Levels 1-10**) are illustrated at low magnification in **Plates 170-179**. High-magnification views of different areas of the brain are shown in **Plates 180-185**. To maximize image size within page space, all of C609's sections are rotated 90° (landscape orientation). The anterior part of each section is on the left (page bottom), and the posterior part of each section is on the right (page top).

C609 is similar to the other GW8 specimens and shows brain maturation in still another perspective. The telencephalic and rhombencephalic *superventricles* are obvious, along with the slit-like diencephalic and mesencephalic superventricles. The parenchyma, the area between the superficial border of the *neuroepithelium (NEP) / subventricular zone (SVZ)* and the pial membrane, is the region where neurons migrate, settle, and differentiate. The thicknesses of the neuroepithelium and the parenchyma are similar to those in C9226 throughout the brain indicating brain maturation in both specimens is similar.

The parenchyma is thick and bordered by a thin NEP in the medulla, pons, and midbrain tegmentum without sur-

rounding dense sojourn zones. Most neurons have been generated here, few are migrating, and most are settled and differentiating. The two exceptions seen in C9226 are also seen in C145. First, presumptive facial motor neurons are clumped near the pontomedullary trench and some are migrating toward their ventral pontine/medullary settling sites. Second, the thicker *precerebellar neuroepithelium* in the medulla is generating predominantly pontine gray neurons; many precerebellar neurons are migrating in the *anterior and posterior extramural migratory streams*. The cerebellar NEP is thicker than that in the pons and medulla, and the cerebellar parenchyma has a dense Purkinje cell sojourn zone below presumptive earlier-generated deep neurons; the *external germinal layer (egl)* is rudimentary. The mesencephalic tectal NEP is thick adjacent to a thin parenchyma that contains dense sojourning and migrating neurons; substantial neurogenesis is ongoing in both the superior and inferior colliculi in the midbrain tectum.

The prominent diencephalic NEP and thick parenchyma filled with dense zones of sojourning and migrating neurons is remarkable in C609. Many migratory streams are visible in the thalamus, and migrating subthalamic nucleus neurons can be followed from the posterior hypothalamic NEP to the subthalamic nucleus. Although many diencephalic neurons have been generated by GW8, most of them are still migrating and few have settled. In spite of that, there are large accumulations of fibers (presumably thalamic axons) in the lateral thalamus and internal capsule, indicating that young thalamic neurons grow axons toward the cerebral cortex before settling and differentiating.

Within the telencephalon, the cerebral cortex has a thick NEP and a very thin parenchyma, indicating that most of its neurons are still not generated. The cerebral cortical NEP is the sole germinal matrix. The *stratified transitional field (STF)* contains *STF1* and *STF5* only in lateral areas. The pronounced anterolateral (thicker) to dorsomedial (thinner) maturation gradient is evident in the cerebral cortex. The basal telencephalic NEP/SVZ and parenchyma are both thick because there are large early-generated neuronal populations (for example, globus pallidus and substantia innominata) and massive late-generated populations (striatal neurons in the caudate and putamen). Most of the neurons settling in the basal telencephalon at GW8 are those of the early-generated populations, while many striatal neurons have not been generated yet.

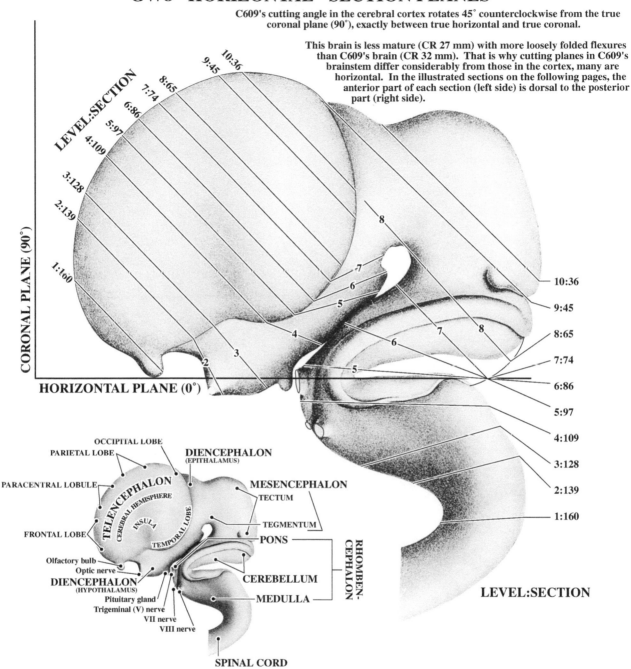

## GW8 "HORIZONTAL" SECTION PLANES

C609's cutting angle in the cerebral cortex rotates 45° counterclockwise from the true coronal plane (90°), exactly between true horizontal and true coronal.

This brain is less mature (CR 27 mm) with more loosely folded flexures than C609's brain (CR 32 mm). That is why cutting planes in C609's brainstem differ considerably from those in the cortex, many are horizontal. In the illustrated sections on the following pages, the anterior part of each section (left side) is dorsal to the posterior part (right side).

LEVEL:SECTION

10:36
9:45
8:65
7:74
6:86
5:97
4:109
3:128
2:139
1:160

CORONAL PLANE (90°)

HORIZONTAL PLANE (0°)

10:36
9:45
8:65
7:74
6:86
5:97
4:109
3:128
2:139
1:160

LEVEL:SECTION

OCCIPITAL LOBE
PARIETAL LOBE
PARACENTRAL LOBULE
DIENCEPHALON (EPITHALAMUS)
TELENCEPHALON
CEREBRAL HEMISPHERE
INSULA
MESENCEPHALON
TECTUM
TEGMENTUM
FRONTAL LOBE
TEMPORAL LOBE
PONS
Olfactory bulb
Optic nerve
DIENCEPHALON (HYPOTHALAMUS)
Pituitary gland
Trigeminal (V) nerve
VII nerve
VIII nerve
CEREBELLUM
MEDULLA
RHOMBEN-CEPHALON
SPINAL CORD

**Figure 8.** The lateral view of the brain and upper cervical spinal cord from a specimen with a crown-rump length of 27 mm (modified from Figure 37, Table VII, Hochstetter, 1919) serves to show the approximate locations and cutting angles of the illustrated sections of C609 in the following pages. The small inset identifies the major structural features. The line in the cerebellum and dorsal edges of the pons and medulla is the cut edge of the medullary velum.

**PLATE 170A**

**GW8 Horizontal**
**CR 32 mm**
**C609**
**Level 1:**
**Section 160**

2 mm

**LAYERS OF THE CORTICAL *STRATIFIED TRANSITIONAL FIELD (STF)***

*STF1*  Superficial fibrous layer with an early developmental stage (*t1*) when many cells are migrating through it, followed by a late stage (*t2*) with sparse cells.  Endures as the subcortical white matter.

*STF5*  Deep cellular layer that is prominent during the first trimester, the first sojourn zone to appear outside the germinal matrix.

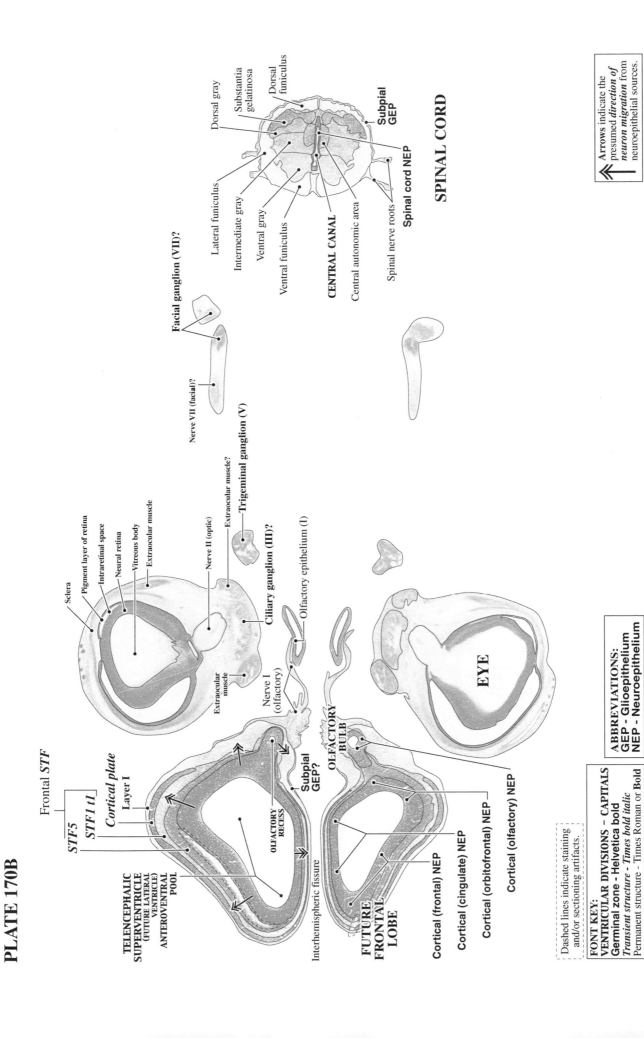

427

**PLATE 170B**

SPINAL CORD

Dorsal gray
Substantia gelatinosa
Dorsal funiculus
**Subpial GEP**
Lateral funiculus
Intermediate gray
Ventral gray
Ventral funiculus
**CENTRAL CANAL**
Central autonomic area
Spinal cord NEP
Spinal nerve roots

**Facial ganglion (VII)?**
Nerve VII (facial)?
**Trigeminal ganglion (V)**
Extraocular muscle?

Sclera
Pigment layer of retina
Intraretinal space
Neural retina
Vitreous body
Extraocular muscle
Nerve II (optic)
**Ciliary ganglion (III)?**
Olfactory epithelium (I)
Extraocular muscle
Nerve I (olfactory)
**Subpial GEP?**
**OLFACTORY BULB**
**EYE**

Frontal *STF*
*STF5*
*STF1t1*
*Cortical plate*
Layer I
**TELENCEPHALIC SUPERVENTRICLE (FUTURE LATERAL VENTRICLE) ANTEROVENTRAL POOL**
**OLFACTORY RECESS**
Interhemispheric fissure
**FUTURE FRONTAL LOBE**
Cortical (frontal) NEP
Cortical (cingulate) NEP
Cortical (orbitofrontal) NEP
Cortical (olfactory) NEP

Arrows indicate the presumed *direction of neuron migration* from neuroepithelial sources.

Dashed lines indicate staining and/or sectioning artifacts.

**FONT KEY:**
VENTRICULAR DIVISIONS - CAPITALS
**Germinal zone - Helvetica bold**
*Transient structure - Times bold italic*
Permanent structure - Times Roman or Bold

**ABBREVIATIONS:**
**GEP** - Glioepithelium
**NEP** - Neuroepithelium

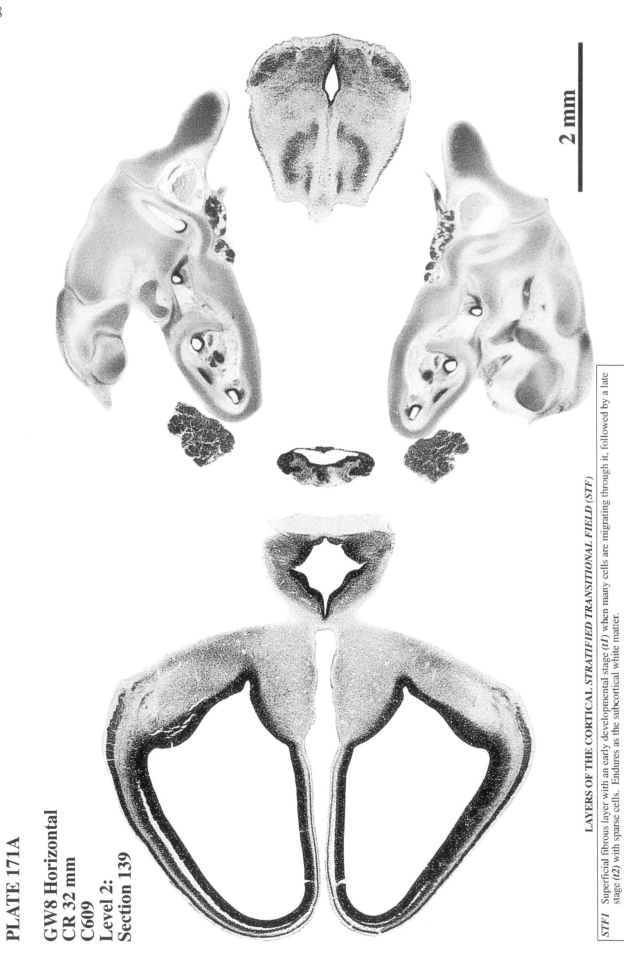

**PLATE 171A**

**GW8 Horizontal**
**CR 32 mm**
**C609**
**Level 2:**
**Section 139**

**LAYERS OF THE CORTICAL *STRATIFIED TRANSITIONAL FIELD* (STF)**

*STF1*  Superficial fibrous layer with an early developmental stage (*t1*) when many cells are migrating through it, followed by a late
stage (*t2*) with sparse cells.  Endures as the subcortical white matter.

*STF5*  Deep cellular layer that is prominent during the first trimester, the first sojourn zone to appear outside the germinal matrix.

2 mm

## PLATE 171B

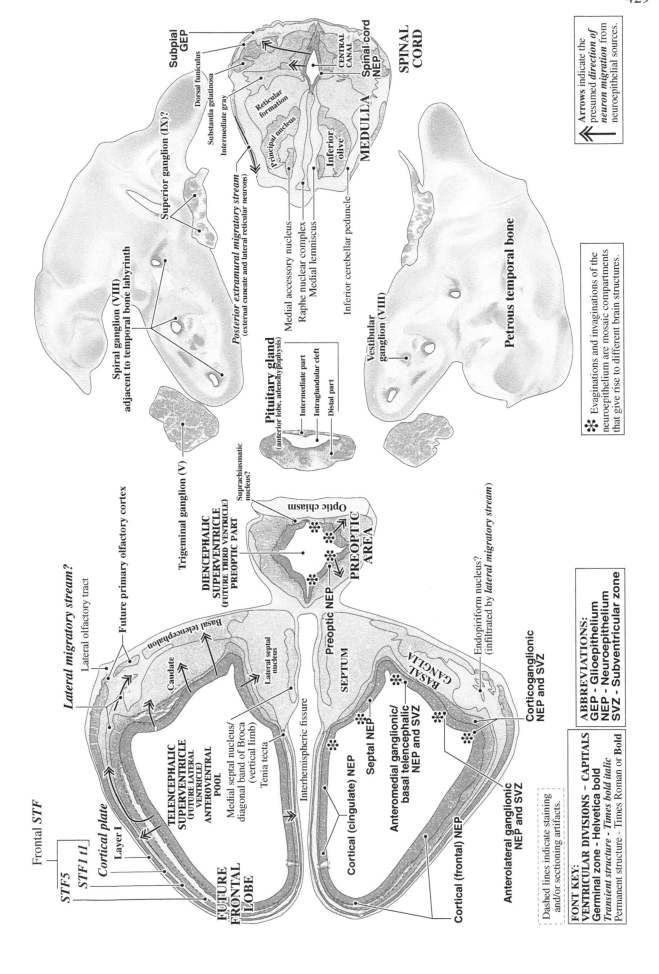

**Frontal STF**

STF5
STF1 t1

*Cortical plate*
Layer I

**FUTURE FRONTAL LOBE**

**TELENCEPHALIC SUPERVENTRICLE** (FUTURE LATERAL VENTRICLE) **ANTEROVENTRAL POOL**

Medial septal nucleus/ diagonal band of Broca (vertical limb)

Tenia tecta

Interhemispheric fissure

Caudate

Basal telencephalon

*Lateral migratory stream?*
Lateral olfactory tract

Future primary olfactory cortex

Trigeminal ganglion (V)

**DIENCEPHALIC SUPERVENTRICLE** (FUTURE THIRD VENTRICLE) **PREOPTIC PART**

Suprachiasmatic nucleus?

Optic chiasm

Preoptic NEP

**PREOPTIC AREA**

**SEPTUM**

Lateral septal nucleus

Cortical (cingulate) NEP

Septal NEP

**Anteromedial ganglionic/ basal telencephalic NEP and SVZ**

**BASAL GANGLIA**

Endopiriform nucleus? (infiltrated by *lateral migratory stream*)

**Corticoganglionic NEP and SVZ**

**Cortical (frontal) NEP**

**Anterolateral ganglionic NEP and SVZ**

Spiral ganglion (VIII) adjacent to temporal bone labyrinth

Superior ganglion (IX)?

Vestibular ganglion (VIII)

**Petrous temporal bone**

**Subpial GEP**
Dorsal funiculus
Dorsal funiculus
Substantia gelatinosa
Intermediate gray

Reticular formation

Principal nucleus

*Posterior extramural migratory stream* (external cuneate and lateral reticular neurons)

Medial accessory nucleus
Raphe nuclear complex
Medial lemniscus

Inferior olive

Inferior cerebellar peduncle

**MEDULLA**

CENTRAL CANAL

Spinal-cord NEP

**SPINAL CORD**

**Pituitary gland** (anterior lobe, adenohypophysis)
Intermediate part
Intraglandular cleft
Distal part

Arrows indicate the presumed *direction of neuron migration* from neuroepithelial sources.

✳ Evaginations and invaginations of the neuroepithelium are mosaic compartments that give rise to different brain structures.

Dashed lines indicate staining and/or sectioning artifacts.

**FONT KEY:**
VENTRICULAR DIVISIONS – CAPITALS
Germinal zone – Helvetica bold
*Transient structure – Times bold italic*
Permanent structure – Times Roman or Bold

**ABBREVIATIONS:**
GEP – Glioepithelium
NEP – Neuroepithelium
SVZ – Subventricular zone

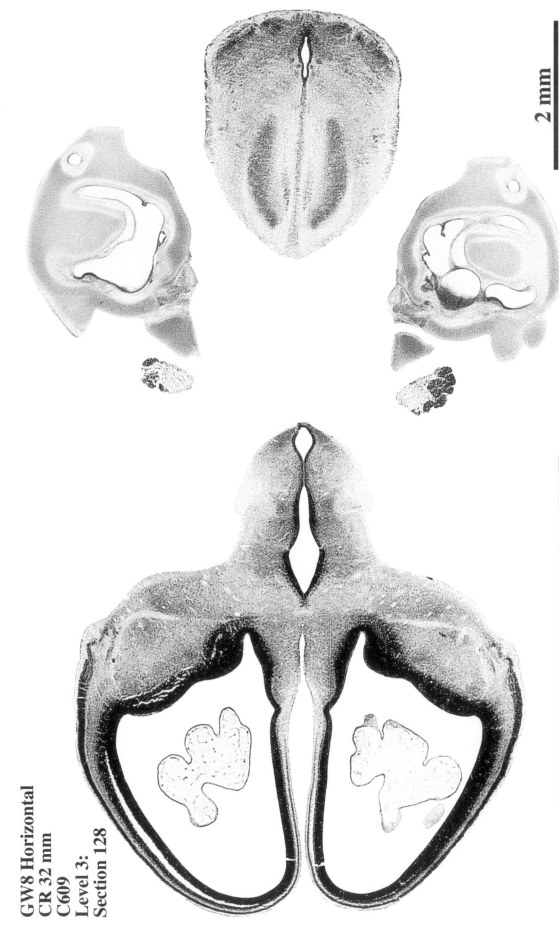

430

**PLATE 172A**

**GW8 Horizontal**
**CR 32 mm**
**C609**
**Level 3:**
**Section 128**

**LAYERS OF THE CORTICAL STRATIFIED TRANSITIONAL FIELD (STF)**

*STF1* Superficial fibrous layer with an early developmental stage (*t1*) when many cells are migrating through it, followed by a late stage (*t2*) with sparse cells. Endures as the subcortical white matter.

*STF5* Deep cellular layer that is prominent during the first trimester, the first sojourn zone to appear outside the germinal matrix.

2 mm

# PLATE 172B

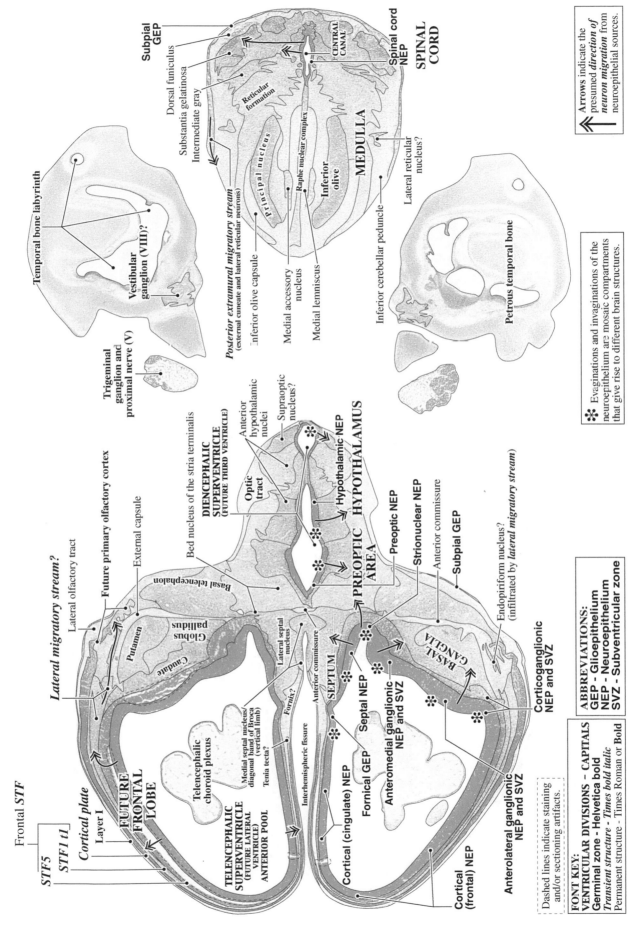

431

**Arrows** indicate the presumed *direction of neuron migration* from neuroepithelial sources.

✻ Evaginations and invaginations of the neuroepithelium are mosaic compartments that give rise to different brain structures.

Dashed lines indicate staining and/or sectioning artifacts.

**FONT KEY:**
VENTRICULAR DIVISIONS - CAPITALS
Germinal zone - Helvetica bold
*Transient structure - Times bold italic*
Permanent structure - Times Roman or Bold

**ABBREVIATIONS:**
GEP - Glioepithelium
NEP - Neuroepithelium
SVZ - Subventricular zone

Subpial GEP
Dorsal funiculus
Substantia gelatinosa
Intermediate gray
Reticular formation
Spinal cord NEP
CENTRAL CANAL
SPINAL CORD
MEDULLA
Principal nucleus
Raphe nuclear complex
Inferior olive
*Posterior extramural migratory stream* (external cuneate and lateral reticular neurons)
Inferior olive capsule
Medial accessory nucleus
Medial lemniscus
Inferior cerebellar peduncle
Lateral reticular nucleus?
Petrous temporal bone

Temporal bone labyrinth
Vestibular ganglion (VIII)?
Trigeminal ganglion and proximal nerve (V)

Frontal *STF*
*STF5*
*STF11t1*
*Cortical plate*
Layer I
**FUTURE FRONTAL LOBE**
*Lateral migratory stream?*
Lateral olfactory tract
Future primary olfactory cortex
External capsule
Bed nucleus of the stria terminalis
Basal telencephalon
Putamen
Globus pallidus
Caudate
Lateral septal nucleus
Fornix?
Anterior commissure
Interhemispheric fissure
Tenia tecta?
Medial septal nucleus/ diagonal band of Broca (vertical limb)
Telencephalic choroid plexus
**TELENCEPHALIC SUPERVENTRICLE** (FUTURE LATERAL VENTRICLE) ANTERIOR POOL
**Cortical (frontal) NEP**
**Anterolateral ganglionic NEP and SVZ**
**Cortical (cingulate) NEP**
**Fornical GEP**
**Septal NEP**
**SEPTUM**
**Anteromedial ganglionic NEP and SVZ**
**BASAL GANGLIA**
**Corticoganglionic NEP and SVZ**
Endopiriform nucleus? (infiltrated by *lateral migratory stream*)
**Subpial GEP**
Anterior commissure
**Strionuclear NEP**
**Preoptic NEP**
**PREOPTIC AREA**
**Hypothalamic NEP**
**HYPOTHALAMUS**
Optic tract
Supraoptic nucleus?
Anterior hypothalamic nuclei
**DIENCEPHALIC SUPERVENTRICLE** (FUTURE THIRD VENTRICLE)

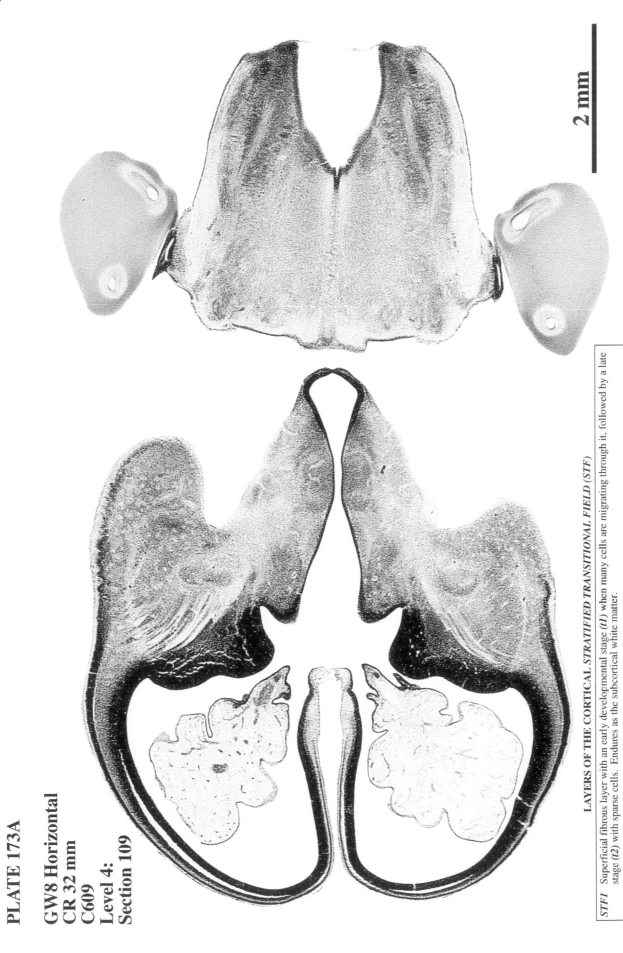

**PLATE 173A**

**GW8 Horizontal**
**CR 32 mm**
**C609**
**Level 4:**
**Section 109**

2 mm

**LAYERS OF THE CORTICAL *STRATIFIED TRANSITIONAL FIELD (STF)***

*STF1*  Superficial fibrous layer with an early developmental stage (*t1*) when many cells are migrating through it, followed by a late stage (*t2*) with sparse cells. Endures as the subcortical white matter.

*STF5*  Deep cellular layer that is prominent during the first trimester, the first sojourn zone to appear outside the germinal matrix.

# PLATE 173B

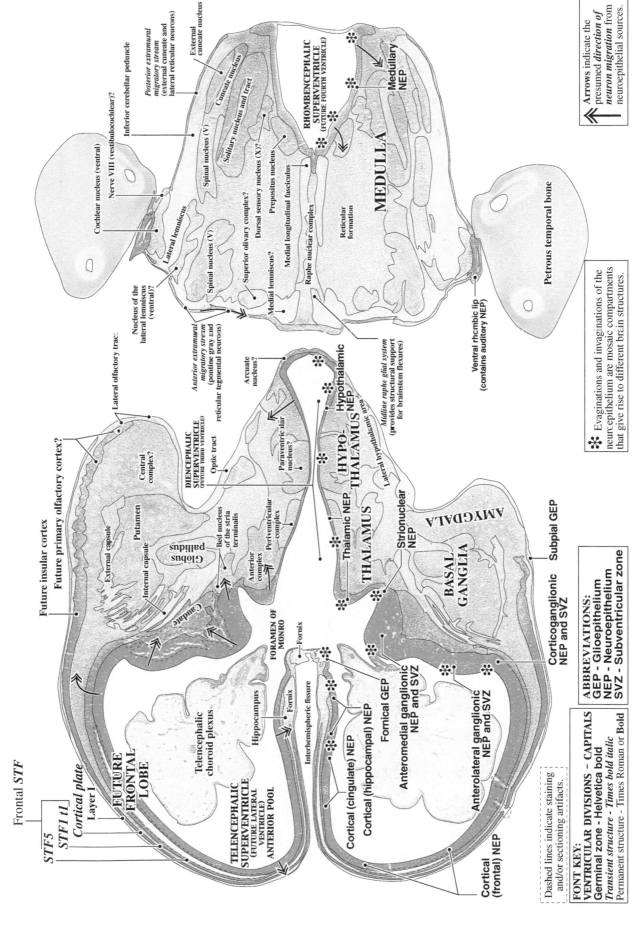

433

Frontal *STF*

*STF5*
*STF1 t1*
*Cortical plate*
Layer I

**FUTURE
FRONTAL
LOBE**

Future insular cortex
Future primary olfactory cortex?

Lateral olfactory tract

External capsule
Putamen
Internal capsule
Central
complex?

Globus pallidus

**DIENCEPHALIC
SUPERVENTRICLE**
(FUTURE THIRD VENTRICLE)
Optic tract

Bed nucleus
of the stria
terminalis

Paraventricular
nucleus?

Arcuate
nucleus?

Anterior
complex
Periventricular
complex

Caudate

**TELENCEPHALIC
SUPERVENTRICLE**
(FUTURE LATERAL
VENTRICLE)
**ANTERIOR POOL**

Telencephalic
choroid plexus

Hippocampus

FORAMEN OF
MONRO
Fornix

Fornix

Fornix

Interhemispheric fissure

Cortical (cingulate) NEP
Cortical (hippocampal) NEP

Fornical GEP

Anteromedial ganglionic
NEP and SVZ

Anterolateral ganglionic
NEP and SVZ

Cortical
(frontal) NEP

Cortical nucleus (ventral)
Nerve VIII (vestibulocochlear)?
Inferior cerebellar peduncle

*Posterior extramural
migratory stream*
(external cuneate and
lateral reticular neurons)

External
cuneate nucleus

Cuneate nucleus

Spinal nucleus (V)

Solitary nucleus and tract

Dorsal sensory nucleus (X)?

Prepositus nucleus

**RHOMBENCEPHALIC
SUPERVENTRICLE**
(FUTURE FOURTH VENTRICLE)

Medullary
NEP

Lateral lemniscus

Spinal nucleus (V)

Superior olivary complex?

Medial lemniscus?

Medial longitudinal fasciculus

Raphe nuclear complex

Reticular
formation

**MEDULLA**

Nucleus of the
lateral lemniscus (ventral)?

*Anterior extramural
migratory stream*
(pontine gray and
reticular tegmental neurons)

*Midline raphe glial system*
(provides structural support
for brainstem flexures)

Ventral rhombic lip
(contains auditory NEP)

Petrous temporal bone

**HYPO-
THALAMUS**

Hypothalamic
NEP

Thalamic NEP

**THALAMUS**

*Lateral hypothalamic area*

Strionuclear
NEP

**AMYGDALA**

**BASAL
GANGLIA**

Subpial GEP

Corticoganglionic
NEP and SVZ

Arrows indicate the
presumed *direction of
neuron migration* from
neuroepithelial sources.

✻ Evaginations and invaginations of the
neurepithelium are mosaic compartments
that give rise to different brain structures.

Dashed lines indicate staining
and/or sectioning artifacts.

**FONT KEY:**
**VENTRICULAR DIVISIONS – CAPITALS**
**Germinal zone - Helvetica bold**
*Transient structure - Times bold italic*
Permanent structure - Times Roman or Bold

**ABBREVIATIONS:**
GEP - Glioepithelium
NEP - Neuroepithelium
SVZ - Subventricular zone

434

**PLATE 174A**

**GW8 Horizontal**
**CR 32 mm**
**C609**
**Level 5:**
**Section 97**

**See a high-magnification view of the diencephalon and basal ganglia in Plates 180A and B.**

2 mm

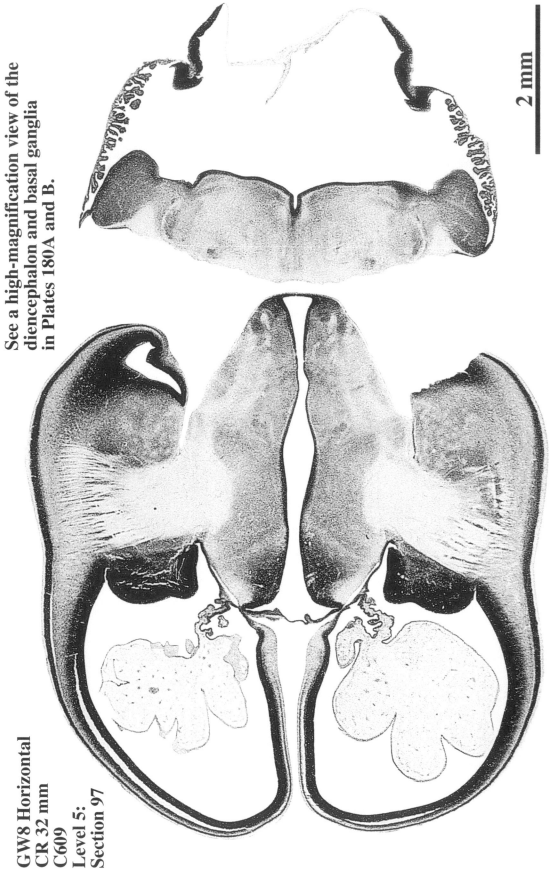

**LAYERS OF THE CORTICAL *STRATIFIED TRANSITIONAL FIELD* (STF)**

*STF1*  Superficial fibrous layer with an early developmental stage (*t1*) when many cells are migrating through it, followed by a late stage (*t2*) with sparse cells.  Endures as the subcortical white matter.

*STF5*  Deep cellular layer that is prominent during the first trimester, the first sojourn zone to appear outside the germinal matrix.

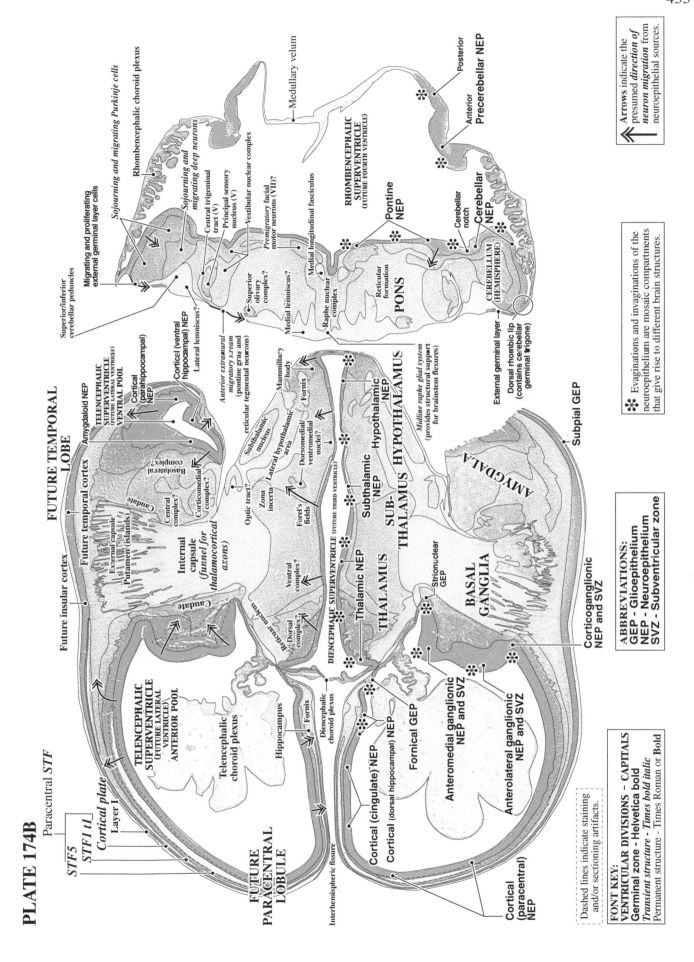

# PLATE 174B

435

Paracentral *STF*

STF5
STF1 t1
*Cortical plate*
Layer I

**FUTURE TEMPORAL LOBE**

Rhombencephalic choroid plexus

*Sojourning and migrating Purkinje cells*

Migrating and proliferating external germinal layer cells

Superior/inferior cerebellar peduncles

*Sojourning and migrating deep neurons*

Central trigeminal tract (V)

Principal sensory nucleus (V)

Vestibular nuclear complex

Premigratory facial motor neurons (VII)?

Medial longitudinal fasciculus

Medullary velum

Posterior

Anterior **Precerebellar NEP**

**RHOMBENCEPHALIC SUPERVENTRICLE** (FUTURE FOURTH VENTRICLE)

**Pontine NEP**

Cerebellar notch

**Cerebellar NEP**

**CEREBELLUM (HEMISPHERE)**

External germinal layer

Dorsal rhombic lip (contains cerebellar germinal trigone)

Superior olivary complex?

Medial lemniscus?

Raphe nuclear complex

Reticular formation **PONS**

Amygdaloid NEP

**TELENCEPHALIC SUPERVENTRICLE** (FUTURE LATERAL VENTRICLE) **VENTRAL POOL**

Future temporal cortex

Future insular cortex

Cortical (parahippocampal) NEP

**Cortical** (ventral hippocampal) NEP

Lateral lemniscus?

*Anterior extramural migratory stream (pontine gray and reticular tegmental neurons)*

Mammillary body

Fornix

**Hypothalamic NEP**

**SUB-HYPOTHALAMUS**

Subthalamic nucleus

Lateral hypothalamic area

Dorsomedial/ ventromedial nuclei?

**Subthalamic NEP**

**SUBTHALAMUS**

**AMYGDALA**

Subpial GEP

Basolateral complex?

Corticomedial complex?

Central complex?

Caudate

External capsule

Putamen (islands)

**Internal capsule** *(funnel for thalamocortical axons)*

Caudate

Reticular nucleus

Optic tract?

Zona incerta

Forel's fields

Ventral complex?

Dorsal complex?

**Thalamic NEP**

**THALAMUS**

Strionuclear GEP

**BASAL GANGLIA**

Corticoganglionic NEP and SVZ

**DIENCEPHALIC SUPERVENTRICLE** (FUTURE THIRD VENTRICLE)

*Midline raphe glial system (provides structural support for brainstem flexures)*

**TELENCEPHALIC SUPERVENTRICLE** (FUTURE LATERAL VENTRICLE) **ANTERIOR POOL**

Telencephalic choroid plexus

Hippocampus

Fornix

Diencephalic choroid plexus

**FUTURE PARACENTRAL LOBULE**

Interhemispheric fissure

Cortical (cingulate) NEP

Cortical (dorsal hippocampal) NEP

Fornical GEP

Anteromedial ganglionic NEP and SVZ

Anterolateral ganglionic NEP and SVZ

**Cortical (paracentral) NEP**

Dashed lines indicate staining and/or sectioning artifacts.

Arrows indicate the presumed *direction of neuron migration* from neuroepithelial sources.

✳ Evaginations and invaginations of the neuroepithelium are mosaic compartments that give rise to different brain structures.

**ABBREVIATIONS:**
GEP - Glioepithelium
NEP - Neuroepithelium
SVZ - Subventricular zone

**FONT KEY:**
VENTRICULAR DIVISIONS - CAPITALS
Germinal zone - Helvetica bold
*Transient structure - Times bold italic*
Permanent structure - Times Roman or Bold

**See high-magnification views of the diencephalon from section 91 in Plates 181A and B, from this section in Plates 182A and B, and of the hypothalamus from section 92 in Plates 185A and B.**

**PLATE 175A**

**GW8 Horizontal**
**CR 32 mm**
**C609**
**Level 6:**
**Section 86**

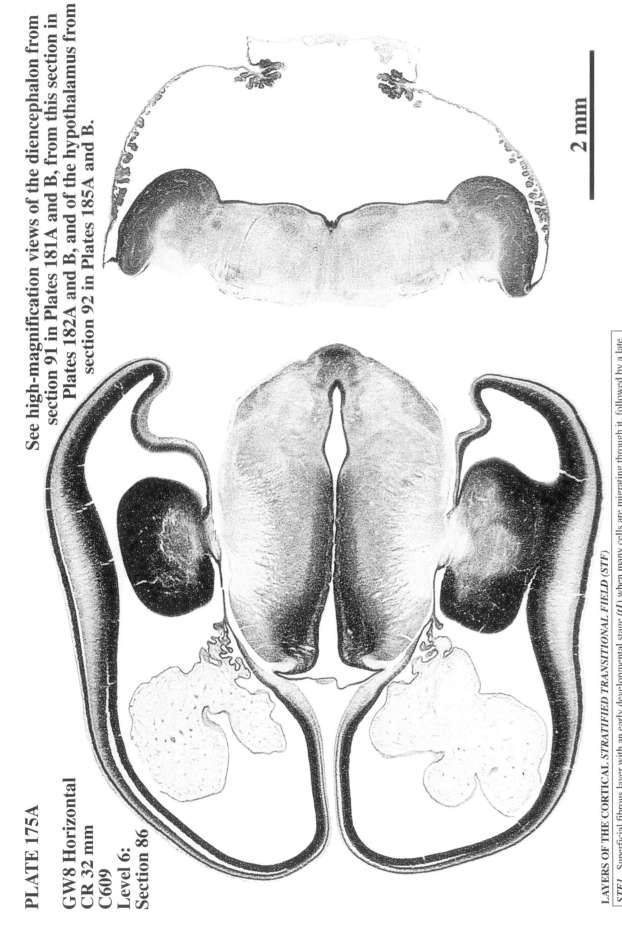

2 mm

*LAYERS OF THE CORTICAL STRATIFIED TRANSITIONAL FIELD (STF)*

*STF1*  Superficial fibrous layer with an early developmental stage (*t1*) when many cells are migrating through it, followed by a late stage (*t2*) with sparse cells.  Endures as the subcortical white matter.

*STF5*  Deep cellular layer that is prominent during the first trimester, the first sojourn zone to appear outside the germinal matrix.

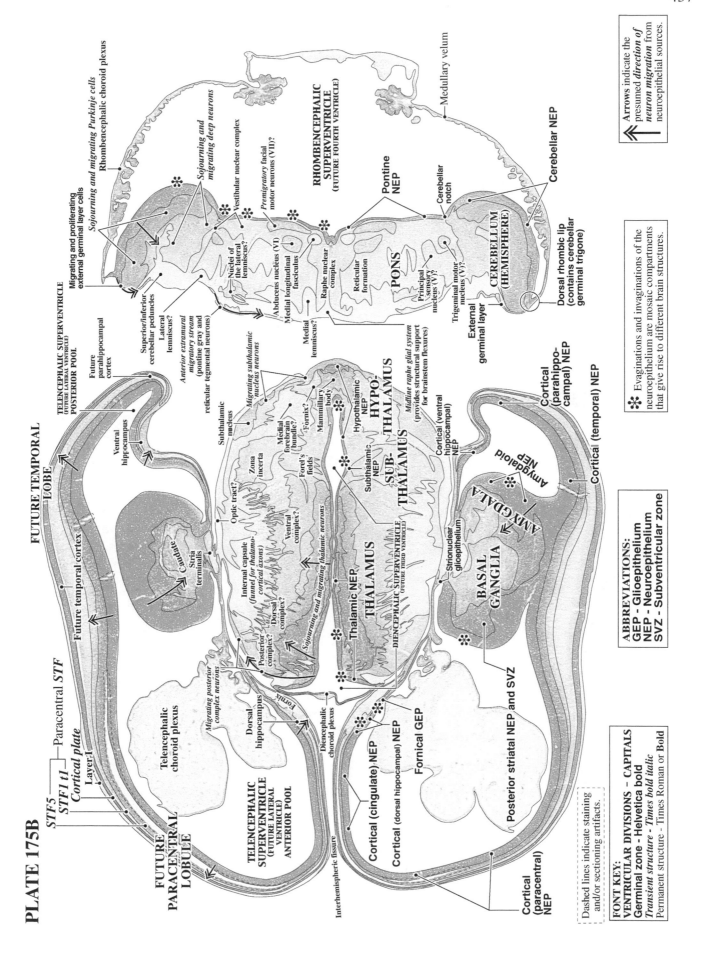

**PLATE 175B**

FUTURE TEMPORAL LOBE

FUTURE PARACENTRAL LOBULE

STF5
STF1 t1 — Paracentral STF
— Paracentral STF
Cortical plate
Layer 1

TELENCEPHALIC SUPERVENTRICLE
(FUTURE LATERAL VENTRICLE)
POSTERIOR POOL

Future parahippocampal cortex

Future temporal cortex

Ventral hippocampus

Caudate
Stria terminalis

Telencephalic choroid plexus

Dorsal hippocampus

TELENCEPHALIC SUPERVENTRICLE
(FUTURE LATERAL VENTRICLE)
ANTERIOR POOL

Interhemispheric fissure

Cortical (cingulate) NEP

Cortical (dorsal hippocampal) NEP

Fornical GEP

Diencephalic choroid plexus

Posterior striatal NEP and SVZ

Cortical (paracentral) NEP

BASAL GANGLIA

Strionuclear glioepithelium

Cortical (ventral hippocampal) NEP

AMYGDALA

Amygdaloid NEP

Cortical (temporal) NEP

Cortical (parahippo-campal) NEP

Migrating posterior complex neurons

Posterior complex?

Dorsal complex?

Internal capsule (funnel for thalamo-cortical axons)

Optic tract?

Zona incerta

Medial forebrain bundle?

Forel's fields

Ventral complex?

Fornix?

Fornix

Subthalamic nucleus

Migrating subthalamic nucleus neurons

Mammillary body

Sojourning and migrating thalamic neurons

Thalamic NEP

THALAMUS

DIENCEPHALIC SUPERVENTRICLE
(FUTURE THIRD VENTRICLE)

Subthalamic NEP

SUB-THALAMUS

Hypothalamic NEP

HYPO-THALAMUS

Midline raphe glial system (provides structural support for brainstem flexures)

TELENCEPHALIC SUPERVENTRICLE
(FUTURE LATERAL VENTRICLE)
POSTERIOR POOL

Migrating and proliferating external germinal layer cells

Sojourning and migrating Purkinje cells

Rhombencephalic choroid plexus

Sojourning and migrating deep neurons

Vestibular nuclear complex

Premigratory facial motor neurons (VII)?

RHOMBENCEPHALIC SUPERVENTRICLE
(FUTURE FOURTH VENTRICLE)

Medullary velum

Superior/inferior cerebellar peduncles

Lateral lemniscus?

Anterior extramural migratory stream (pontine gray and reticular tegmental neurons)

Nuclei of the lateral lemniscus?

Abducens nucleus (VI)

Medial longitudinal fasciculus

Raphe nuclear complex

Reticular formation

Medial lemniscus?

Principal sensory nucleus (V)?

Trigeminal motor nucleus (V)?

External germinal layer

Pontine NEP

PONS

Cerebellar notch

CEREBELLUM (HEMISPHERE)

Cerebellar NEP

Dorsal rhombic lip (contains cerebellar trigone)

Cerebellar NEP

Arrows indicate the presumed *direction of neuron migration* from neuroepithelial sources.

✻ Evaginations and invaginations of the neuroepithelium are mosaic compartments that give rise to different brain structures.

**ABBREVIATIONS:**
GEP - Glioepithelium
NEP - Neuroepithelium
SVZ - Subventricular zone

Dashed lines indicate staining and/or sectioning artifacts.

**FONT KEY:**
VENTRICULAR DIVISIONS – CAPITALS
Germinal zone - Helvetica bold
*Transient structure - Times bold italic*
Permanent structure - Times Roman or Bold

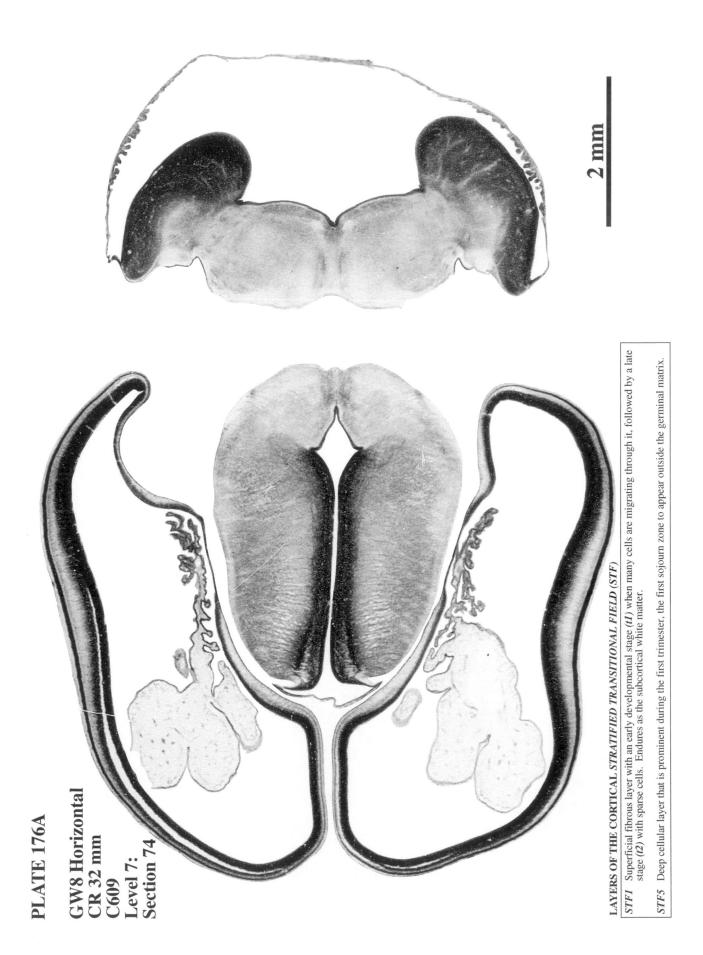

**PLATE 176A**

**GW8 Horizontal**
**CR 32 mm**
**C609**
**Level 7:**
**Section 74**

2 mm

*LAYERS OF THE CORTICAL STRATIFIED TRANSITIONAL FIELD (STF)*

*STF1*  Superficial fibrous layer with an early developmental stage (*t1*) when many cells are migrating through it, followed by a late
stage (*t2*) with sparse cells.  Endures as the subcortical white matter.

*STF5*  Deep cellular layer that is prominent during the first trimester, the first sojourn zone to appear outside the germinal matrix.

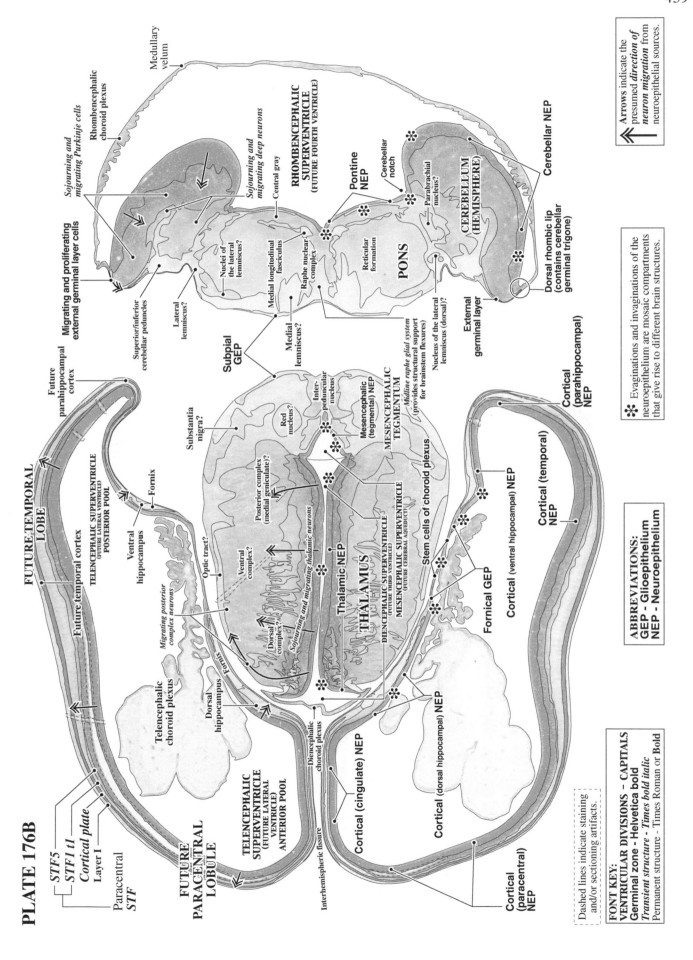

PLATE 176B

STF5
STF1 t1
Cortical plate
Layer I

Paracentral STF

FUTURE TEMPORAL LOBE

Future parahippocampal cortex

Future temporal cortex

Ventral hippocampus

Telencephalic choroid plexus

Dorsal hippocampus

FUTURE PARACENTRAL LOBULE

TELENCEPHALIC SUPERVENTRICLE (FUTURE LATERAL VENTRICLE) ANTERIOR POOL

TELENCEPHALIC SUPERVENTRICLE (FUTURE LATERAL VENTRICLE) POSTERIOR POOL

Fornix

Diencephalic choroid plexus

Interhemispheric fissure

Optic tract?

Migrating posterior complex neurons

Fornix

Dorsal complex?

Ventral complex?

Sojourning and migrating thalamic neurons

Cortical (cingulate) NEP

Cortical (dorsal hippocampal) NEP

Fornical GEP

Cortical (ventral hippocampal) NEP

Cortical (paracentral) NEP

Cortical (temporal) NEP

Cortical (parahippocampal) NEP

Substantia nigra?

Posterior complex (medial geniculate)?

Red nucleus?

Inter-peduncular nucleus

Mesencephalic (tegmental) NEP

MESENCEPHALIC TEGMENTUM

Midline raphe glial system (provides structural support for brainstem flexures)

MESENCEPHALIC SUPERVENTRICLE (FUTURE CEREBRAL AQUEDUCT)

DIENCEPHALIC SUPERVENTRICLE (FUTURE THIRD VENTRICLE)

Thalamic NEP

THALAMUS

Stem cells of choroid plexus

Medullary velum

Rhombencephalic choroid plexus

Sojourning and migrating Purkinje cells

Sojourning and migrating deep neurons

Migrating and proliferating external germinal layer cells

Superior/inferior cerebellar peduncles

Lateral lemniscus?

Nuclei of the lateral lemniscus?

Medial longitudinal fasciculus

Central gray

RHOMBENCEPHALIC SUPERVENTRICLE (FUTURE FOURTH VENTRICLE)

Raphe nuclear complex

Reticular formation

PONS

Pontine NEP

Cerebellar notch

Parabrachial nucleus?

CEREBELLUM (HEMISPHERE)

Cerebellar NEP

Dorsal rhombic lip (contains cerebellar germinal trigone)

External germinal layer

Nucleus of the lateral lemniscus (dorsal)?

Subpial GEP

Medial lemniscus?

439

Arrows indicate the presumed direction of neuron migration from neuroepithelial sources.

✳ Evaginations and invaginations of the neuroepithelium are mosaic compartments that give rise to different brain structures.

ABBREVIATIONS:
GEP - Glioepithelium
NEP - Neuroepithelium

FONT KEY:
VENTRICULAR DIVISIONS - CAPITALS
Germinal zone - Helvetica bold
Transient structure - Times bold italic
Permanent structure - Times Roman or Bold

Dashed lines indicate staining and/or sectioning artifacts.

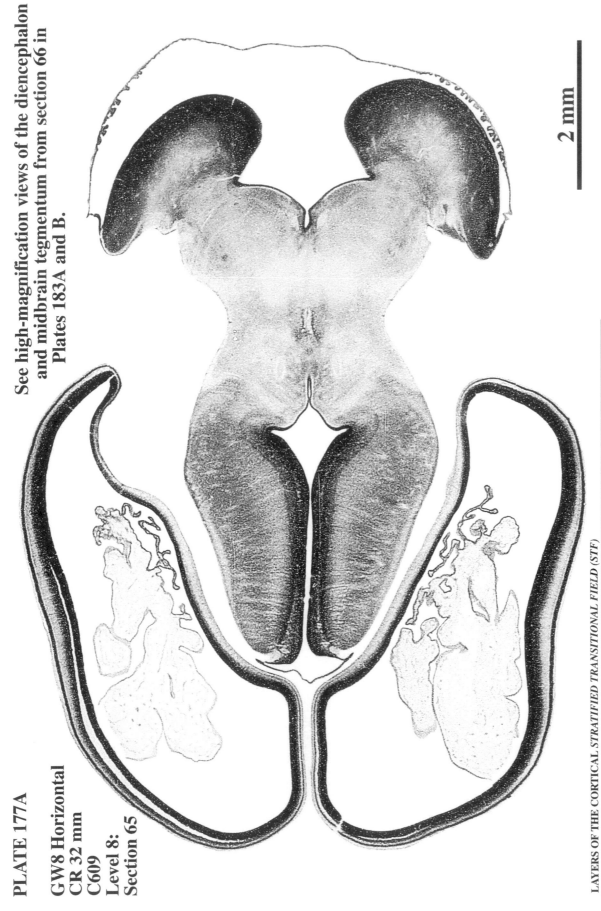

PLATE 177A

GW8 Horizontal
CR 32 mm
C609
Level 8:
Section 65

See high-magnification views of the diencephalon
and midbrain tegmentum from section 66 in
Plates 183A and B.

**LAYERS OF THE CORTICAL STRATIFIED TRANSITIONAL FIELD (STF)**

*STF1* Superficial fibrous layer with an early developmental stage (*t1*) when many cells are migrating through it, followed by a late
stage (*t2*) with sparse cells. Endures as the subcortical white matter.

*STF5* Deep cellular layer that is prominent during the first trimester, the first sojourn zone to appear outside the germinal matrix.

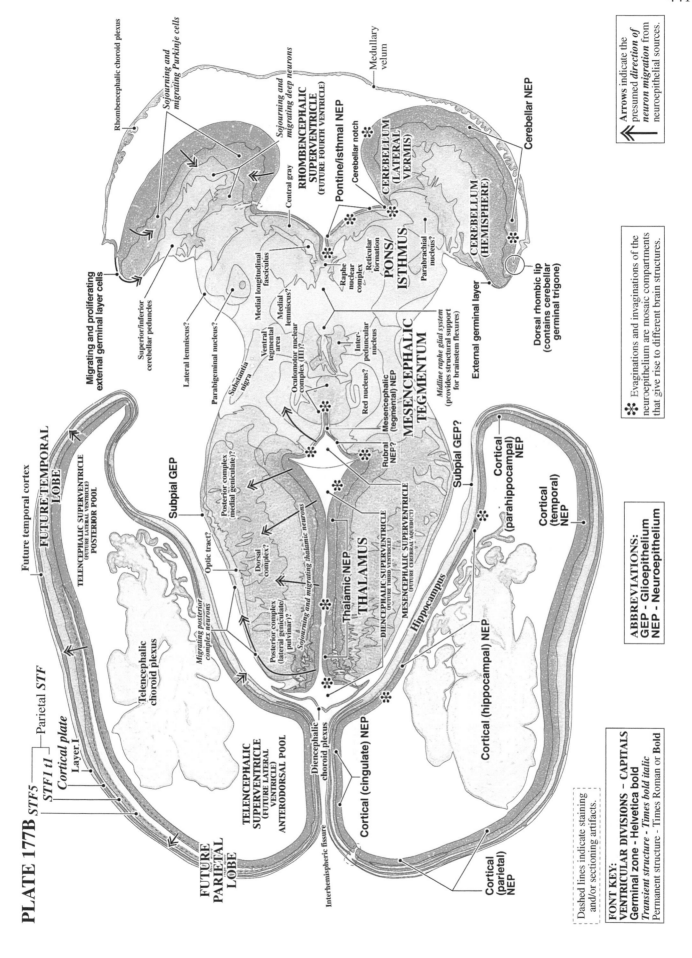

**PLATE 177B**

Arrows indicate the presumed *direction of neuron migration* from neuroepithelial sources.

✳ Evaginations and invaginations of the neuroepithelium are mosaic compartments that give rise to different brain structures.

**ABBREVIATIONS:**
GEP - Glioepithelium
NEP - Neuroepithelium

Dashed lines indicate staining and/or sectioning artifacts.

**FONT KEY:**
VENTRICULAR DIVISIONS - CAPITALS
Germinal zone - Helvetica bold
*Transient structure - Times bold italic*
Permanent structure - Times Roman or Bold

Rhombencephalic choroid plexus

*Sojourning and migrating Purkinje cells*

*Sojourning and migrating deep neurons*

Medullary velum

Central gray

**RHOMBENCEPHALIC SUPERVENTRICLE (FUTURE FOURTH VENTRICLE)**

Pontine/isthmal NEP

Cerebellar notch

Cerebellar NEP

**CEREBELLUM (LATERAL VERMIS)**

**CEREBELLUM (HEMISPHERE)**

**Migrating and proliferating external germinal layer cells**

Superior/inferior cerebellar peduncles

Lateral lemniscus?

Parabigeminal nucleus?

Medial longitudinal fasciculus

Medial lemniscus?

Raphe nuclear complex

Reticular formation

Parabrachial nucleus?

**PONS/ ISTHMUS**

Substantia nigra

Ventral tegmental area

Oculomotor nuclear complex (III)?

Inter- peduncular nucleus

Red nucleus?

Mesencephalic (tegmental) NEP

**MESENCEPHALIC TEGMENTUM**

*Midline raphe glial system (provides structural support for brainstem flexures)*

External germinal layer

Dorsal rhombic lip (contains cerebellar germinal trigone)

Future temporal cortex

**FUTURE TEMPORAL LOBE**

**TELENCEPHALIC SUPERVENTRICLE (FUTURE LATERAL VENTRICLE) POSTERIOR POOL**

Parietal *STF*

*STF1 t1*
*STF5*

*Cortical plate*
Layer I

**Subpial GEP**

Posterior complex (medial geniculate)?

Optic tract?

Dorsal complex?

Posterior complex (lateral geniculate/ pulvinar)?

*Migrating posterior complex neurons*

*Sojourning and migrating thalamic neurons*

**Thalamic NEP**

**THALAMUS**

Rubral NEP?

Subpial GEP?

**DIENCEPHALIC SUPERVENTRICLE (FUTURE THIRD VENTRICLE)**

**MESENCEPHALIC SUPERVENTRICLE (FUTURE CEREBRAL AQUEDUCT)**

Telencephalic choroid plexus

**FUTURE PARIETAL LOBE**

**TELENCEPHALIC SUPERVENTRICLE (FUTURE LATERAL VENTRICLE) ANTERODORSAL POOL**

Interhemispheric fissure

Diencephalic choroid plexus

**Cortical (cingulate) NEP**

Hippocampus

**Cortical (parahippocampal) NEP**

**Cortical (temporal) NEP**

**Cortical (hippocampal) NEP**

**Cortical (parietal) NEP**

**PLATE 178A**

**GW8 Horizontal
CR 32 mm
C609
Level 9:
Section 45**

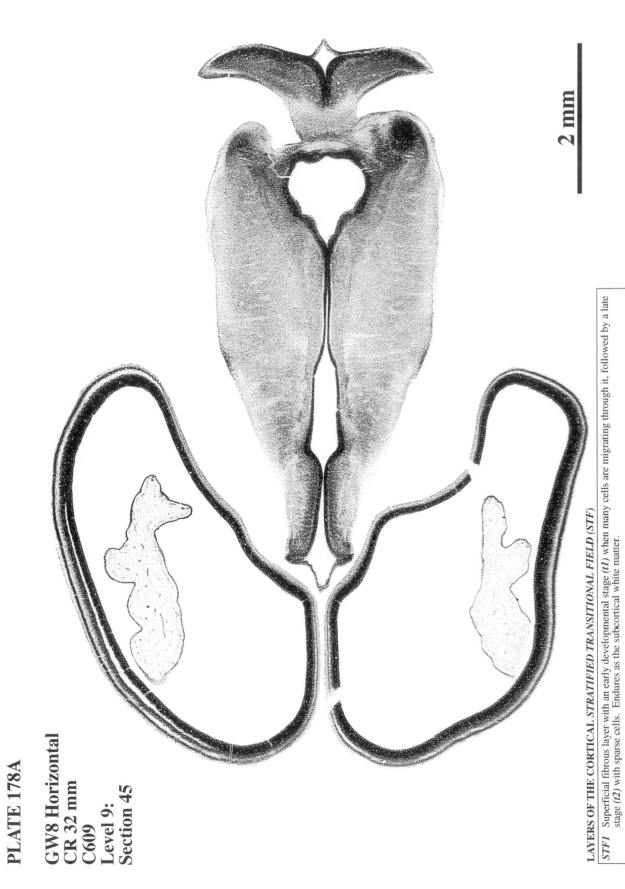

**2 mm**

*LAYERS OF THE CORTICAL STRATIFIED TRANSITIONAL FIELD (STF)*

*STF1*  Superficial fibrous layer with an early developmental stage (*t1*) when many cells are migrating through it, followed by a late stage (*t2*) with sparse cells.  Endures as the subcortical white matter.

*STF5*  Deep cellular layer that is prominent during the first trimester, the first sojourn zone to appear outside the germinal matrix.

## PLATE 178B

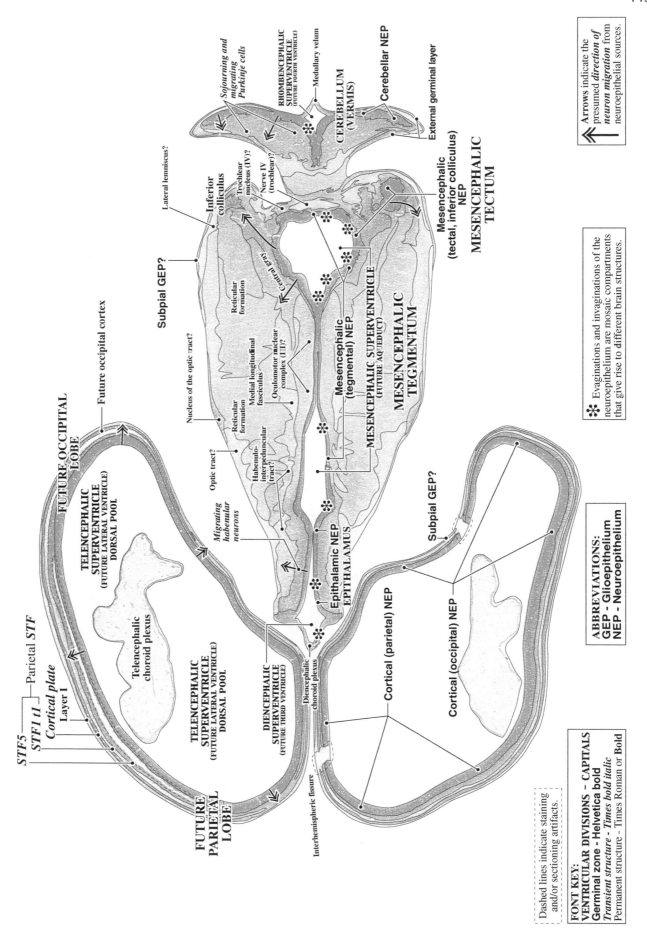

STF5
STF1 t1 —— Parietal STF
*Cortical plate*
*Layer I*

**FUTURE PARIETAL LOBE**

**FUTURE OCCIPITAL LOBE**

Future occipital cortex

**TELENCEPHALIC SUPERVENTRICLE** (FUTURE LATERAL VENTRICLE) **DORSAL POOL**

Telencephalic choroid plexus

**TELENCEPHALIC SUPERVENTRICLE** (FUTURE LATERAL VENTRICLE) **DORSAL POOL**

**DIENCEPHALIC SUPERVENTRICLE** (FUTURE THIRD VENTRICLE)

Diencephalic choroid plexus

Interhemispheric fissure

Subpial GEP?

*Migrating habenular neurons*

Epithalamic NEP
**EPITHALAMUS**

Optic tract?

Habenulo-interpeduncular tract?

Oculomotor nuclear complex (III)?

Medial longitudinal fasciculus

Reticular formation

Nucleus of the optic tract?

**Subpial GEP?**

Lateral lemniscus?

Inferior colliculus

Reticular formation

Central gray?

Trochlear nucleus (IV)?

Nerve IV (trochlear)?

Medullary velum

*Sojourning and migrating Purkinje cells*

**RHOMBENCEPHALIC SUPERVENTRICLE** (FUTURE FOURTH VENTRICLE)

**CEREBELLUM (VERMIS)**

**Cerebellar NEP**

External germinal layer

Mesencephalic (tectal, inferior colliculus) NEP

**MESENCEPHALIC TECTUM**

Mesencephalic (tegmental) NEP

**MESENCEPHALIC SUPERVENTRICLE** (FUTURE AQUEDUCT)

**MESENCEPHALIC TEGMENTUM**

**Cortical (parietal) NEP**

Subpial GEP?

**Cortical (occipital) NEP**

Arrows indicate the presumed *direction of neuron migration* from neuroepithelial sources.

* Evaginations and invaginations of the neuroepithelium are mosaic compartments that give rise to different brain structures.

**ABBREVIATIONS:**
GEP - Glioepithelium
NEP - Neuroepithelium

**FONT KEY:**
VENTRICULAR DIVISIONS - CAPITALS
Germinal zone - **Helvetica bold**
*Transient structure - Times bold italic*
Permanent structure - Times Roman or Bold

Dashed lines indicate staining and/or sectioning artifacts.

444

**PLATE 179A**

**GW8 Horizontal**
**CR 32 mm**
**C609**
**Level 10:**
**Section 36**

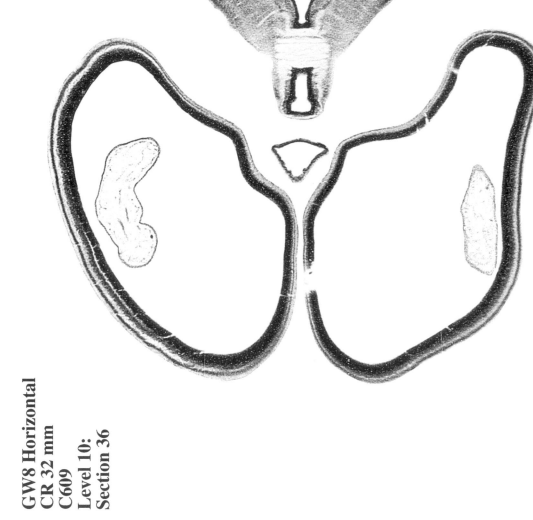

2 mm

*LAYERS OF THE CORTICAL STRATIFIED TRANSITIONAL FIELD (STF)*

*STF1* Superficial fibrous layer with an early developmental stage (*t1*) when many cells are migrating through it, followed by a late stage (*t2*) with sparse cells. Endures as the subcortical white matter.

*STF5* Deep cellular layer that is prominent during the first trimester, the first sojourn zone to appear outside the germinal matrix.

# PLATE 179B

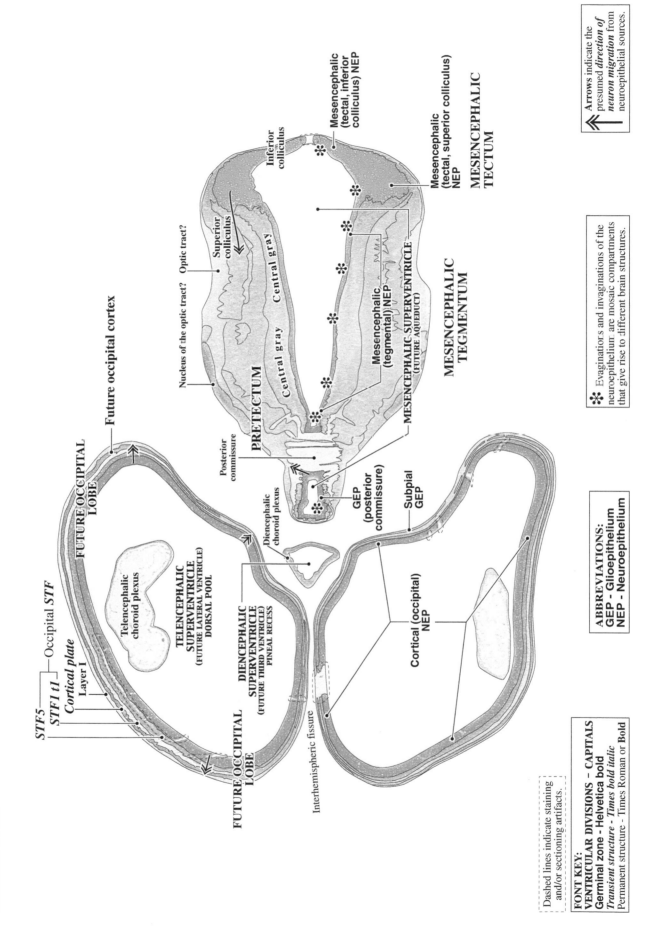

445

STF5
STF1 t1 ⎤— Occipital STF
Cortical plate ⎦
Layer I

Future occipital cortex

FUTURE OCCIPITAL LOBE

Telencephalic choroid plexus

TELENCEPHALIC SUPERVENTRICLE
(FUTURE LATERAL VENTRICLE)
DORSAL POOL

DIENCEPHALIC SUPERVENTRICLE
(FUTURE THIRD VENTRICLE)
PINEAL RECESS

Diencephalic choroid plexus

FUTURE OCCIPITAL LOBE

Interhemispheric fissure

Cortical (occipital) NEP

GEP (posterior commissure)

Subpial GEP

Posterior commissure

PRETECTUM

Nucleus of the optic tract?    Optic tract?

Superior colliculus

Central gray    Central gray

Inferior colliculus

Superior colliculus

Mesencephalic (tectal, inferior colliculus) NEP

Mesencephalic (tectal, superior colliculus) NEP

MESENCEPHALIC TECTUM

Mesencephalic (tegmental) NEP

MESENCEPHALIC SUPERVENTRICLE
(FUTURE AQUEDUCT)

MESENCEPHALIC TEGMENTUM

Arrows indicate the presumed *direction of neuron migration* from neuroepithelial sources.

✳ Evaginations and invaginations of the neuroepithelium are mosaic compartments that give rise to different brain structures.

FONT KEY:
VENTRICULAR DIVISIONS – CAPITALS
Germinal zone - Helvetica bold
*Transient structure - Times bold italic*
Permanent structure - Times Roman or Bold

ABBREVIATIONS:
GEP - Glioepithelium
NEP - Neuroepithelium

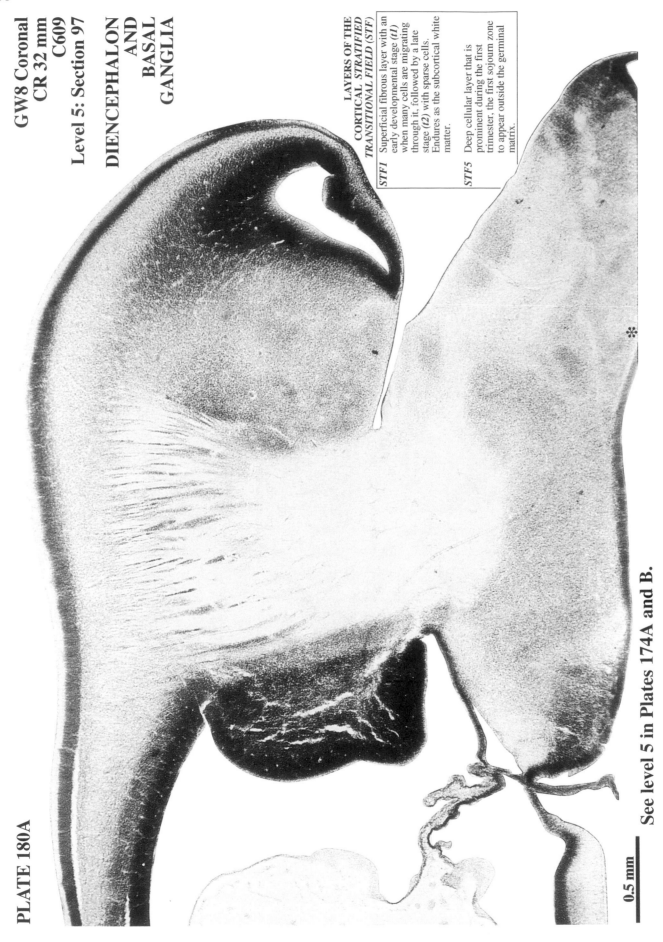

PLATE 180A

GW8 Coronal
CR 32 mm
C609
Level 5: Section 97

DIENCEPHALON
AND
BASAL
GANGLIA

**LAYERS OF THE
CORTICAL *STRATIFIED
TRANSITIONAL FIELD (STF)***

*STF1*  Superficial fibrous layer with an
early developmental stage (*t1*)
when many cells are migrating
through it, followed by a late
stage (*t2*) with sparse cells.
Endures as the subcortical white
matter.

*STF5*  Deep cellular layer that is
prominent during the first
trimester, the first sojourn zone
to appear outside the germinal
matrix.

0.5 mm

See level 5 in Plates 174A and B.

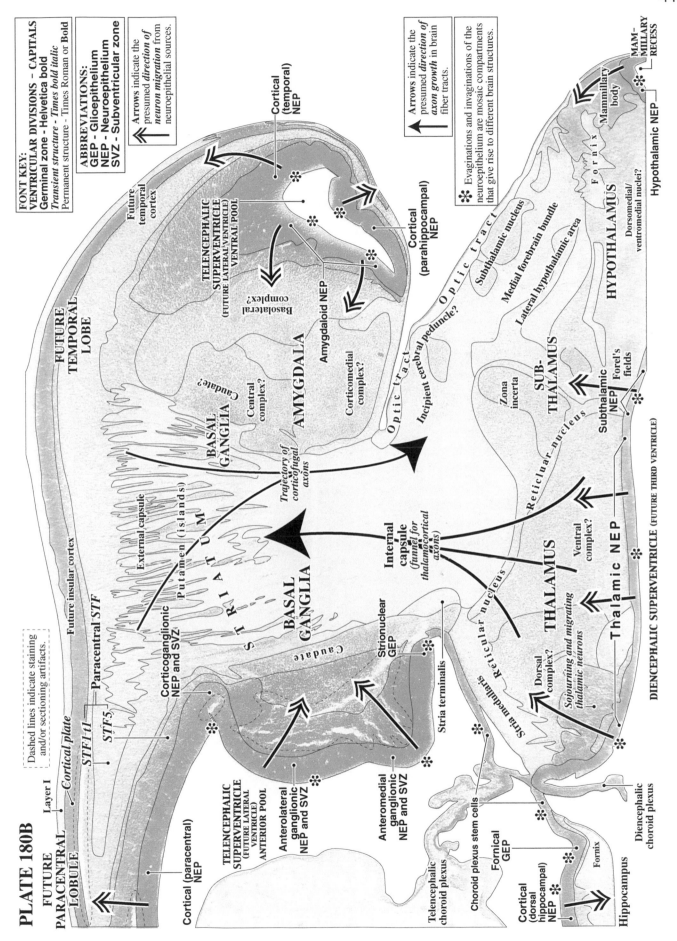

447

PLATE 180B

GW8 Coronal
CR 32 mm
C609
Between levels 5 and 6:
Section 91

DIENCEPHALON

See levels 5 and 6
in Plates 174A and B
to 175A and B.

0.5 mm

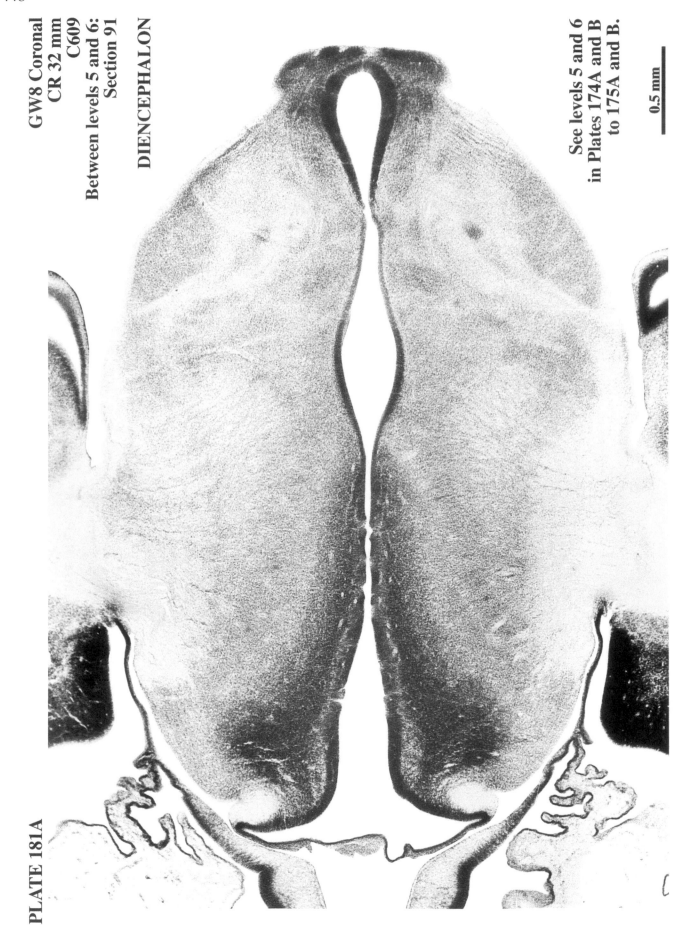

PLATE 181A

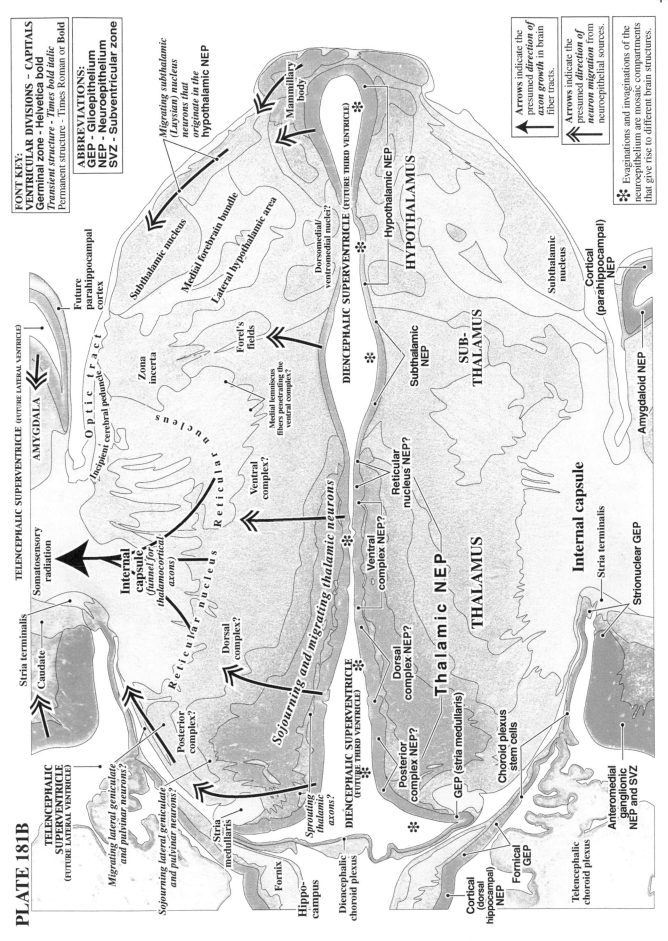

PLATE 181B

FONT KEY:
VENTRICULAR DIVISIONS – CAPITALS
Germinal zone – Helvetica bold
*Transient structure - Times bold italic*
Permanent structure - Times Roman or Bold

ABBREVIATIONS:
GEP - Glioepithelium
NEP - Neuroepithelium
SVZ - Subventricular zone

Arrows indicate the presumed *direction of axon growth* in brain fiber tracts.

Arrows indicate the presumed *direction of neuron migration* from neuroepithelial sources.

✳ Evaginations and invaginations of the neuroepithelium are mosaic compartments that give rise to different brain structures.

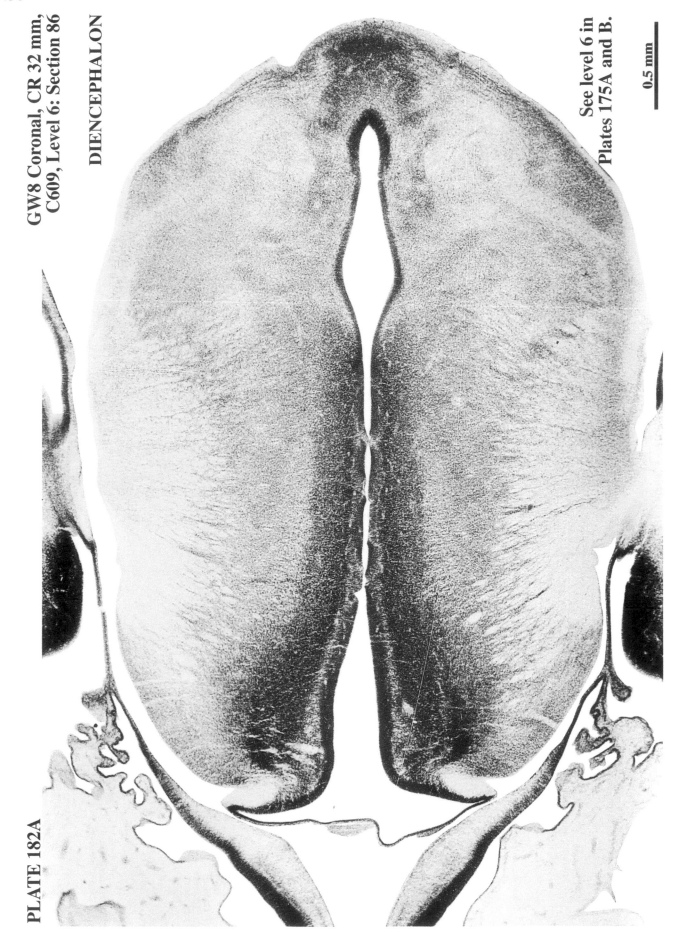

See level 6 in
Plates 175A and B.

0.5 mm

PLATE 182A

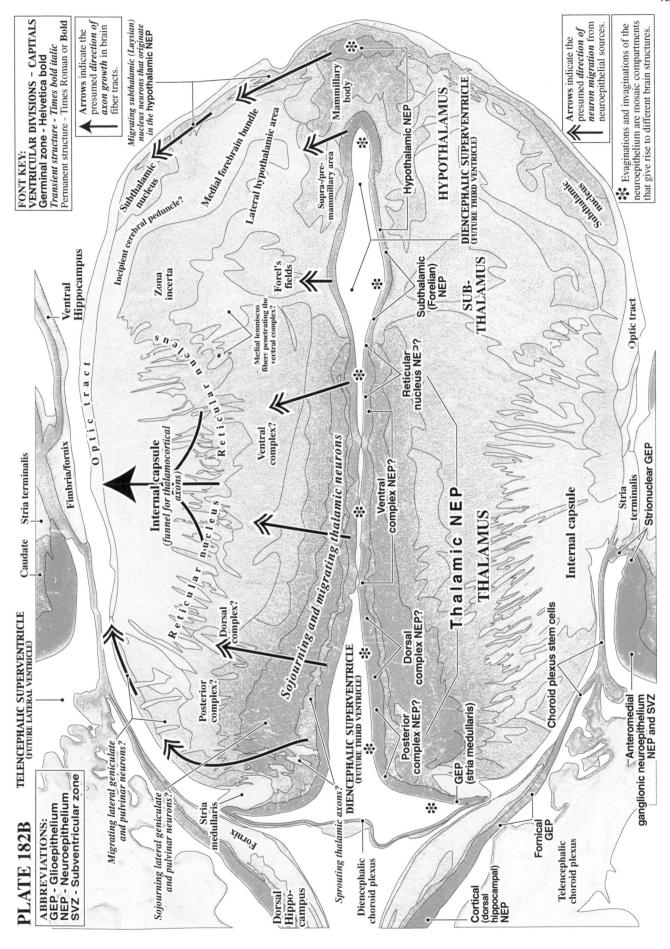

451

**PLATE 182B**

TELENCEPHALIC SUPERVENTRICLE
(FUTURE LATERAL VENTRICLE)

**ABBREVIATIONS:**
GEP - Glioepithelium
NEP - Neuroepithelium
SVZ - Subventricular zone

FONT KEY:
VENTRICULAR DIVISIONS – CAPITALS
Germinal zone - Helvetica bold
*Transient structure - Times bold italic*
Permanent structure - Times Roman or Bold

Arrows indicate the
presumed *direction of
axon growth* in brain
fiber tracts.

*Migrating subthalamic (Luysian)
nucleus neurons that originate
in the hypothalamic NEP*

Arrows indicate the
presumed *direction of
neuron migration* from
neuroepithelial sources.

✳ Evaginations and invaginations of the
neuroepithelium are mosaic compartments
that give rise to different brain structures.

Mammillary
body

Supra-/pre-
mammillary area

*Lateral hypothalamic area*

*Medial forebrain bundle*

Subthalamic
nucleus

*Incipient cerebral peduncle?*

Zona
incerta

Forel's
fields

*Medial lemniscus
fibers penetrating the
ventral complex?*

Hypothalamic NEP

**HYPOTHALAMUS**

DIENCEPHALIC SUPERVENTRICLE
(FUTURE THIRD VENTRICLE)

*Subthalamic
nuclei*

Subthalamic
(Forelian)
NEP

**SUB-
THALAMUS**

Reticular
nucleus NE??

Ventral
complex?

*Reticular nucleus*

*Reticular nucleus*

**Ventral
complex NEP?**

*Ventral
complex?*

Optic tract

Ventral
Hippocampus

Fimbria/fornix

Stria terminalis

Caudate

**Internal capsule**
(funnel for thalamocortical
axons)

O p t i c   t r a c t

*Reticular nucleus*

*Dorsal
complex?*

**Sojourning and migrating thalamic neurons**

*Posterior
complex?*

**Dorsal
complex NEP?**

**Posterior
complex NEP?**

**GEP
(stria medullaris)**

**T h a l a m i c   N E P**

**THALAMUS**

Internal capsule

Choroid plexus stem cells

Stria
terminalis

Strionuclear GEP

**Anteromedial
ganglionic neuroepithelium
NEP and SVZ**

Optic tract

*Migrating lateral geniculate
and pulvinar neurons?*

*Sojourning lateral geniculate
and pulvinar neurons?*

Stria
medullaris

Fornix

*Sprouting thalamic axons?*

DIENCEPHALIC SUPERVENTRICLE
(FUTURE THIRD VENTRICLE)

Diencephalic
choroid plexus

Telencephalic
choroid plexus

**Cortical
(dorsal
hippocampal) NEP**

**Fornical
GEP**

Dorsal
Hippo-
campus

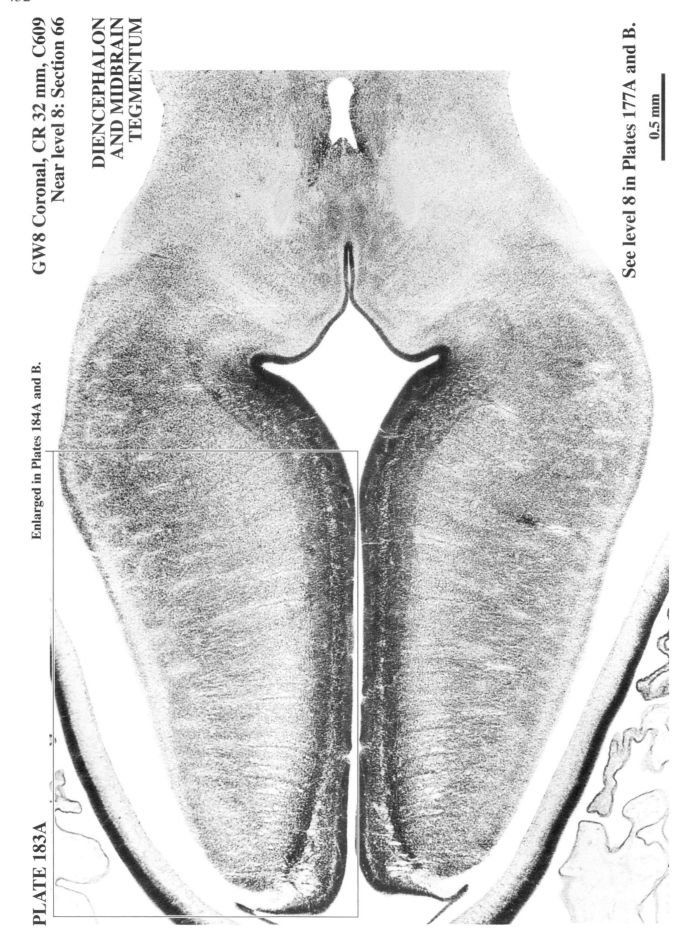

GW8 Coronal, CR 32 mm, C609
Near level 8: Section 66

DIENCEPHALON
AND MIDBRAIN
TEGMENTUM

See level 8 in Plates 177A and B.

0.5 mm

Enlarged in Plates 184A and B.

PLATE 183A

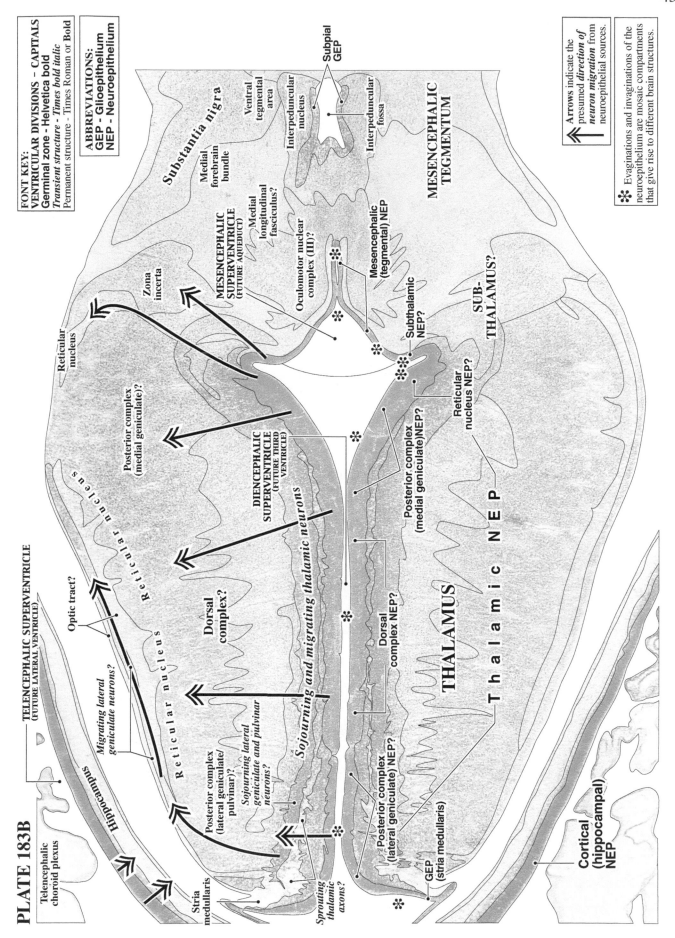

PLATE 183B

FONT KEY:
VENTRICULAR DIVISIONS – CAPITALS
Germinal zone – Helvetica bold
*Transient structure – Times bold italic*
Permanent structure – Times Roman or Bold

ABBREVIATIONS:
GEP – Glioepithelium
NEP – Neuroepithelium

Arrows indicate the presumed *direction of neuron migration* from neuroepithelial sources.

✱ Evaginations and invaginations of the neuroepithelium are mosaic compartments that give rise to different brain structures.

Telencephalic choroid plexus

TELENCEPHALIC SUPERVENTRICLE
(FUTURE LATERAL VENTRICLE)

*Optic tract?*

Hippocampus

*Migrating lateral geniculate neurons?*

*Reticular nucleus*

*Sojourning lateral geniculate and pulvinar neurons?*

*Posterior complex (lateral geniculate/pulvinar)?*

*Stria medullaris*

*Sprouting thalamic axons?*

Reticular nucleus

Dorsal complex?

DIENCEPHALIC SUPERVENTRICLE
(FUTURE THIRD VENTRICLE)

*Sojourning and migrating thalamic neurons*

Posterior complex (medial geniculate)?

Zona incerta

Reticular nucleus

MESENCEPHALIC SUPERVENTRICLE
(FUTURE AQUEDUCT)

*Medial longitudinal fasciculus?*

Oculomotor nuclear complex (III)?

Substantia nigra

Medial forebrain bundle

Ventral tegmental area

Interpeduncular nucleus

Interpeduncular fossa

Subpial GEP

Mesencephalic (tegmental) NEP

Subthalamic NEP?

SUB-THALAMUS?

MESENCEPHALIC TEGMENTUM

Reticular nucleus NEP?

Posterior complex (medial geniculate) NEP?

Dorsal complex NEP?

THALAMUS

Thalamic NEP

Posterior complex (lateral geniculate) NEP?

GEP (stria medullaris)

Cortical (hippocampal) NEP

PLATE 184A

GW8 Coronal, CR 32 mm, C609, Near level 8: Section 66

THALAMUS
(TRANSIENT
DEVELOPMENTAL LAYERS)

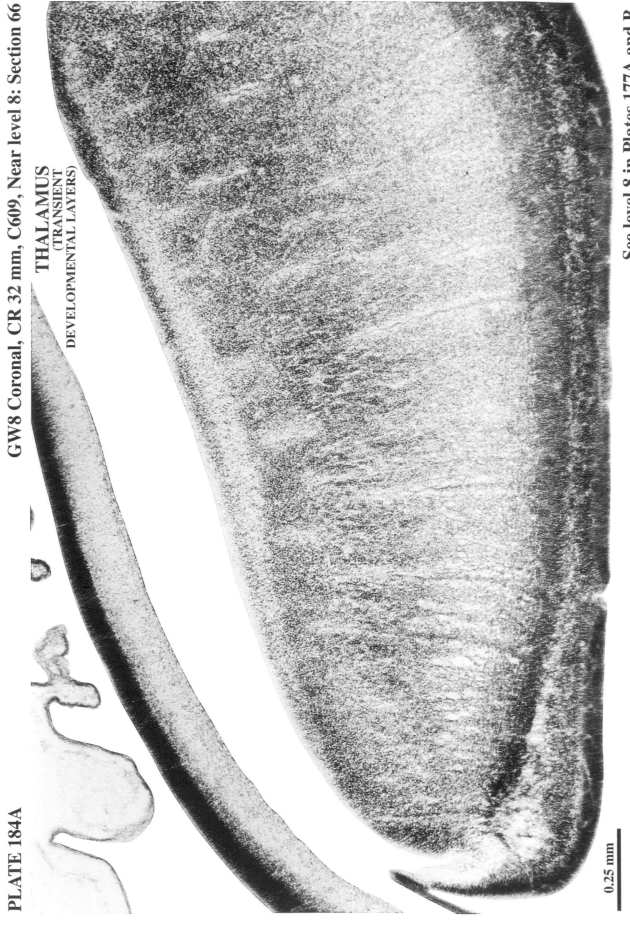

See level 8 in Plates 177A and B.

0.25 mm

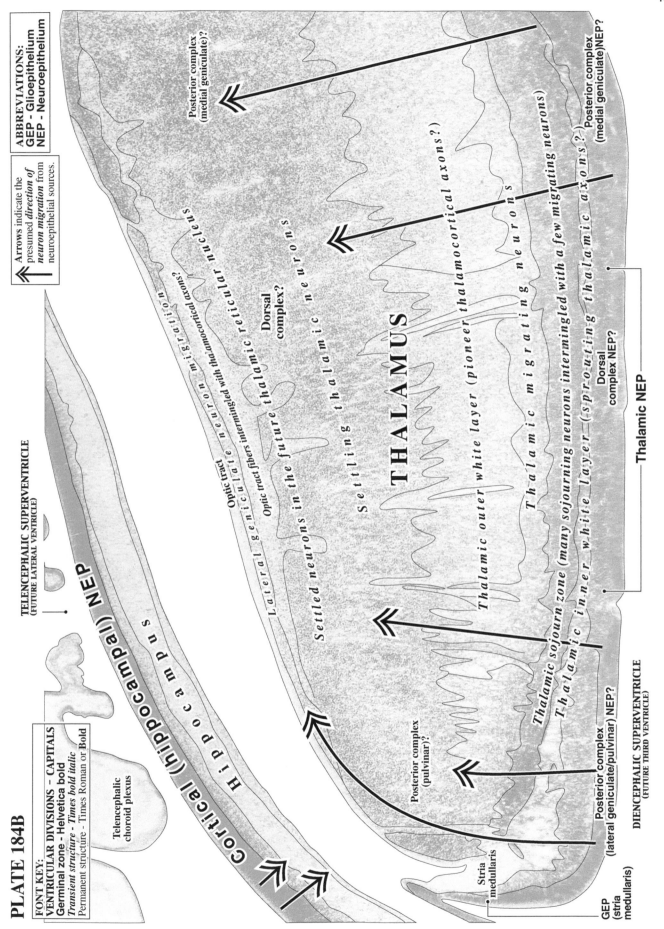

455

PLATE 184B

FONT KEY:
VENTRICULAR DIVISIONS – CAPITALS
Germinal zone - **Helvetica bold**
*Transient structure - Times bold italic*
Permanent structure - Times Roman or **Bold**

Telencephalic choroid plexus

**Cortical (hippocampal) NEP**

C o r t i c a l   H i p p o c a m p u s

TELENCEPHALIC SUPERVENTRICLE
(FUTURE LATERAL VENTRICLE)

↞ Arrows indicate the presumed *direction of neuron migration* from neuroepithelial sources.

ABBREVIATIONS:
GEP - Glioepithelium
NEP - Neuroepithelium

Posterior complex (medial geniculate)?

Posterior complex (medial geniculate)NEP?

*O p t i c   t r a c t   n e u r o n   m i g r a t i o n*

Lateral geniculate neuron intermingled with thalamocortical nucleus

*Optic tract fibers intermingled with thalamic reticular axons?*

*L a t e r a l   g e n i c u l a t e   f u t u r e   t h a l a m i c   n e u r o n s*

Dorsal complex?

*S e t t l e d   n e u r o n s   i n   t h e   f u t u r e   t h a l a m i c   n e u r o n s*

*S e t t l i n g   t h a l a m i c*

**THALAMUS**

*Thalamic outer white layer (pioneer thalamocortical axons?)*

*T h a l a m i c   m i g r a t i n g   n e u r o n s*

*Thalamic sojourn zone (many sojourning neurons intermingled with a few migrating neurons)*

*T h a l a m i c   i n n e r   w h i t e   l a y e r   (s p r o u t i n g   t h a l a m i c   a x o n s ?)*

Dorsal complex NEP?

**Thalamic NEP**

Posterior complex (pulvinar)?

Posterior complex (lateral geniculate/pulvinar) NEP?

Stria medullaris

GEP (stria medullaris)

DIENCEPHALIC SUPERVENTRICLE
(FUTURE THIRD VENTRICLE)

**PLATE 185A**

GW8 Coronal, CR 32 mm, C609
Between levels 5 and 6: Section 92
HYPOTHALAMUS AND SUBTHALAMUS

0.25 mm

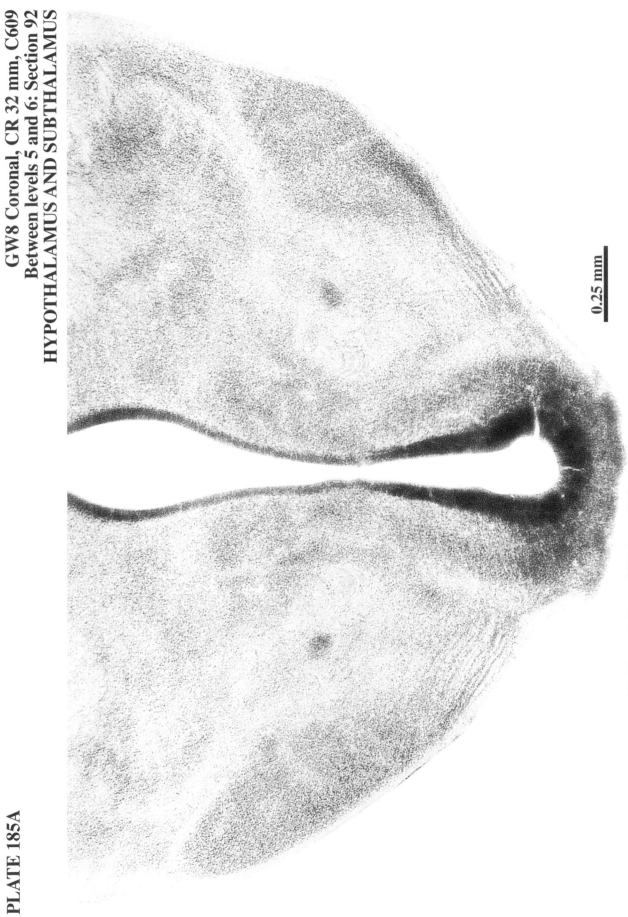

See levels 5 and 6 in Plates 174A and B to 175A and B.

**PLATE 185B**

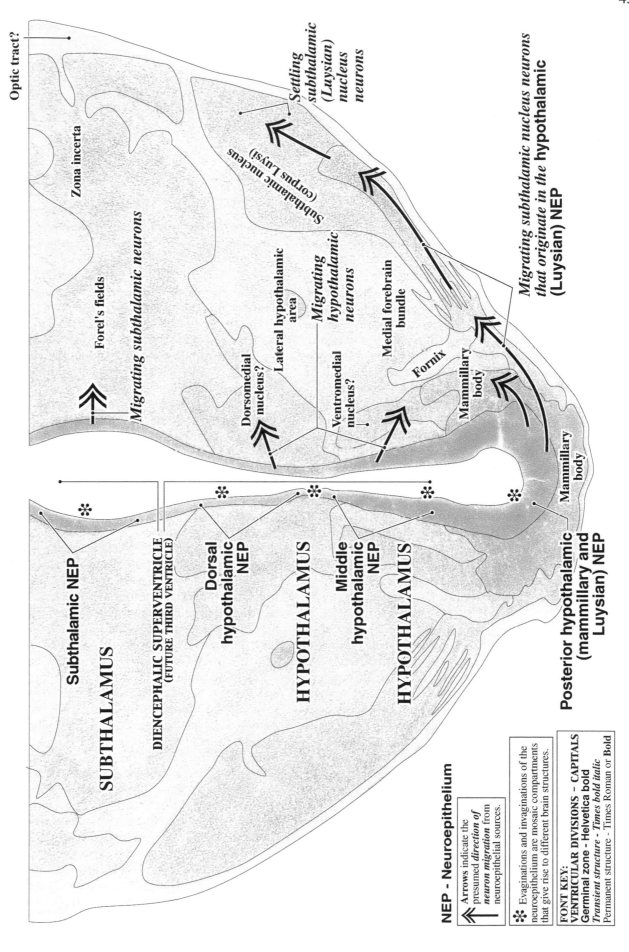

Optic tract?

Zona incerta

Forel's fields

*Migrating subthalamic neurons*

Dorsomedial nucleus?

Lateral hypothalamic area

*Migrating hypothalamic neurons*

Ventromedial nucleus?

Medial forebrain bundle

Fornix

Mammillary body

Subthalamic nucleus (corpus Luysi)

*Settling subthalamic (Luysian) nucleus neurons*

*Migrating subthalamic nucleus neurons that originate in the hypothalamic (Luysian) NEP*

Subthalamic NEP

**SUBTHALAMUS**

DIENCEPHALIC SUPERVENTRICLE
(FUTURE THIRD VENTRICLE)

Dorsal hypothalamic NEP

**HYPOTHALAMUS**

Middle hypothalamic NEP

**HYPOTHALAMUS**

Mammillary body

Posterior hypothalamic (mammillary and Luysian) NEP

**NEP – Neuroepithelium**

Arrows indicate the presumed *direction of neuron migration* from neuroepithelial sources.

✳ Evaginations and invaginations of the neuroepithelium are mosaic compartments that give rise to different brain structures.

FONT KEY:
VENTRICULAR DIVISIONS – CAPITALS
Germinal zone - Helvetica bold
*Transient structure - Times bold italic*
Permanent structure - Times Roman or Bold

# PART X: GW7.5 CORONAL

This is specimen number 966 in the Carnegie Collection, designated here as C966. A normal female fetus with a crown-rump length (CR) of 23 mm was collected in 1914. The fetus is estimated to be at gestational week (GW) 7.5. The entire fetus was fixed in bichloric acetic acid, embedded in celloidin, cut transversely in 40 μm sections, and was stained with aluminum cochineal. The histological preservation of this specimen is excellent, and the sections are cut nearly perfectly bilaterally. Several years ago, an excellent 3-D reconstruction of the brain and upper cervical spinal cord was done by piecing together cardboard cutouts of the brain outlines in each section and then gluing them together; the rhombencephalic superventricle was not included in that reconstruction. A photograph of that model (which is still part of the Carnegie Collection today) shows us the exact location and cutting plane of C966's sections (**Figure 9**). Like most of the specimens in this Volume, the sections are not cut exactly in one plane, but C966's sections are much closer to the coronal than the horizontal plane. Photographs of 21 sections (**Levels 1-21**) of the brain in the head are shown in **Plates 186-206**. Our computer-aided three-dimensional reconstructions of the brain, ventricles (including the rhombencephalic superventricle), and selected neuroepithelial components are shown on the cover and in **Figures 10-19**.

C966 is considerably less mature than the GW8 specimens. The *superventricles* are large in the centers of all brain structures, especially in the telencephalon and rhombencephon. Even though the diencephalic superventricle is approaching a slit-like shape, it is wider than that in the GW8 specimens, and the mesencephalic superventricle forms a more balloon-like expansion beneath the rudimentary tectum. Like the GW8 specimens, the respective thicknesses of the neuroepithelium (NEP) and parenchyma are keys to determining the degree of maturation of various brain structures.

The parenchyma is thick and bordered by a thin NEP in the medulla and pons, indicating that many neurons have been generated here, but the production of late-generated neurons continues. There are layers of dense cells adjacent to the lateral pontine NEP where vestibular nuclear neurons and trigeminal nuclear neurons may be sojourning prior to migration and settling. There is a larger accumulation of presumptive facial motor neurons near the midline NEP in the pontomedullary trench and some are migrating toward the indistinct facial motor nucleus. The *precerebellar neuroepithelium* in the medulla is thicker and generating more precerebellar neurons than at GW8; many neurons are entering the inferior olive after migrating in the *posterior intramural migratory stream*, but the anterior and posterior extramural migratory streams are absent; that confirms neurogenetic data in rats that the inferior olive contains the earliest-generated precerebellar neurons. The cerebellar NEP is thick and difficult to distinguish from an adjacent dense sojourn zone in the cerebellar parenchyma, called *cerebellar transitional field (CTF) 6*. The remaining *CTF* has alternating layers of cells and fibers (*CTF1-5*). The *external germinal layer (egl)* is completely absent. If one can extrapolate from data on cerebellar neurogenesis in rats, the human cerebellar NEP by GW7.5 has generated all of the deep neurons and is now producing "middle-aged" Purkinje cells; the oldest Purkinje cells are sojourning in *CTF6*. Both the mesencephalic tegmental NEP and the isthmal NEP are nearly the same thickness as the pontine and medullary NEPs, but dense sojourn zones of young neurons are more obvious in the adjacent parenchyma. The superficial border of a thick mesencephalic tectal NEP is difficult to distinguish from dense wavefronts of young neurons extending into a thin parenchyma. The majority of neurons in both the superior and inferior colliculi have not yet been generated by GW7.5.

The diencephalic NEP is thicker at GW7.5 than at GW8, and the thin parenchyma is filled with dense zones of sojourning and migrating neurons. There is no internal capsule, but some pioneer axons are accumulating in inner and outer white layers in the thalamic parenchyma. The telencephalic NEP is thick in all areas, and the oldest basal telencephalic and basal ganglionic neurons are settling in the thick parenchyma. Most neurons in the septum and striatum have yet to be generated. In the cerebral cortex, the NEP is bordered by a thin primordial plexiform layer that contains the oldest cortical neurons (Cajal-Retzius cells) and subplate neurons; there is no cortical plate. The cortical NEP is expanding and increasing its number of neural stem cells as the telencephalic superventricle grows; nearly all cortical neurons in layers II-VI have still to be generated.

# GW7.5 CORONAL SECTION PLANES

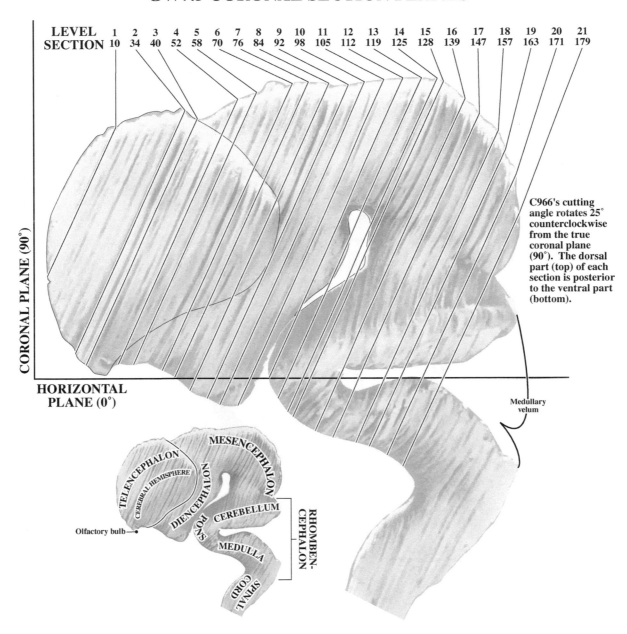

LEVEL
SECTION

| 1 | 2 | 3 | 4 | 5 | 6 | 7 | 8 | 9 | 10 | 11 | 12 | 13 | 14 | 15 | 16 | 17 | 18 | 19 | 20 | 21 |
| 10 | 34 | 40 | 52 | 58 | 70 | 76 | 84 | 92 | 98 | 105 | 112 | 119 | 125 | 128 | 139 | 147 | 157 | 163 | 171 | 179 |

CORONAL PLANE (90°)

HORIZONTAL
PLANE (0°)

C966's cutting
angle rotates 25°
counterclockwise
from the true
coronal plane
(90°). The dorsal
part (top) of each
section is posterior
to the ventral part
(bottom).

Medullary
velum

MESENCEPHALON

TELENCEPHALON
CEREBRAL HEMISPHERE

DIENCEPHALON

PONS

CEREBELLUM

MEDULLA

RHOMBEN-
CEPHALON

SPINAL CORD

Olfactory bulb

**Figure 9.** The lateral view of a 3-D model of C966's brain and upper cervical spinal cord (part of the Carnegie Collection at the National Museum of Health and Medicine) shows the exact locations and cutting angles of the illustrated sections of C966 in the following pages. The small inset identifies the major structural features. The medullary velum was not reconstructed so that the rhombencephalic superventricle appears as an open gap beneath the cerebellum.

**PLATE 186A**

**GW7.5  Coronal/horizontal**
**CR 23 mm**
**C966**
**Level 1: Section 10**

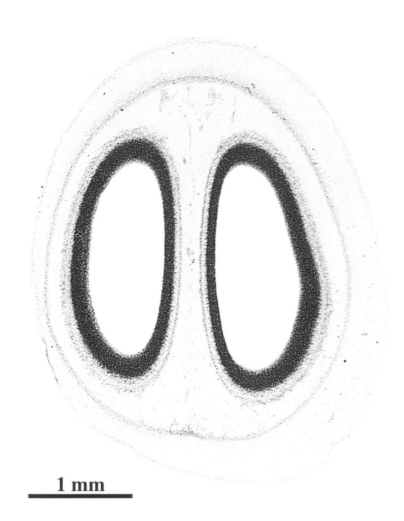

1 mm

FONT KEY:
**VENTRICULAR DIVISIONS – CAPITALS**
**Germinal zone - Helvetica bold**
*Transient structure - Times bold italic*
Permanent structure - Times Roman or **Bold**

**NEP - Neuroepithelium**

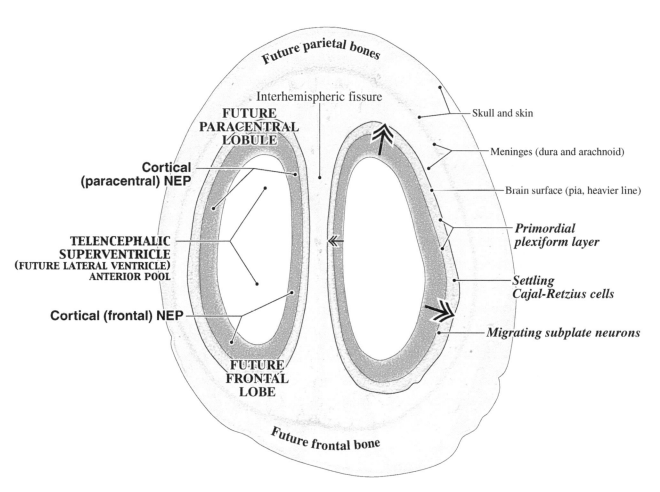

Future parietal bones

Interhemispheric fissure

**FUTURE PARACENTRAL LOBULE**

Skull and skin

Meninges (dura and arachnoid)

**Cortical (paracentral) NEP**

Brain surface (pia, heavier line)

*Primordial plexiform layer*

**TELENCEPHALIC SUPERVENTRICLE (FUTURE LATERAL VENTRICLE) ANTERIOR POOL**

*Settling Cajal-Retzius cells*

**Cortical (frontal) NEP**

*Migrating subplate neurons*

**FUTURE FRONTAL LOBE**

Future frontal bone

The cerebral cortex in this specimen does not have a ***cortical plate*** and a ***stratified transitional field*** outside the cortical neuroepithelium. Instead, there is a ***primordial plexiform layer*** composed of early-generated Cajal-Retzius cells and the slightly later-generated subplate neurons.

The large Cajal-Retzius cells settle subjacent to the pia meninx and remain superficial in cortical Layer I throughout later development. The deep border of Layer I is not defined until a cortical plate appears.

Subplate neurons accumulate in a loosely defined network beneath the Cajal-Retzius cells. In later development they become radially aligned and are the pioneer neurons in the cortical plate. Eventually, they delaminate and settle in the subplate (cortical Layer VII).

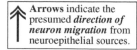

**Arrows** indicate the presumed ***direction of neuron migration*** from neuroepithelial sources.

**PLATE 187A**

**GW7.5 Coronal/horizontal**
**CR 23 mm**
**C966**
**Level 2: Section 34**

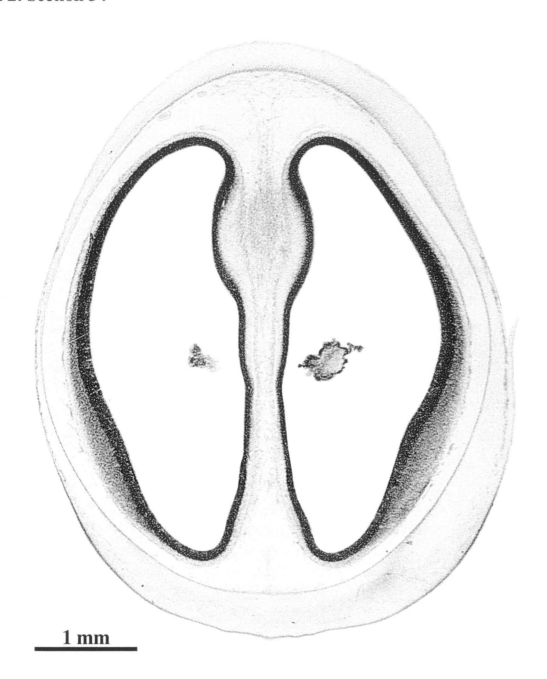

1 mm

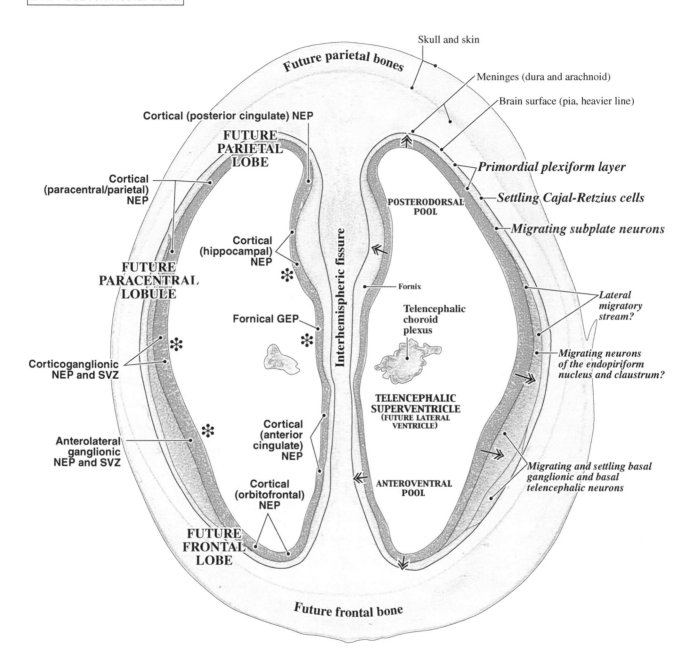

Skull and skin

*Future parietal bones*

Meninges (dura and arachnoid)

Brain surface (pia, heavier line)

Cortical (posterior cingulate) NEP

**FUTURE PARIETAL LOBE**

Cortical (paracentral/parietal) NEP

*Primordial plexiform layer*

**POSTERODORSAL POOL**

*Settling Cajal-Retzius cells*

*Migrating subplate neurons*

Cortical (hippocampal) NEP

**FUTURE PARACENTRAL LOBULE**

Fornix

*Lateral migratory stream?*

Fornical GEP

**Telencephalic choroid plexus**

*Migrating neurons of the endopiriform nucleus and claustrum?*

Corticoganglionic NEP and SVZ

**TELENCEPHALIC SUPERVENTRICLE**
**(FUTURE LATERAL VENTRICLE)**

Interhemispheric fissure

Cortical (anterior cingulate) NEP

Anterolateral ganglionic NEP and SVZ

*Migrating and settling basal ganglionic and basal telencephalic neurons*

**ANTEROVENTRAL POOL**

Cortical (orbitofrontal) NEP

**FUTURE FRONTAL LOBE**

*Future frontal bone*

**Arrows** indicate the presumed *direction of neuron migration* from neuroepithelial sources.

✴ Evaginations and invaginations of the neuroepithelium are mosaic compartments that give rise to different brain structures.

**PLATE 188A**

**GW7.5 Coronal/horizontal**
**CR 23 mm**
**C966**
**Level 3: Section 40**

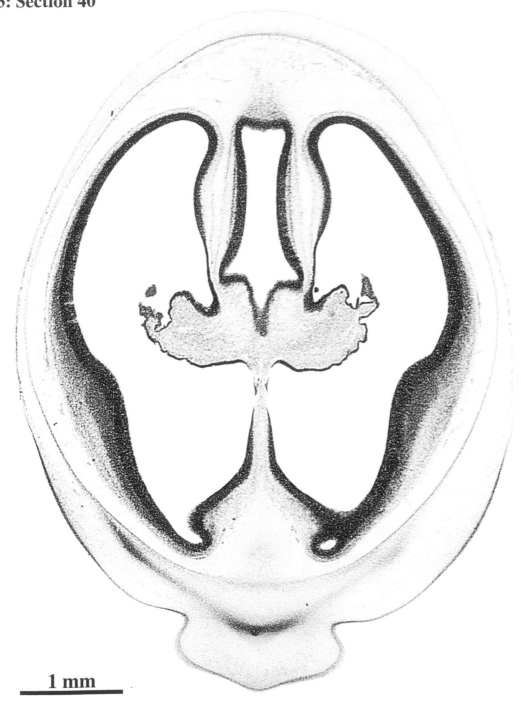

1 mm

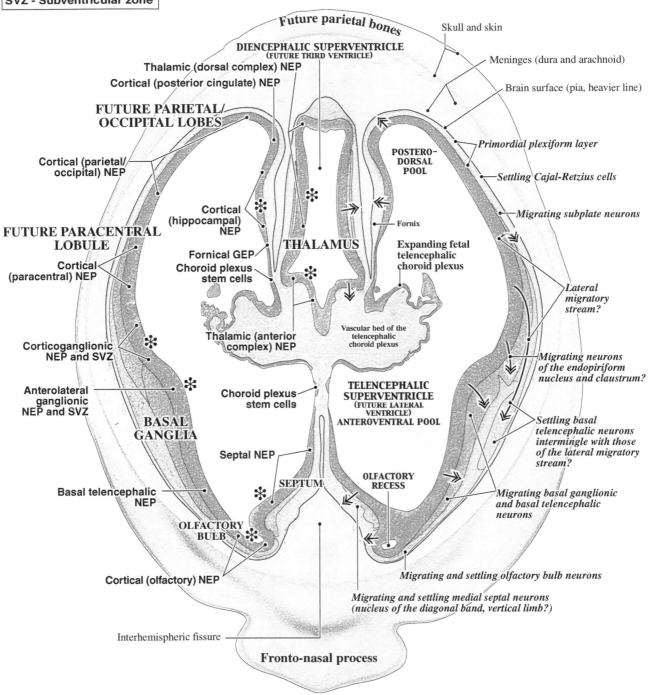

**FONT KEY:**
VENTRICULAR DIVISIONS – CAPITALS
Germinal zone - Helvetica bold
*Transient structure - Times bold italic*
Permanent structure - Times Roman or **Bold**

**ABBREVIATIONS:**
GEP - Glioepithelium
NEP - Neuroepithelium
SVZ - Subventricular zone

Future parietal bones

Skull and skin

DIENCEPHALIC SUPERVENTRICLE
(FUTURE THIRD VENTRICLE)

Meninges (dura and arachnoid)

Thalamic (dorsal complex) NEP

Cortical (posterior cingulate) NEP

Brain surface (pia, heavier line)

**FUTURE PARIETAL/
OCCIPITAL LOBES**

**POSTERO-
DORSAL
POOL**

*Primordial plexiform layer*

Cortical (parietal/
occipital) NEP

*Settling Cajal-Retzius cells*

*Migrating subplate neurons*

Cortical
(hippocampal)
NEP

Fornix

**FUTURE PARACENTRAL
LOBULE**

**THALAMUS**

Cortical
(paracentral) NEP

Fornical GEP
Choroid plexus
stem cells

***Expanding fetal
telencephalic
choroid plexus***

*Lateral
migratory
stream?*

Corticoganglionic
NEP and SVZ

Thalamic (anterior
complex) NEP

*Vascular bed of the
telencephalic
choroid plexus*

*Migrating neurons
of the endopiriform
nucleus and claustrum?*

Anterolateral
ganglionic
NEP and SVZ

Choroid plexus
stem cells

**TELENCEPHALIC
SUPERVENTRICLE
(FUTURE LATERAL
VENTRICLE)
ANTEROVENTRAL POOL**

*Settling basal
telencephalic neurons
intermingle with those
of the lateral migratory
stream?*

**BASAL
GANGLIA**

Septal NEP

Basal telencephalic
NEP

***OLFACTORY
RECESS***

*Migrating basal ganglionic
and basal telencephalic
neurons*

**SEPTUM**

**OLFACTORY
BULB**

*Migrating and settling olfactory bulb neurons*

Cortical (olfactory) NEP

*Migrating and settling medial septal neurons
(nucleus of the diagonal band, vertical limb?)*

Interhemispheric fissure

**Fronto-nasal process**

**Arrows** indicate the
presumed *direction of
neuron migration* from
neuroepithelial sources.

✳ Evaginations and invaginations of the
neuroepithelium are mosaic compartments
that give rise to different brain structures.

**PLATE 189A**

**GW7.5 Coronal/horizontal**
**CR 23 mm**
**C966**
**Level 4: Section 52**

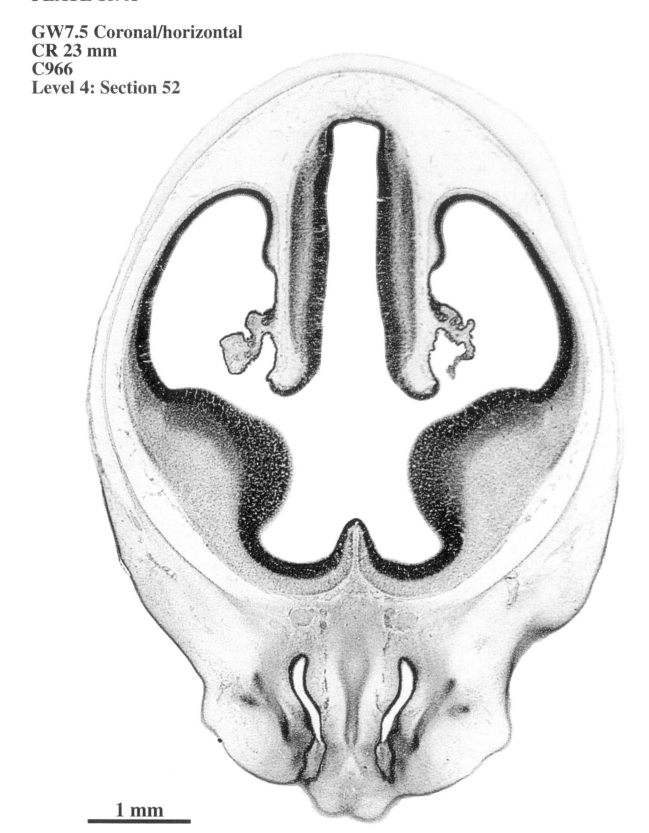

1 mm

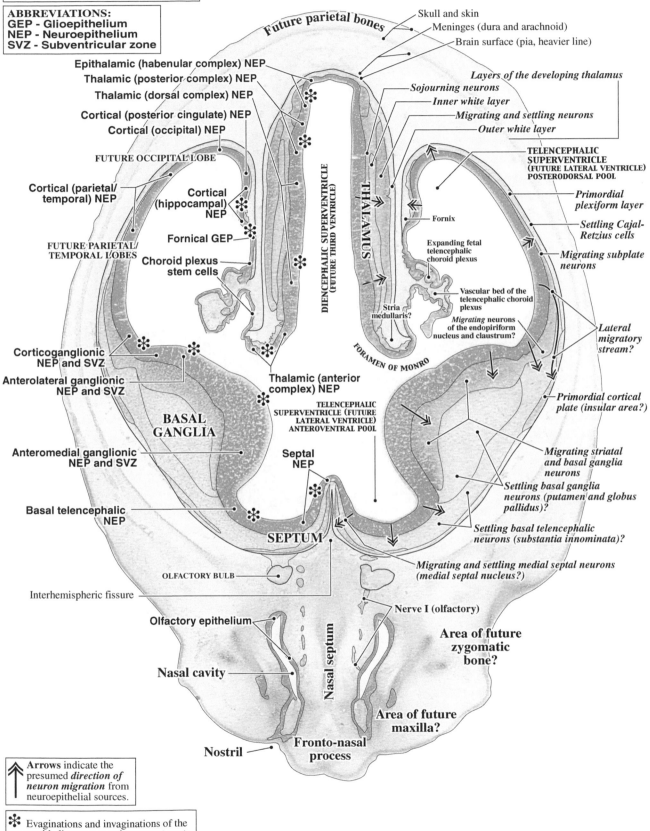

**FONT KEY:**
**VENTRICULAR DIVISIONS – CAPITALS**
**Germinal zone - Helvetica bold**
*Transient structure - Times bold italic*
Permanent structure - Times Roman or **Bold**

**ABBREVIATIONS:**
**GEP - Glioepithelium**
**NEP - Neuroepithelium**
**SVZ - Subventricular zone**

Future parietal bones

Epithalamic (habenular complex) NEP
Thalamic (posterior complex) NEP
Thalamic (dorsal complex) NEP
Cortical (posterior cingulate) NEP
Cortical (occipital) NEP
FUTURE OCCIPITAL LOBE
Cortical (parietal/temporal) NEP
Cortical (hippocampal) NEP
FUTURE PARIETAL/TEMPORAL LOBES
Fornical GEP
Choroid plexus stem cells

Corticoganglionic NEP and SVZ
Anterolateral ganglionic NEP and SVZ
Thalamic (anterior complex) NEP

BASAL GANGLIA

Anteromedial ganglionic NEP and SVZ

Basal telencephalic NEP

Septal NEP

TELENCEPHALIC SUPERVENTRICLE (FUTURE LATERAL VENTRICLE) ANTEROVENTRAL POOL

SEPTUM

OLFACTORY BULB

Interhemispheric fissure

Olfactory epithelium

Nasal cavity

Nostril

Fronto-nasal process

Skull and skin
Meninges (dura and arachnoid)
Brain surface (pia, heavier line)
*Layers of the developing thalamus*
*Sojourning neurons*
*Inner white layer*
*Migrating and settling neurons*
*Outer white layer*
**TELENCEPHALIC SUPERVENTRICLE (FUTURE LATERAL VENTRICLE) POSTERODORSAL POOL**
*Primordial plexiform layer*
*Settling Cajal-Retzius cells*
*Migrating subplate neurons*

DIENCEPHALIC SUPERVENTRICLE (FUTURE THIRD VENTRICLE)
THALAMUS
Fornix
Expanding fetal telencephalic choroid plexus
Vascular bed of the telencephalic choroid plexus
*Migrating neurons of the endopiriform nucleus and claustrum?*
Stria medullaris?
FORAMEN OF MONRO

*Lateral migratory stream?*
*Primordial cortical plate (insular area?)*
*Migrating striatal and basal ganglia neurons*
*Settling basal ganglia neurons (putamen and globus pallidus)?*
*Settling basal telencephalic neurons (substantia innominata)?*
*Migrating and settling medial septal neurons (medial septal nucleus?)*

Nerve I (olfactory)

**Area of future zygomatic bone?**

Nasal septum

**Area of future maxilla?**

**Arrows** indicate the presumed *direction of neuron migration* from neuroepithelial sources.

✷ Evaginations and invaginations of the neuroepithelium are mosaic compartments that give rise to different brain structures.

**PLATE 190A**

**GW7.5 Coronal/horizontal**
**CR 23 mm**
**C966**
**Level 5: Section 58**

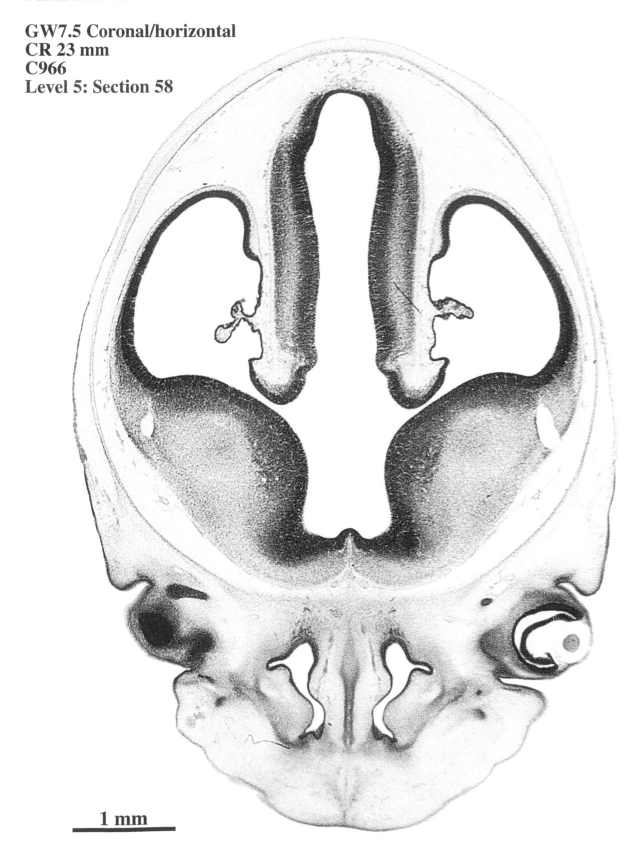

1 mm

FONT KEY:
VENTRICULAR DIVISIONS – CAPITALS
Germinal zone - Helvetica bold
*Transient structure - Times bold italic*
Permanent structure - Times Roman or **Bold**

ABBREVIATIONS:
GEP - Glioepithelium
NEP - Neuroepithelium
SVZ - Subventricular zone

Skull and skin
Meninges (dura and arachnoid)
Brain surface (pia, heavier line)

Future parietal bones

Epithalamic (habenular complex) NEP
Thalamic (posterior complex) NEP
Thalamic (dorsal complex) NEP

Cortical (parahippocampal) NEP

FUTURE OCCIPITAL/
TEMPORAL LOBES

Cortical
(hippo-
campal)
NEP

Fornical GEP

Cortical (occipital/
temporal) NEP

Choroid plexus
stem cells

Thalamic
(reticular
protuberance)
NEP

Thalamic (anterior
complex) NEP

Corticoganglionic
NEP and SVZ

Anterolateral ganglionic
NEP and SVZ

BASAL
GANGLIA

Anteromedial/posterior
ganglionic NEP and SVZ?

*Strionuclear
NEP?*

*Basal
telencephalic
NEP?*

*Septal
NEP*

Future orbito-sphenoid process?

SEPTUM

Interhemispheric
fissure

Nerve I (olfactory)

Pigment layer of retina

Olfactory epithelium

Nasal cavity

Nasal septum

Fronto-nasal
process

*Layers of the developing thalamus*
*Sojourning neurons*
*Inner white layer?*
*Migrating and settling neurons*
*Outer white layer*

DIENCEPHALIC SUPERVENTRICLE
(FUTURE THIRD VENTRICLE, THALAMIC POOL)

THALAMUS

TELENCEPHALIC
SUPERVENTRICLE
(FUTURE LATERAL VENTRICLE)
POSTERODORSAL POOL

*Primordial
plexiform layer*

Fornix

*Settling Cajal-
Retzius cells*

Telencephalic
choroid
plexus

*Migrating subplate
neurons*

Vascular bed of the
telencephalic choroid
plexus

Stria
medullaris?

*Migrating neurons
of the endopiriform
nucleus and claustrum?*

*Lateral
migratory
stream?*

FORAMEN
OF MONRO

*Primordial cortical
plate (insular area?)*

FUTURE
LATERAL VENTRICLE
ANTERO-
VENTRAL POOL

*Channels for
internal capsule?*

*Migrating basal ganglia
neurons*

*Settling basal ganglia
neurons (putamen and globus
pallidus)?*

*Settling basal telencephalic
neurons (substantia innominata)?*

*Migrating and settling
bed nucleus of the stria terminalis neurons?*

*Migrating and settling
medial septal nucleus
neurons?*

Vitreous body
Cornea
Lens
Neural retina
Intraretinal space
Pigment layer of retina
Sclera
Eyelid
Eye

Nasal
conchae

Maxilla?

Area of future
zygomatic
bone?

**Arrows** indicate the
presumed *direction of
neuron migration* from
neuroepithelial sources.

✳ Evaginations and invaginations of the
neuroepithelium are mosaic compartments
that give rise to different brain structures.

**PLATE 191A**

GW7.5 Coronal/horizontal
CR 23 mm
C966
Level 6: Section 70

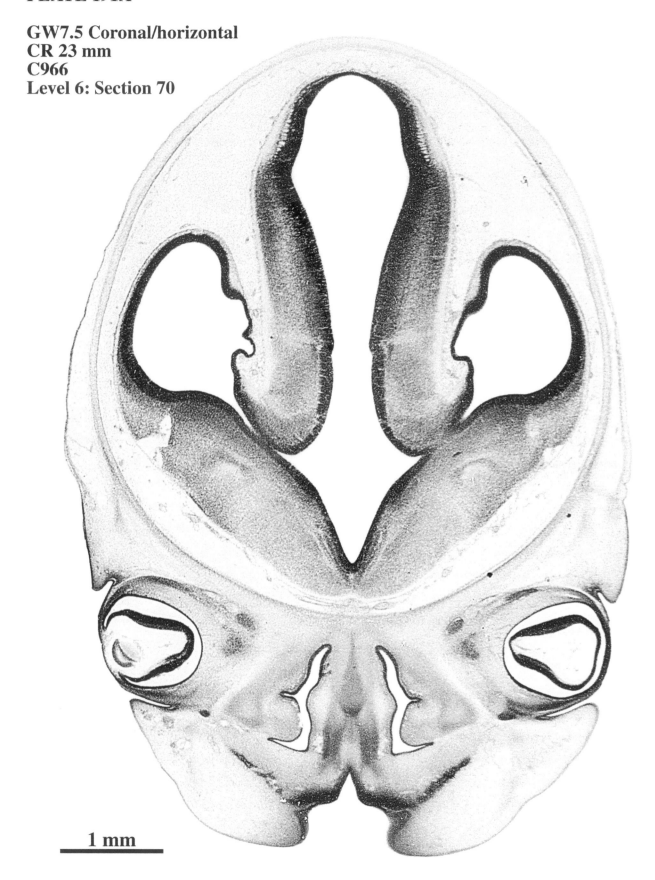

1 mm

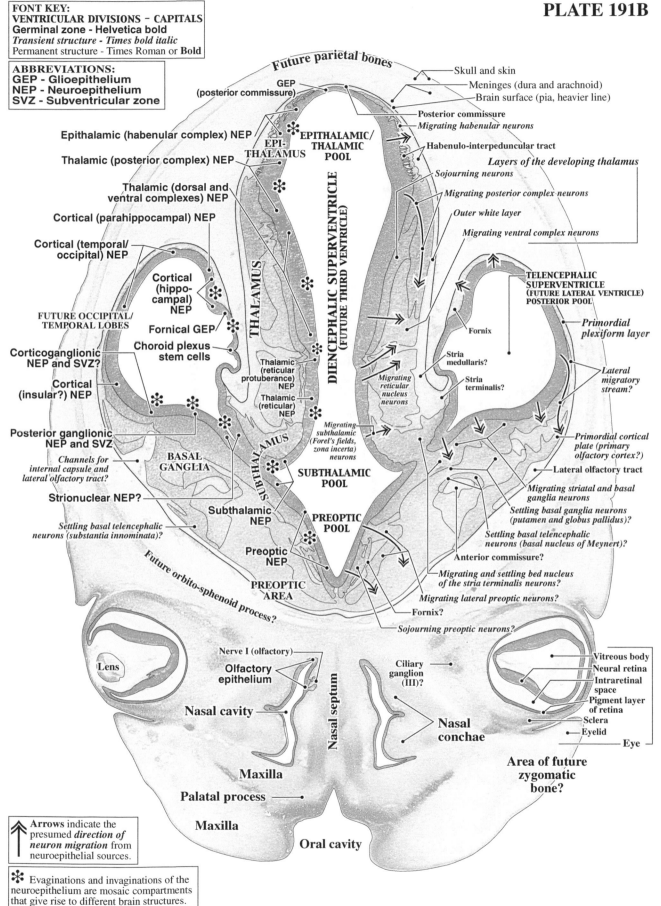

**PLATE 192A**

**GW7.5 Coronal/horizontal**
**CR 23 mm**
**C966**
**Level 7: Section 76**

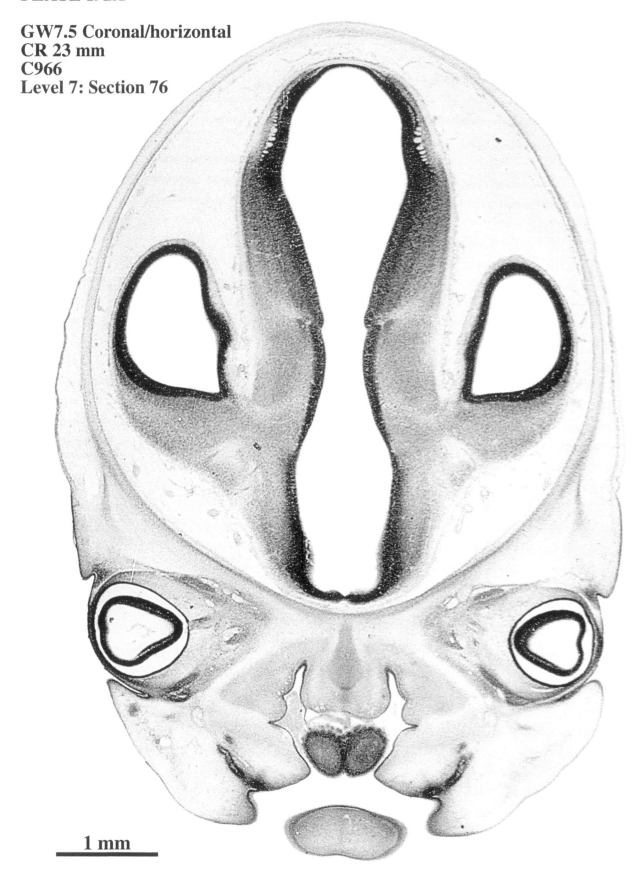

1 mm

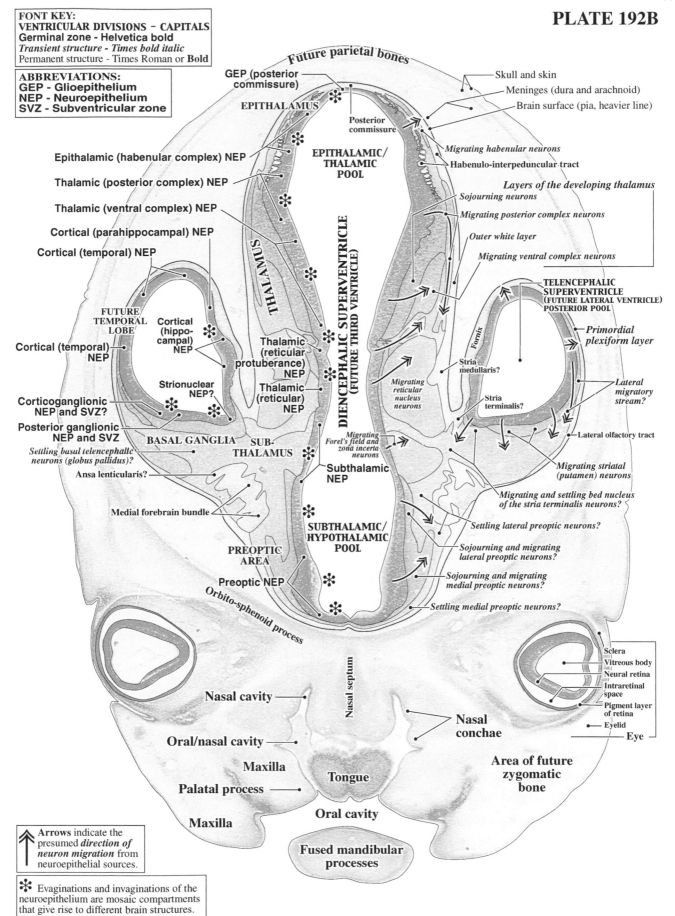

**FONT KEY:**
VENTRICULAR DIVISIONS – CAPITALS
Germinal zone - Helvetica bold
*Transient structure - Times bold italic*
Permanent structure - Times Roman or **Bold**

**ABBREVIATIONS:**
GEP - Glioepithelium
NEP - Neuroepithelium
SVZ - Subventricular zone

Future parietal bones

GEP (posterior commissure)

EPITHALAMUS

Epithalamic (habenular complex) NEP

Thalamic (posterior complex) NEP

Thalamic (ventral complex) NEP

Cortical (parahippocampal) NEP

Cortical (temporal) NEP

FUTURE TEMPORAL LOBE

Cortical (temporal) NEP

Cortical (hippo-campal) NEP

Strionuclear NEP?

Corticoganglionic NEP and SVZ?

Posterior ganglionic NEP and SVZ

*Settling basal telencephalic neurons (globus pallidus)?*

Ansa lenticularis?

Medial forebrain bundle

BASAL GANGLIA

SUB-THALAMUS

PREOPTIC AREA

Preoptic NEP

*Orbito-sphenoid process*

THALAMUS

Thalamic (reticular protuberance) NEP

Thalamic (reticular) NEP

DIENCEPHALIC SUPERVENTRICLE (FUTURE THIRD VENTRICLE)

EPITHALAMIC/ THALAMIC POOL

Posterior commissure

*Migrating habenular neurons*

Habenulo-interpeduncular tract

*Layers of the developing thalamus*
*Sojourning neurons*
*Migrating posterior complex neurons*
*Outer white layer*
*Migrating ventral complex neurons*

Skull and skin

Meninges (dura and arachnoid)

Brain surface (pia, heavier line)

TELENCEPHALIC SUPERVENTRICLE (FUTURE LATERAL VENTRICLE) POSTERIOR POOL

*Primordial plexiform layer*

Fornix

*Migrating reticular nucleus neurons*

Stria medullaris?

Stria terminalis?

*Lateral migratory stream?*

*Lateral olfactory tract*

*Migrating striatal (putamen) neurons*

*Migrating and settling bed nucleus of the stria terminalis neurons?*

*Migrating Forel's field and zona incerta neurons*

Subthalamic NEP

SUBTHALAMIC/ HYPOTHALAMIC POOL

*Settling lateral preoptic neurons?*

*Sojourning and migrating lateral preoptic neurons?*

*Sojourning and migrating medial preoptic neurons?*

*Settling medial preoptic neurons?*

Nasal cavity

Oral/nasal cavity

Maxilla

Palatal process

Maxilla

Nasal septum

Tongue

Oral cavity

Fused mandibular processes

Nasal conchae

Area of future zygomatic bone

Sclera
Vitreous body
Neural retina
Intraretinal space
Pigment layer of retina
Eyelid
Eye

**Arrows** indicate the presumed *direction of neuron migration* from neuroepithelial sources.

✳ Evaginations and invaginations of the neuroepithelium are mosaic compartments that give rise to different brain structures.

**PLATE 193A**

GW7.5 Coronal/horizontal
CR 23 mm
C966
Level 8: Section 84

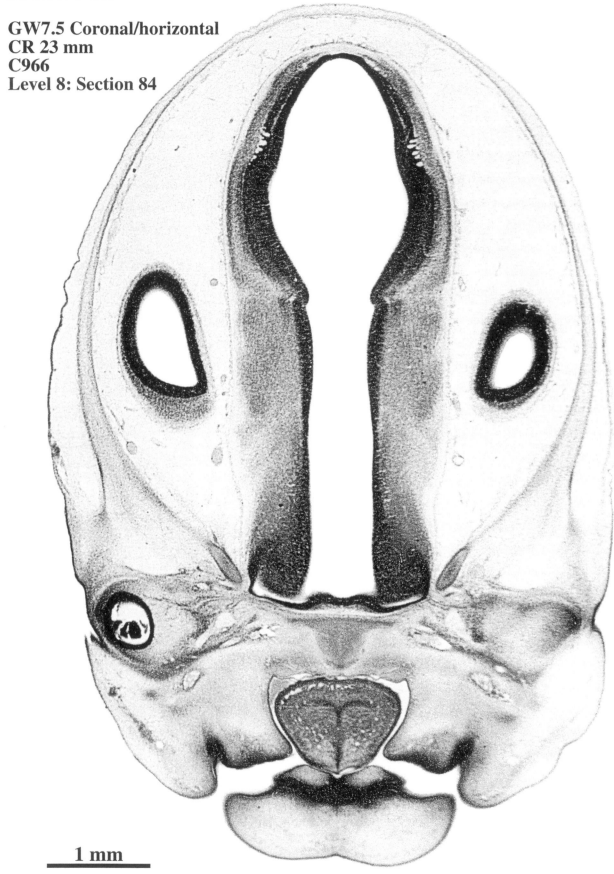

<u>1 mm</u>

**FONT KEY:**
**VENTRICULAR DIVISIONS – CAPITALS**
**Germinal zone - Helvetica bold**
*Transient structure - Times bold italic*
Permanent structure - Times Roman or **Bold**

**ABBREVIATIONS:**
**GEP** - Glioepithelium
**NEP** - Neuroepithelium
**SVZ** - Subventricular zone

Future parietal bones

GEP (posterior commissure)

Epithalamic (habenular complex) NEP

Future parietal bone

**EPI-THALAMUS**

**EPITHALAMIC POOL**

Posterior commissure

Thalamic (posterior complex) NEP

Cortical (temporal) NEP

**THALAMUS**

**FUTURE TEMPORAL LOBE**

**DIENCEPHALIC SUPERVENTRICLE (FUTURE THIRD VENTRICLE)**

Thalamic (reticular proluberance) NEP

Thalamic (reticular) NEP

Amygdaloid NEP and SVZ

*AMYGDALA*

**SUB-THALAMUS**

Sub-thalamic NEP

Medial forebrain bundle?

**PREOPTIC AREA**

**SUB-THALAMIC/ PREOPTIC POOL**

Preoptic NEP

Optic chiasm GEP?

Sclera
Pigment layer of retina
Intraretinal space
Neural retina
Eyelid
Eye

*Orbito-sphenoid process*

Optic nerve and chiasm GEP

Sphenoid

**Maxilla**

**Palatal process**

**Maxilla**

**Tongue**

**Oral cavity**

**Area of future zygomatic bone**

**Fused mandibular processes**

Skull and skin
Meninges (dura and arachnoid)
Brain surface (pia, heavier line)

*Migrating habenular neurons*
Habenulo-interpeduncular tract

*Layers of the developing thalamus*

*Sojourning neurons (posterior complex)*

*Migrating posterior complex neurons (pulvinar and lateral geniculate body)*

*Outer white layer*

**THALAMIC POOL**

Ventral lateral geniculate body?

*Migrating reticular nucleus neurons*

*Primordial plexiform layer*

**TELENCEPHALIC SUPERVENTRICLE (FUTURE LATERAL VENTRICLE) POSTERIOR POOL**

*Migrating amygdaloid neurons*

*Settling subthalamic neurons (zona incerta?)*

*Migrating subthalamic (Forel's field) neurons*

*Settling lateral preoptic neurons?*

*Sojourning and migrating lateral preoptic neurons?*

*Optic tract?*

*Sojourning and migrating medial preoptic neurons?*

*Settling medial preoptic neurons?*

**OPTIC RECESS**

*Migrating optic nerve and chiasm glia?*

**Arrows** indicate the presumed *direction of neuron migration* from neuroepithelial sources.

✱ Evaginations and invaginations of the neuroepithelium are mosaic compartments that give rise to different brain structures.

**PLATE 194A**

GW7.5
Coronal/horizontal
CR 23 mm
C966
Level 9:
Section 92

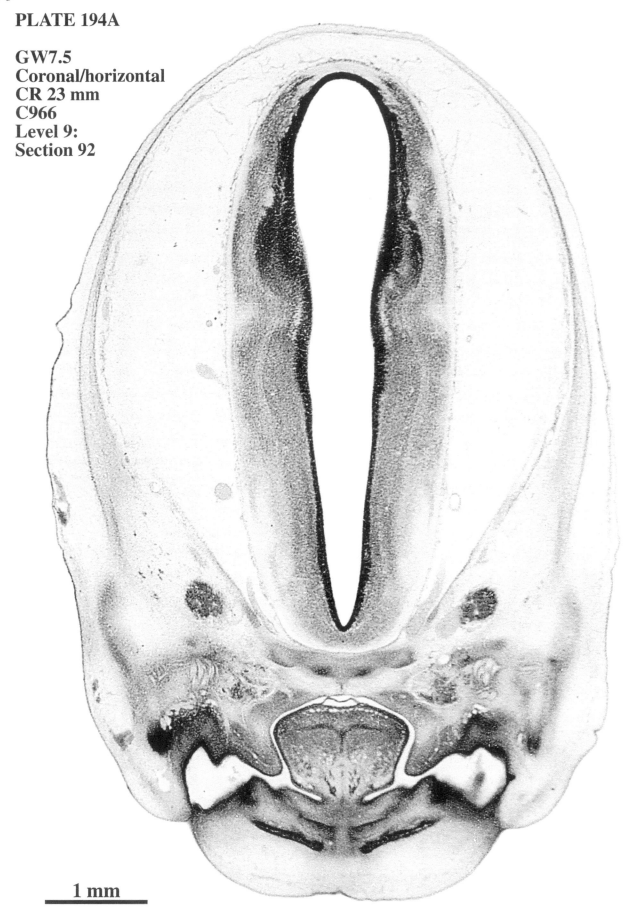

1 mm

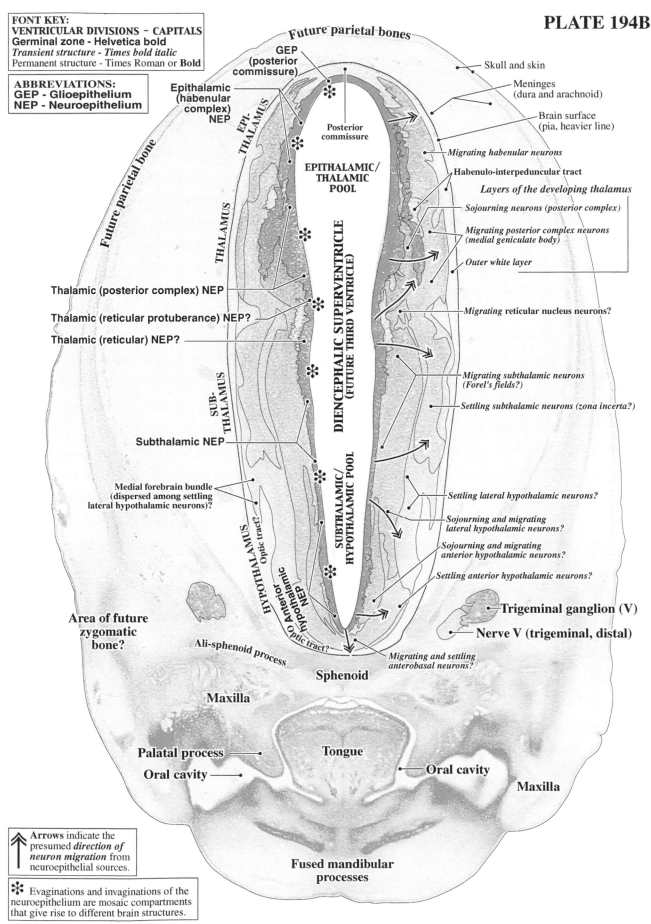

477

PLATE 194B

FONT KEY:
VENTRICULAR DIVISIONS – CAPITALS
Germinal zone - Helvetica bold
*Transient structure - Times bold italic*
Permanent structure - Times Roman or **Bold**

ABBREVIATIONS:
GEP - Glioepithelium
NEP - Neuroepithelium

Future parietal bones

GEP
(posterior
commissure)

Epithalamic
(habenular
complex)
NEP

EPI-
THALAMUS

Posterior
commissure

EPITHALAMIC/
THALAMIC
POOL

THALAMUS

Future parietal bone

Thalamic (posterior complex) NEP

Thalamic (reticular protuberance) NEP?

Thalamic (reticular) NEP?

DIENCEPHALIC SUPERVENTRICLE
(FUTURE THIRD VENTRICLE)

SUB-
THALAMUS

Subthalamic NEP

SUBTHALAMIC/
HYPOTHALAMIC POOL

Medial forebrain bundle
(dispersed among settling
lateral hypothalamic neurons)?

HYPOTHALAMUS

*Optic tract?*

O Anterior
hypothalamic NEP

Skull and skin

Meninges
(dura and arachnoid)

Brain surface
(pia, heavier line)

*Migrating habenular neurons*

Habenulo-interpeduncular tract

*Layers of the developing thalamus*

*Sojourning neurons (posterior complex)*

*Migrating posterior complex neurons
(medial geniculate body)*

*Outer white layer*

*Migrating reticular nucleus neurons?*

*Migrating subthalamic neurons
(Forel's fields?)*

*Settling subthalamic neurons (zona incerta?)*

*Settling lateral hypothalamic neurons?*

*Sojourning and migrating
lateral hypothalamic neurons?*

*Sojourning and migrating
anterior hypothalamic neurons?*

*Settling anterior hypothalamic neurons?*

**Trigeminal ganglion (V)**

**Nerve V (trigeminal, distal)**

Area of future
zygomatic
bone?

*Optic tract?*

Ali-sphenoid process

Sphenoid

*Migrating and settling
anterobasal neurons?*

Maxilla

Palatal process

Tongue

Oral cavity

Oral cavity

Maxilla

Fused mandibular
processes

↟ **Arrows** indicate the
presumed *direction of
neuron migration* from
neuroepithelial sources.

✱ Evaginations and invaginations of the
neuroepithelium are mosaic compartments
that give rise to different brain structures.

**PLATE 195A**

**GW7.5**
**Coronal/horizontal**
**CR 23 mm, C966**
**Level 10:**
**Section 98**

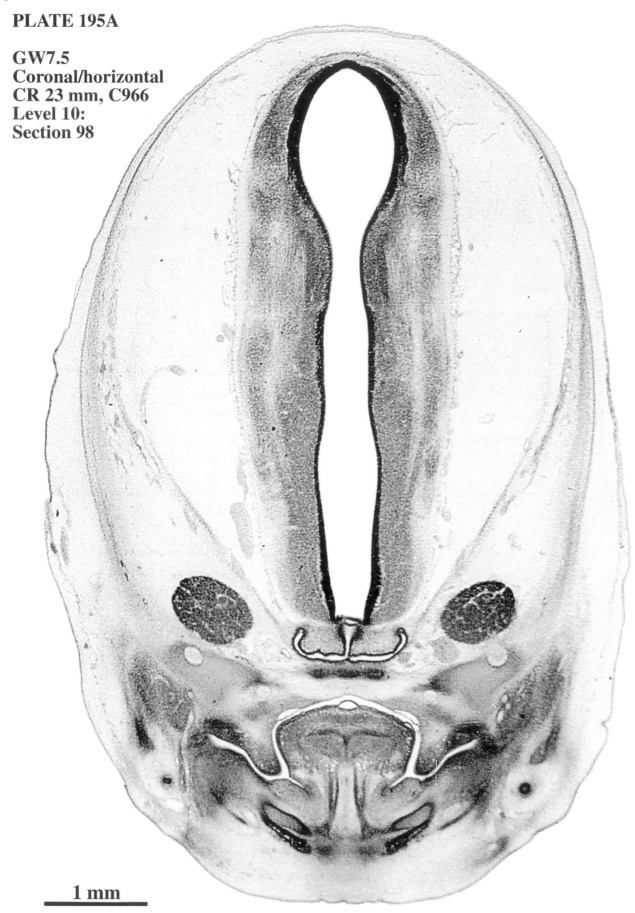

1 mm

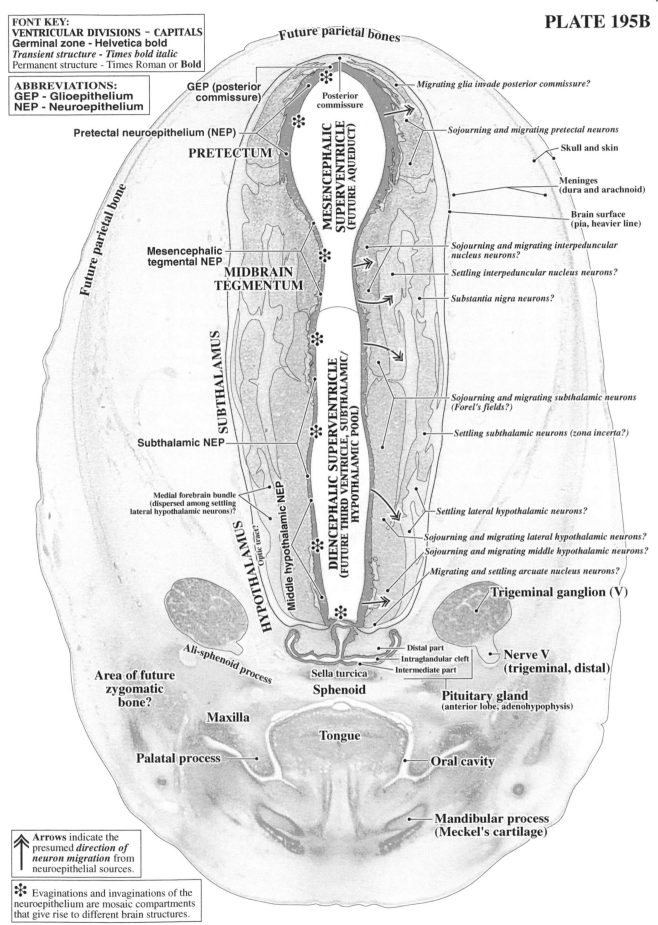

FONT KEY:
VENTRICULAR DIVISIONS – CAPITALS
Germinal zone - Helvetica bold
*Transient structure - Times bold italic*
Permanent structure - Times Roman or **Bold**

ABBREVIATIONS:
GEP - Glioepithelium
NEP - Neuroepithelium

Future parietal bones

*Migrating glia invade posterior commissure?*

GEP (posterior commissure)

Posterior commissure

Pretectal neuroepithelium (NEP)

*Sojourning and migrating pretectal neurons*

**Skull and skin**

**PRETECTUM**

**MESENCEPHALIC SUPERVENTRICLE (FUTURE AQUEDUCT)**

**Meninges (dura and arachnoid)**

**Brain surface (pia, heavier line)**

Mesencephalic tegmental NEP

*Sojourning and migrating interpeduncular nucleus neurons?*

**Future parietal bone**

**MIDBRAIN TEGMENTUM**

*Settling interpeduncular nucleus neurons?*

*Substantia nigra neurons?*

**SUBTHALAMUS**

*Sojourning and migrating subthalamic neurons (Forel's fields?)*

Subthalamic NEP

*Settling subthalamic neurons (zona incerta?)*

**DIENCEPHALIC SUPERVENTRICLE (FUTURE THIRD VENTRICLE, SUBTHALAMIC/HYPOTHALAMIC POOL)**

*Medial forebrain bundle (dispersed among settling lateral hypothalamic neurons?)*

*Settling lateral hypothalamic neurons?*

*Optic tract?*

Middle hypothalamic NEP

*Sojourning and migrating lateral hypothalamic neurons?*

**HYPOTHALAMUS**

*Sojourning and migrating middle hypothalamic neurons?*

*Migrating and settling arcuate nucleus neurons?*

**Trigeminal ganglion (V)**

*Ali-sphenoid process*

Distal part
Intraglandular cleft
Intermediate part

**Nerve V (trigeminal, distal)**

**Area of future zygomatic bone?**

Sella turcica

**Sphenoid**

**Pituitary gland (anterior lobe, adenohypophysis)**

**Maxilla**

**Tongue**

**Palatal process**

**Oral cavity**

**Mandibular process (Meckel's cartilage)**

**Arrows** indicate the presumed *direction of neuron migration* from neuroepithelial sources.

✱ Evaginations and invaginations of the neuroepithelium are mosaic compartments that give rise to different brain structures.

**PLATE 196A**

GW7.5
Coronal/horizontal
CR 23 mm, C966
Level 11:
Section 105

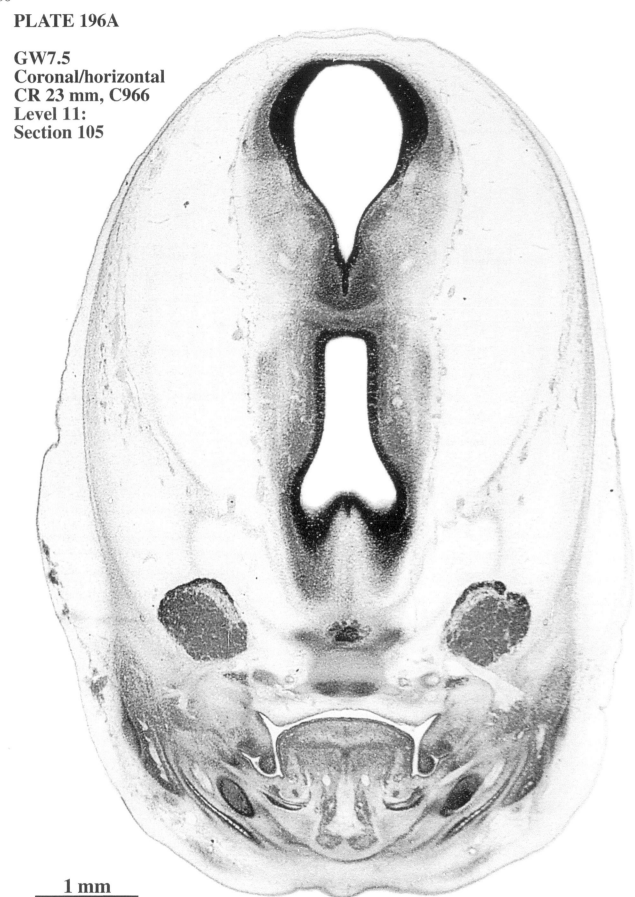

1 mm

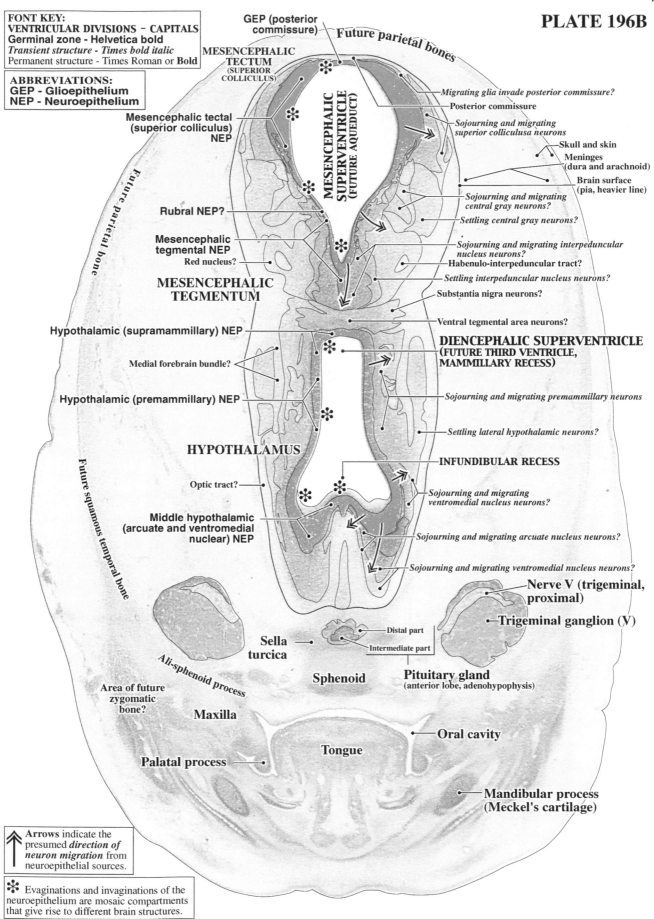

**FONT KEY:**
**VENTRICULAR DIVISIONS – CAPITALS**
**Germinal zone - Helvetica bold**
*Transient structure - Times bold italic*
Permanent structure - Times Roman or **Bold**

**ABBREVIATIONS:**
**GEP - Glioepithelium**
**NEP - Neuroepithelium**

GEP (posterior commissure)

Future parietal bones

**MESENCEPHALIC TECTUM (SUPERIOR COLLICULUS)**

Mesencephalic tectal (superior colliculus) **NEP**

**MESENCEPHALIC SUPERVENTRICLE (FUTURE AQUEDUCT)**

*Migrating glia invade posterior commissure?*
Posterior commissure
*Sojourning and migrating superior colliculusa neurons*
Skull and skin
Meninges (dura and arachnoid)
Brain surface (pia, heavier line)

Rubral **NEP?**

Mesencephalic tegmental **NEP**
Red nucleus?

*Sojourning and migrating central gray neurons?*
*Settling central gray neurons?*

**MESENCEPHALIC TEGMENTUM**

*Sojourning and migrating interpeduncular nucleus neurons?*
Habenulo-interpeduncular tract?
*Settling interpeduncular nucleus neurons?*
Substantia nigra neurons?
Ventral tegmental area neurons?

Hypothalamic (supramammillary) **NEP**

Medial forebrain bundle?

Hypothalamic (premammillary) **NEP**

**DIENCEPHALIC SUPERVENTRICLE (FUTURE THIRD VENTRICLE, MAMMILLARY RECESS)**

*Sojourning and migrating premammillary neurons*

*Settling lateral hypothalamic neurons?*

**HYPOTHALAMUS**

**INFUNDIBULAR RECESS**

Optic tract?

*Sojourning and migrating ventromedial nucleus neurons?*

Middle hypothalamic (arcuate and ventromedial nuclear) **NEP**

*Sojourning and migrating arcuate nucleus neurons?*

*Sojourning and migrating ventromedial nucleus neurons?*

**Nerve V (trigeminal, proximal)**

**Trigeminal ganglion (V)**

*Ali-sphenoid process*

**Sella turcica**

Distal part
Intermediate part

**Sphenoid**

**Pituitary gland** (anterior lobe, adenohypophysis)

Area of future zygomatic bone?

**Maxilla**

**Oral cavity**

**Tongue**

**Palatal process**

**Mandibular process (Meckel's cartilage)**

*Future parietal bone*

*Future squamous temporal bone*

**Arrows** indicate the presumed *direction of neuron migration* from neuroepithelial sources.

✱ Evaginations and invaginations of the neuroepithelium are mosaic compartments that give rise to different brain structures.

**PLATE 197A**

**GW7.5**
**Coronal/horizontal**
**CR 23 mm, C966**
**Level 12:**
**Section 112**

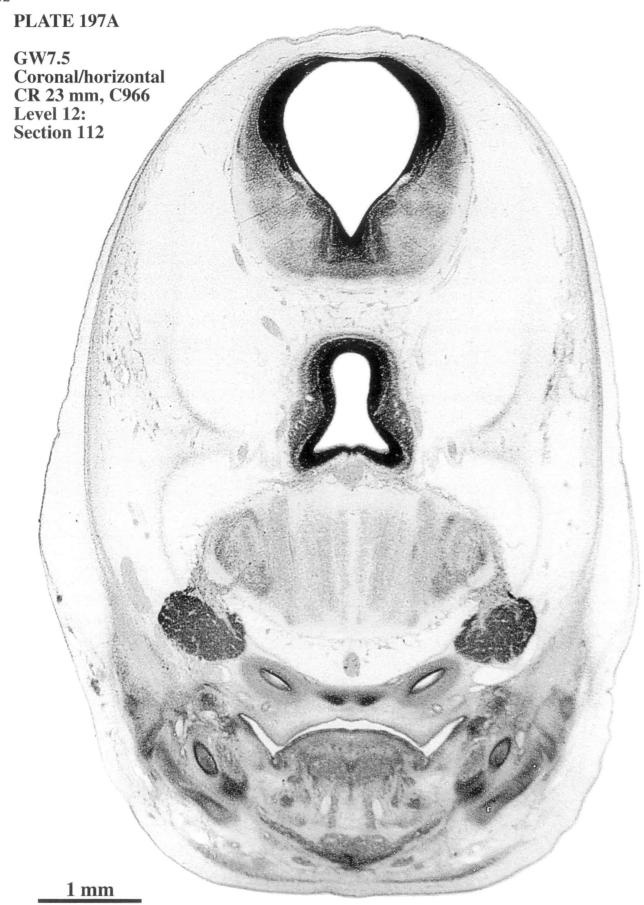

1 mm

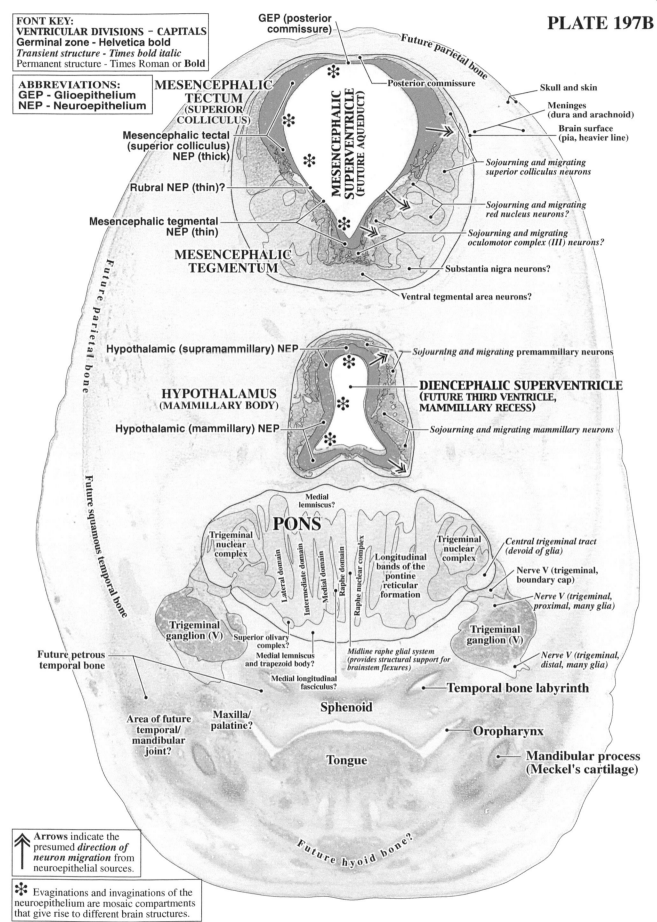

FONT KEY:
VENTRICULAR DIVISIONS – CAPITALS
Germinal zone - Helvetica bold
*Transient structure - Times bold italic*
Permanent structure - Times Roman or **Bold**

ABBREVIATIONS:
GEP - Glioepithelium
NEP - Neuroepithelium

GEP (posterior commissure)

Future parietal bone

Posterior commissure

Skull and skin

Meninges (dura and arachnoid)

Brain surface (pia, heavier line)

MESENCEPHALIC TECTUM (SUPERIOR COLLICULUS)

MESENCEPHALIC SUPERVENTRICLE (FUTURE AQUEDUCT)

Mesencephalic tectal (superior colliculus) NEP (thick)

*Sojourning and migrating superior colliculus neurons*

Rubral NEP (thin)?

*Sojourning and migrating red nucleus neurons?*

Mesencephalic tegmental NEP (thin)

*Sojourning and migrating oculomotor complex (III) neurons?*

MESENCEPHALIC TEGMENTUM

Substantia nigra neurons?

Ventral tegmental area neurons?

Future parietal bone

Hypothalamic (supramammillary) NEP

*Sojourning and migrating premammillary neurons*

HYPOTHALAMUS (MAMMILLARY BODY)

DIENCEPHALIC SUPERVENTRICLE (FUTURE THIRD VENTRICLE, MAMMILLARY RECESS)

Hypothalamic (mammillary) NEP

*Sojourning and migrating mammillary neurons*

Future squamous temporal bone

Medial lemniscus?

PONS

Trigeminal nuclear complex

Trigeminal nuclear complex

*Central trigeminal tract (devoid of glia)*

Lateral domain

Intermediate domain

Medial domain

Raphe domain

Raphe nuclear complex

Longitudinal bands of the pontine reticular formation

Nerve V (trigeminal, boundary cap)

*Nerve V (trigeminal, proximal, many glia)*

Trigeminal ganglion (V)

Trigeminal ganglion (V)

Superior olivary complex?

Medial lemniscus and trapezoid body?

*Midline raphe glial system (provides structural support for brainstem flexures)*

*Nerve V (trigeminal, distal, many glia)*

Future petrous temporal bone

Medial longitudinal fasciculus?

Temporal bone labyrinth

Area of future temporal/ mandibular joint?

Maxilla/ palatine?

Sphenoid

Oropharynx

Tongue

Mandibular process (Meckel's cartilage)

Future hyoid bone?

Arrows indicate the presumed *direction of neuron migration* from neuroepithelial sources.

✳ Evaginations and invaginations of the neuroepithelium are mosaic compartments that give rise to different brain structures.

**PLATE 198A**

**GW7.5**
**Coronal/horizontal**
**CR 23 mm, C966**
**Level 13:**
**Section 119**

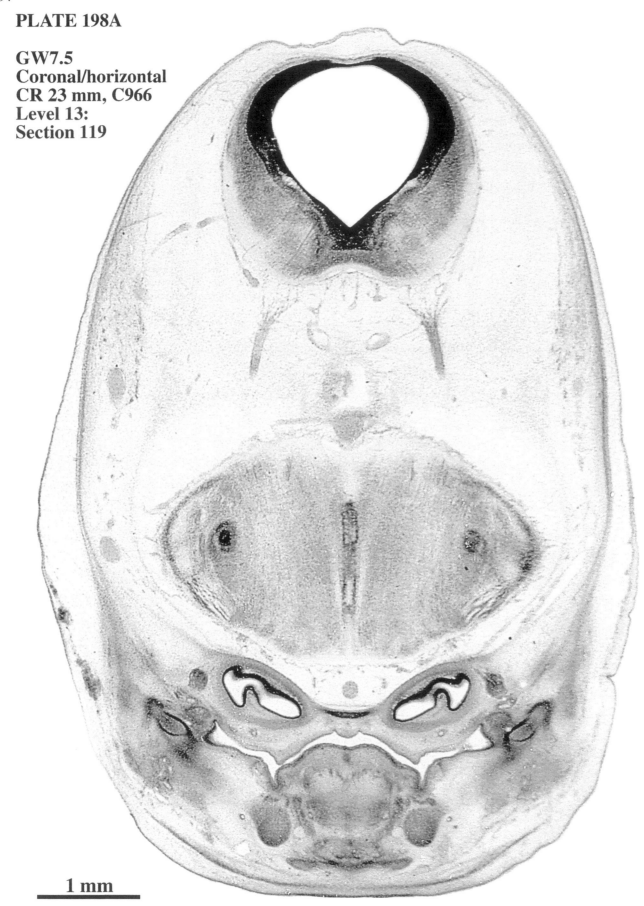

1 mm

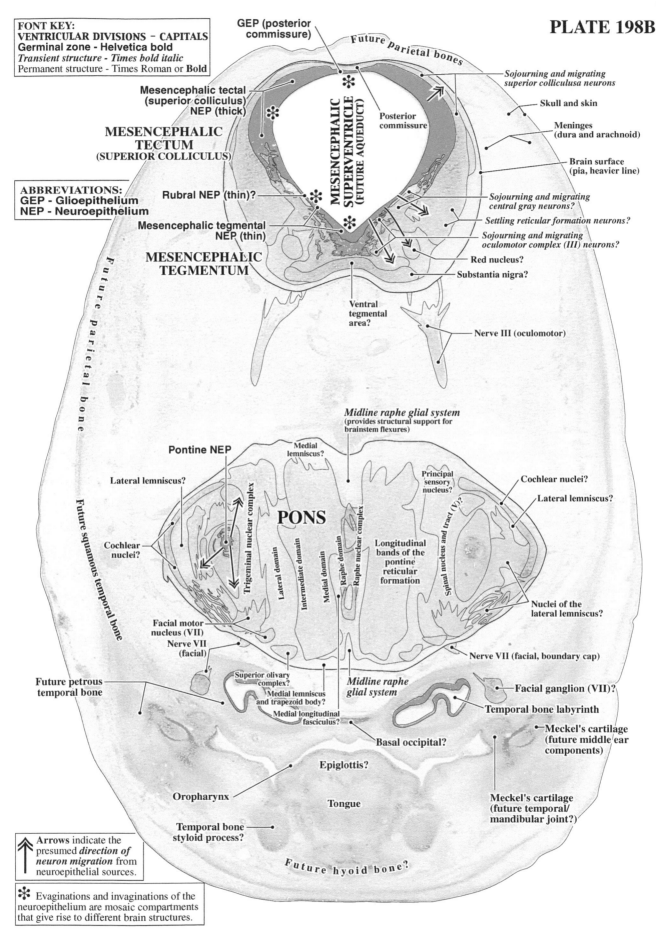

GEP (posterior commissure)

*Future parietal bones*

Mesencephalic tectal (superior colliculus) NEP (thick)

**MESENCEPHALIC TECTUM (SUPERIOR COLLICULUS)**

**MESENCEPHALIC SUPERVENTRICLE (FUTURE AQUEDUCT)**

Posterior commissure

*Sojourning and migrating superior colliculusa neurons*

Skull and skin

Meninges (dura and arachnoid)

Brain surface (pia, heavier line)

Rubral NEP (thin)?

Mesencephalic tegmental NEP (thin)

**MESENCEPHALIC TEGMENTUM**

*Sojourning and migrating central gray neurons?*

*Settling reticular formation neurons?*

*Sojourning and migrating oculomotor complex (III) neurons?*

Red nucleus?

Substantia nigra?

Ventral tegmental area?

Nerve III (oculomotor)

*Future parietal bone*

*Future squamous temporal bone*

*Midline raphe glial system (provides structural support for brainstem flexures)*

**Pontine NEP**

Medial lemniscus?

Principal sensory nucleus?

Cochlear nuclei?

**Lateral lemniscus?**

Lateral lemniscus?

**PONS**

Trigeminal nuclear complex

Lateral domain

Intermediate domain

Medial domain

Raphe domain

Raphe nuclear complex

Longitudinal bands of the pontine reticular formation

Spinal nucleus and tract (V)?

**Cochlear nuclei?**

Nuclei of the lateral lemniscus?

Facial motor nucleus (VII)

Nerve VII (facial)

Superior olivary complex?

Medial lemniscus and trapezoid body?

Medial longitudinal fasciculus?

*Midline raphe glial system*

Nerve VII (facial, boundary cap)

**Facial ganglion (VII)?**

Temporal bone labyrinth

**Future petrous temporal bone**

Basal occipital?

Meckel's cartilage (future middle ear components)

Meckel's cartilage (future temporal/ mandibular joint?)

Epiglottis?

Oropharynx

Tongue

Temporal bone styloid process?

*Future hyoid bone?*

**PLATE 199A**

**GW7.5 Coronal/horizontal**
**CR 23 mm, C966**
**Level 14:**
**Section 125**

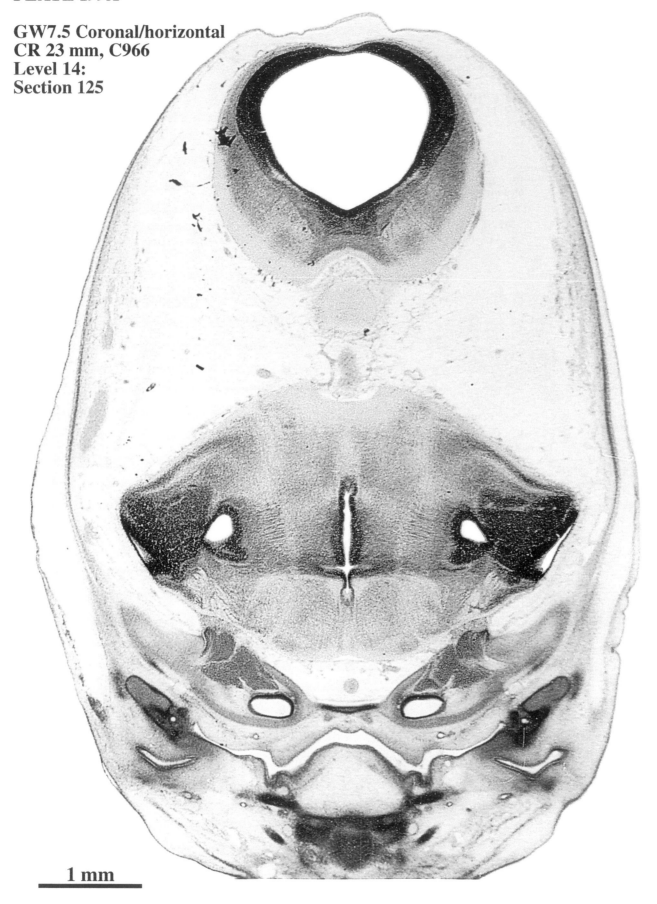

1 mm

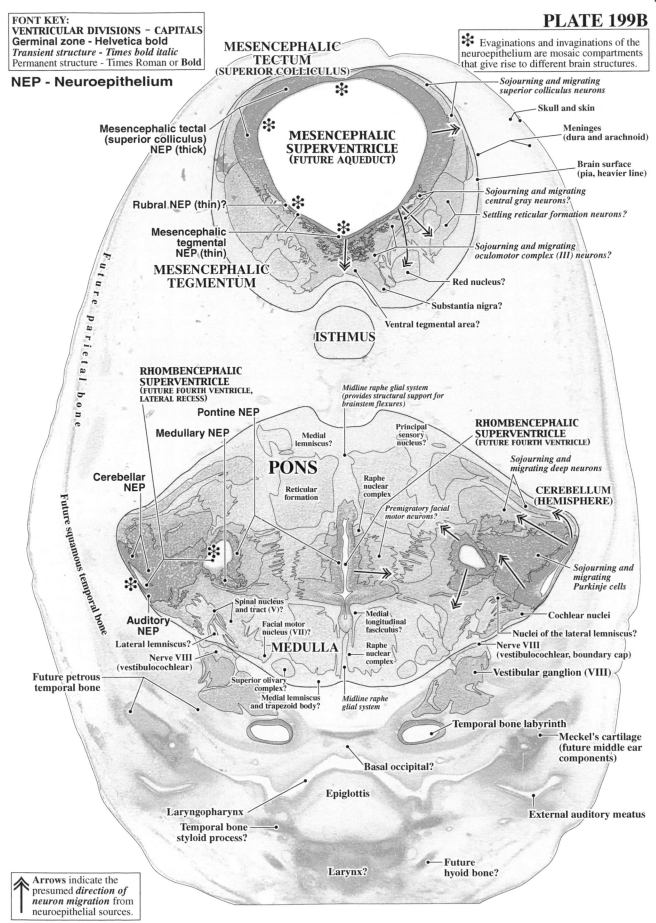

**FONT KEY:**
VENTRICULAR DIVISIONS – CAPITALS
Germinal zone - Helvetica bold
*Transient structure - Times bold italic*
Permanent structure - Times Roman or **Bold**

**NEP - Neuroepithelium**

✱ Evaginations and invaginations of the neuroepithelium are mosaic compartments that give rise to different brain structures.

MESENCEPHALIC
TECTUM
(SUPERIOR COLLICULUS)

*Sojourning and migrating
superior colliculus neurons*

Mesencephalic tectal
(superior colliculus)
NEP (thick)

Skull and skin

Meninges
(dura and arachnoid)

MESENCEPHALIC
SUPERVENTRICLE
(FUTURE AQUEDUCT)

Brain surface
(pia, heavier line)

Rubral NEP (thin)?

*Sojourning and migrating
central gray neurons?*

*Settling reticular formation neurons?*

Mesencephalic
tegmental
NEP (thin)

*Sojourning and migrating
oculomotor complex (III) neurons?*

MESENCEPHALIC
TEGMENTUM

Red nucleus?

Substantia nigra?

Ventral tegmental area?

ISTHMUS

*Future parietal bone*

RHOMBENCEPHALIC
SUPERVENTRICLE
(FUTURE FOURTH VENTRICLE,
LATERAL RECESS)

*Midline raphe glial system
(provides structural support for
brainstem flexures)*

Pontine NEP

Medial
lemniscus?

Principal
sensory
nucleus?

RHOMBENCEPHALIC
SUPERVENTRICLE
(FUTURE FOURTH VENTRICLE)

Medullary NEP

PONS

*Sojourning and
migrating deep neurons*

Cerebellar
NEP

Raphe
nuclear
complex

Reticular
formation

CEREBELLUM
(HEMISPHERE)

*Premigratory facial
motor neurons?*

*Future squamous temporal bone*

*Sojourning and
migrating
Purkinje cells*

Spinal nucleus
and tract (V)?

Medial
longitudinal
fasciculus?

Cochlear nuclei

Facial motor
nucleus (VII)?

Nuclei of the lateral lemniscus?

Auditory
NEP

MEDULLA

Raphe
nuclear
complex

Nerve VIII
(vestibulocochlear, boundary cap)

Lateral lemniscus?

Nerve VIII
(vestibulocochlear)

Vestibular ganglion (VIII)

Superior olivary
complex?

*Midline raphe
glial system*

Future petrous
temporal bone

Medial lemniscus
and trapezoid body?

Temporal bone labyrinth

Meckel's cartilage
(future middle ear
components)

Basal occipital?

Epiglottis

External auditory meatus

Laryngopharynx

Temporal bone
styloid process?

Future
hyoid bone?

Larynx?

↑ **Arrows** indicate the
presumed *direction of
neuron migration* from
neuroepithelial sources.

**PLATE 200A**

**GW7.5 Coronal/horizontal
CR 23 mm, C966
Level 15:
Section 128**

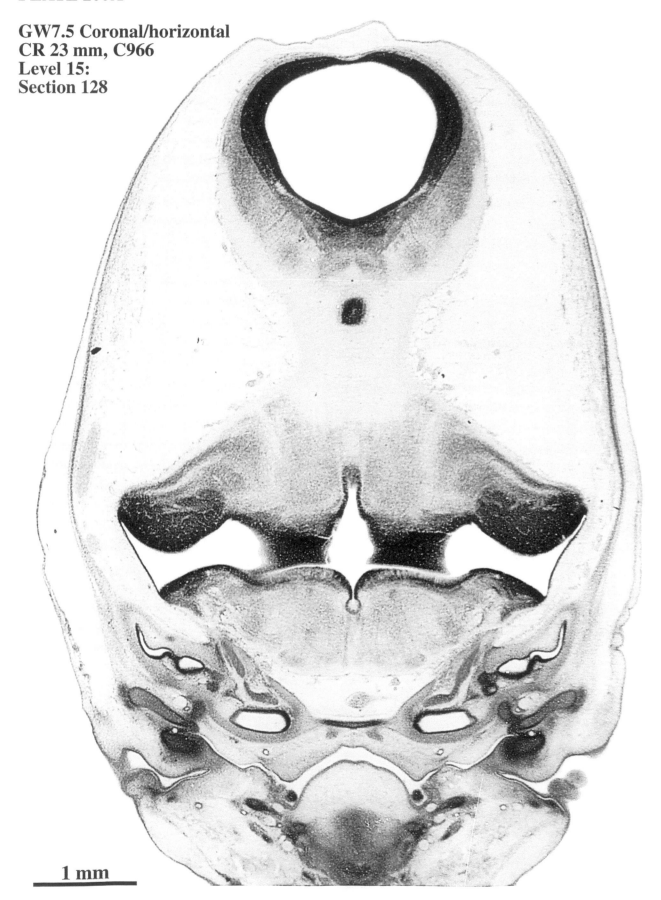

1 mm

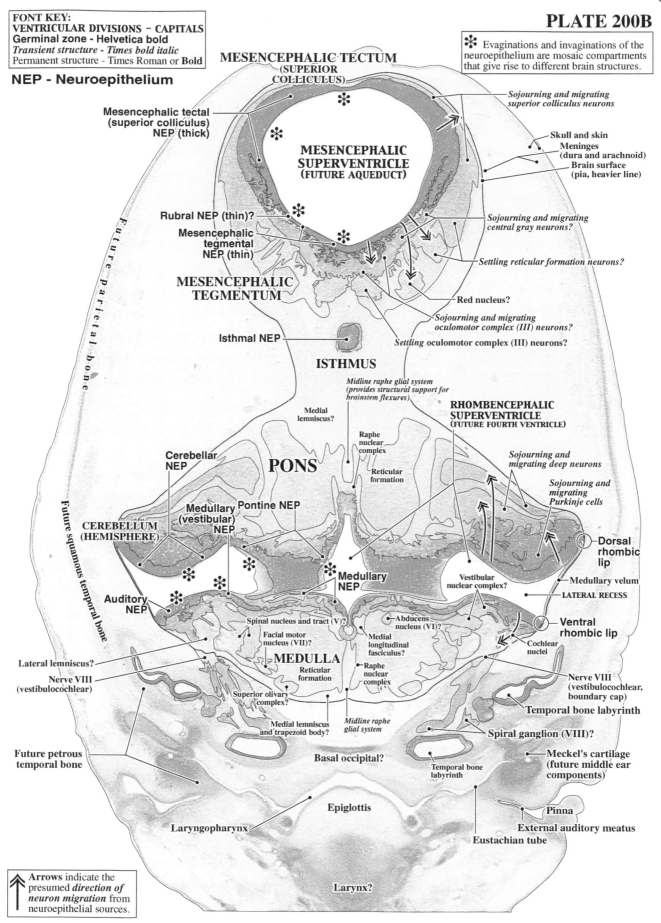

**FONT KEY:**
VENTRICULAR DIVISIONS – CAPITALS
Germinal zone - Helvetica bold
*Transient structure - Times bold italic*
Permanent structure - Times Roman or **Bold**

**NEP - Neuroepithelium**

✳ Evaginations and invaginations of the neuroepithelium are mosaic compartments that give rise to different brain structures.

MESENCEPHALIC TECTUM
(SUPERIOR COLLICULUS)

Mesencephalic tectal (superior colliculus) NEP (thick)

*Sojourning and migrating superior colliculus neurons*

Skull and skin
Meninges (dura and arachnoid)
Brain surface (pia, heavier line)

MESENCEPHALIC SUPERVENTRICLE
(FUTURE AQUEDUCT)

Rubral NEP (thin)?

Mesencephalic tegmental NEP (thin)

*Sojourning and migrating central gray neurons?*

*Settling reticular formation neurons?*

MESENCEPHALIC TEGMENTUM

Red nucleus?

*Sojourning and migrating oculomotor complex (III) neurons?*

Isthmal NEP

*Settling oculomotor complex (III) neurons?*

ISTHMUS

*Midline raphe glial system (provides structural support for brainstem flexures)*

Medial lemniscus?

RHOMBENCEPHALIC SUPERVENTRICLE
(FUTURE FOURTH VENTRICLE)

Raphe nuclear complex

Cerebellar NEP

Reticular formation

*Sojourning and migrating deep neurons*

PONS

*Sojourning and migrating Purkinje cells*

Medullary (vestibular) NEP

Pontine NEP

CEREBELLUM (HEMISPHERE)

Medullary NEP

Dorsal rhombic lip

Medullary velum
LATERAL RECESS

Vestibular nuclear complex?

Auditory NEP

Spinal nucleus and tract (V)?

Abducens nucleus (VI)?

Ventral rhombic lip

Facial motor nucleus (VII)?

Medial longitudinal fasciculus?

Cochlear nuclei

Lateral lemniscus?

MEDULLA

Nerve VIII (vestibulocochlear)

Reticular formation

Raphe nuclear complex

Nerve VIII (vestibulocochlear, boundary cap)

Superior olivary complex?

Temporal bone labyrinth

Medial lemniscus and trapezoid body?

*Midline raphe glial system*

Spiral ganglion (VIII)?

Future petrous temporal bone

Basal occipital?

Meckel's cartilage (future middle ear components)

Temporal bone labyrinth

Epiglottis

Pinna

Laryngopharynx

External auditory meatus

Eustachian tube

Larynx?

*Future parietal bone*

*Future squamous temporal bone*

↑ **Arrows** indicate the presumed *direction of neuron migration* from neuroepithelial sources.

**PLATE 201A**

GW7.5 Coronal/horizontal
CR 23 mm, C966
Level 16:
Section 139

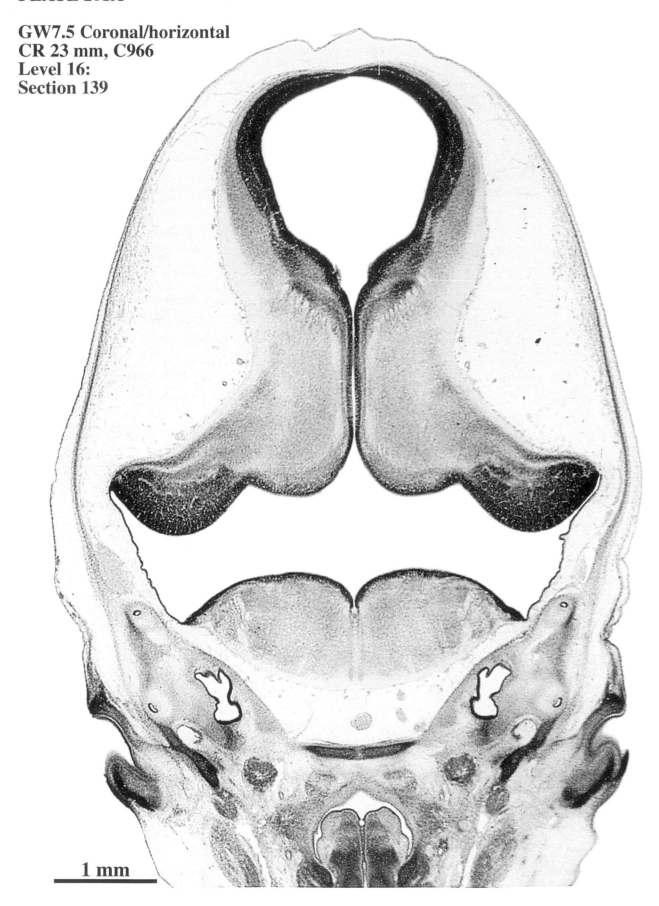

1 mm

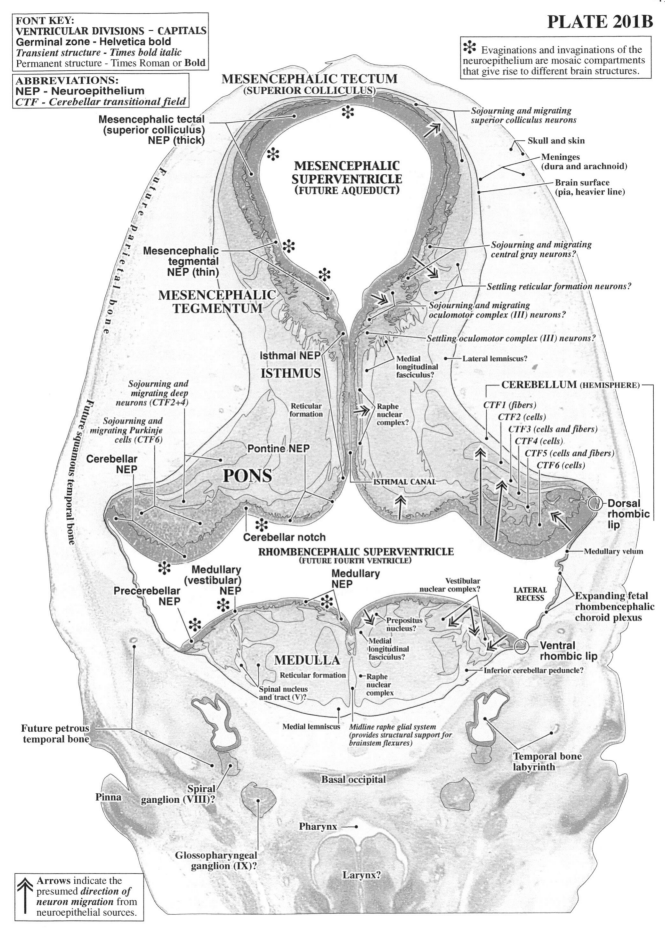

FONT KEY:
VENTRICULAR DIVISIONS – CAPITALS
Germinal zone - Helvetica bold
*Transient structure - Times bold italic*
Permanent structure - Times Roman or **Bold**

ABBREVIATIONS:
NEP - Neuroepithelium
*CTF - Cerebellar transitional field*

❋ Evaginations and invaginations of the neuroepithelium are mosaic compartments that give rise to different brain structures.

MESENCEPHALIC TECTUM
(SUPERIOR COLLICULUS)

Mesencephalic tectal
(superior colliculus)
NEP (thick)

*Sojourning and migrating superior colliculus neurons*

Skull and skin

MESENCEPHALIC
SUPERVENTRICLE
(FUTURE AQUEDUCT)

Meninges
(dura and arachnoid)

Brain surface
(pia, heavier line)

Mesencephalic
tegmental
NEP (thin)

MESENCEPHALIC
TEGMENTUM

*Sojourning and migrating central gray neurons?*

*Settling reticular formation neurons?*

*Sojourning and migrating oculomotor complex (III) neurons?*

*Settling oculomotor complex (III) neurons?*

Isthmal NEP
ISTHMUS

Medial
longitudinal
fasciculus?

Lateral lemniscus?

CEREBELLUM (HEMISPHERE)

Reticular
formation

Raphe
nuclear
complex?

CTF1 *(fibers)*
CTF2 *(cells)*
CTF3 *(cells and fibers)*
CTF4 *(cells)*
CTF5 *(cells and fibers)*
CTF6 *(cells)*

*Sojourning and migrating deep neurons (CTF2+4)*

*Sojourning and migrating Purkinje cells (CTF6)*

Cerebellar
NEP

Pontine NEP

PONS

ISTHMAL CANAL

Dorsal
rhombic
lip

Cerebellar notch

Medullary velum

RHOMBENCEPHALIC SUPERVENTRICLE
(FUTURE FOURTH VENTRICLE)

Medullary
(vestibular)
NEP

Medullary
NEP

Vestibular
nuclear complex?

LATERAL
RECESS

Expanding fetal
rhombencephalic
choroid plexus

Precerebellar
NEP

Prepositus
nucleus?

Medial
longitudinal
fasciculus?

Ventral
rhombic
lip

MEDULLA

Reticular formation

Raphe
nuclear
complex

Inferior cerebellar peduncle?

Spinal nucleus
and tract (V)?

Future petrous
temporal bone

Medial lemniscus

*Midline raphe glial system
(provides structural support for
brainstem flexures)*

Temporal bone
labyrinth

Pinna

Spiral
ganglion (VIII)?

Basal occipital

Glossopharyngeal
ganglion (IX)?

Pharynx

Larynx?

**Arrows** indicate the
presumed *direction of
neuron migration* from
neuroepithelial sources.

**PLATE 202A**

**GW7.5 Coronal/horizontal**
**CR 23 mm, C966**
**Level 17:**
**Section 147**

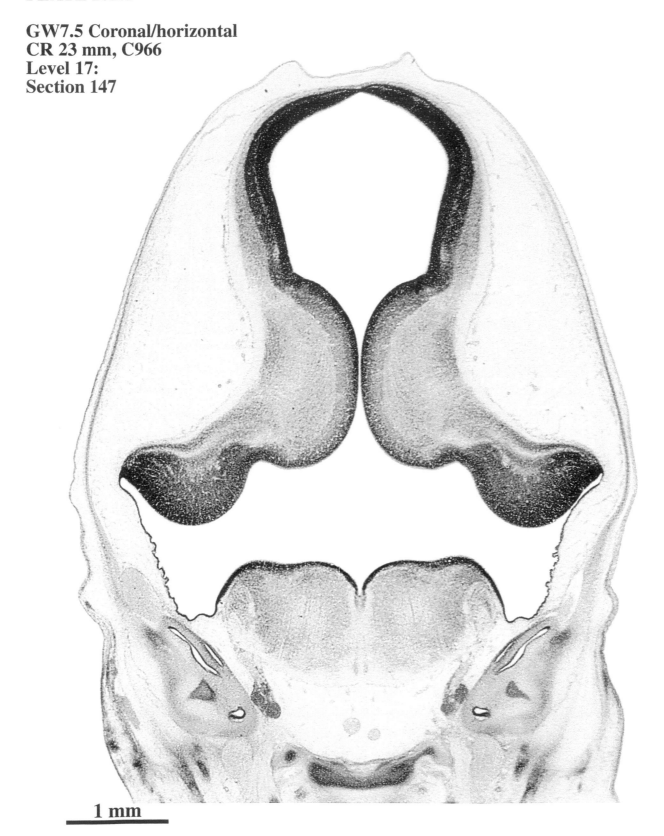

1 mm

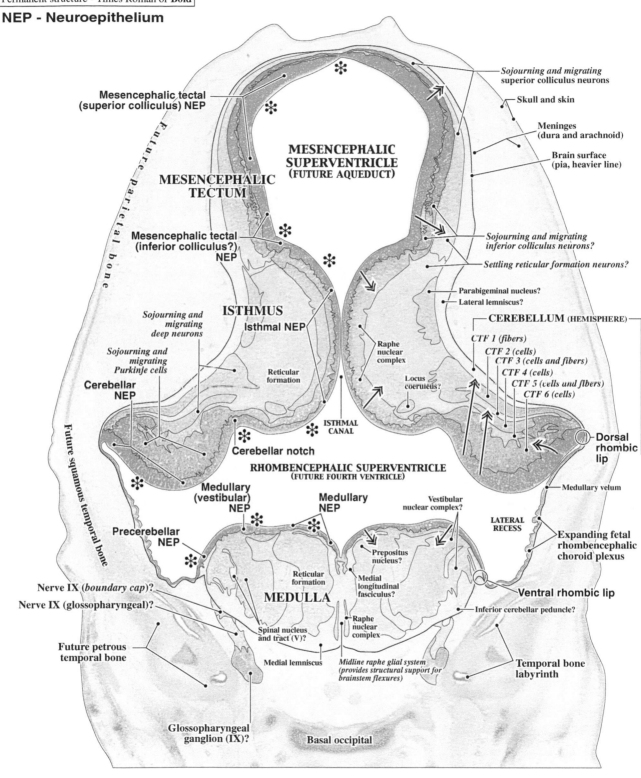

FONT KEY:
**VENTRICULAR DIVISIONS – CAPITALS**
**Germinal zone - Helvetica bold**
*Transient structure - Times bold italic*
Permanent structure - Times Roman or **Bold**

**NEP - Neuroepithelium**

Mesencephalic tectal
(superior colliculus) NEP

*Sojourning and migrating
superior colliculus neurons*

Skull and skin

Meninges
(dura and arachnoid)

**MESENCEPHALIC
SUPERVENTRICLE
(FUTURE AQUEDUCT)**

Brain surface
(pia, heavier line)

**MESENCEPHALIC
TECTUM**

Mesencephalic tectal
(inferior colliculus?)
NEP

*Sojourning and migrating
inferior colliculus neurons?*

*Settling reticular formation neurons?*

Parabigeminal nucleus?
Lateral lemniscus?

**ISTHMUS**

*Sojourning and
migrating
deep neurons*

**Isthmal NEP**

**CEREBELLUM** (HEMISPHERE)

*Sojourning and
migrating
Purkinje cells*

Raphe
nuclear
complex

*CTF 1 (fibers)*
*CTF 2 (cells)*
*CTF 3 (cells and fibers)*
*CTF 4 (cells)*
*CTF 5 (cells and fibers)*
*CTF 6 (cells)*

**Cerebellar
NEP**

Reticular
formation

Locus
coeruleus?

**ISTHMAL
CANAL**

**Dorsal
rhombic
lip**

Cerebellar notch

**RHOMBENCEPHALIC SUPERVENTRICLE**
**(FUTURE FOURTH VENTRICLE)**

Medullary velum

**Medullary
(vestibular)
NEP**

**Medullary
NEP**

Vestibular
nuclear complex?

**LATERAL
RECESS**

**Precerebellar
NEP**

Prepositus
nucleus?

**Expanding fetal
rhombencephalic
choroid plexus**

Reticular
formation

Medial
longitudinal
fasciculus?

**Ventral rhombic lip**

Nerve IX (*boundary cap*)?

**MEDULLA**

Nerve IX (glossopharyngeal)?

Inferior cerebellar peduncle?

Spinal nucleus
and tract (V)?

Raphe
nuclear
complex

**Temporal bone
labyrinth**

**Future petrous
temporal bone**

Medial lemniscus

*Midline raphe glial system
(provides structural support for
brainstem flexures)*

Glossopharyngeal
ganglion (IX)?

**Basal occipital**

*Future parietal bone*

*Future squamous temporal bone*

**Arrows** indicate the
presumed *direction of
neuron migration* from
neuroepithelial sources.

✳ Evaginations and invaginations of the
neuroepithelium are mosaic compartments
that give rise to different brain structures.

**PLATE 203A**

**GW7.5 Coronal/horizontal**
**CR 23 mm, C966**
**Level 18:**
**Section 157**

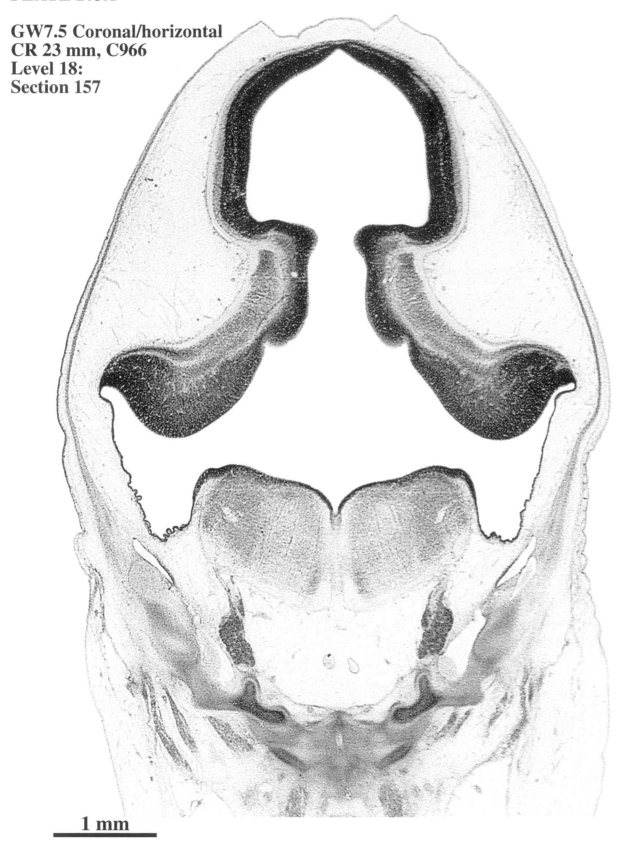

1 mm

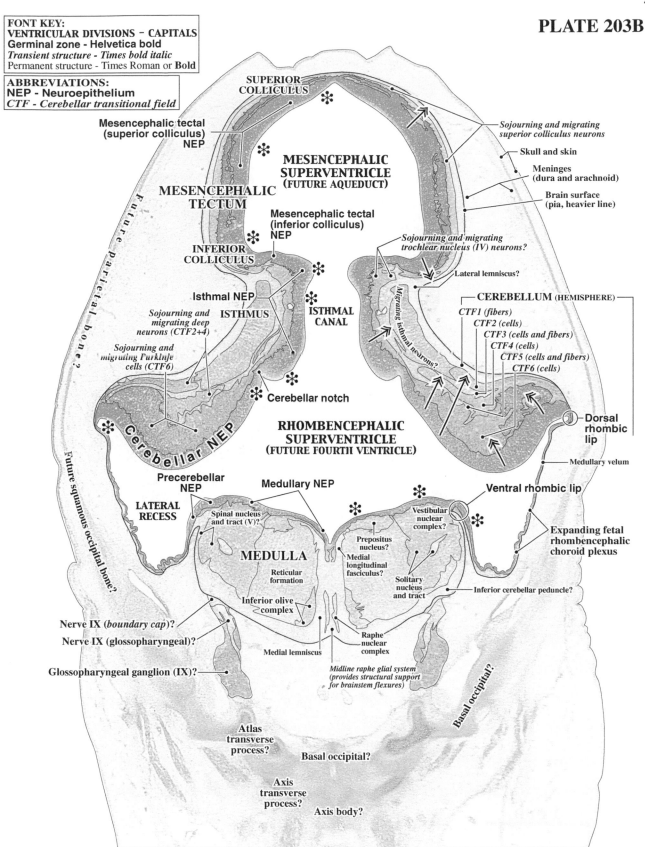

FONT KEY:
VENTRICULAR DIVISIONS – CAPITALS
Germinal zone - Helvetica bold
*Transient structure - Times bold italic*
Permanent structure - Times Roman or **Bold**

ABBREVIATIONS:
NEP - Neuroepithelium
*CTF - Cerebellar transitional field*

Mesencephalic tectal
(superior colliculus)
NEP

SUPERIOR
COLLICULUS

*Sojourning and migrating
superior colliculus neurons*

Skull and skin

Meninges
(dura and arachnoid)

MESENCEPHALIC
SUPERVENTRICLE
(FUTURE AQUEDUCT)

MESENCEPHALIC
TECTUM

Brain surface
(pia, heavier line)

Mesencephalic tectal
(inferior colliculus)
NEP

INFERIOR
COLLICULUS

*Sojourning and migrating
trochlear nucleus (IV) neurons?*

Lateral lemniscus?

*Future parietal bone?*

Isthmal NEP

*Sojourning and
migrating deep
neurons (CTF2+4)*

ISTHMUS

ISTHMAL
CANAL

*Migrating isthmal neurons?*

CEREBELLUM (HEMISPHERE)

*CTF1 (fibers)*
*CTF2 (cells)*
*CTF3 (cells and fibers)*
*CTF4 (cells)*
*CTF5 (cells and fibers)*
*CTF6 (cells)*

*Sojourning and
migrating Purkinje
cells (CTF6)*

Cerebellar notch

Cerebellar NEP

RHOMBENCEPHALIC
SUPERVENTRICLE
(FUTURE FOURTH VENTRICLE)

Dorsal
rhombic
lip

Medullary velum

Precerebellar
NEP

Medullary NEP

Ventral rhombic lip

LATERAL
RECESS

Spinal nucleus
and tract (V)?

Vestibular
nuclear
complex?

Expanding fetal
rhombencephalic
choroid plexus

*Future squamous occipital bone?*

Prepositus
nucleus?

MEDULLA

Reticular
formation

Medial
longitudinal
fasciculus?

Solitary
nucleus
and tract

Inferior cerebellar peduncle?

Inferior olive
complex

Nerve IX (*boundary cap*)?

Nerve IX (glossopharyngeal)?

Raphe
nuclear
complex

Medial lemniscus

Glossopharyngeal ganglion (IX)?

*Midline raphe glial system
(provides structural support
for brainstem flexures)*

*Basal occipital?*

Atlas
transverse
process?

Basal occipital?

Axis
transverse
process?

Axis body?

**Arrows** indicate the
presumed *direction of
neuron migration* from
neuroepithelial sources.

❊ Evaginations and invaginations of the
neuroepithelium are mosaic compartments
that give rise to different brain structures.

496

**GW7.5 Coronal/horizontal**
**CR 23 mm, C966**
**Level 19:**
**Section 163**

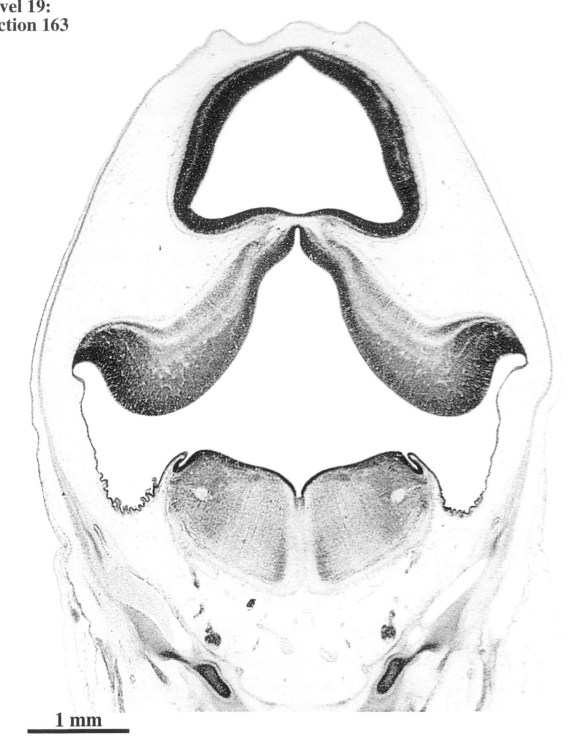

1 mm

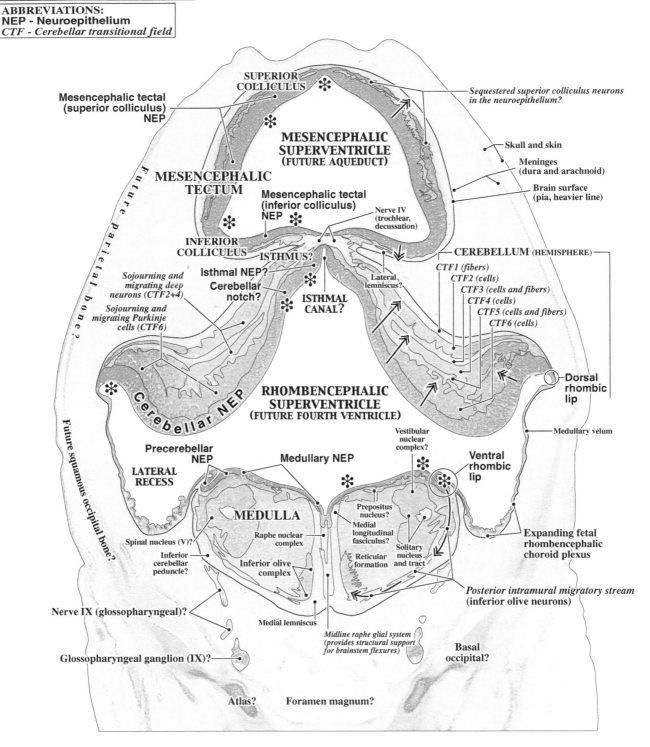

**FONT KEY:**
**VENTRICULAR DIVISIONS – CAPITALS**
**Germinal zone - Helvetica bold**
*Transient structure - Times bold italic*
Permanent structure - Times Roman or **Bold**

**ABBREVIATIONS:**
**NEP** - Neuroepithelium
*CTF - Cerebellar transitional field*

**SUPERIOR COLLICULUS** ✳

Mesencephalic tectal (superior colliculus) NEP

*Sequestered superior colliculus neurons in the neuroepithelium?*

**MESENCEPHALIC SUPERVENTRICLE (FUTURE AQUEDUCT)**

Skull and skin

Meninges (dura and arachnoid)

Brain surface (pia, heavier line)

**MESENCEPHALIC TECTUM**

*Future parietal bone?*

Mesencephalic tectal (inferior colliculus) NEP

Nerve IV (trochlear, decussation)

**CEREBELLUM** (HEMISPHERE)

**INFERIOR COLLICULUS**

ISTHMUS?

*Isthmal NEP?*

*CTF1 (fibers)*
*CTF2 (cells)*
*CTF3 (cells and fibers)*
*CTF4 (cells)*
*CTF5 (cells and fibers)*
*CTF6 (cells)*

*Sojourning and migrating deep neurons (CTF2+4)*

*Cerebellar notch?*

Lateral lemniscus?

*Sojourning and migrating Purkinje cells (CTF6)*

**ISTHMAL CANAL?**

**Cerebellar NEP**

**RHOMBENCEPHALIC SUPERVENTRICLE (FUTURE FOURTH VENTRICLE)**

Dorsal rhombic lip

Medullary velum

*Future squamous occipital bone?*

**Precerebellar NEP**

**Medullary NEP**

Vestibular nuclear complex?

Ventral rhombic lip

**LATERAL RECESS**

**MEDULLA**

Prepositus nucleus?

Medial longitudinal fasciculus?

Expanding fetal rhombencephalic choroid plexus

Spinal nucleus (V)?

Raphe nuclear complex

Solitary nucleus and tract

Reticular formation

Inferior cerebellar peduncle?

Inferior olive complex

*Posterior intramural migratory stream (inferior olive neurons)*

Nerve IX (glossopharyngeal)?

Medial lemniscus

*Midline raphe glial system (provides structural support for brainstem flexures)*

Basal occipital?

Glossopharyngeal ganglion (IX)?

Atlas?        Foramen magnum?

**Arrows** indicate the presumed *direction of neuron migration* from neuroepithelial sources.

✳ Evaginations and invaginations of the neuroepithelium are mosaic compartments that give rise to different brain structures.

**PLATE 205A**

**GW7.5 Coronal/horizontal**
**CR 23 mm, C966**
**Level 20:**
**Section 171**

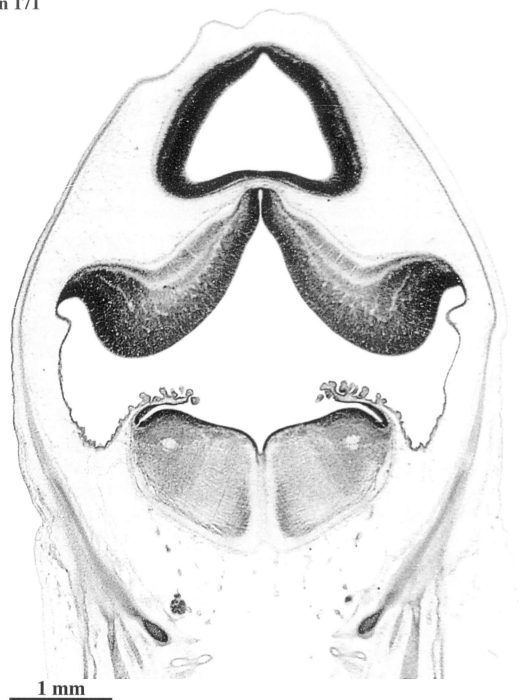

1 mm

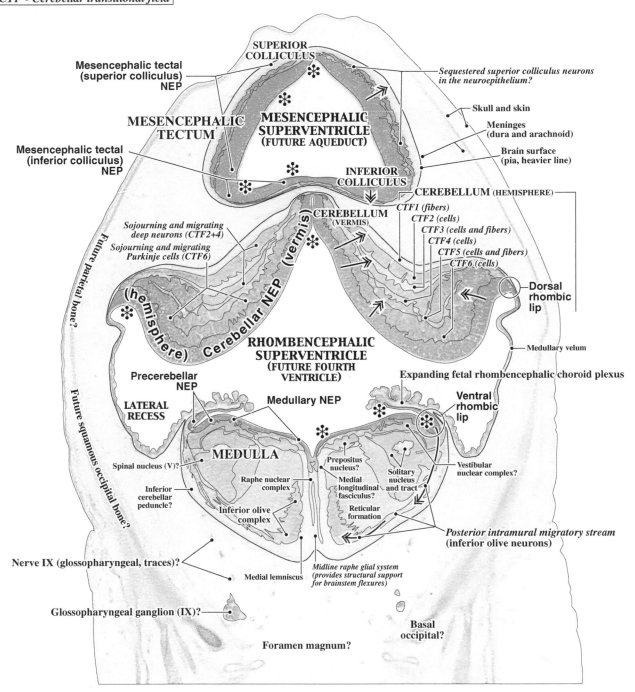

FONT KEY:
**VENTRICULAR DIVISIONS – CAPITALS**
**Germinal zone - Helvetica bold**
***Transient structure - Times bold italic***
Permanent structure - Times Roman or **Bold**

ABBREVIATIONS:
**NEP - Neuroepithelium**
*CTF - Cerebellar transitional field*

**SUPERIOR COLLICULUS**

Mesencephalic tectal (superior colliculus) NEP

*Sequestered superior colliculus neurons in the neuroepithelium?*

**MESENCEPHALIC SUPERVENTRICLE (FUTURE AQUEDUCT)**

**MESENCEPHALIC TECTUM**

Skull and skin

Meninges (dura and arachnoid)

Brain surface (pia, heavier line)

Mesencephalic tectal (inferior colliculus) NEP

**INFERIOR COLLICULUS**

**CEREBELLUM** (HEMISPHERE)

*CTF1 (fibers)*
*CTF2 (cells)*
*CTF3 (cells and fibers)*
*CTF4 (cells)*
*CTF5 (cells and fibers)*
*CTF6 (cells)*

**CEREBELLUM (VERMIS)**

*Sojourning and migrating deep neurons (CTF2+4)*

*Sojourning and migrating Purkinje cells (CTF6)*

**Cerebellar NEP (vermis)**

Dorsal rhombic lip

**(hemisphere)**

*Future parietal bone?*

*Future squamous occipital bone?*

**RHOMBENCEPHALIC SUPERVENTRICLE (FUTURE FOURTH VENTRICLE)**

Medullary velum

Expanding fetal rhombencephalic choroid plexus

**Precerebellar NEP**

**Medullary NEP**

**LATERAL RECESS**

Ventral rhombic lip

**MEDULLA**

Spinal nucleus (V)?

Prepositus nucleus?

Solitary nucleus and tract

Vestibular nuclear complex?

Raphe nuclear complex

Inferior cerebellar peduncle?

Medial longitudinal fasciculus?

Inferior olive complex

Reticular formation

*Posterior intramural migratory stream (inferior olive neurons)*

Nerve IX (glossopharyngeal, traces)?

Medial lemniscus

*Midline raphe glial system (provides structural support for brainstem flexures)*

Glossopharyngeal ganglion (IX)?

Basal occipital?

Foramen magnum?

**Arrows** indicate the presumed *direction of neuron migration* from neuroepithelial sources.

✳ Evaginations and invaginations of the neuroepithelium are mosaic compartments that give rise to different brain structures.

**PLATE 206A**

**GW7.5 Coronal/horizontal**
**CR 23 mm, C966**
**Level 21:**
**Section 179**

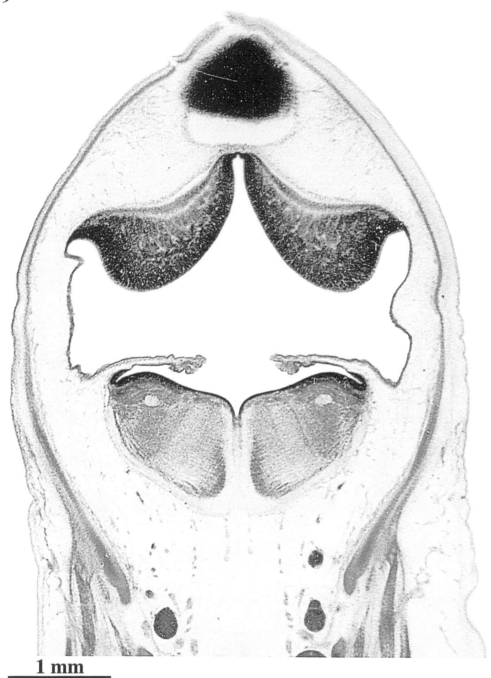

1 mm

# PLATE 206B

FONT KEY:
VENTRICULAR DIVISIONS - CAPITALS
**Germinal zone - Helvetica bold**
***Transient structure - Times bold italic***
Permanent structure - Times Roman or **Bold**

ABBREVIATIONS:
**NEP - Neuroepithelium**
***CTF - Cerebellar transitional field***

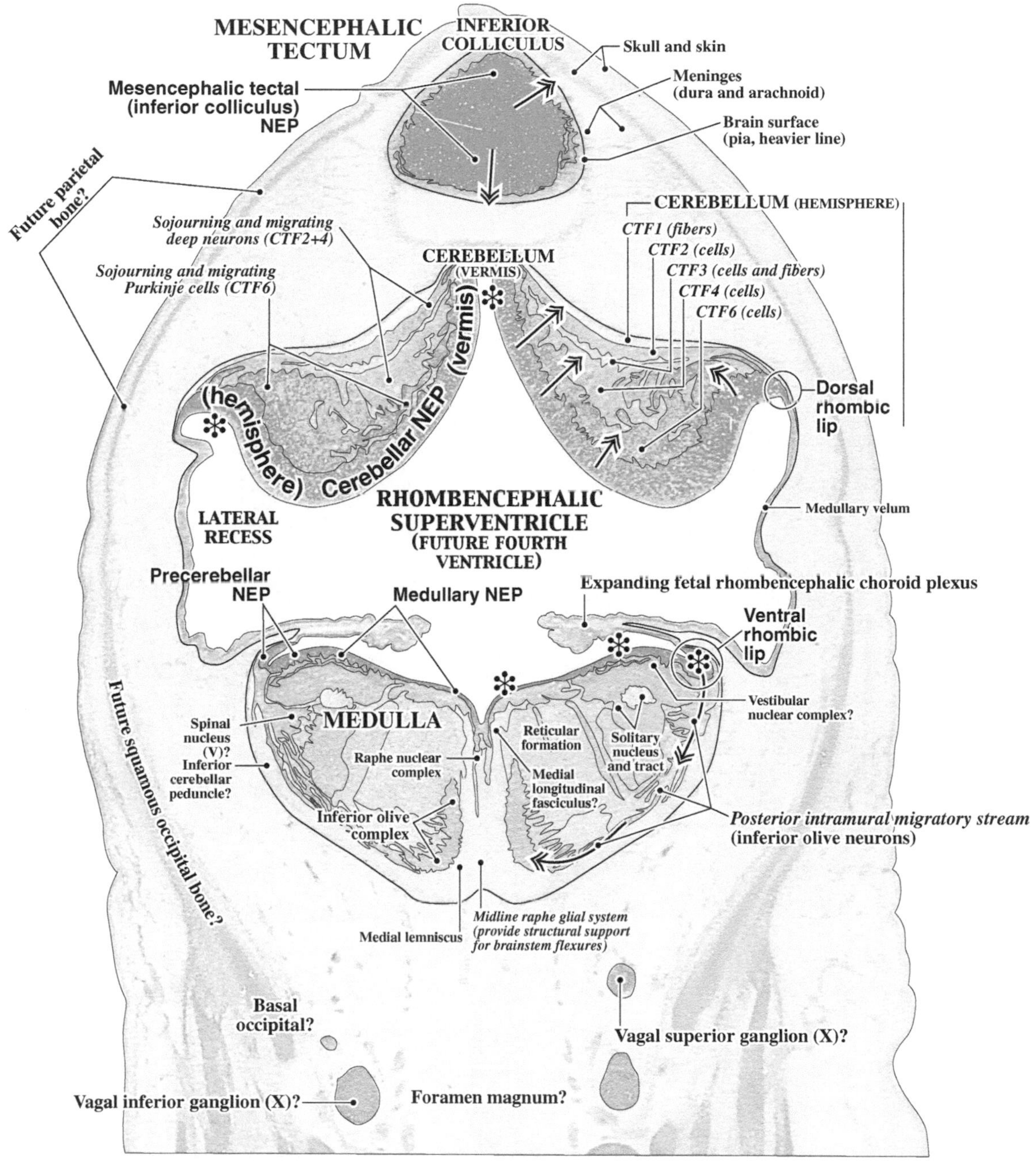
MESENCEPHALIC TECTUM
INFERIOR COLLICULUS
Skull and skin
Mesencephalic tectal (inferior colliculus) NEP
Meninges (dura and arachnoid)
Brain surface (pia, heavier line)
Future parietal bone?
CEREBELLUM (HEMISPHERE)
CTF1 (fibers)
Sojourning and migrating deep neurons (CTF2+4)
CTF2 (cells)
CEREBELLUM (VERMIS)
CTF3 (cells and fibers)
Sojourning and migrating Purkinje cells (CTF6)
CTF4 (cells)
CTF6 (cells)
(hemisphere) Cerebellar NEP (vermis)
Dorsal rhombic lip
RHOMBENCEPHALIC SUPERVENTRICLE (FUTURE FOURTH VENTRICLE)
LATERAL RECESS
Medullary velum
Precerebellar NEP
Medullary NEP
Expanding fetal rhombencephalic choroid plexus
Ventral rhombic lip
Vestibular nuclear complex?
MEDULLA
Spinal nucleus (V)?
Inferior cerebellar peduncle?
Reticular formation
Solitary nucleus and tract
Raphe nuclear complex
Medial longitudinal fasciculus?
Inferior olive complex
Posterior intramural migratory stream (inferior olive neurons)
Future squamous occipital bone?
Midline raphe glial system (provide structural support for brainstem flexures)
Medial lemniscus
Basal occipital?
Vagal superior ganglion (X)?
Vagal inferior ganglion (X)?
Foramen magnum?

**Arrows** indicate the presumed ***direction of neuron migration*** from neuroepithelial sources.

✻ Evaginations and invaginations of the neuroepithelium are mosaic compartments that give rise to different brain structures.

# FIGURE 10

# GW7.5, CR23 mm, C966, COMPUTER-AIDED 3-D RECONSTRUCTION

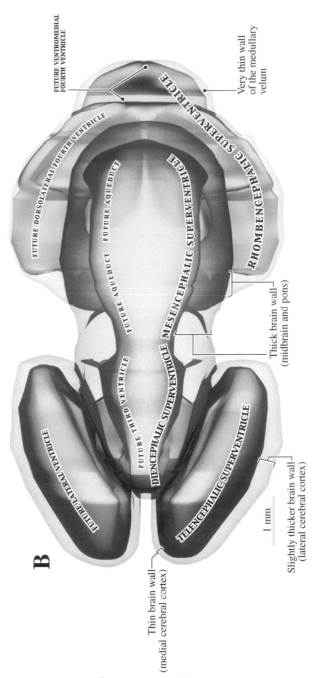

**A**

Medullary velum

MEDULLA

HEMISPHERE VERMIS CEREBELLUM

SUPERIOR COLLICULUS

TECTUM

MESENCEPHALON

PRE-TECTUM

EPI-THALAMUS

THALAMUS

Future cingulate gyrus

DIENCEPHALON

PONS

TELENCEPHALON

FUTURE PARIETAL LOBE

FUTURE OCCIPITAL LOBE

FUTURE PARACENTRAL LOBULE

FUTURE FRONTAL LOBE

1 mm

## A. TOP VIEW OF THE BRAIN SURFACE

The observer is looking straight down on the top of the brain. Note the small size of the telencephalon. In a mature brain, the only structure visible from the top is the telencephalon. At GW 7.5, the telencephalon does not yet cover the dorsal surface of the diencephalon. All of the mesencephalic tectum is visible. The cerebellum appears as a wide ledge forming a horseshoe-shaped understory beneath the tectum. The pons is only partly visible connecting to the anteroventral edge of the cerebellum. The medullary velum is all that is visible of the medulla.

**B**

FUTURE VENTROMEDIAL FOURTH VENTRICLE

FUTURE DORSOLATERAL FOURTH VENTRICLE

FUTURE AQUEDUCT

FUTURE AQUEDUCT

MESENCEPHALIC SUPERVENTRICLE

RHOMBENCEPHALIC SUPERVENTRICLE

Very thin wall of the medullary velum

Thick brain wall (midbrain and pons)

FUTURE LATERAL VENTRICLE

FUTURE THIRD VENTRICLE

DIENCEPHALIC SUPERVENTRICLE

TELENCEPHALIC SUPERVENTRICLE

Thin brain wall (medial cerebral cortex)

Slightly thicker brain wall (lateral cerebral cortex)

1 mm

## B. TOP VIEW SHOWING THE SUPERVENTRICLES

A substantial portion of the brain's volume is occupied by the ventricles. In a mature brain, the ventricles are small, central cavities. At GW 7.5, the diencephalic and mesencephalic superventricles are already narrowing to resemble their adult shapes, and the brain wall is thickest in that region. Most of the telencephalon is filled with the paired telencephalic superventricles, but note that the brain wall is thicker laterally than medially in accordance with the ventrolateral (earlier maturing) to dorsomedial (later maturing) developmental gradient. The dark areas within the brain wall are caused by looking through more than one thickness.

# FIGURE 11

## GW7.5, CR23 mm, C966, COMPUTER-AIDED 3-D RECONSTRUCTION

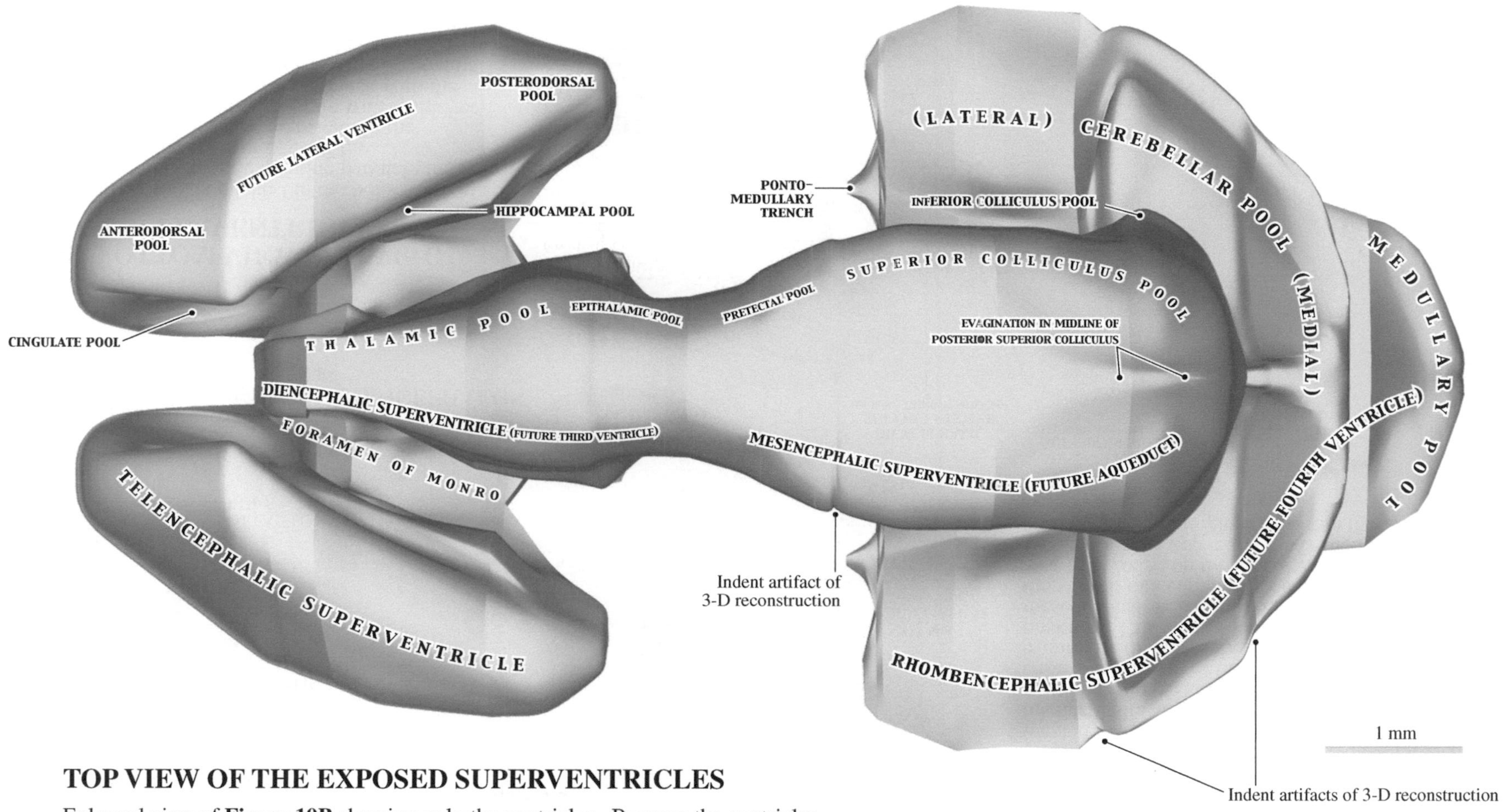

MEDULLARY POOL

CEREBELLAR POOL (MEDIAL)

(LATERAL)

INFERIOR COLLICULUS POOL

SUPERIOR COLLICULUS POOL

EVAGINATION IN MIDLINE OF POSTERIOR SUPERIOR COLLICULUS

SUPRAVENTRICLE (FUTURE FOURTH VENTRICLE)

RHOMBENCEPHALIC SUPER

MESENCEPHALIC SUPERVENTRICLE (FUTURE AQUEDUCT)

PONTO–MEDULLARY TRENCH

PREFECTAL POOL

EPITHALAMIC POOL

THALAMIC POOL

DIENCEPHALIC SUPERVENTRICLE (FUTURE THIRD VENTRICLE)

FORAMEN OF MONRO

TELENCEPHALIC SUPERVENTRICLE

POSTERODORSAL POOL

HIPPOCAMPAL POOL

FUTURE LATERAL VENTRICLE

ANTERODORSAL POOL

CINGULATE POOL

Indent artifact of 3-D reconstruction

Indent artifacts of 3-D reconstruction

1 mm

## TOP VIEW OF THE EXPOSED SUPERVENTRICLES

Enlarged view of **Figure 10B** showing only the ventricles. Because the ventricles contain fluid, subdivisions are called "pools." The various pools are named according to adjacent structures, which often produce evaginations into the ventricle (such as the cingulate and hippocampal pools). The complex shape of the brain ventricles is primarily a result of (1) localized differential proliferation in the adjacent neuroepithelium and (2) flexures in the diencephalon, midbrain, pons, and medulla during development.

# FIGURE 12

## GW7.5, CR23 mm, C966, COMPUTER-AIDED 3-D RECONSTRUCTION

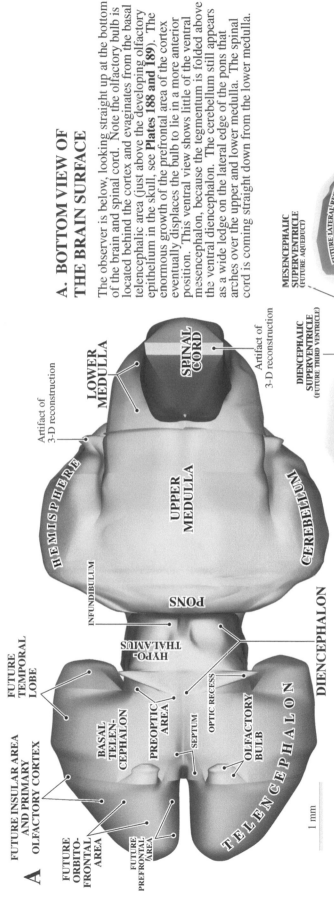

### A. BOTTOM VIEW OF THE BRAIN SURFACE

The observer is below, looking straight up at the bottom of the brain and spinal cord. Note the olfactory bulb is located behind the cortex and evaginates from the basal telencephalic area (just above the developing olfactory epithelium in the skull, see **Plates 188 and 189**). The enormous growth of the prefrontal area of the cortex eventually displaces the bulb to lie in a more anterior position. This ventral view shows little of the ventral mesencephalon, because the tegmentum is folded above the ventral diencephalon. The cerebellum still appears as a wide ledge on the lateral edge of the pons that arches over the upper and lower medulla. The spinal cord is coming straight down from the lower medulla.

### B. BOTTOM VIEW SHOWING THE SUPERVENTRICLES

The brain wall is thickest in the pons from this viewpoint because the pontine parenchyma is in front of the anterior edge of the rhombencephalic superventricle. In the telencephalon, the lateral brain wall is thicker than the medial wall in accordance with the ventrolateral (earlier maturing) to dorsomedial (later maturing) developmental gradient. As in **Figure 10B**, the dark areas within the brain wall are caused by looking through more than one thickness.

# FIGURE 13

## GW7.5, CR23 mm, C966, COMPUTER-AIDED 3-D RECONSTRUCTION

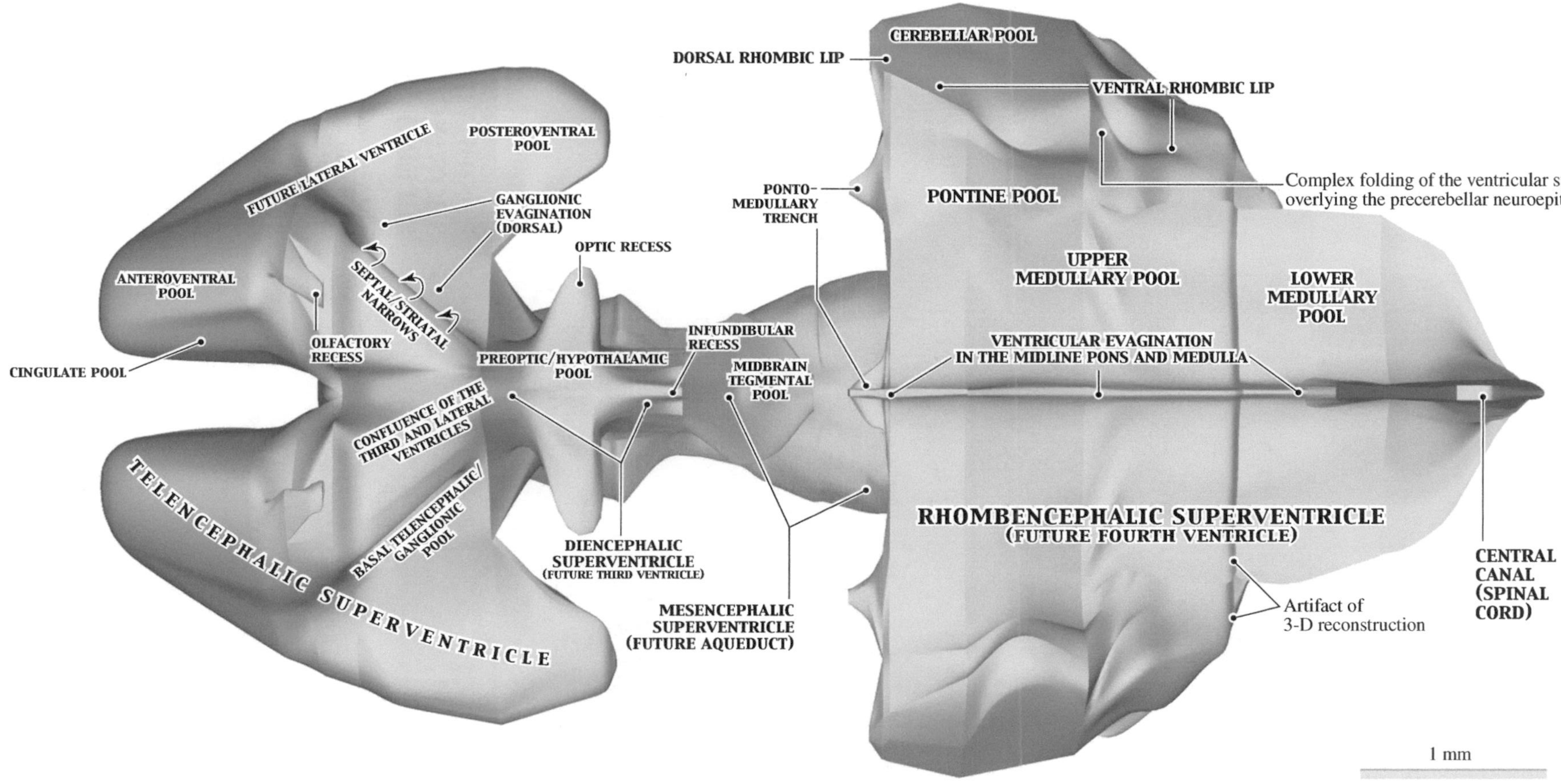

Complex folding of the ventricular surface overlying the precerebellar neuroepithelium

CEREBELLAR POOL

VENTRAL RHOMBIC LIP

DORSAL RHOMBIC LIP

PONTINE POOL

UPPER MEDULLARY POOL

LOWER MEDULLARY POOL

VENTRICULAR EVAGINATION IN THE MIDLINE PONS AND MEDULLA

RHOMBENCEPHALIC SUPERVENTRICLE (FUTURE FOURTH VENTRICLE)

Artifact of 3-D reconstruction

CENTRAL CANAL (SPINAL CORD)

PONTO–MEDULLARY TRENCH

INFUNDIBULAR RECESS

MIDBRAIN TEGMENTAL POOL

OPTIC RECESS

PREOPTIC/HYPOTHALAMIC POOL

DIENCEPHALIC SUPERVENTRICLE (FUTURE THIRD VENTRICLE)

MESENCEPHALIC SUPERVENTRICLE (FUTURE AQUEDUCT)

POSTEROVENTRAL POOL

GANGLIONIC EVAGINATION (DORSAL)

FUTURE LATERAL VENTRICLE

SEPTAL/STRIATAL NARROWS

OLFACTORY RECESS

CONFLUENCE OF THE THIRD AND LATERAL VENTRICLES

BASAL GANGLIONIC POOL

ANTEROVENTRAL POOL

CINGULATE POOL

TELENCEPHALIC SUPERVENTRICLE

1 mm

## BOTTOM VIEW OF THE EXPOSED SUPERVENTRICLES

Enlarged view of **Figure 12B** showing only the ventricles. The complex evaginations of the ventricular surface indicate the heterogeneous nature of the neuroepithelium that lines different ventricular divisions.

# FIGURE 14

## A. SIDE VIEW OF THE BRAIN SURFACE

The observer is beside the model, looking at the lateral surface of the brain and upper cervical spinal cord. Note the posterior-facing olfactory bulb beneath the basal telencephaon. This lateral view shows most clearly all of the flexures in the brainstem.

## B. SIDE VIEW SHOWING THE SUPERVENTRICLES

The varying thicknesses of the brain wall mirror a maturation gradient: the medulla has the thickest wall and is most mature, while the thin wall of the cerebral cortex is one of the least mature brain areas. As in **Figures 10B and 12B**, the dark areas within the brain wall are caused by looking through more than one thickness.

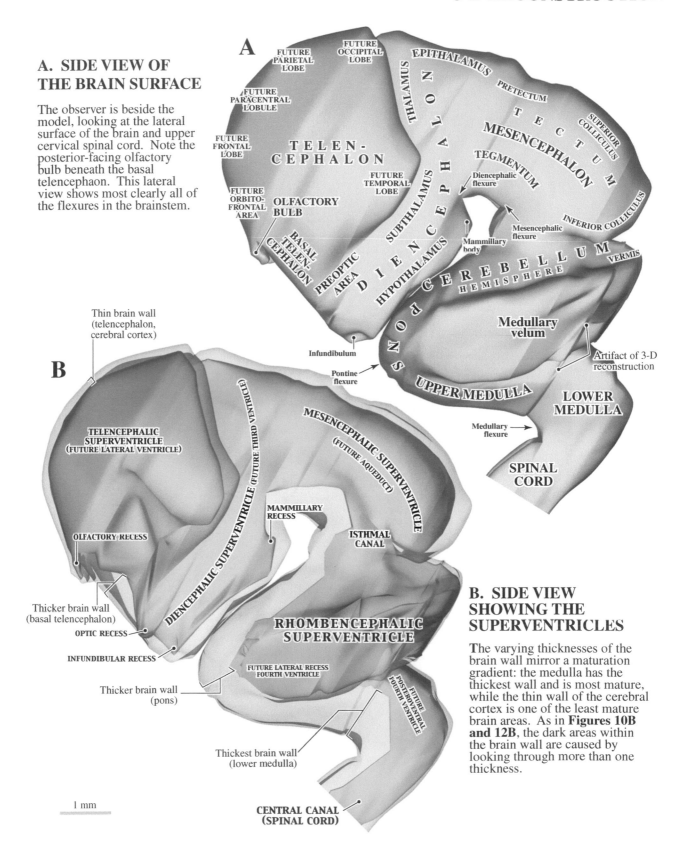

1 mm

# GW7.5, CR23 mm, C966, COMPUTER-AIDED 3-D RECONSTRUCTION

**FIGURE 15**

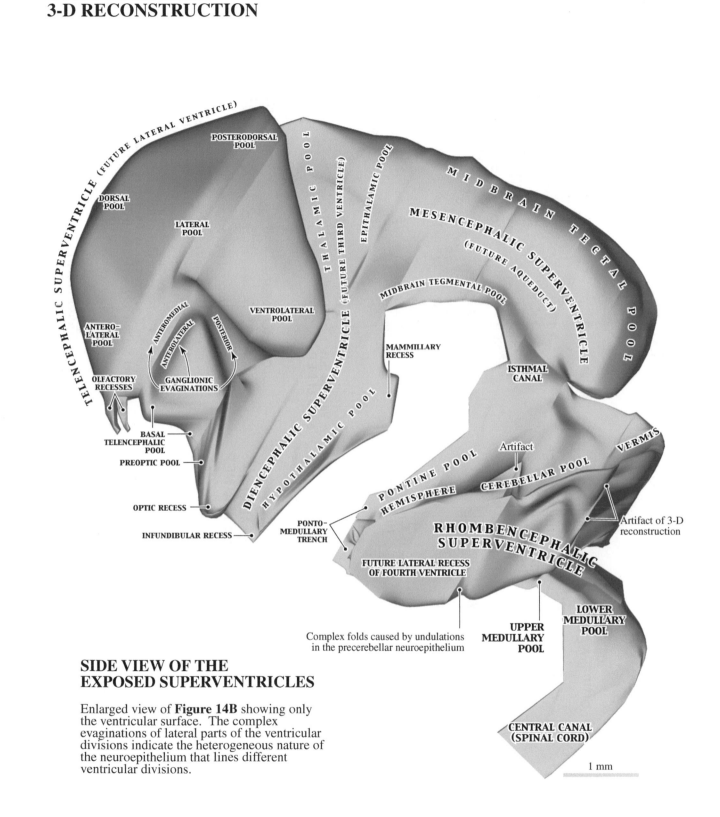

## SIDE VIEW OF THE EXPOSED SUPERVENTRICLES

Enlarged view of **Figure 14B** showing only the ventricular surface. The complex evaginations of lateral parts of the ventricular divisions indicate the heterogeneous nature of the neuroepithelium that lines different ventricular divisions.

1 mm

# FIGURE 16

## GW7.5, CR23 mm, C966, COMPUTER-AIDED 3-D RECONSTRUCTION OF THE LATERAL TELENCEPHALIC NEUROEPITHELIUM

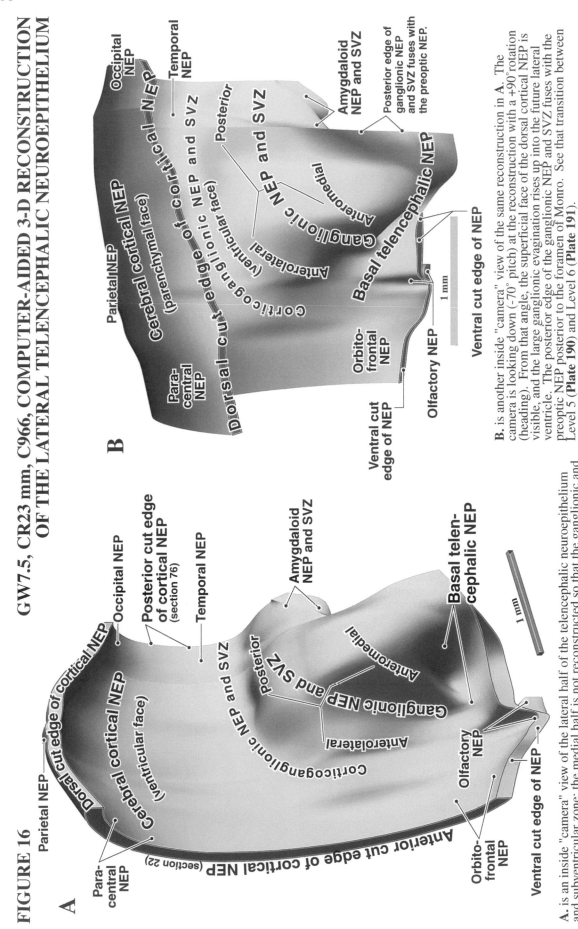

**A.** is an inside "camera" view of the lateral half of the telencephalic neuroepithelium and subventricular zone; the medial half is not reconstructed so that the ganglionic and olfactory evaginations are visible. The reconstruction includes sections 22 (between **Levels 1 and 2, Plates 187 and 188**) through 76 (**Level 7, Plate 192**). The camera is looking straight (0° pitch) at the reconstruction with a +60° rotation (heading). The large ganglionic evagination is evident in the ventrolateral NEP and SVZ.

**B.** is another inside "camera" view of the same reconstruction in **A.** The camera is looking down (-70° pitch) at the reconstruction with a +90° rotation (heading). From that angle, the superficial face of the dorsal cortical NEP is visible, and the large ganglionic evagination rises up into the future lateral ventricle. The posterior edge of the ganglionic NEP and SVZ fuses with the preoptic NEP posterior to the foramen of Monro. See that transition between Level 5 (**Plate 190**) and Level 6 (**Plate 191**).

ABBREVIATIONS:
NEP - Neuroepithelium
SVZ - Subventricular zone

Neuroepithelial compartments labeled in Helvetica Bold.

509

# GW7.5, CR23 mm, C966, COMPUTER-AIDED 3-D RECONSTRUCTION OF THE DIENCEPHALIC NEUROEPITHELIUM

Two inside "camera" views of the right half of the diencephalic neuroepithelium that includes section 44 (near **Level 3, Plate 188**) through section 112 (**Level 12, Plate 197**); the NEP in all sections has been cut where it bridges the midline dorsally and ventrally so that its folds and undulations can be observed.

**A.** The camera views the front of the reconstruction with +45°heading and -25° pitch. From this angle, we see the ventricular face of the NEP.

**B.** The camera views the back of the reconstruction with 180° heading and -10° pitch. From this angle, we see some of the parenchymal face of the NEP. Note the prominent folding in the anterior complex NEP, the thick posterior complex NEP, and the back wall of the NEP lining the optic recess.

It is postulated that the multiple evaginations and invaginations of the diencephalic neuroepithelium are mosaic compartments that give rise to different thalamic, hypothalamic, and preoptic nuclei.

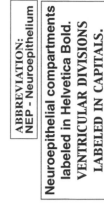

ABBREVIATION:
NEP - Neuroepithelium

Neuroepithelial compartments labeled in Helvetica Bold. **VENTRICULAR DIVISIONS LABELED IN CAPITALS.**

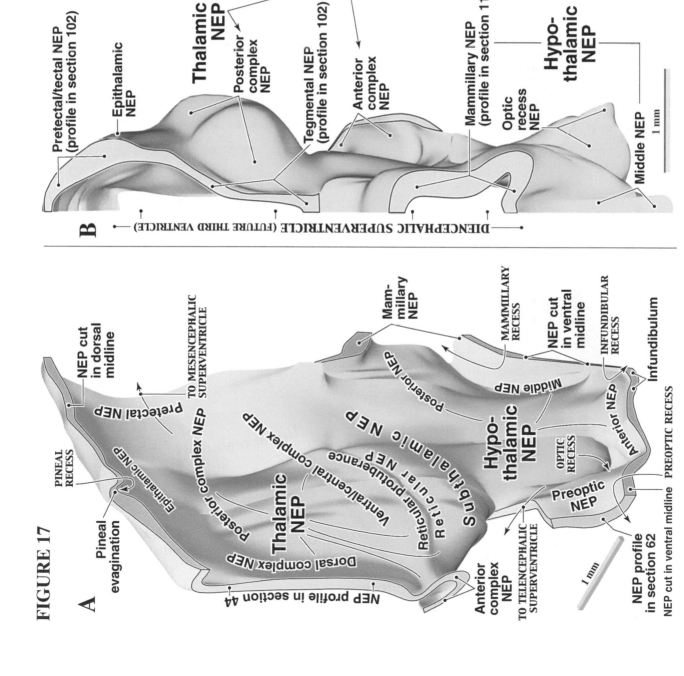

**FIGURE 17**

510

# FIGURE 18
## GW7.5, CR23 mm, C966, COMPUTER-AIDED 3-D RECONSTRUCTION OF THE RHOMBENCEPHALIC NEUROEPITHELIUM

The reconstruction includes section 128 (**Level 15, Plate 200**) through section 193 (posterior to **Level 21, Plate 179**); the NEP in all sections has been cut where it bridges the midline dorsally and ventrally, and only the right half is shown. The camera is looking at the front of the NEP (0° heading, -10° pitch).

**A.** shows both the roof and floor plates of the rhombencephalic NEP with orientation labels.

**B.** shows the parenchymal face of the roof NEP bordering the differentiating zones of the pons, isthmus, and cerebellum. Since the edges between *CTF6* (Purkinje cell sojourn zone) and the **cerebellar NEP** are virtually indistinguishable, both are included in the reconstruction.

**C.** shows the ventricular face of the floor NEP overlying the upper (adjacent to pons) and lower medulla (adjacent to spinal cord).

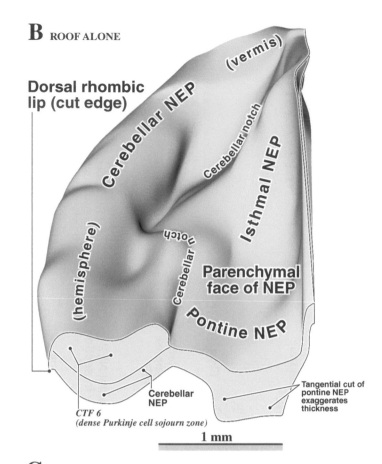

**B** ROOF ALONE

Dorsal rhombic lip (cut edge)

Cerebellar NEP

(vermis)

Cerebellar notch

(hemisphere)

Cerebellar notch

Isthmal NEP

Parenchymal face of NEP

Pontine NEP

Cerebellar NEP

*CTF 6 (dense Purkinje cell sojourn zone)*

Tangential cut of pontine NEP exaggerates thickness

1 mm

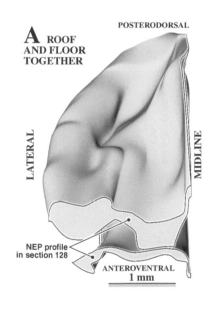

**A** ROOF AND FLOOR TOGETHER

POSTERODORSAL

LATERAL

MIDLINE

NEP profile in section 128

ANTEROVENTRAL

1 mm

> **ABBREVIATIONS:**
> *CTF - Cerebellar transitional field*
> NEP - Neuroepithelium

> **Neuroepithelial compartments labeled in Helvetica Bold.**
> *Transient developmental structures labeled in Times Bold Italic*

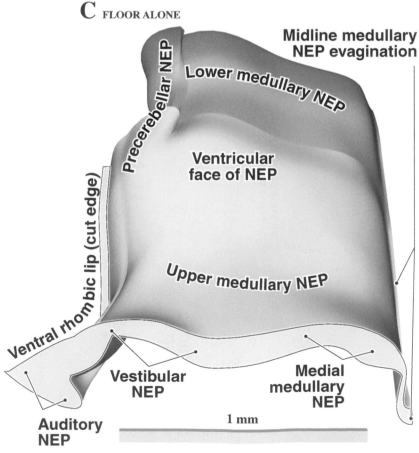

**C** FLOOR ALONE

Midline medullary NEP evagination

Precerebellar NEP

Lower medullary NEP

Ventricular face of NEP

Upper medullary NEP

Ventral rhombic lip (cut edge)

Vestibular NEP

Medial medullary NEP

Auditory NEP

1 mm

# FIGURE 19
## GW7.5, CR23 mm, C966, COMPUTER-AIDED 3-D RECONSTRUCTION OF THE RHOMBENCEPHALIC NEUROEPITHELIUM

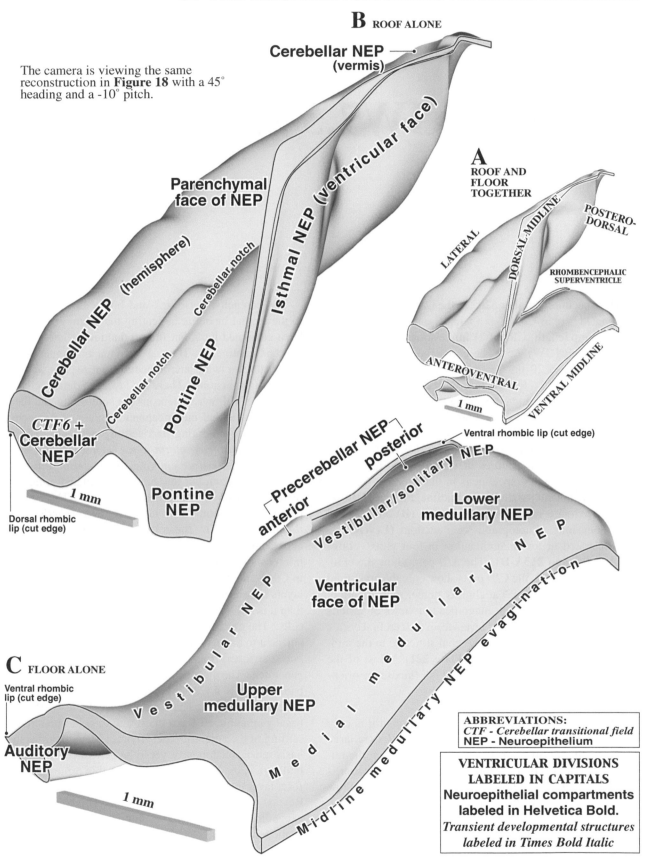

**B** ROOF ALONE

Cerebellar NEP (vermis)

The camera is viewing the same reconstruction in **Figure 18** with a 45° heading and a -10° pitch.

Parenchymal face of NEP

Isthmal NEP (ventricular face)

**A** ROOF AND FLOOR TOGETHER

LATERAL
DORSAL MIDLINE
POSTERO-DORSAL
RHOMBENCEPHALIC SUPERVENTRICLE
ANTEROVENTRAL
VENTRAL MIDLINE
1 mm

Cerebellar NEP (hemisphere)
Cerebellar notch
Cerebellar notch
Pontine NEP

*CTF6 +* Cerebellar NEP

1 mm

Pontine NEP

Dorsal rhombic lip (cut edge)

Precerebellar NEP
posterior
anterior
Ventral rhombic lip (cut edge)

Vestibular/solitary NEP

Lower medullary NEP

Ventricular face of NEP

Vestibular NEP
Medial medullary NEP
Upper medullary NEP

Midline medullary NEP evagination

**C** FLOOR ALONE

Ventral rhombic lip (cut edge)

Auditory NEP

1 mm

**ABBREVIATIONS:**
*CTF - Cerebellar transitional field*
*NEP - Neuroepithelium*

**VENTRICULAR DIVISIONS LABELED IN CAPITALS**
**Neuroepithelial compartments labeled in Helvetica Bold.**
*Transient developmental structures labeled in Times Bold Italic*

# PART XI: GW7.5 SAGITTAL

This specimen number 6202 in the Carnegie Collection, designated here as C6202, is a fetus of unknown sex with a crown-rump length (CR) of 21 mm estimated to be at gestational week (GW) 7.5. The entire fetus was cut in the sagittal plane in 20 μm sections and stained with hematoxylin and eosin. Information on the date of specimen collection, fixative, and embedding medium was not available to us. Since there is no photograph of this specimen before histological processing, a specimen from Hochstetter (1919) that is comparable in age to C6202 is used to show external brain features at GW7.5 (**A, Figure 20**). Like all other sagittal specimens in this Volume, C6202's sections are not cut parallel to the midline; **Figure 20** shows the approximate rotations in horizontal (**B**) and vertical (**C**) dimensions. Photographs of 7 sections (**Levels 1-7**) are illustrated at low magnification in four parts (**Plates 207A-D** through **213A-D**). The **A/B** parts show the brain in place in the skull; the **C/D** parts show only the brain (and some peripheral ganglia) at slightly higher magnification. **Plates 214-221** show high-magnification views of various parts of the brain at different levels from the cerebral cortex (**Plate 214**) to the midline raphe glial structure in the midbrain and cervical spinal cord (**Plate 221**). Most of the high-magnification plates are rotated 90° (landscape orientation) to more efficiently use page space.

C6202 is less mature than C966 even though it is only 2 mm shorter in CR length. The *superventricles* are large in the centers of all brain structures, especially in the telencephalon and rhombencephon, but the telencephalon is smaller than it is in C966. Sections near the midline show the enormous size of the diencephalic and mesencephalic superventricles. The respective thicknesses of the *neuroepithelium* (NEP) and parenchyma are keys to determining the degree of maturation of various brain structures.

The parenchyma is thick and bordered by a thin NEP in the medial medulla, indicating that many neurons have been generated here, but the production of late-generated neurons continues. The parenchyma is thinner and the NEP is thicker in the lateral medulla, entire pons, and entire midbrain tegmentum. There are layers of dense cells adjacent to the NEP where neurons are sojourning and the rest of the pontine and mesencephalic tegmental parenchyma is filled with migrating and settling neurons. The *precerebellar neuroepithelium* in the lateral medulla is thick and many neurons are entering the *posterior intramural migratory stream*, but few have accumulated in the indistinct inferior olive in the medial medulla. The cerebellar parenchyma has only four layers in the *cerebellar transitional field (CTF)*. The cells in *CTF2*, *3*, and *4* are probably deep neurons that will eventually settle in the dentate, interpositus, and fastigial nuclei. *CTF5* and *6* are present in C966 and presumably contain Purkinje cells, but these layers are absent in C6202. The cerebellar NEP is thicker than it is in C966 and is now producing the oldest Purkinje cells. The mesencephalic tectal NEP is very thick and lies adjacent to a thin parenchyma. The majority of neurons in both the superior and inferior colliculi have not yet been generated. The diencephalic NEP is thick and the adjacent parenchyma is filled with dense zones of sojourning and migrating neurons. The basal telencephalic NEP and the basal ganglionic NEP are thick, and the oldest neurons are settling in the adjacent parenchyma. Many neurons in the basal telencephalon have yet to be generated. In the cerebral cortex, the NEP is bordered by a thin primordial plexiform layer that contains the oldest cortical neurons (Cajal-Retzius cells) and subplate neurons. The cortical NEP is expanding and increasing its number of neural stem cells as the telencephalic superventricle grows; nearly all cortical neurons in layers II-VI have still to be generated.

## GW7.5 SAGITTAL

A perfect sagittal cut through the brain bisects the cerebral cortex into two separate hemispheres by passing through the interhemispheric fissure, and does the same in the brainstem by passing through the midline of the ventricles.

Sections of C6202's brain rotate 10.5° counterclockwise from the horizontal midline running through the cerebral cortex and midbrain tectum (top view). C6202's sections are quite close to the vertical midline, rotating only 2.8° counterclockwise (back view). In the sections illustrated on the following pages, the telencephalon and diencephalon (top) are tilted away from the observer, while the medulla and upper spinal cord (bottom) are tilted toward the observer.

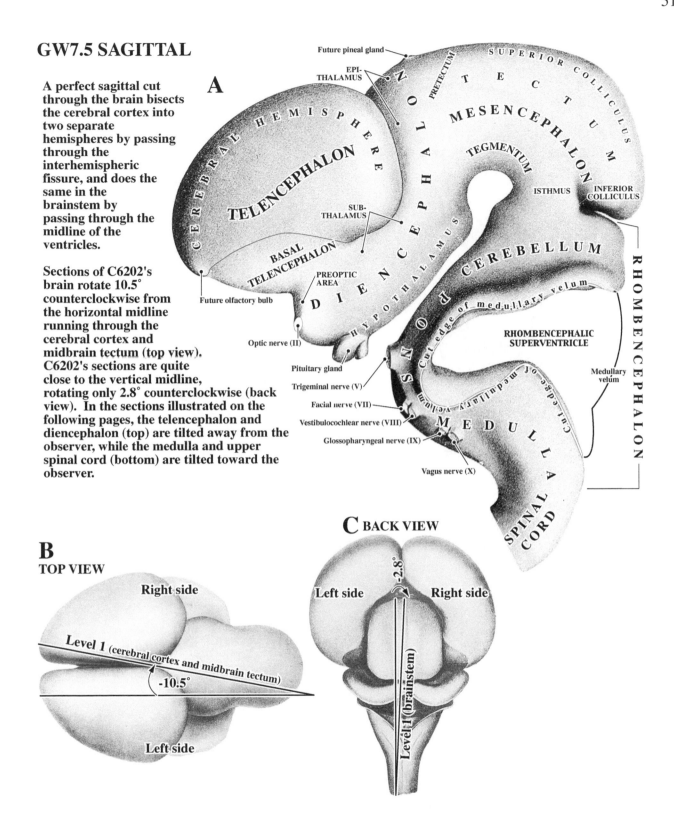

**Figure 20. A**, The lateral view of the brain and upper cervical spinal cord from a specimen with a crown-rump length of 19.4 mm (modified from Figure 35, Table VI, Hochstetter, 1919) identifies external features of a brain similar to C6202 (CR 21 mm). **B**, Top view of the brain in **A** (modified from Figure 37, Table VI, Hochstetter, 1919) shows how C6202's sections rotate from a line parallel to the horizontal midline in the interhemispheric fissure and midbrain tectum. **C**, Back view of the brain with a crown-rump length of 38 mm (modified from Figure 44, Table VIII, Hochstetter, 1919) shows how C6202's sections rotate from a line parallel to the vertical midline in the brainstem and upper cervical spinal cord.

**PLATE 207A**

**GW7.5 Sagittal**
**CR 21 mm, C6202**
**Level 1: Slide 30, Section 2**
**SKULL, MAJOR BRAIN**
**STRUCTURES, AND**
**VENTRICULAR**
**DIVISIONS**

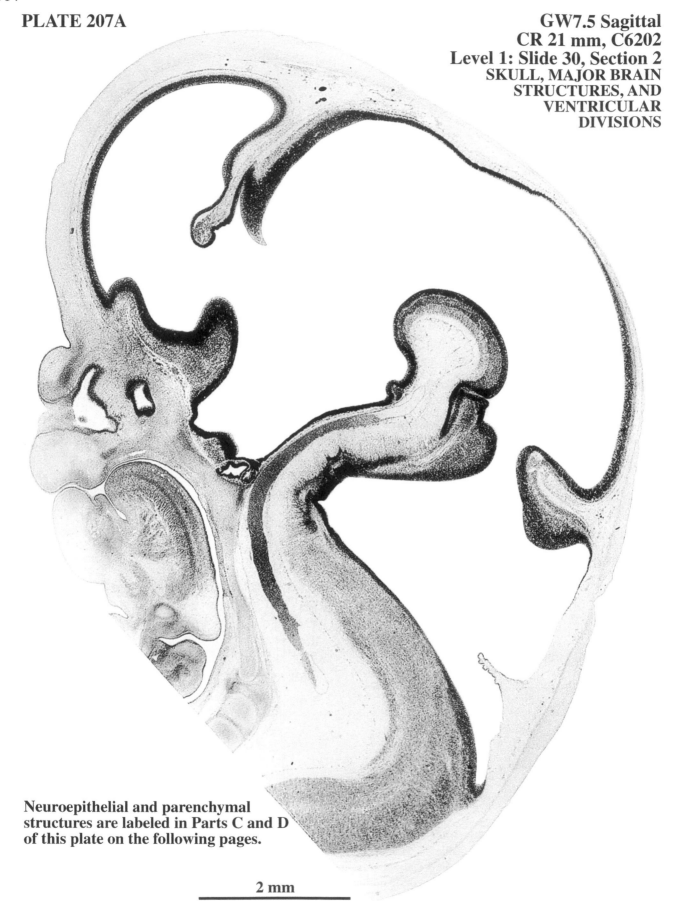

**Neuroepithelial and parenchymal
structures are labeled in Parts C and D
of this plate on the following pages.**

2 mm

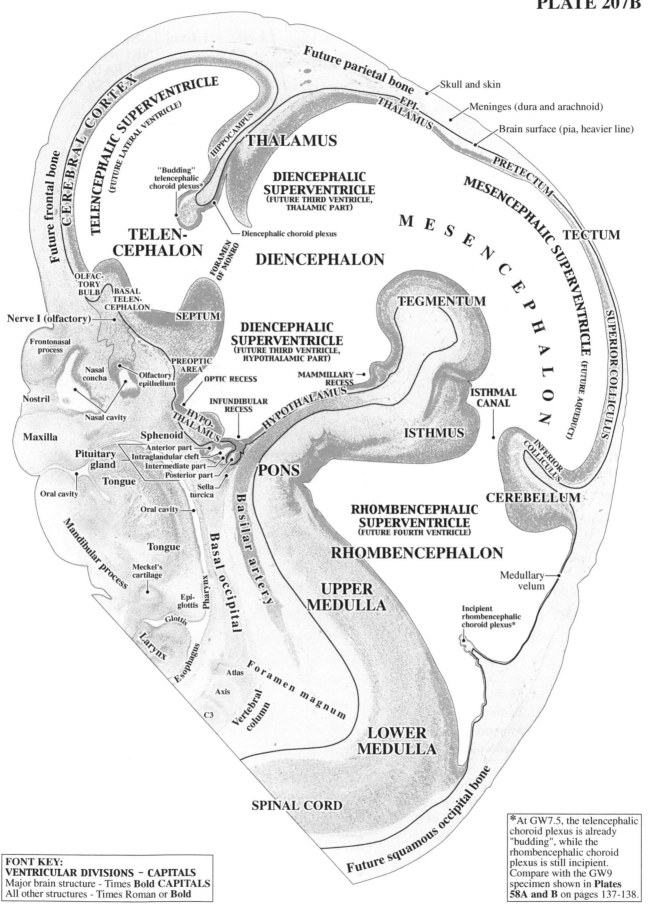

Skull and skin

Meninges (dura and arachnoid)

Brain surface (pia, heavier line)

Future parietal bone

CEREBRAL CORTEX

TELENCEPHALIC SUPERVENTRICLE
(FUTURE LATERAL VENTRICLE)

Future frontal bone

HIPPOCAMPUS

THALAMUS

EPI-THALAMUS

PRETECTUM

MESENCEPHALIC SUPERVENTRICLE

MESENCEPHALON

DIENCEPHALIC SUPERVENTRICLE
(FUTURE THIRD VENTRICLE, THALAMIC PART)

"Budding" telencephalic choroid plexus*

Diencephalic choroid plexus

MESENCEPHALIC TECTUM

TELEN-CEPHALON

DIENCEPHALON

FORAMEN OF MONRO

OLFAC-TORY BULB

BASAL TELEN-CEPHALON

Nerve I (olfactory)

SEPTUM

Frontonasal process

Nasal concha

Olfactory epithelium

PREOPTIC AREA

DIENCEPHALIC SUPERVENTRICLE
(FUTURE THIRD VENTRICLE, HYPOTHALAMIC PART)

TEGMENTUM

SUPERIOR COLLICULUS (FUTURE AQUEDUCT)

Nostril

OPTIC RECESS

MAMMILLARY RECESS

ISTHMAL CANAL

INFERIOR COLLICULUS

Maxilla

Nasal cavity

INFUNDIBULAR RECESS

HYPO-THALAMUS

HYPOTHALAMUS

ISTHMUS

Pituitary gland

Sphenoid

Anterior part

Intraglandular cleft

Intermediate part

Posterior part

PONS

CEREBELLUM

Tongue

Sella turcica

Oral cavity

Oral cavity

RHOMBENCEPHALIC SUPERVENTRICLE
(FUTURE FOURTH VENTRICLE)

Tongue

Basal occipital

RHOMBENCEPHALON

Medullary velum

Mandibular process

Meckel's cartilage

Basal artery

UPPER MEDULLA

Incipient rhombencephalic choroid plexus*

Epi-glottis

Pharynx

Glottis

Larynx

Esophagus

Atlas

Axis

C3

Foramen magnum

Vertebral column

LOWER MEDULLA

SPINAL CORD

Future squamous occipital bone

*At GW7.5, the telencephalic choroid plexus is already "budding", while the rhombencephalic choroid plexus is still incipient. Compare with the GW9 specimen shown in **Plates 58A and B** on pages 137-138.

**FONT KEY:**
**VENTRICULAR DIVISIONS – CAPITALS**
Major brain structure - Times **Bold CAPITALS**
All other structures - Times Roman or **Bold**

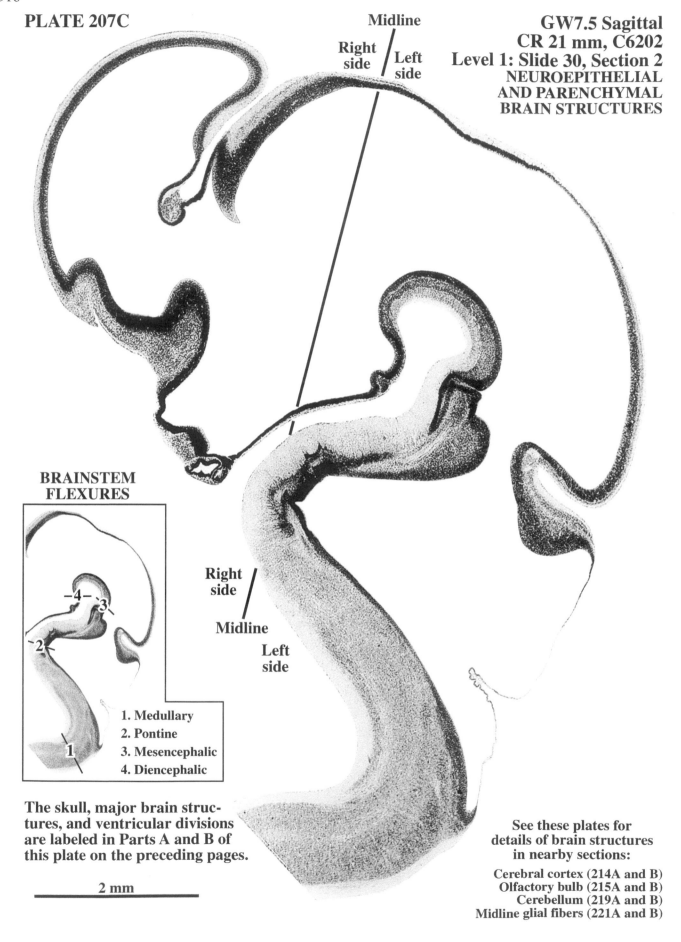

**PLATE 207C**

Midline

Right side / Left side

**GW7.5 Sagittal
CR 21 mm, C6202
Level 1: Slide 30, Section 2
NEUROEPITHELIAL
AND PARENCHYMAL
BRAIN STRUCTURES**

**BRAINSTEM
FLEXURES**

−4−
−3
−2

Right side

Midline

Left side

1. Medullary
2. Pontine
3. Mesencephalic
4. Diencephalic

The skull, major brain structures, and ventricular divisions
are labeled in Parts A and B of
this plate on the preceding pages.

2 mm

See these plates for
details of brain structures
in nearby sections:

Cerebral cortex (214A and B)
Olfactory bulb (215A and B)
Cerebellum (219A and B)
Midline glial fibers (221A and B)

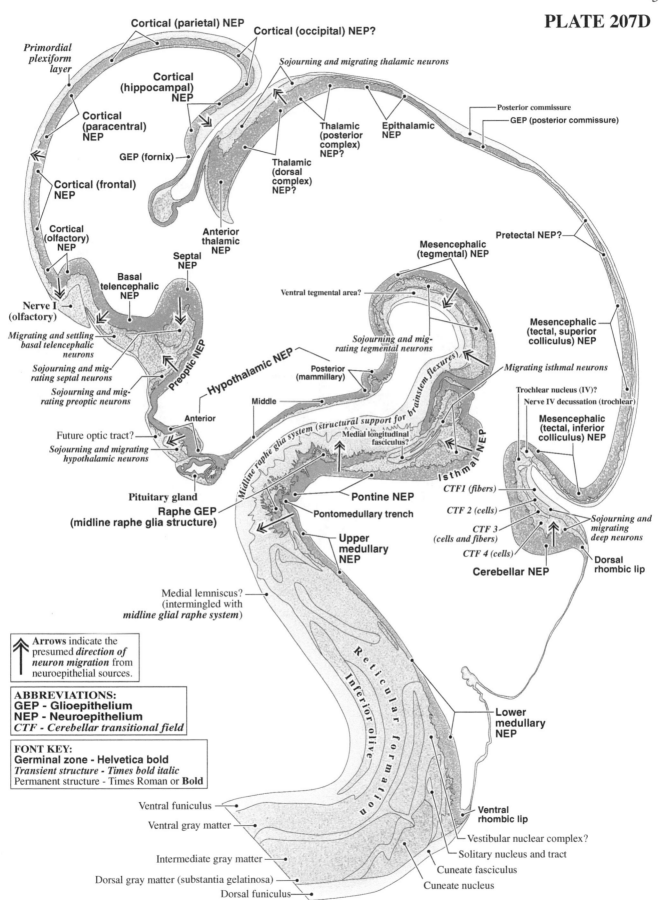

Cortical (parietal) NEP

Cortical (occipital) NEP?

*Primordial plexiform layer*

*Sojourning and migrating thalamic neurons*

**Cortical (hippocampal) NEP**

Posterior commissure

GEP (posterior commissure)

**Cortical (paracentral) NEP**

GEP (fornix)

Thalamic (posterior complex) NEP?

**Epithalamic NEP**

**Cortical (frontal) NEP**

Thalamic (dorsal complex) NEP?

**Cortical (olfactory) NEP**

Pretectal NEP?

**Anterior thalamic NEP**

**Mesencephalic (tegmental) NEP**

**Septal NEP**

**Nerve I (olfactory)**

**Basal telencephalic NEP**

*Ventral tegmental area?*

**Mesencephalic (tectal, superior colliculus) NEP**

*Migrating and settling basal telencephalic neurons*

**Preoptic NEP**

**Hypothalamic NEP**

Posterior (mammillary)

*Sojourning and mig- rating tegmental neurons*

*Migrating isthmal neurons*

*Sojourning and mig- rating septal neurons*

Trochlear nucleus (IV)?

Nerve IV decussation (trochlear)

*Sojourning and mig- rating preoptic neurons*

Middle

**Mesencephalic (tectal, inferior colliculus) NEP**

Anterior

*Future optic tract?*

Medial longitudinal fasciculus?

*Sojourning and migrating hypothalamic neurons*

*Midline raphe glia system (structural support for brainstem flexures)*

**Isthmal NEP**

CTF1 (fibers)

**Pituitary gland**

CTF 2 (cells)

*Sojourning and migrating deep neurons*

**Raphe GEP (midline raphe glia structure)**

**Pontine NEP**

CTF 3 (cells and fibers)

Pontomedullary trench

CTF 4 (cells)

Dorsal rhombic lip

**Upper medullary NEP**

**Cerebellar NEP**

Medial lemniscus? (intermingled with *midline glial raphe system*)

**Lower medullary NEP**

**Arrows** indicate the presumed *direction of neuron migration* from neuroepithelial sources.

**ABBREVIATIONS:**
**GEP** - Glioepithelium
**NEP** - Neuroepithelium
*CTF - Cerebellar transitional field*

**FONT KEY:**
Germinal zone - Helvetica bold
*Transient structure - Times bold italic*
Permanent structure - Times Roman or **Bold**

*Reticular formation*

*Inferior olive*

**Ventral rhombic lip**

Ventral funiculus

Ventral gray matter

Vestibular nuclear complex?

Intermediate gray matter

Solitary nucleus and tract

Cuneate fasciculus

Dorsal gray matter (substantia gelatinosa)

Cuneate nucleus

Dorsal funiculus

**PLATE 208A**

GW7.5 Sagittal
CR 21 mm, C6202
Level 2: Slide 26, Section 2
SKULL, MAJOR BRAIN
STRUCTURES, AND
VENTRICULAR
DIVISIONS

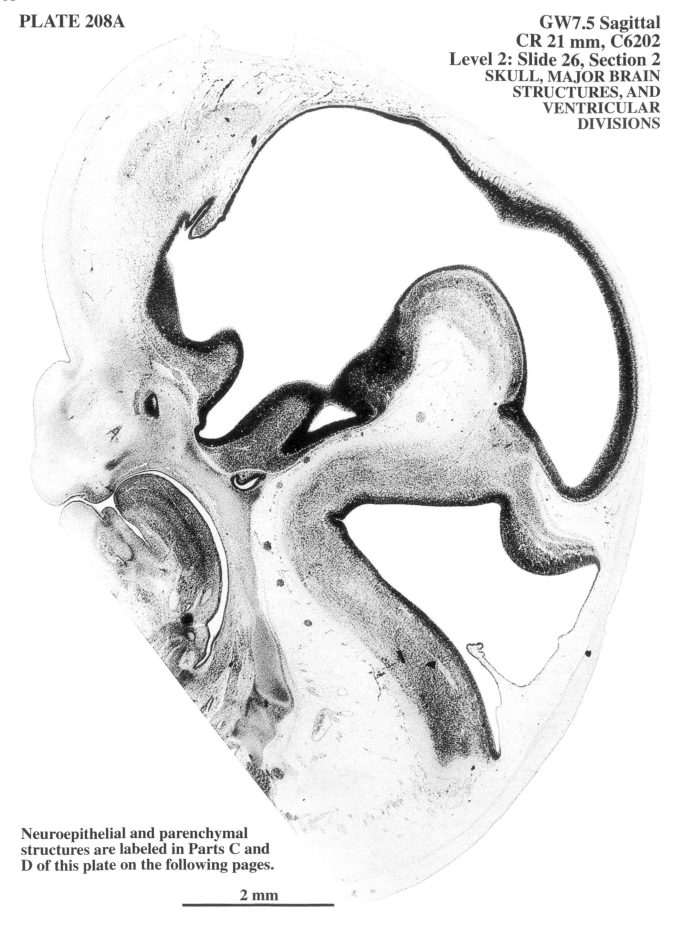

Neuroepithelial and parenchymal
structures are labeled in Parts C and
D of this plate on the following pages.

2 mm

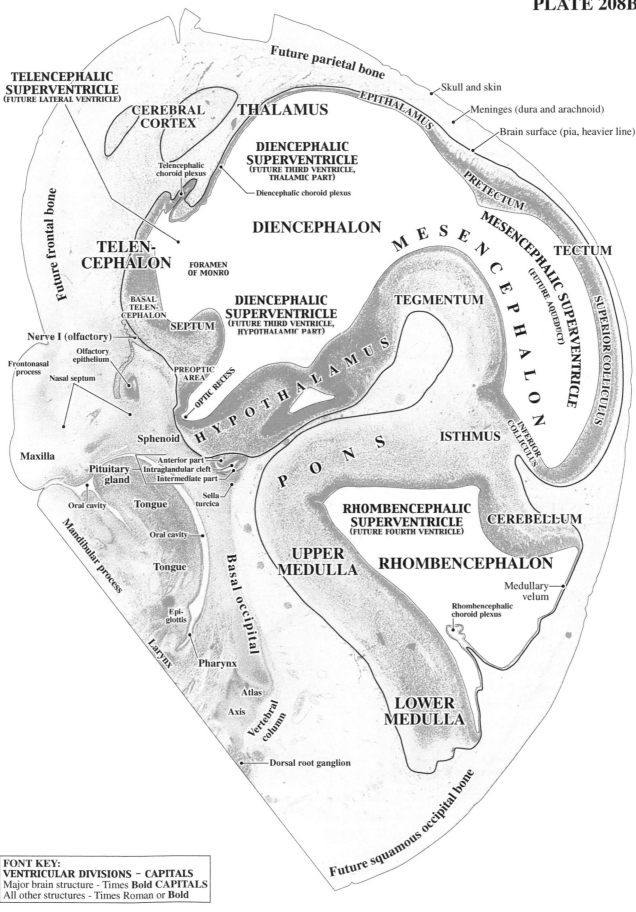

TELENCEPHALIC
SUPERVENTRICLE
(FUTURE LATERAL VENTRICLE)

Future parietal bone

Skull and skin

Meninges (dura and arachnoid)

Brain surface (pia, heavier line)

CEREBRAL
CORTEX

THALAMUS

EPITHALAMUS

DIENCEPHALIC
SUPERVENTRICLE
(FUTURE THIRD VENTRICLE,
THALAMIC PART)

Telencephalic
choroid plexus

Diencephalic choroid plexus

DIENCEPHALON

PRETECTUM

MESENCEPHALIC
SUPERVENTRICLE
(FUTURE AQUEDUCT)

TECTUM

Future frontal bone

TELEN-
CEPHALON

FORAMEN
OF MONRO

M E S E N C E P H A L O N

SUPERIOR COLLICULUS

BASAL
TELEN-
CEPHALON

SEPTUM

DIENCEPHALIC
SUPERVENTRICLE
(FUTURE THIRD VENTRICLE,
HYPOTHALAMIC PART)

TEGMENTUM

Nerve I (olfactory)

Olfactory
epithelium

Frontonasal
process

Nasal septum

PREOPTIC
AREA

OPTIC RECESS

H Y P O T H A L A M U S

ISTHMUS

INFERIOR COLLICULUS

Sphenoid

P O N S

Maxilla

Pituitary
gland

Anterior part
Intraglandular cleft
Intermediate part

Sella
turcica

RHOMBENCEPHALIC
SUPERVENTRICLE
(FUTURE FOURTH VENTRICLE)

CEREBELLUM

Oral cavity

Tongue

Oral cavity

UPPER
MEDULLA

RHOMBENCEPHALON

Tongue

Medullary
velum

Epi-
glottis

Basal occipital

Rhombencephalic
choroid plexus

Larynx

Pharynx

Atlas

Axis

Vertebral column

LOWER
MEDULLA

Mandibular process

Dorsal root ganglion

Future squamous occipital bone

FONT KEY:
**VENTRICULAR DIVISIONS – CAPITALS**
Major brain structure - Times **Bold CAPITALS**
All other structures - Times Roman or **Bold**

**PLATE 208C**

GW7.5 Sagittal
CR 21 mm, C6202
Level 2: Slide 26, Section 2
NEUROEPITHELIAL
AND PARENCHYMAL
BRAIN STRUCTURES

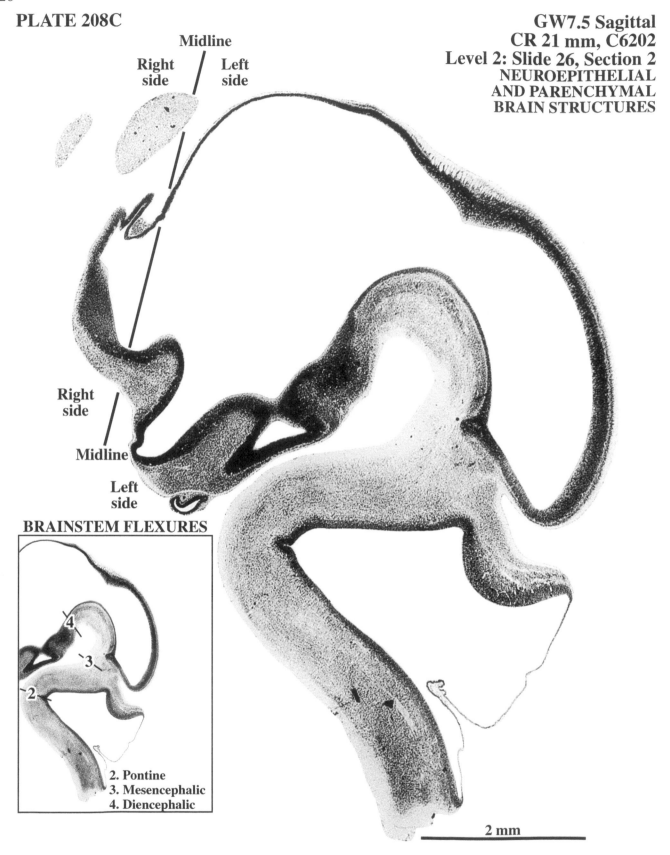

Midline

Right
side

Left
side

Right
side

Midline

Left
side

**BRAINSTEM FLEXURES**

4

3

2

2. Pontine
3. Mesencephalic
4. Diencephalic

2 mm

The skull, major brain structures,
and ventricular divisions are
labeled in Parts A and B of this
plate on the preceding pages.

See Plates 210A and B for details of the
anterior and middle hypothalamus, Plates
218A and B for details of the midbrain
tegmentum in nearby sections.

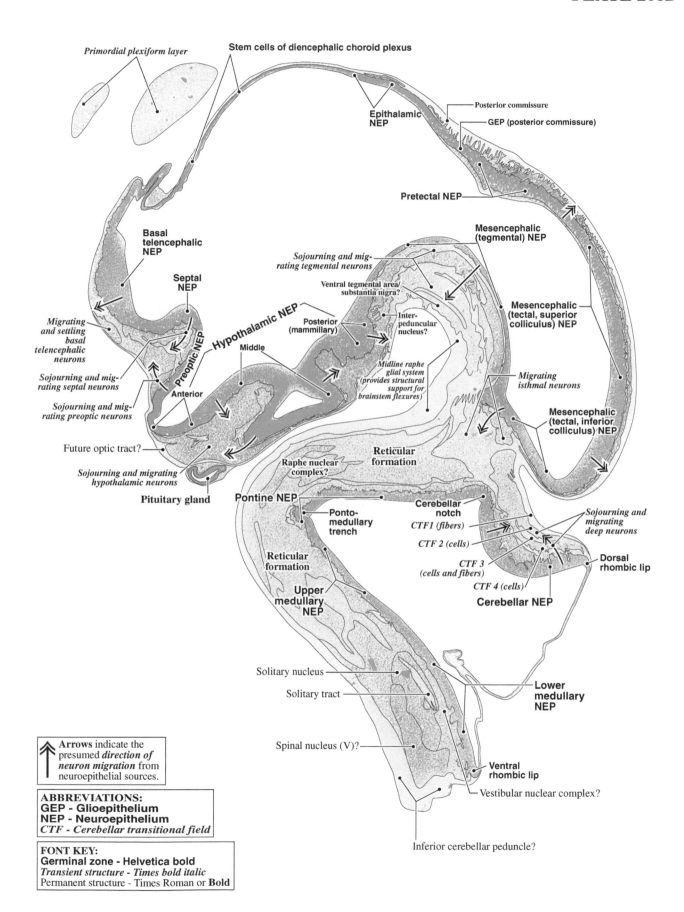

*Primordial plexiform layer*

Stem cells of diencephalic choroid plexus

**Epithalamic
NEP**

Posterior commissure

GEP (posterior commissure)

**Pretectal NEP**

**Mesencephalic
(tegmental) NEP**

**Basal
telencephalic
NEP**

**Septal
NEP**

*Sojourning and mig-
rating tegmental neurons*

Ventral tegmental area/
substantia nigra?

**Mesencephalic
(tectal, superior
colliculus) NEP**

**Hypothalamic NEP**

Posterior
(mammillary)

Inter-
peduncular
nucleus?

*Migrating
and settling
basal
telencephalic
neurons*

**Preoptic NEP**

Middle

*Midline raphe
glial system
(provides structural
support for
brainstem flexures)*

*Migrating
isthmal neurons*

*Sojourning and mig-
rating septal neurons*

Anterior

**Mesencephalic
(tectal, inferior
colliculus) NEP**

*Sojourning and mig-
rating preoptic neurons*

Future optic tract?

**Reticular
formation**

*Sojourning and migrating
hypothalamic neurons*

Raphe nuclear
complex?

**Reticular
formation**

**Pituitary gland**

**Pontine NEP**

Ponto-
medullary
trench

**Cerebellar
notch**

*CTF1 (fibers)*

*Sojourning and
migrating
deep neurons*

*CTF 2 (cells)*

*CTF 3
(cells and fibers)*

**Dorsal
rhombic lip**

**Reticular
formation**

*CTF 4 (cells)*

**Upper
medullary
NEP**

**Cerebellar NEP**

Solitary nucleus

**Lower
medullary
NEP**

Solitary tract

Spinal nucleus (V)?

**Ventral
rhombic lip**

Vestibular nuclear complex?

Inferior cerebellar peduncle?

**Arrows** indicate the
presumed *direction of
neuron migration* from
neuroepithelial sources.

**ABBREVIATIONS:**
**GEP - Glioepithelium**
**NEP - Neuroepithelium**
*CTF - Cerebellar transitional field*

**FONT KEY:**
**Germinal zone - Helvetica bold**
*Transient structure - Times bold italic*
Permanent structure - Times Roman or **Bold**

**GW7.5 Sagittal
CR 21 mm, C6202
Level 3: Slide 24, Section 2**
Left side of brain
**SKULL, MAJOR BRAIN
STRUCTURES, AND
VENTRICULAR
DIVISIONS**

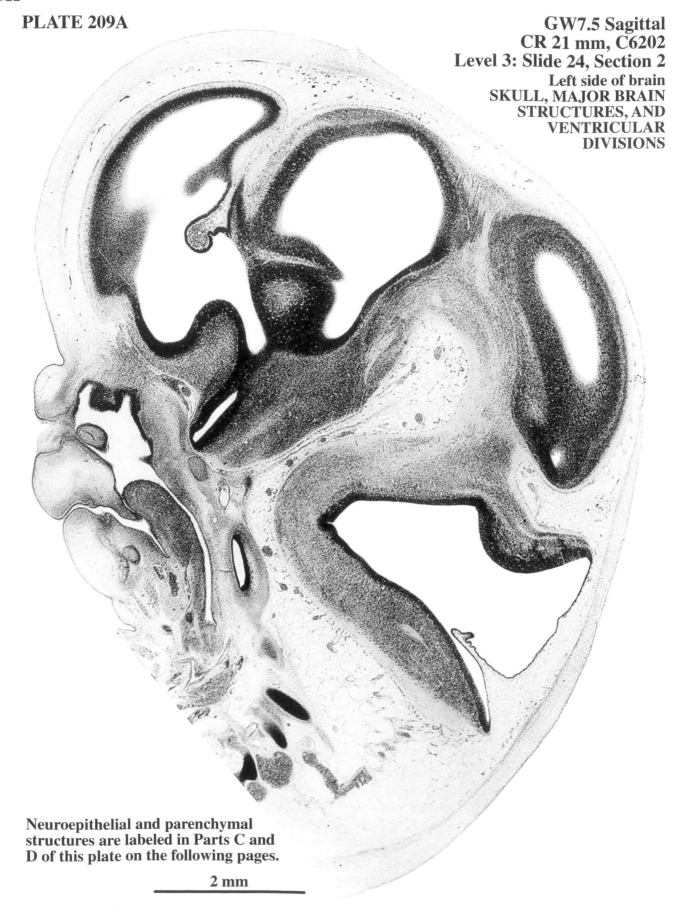

**Neuroepithelial and parenchymal
structures are labeled in Parts C and
D of this plate on the following pages.**

2 mm

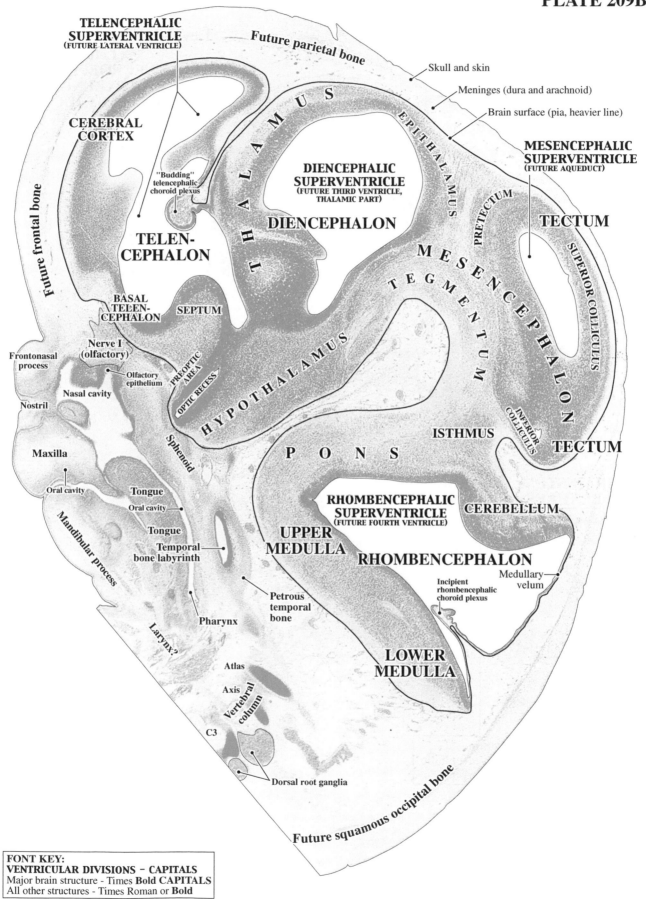

**TELENCEPHALIC SUPERVENTRICLE**
(FUTURE LATERAL VENTRICLE)

Future parietal bone

Skull and skin

Meninges (dura and arachnoid)

Brain surface (pia, heavier line)

**CEREBRAL CORTEX**

T H A L A M U S

**DIENCEPHALIC SUPERVENTRICLE**
(FUTURE THIRD VENTRICLE, THALAMIC PART)

**DIENCEPHALON**

EPITHALAMUS

**MESENCEPHALIC SUPERVENTRICLE**
(FUTURE AQUEDUCT)

PRETECTUM

**TECTUM**

"Budding" telencephalic choroid plexus

**TELEN-CEPHALON**

M E S E N C E P H A L O N

T E G M E N T U M

SUPERIOR COLLICULUS

Future frontal bone

**BASAL TELEN-CEPHALON**

**SEPTUM**

Nerve I (olfactory)

Frontonasal process

PREOPTIC AREA

H Y P O T H A L A M U S

Olfactory epithelium

OPTIC RECESS

Nasal cavity

Nostril

INFERIOR COLLICULUS

**ISTHMUS**

**TECTUM**

Maxilla

Sphenoid

P O N S

Oral cavity

**Tongue**

Oral cavity

**Tongue**

**RHOMBENCEPHALIC SUPERVENTRICLE**
(FUTURE FOURTH VENTRICLE)

**CEREBELLUM**

Mandibular process

Temporal bone labyrinth

**UPPER MEDULLA**

**RHOMBENCEPHALON**

Medullary velum

Petrous temporal bone

Incipient rhombencephalic choroid plexus

Pharynx

**LOWER MEDULLA**

Larynx?

Atlas

Axis

Vertebral column

C3

Dorsal root ganglia

Future squamous occipital bone

**FONT KEY:**
**VENTRICULAR DIVISIONS – CAPITALS**
Major brain structure - Times **Bold CAPITALS**
All other structures - Times Roman or **Bold**

**PLATE 209C**

GW7.5 Sagittal
CR 21 mm, C6202
Level 3: Slide 24, Section 2
Left side of brain
NEUROEPITHELIAL
AND PARENCHYMAL
BRAIN STRUCTURES

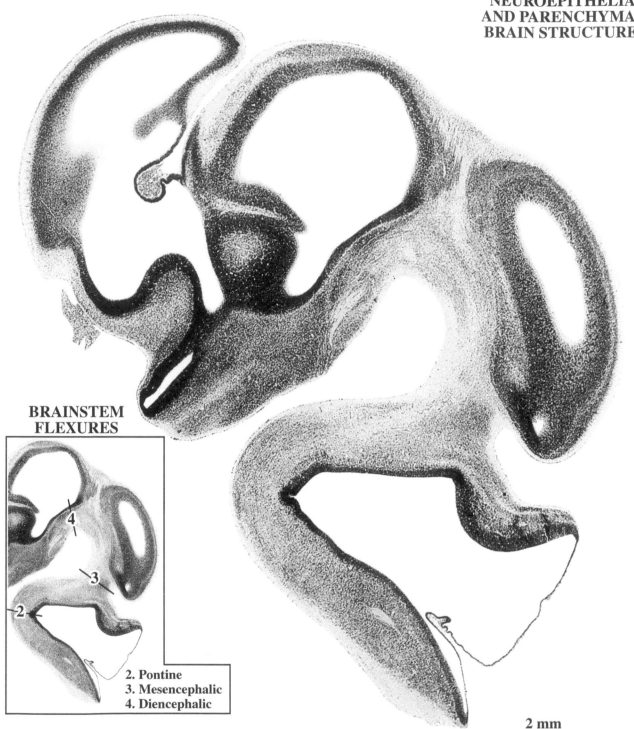

**BRAINSTEM
FLEXURES**

2. Pontine
3. Mesencephalic
4. Diencephalic

2 mm

The skull, major brain structures,
and ventricular divisions are
labeled in Parts A and B of this
plate on the preceding pages.

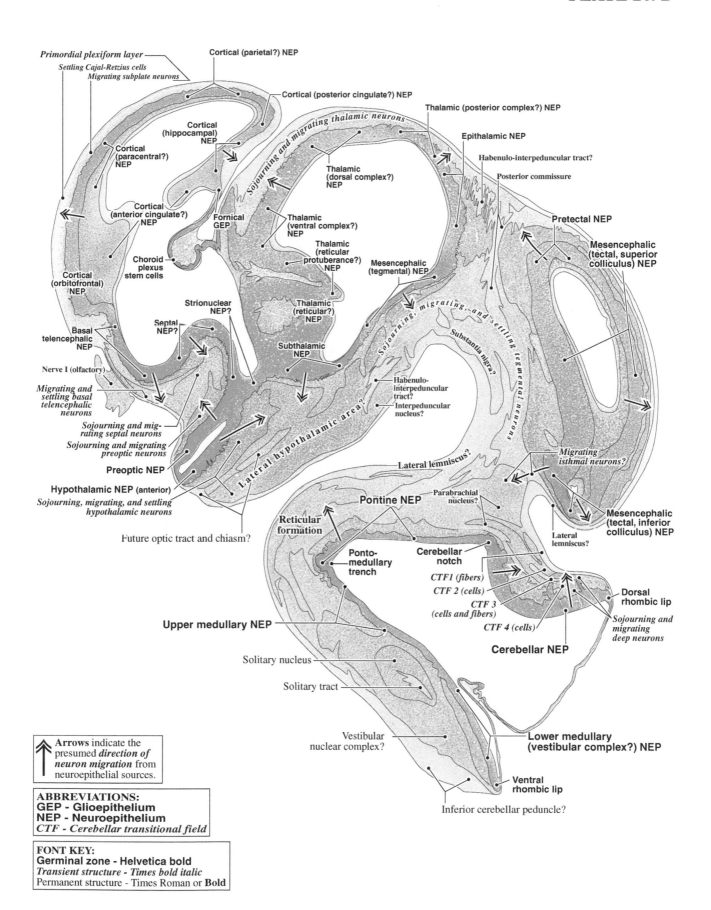

*Primordial plexiform layer*
*Settling Cajal-Retzius cells*
*Migrating subplate neurons*

Cortical (parietal?) NEP

Cortical (posterior cingulate?) NEP

Cortical (hippocampal) NEP

Cortical (paracentral?) NEP

*Sojourning and migrating thalamic neurons*

Thalamic (posterior complex?) NEP

Epithalamic NEP

Habenulo-interpeduncular tract?

Posterior commissure

Thalamic (dorsal complex?) NEP

Cortical (anterior cingulate?) NEP

Fornical GEP

Thalamic (ventral complex?) NEP

Pretectal NEP

Choroid plexus stem cells

Thalamic (reticular protuberance?) NEP

Mesencephalic (tegmental) NEP

Mesencephalic (tectal, superior colliculus) NEP

Cortical (orbitofrontal) NEP

Strionuclear NEP?

Thalamic (reticular?) NEP

*Sojourning, migrating, and settling tegmental neurons*

Basal telencephalic NEP

Septal NEP?

Subthalamic NEP

*Substantia nigra?*

Nerve I (olfactory)

Habenulo-interpeduncular tract?

Interpeduncular nucleus?

*Migrating and settling basal telencephalic neurons*

*Lateral hypothalamic area?*

Lateral lemniscus?

*Migrating isthmal neurons?*

*Sojourning and migrating septal neurons*

*Sojourning and migrating preoptic neurons*

**Preoptic NEP**

Mesencephalic (tectal, inferior colliculus) NEP

**Hypothalamic NEP (anterior)**

*Sojourning, migrating, and settling hypothalamic neurons*

Parabrachial nucleus?

Lateral lemniscus?

**Pontine NEP**

Future optic tract and chiasm?

**Reticular formation**

**Ponto-medullary trench**

Cerebellar notch

CTF1 (fibers)

CTF 2 (cells)

CTF 3 (cells and fibers)

Dorsal rhombic lip

*Sojourning and migrating deep neurons*

CTF 4 (cells)

**Upper medullary NEP**

**Cerebellar NEP**

Solitary nucleus

Solitary tract

Vestibular nuclear complex?

**Lower medullary (vestibular complex?) NEP**

Ventral rhombic lip

Inferior cerebellar peduncle?

↑ **Arrows** indicate the presumed *direction of neuron migration* from neuroepithelial sources.

**ABBREVIATIONS:**
**GEP - Glioepithelium**
**NEP - Neuroepithelium**
*CTF - Cerebellar transitional field*

**FONT KEY:**
**Germinal zone - Helvetica bold**
*Transient structure - Times bold italic*
Permanent structure - Times Roman or **Bold**

**PLATE 210A**

**GW7.5 Sagittal**
**CR 21 mm, C6202**
**Level 4: Slide 22, Section 2**
Left side of brain
**SKULL, MAJOR BRAIN**
**STRUCTURES, AND**
**VENTRICULAR**
**DIVISIONS**

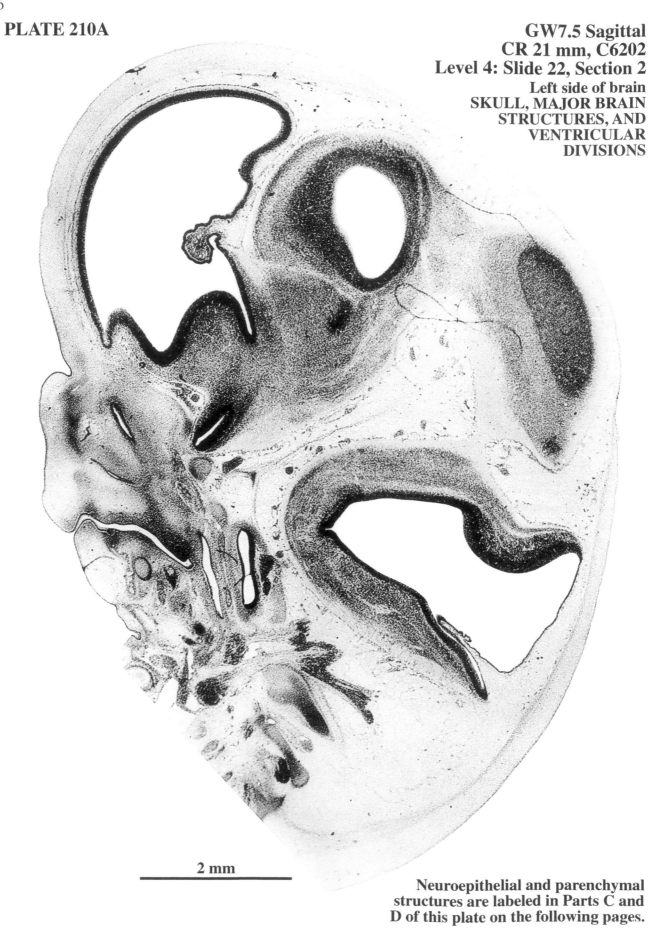

2 mm

**Neuroepithelial and parenchymal**
**structures are labeled in Parts C and**
**D of this plate on the following pages.**

**DIENCEPHALIC SUPERVENTRICLE**
(FUTURE THIRD VENTRICLE, THALAMIC PART)

Future parietal bone

Skull and skin

Meninges (dura and arachnoid)

Brain surface (pia, heavier line)

**CEREBRAL CORTEX**

HIPPOCAMPUS

**TELENCEPHALIC SUPERVENTRICLE**
(FUTURE LATERAL VENTRICLE)

"Budding" telencephalic choroid plexus

Future frontal bone

**TELEN-CEPHALON**

T H A L A M U S

EPITHALAMUS

T E C T U M

PRETECTUM?

**DIENCEPHALON**

M E S E N C E P H A L O N

SUPERIOR COLLICULUS

OLFACTORY BULB

**BASAL GANGLIA**

BASAL TELENCEPHALON

SUBTHALAMUS

TEGMENTUM?

INFERIOR COLLICULUS

T E C T U M

Nerve I (olfactory)

Frontonasal process

Olfactory epithelium

PREOPTIC AREA

**HYPO-THALAMUS**

Nasal cavity

OPTIC RECESS

Nerve VIII (vestibulocochlear)

**PONS**

Vestibular ganglion (VIII)

Nerve VIII (vestibulocochlear, *boundary cap*)

**Maxilla**

Sphenoid

**PONS**

**CEREBELLUM**

Petrous temporal bone

**RHOMBENCEPHALIC SUPERVENTRICLE**
(FUTURE FOURTH VENTRICLE)

Oral cavity

Incipient rhombencephalic choroid plexus

**Mandibular process**

UPPER MEDULLA

**RHOMBENCEPHALON**

Medullary velum

Meckel's cartilage

Eustachian tube?

Nerve IX (glossopharyngeal)

LOWER MEDULLA

Temporal bone labyrinth

Superior ganglion (IX)

Superior ganglion (X)

Inferior ganglion (IX)

Nerve X (vagus)

Inferior ganglion (X)

Atlas

Nerve X (vagus)

Vertebral column

Dorsal root ganglion

Axis

C3

Future squamous occipital bone

**FONT KEY:**
**VENTRICULAR DIVISIONS – CAPITALS**
Major brain structure - Times **Bold CAPITALS**
All other structures - Times Roman or **Bold**

**PLATE 210C**

**GW7.5 Sagittal**
**CR 21 mm, C6202**
**Level 4: Slide 22, Section 2**
Left side of brain
**NEUROEPITHELIAL**
**AND PARENCHYMAL**
**BRAIN STRUCTURES,**
**PERIPHERAL GANGLIA**

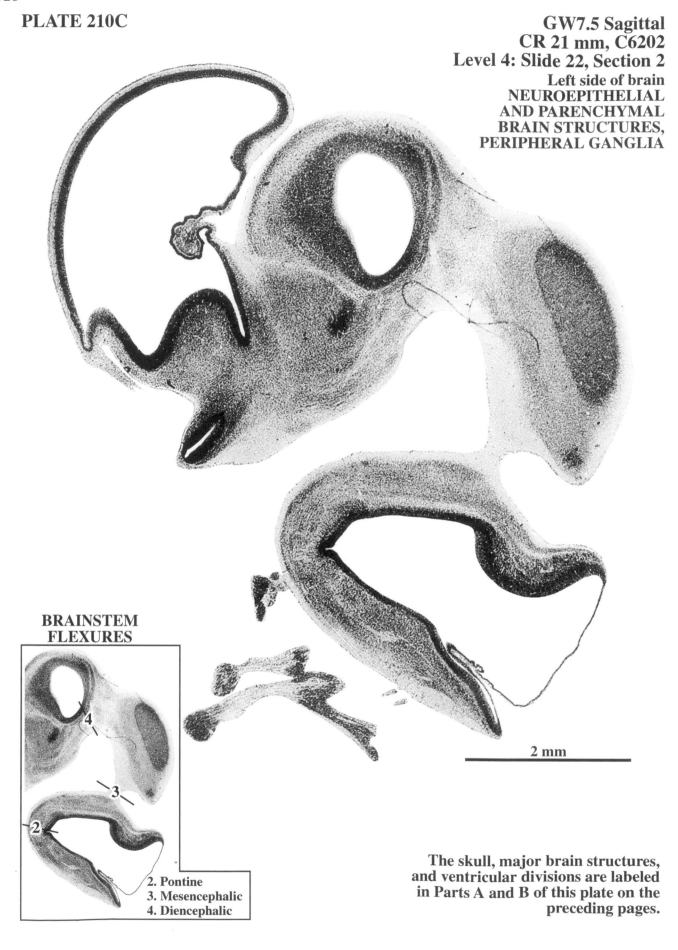

**BRAINSTEM**
**FLEXURES**

2. Pontine
3. Mesencephalic
4. Diencephalic

2 mm

The skull, major brain structures,
and ventricular divisions are labeled
in Parts A and B of this plate on the
preceding pages.

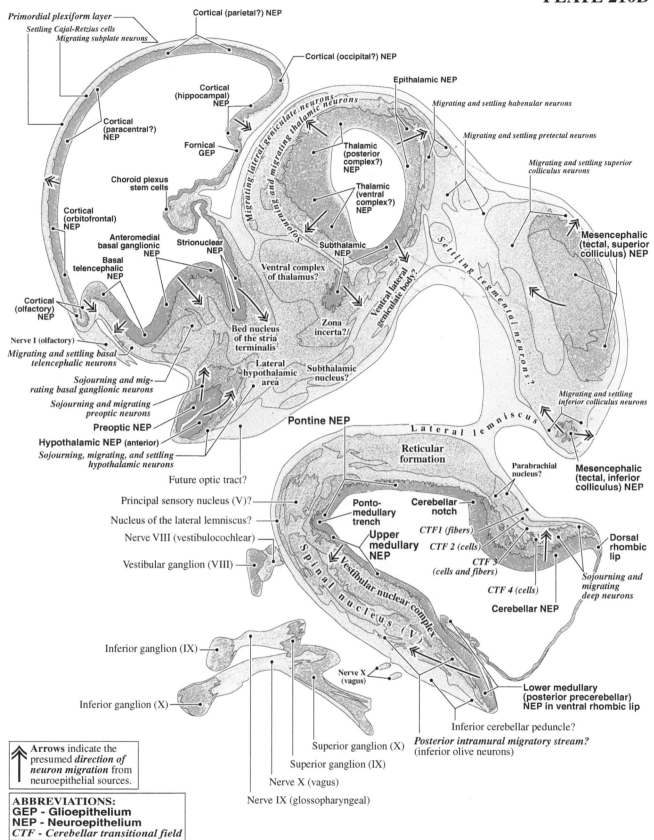

*Primordial plexiform layer*
*Settling Cajal-Retzius cells*
*Migrating subplate neurons*
Cortical (parietal?) NEP
Cortical (occipital?) NEP
Epithalamic NEP
*Migrating and settling habenular neurons*
Cortical (hippocampal) NEP
*Migrating lateral geniculate neurons*
*Migrating and settling thalamic neurons*
Thalamic (posterior complex?) NEP
*Migrating and settling pretectal neurons*
Cortical (paracentral?) NEP
Fornical GEP
*Migrating and settling superior colliculus neurons*
Cortical (orbitofrontal) NEP
Choroid plexus stem cells
Thalamic (ventral complex?) NEP
*Sojourning and migrating thalamic neurons*
Anteromedial basal ganglionic NEP
Strionuclear NEP
Subthalamic NEP
**Mesencephalic (tectal, superior colliculus) NEP**
Basal telencephalic NEP
Ventral complex of thalamus?
*Settling tegmental neurons?*
Cortical (olfactory) NEP
Bed nucleus of the stria terminalis
Zona incerta?
*Ventral lateral geniculate body?*
*Nerve I (olfactory)*
*Migrating and settling basal telencephalic neurons*
Lateral hypothalamic area
Subthalamic nucleus?
*Sojourning and migrating basal ganglionic neurons*
*Migrating and settling inferior colliculus neurons*
*Sojourning and migrating preoptic neurons*
*Lateral lemniscus*
**Preoptic NEP**
**Hypothalamic NEP (anterior)**
**Pontine NEP**
Reticular formation
*Sojourning, migrating, and settling hypothalamic neurons*
**Mesencephalic (tectal, inferior colliculus) NEP**
*Future optic tract?*
Parabrachial nucleus?
Principal sensory nucleus (V)?
Ponto-medullary trench
Cerebellar notch
*CTF1 (fibers)*
Nucleus of the lateral lemniscus?
**Upper medullary NEP**
*CTF 2 (cells)*
**Dorsal rhombic lip**
Nerve VIII (vestibulocochlear)
*CTF 3 (cells and fibers)*
Vestibular ganglion (VIII)
*Spinal nucleus*
*Vestibular nuclear complex (V)*
*CTF 4 (cells)*
*Sojourning and migrating deep neurons*
**Cerebellar NEP**
Inferior ganglion (IX)
Nerve X (vagus)
**Lower medullary (posterior precerebellar) NEP in ventral rhombic lip**
Inferior ganglion (X)
Inferior cerebellar peduncle?
*Posterior intramural migratory stream?* (inferior olive neurons)
Superior ganglion (X)
Superior ganglion (IX)
Nerve X (vagus)
Nerve IX (glossopharyngeal)

**Arrows** indicate the presumed *direction of neuron migration* from neuroepithelial sources.

**ABBREVIATIONS:**
**GEP - Glioepithelium**
**NEP - Neuroepithelium**
*CTF - Cerebellar transitional field*

**FONT KEY:**
**Germinal zone - Helvetica bold**
*Transient structure - Times bold italic*
Permanent structure - Times Roman or **Bold**

**PLATE 211A**

GW7.5 Sagittal
CR 21 mm, C6202
Level 5: Slide 21, Section 2
Left side of brain
SKULL, MAJOR BRAIN
STRUCTURES, AND
VENTRICULAR
DIVISIONS

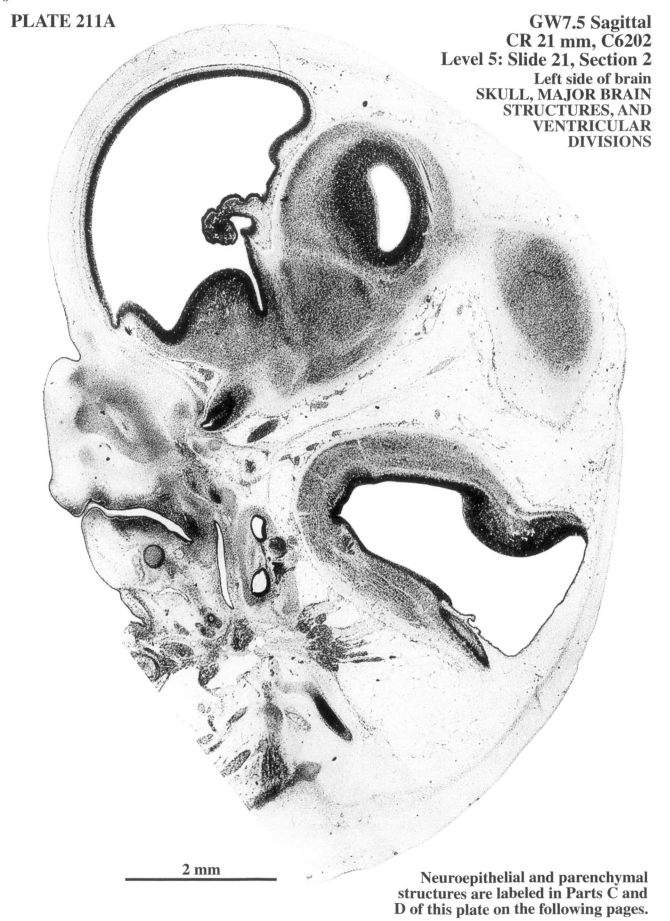

2 mm

Neuroepithelial and parenchymal
structures are labeled in Parts C and
D of this plate on the following pages.

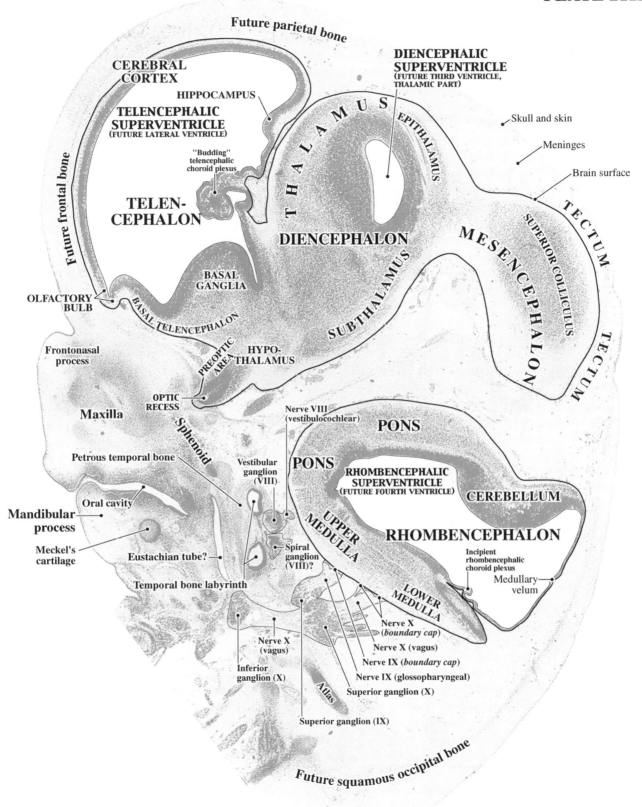

Future parietal bone

**CEREBRAL CORTEX**

HIPPOCAMPUS

**DIENCEPHALIC SUPERVENTRICLE**
(FUTURE THIRD VENTRICLE, THALAMIC PART)

**TELENCEPHALIC SUPERVENTRICLE**
(FUTURE LATERAL VENTRICLE)

T H A L A M U S

EPITHALAMUS

Skull and skin

Meninges

Brain surface

"Budding" telencephalic choroid plexus

**TELEN- CEPHALON**

**DIENCEPHALON**

*T E C T U M*

*M E S E N C E P H A L O N*

*SUPERIOR COLLICULUS*

Future frontal bone

**BASAL GANGLIA**

OLFACTORY BULB

*BASAL TELENCEPHALON*

*SUBTHALAMUS*

*T E C T U M*

Frontonasal process

*PREOPTIC AREA*

**HYPO- THALAMUS**

OPTIC RECESS

Maxilla

Nerve VIII (vestibulocochlear)

**PONS**

Sphenoid

Petrous temporal bone

Vestibular ganglion (VIII)

**PONS**

**RHOMBENCEPHALIC SUPERVENTRICLE**
(FUTURE FOURTH VENTRICLE)

**CEREBELLUM**

Oral cavity

**Mandibular process**

Meckel's cartilage

Eustachian tube?

Spiral ganglion (VIII)?

**UPPER MEDULLA**

**RHOMBENCEPHALON**

Incipient rhombencephalic choroid plexus

Medullary velum

Temporal bone labyrinth

**LOWER MEDULLA**

Nerve X (*boundary cap*)

Nerve X (vagus)

Nerve X (vagus)

Nerve IX (*boundary cap*)

Inferior ganglion (X)

Nerve IX (glossopharyngeal)

Superior ganglion (X)

*Atlas*

Superior ganglion (IX)

Future squamous occipital bone

**FONT KEY:**
**VENTRICULAR DIVISIONS – CAPITALS**
Major brain structure - Times **Bold CAPITALS**
All other structures - Times Roman or **Bold**

**PLATE 211C**

**GW7.5 Sagittal**
**CR 21 mm, C6202**
**Level 5: Slide 21, Section 2**
Left side of brain
**NEUROEPITHELIAL**
**AND PARENCHYMAL**
**BRAIN STRUCTURES,**
**PERIPHERAL GANGLIA**

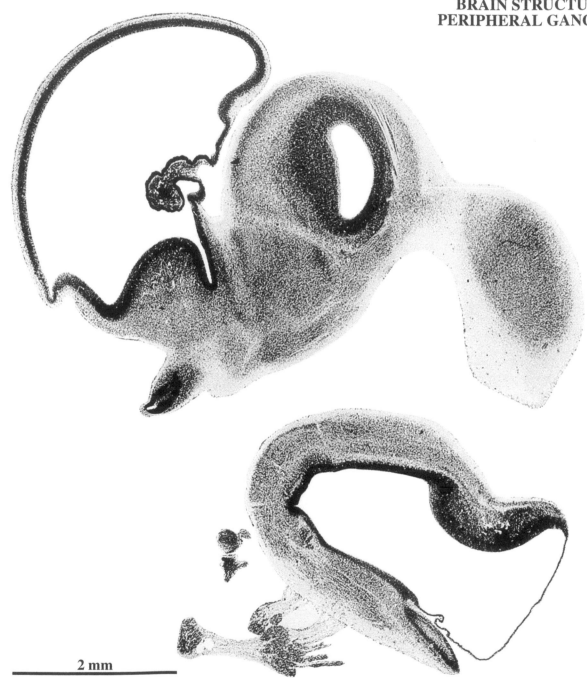

2 mm

The skull, major brain structures,
and ventricular divisions are
labeled in Parts A and B of this
plate on the preceding pages.

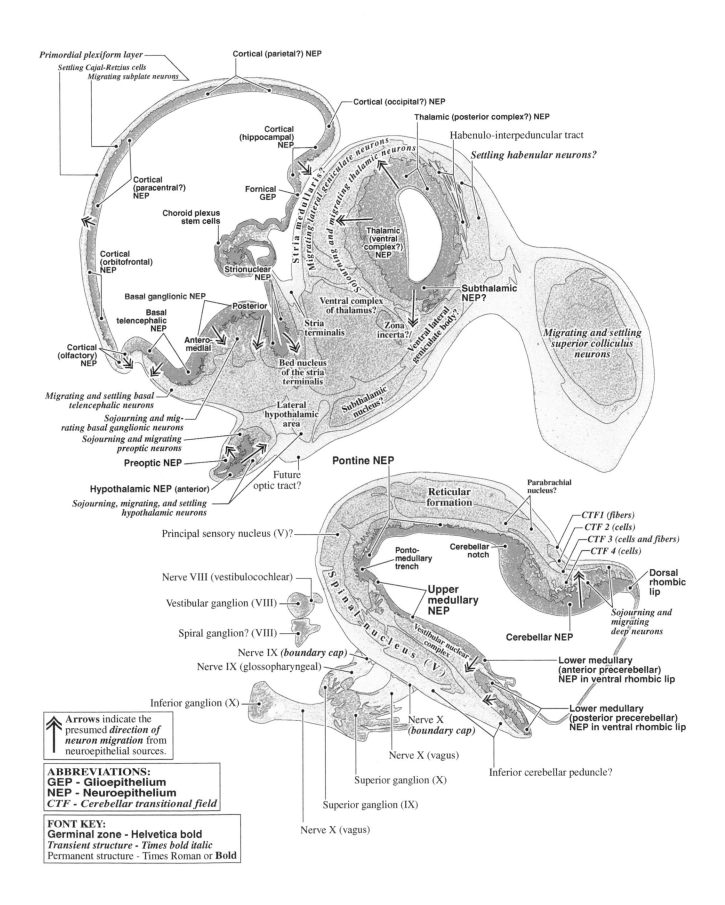

*Primordial plexiform layer* — Cortical (parietal?) NEP

*Settling Cajal-Retzius cells*
*Migrating subplate neurons*

Cortical (occipital?) NEP

Thalamic (posterior complex?) NEP

Habenulo-interpeduncular tract

Cortical
(hippocampal)
NEP

*Settling habenular neurons?*

Cortical
(paracentral?)
NEP

Fornical
GEP

*Stria medullaris?*

*Migrating lateral geniculate neurons*

*Sojourning and migrating thalamic neurons*

Choroid plexus
stem cells

Thalamic
(ventral
complex?)
NEP

Cortical
(orbitofrontal)
NEP

Strionuclear
NEP

**Subthalamic
NEP?**

Basal ganglionic NEP

Posterior

Ventral complex
of thalamus?

Basal
telencephalic
NEP

Stria
terminalis

Zona
incerta?

*Ventral lateral
geniculate body?*

*Migrating and settling
superior colliculus
neurons*

Antero-
medial

Cortical
(olfactory)
NEP

Bed nucleus
of the stria
terminalis

*Migrating and settling basal
telencephalic neurons*

Lateral
hypothalamic
area

*Subthalamic
nucleus?*

*Sojourning and mig-
rating basal ganglionic neurons*

*Sojourning and migrating
preoptic neurons*

**Preoptic NEP**

**Pontine NEP**

**Reticular
formation**

*Parabrachial
nucleus?*

Future
optic tract?

**Hypothalamic NEP (anterior)**

*Sojourning, migrating, and settling
hypothalamic neurons*

*CTF1 (fibers)*
*CTF 2 (cells)*
*CTF 3 (cells and fibers)*
*CTF 4 (cells)*

Principal sensory nucleus (V)?

Ponto-
medullary
trench

Cerebellar
notch

Nerve VIII (vestibulocochlear)

*Spinal nucleus (V)*

**Upper
medullary
NEP**

**Dorsal
rhombic
lip**

Vestibular ganglion (VIII)

*Sojourning
and
migrating
deep neurons*

Spiral ganglion? (VIII)

*Vestibular nuclear
complex*

**Cerebellar NEP**

Nerve IX *(boundary cap)*

Nerve IX (glossopharyngeal)

Lower medullary
(anterior precerebellar)
NEP in ventral rhombic lip

Inferior ganglion (X)

Nerve X
*(boundary cap)*

Lower medullary
(posterior precerebellar)
NEP in ventral rhombic lip

Nerve X (vagus)

Superior ganglion (X)

Inferior cerebellar peduncle?

Superior ganglion (IX)

Nerve X (vagus)

**Arrows** indicate the
presumed *direction of
neuron migration* from
neuroepithelial sources.

**ABBREVIATIONS:**
**GEP - Glioepithelium**
**NEP - Neuroepithelium**
*CTF - Cerebellar transitional field*

**FONT KEY:**
**Germinal zone - Helvetica bold**
*Transient structure - Times bold italic*
Permanent structure - Times Roman or **Bold**

**PLATE 212A**

**GW7.5 Sagittal**
**CR 21 mm, C6202**
**Level 6: Slide 18, Section 2**
Left side of brain
**SKULL, MAJOR BRAIN**
**STRUCTURES, AND**
**VENTRICULAR**
**DIVISIONS**

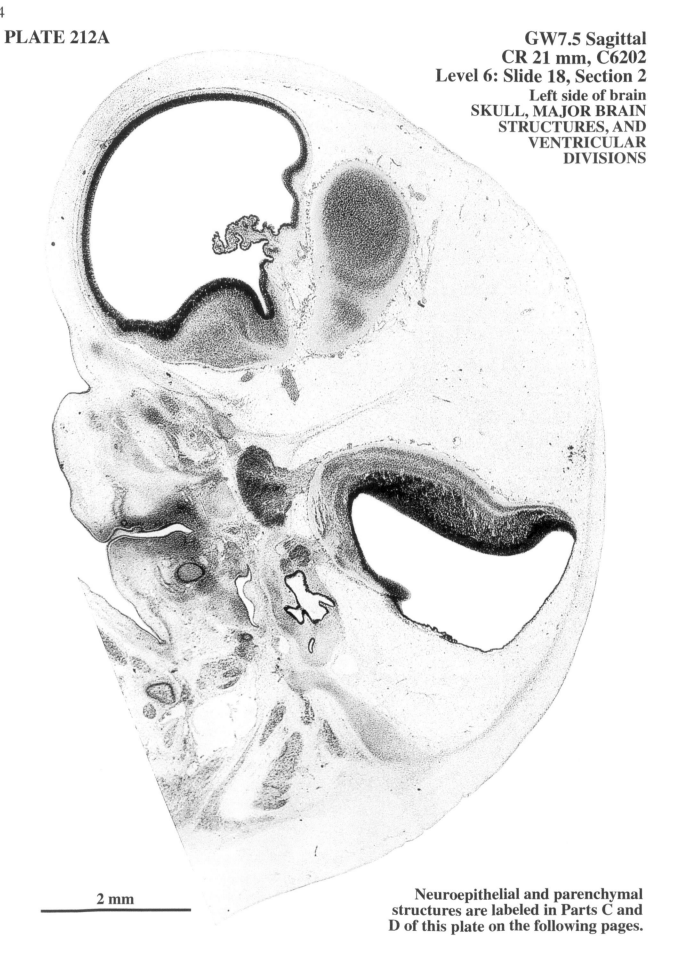

2 mm

**Neuroepithelial and parenchymal**
**structures are labeled in Parts C and**
**D of this plate on the following pages.**

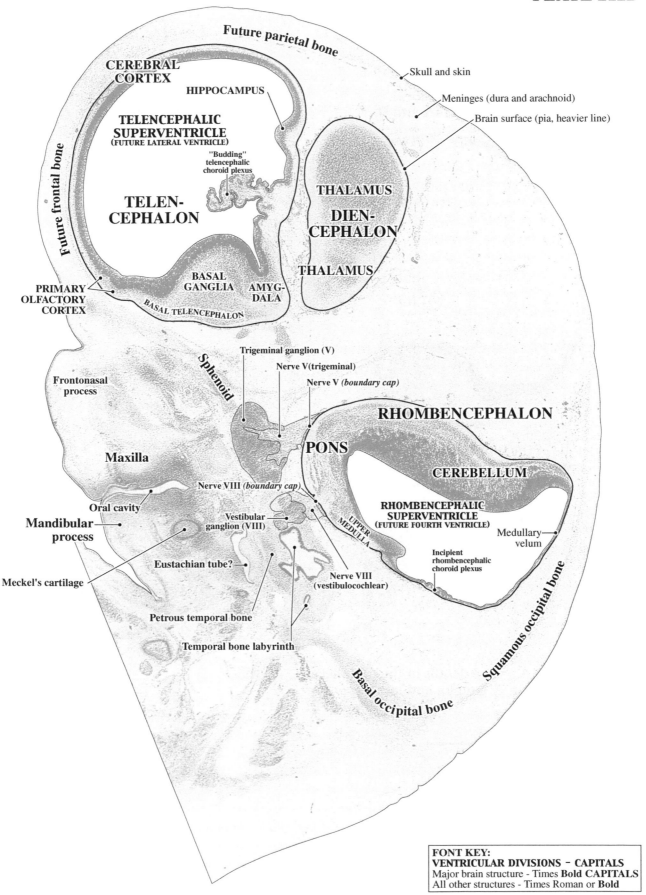

Future parietal bone

**CEREBRAL CORTEX**

HIPPOCAMPUS

Skull and skin

Meninges (dura and arachnoid)

Brain surface (pia, heavier line)

**TELENCEPHALIC SUPERVENTRICLE**
(FUTURE LATERAL VENTRICLE)

Future frontal bone

"Budding" telencephalic choroid plexus

**THALAMUS**

**DIEN-CEPHALON**

**TELEN-CEPHALON**

**THALAMUS**

PRIMARY OLFACTORY CORTEX

**BASAL GANGLIA**

**AMYG-DALA**

BASAL TELENCEPHALON

Trigeminal ganglion (V)

Nerve V (trigeminal)

Nerve V (*boundary cap*)

Frontonasal process

Sphenoid

**RHOMBENCEPHALON**

**PONS**

Maxilla

**CEREBELLUM**

Nerve VIII (*boundary cap*)

**RHOMBENCEPHALIC SUPERVENTRICLE**
(FUTURE FOURTH VENTRICLE)

**Mandibular process**

Oral cavity

Vestibular ganglion (VIII)

UPPER MEDULLA

Medullary velum

Meckel's cartilage

Eustachian tube?

Nerve VIII (vestibulocochlear)

Incipient rhombencephalic choroid plexus

Petrous temporal bone

Temporal bone labyrinth

Basal occipital bone

Squamous occipital bone

FONT KEY:
**VENTRICULAR DIVISIONS - CAPITALS**
Major brain structure - Times **Bold CAPITALS**
All other structures - Times Roman or **Bold**

**PLATE 212C**

**Left side of brain
NEUROEPITHELIAL
AND PARENCHYMAL
BRAIN STRUCTURES;
PERIPHERAL GANGLIA**

**See details of the cerebral
cortex in Plates 214A and B.**

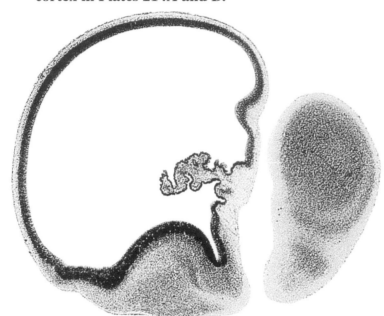

**See details of the hippocampus,
basal ganglia, and amygdala in
Plates 216A and B.**

**See details of the
cerebellum in Plates
219A and B.**

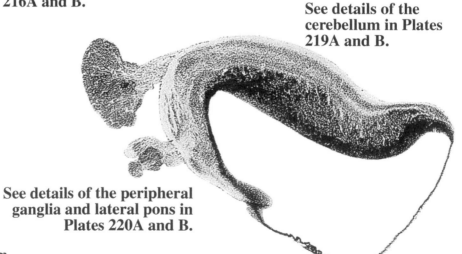

**See details of the peripheral
ganglia and lateral pons in
Plates 220A and B.**

2 mm

**The skull, major brain structures,
and ventricular divisions are
labeled in Parts A and B of this
plate on the preceding pages.**

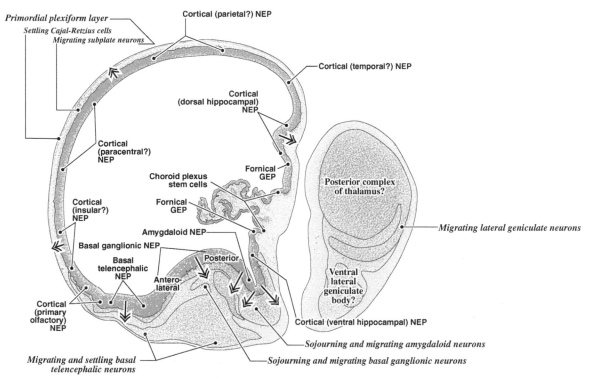

*Primordial plexiform layer*
*Settling Cajal-Retzius cells*
*Migrating subplate neurons*

Cortical (parietal?) NEP

Cortical (temporal?) NEP

Cortical
(dorsal hippocampal)
NEP

Fornical
GEP

Choroid plexus
stem cells

Fornical
GEP

Posterior complex
of thalamus?

*Migrating lateral geniculate neurons*

Cortical
(paracentral?)
NEP

Cortical
(insular?)
NEP

Amygdaloid NEP

Basal ganglionic NEP

Posterior

Basal
telencephalic
NEP

Antero-
lateral

Ventral
lateral
geniculate
body?

Cortical
(primary
olfactory)
NEP

Cortical (ventral hippocampal) NEP

*Migrating and settling basal
telencephalic neurons*

*Sojourning and migrating amygdaloid neurons*

*Sojourning and migrating basal ganglionic neurons*

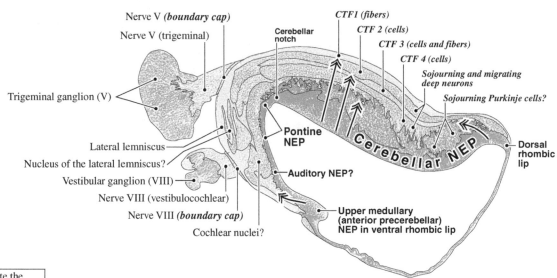

Nerve V *(boundary cap)*

Nerve V (trigeminal)

Cerebellar
notch

*CTF1 (fibers)*

*CTF 2 (cells)*

*CTF 3 (cells and fibers)*

*CTF 4 (cells)*

*Sojourning and migrating
deep neurons*

*Sojourning Purkinje cells?*

Trigeminal ganglion (V)

**Pontine
NEP**

*Cerebellar NEP*

Dorsal
rhombic
lip

Lateral lemniscus

Nucleus of the lateral lemniscus?

Vestibular ganglion (VIII)

Nerve VIII (vestibulocochlear)

Nerve VIII *(boundary cap)*

Cochlear nuclei?

**Auditory NEP?**

**Upper medullary
(anterior precerebellar)
NEP in ventral rhombic lip**

**Arrows** indicate the
presumed *direction of
neuron migration* from
neuroepithelial sources.

**ABBREVIATIONS:
GEP - Glioepithelium
NEP - Neuroepithelium**
*CTF - Cerebellar transitional field*

**FONT KEY:
Germinal zone - Helvetica bold**
*Transient structure - Times bold italic*
Permanent structure - Times Roman or **Bold**

**PLATE 213A**

GW7.5 Sagittal
CR 21 mm, C6202
Level 7: Slide 17, Section 2
Left side of brain
SKULL, MAJOR BRAIN
STRUCTURES, AND
VENTRICULAR
DIVISIONS

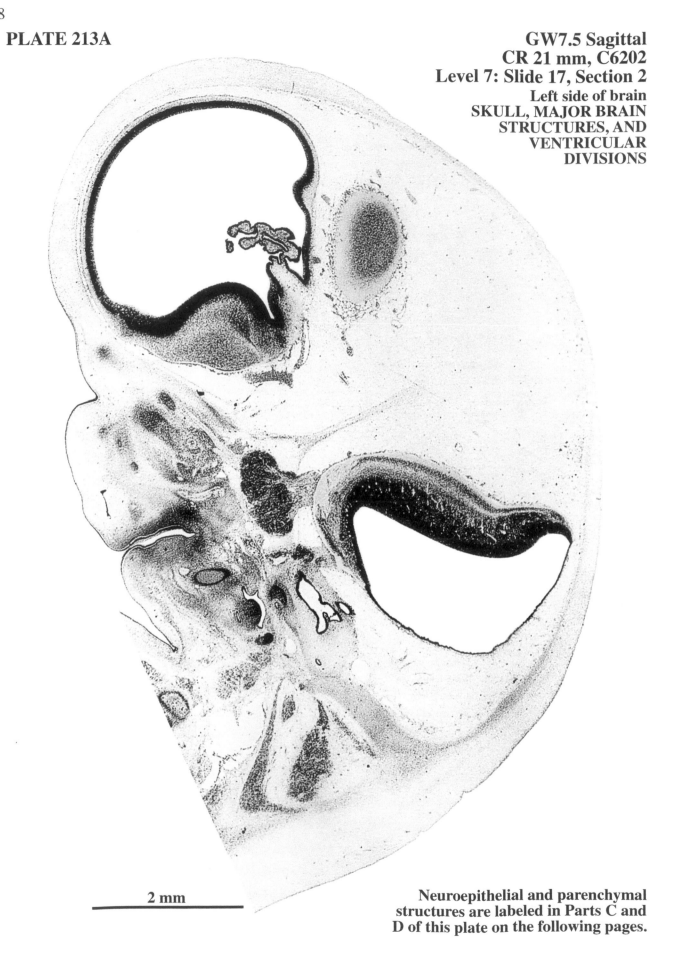

2 mm

Neuroepithelial and parenchymal
structures are labeled in Parts C and
D of this plate on the following pages.

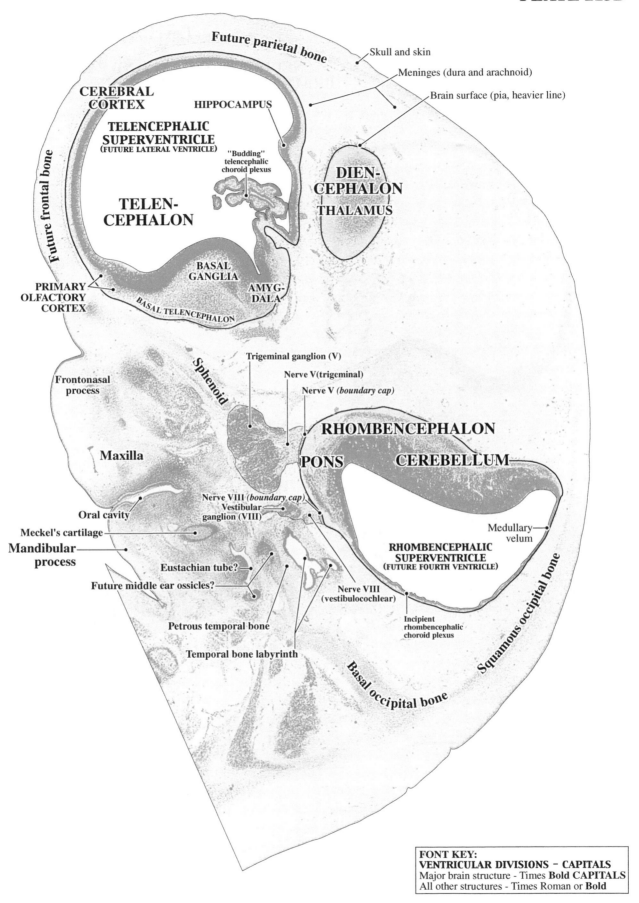

Future parietal bone

Skull and skin

Meninges (dura and arachnoid)

Brain surface (pia, heavier line)

**CEREBRAL CORTEX**

HIPPOCAMPUS

**TELENCEPHALIC SUPERVENTRICLE**
(FUTURE LATERAL VENTRICLE)

"Budding" telencephalic choroid plexus

**DIEN-CEPHALON**

**THALAMUS**

**TELEN-CEPHALON**

Future frontal bone

**BASAL GANGLIA**

**AMYG-DALA**

PRIMARY OLFACTORY CORTEX

*BASAL TELENCEPHALON*

Trigeminal ganglion (V)

Nerve V (trigeminal)

Nerve V *(boundary cap)*

Frontonasal process

*Sphenoid*

**RHOMBENCEPHALON**

**Maxilla**

**PONS**

**CEREBELLUM**

Nerve VIII *(boundary cap)*

Vestibular ganglion (VIII)

Oral cavity

Medullary velum

Meckel's cartilage

**Mandibular process**

**RHOMBENCEPHALIC SUPERVENTRICLE**
(FUTURE FOURTH VENTRICLE)

Eustachian tube?

Future middle ear ossicles?

Nerve VIII (vestibulocochlear)

Incipient rhombencephalic choroid plexus

Petrous temporal bone

*Basal occipital bone*

*Squamous occipital bone*

Temporal bone labyrinth

FONT KEY:
**VENTRICULAR DIVISIONS – CAPITALS**
Major brain structure - Times **Bold CAPITALS**
All other structures - Times Roman or **Bold**

**PLATE 213C**

GW7.5 Sagittal
CR 21 mm, C6202
Level 7: Slide 17, Section 2
Left side of brain
NEUROEPITHELIAL AND PARENCHYMAL
BRAIN STRUCTURES, PERIPHERAL GANGLIA

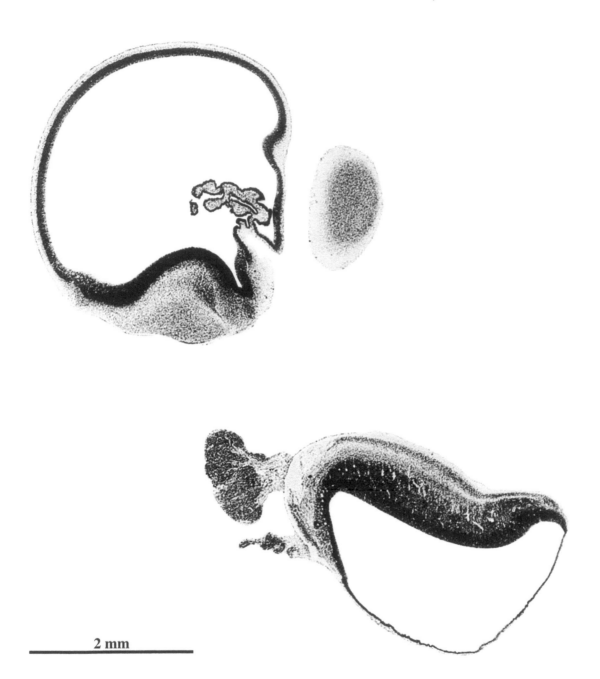

2 mm

The skull, major brain structures,
and ventricular divisions are
labeled in Parts A and B of this
plate on the preceding pages.

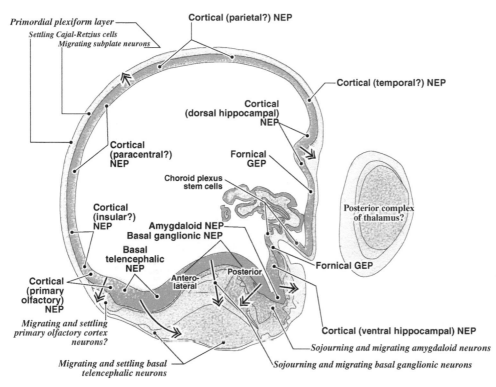

*Primordial plexiform layer*
*Settling Cajal-Retzius cells*
*Migrating subplate neurons*

Cortical (parietal?) NEP

Cortical (temporal?) NEP

Cortical
(dorsal hippocampal)
NEP

Fornical
GEP

Cortical
(paracentral?)
NEP

Choroid plexus
stem cells

Posterior complex
of thalamus?

Cortical
(insular?)
NEP

Amygdaloid NEP
Basal ganglionic NEP

Basal
telencephalic
NEP

Antero-
lateral

Posterior

Fornical GEP

Cortical
(primary
olfactory)
NEP

Cortical (ventral hippocampal) NEP

*Migrating and settling*
*primary olfactory cortex*
*neurons?*

*Sojourning and migrating amygdaloid neurons*

*Migrating and settling basal*
*telencephalic neurons*

*Sojourning and migrating basal ganglionic neurons*

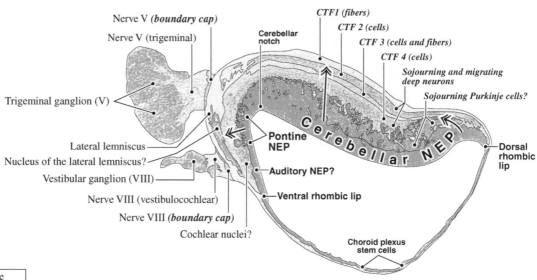

Nerve V *(boundary cap)*

Nerve V (trigeminal)

Cerebellar
notch

*CTF1 (fibers)*
*CTF 2 (cells)*
*CTF 3 (cells and fibers)*
*CTF 4 (cells)*
*Sojourning and migrating*
*deep neurons*
*Sojourning Purkinje cells?*

Trigeminal ganglion (V)

Pontine
NEP

Cerebellar NEP

Dorsal
rhombic
lip

Lateral lemniscus

Nucleus of the lateral lemniscus?

Vestibular ganglion (VIII)

Auditory NEP?

Nerve VIII (vestibulocochlear)

Ventral rhombic lip

Nerve VIII *(boundary cap)*

Cochlear nuclei?

Choroid plexus
stem cells

**Arrows** indicate the
presumed *direction of*
*neuron migration* from
neuroepithelial sources.

ABBREVIATIONS:
**GEP - Glioepithelium**
**NEP - Neuroepithelium**
*CTF - Cerebellar transitional field*

FONT KEY:
**Germinal zone - Helvetica bold**
*Transient structure - Times bold italic*
Permanent structure - Times Roman or **Bold**

542

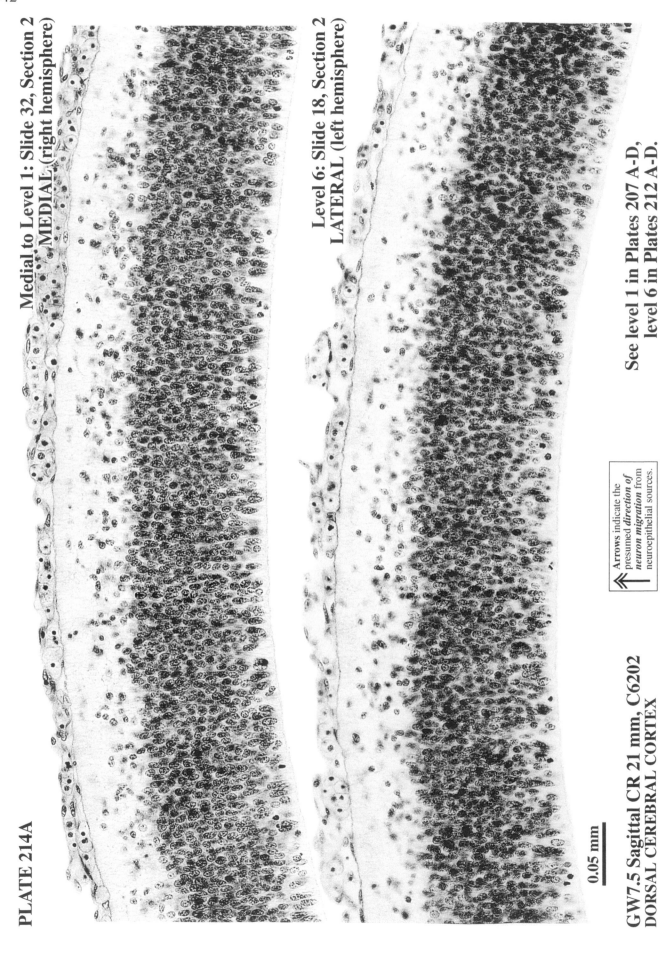

PLATE 214A

**Medial to Level 1: Slide 32, Section 2**
**MEDIAL (right hemisphere)**

**Level 6: Slide 18, Section 2**
**LATERAL (left hemisphere)**

See level 1 in Plates 207 A-D,
level 6 in Plates 212 A-D.

Arrows indicate the
presumed *direction of
neuron migration* from
neuroepithelial sources.

0.05 mm

GW7.5 Sagittal CR 21 mm, C6202
DORSAL CEREBRAL CORTEX

PLATE 214B

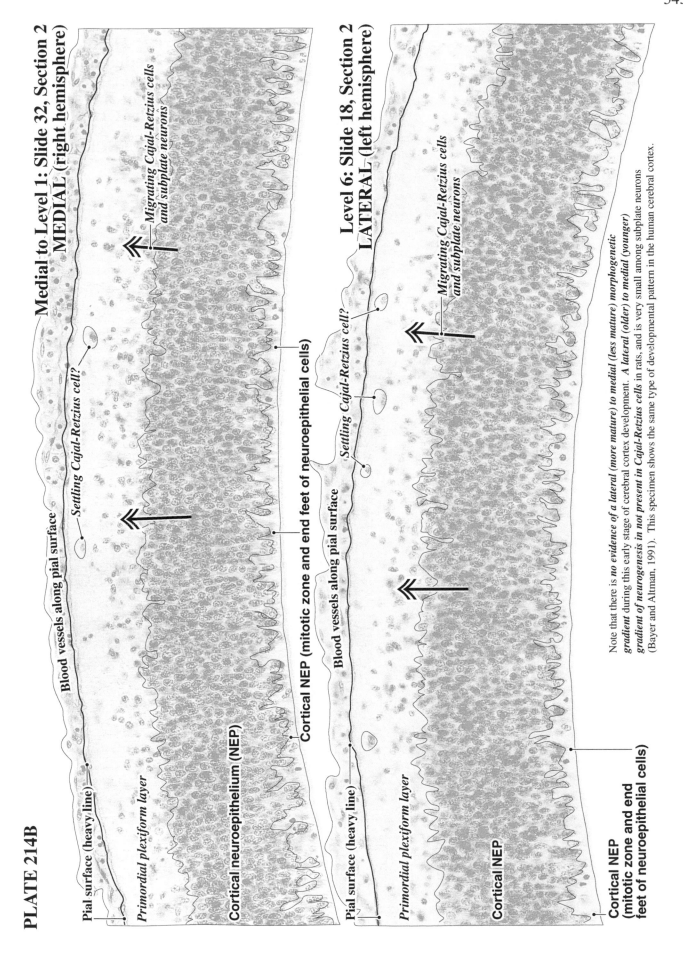

**Medial-to Level 1: Slide 32, Section 2**
**MEDIAL (right hemisphere)**

Pial surface (heavy line)

Blood vessels along pial surface

*Settling Cajal-Retzius cell?*

*Primordial plexiform layer*

*Migrating Cajal-Retzius cells and subplate neurons*

**Cortical neuroepithelium (NEP)**

Cortical NEP (mitotic zone and end feet of neuroepithelial cells)

**Level 6: Slide 18, Section 2**
**LATERAL (left hemisphere)**

Pial surface (heavy line)

Blood vessels along pial surface

*Settling Cajal-Retzius cell?*

*Primordial plexiform layer*

*Migrating Cajal-Retzius cells and subplate neurons*

**Cortical NEP**

**Cortical NEP (mitotic zone and end feet of neuroepithelial cells)**

Note that there is *no evidence of a lateral (more mature) to medial (less mature) morphogenetic gradient* during this early stage of cerebral cortex development. *A lateral (older) to medial (younger) gradient of neurogenesis in not present in Cajal-Retzius cells* in rats, and is very small among subplate neurons (Bayer and Altman, 1991). This specimen shows the same type of developmental pattern in the human cerebral cortex.

**PLATE 215A**

GW7.5 Sagittal
CR 21 mm, C6202
Medial to Level 1:
Slide 31, Section 2
Right side of brain
OLFACTORY BULB
AND BASAL
TELENCEPHALON

See level 1 in
Plates 207A-D.

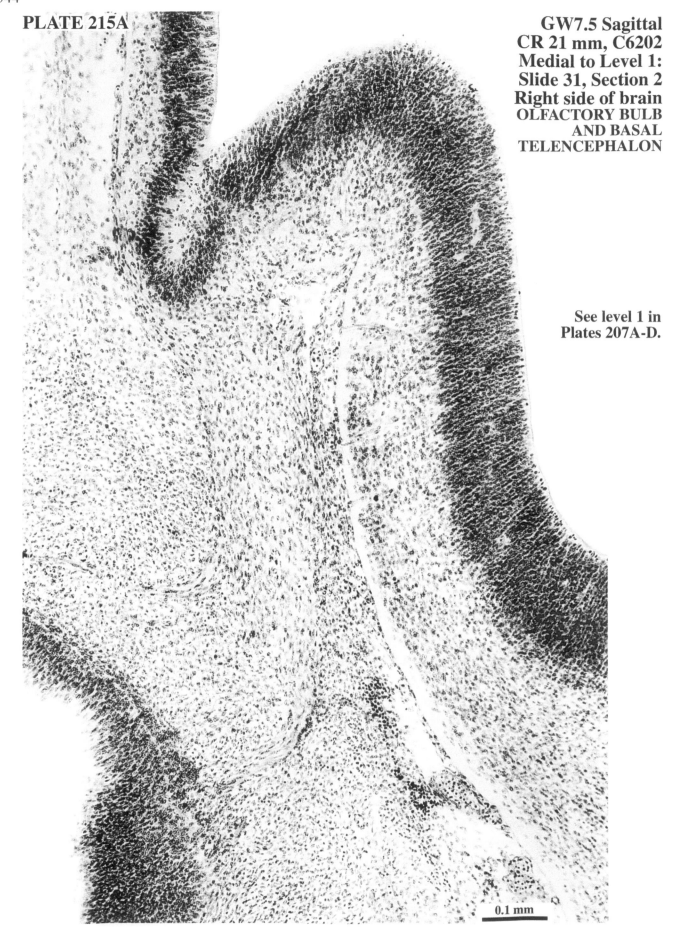

0.1 mm

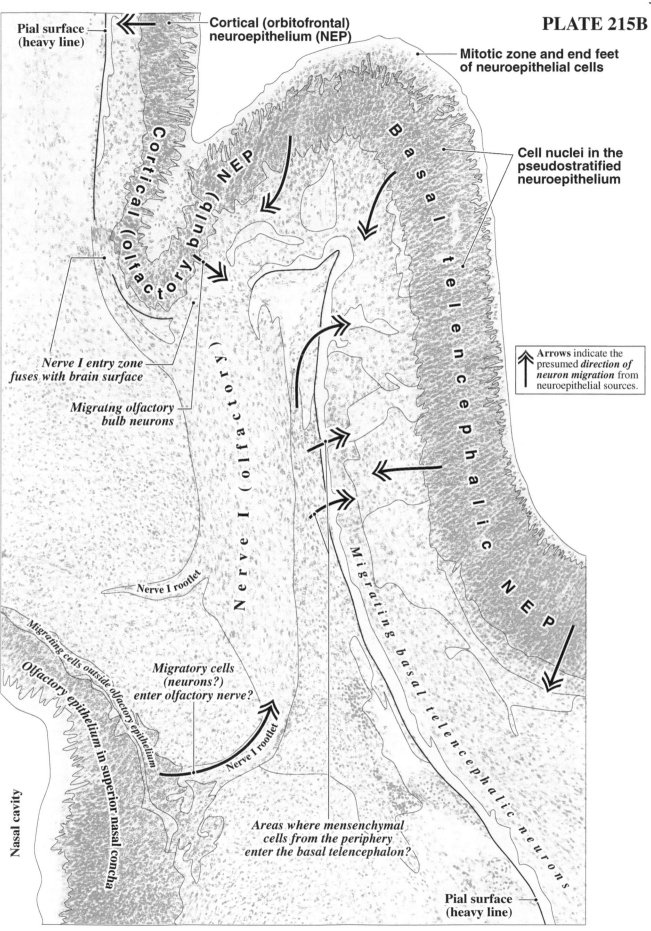

**PLATE 215B**

Pial surface
(heavy line)

Cortical (orbitofrontal)
neuroepithelium (NEP)

Mitotic zone and end feet
of neuroepithelial cells

Cell nuclei in the
pseudostratified
neuroepithelium

Cortical (olfactory bulb) NEP

Basal telencephalic NEP

Arrows indicate the
presumed *direction of
neuron migration* from
neuroepithelial sources.

Nerve I entry zone
*fuses with brain surface*

*Migratng olfactory
bulb neurons*

Nerve I (olfactory)

Nerve I rootlet

*Migrating cells outside olfactory epithelium*

Olfactory epithelium in superior nasal concha

Nasal cavity

*Migratory cells
(neurons?)
enter olfactory nerve?*

Nerve I rootlet

*Migrating basal telencephalic neurons*

*Areas where mensenchymal
cells from the periphery
enter the basal telencephalon?*

Pial surface
(heavy line)

**PLATE 216A**

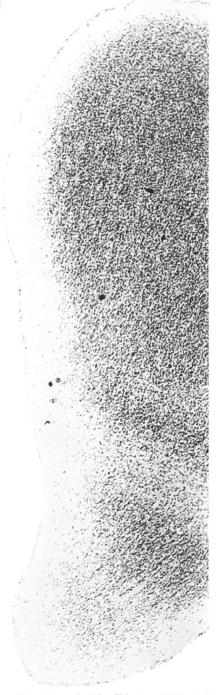

GW7.5 Sagittal
CR 21 mm, C6202
Level 6: Slide 18, Section 2
Left side of brain
HIPPOCAMPUS, BASAL
GANGLIA, AND AMYGDALA

See Level 6 in Plates 212 A-D.

0.5 mm

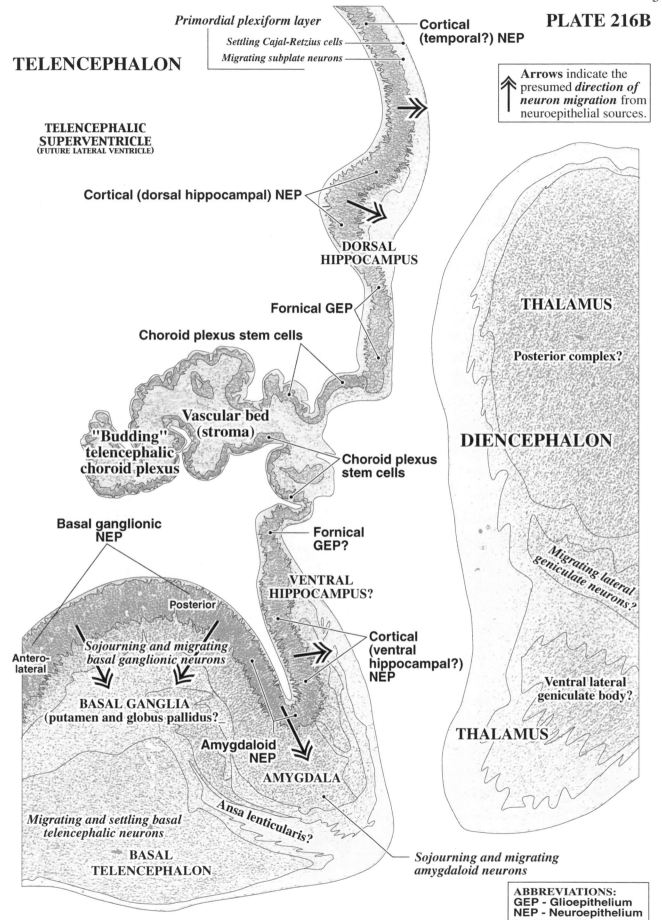

547

PLATE 216B

*Primordial plexiform layer* — Cortical (temporal?) NEP

*Settling Cajal-Retzius cells*

*Migrating subplate neurons*

**TELENCEPHALON**

Arrows indicate the presumed *direction of neuron migration* from neuroepithelial sources.

**TELENCEPHALIC SUPERVENTRICLE** (FUTURE LATERAL VENTRICLE)

Cortical (dorsal hippocampal) NEP

**DORSAL HIPPOCAMPUS**

Fornical GEP

**THALAMUS**

Posterior complex?

Choroid plexus stem cells

Vascular bed (stroma)

**DIENCEPHALON**

"Budding" telencephalic choroid plexus

Choroid plexus stem cells

*Migrating lateral geniculate neurons?*

Basal ganglionic NEP

Fornical GEP?

**VENTRAL HIPPOCAMPUS?**

Posterior

*Sojourning and migrating basal ganglionic neurons*

Antero-lateral

Cortical (ventral hippocampal?) NEP

**Ventral lateral geniculate body?**

**BASAL GANGLIA** (putamen and globus pallidus?)

**THALAMUS**

Amygdaloid NEP

**AMYGDALA**

*Ansa lenticularis?*

*Migrating and settling basal telencephalic neurons*

**BASAL TELENCEPHALON**

*Sojourning and migrating amygdaloid neurons*

**ABBREVIATIONS:**
GEP - Glioepithelium
NEP - Neuroepithelium

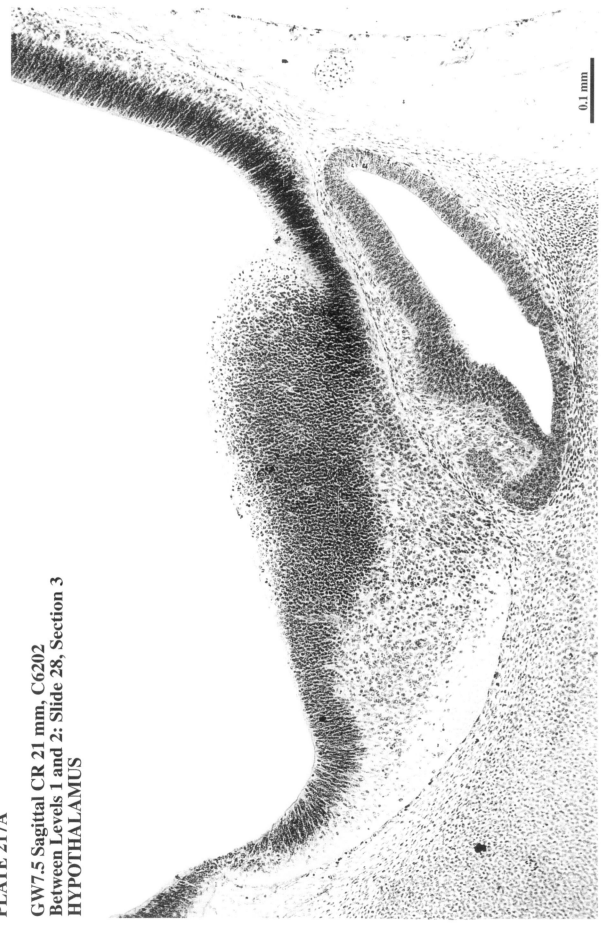

**PLATE 217A**

**GW7.5 Sagittal CR 21 mm, C6202**
**Between Levels 1 and 2: Slide 28, Section 3**
**HYPOTHALAMUS**

0.1 mm

See level 1 in Plates 207A-D, level 2 in Plates 208 A-D.

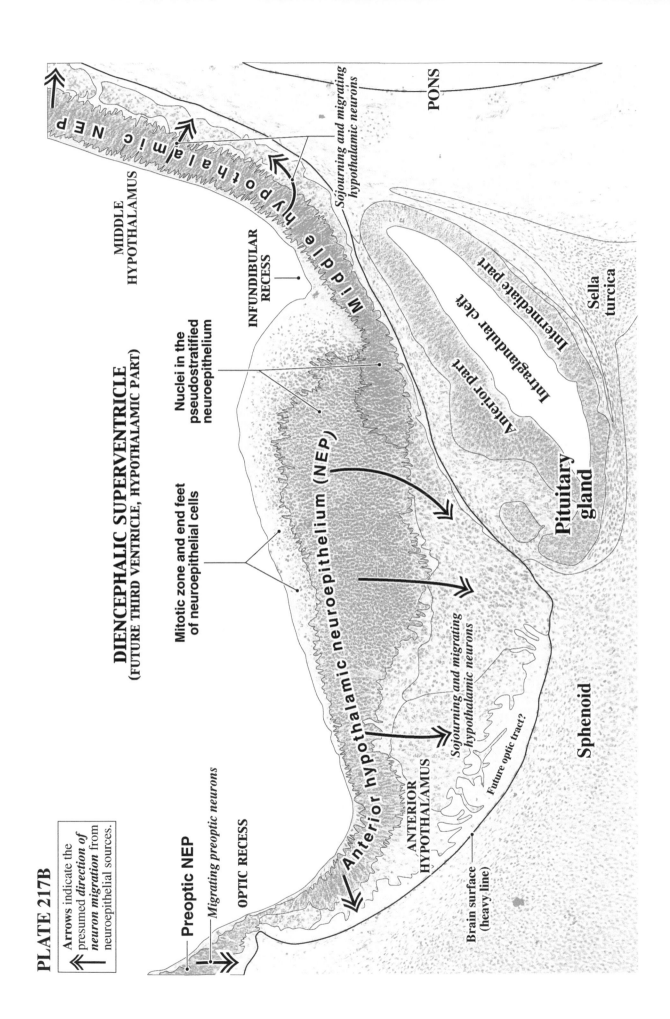

549

PLATE 217B

Arrows indicate the presumed *direction of neuron migration* from neuroepithelial sources.

**DIENCEPHALIC SUPERVENTRICLE**
(FUTURE THIRD VENTRICLE, HYPOTHALAMIC PART)

MIDDLE HYPOTHALAMUS

**Middle hypothalamic NEP**

*Sojourning and migrating hypothalamic neurons*

PONS

INFUNDIBULAR RECESS

Nuclei in the pseudostratified neuroepithelium

Intermediate part

Intraglandular cleft

Anterior part

Sella turcica

**Pituitary gland**

Mitotic zone and end feet of neuroepithelial cells

**Anterior hypothalamic neuroepithelium (NEP)**

*Sojourning and migrating hypothalamic neurons*

**ANTERIOR HYPOTHALAMUS**

Future optic tract?

**Sphenoid**

Brain surface (heavy line)

**Preoptic NEP**

*Migrating preoptic neurons*

OPTIC RECESS

550

PLATE 218A

GW7.5 Sagittal CR 21 mm, C6202
Between Levels 1 and 2:
Slide 28, Section 3
MIDBRAIN
TEGMENTUM

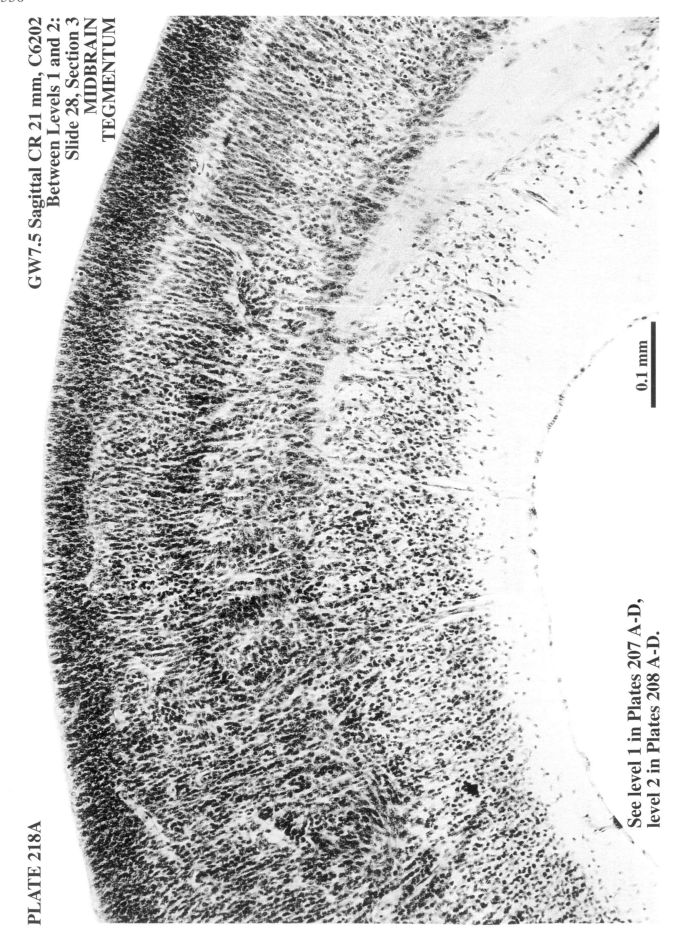

0.1 mm

See level 1 in Plates 207 A-D,
level 2 in Plates 208 A-D.

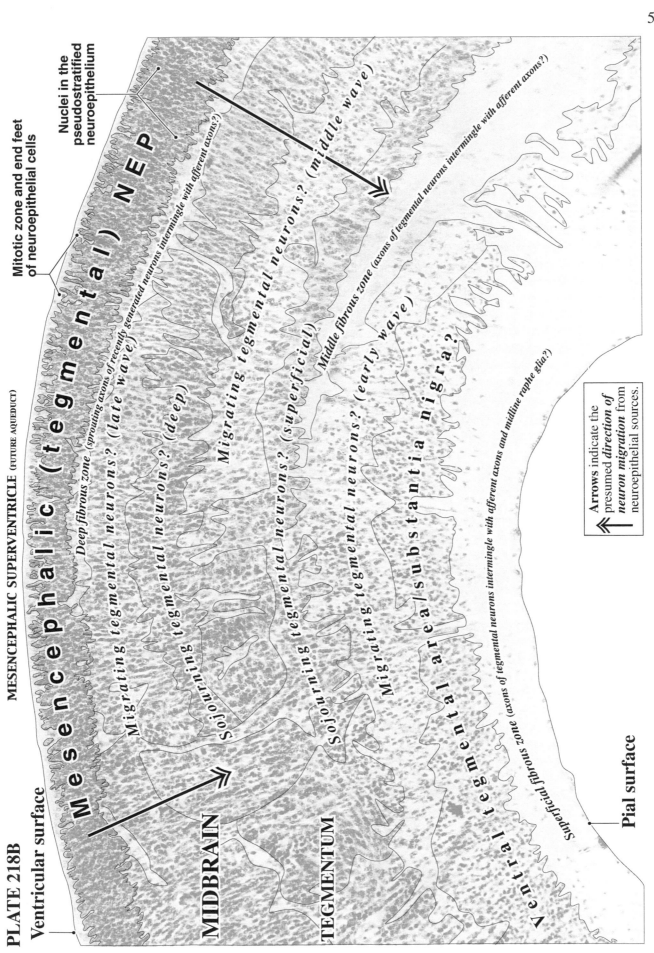

551

PLATE 218B

MESENCEPHALIC SUPERVENTRICLE (FUTURE AQUEDUCT)

Ventricular surface

Nuclei in the pseudostratified neuroepithelium

Mitotic zone and end feet of neuroepithelial cells

Mesencephalic (tegmental) NEP

Deep fibrous zone (sprouting axons of recently generated neurons intermingle with afferent axons?)

Migrating tegmental neurons? (late wave)

Sojourning tegmental neurons? (deep)

Migrating tegmental neurons? (superficial)

Middle fibrous zone (axons of tegmental neurons intermingle with afferent axons?)

Migrating tegmental neurons? (middle wave)

Sojourning tegmental neurons?

Migrating tegmental neurons? (early wave)

Ventral tegmental area/substantia nigra?

Superficial fibrous zone (axons of tegmental neurons intermingle with afferent axons and midline raphe glia?)

MIDBRAIN

TEGMENTUM

Pial surface

Arrows indicate the presumed direction of neuron migration from neuroepithelial sources.

PLATE 219A

GW7.5 Sagittal CR 21 mm, C6202
CEREBELLUM

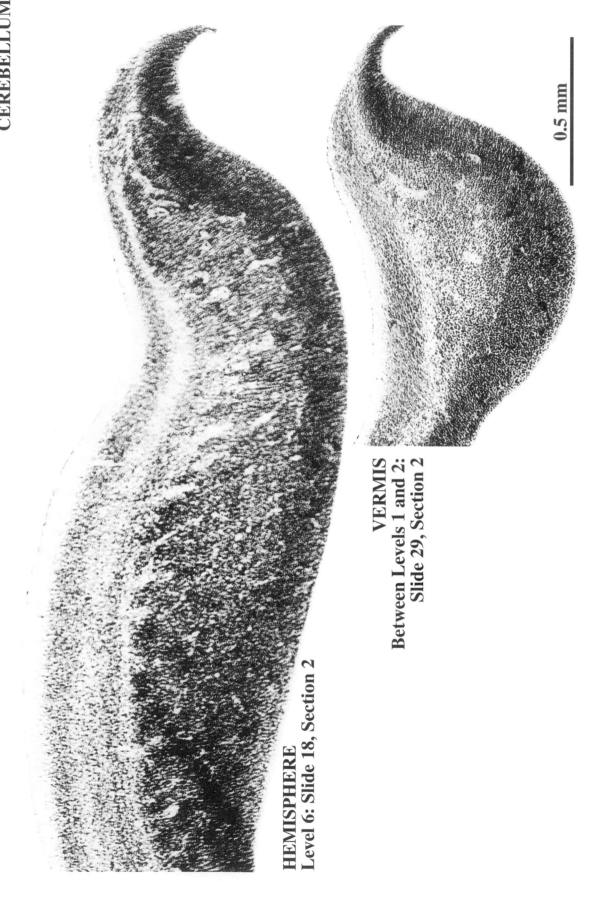

HEMISPHERE
Level 6: Slide 18, Section 2

VERMIS
Between Levels 1 and 2:
Slide 29, Section 2

0.5 mm

See level 1 in Plates 207 A-D, level 2 in Plates 208 A-D, and level 6 in Plates 212 A-D.

**PLATE 219B**

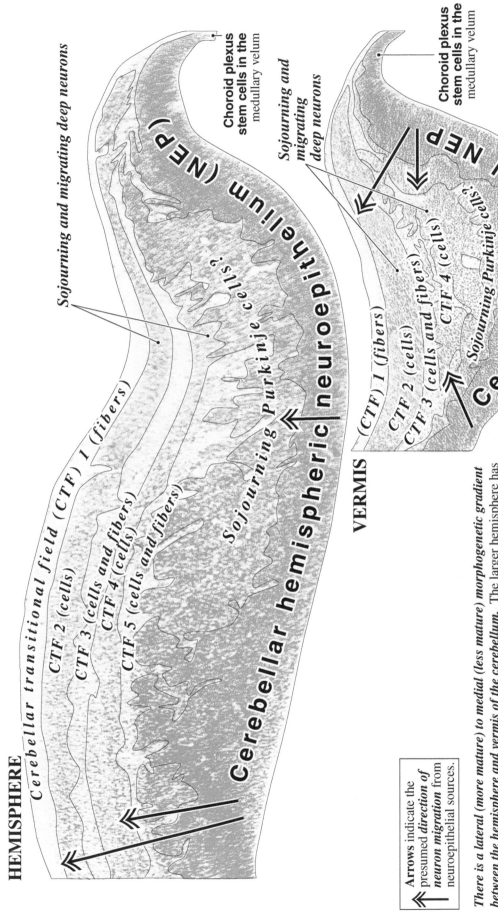

**HEMISPHERE**

*Cerebellar transitional field (CTF) 1 (fibers)*

CTF 2 (cells)

CTF 3 (cells and fibers)

CTF 4 (cells)

CTF 5 (cells and fibers)

*Sojourning Purkinje cells?*

**Cerebellar hemispheric neuroepithelium (NEP)**

*Sojourning and migrating deep neurons*

**Choroid plexus stem cells in the** medullary velum

**VERMIS**

*(CTF) 1 (fibers)*

CTF 2 (cells)

CTF 3 (cells and fibers)

CTF 4 (cells)

*Sojourning Purkinje cells*

**Cerebellar vermal NEP**

*Sojourning and migrating deep neurons*

**Choroid plexus stem cells in the** medullary velum

Arrows indicate the presumed *direction of neuron migration* from neuroepithelial sources.

*There is a lateral (more mature) to medial (less mature) morphogenetic gradient between the hemisphere and vermis of the cerebellum.* The larger hemisphere has better defined and more transitional layers than the smaller vermis. Deep neurons predominate in the transitional field outside the germinal matrix throughout the cerebellum, indicating that they are earlier-generated neurons. Within both the hemispheric and vermal parts of the **cerebellar neuroepithelium**, more Purkinje cells and fewer deep neurons are being generated. Note that the **external germinal layer** is not present at this stage of cerebellar development.

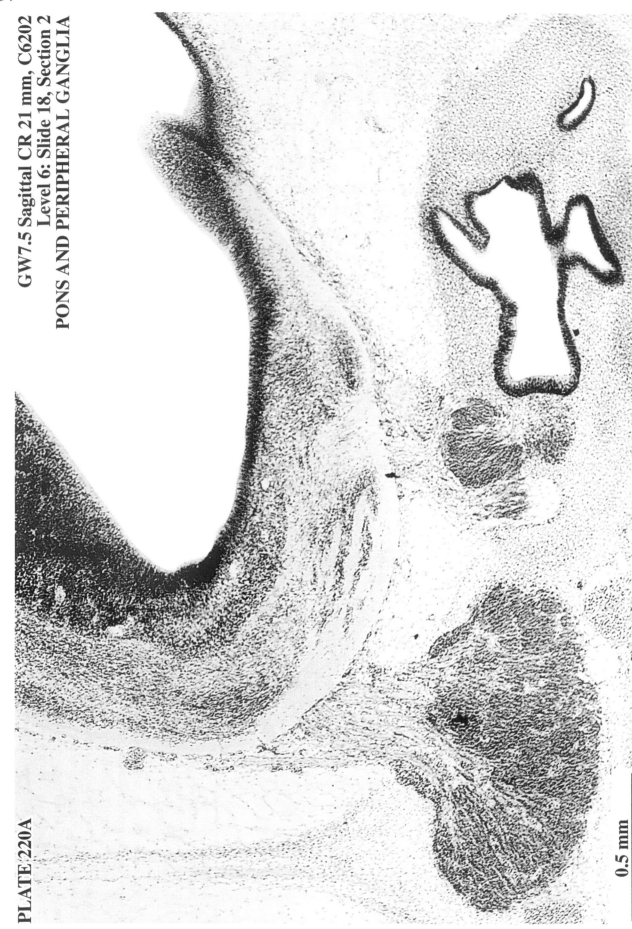

GW7.5 Sagittal CR 21 mm, C6202
Level 6: Slide 18, Section 2
PONS AND PERIPHERAL GANGLIA

PLATE 220A

0.5 mm

See level 6 in Plates 212 A-D.

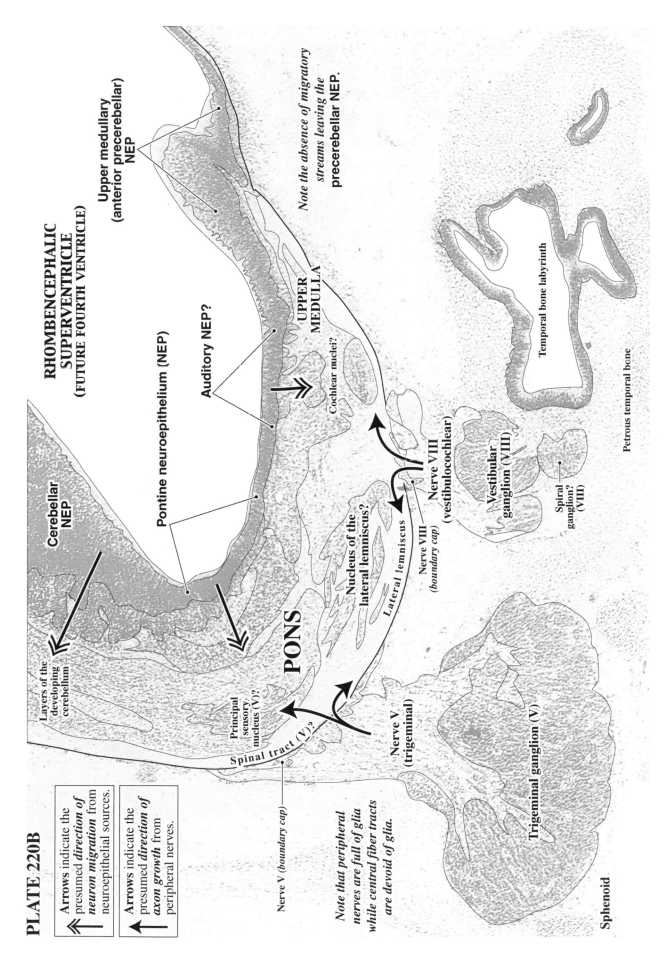

555

## PLATE 220B

⇐ **Arrows** indicate the presumed *direction of neuron migration* from neuroepithelial sources.

← **Arrows** indicate the presumed *direction of axon growth* from peripheral nerves.

**RHOMBENCEPHALIC SUPERVENTRICLE (FUTURE FOURTH VENTRICLE)**

Upper medullary (anterior precerebellar) NEP

*Note the absence of migratory streams leaving the precerebellar NEP.*

Pontine neuroepithelium (NEP)

Auditory NEP?

**UPPER MEDULLA**

Cochlear nuclei?

Cerebellar NEP

Layers of the developing cerebellum

Principal sensory nucleus (V)?

**PONS**

Nucleus of the lateral lemniscus?

Lateral lemniscus

Nerve VIII (vestibulocochlear)

Nerve VIII (boundary cap)

Vestibular ganglion (VIII)

Spiral ganglion? (VIII)

Temporal bone labyrinth

Petrous temporal bone

Spinal tract (V)?

Nerve V (boundary cap)

*Note that peripheral nerves are full of glia while central fiber tracts are devoid of glia.*

Nerve V (trigeminal)

Trigeminal ganglion (V)

Sphenoid

GW7.5 Sagittal CR 21 mm, C6202
Medial to Level 1
**MIDLINE RAPHE
GLIAL STRUCTURE**

**Pons adjacent to
the pontine
flexure in slide 31,
section 2.**

**Cervical spinal
cord adjacent to
the medullary
flexure in slide 32,
section 2.**

0.1 mm

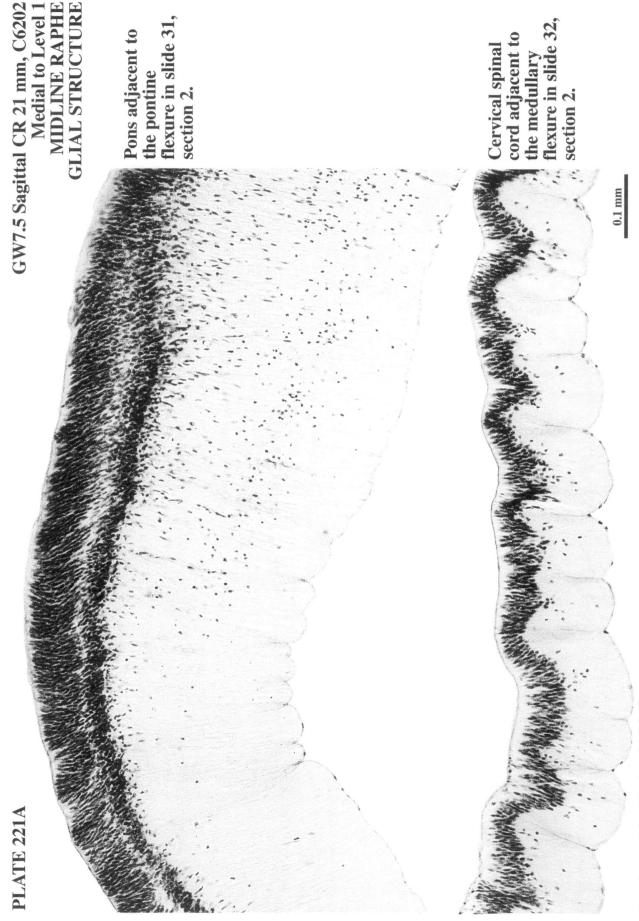

**PLATE 221A**

**See level 1 in Plates 207A-D.**

# PLATE 221B

## PONS

The **midline raphe glial system** is prominent in regions where the shape of the brain and spinal cord sharply change curvature. Van Hartesveldt et al. (1986) described this in rats, and it is virtually identical in man. The strong fibrous palisades may provide structural stability in the region of these curvatures. Consequently, we call the specialized glia MORPHOCYTES.

There is structural variability between the pons and spinal cord. In the PONS (top panel) there are two cell-dense layers near the ventricular surface. Since mitotic figures are rare in the layer adjacent to the ventricle, it may not be an active germinal zone generating glia. However, at earlier stages of development (to be described in Volume 5), this layer is full of stem cells generating midline raphe glia. The second layer is most likely the densely packed cell bodies that have long fibrous processes extending to the pial surface rather than a premigratory sojourn zone. Morphocytes may be predominantly nonmigratory cells that differentiate at the site of their generation (similar to the ependymal cells that eventually line the ventricular system).

In the SPINAL CORD (bottom panel) there is one cell-dense layer adjacent to the central canal. These are the cell bodies with relatively short (compared to the pons) fibrous processes that extend to the pial surface.

In both regions, there are widely scattered cells dispersed between the fibers. These are most likely other types of glial cells that do not play a structural role (nonmorphocytes) in the raphe system.

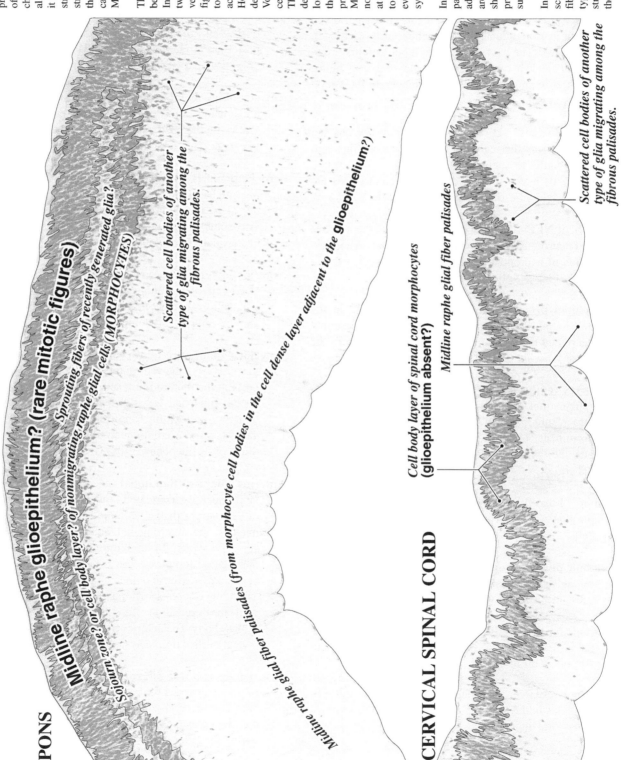

**Midline raphe glioepithelium? (rare mitotic figures)**

*Sojourn zone? or cell body layer? of nonmigrating raphe glial cells (MORPHOCYTES)*

*Sprouting fibers of recently generated glia?*

*Scattered cell bodies of another type of glia migrating among the fibrous palisades.*

*Midline raphe glial fiber palisades (from morphocyte cell bodies in the cell dense layer adjacent to the glioepithelium.?)*

## CERVICAL SPINAL CORD

*Cell body layer of spinal cord morphocytes (glioepithelium absent?)*

*Midline raphe glial fiber palisades*

*Scattered cell bodies of another type of glia migrating among the fibrous palisades.*

# GLOSSARY

## A

**Abducens nucleus (VI)** – An aggregate of cranial nerve motor neurons situated beneath the fourth ventricle in the *pons. The nucleus receives input from the *vestibular nuclear complex and is the source of motor fibers of cranial *nerve VI that innervate the lateral rectus muscle of the eye.

**Accessory nucleus (XI)** – A column of motor neurons that extends from the region of the *nucleus ambiguus in the medulla to segments 5-6 of the cervical spinal cord. Its axons form *nerve XI that innervates the sternocleidomastoid and trapezius muscles.

**Accumbent NEP** – Neuroepithelial source of the neurons and neuroglia of the *nucleus accumbens. After cessation of neurogenesis, this germinal matrix is transformed into a *glioepithelium.

**Alisphenoid process** – A cartilaginous structure in the developing skull that surrounds part of the trigeminal nerve and the facial nerve; eventually it becomes the posterior wing of the *sphenoid bone.

**Allocortex (embryonic)** – Portion of the cerebral hemispheres that, in contrast to the *neocortex, develops into a "three-layered" (oligolaminar) cortex. Among prominent allocortical regions are the *hippocampus, the *primary olfactory cortex, and the *entorhinal cortex.

**Ammon's horn (hippocampus)** – Part of the *hippocampus that contains a prominent layer of large pyramidal cells.

**Ammonic migration and sojourn zone** – Migrating pyramidal cells originating in the *ammonic NEP that settle in *Ammon's horn.

**Ammonic NEP** – Neuroepithelial patch of the *hippocampal NEP, the putative source of the pyramidal cells of Ammon's horn.

**Amygdala** – A large subcortical structure with several subdivisions in the temporal lobe, implicated in adults in behavioral aggression.

**Amygdaloid NEP** – Neuroepithelial lining of the anterior wall of the posteroventral *telencephalic superventricle, the presumptive site of origin of neurons and glia of the *amygdala. It is continuous rostrally with the posterior *striatal NEP and caudally with the ventral *hippocampal NEP.

**Amygdalohippocampal area** – Also called the cortical amygdaloid transition area, connected in the mature brain with the *bed nucleus of the stria terminalis and the *ventromedial hypothalamic nucleus.

**Ansa lenticularis** – Fiber tract that originates in the internal (medial) segment of the *globus pallidus, courses dorsal to the *zona incerta, and terminates in the *thalamus, in particular the *ventral complex and the *centromedian nucleus.

**Anterior amygdaloid area** – A region of small to medium-sized cells in the mature *amygdala that represents a transition zone between the *substantia innominata and the amygdaloid complex proper.

**Anterior commissure (embryonic)** – Large fiber bundle that crosses in the ventral telencephalon and interconnects in the mature brain several forebrain structures on the right and left sides, including the *primary olfactory cortex, the *entorhinal cortex, the *amygdala, and some components of the *temporal lobe. Some of its fibers have crossed the midline by GW9.

**Anterior complex (thalamus)** – A group of anterior thalamic nuclei with related connections. Components of the anterior thalamic complex are the anterodorsal nucleus, the anteromedial nucleus, and the anteroventral nucleus. In he mature brain, the afferents of the anterior complex come principally from the *hippocampal region and the *mammillary body. The ascending efferents terminate in the *cingulate gyrus, while the descending efferents terminate in the *mammillary body. The anterior thalamic complex is a component of the "limbic system."

**Anterior extramural migratory stream** – Large stream of young neurons migrating anteroventrally from the *anterior precerebellar NEP to the *pontine gray and the *reticular tegmental nucleus. The stream forms during the latter part of the late first trimester.

**Anterior pituitary gland (embryonic)** – The anterior lobe of the pituitary gland, also known as the adenohypophysis. It is derived from Rathke's pouch during embryonic development.

**Anterior precerebellar NEP** – Neuroepithelial source of neurons of the *pontine gray and *reticular tegmental nucleus. It lines the *rhombencephalic superventricle dorsally and the young neurons migrate to their target structures by way of the *anterior extramural migratory stream.

**Aqueduct (embryonic)** – *See* **Mesencephalic superventricle**

**Arcuate nucleus (hypothalamus)** – A small-celled nucleus that surrounds the base of the third ventricle posteriorly. It contains releasing hormones and is involved in the central nervous regulation of the anterior pituitary gland.

**Atlas** – The first cervical vertebra.

**Auditory germinal trigone** – Proliferative site in the *rhombencephalon that contains the NEP and the external germinal layer of the *cochlear nuclei and contributes stem cells to the rhombencephalic *choroid plexus.

**Axis** – The second cervical vertebra.

# B

**Basal ganglia** – A broad term that includes three large ganglionic (subcortical) components of the telencephalon, the *caudate nucleus, the *putamen, and the *globus pallidus. The latter two are also referred to as the *striatum. Basal ganglia pathologies have been linked to Parkinsonism, Huntington's disease, and other motor abnormalities.

**Basal ganglionic NEP and SVZ** – Composed of three large hillocks or eminences protruding into the *telencephalic superventricle – the anterolateral, the anteromedial, and the posterior – and contain the neuroepithelial and subventricular stem cells that furnish neurons and neuroglia to the *basal ganglia. A fourth component is the cortico-ganglionic NEP/SVZ. The SVZ is far more prominent in the basal ganglia than in the developing *cerebral cortex.

**Basal nucleus (amygdala)** – The largest nucleus in the amygdala that forms a major part of the *basolateral complex. It is separated from the *lateral nucleus by a thin fibrous band.

**Basal nucleus of Meynert** – Large-celled component of the *substantia innominata that provides in the mature brain cholinergic input to the *cerebral cortex.

**Basal occipital** – The occipital bone surrounding the *foramen magnum.

**Basal telencephalic NEP and SVZ** – Putative source of neurons and neuroglia of the *basal nucleus of Meynert and the *substantia innominata.

**Basolateral complex (amygdala)** – The largest and best differentiated part of the amygdala in humans. Its principal components are the *basal nucleus and the *lateral nucleus.

**Bed nucleus of the stria terminalis** – A large subcortical telencephalic field with indistinct boundaries. It is situated medial to the *globus pallidus, lateral to the *septum, and is transected by the *anterior commissure; a thin portion extends back to the *amygdala adjacent to the *stria terminalis. It has its own germinal source, the *strionuclear NEP and GEP.

**Boundary cap** – A thin ring of proliferative cells surrounding the spinal and cranial nerves at the site they enter the central nervous system.

**Brachium of the inferior colliculus** – A fiber tract situated superficially in the fibrous layer covering of the *inferior colliculus. It is composed of ascending auditory fibers from the inferior colliculus and auditory nuclei in the *pons to the *medial geniculate body.

# C

**Calcarine sulcus (embryonic)** – Cortical fissure in the occipital lobe; its wall is the target of the visual fibers from the *lateral geniculate body. It is recognizable in incipient form by the end of the first trimester.

**Caudate nucleus** – Elongated and arched component of the *basal ganglia beneath the *cerebral cortex. It abuts the lateral ventricle and extends from anterodorsal (its "head") to posteroventral (its "tail").

**Cajal-Retzius cells** – The earliest-generated *cerebral cortex neurons. Their perikarya are oriented parallel to the pial surface in *layer I.

**Central autonomic area (spinal cord)** – Region of the spinal cord that surrounds the *central canal and is implicated in nociceptive and autonomic functions. It may be continuous rostrally with the periaqueductal *central gray.

**Central canal (embryonic)** – Portion of the ventricular system that extends caudally from the cervical to the sacral segments of the *spinal cord. During embryonic development, the proliferative *neuroepithelium lining this canal is the source of neurons and neuroglia of the spinal cord. After the cessation of neurogenesis, the shrunken central canal is lined by the *ependyma.

**Central complex (thalamus)** – A group of contiguous central thalamic nuclei, including the *centromedian, central lateral, and paracentral nuclei.

**Central gray (periaqueductal)** – Oval shaped region in the core of the mesencephalon that surrounds the *aqueduct and is capped by the *superior colliculus and the *inferior colliculus.

**Central nucleus (amygdala)** – Part of the *corticomedial complex that is sometimes put in a class by itself. A large nucleus situated lateral to the medial nucleus that extends from the *anterior amygdaloid area to the caudal pole of the amygdala where it blends with the *putamen and the tail of the *caudate nucleus.

**Central nucleus (inferior colliculus)** – Laminated core of the *inferior colliculus where auditory fibers of the *lateral lemniscus terminate in a tonotopic order.

**Centromedian nucleus (thalamus, embryonic)** – Large spherical structure surrounded by fibers of the internal medullary lamina, classified with the *central complex of the thalamus. It is a paleothalamic structure that has extensive connections with the *striatum and the *reticular formation of the midbrain and is prominent by the end of the first trimester.

**Cerebellar NEP** – An extensive neuroepithelial matrix that lines the *rhombencephalic superventricle lateral to the *dorsal rhombic lip. It is the direct source of the neurons of the *cerebellar deep nuclei and the *Purkinje cells, and an indirect source of its basket, stellate and granule cells, by way of its secondary proliferative matrix, the *external germinal layer.

**Cerebellar transitional field (CTF)** – Transient cellular and fibrous layers, composed of migrating deep neurons and Purkinje cells, and of exiting and entering fiber tracts, prior to the formation of the cerebellar cortex during the second trimester.

**Cerebellum (deep nuclei)** – Three pairs of ganglionic structures in the depth of the cerebellum: the medial *fastigial nucleus; the intermediate *interpositus nucleus, and the lateral *dentate nucleus. The efferent fibers of cerebellar *Purkinje cells synapse with the neurons of the cerebellar deep nuclei which, in turn, are the source of cerebellofugal fibers that terminate in structures outside the cerebellum. The early-generated deep nuclei neurons are mostly situated superficially (above the later generated Purkinje cells) during the late first trimester.

**Cerebellum (hemisphere)** – Portion of the cerebellum situated on either side of the midline *vermis. It is prominent in the cerebellum in higher mammals and man.

**Cerebellum (vermis)** – Midline portion of the cerebellar cortex. It is a relatively small part of the cerebellum in higher mammals and man.

**Cerebral cortex (embryonic)** – The expanding and differentiating bilateral brain tissue covering the lateral, dorsal and medial aspects of the telencephalic superventricles. It has two major components, the *neocortex and the *allocortex. The *stratified transitional field and the *cortical plate begin to form during the late first trimester.

**Cerebral peduncle (embryonic)** – Fibrous region along the ventrolateral aspect of the diencephalon and mesencephalon, containing fibers of the *corticofugal tract. It begins to form by the end of the first trimester and some of its fibers reach the formative *pontine gray.

**Choroid plexus (embryonic)** – Glycogen-rich epithelial tissue that forms and begins to expand during the later first trimester in the *rhombencephalic and *telencephalic superventricles. It is formed by proliferative stem cells associated with the cerebellar *germinal trigone and an analogous germinal site in the *hippocampus. The fetal choroid plexus may play a role in the anaerobic metabolism of the early developing brain. During late-fetal development, it becomes gradually transformed into the mature choroid plexus of the shrunken fourth ventricle, with a different cellular composition and perhaps a different function.

**Cingulate cortex** – A rostrocaudally extending medial region of the cerebral cortex with an *allocortical organization. By virtue of its structure and principal connections, it is considered a component of the "limbic system."

**Cingulate NEP** – Long neuroepithelial stretch along the medial bank of the *telencephalic superventricle that generates the neurons and glia of the *cingulate cortex.

**Claustrum (embryonic)** – Subcortical gray matter adjacent to the *insula. During embryonic development, it is in the path of the *lateral migratory stream.

**Cochlear NEP** – Neuroepithelial patch in the vicinity of the *ventral rhombic lip, the putative source of neurons and glia of the ventral and dorsal *cochlear nuclei.

**Cochlear nuclei** – The dorsal and ventral auditory nuclei are targets of *nerve VIII auditory fibers. Both nuclei contribute axons to the *lateral lemniscus that terminate in the *nuclei of the lateral lemniscus, the *inferior colliculus, and the *medial geniculate body.

**Cortical NEP** – Extensive and continuous neuroepithelial lining of the lateral, dorsal and medial banks of the *telencephalic superventricle. It is the sole constituent of the *cerebral cortex during the early embryonic period but, following a strict timetable and spatial gradient, it expands and then shrinks as a class of differentiating neurons and glia leave it to enter the *stratified transitional field and migrate to the *cortical plate. The cortical NEP is also the source of a secondary proliferative matrix, the *subventricular zone, and of a fate-restricted *glioepithelium and the cells that line the enduring *ependyma.

**Cortical nucleus (amygdala)** – Also called the periamygdaloid cortex, part of the *corticomedial complex in the superficial amygdala.

**Cortical plate** – The densely packed cellular band in the embryonic and fetal *cerebral cortex that later becomes the laminated *gray matter. It is situated between the future molecular layer, or *layer I and the subplate (the future layer VII).

**Corticofugal fibers (embryonic)** – Collective term for the efferent fiber system (the traditional pyramidal tract) that originates in the cerebral cortex and terminates in subcortical structures. It is known by different names along its path from rostral to caudal: *internal capsule, *cerebral peduncle, *transpontine corticofugal tract, and corticospinal tract. The corticofugal tract begins to form during the end of the late first trimester but its fibers do not reach the spinal cord before the end of the second trimester.

**Corticoganglionic NEP** – A transitional neuroepithelium at the junction between the *cortical NEP and the *basal ganglionic NEP and SVZ.

**Corticomedial complex (amygdala)** – Portion of the *amygdala that includes the *anterior amygdaloid area, the *central nucleus, and the *cortical nucleus.

**Cricoid cartilage** – A ring-shaped cartilage that forms the base of the larynx.

**Cuneate fasciculus** – A large fiber tract in the dorsolateral *spinal cord. It is composed of the ascending branch of the primary sensory axons of dorsal root ganglion cells that terminate topographically in the *cuneate nucleus.

**Cuneate nucleus (embryonic)** – The target of the *cuneate fasciculus in the lower *medulla and the source of second-order somatosensory fibers that cross the midline and enter the contralateral *medial lemniscus. It is recognizable by GW8.

## D

**Dentate gyrus** – Curved small-celled component of the *hippocampus, interlocked with the large-celled *Ammon's horn. Although the *dentate migration is recognizable by the end of the first trimester, the blades of the dentate gyrus do not form until the second trimester.

**Dentate migration** – Precursors of granule cells of the *dentate gyrus that migrate from a patch of the *hippocampal NEP to form what will later become the secondary germinal matrix of the hippocampus, the *subgranular zone.

**Dentate nucleus (embryonic)** – Lobulated and largest of the *cerebellar deep nuclei in the core of the cerebellar hemispheres, it is the principal source of efferent fibers of the *superior cerebellar peduncle. In the embryonic cerebellum, the early-generated neurons of the dentate nucleus are situated superficially and do not descend until after the late-generated *Purkinje cells migrate toward the surface to form the cerebellar cortex toward the early fetal period.

**Diagonal band of Broca** – Oblique nucleus situated ventral to the medial *septum. It is subdivided into a vertical limb dorsally and a horizontal limb ventrally.

**Diencephalic NEP** – The extensive neuroepithelial lining of the *diencephalic superventricle. Its different mosaic components, distinguished by their bilaterally symmetrical variable thicknesses and evaginations or invaginations into the ventricle, are the source of neurons and glia of the different nuclei of the *diencephalon.

**Diencephalic superventricle** – Large midline component of the embryonic ventricular system that is later reduced to the small and narrow third ventricle. It is confluent laterally, by way of the foramen of Monro, with the *telencephalic superventricle, and caudally with the *mesencephalic superventricle. Its lining, the *diencephalic NEP, is the source of all the neurons and glia of the *diencephalon.

**Diencephalon (embryonic)** – Extensive forebrain region flanked laterally by the telencephalon and continuous caudally with the mesencephalon. Among its larger components are the *epithalamus and the *thalamus dorsally; the *preoptic area, *hypothalamus, and *subthalamus ventrally; and the *pretectum caudally. Its development precedes that of dorsal *telencephalon.

**Dorsal complex (thalamus)** – Collective term for two structurally and functionally related dorsally situated thalamic regions, the *dorsomedial nucleus and the dorsolateral nucleus.

**Dorsal funiculus** – *See* **Dorsal white matter**

**Dorsal gray matter (spinal cord)** – Wing-shaped region of the spinal gray matter, the target of the local collaterals dorsal root afferent fibers. Its principal component is the small-celled substantia gelatinosa. The neurons of the dorsal gray matter originate in the *neuroepithelium flanking the transient dorsal spinal canal.

**Dorsal hippocampus (embryonic)** – The portion of the *hippocampus located above the thalamus during the early stages of embryonic development. In the late second trimester, this region is displaced by the expanding *cerebral cortex laterally and then ventrally.

**Dorsal motor nucleus (X) (embryonic)** – An early-forming column of parasympathetic preganglionic motor neurons dorsolateral to the *hypoglossal nucleus. Their axons leave the brain in cranial *nerve X and terminate in the parasympathetic ganglia supplying the viscera of the thoracic, pericardial, and abdominal cavities. *See also* **Nerve X**.

**Dorsal rhombic lip** – The dorsolateral anchoring point of the *medullary velum that covers the expanding *rhombencephalic superventricle. Following the formation of the cerebellar *external germinal layer by the end of the first trimester, it is recognized as the *germinal trigone. The *ventral rhombic lip forms the ventrolateral anchoring point of the medullary velum.

**Dorsal sensory nucleus (X) (embryonic)** – A medial nucleus in the early-forming *solitary nuclear complex of the *medulla that lies dorsolateral to the *dorsal motor nucleus (X). *See also* **Nerve X**.

**Dorsal tegmental nucleus** – Situated in the central gray dorsal to the *trochlear nucleus and extending caudally into the pons. It is targeted by fibers of the *mammillotegmental tract.

**Dorsal white matter (spinal cord)** – Medial fibrous component of the white matter situated between the wings of the *dorsal gray matter; also known as the dorsal column or the dorsal funiculus. It contains ascending somatosensory and proprioceptive fibers that terminate in the dorsal column nuclei of the medulla. In the upper spinal cord it has two distinguishable parts, the *gracile fasciculus medially, and the *cuneate fasciculus laterally.

**Dorsomedial nucleus (hypothalamus)** – Area situated above the more distinct *ventromedial nucleus of the hypothalamus. Its principal connections are with the *bed nucleus of the stria terminalis and the *septum.

**Dorsomedial nucleus (thalamus)** – Also known as the medial dorsal nucleus, this component of the *dorsal complex is situated between the internal medullary lamina and the periventricular gray. Its principal connections are with the *amygdala, the *hypothalamus, the *olfactory tubercle, and the *orbital gyrus. It is considered a paleothalamic component of the limbic system rather than as a neothalamic relay nucleus to the neocortex.

# E

**Endopiriform nucleus (embryonic)** – Small telencephalic nucleus deep to the *primary olfactory cortex and ventral to the *claustrum. The *lateral migratory stream percolates through this nucleus during development.

**Entorhinal cortex** – Allocortical component of the *parahippocampal gyrus. It is bordered internally by the *subicular complex and is separated from the *neocortex in the maturing brain by the rhinal sulcus. It is the source of the perforant pathway to the *hippocampus.

**Ependyma** – Layer of cuboidal cells that line the lumen of the permanent brain *ventricles and *central canal after dissolution of the proliferative *neuroepithelium.

**Epithalamic NEP** – Neuroepithelial patch of the *diencephalic NEP, the putative source of neurons and glia of the *epithalamus.

**Epithalamus** – Collective term for the region of the dorsal diencephalon consisting of the *habenular nuclei, the stria medullaris, and the *habenulo-interpeduncular tract.

**Ethmoid** – A cranial bone that lies beneath the olfactory bult and forms part of the nasal septum.

**Eustachian tube** – A canal between the oral cavity and the middle ear.

**External capsule** – Slender fiber band situated between the *claustrum and the *putamen.

**External cuneate nucleus** – Situated lateral to the *cuneate nucleus in the *medulla, also known as the accessory cuneate nucleus, it is the source of the cuneocerebellar tract. The nucleus relays somesthetic and proprioceptive information from anterior regions of the body to the *cerebellum.

**External germinal layer (cerebellum)** – Subpial, secondary germinal matrix of the cerebellar cortex, the source of its late-differentiating granule, stellate, and basket cells. It begins to form at the end of the first trimester and persists as a source of neurons over the surface of the human cerebellar cortex until the end of the second year of postnatal life.

# F

**Facial motor nucleus** – A large aggregate of somatic motor neurons in the ventrolateral pons dorsal to the *superior olive complex. It is the source of the motor fibers of *nerve VII that innervate the facial mimetic muscles. Subdivisions of this nucleus innervate different facial muscles.

**Fastigial nucleus (cerebellum)** – A deep nucleus of the *cerebellum, also known as the medial cerebellar nucleus. It is the target of Purkinje cell axons that originate in the *vermis. Its axons contribute to the large efferent system that leaves the cerebellum.

**Foramen magnum** – The large opening in the occipital bone that surrounds the spinal cord.

**Foramen of Monro (embryonic)** – Large bilateral channels that connect the paired *telencephalic superventricles with the midline *diencephalic superventricle.

**Forel's fields (embryonic)** – Early differentiating subthalamic tegmental field (H1 and H2) traversed by fibers of the *ansa lenticularis, lenticular fasciculus, and subthalamic fasciculus.

**Fornical GEP** – Fate-restricted germinal extension of the hippocampal NEP, the glioepithelium that surrounds the fornix. It may be the germinal source of the oligodendrocytes of the fornix.

**Fornix** – An early forming fiber tract of the *hippocampus that distributes fibers in the mature brain to the *septum and the thalamic *anterior complex, and terminates in the *mammillary body.

**Fourth ventricle (embryonic)** – *See* **Rhombencephalic superventricle**

**Frontal bone** – Part of the skull that lies over the frontal lobes of the cerebral cortex.

**Frontal lobe** or **cortex (embryonic)** – Region of the developing *cerebral cortex that will grow into the large frontal lobe anterior to the central sulcus. The *orbital cortex is sometimes distinguished as a separate component.

**Frontal NEP** – Long stretch of the cortical neuroepithelium that is presumed to generate the cells of the *frontal lobe. Before settling in the *cortical plate, the migrating and sojourning neurons form a site-specific *stratified transitional field.

**Frontal nasal process** – A primordium in the immature skull of the future frontal and nasal bones.

# G

**Ganglionic narrows** – Ventricular region where the *basal ganglionic eminences protrude into the lumen of the *telencephalic superventricle.

**Germinal trigone (cerebellum)** – Proliferative germinal matrix of the *dorsal rhombic lip with three prongs: the *cerebellar NEP, the *external germinal layer, and the stem cells of the rhombencephalic *choroid plexus.

**Glioepithelium (GEP)** – Fate-restricted transient germinal matrix in the developing brain, the presumed source of neuroglia (astrocytes and oligodendroglia) precursors. There are two types of glioepithelia, the *perifascicular GEP that surround fiber tracts, such as the *fornical GEP, and another that covers the surface of the brain, the *subpial granular layer. Glioepithelia are easiest to recognize without special glial markers at sites of considerable distance from neuronal aggregates or their migratory routes.

**Glioepithelium/Ependyma (G/EP)** – Transient proliferative linings of the ventricle that endure into adulthood and give rise to both glia and cells of the *ependyma.

**Globus pallidus** – Component of the *striatum, situated medial to the *putamen, with an external (lateral) and internal (medial) segment. Pallidal fibers are the principal efferents from the striatum to the *thalamus, the *subthalamus, and the *tegmentum.

**Gracile fasciculus** – A large fiber tract in the dorsomedial *spinal cord. It is composed of the branch of axons of dorsal root ganglia that convey somatosensory input from lower parts of the body. The fibers terminate topographically in the *gracile nucleus.

**Gracile nucleus (embryonic)** – Gray mass in the core of the *gracile fasciculus that receives input from that fiber tract. Axons of these neurons cross the midline in the *medulla, enter the contralateral *medial lemniscus. It is recognizable by GW8.

**Gray matter (embryonic)** – General term for the laminated component of the mature cerebral cortex with a high concentration of neuronal cell bodies and nerve processes but few myelinated fibers. In the embryonic cerebral cortex, the unlaminated cortical gray matter is known as the *cortical plate.

# H

**Habenular nuclei (complex)** – Mediodorsally situated nuclei in the posterodorsal *thalamus; sometimes distinguished from the thalamus proper as the *epithalamus. Its two distinct components are the medial and the lateral habenular nuclei. Habenular afferents come principally from the *septum and the *hypothalamus, and its efferents form the *habenulo-interpeduncular tract.

**Habenulo-interpeduncular tract (embryonic)** – An early forming fiber bundle, also known as the fasciculus retroflexus, that originates in the *habenular nuclei. It courses through the posterior *thalamus and terminates in the *interpeduncular nucleus of the midbrain.

**Hippocampal NEP** – Long medial stretch of the *cortical NEP, the putative neuroepithelial source of neurons and glia of the hippocampus. It has three distinctive parts, the *ammonic NEP, the *dentate migration, and the *fornical GEP.

**Hippocampal region** – An inclusive term (also called the hippocampal formation) that includes not only the *hippocampus proper but also the *subicular complex and other components of the *parahippocampal cortex.

**Hippocampus** – A distinctive allocortical (three-layered or oligolaminar) region formed by the interlocking *dentate gyrus and *Ammon's horn. The hippocampus is continuous with the *subicular complex. The principal afferents of the hippocampus travel in the alveolar and perforant paths; its efferents leave by way of the *fimbria that join the *fornix.

**Honeycomb matrix** – Strands of radially oriented fibers surrounded by migrating cells that may be present by the end of the first trimester in the *superior colliculus

and forms during the second trimester in the *stratified transitional field of the sensory areas of the developing *cerebral cortex.

**Hook bundle (embryonic)** – Intracerebellar decussating fiber tract that originates in the deep nuclei of the cerebellum and presumed to leave it contralaterally as the uncinate fasciculus. A prominent fibrous region identified at the base of the cerebellum in GW8-GW9 specimens may be the sprouting fibers of this tract.

**Hypoglossal nucleus (XII)** – A column of somatic motor neurons near the floor of the *fourth ventricle in the caudal medulla. Their axons form cranial *nerve XII that innervate the intrinsic and extrinsic muscles of the tongue.

**Hyoid bone** – A small bone embedded in the throat muscles that lies below the tongue and above the larynx.

**Hypothalamic NEP** – Stretch of the *diencephalic NEP, situated posterior to the preoptic NEP and ventral to the subthalamic NEP. It is the source of neurons and glia of the nuclei of the *hypothalamus.

**Hypothalamus (embryonic)** – Early differentiating large diencephalic region that surrounds the ventral (or hypothalamic) portion of *diencephalic superventricle. It is continuous anteriorly with the *preoptic area and merges caudally with the midbrain *tegmentum. The hypothalamus contains a large number of discrete nuclei, among them the *suprachiasmatic nucleus, the *supraoptic nucleus, the *paraventricular nucleus, the *arcuate nucleus, the *ventromedial nucleus, the *dorsomedial nucleus, the *lateral tuberal nucleus, and the *mammillary body.

# I

**Inferior cerebellar peduncle** – Large fiber tract, also known as the restiform body. It contains ascending afferents to the cerebellum from the *spinal cord, the *external cuneate nucleus, the *inferior olive, and the *lateral reticular nucleus.

**Inferior collicular NEP** – Distinctive neuroepithelial component of the *tectal NEP surrounding the posterior pool of the mesencephalic syperventricle and the source of neurons of the *inferior colliculus.

**Inferior colliculus** – Paired inferior hillocks of the midbrain *tectum that receive primary, secondary, and higher order auditory afferents. The output of the inferior colliculus is mainly to the *medial geniculate body in the thalamus, but some axons extend to the primary auditory cortex in the *temporal lobe.

**Inferior olive** – A distinctive convoluted region in the ventrolateral *medulla, with the large principal nucleus and the smaller accessory nuclei. Its axons join the contralateral *inferior cerebellar peduncle and terminate in the *cerebellar cortex as climbing fibers.

**Inferior olive (embryonic)** – A compact cell aggregate in the lower *medulla, formed by neurons of the *posterior intramural migratory stream during the late first trimester. Its lamination does not begin until the end of the second trimester.

**Inferior vestibular nucleus** – This nucleus begins caudally in the *medulla near the *external cuneate nucleus and extends rostrally along the medial border of the *inferior cerebellar peduncle.

**Infundibular recess** – Small recess of the third ventricle that evaginates into the *infundibulum.

**Infundibulum** – Stalk extending from the ventral *hypothalamus that forms a link with the pituitary gland.

**Insula (insular cortex)** – Large buried neocortical region that is continuous internally with the *frontal, *parietal and *temporal lobes.

**Interhemispheric fissure** – Longitudinal cleft that separates the two cerebral hemispheres. The corpus callosum that traversed it in the maturing brain has not yet started to form in the first trimester.

**Internal capsule (embryonic)** – Massive fiber tract between the *thalamus and *striatum, composed of *thalamocortical and *corticofugal fibers. It is continuous with the incipient *cerebral peduncle caudally.

**Interpeduncular nucleus** – Midline mesencephalic structure above the interpeduncular fossa and between the *cerebral peduncles. It is the target of fibers of the *habenulo-interpeduncular tract.

**Interpositus nucleus (cerebellum)** – A deep cerebellar nucleus located between the *dentate nucleus and the *fastigial nucleus. It contains a lateral and a medial group of neurons, the emboliform near the *dentate nucleus, and the globosus near the *fastigial nucleus.

**Isthmal canal** – Channel that interconnects the *mesencephalic and *rhombencephalic superventricles.

**Isthmal NEP** – The putative source of neurons and glia of the *isthmus.

**Isthmus** – Transitional brain region between the *midbrain tectum, the *pons, and the *cerebellum.

# L

**Lateral geniculate body** – Large posteroventral thalamic region, composed in the maturing brain of the laminated dorsal lateral geniculate nucleus and its capsule of ipsi- and contralateral *optic tract fibers. Its axons project, by way of the *visual radiation, to the ipsilateral *occipital lobe.

**Lateral geniculate nucleus (embryonic)** – The neurons of this prominent thalamic nucleus appear to be generated dorsally in a distinct neuroepithelial locus and migrate ventrolaterally where they meet the incoming *optic tract fibers.

**Lateral hypothalamic area** – An ill-defined fibrous region of the *hypothalamus with scattered neurons medial to the cerebral peduncle. It is traversed by many fiber tracts, including the *medial forebrain bundle.

**Lateral lemniscus** – The fiber tract on the lateral surface of the *pons that contains secondary auditory fibers from the dorsal and ventral *cochlear nuclei and higher-order auditory fibers from the *superior olivary complex. The dorsal and ventral *nuclei of the lateral lemniscus are embedded within the fiber tract.

**Lateral migratory stream (cortical)** – Tangentially migrating neurons and glia in the developing *cerebral cortex that leave dorsal *cortical NEP and migrate laterally and ventrally to the *insula, the *temporal lobe, and other telencephalic structures that lack a nearby germinal matrix. The bulk of the lateral migratory stream follows a trajectory outlined by the receding *subventricular zone between the *basal ganglia and the lateral cortex.

**Lateral nucleus (amygdala)** – The most lateral nucleus in the *basolateral complex of the *amygdala. Cells from the *lateral migratory stream appear to enter the nucleus at the distinctive saw-toothed lateral edge.

**Lateral olfactory tract** – *See* **Olfactory tract**

**Lateral preoptic area** – An anterior continuation of the *lateral hypothalamic area in the *preoptic area.

**Lateral reticular nucleus** – A relatively discrete group of neurons in the caudal medulla, dorsolateral to the *inferior olive. The neurons of this precerebellar relay nucleus receive topographic exteroceptive and proprioceptive afferents from the *spinal cord and project ipsilaterally to the *cerebellum via the *inferior cerebellar peduncle.

**Lateral septal nucleus** – An indistinct gray mass in the lateral *septum that is closely associated with the *fornix, which provides input to these neurons.

**Lateral tuberal nucleus (hypothalamus)** – Two or three distinct spherical masses near the inferior surface of the *lateral hypothalamic area.

**Lateral ventricle**s – *See* <u>**Telencephalic superventricle**</u>

**Lateral vestibular nucleus** – Also called Deiter's nucleus. A collection of large neurons lying dorsolaterally along the wall of the *fourth ventricle. It receives primary sensory input from the vestibular ganglion via cranial *nerve VIII and its large neurons are the source of the vestibulospinal tract.

<u>**Layer I (embryonic)**</u> – Cell sparse layer beneath the pia throughout the *cerebral cortex. This is the first cortical layer to develop and contains the earliest generated cortical neurons, including the *Cajal-Retzius cells.

**Locus coeruleus** – Aggregate of large pigmented cells in the *pons. It is the major source of ascending and descending noradrenergic fibers that are widely distributed throughout the central nervous system.

<u>**Luysian NEP**</u> – Putative source of neurons of the so-called *subthalamic nucleus (corpus Luysi) in the *hypothalamic NEP. The intramural migration of these neurons has been traced from the region of the formative *mammillary body to the level of the subthalamus dorsolaterally.

# M

**Mammillary body** – Distinctive region in the posteroventral hypothalamus, composed of the medial and lateral mammillary nuclei. Its principal afferents are from the *septum and *subiculum that course in the *fornix; its efferents form the *mammillothalamic and *mammillotegmental tracts.

**Mammillotegmental tract** – Descending fiber bundle containing *mammillary body efferents to the brain stem, including the *dorsal tegmental nucleus.

**Mammillothalamic tract** – Ascending fiber bundle containing efferents of the *mammillary body that terminate in the thalamic *anterior complex.

**Mandible** – The lower jaw bone.

**Maxilla** – The upper jaw bone.

<u>**Meckel's cartilage**</u> – Cartilage of the first pharyngeal arch that is associated with the formation of the mandible (lower jaw bone) and the ossicles of the middle ear.

**Medial accessory olive** – A small nucleus in the *inferior olive complex that contains densely packed neurons along the lateral border of the medial lemniscus. It receives proprioceptive input from the spinal cord and its efferents reach the contralateral cerebellum (mainly the vermis) by way of the *inferior cerebellar peduncle.

**Medial forebrain bundle** – A diffuse fiber tract that extends from the *olfactory tubercle, through the *lateral hypothalamic area, to the *substantia nigra in the midbrain *tegmentum.

**Medial geniculate body** – Principal thalamic relay station in the auditory pathway to the *cerebral cortex. Its afferents originate in the *trapezoid body, the *superior olivary complex, the * nuclei of the lateral lemniscus, and the *inferior colliculus. Its efferents form the auditory radiation that terminates in the *temporal lobe.

**Medial lemniscus** – Large fiber bundle conveying tactile and other somatosensory input to the thalamus. It originates in the *gracile and *cuneate nuclei in the *medulla, crosses to the opposite side, ascends through the *pons and *midbrain, and terminates in the *ventral posterolateral and *ventral posteromedial nuclei of the thalamus.

**Medial lemniscus (decussation)** – Also known as the sensory decussation, it is composed of ascending fibers of the medial lemniscus that cross to the opposite side in the medulla.

**Medial longitudinal fasciculus** – A dorsomedial tract in the *midbrain, *pons, and *medulla that contains ascending and descending vestibular fibers coursing beneath the *oculomotor nuclear complex, the *trochlear nucleus, the *abducens nucleus, and the *hypoglossal nucleus. It turns ventrally in the posterior medulla and extends into the ventral funiculus of the cervical spinal cord.

**Medial preoptic area** – A rounded mass of neurons implicated in reproductive functions.

**Medial septal nucleus** – An indistinct nucleus in the midline septum that is continuous with the vertical limb of the *diagonal band of Broca.

**Medial vestibular nucleus** – Component of the *vestibular nuclear complex situated underneath the *fourth ventricle medial to the other vestibular nuclei. Its neurons receive primary sensory input from the vestibular ganglion via cranial *nerve VIII and project to the cerebellum.

**Median preoptic nucleus** – A small nucleus in the *preoptic area that forms a narrow cap around the *anterior commissure in the midline.

**Medulla (embryonic)** – Region of the neuraxis, also known as the medulla oblongata, surrounding the posterior *rhombencephalic superventricle and bounded by the *pons rostrally and the *spinal cord caudally. An extremely heterogeneous region containing sensory, somatic motor, and visceral motor nuclei as well as several ascending, descending, and decussating fiber tracts.

**Medullary NEP** – Extensive neuroepithelial site that lines the variegated caudal bank of the *rhombencephalic superventricle. Its several subdivisions are the source of neurons and glia of the different sensory, relay and motor nuclei of the medulla.

**Medullary velum** – Membranous roof of the expanding *rhombencephalic superventricle. Its lateral anchor to the *rhombencephalon is known as the *dorsal rhombic lip and the *ventral rhombic lip, and its inner surface serves as a substratum for the expanding rhombencephalic choroid plexus.

**Mesencephalic NEP** – The extensive neuroepithelium that lines the large *mesencephalic superventricle. Its major divisions are the anterior *pretectal NEP, the dorsal *tectal NEP, the ventral *tegmental NEP, and the posterior *isthmal NEP. Subdivisions are the source, among others, of neurons and glia of the *superior colliculus and *inferior colliculus dorsally, and of the *red nucleus, the *oculomotor nuclear complex, and several tegmental nuclei ventrally.

**Mesencephalic nucleus (V)** – Large neurons scattered along the lateral border of the *central gray of the midbrain and pons. They are primary sensory neurons that enter the brain from the periphery early in development relaying proprioceptive information from the muscles of mastication.

**Mesencephalic superventricle** – Greatly inflated lumen of the embryonic *mesencephalon, situated between the *diencephalic superventricle rostrally and the *rhombencephalic superventricle caudally. The connection with the latter is by way of the *isthmal canal. It shrinks in the maturing brain into the narrow aqueduct.

**Mesencephalon (embryonic)** – Anterior part of the developing brainstem surrounding the *mesencephalic superventricle consisting of the *pretectum and *tectum dorsally, and of the *tegmentum ventrally. Among the intermediate components of the mesencephalon are the *central gray, the *oculomotor nuclear complex, the *red nucleus and the *reticular formation.

**Meyer's loop** – Part of the *visual radiation that takes a sharp curve in the *temporal lobe as it proceeds to the *occipital lobe.

**Midbrain** – *See* **Mesencephalon**

**Middle cerebellar peduncle** – Massive tract of *pontocerebellar fibers that originate in the *pontine gray and enter the cerebellum posterolateral to the *inferior cerebellar peduncle. The earliest fibers of this system emerge at the end of the first trimester.

**Motor nucleus, V** – *See* **Trigeminal, motor nucleus**

**Motor nucleus, VII** – *See* **Facial, motor nucleus**

# N

**Nasal conchae** – Scroll-like processes of the *ethmoid and *maxilla that project into the lateral nasal cavity.

**Neocortex (embryonic)** – Portion of the cerebral hemispheres that develops a "six-layered" (multilaminar) cortical *gray matter. Neocortical development begins with the expansion of the primordial *cortical NEP devoid of differentiated neurons. It is followed by the formation of the *cortical plate, the *subventricular zone, and the different layers of the *stratified transitional field, in association with the ingrowth of *thalamocortical fibers, the outgrowth of *corticofugal fibers and the onset of the formation of intracortical connections. The principal divisions of the neocortex are the *frontal lobe, the *paracentral lobule, the *parietal lobe, the *temporal lobe, and the *occipital lobe.

**Neocortical NEP** – Extensive *neuroepithelium that lines the lateral and dorsal aspects the *telencephalic superventricles. The proliferating neocortical NEP cells are the source of neurons and glia that migrate to the nearby *cortical plate by way of the *stratified transitional field. Some of its cells may move to more distant sites by way of the *lateral migratory stream.

**Nerve I** – *See* **Olfactory nerve**

**Nerve II** – *See* **Optic nerve**

**Nerve III (oculomotor)** – Cranial motor nerve originating in the *oculomotor nuclear complex. It innervates all the extraocular muscles – except the lateral rectus and superior oblique – and the skeletal muscles of the eyelid, the smooth sphincter muscles of the iris, and the ciliary muscles of the lens.

**Nerve IV (trochlear)** – Cranial motor nerve composed of axons of the *trochlear nucleus that innervates the superior oblique muscle of the eye. This nerve is unique because it exits from the dorsal surface of the *midbrain behind the *inferior colliculus.

**Nerve V and ganglia (trigeminal)**– A mixed sensory and motor cranial nerve that has three peripheral branches, the ophthalmic, the maxillary, and the mandibular. All three branches contain peripheral sensory fibers from the trigeminal ganglion that terminate in the *trigeminal principal sensory nucleus, the *trigeminal spinal nucleus, and the substantia gelatinosa in upper cervical segments of the *spinal cord. A bundle of fibers in the mandibular branch, originating in the *trigeminal motor nucleus, innervates the muscles of mastication.

**Nerve VI (abducens)** – A motor cranial nerve that originates in the *abducens nucleus and emerges near the midline at the caudal border of the *pons. The fibers innervate the lateral rectus muscle of the eye.

**Nerve VII and ganglion (facial)** – A mixed sensory and motor nerve, the facial nerve has three components. Primary sensory gustatory fibers from the geniculate ganglion enter the *solitary tract and nucleus. Somatic motor fibers from the *facial motor nucleus innervate the mimetic muscles. Visceral motor (parasympathetic) fibers from preganglionic neurons of the indistinct salivatory nucleus target the pterygopalatine and submandibular ganglia.

**Nerve VIII and ganglia (cochlear, vestibular)** – A sensory cranial nerve that contains primary auditory afferents from the spiral ganglion in the cochlea and primary vestibular afferents from the vestibular (Scarpa's) ganglion. The auditory afferents terminate in the dorsal and ventral *cochlear nuclei; the vestibular afferents terminate in the nuclei of the *vestibular nuclear complex and some reach the *cerebellum.

**Nerve IX and ganglia (glossopharyngeal)** – A mixed sensory and motor cranial nerve. The sensory part of nerve IX originates in the superior and inferior ganglia, and relays gustatory input from the posterior third of the tongue and visceral sensory input from the tonsils, the Eustachian tube, and the carotid sinus. These fibers enter the solitary tract and terminate in the *solitary nucleus. The somatic motor part of nerve IX originates in the *nucleus ambiguus and innervates the pharyngeal and laryngeal muscles. The

visceral motor fibers from parasympathetic preganglionic neurons in the salivatory nucleus terminate in the otic ganglion.

**Nerve X and ganglia (vagus)** – A mixed sensory and motor nerve, with some somatic and many visceral afferents and efferents associated with the craniosacral parasympathetic ganglia. The sensory fibers originate peripherally in the superior and inferior ganglia and are widely distributed throughout the body, including the pharynx, larynx, trachea, esophagus, and all the thoracic and abdominal viscera. They terminate centrally in the *solitary nucleus and at other medullary sites. Most of its preganglionic motor neurons are located in the *dorsal motor nucleus (X).

**Nerve XI (accessory)** – This motor nerve has a cranial and a spinal component. The cranial fibers originate in the *nucleus ambiguus and innervate the muscles of the larynx and pharynx. The spinal motor fibers originate in a motor column of the cervical *spinal cord and innervate the sternocleidomastoid and upper trapezius muscles.

**Nerve XII (hypoglossal)** – A somatic motor cranial nerve that originates in the *hypoglossal nucleus and innervates the intrinsic and extrinsic muscles of the tongue.

**Neuroepithelium (NEP)** – Pluripotential pseudostratified tissue of neural stem cells that extends from the frontal pole rostrally to the last segment of the spinal cord caudally and is the source of all neurons and neuroglia of the developing central nervous system. The NEP cells initially form the neural plate, then fold dorsally and fuse to form the neural tube (future spinal cord) caudally and the variegated cephalic vesicles (the future brain) rostrally. After that closure, the lumen of the cephalic vesicles expands enormously to form the *rhombencephalic, *mesencephalic, *diencephalic, and *telencephalic superventricles. This expansion provides the space for the mitotic division of NEP cell nuclei that have to shuttle to the fluid-filled lumen to undergo mitosis. The continuous but variegated cephalic NEP lining the ventricles has a mosaic organization, being composed of bilaterally symmetrical long and intermediate stretches, and short patches that give rise to neurons and glia of different brain regions, distinct structures, and specific cell types. Examples of long stretches are the *cortical NEP and the *cerebellar NEP; of intermediate stretches, the *thalamic NEP and the *hypothalamic NEP of the inclusive *diencephalic NEP; and of short patches, the *ammonic NEP and *dentate NEP of the inclusive *hippocampal NEP. The primary NEP is also the source of several *secondary germinal matrices that generate microneurons with locally arborizing axons. Finally, as neurogenesis winds down the pluripotential NEP is transformed at many sites into a *glioepithelium, such as the *fornical GEP, or into the *ependyma that lines the enduring ventricles.

**Nuclei of the lateral lemniscus** – Both components of this system, the dorsal nucleus and the ventral nucleus, receive their major input from the *cochlear nuclei and the *superior olivary complex by way of the *lateral lemniscus. Both nuclei are connected with the *inferior colliculus and the *medial geniculate body.

**Nucleus accumbens** – Ganglionic component of the ventral telencephalon ventromedial to the *striatum. It is distinguished from the striatum by its cellular organization, molecular composition, and intimate connections with the *hypothalamus, *amygdala, and other regions of the limbic system.

**Nucleus ambiguus** – Aggregate of somatic motor neurons that form a thin column in the ventrolateral medulla. Its axons innervate the muscles of the larynx and pharynx via *nerve IX.

# O

**Occipital lobe** or **cortex (embryonic)** – Posterior region of the developing cerebral cortex that will be the target of *visual radiation fibers from the *lateral geniculate body and the *pulvinar.

**Occipital NEP** – Putative neuroepithelial source of the neurons and glia of the occipital lobe. It is flanked in the fetal neocortex by the occipital *subventricular zone and the *stratified transitional field.

**Oculomotor nerve** – *See* **Nerve III**

**Oculomotor nuclear complex** – Situated at the base of the periaqueductal *central gray, the cell columns of this complex extend from the anterior pole of the *superior colliculus rostrally to the *trochlear nucleus caudally. Its somatic motor nuclei innervate the medial rectus, inferior rectus, superior rectus and inferior oblique muscles of the eye, and are associated with the fibers of the *medial longitudinal fasciculus. Most prominent of its autonomic (preganglionic) components is the dorsally located Edinger-Westphal nucleus.

**Odontoid process** – A tooth-like projection on the superior surface of the *axis that articulates with the *atlas.

**Olfactory bulb** – Laminated brain structure where the first-order fibers of the *olfactory nerve terminate and the second-order fibers of the *olfactory tract orig-

inate. It is composed of three classes of neurons: large mitral cells, the intermediate tufted cells, and the small granule cells.

**Olfactory NEP** – Putative source of the earlier-generated mitral and tufted cells of the olfactory bulb that lines the olfactory recess of the *telencephalic superventricle. The later-generated granule cells are supplied by the *rostral migratory stream.

**Olfactory nerve (embryonic)** – Composed of the fine axons of bipolar cells in the olfactory epithelium that terminate in the *olfactory bulb. The nerve is recognizable in GW7.5 specimens.

**Olfactory tract** – Large fiber bundle of second-order fibers that originate in the *olfactory bulb with two parts, the larger lateral olfactory stria and the smaller medial stria. The fibers of the lateral stria terminate in the *olfactory tubercle, the *primary olfactory cortex, and *corticomedial complex of the amygdala.

**Olfactory tubercle** – Allocortical area in the ventral telencephalon between the *diagonal band of Broca and the *nucleus accumbens. Its input comes mainly from the lateral *olfactory tract.

**Optic chiasm (embryonic)** – Site of crossing of fibers of the *optic nerve. Fibers from the nasal half of each retina cross here to the opposite side while those from temporal half proceed uncrossed. The earliest crossing fibers appear in GW7.5 specimens.

**Optic nerve** – This large fiber tract contains the axons of retinal ganglion cells. Beyond the *optic chiasm it is called the *optic tract.

**Optic tract** – Large bundle of crossed and uncrossed retinal afferent fibers. In the human brain the majority of the fibers terminate in the *lateral geniculate body; others proceed to the *superior colliculus, the *suprachiasmatic nucleus, the *pretectum, and some other structures.

**Orbital cortex** – Ventromedial region of the *frontal lobe with afferents from the thalamic dorsomedial nucleus and efferents to the *preoptic area and the *hypothalamus.

**Orbitofrontal NEP** – Putative source of neurons and glia of the orbitofrontal cortex.

**Orbitosphenoid** – A cartilaginous structure in the developing skull that surrounds the *optic nerve and eventually becomes the lesser wing of the sphenoid bone.

**P**

**Palatal process** – Part of the *maxilla that forms the hard palate in the roof of the mouth.

**Pallidum** – *See* **Globus pallidus**

**Parabrachial nucleus** – Dorsolateral pontine structure with indistinct boundaries that surrounds the *superior cerebellar peduncle. Its principal input comes from the *solitary nucleus and its efferents target the *ventral posteromedial nucleus of the thalamus, the *amygdala, and the *insular cortex.

**Paracentral lobule (embryonic)** – Incipient cortical region that becomes later divided by the central sulcus into the precentral gyrus and the postcentral gyrus. The term is used to distinguish this site of the presumptive motor and sensory projection areas from the *frontal lobe anteriorly and the *parietal lobe posteriorly.

**Paracentral NEP** – Putative neuroepithelium of the *paracentral lobule in the developing neocortex. It is flanked by the paracentral *subventricular zone and the paracentral *stratified transitional field.

**Parahippocampal cortex (embryonic)** – Allocortical and neocortical region between the *hippocampus and the *temporal lobe that will become the parahippocampal gyrus. Its subdivisions are the *subicular complex and the *entorhinal cortex.

**Parahippocampal NEP** – Putative source of the neurons and glia of the parahippocampal gyrus. It is flanked by the parahippocampal *subventricular zone and the parahippocampal *stratified transitional field.

**Paraventricular nucleus (hypothalamus)** – Prominent neuroendocrine structure abutting the third ventricle with a magnocellular and a parvocellular division. The large neurons of the paraventricular nucleus are the source of oxytocin and vasopressin that reach the posterior pituitary gland by axoplasmic flow. The small neurons of the nucleus are the source of releasing hormones conveyed to the portal vessels of the median eminence.

**Parietal bone** – A cranial bone in the skull that eventually lies over the parietal lobe of the cerebral cortex.

**Parietal lobe of cortex (embryonic)** – Region of the developing neocortex bounded anteriorly by the *paracentral lobule and posteriorly by the *occipital lobe.

**Parietal NEP** – Long stretch of the cortical neuroepithelium containing the neural stem cells of the *parietal lobe. It is flanked by the parietal *subventricular zone and *stratified transitional field.

**Perifascicular GEP** – Fate-restricted glioepithelium, the presumed source of oligodendrocytes that surround a fiber tract, such as the *fornical GEP.

**Periventricular complex (thalamus)** – The thalamic region surrounding the *mesencephalic superventricle that will become partitioned as the paracentral, parafascicular, paratenial, paraventricular, and reuniens nuclei. Its principal connections are with limbic system; connections with the *neocortex are sparse.

**Petrous temporal bone** – Part of the temporal bone that contains the internal ear and semicircular canals.

**Pineal gland** – Midline endocrine gland connected by its stalk to the pineal recess of the dorsal *mesencephalic superventricle. It secretes melatonin and other indoleamines. It is believed to receive indirect visual input from the retina.

**Piriform cortex** – *See* **Primary olfactory cortex**

**Pituitary gland** – *See* **Anterior pituitary gland**; **Posterior pituitary gland**

**Pons (embryonic)** – Developing brainstem region, situated between the *isthmus and the *medulla, that surrounds the anterior part of the *rhombencephalic superventricle. It contains some early ascending, descending and decussating fiber tracts, the sensory and motor nuclei of some of the cranial nerves, and the *reticular formation.

**Pontine gray (embryonic)** – This massive basal region of the *pons is just beginning to form as neurons of the *anterior extramural migratory stream start to settle and the earliest descending *corticofugal fibers reach the site. Corticofugal axons that collateralize here are the principal afferents of the pontine gray neurons that are, in turn, the source of the pontocerebellar fibers of the *middle cerebellar peduncle.

**Posterior commissure (embryonic)** – Early forming decussating fiber tract in the dorsal *mesencephalon that interconnects several nuclei in the *pretectum and *tectum.

**Posterior complex (thalamus)** – Division of the thalamus that includes the *lateral geniculate body, the *pulvinar, and the *medial geniculate body. The neurons of the lateral geniculate body and pulvinar appear to originate in a distant source and migrate from dorsal to ventral over an extended period.

**Posterior intramural migratory stream** – Stream of young neurons that migrate inside the parenchyma from their source in the posterior *precerebellar NEP dorsally, to form the *inferior olive in the ventral medulla.

**Posterior extramural migratory stream** – Stream of young neurons that originate in the posterior *precerebellar NEP, migrate outside the parenchyma, cross the midline ventrally, and settle on the opposite side to form the contralateral *external cuneate nucleus and *lateral reticular nucleus.

**Posterior pituitary gland** – The posterior lobe of the pituitary gland, also known as the neurohypophysis, is the terminal and storage site of hypothalamic neurosecretory cells.

**Precerebellar NEP** – Dorsally situated neuroepithelium that lines the *rhombencephalic superventricle in the vicinity of the ventral rhombic lip and is the source of neurons of the *precerebellar nuclei. Neurons of its rostral division migrate in the *anterior extramural migratory stream and settle in the *pontine gray and the *reticular tegmental nucleus. Neurons of its posterior division form two migratory streams, the *posterior intramural migratory stream that forms the *inferior olive, and the *posterior extramural migratory stream that crosses to the opposite side and forms the *lateral reticular nucleus and the *external cuneate nucleus.

**Precerebellar nuclei** – A series of nuclei in the *medulla and *pons that provide massive higher-order input to the *cerebellum, including the *inferior olive, the *external cuneate nucleus, the *lateral reticular nucleus, and the *pontine gray.

**Premammillary area** – Region with ill-defined boundaries anterior to the *mammillary body in the *hypothalamus.

**Preoptic area (embryonic)** – Early developing midline region surrounding the preoptic recess of the *diencephalic superventricle. It is contiguous anteriorly with the ventral telencephalon and blends posteriorly with the anterior *hypothalamus. It is implicated in the regulation of sexual behavior and reproductive functions.

**Prepositus nucleus** – Situated in the dorsomedial *medulla, it extends from the anterior part of the *hypoglossal nucleus to the posterior part of the *abducens nucleus.

**Presubiculum** – *Allocortical component of the *parahippocampal gyrus between the *subiculum and the *parasubiculum.

**Pretectum** – Dorsal area between the posterior *thalamus and the *superior colliculus with an early forming fiber system, the *posterior commissure.

**Primary olfactory cortex** – Allocortical region, also called the piriform lobe, where fibers of the lateral olfactory tract terminate. It is situated rostral to the *entorhinal cortex and includes the prepiriform area along the rhinal fissure and the periamygdaloid area.

**Primordial plexiform layer (cortical)** – The first transitional layer to appear outside of the *cerebral cortical NEP that contains *Cajal-Retzius cells and *subplate neurons. The *cortical plate forms within its boundaries later on in development.

**Principal sensory nucleus (V)** – *See* **Trigeminal, principal sensory nucleus**

**Pulvinar (thalamus)** – Large nucleus of the thalamic *posterior complex. Its subdivisions send fibers to various regions of the *parietal lobe, *occipital lobe, *temporal lobe, and frontal lobe It has been implicated in multisensory integration.

**Purkinje cells (embryonic)** – These neurons, which form a monolayer in the maturing cerebellar cortex, are generated in the *cerebellar NEP towards the end of the first trimester, subsequent to the production of the *cerebellar deep nuclei neurons. Hence they are initially situated beneath the layer of deep neurons adjacent to the cerebellar NEP. Later they migrate toward the surface of the formative cerebellar cortex as the *external germinal layer forms there.

**Putamen** – Lateral component of the *striatum. It lies between the *external capsule and the *globus pallidus. It is the major source of striatal efferents to the *thalamus, *subthalamic nucleus, *substantia nigra, and *tegmentum.

# R

**Raphe migration** – Streams of cells that originate in the dorsal *medullary NEP and are later distributed in the midline of the ventral medulla.

**Raphe nuclear complex** – Several smaller and some larger cell aggregates that extend in and near the midline from the *midbrain rostrally to the *medulla caudally. The raphe cells are the principal source of serotonin-containing fibers distributed along the entire neuraxis from the forebrain to the *spinal cord. They are involved, as neuromodulators, in the regulation of sleep, wakefulness and emotional arousal.

**Red nucleus (embryonic)** – A prominent nucleus in the maturing brain with a small-celled (parvocellular) and a large-celled (magnocellular) division. It is recognizable during the first trimester in the vicinity of a germinal patch, identified as the *rubral NEP. The associated fibers may be early elements of the *superior cerebellar peduncle.

**Reticular belt (thalamus)** – Distinctive component of the thalamus; it is coextensive with the thalamic *reticular nucleus.

**Reticular formation** – A large collection of scattered neurons, enmeshed in a complex network of fibers, in the core of the *medulla, *pons, and *mesencephalon.

**Reticular tegmental nucleus (embryonic)** – Situated dorsal to the *pontine gray, this *precerebellar nucleus, also known as the nucleus reticularis tegmenti pontis, begins to form toward the end of the first trimester ahead of the pontine gray.

**Reticular nucleus (thalamus, embryonic)** – An early formong thin belt of cells and fibers between the wall of the *thalamus and the *internal capsule. Virtually all fibers that interconnect the thalamus and the cerebral cortex traverse the thalamic reticular nucleus.

**Rhombencephalic superventricle** – The greatly inflated NEP-lined lumen of the *rhombencephalon, situated between the *isthmal canal of the *mesencephalic superventricle rostrally and the central canal of the *spinal cord caudally.

**Rhombencephalon (embryonic)** – An extremely heterogeneous hindbrain region lining the *rhombencephalic superventricle, that includes the developing *cerebellum, *pons, and *medulla.

**Rostral migratory stream** – A large stream of mitotic and postmitotic cells in the forebrain extending from the cerebral *subventricular zone to the *olfactory bulb. It is a source of late generated neurons and persists after the NEP has receded or disappeared.

**Rubral NEP** – A distinctive neuroepithelial patch lining the *mesencephalic superventricle and situated between the *tectal NEP and the *tegmental NEP. It is the putative source of neurons of the early generated neurons of the *red nucleus.

**S**

**Secondary germinal matrix** – Layer or field of proliferative precursors of neurons and glia abutting or some distance from the primary *neuroepithelium. These fate-restricted stem cells are progeny of the NEP and persist for varying periods postnatally (some into adulthood). Examples of secondary germinal matrices are the *external germinal layer of the cerebellum, the *subventricular zone of the neocortex, the *subgranular zone of the dentate gyrus, and the *striatal subventricular zone. Typically, the secondary germinal matrices are the source of late-generated short-axoned interneurons, or microneurons.

**Sella turcica** – Part of the *sphenoid bone that surrounds the pituitary gland.

**Septum** – Midline telencephalic structure with two components, the *medial and *lateral nuclei. Its principal connections are with the *hippocampus and the *hypothalamus by way of the *fornix. The septum is a focal component of the limbic system.

**Sojourn zone** – Transient cellular layers formed by neurons that halt their migration for varying periods before they proceed to their final destination. They have been recognized at various sites, among them the *stratified transitional field of the cerebral cortex and the *cerebellar transitional field. It is hypothesized that the sojourn zones are sites where transient or enduring connections are made as the coarse circuitry of a brain region is established.

**Solitary tract and nucleus (embryonic)** – The solitary tract is an early forming medullary fiber system that may contain the primary sensory fibers of cranial *nerves VII, IX, and X that convey gustatory (VII and IX) and visceral-sensory information (IX and X) to the solitary nucleus. The *dorsal sensory nucleus of nerve X and the *commissural nucleus of nerve X are part of this nuclear complex.

**Sphenoid** – The skull bone that lies mainly beneath the hypothalamus, basal telencephalon, and mesencephalon. It contains the *alisphenoid and *orbitosphenoid processes and the *sella turcica around the *pituitary gland.

**Spinal cord** – Caudal tubular component of the central nervous system that surrounds the *central canal. Its core of gray matter (the dorsal horn, intermediate gray and ventral horn) and surrounding white matter (the dorsal, lateral and ventral funiculi) blend rostrally with the lower medulla.

**Spinal nucleus (V)** – *See* **Trigeminal, spinal nucleus**

**Spinal tract (V)** – *See* **Trigeminal, spinal tract**

**Squamous temporal bone** – The flat part of the *temporal bone that covers the *temporal lobe of the cerebral cortex.

**Stratified transitional field (STF)** – Transient component of the fetal *cerebral cortex, sandwiched between the *neuroepithelium and the *cortical plate. By the second trimester (GW13), illustrated in Volume 3, it has six layers of alternating cells and fibers (STF1 to STF6) that vary in their configuration in different lobes of the cerebral cortex. In the oldest specimens illustrated in this Volume (GW11), the STF has only two layers (STF1, 4/5) or at most three layers (STF1, 4, 5). The cell-rich STF4/5 is composed of sojourning neurons, and the cell-sparse STF1 of incoming *thalamocortical fibers with possibly some outgoing *corticofugal fibers.

**Stria medullaris (thalamus)** – Mediodorsal fiber bundle in the diencephalon coursing in an anteroposterior direction and terminating in the *habenular nuclei.

**Stria terminalis** – Arched fiber bundle that originates in the *amygdala, courses along the medial surface of the *caudate nucleus, and terminates in the *bed nucleus of the stria terminalis, the anterior *hypothalamus, and the *preoptic area.

**Striatal NEP** – Primary germinal source of neurons of the *caudate nucleus, *putamen, and *globus pallidus. It has a large anterolateral and anteromedial division, also known as the lateral and medial eminences, and a small posterior division that generates the neurons of the tail of the caudate nucleus. The posterior striatal NEP is continuous with the *amygdaloid NEP

**Striatal subventricular zone (SVZ)** – A massive *secondary germinal matrix flanking the striatal NEP. It generates the bulk of the neurons of *caudate nucleus, *putamen and *globus pallidus. It may also be the source of some cortical neurons.

**Striatum** – Term used for two components of the *basal ganglia: the *caudate nucleus and the *putamen.

**Strionuclear GEP** – Fate-restricted glioepithelium, the putative source of the glia of the stria terminalis, stria medullaris, and possibly other nearby fiber tracts.

**Strionuclear NEP** – Putative neuroepithelial source of the neurons and glia of the *bed nucleus of the stria terminalis. It is situated beneath the *striatal NEP in a notch of the *lateral ventricle near the *foramen of Monro.

**Subcommissural organ** – A highly vascularized circumventricular neuroendocrine organ located beneath the *posterior commissure in the roof of the *mesencephalon.

**Subgranular zone (hippocampus)** – A long-persisting secondary germinal matrix beneath the *granular layer of the dentate gyrus, the source of late generated dentate granule cells. It is recognizable in incipient form by the end of the first trimester.

**Subicular complex** – Collective term for the *parasubiculum, the *presubiculum, and the *subiculum in the *parahippocampal cortex.

**Subpial granular layer** – Transient cellular layer between *layer I and the pia in some regions of the developing *cerebral cortex. It may be a source of cortical astrocytes.

**Subplate** – A poorly-defined layer beneath the *cortical plate. It contains neurons that are the transient pioneer residents of the cortical plate.

**Substantia innominata** – Extensive telencephalic area with indistinct boundaries beneath the *globus pallidus. A prominent component of the substantia innominata is the *basal nucleus of Meynert.

**Substantia nigra (embryonic)** – A pigmented *midbrain tegmental structure in the maturing brain at the base of the *cerebral peduncle. It has two components, the dopaminergic pars compacta, and the GABAergic pars reticulata.

**Subthalamic NEP** – Neuroepithelial patch between the *thalamic NEP and the *hypothalamic NEP. It is also identified as the Forelian NEP to distinguish it from the *Luysian NEP.

**Subthalamic nucleus** – Biconvex diencephalic structure, also known as corpus Luysii, situated above the substantia nigra between the *zona incerta and the base of the *internal capsule. It has extensive reciprocal connections with the *globus pallidus, hence it is considered a component of the *basal ganglia circuitry. Subthalamic lesions produce persistent choroid movements (hemiballism) in the arms, legs and face.

**Subthalamus (embryonic)** – Diencephalic region situated between the *thalamus dorsally and the *hypothalamus ventrally. Its major components, *Forel's fields and the *zona incerta are recognizable in late first trimester fetuses.

**Subthalamic nucleus NEP** – *See* **Luysian NEP**

**Subventricular zone (SVZ)** – Secondary germinal matrix, derived from the primary *neuroepithelium. The SVZ flanks the NEP during early development and then abuts the ependyma when the NEP dissolves. The nuclei of proliferative SVZ cells, unlike the nuclei of NEP cells, do not shuttle to the lumen of the ventricle during mitosis. Prominent SVZs in the telencephalon are found in the *cerebral cortex and the *striatum.

**Superior cerebellar peduncle (embryonic)** –A large fiber tract that originates mainly in the *dentate nucleus and *interpositus nucleus. It is present by the end of the first trimester in the formative cerebellum and appears to reach the level of the *red nucleus.

**Superior colliculus (embryonic)** – Anterior component of the *tectum (known in lower vertebrates as the optic lobe) is a direct target of a small complement of optic nerve fibers. Several waves of migrating cells suggest its imminent lamination by the end of the first trimester. There are indications that the entering optic fibers form a *honeycomb matrix superficially.

**Superior olivary complex** – A group of neurons in the ventrolateral posterior *pons that receive auditory input from the dorsal and ventral *cochlear nuclei. Ipsilateral and contralateral fibers of the complex join the *lateral lemniscus and terminate in the *inferior colliculus and in the *medial geniculate body.

**Superventricles** – The small ventricles of the mature central nervous system – known as the lateral and third ventricles in the forebrain, the aqueduct in the midbrain, the fourth ventricle in the pons and medulla, and the central canal in the spinal cord – are a continuous system of narrow channels lined by an enduring *ependyma and filled with cerebrospinal fluid. In contrast, the ventricles of the embryonic and early fetal brain are greatly inflated, balloon-like cisterns, hence the term superventricles, lined by a continuously changing germinal matrix, the *neuroepithelium and a fetal *choroid plexus that differs from its mature counterpart. Its large and variegated components are distinguished as the *telencephalic, the *diencephalic, the *mesencephalic, and the *rhombencephalic superventricles.

**Suprachiasmatic nucleus (hypothalamus)** – Small, paired midline structure above the *optic chiasm. It is implicated in the photic entrainment of the circadian rhythm.

**Supramammillary area (hypothalamus)** – Hypothalamic region that caps the *mammillary body. Experimental studies in animals indicate that its cells project to the *dentate gyrus of the hippocampus.

**Supraoptic nucleus (hypothalamus)** – Located above the optic tract lateral to the optic chiasm. The large secretory neurons of this nucleus produce arginine vasopressin and oxytocin that are conveyed by axoplasmic flow to the posterior lobe of the pituitary gland.

**T**

**Tectal NEP** – Extensive mesencephalic, smooth-surfaced NEP that lines the dorsal bank of the *mesencephalic superventricle. Its larger anterior part generates the neurons and glia of the *superior colliculus, its smaller posterior part that of the *inferior colliculus.

**Tectum (embryonic)** – Dorsal region of the *mesencephalon, consisting of the earlier developing *superior colliculus anteriorly and the later developing *inferior colliculus posteriorly.

**Tegmental NEP** – The variegated ventral stretch of the *mesencephalic NEP that contains shorter patches that produce neurons and glia for various tegmental nuclei, such as the *substantia nigra and the *ventral tegmental area.

**Tegmentum (embryonic)** – Ventral and ventrolateral region of the *mesencephalon and *rhombencephalon with indistinct boundaries. In addition to several brainstem nuclei, it contains many early-forming ascending, decussating, and descending fiber tracts. Some tegmental nuclei have been implicated in somatomotor and visceromotor functions. The onset of development of some components of the tegmentum appear to antedate that of the *tectum.

**Telencephalic superventricle** – The largest component of the superventricles that begins to expand during the early first trimester and shrinks during the third trimester. It is bounded laterally, dorsally and dorsomedially by the long stretch of the *cortical NEP, and ventromedially and ventrally by the shorter *olfactory, *septal, *striatal, *hippocampal and *amygdaloid NEPs. A large portion of its lumen is occupied by the fetal telencephalic *choroid plexus.

**Telencephalon (embryonic)** – Extensive forebrain region consisting of both cortical and subcortical components whose neurons and glia are produced by NEP stretches and patches lining the *telencephalic superventricle.

**Temporal bone** – A cranial bone in the skull that covers the *temporal lobe of the cerebral cortex and contains the internal ear and semicircular canals.

**Temporal lobe** or **cortex (embryonic)** – Lateral and ventral portion of the developing cerebral cortex that will later become separated from much of the rest of the cerebral hemisphere by the lateral fissure.

**Temporal NEP** – Putative source of neurons and glia of the *temporal lobe. It is flanked during fetal development by the temporal *subventricular zone and the temporal *stratified transitional field.

**Tenia tecta** – Components of the *cerebral cortex that extend into the dorsomedial *septum (dorsal tenia tecta) and medial olfactory peduncle (ventral tenia tecta).

**Thalamic NEP** – Stretch of the *diencephalic NEP between the *epithalamic and the *hypothalamic NEPs. Its mosaic divisions are the putative source of neurons and glia of the diverse nuclei of the *thalamus.

**Thalamocortical fibers (embryonic)** – Collective term for the large afferent tracts that proceed from relay nuclei in the thalamus, by way of the *internal capsule, to the *cerebral cortex. These nuclei include the *ventral posterolateral and *ventral posteromedial nuclei, and the *lateral geniculate and *medial geniculate bodies.

**Thalamus** – Massive dorsal diencephalic structure with several distinct and some indistinct nuclei. As a convenience, the thalamus is divided into the following nuclear regions: the *anterior complex, the *central complex, the *dorsal complex, the *periventricular complex, the *posterior complex, the *ventral complex, and the *reticular belt.

**Third ventricle** – *See* **Diencephalic superventricle**

**Thyroid cartilage** – A shield-shaped cartilage in the larynx.

**Transpontine corticofugal tract (embryonic)** – Portion of the large descending fiber tract in the maturing brain that traverses the *pontine gray and gives off collaterals there. Pioneering fibers of this tract are present by the end of the first trimester.

**Trapezoid body** – A fiber tract extending from the ventral *cochlear nucleus to the contralateral *superior olivary complex. It contains second- and higher-order auditory fibers.

**Trigeminal, motor nucleus (embryonic)** – Aggregate of trigeminal somatic motor neurons situated medial to the *trigeminal principal sensory nucleus. It is recognizable in late first trimester embryos.

**Trigeminal, principal sensory nucleus (embryonic)** – The second-order sensory neurons in the trigeminal system located dorsal and lateral to the incoming sensory root of cranial *nerve V. It receives topographic somatosensory input from the face and mouth, and its efferents cross the midline in the pons and proceed to the thalamic *ventral complex in close association with the *medial lemniscus. The nucleus is prominent by the late first trimester.

**Trigeminal, spinal nucleus (embryonic)** – A continuation of the *trigeminal principal sensory nucleus that extends caudally through the *medulla to the second cervical level of the *spinal cord. It is prominent by the late first trimester.

**Trigeminal, spinal tract** – Primary sensory fibers of the trigeminal ganglion that convey touch and pressure information from the face. The axons enter the brain in the pons and proceed caudally, forming a lateral cap around the *trigeminal spinal nucleus where they terminate in a topographic order.

**Trochlear nucleus** – Aggregate of somatic motor neurons located posterior to the *oculomotor nuclear complex that innervate the superior oblique muscle of the eye by way of cranial *nerve IV.

# V

**Ventral anterior nucleus (thalamus, embryonic)** – The ventral anterior nucleus, to be distinguished from the *anteroventral nucleus (which is part of the *anterior complex of the thalamus) is the most rostral component of the thalamic *ventral complex. Its afferents come mostly from the *globus pallidus and the *substantia nigra, and its efferents terminate in the *paracentral lobule of the neocortex. The nucleus is recognizable during the late first trimester as a developing structure with putative migrating and sojourning neurons and sprouting fiber bundles.

**Ventral complex (thalamus, embryonic)** – A group of structurally and functionally related ventrolateral and ventral nuclei of the thalamus, including the *ventral anterior, *ventral lateral, *ventral posterolateral, and *ventral posteromedial nuclei. The ventral thalamic complex is the principal topographically organized relay system of the direct (lemniscal) and indirect (cerebellar and striatal) somatosensory and proprioceptive input system to the sensory and motor areas of the *neocortex. Its components are recognizable during the late first trimester as formative structures with putative migrating and sojourning neurons and sprouting fiber bundles.

**Ventral lateral nucleus (thalamus, embryonic)** – Situated caudal to the *ventral anterior nucleus, this component of the thalamic *ventral complex is the target of input from the *superior cerebellar peduncle and the *red nucleus. Its somatotopically organized fibers terminate in the motor cortex and adjacent areas. The ventral lateral nucleus may be the principal relay from the *cerebellum to the *neocortex. The nucleus is recognizable during the late first trimester as a developing structure with putative migrating and sojourning neurons and sprouting fiber bundles.

**Ventral posterolateral nucleus (thalamus, embryonic)** – Situated caudal to the *ventral lateral nucleus, this region of the thalamic *ventral complex is the target of fibers of the *medial lemniscus that originate in the *cuneate nucleus and the *gracile nucleus and convey somatosensory information from the trunk and the extremities. Its efferents form the somesthetic radiation that terminates in a precise topographic order in the medial part of the *postcentral gyrus. The ventral posterolateral nucleus is the principal thalamic relay of somesthetic input from the trunk and limbs to the *neocortex. The nucleus is recognizable during the late first trimester as a developing structure with putative migrating and sojourning neurons and sprouting fiber bundles.

**Ventral posteromedial nucleus (thalamus, embryonic)** – Situated between the *ventral posterolateral nucleus and the *centromedian nucleus, this nucleus receives afferents from the *trigeminal sensory nuclei and the *parabrachial nucleus that convey sensory information from the face, the tongue, the oral cavity, and the neck. The efferents of this nucleus terminate in a precise topographic order in the lateral part of the *postcentral gyrus. The ventral posteromedial nucleus is the principal thalamic relay of somatosensory and gustatory input from the neck, head, and mouth to the *neocortex. The nucleus is recognizable during the late first trimester as a developing structure with putative migrating and sojourning neurons and sprouting fiber bundles.

**Ventral rhombic lip** – *see* **Precerebellar NEP**

**Ventral tegmental area (embryonic)** – Medial area flanking the *substantia nigra and containing a high concentration of dopaminergic neurons, much like the substantia nigra, pars compacta. It is present as a compact cell mass in the mesencephalon of late first trimester fetuses.

**Ventricles** – *See* **Superventricles**

**Ventromedial nucleus (hypothalamus)** – Large spherical nucleus that flanks the third ventricle and is surrounded by a fibrous shell. It has reciprocal connections with the *amygdala, the *bed nucleus of the stria terminalis, the *septum, and the *subiculum. It has been implicated in motivational functions related to feeding and sexual behavior.

**Vermis** – *See* **Cerebellum (vermis)**

**Vestibular nuclear complex** – A large area in the dorsal medulla, composed of the *medial, the *lateral, the *superior, and the *inferior vestibular nuclei. These nuclei get primary sensory input from the vestibular ganglion; their efferents join the *medial longitudinal fasciculus and form the vestibulospinal tract.

**Visual radiation (embryonic)** – Thalamocortical fibers that originate in the *lateral geniculate body and terminate in the striate cortex of the *occipital lobe. The identification of *Meyer's loop at GW11 suggests that this tract may reach the occipital lobe by the end of the first trimester.

## W

**White matter** – General term for extensive regions in the brain and spinal cord composed of myelinated fiber tracts but few or no neuronal cell bodies. In histological preparations with myelin stains, the white matter appears black. In laminated brain regions, as in the *cerebral cortex, the white matter is called the medullary layer.

## Z

**Zygomatic bone** – A facial bone in the cheek bone.

**Zona incerta (embryonic)** – Region in the *subthalamus with uncertain boundaries with *Forel's fields. It is a prominent area in the late first trimester *diencephalon.

---

**An asterisk in front of a term indicates that it is a separate entry in the Glossary with additional information. Terms referring to transient developmental structures are underlined.**

T - #0545 - 071024 - C592 - 279/216/28 - PB - 9780367390945 - Gloss Lamination